Handbook of Optical Sensors

Handbook of
Optical Sensors

Edited by

José Luís Santos
Faramarz Farahi

CRC Press
Taylor & Francis Group
Boca Raton London New York

CRC Press is an imprint of the
Taylor & Francis Group, an **informa** business

CRC Press
Taylor & Francis Group
6000 Broken Sound Parkway NW, Suite 300
Boca Raton, FL 33487-2742

First issued in paperback 2017

ISBN-13: 978-1-4398-6685-6 (hbk)
ISBN-13: 978-1-138-19866-1 (pbk)

Visit the Taylor & Francis Web site at
http://www.taylorandfrancis.com

and the CRC Press Web site at
http://www.crcpress.com

To Our Families

Contents

SECTION III Fiber-Optic Sensors

SECTION IV The Dynamic Field of Optical Sensing

Preface

The progress of science and technology has led to the development of complex structures with increasing sophistication, but still nothing comparable to what we observe in biological systems. Complex biological systems have three key characteristics associated with their operation: (1) Energy flow inside the system and energy exchange with its surrounding environment are coordinated to achieve specific functionalities. (2) Such flow of energy derives from analysis of the system about its internal state, also conditioned by the external environment, and the consequent identification of actions that must be performed to restore the balances necessary for adequate functioning of the system. (3) This analysis is fed by a constant stream of information from the internal state and external surroundings of the system, which increases markedly with the system's complexity.

This last characteristic relies on a sensing mechanism, where sensors are designed for specific quantities according to natural laws that determine the quantitative relationship between different variables. Therefore, distinct phenomena and representation can be related to the measurement of certain quantities such as mechanical, thermodynamic, electrical, magnetic, optical, and other properties, each of which offers specific advantages and unique representation, where the natural choice depends on a number of factors with emphasis on the intended application. Electromagnetic radiation described by its amplitude, frequency (wavelength), polarization, and phase provides a range of options to design sensors for a wide range of parameters with performances that can be adjusted according to the needs. It is not surprising that electromagnetism has long been regarded as an important phenomenon in the field of sensors. Because of the central role of vision as a sensory mechanism of living beings, the optical band of the electromagnetic spectrum is of particular importance. Optical sensors refer to sensing mechanisms that use electromagnetic radiation in the optical band, although recently longer (terahertz) and shorter (x-ray) wavelengths have been used within the context of optical systems. Since one important means of communication of humans with the environment is within the optical band of the electromagnetic spectrum, it seems natural for us to think of optics as a branch of science and technology for sensing. Over time, technological progress has increased our abilities to control the properties of light, and, with that, the construction of better sensor devices and optical instruments, enabling rapid progress in the field of optical sensing.

The importance of optical sensing is now recognized and its potential in future technological advances is immeasurable. But the truth is, if we ignore the sensing systems associated with vision, it was only in the second half of the twentieth century that the designation of the *optical sensor* became common. A survey of the publications of the Optical Society of America (OSA) shows that in 1961 the word *optical sensor* appeared in an OSA journal for the first time. The reason for this relatively late bloom is the level of technological demands it requires, which is not only due to the difficulties associated with sensor fabrication, but also the requirement for advanced optical sources, detectors, modulators, and waveguides that allow light propagation to and from the measurement region. Remarkable achievements in all these areas in recent decades allowed for the rapid progress in the field of optical sensing. In particular, optical fibers that play a dual role as light transporters and communication channels enable the relatively straightforward implementation of the sensors network.

The accelerated growth recently seen in the field of optical sensors is driven by two factors: the demands introduced by the applications and the progress of science and technology in other fields such as materials science, microelectronics, metamaterials, and quantum physics. This interesting dynamic has motivated us to seek help from many experts in this field and coordinate an effort with the objective of providing a comprehensive and integrated view of optical sensors, where each specific area is presented in detail, but within a general context. With this approach, our intention has been to offer a source of information to practitioners and those seeking optical solutions for their specific needs, and a reference for the students and investigators who are the intellectual driving force of this exciting field.

In pursuit of this objective, the book is organized into four sections, with a total of 21 chapters. In the first section, Optical Sensing and Measurement, the basic aspects of optical sensors and the principles of optical metrology are discussed in two chapters and a brief historical review of the subject is presented. The second section, Optical Measurement Principles and Techniques, starts with a chapter focused on the role of optical waveguides in sensing, followed by four chapters that discuss sensor technologies based on intensity modulation, phase modulation (interferometric), fluorescence, and plasmonic waves, followed by three chapters describing wavefront sensing, multiphoton microscopy, and imaging based on optical coherence tomography.

The third section, Fiber-Optic Sensors, explores the specific but increasingly important field of optical fiber sensing, which includes ten chapters starting with a historical overview of this subject, and a chapter specifically on light guiding in standard and microstructured optical fibers. Sensor technologies based on fiber optics benefit from techniques such as the modulation of intensity and phase of light, therefore the next two chapters address sensing supported by these modulation formats. This is followed by chapters on sensor multiplexing and distributed sensing. A chapter is devoted to fiber Bragg grating because of its importance in the field of fiber sensors. The next chapters focus on the application of fiber sensors for the detection of chemicals, with a chapter that can be considered a case study on the impact of the fiber-optic sensing approach through analysis and comparison with electrical- and optical-based strain gages. This part concludes with a discussion of the important issue of standardization in optical sensors.

The fourth section, The Dynamic Field of Optical Sensing, is composed of one chapter that intends to provide a broad perspective of the field and to identify possible new trends that could shape the future of this area. It starts by discussing some of the well-established sensing technologies such as wavelength-encoded sensors and distributed sensing, which are expected to experience significant progress in the near future considering the high level of interest and R&D resources they have attracted. Then, the chapter addresses some of the recent developments in optics that could potentially impact the field of optical sensors beyond what can be envisioned today. Metamaterials could certainly provide a new vehicle for the design of novel sensing structures and on the horizon the enticing idea of entangled quantum states of light offers exciting possibilities in optical sensing offers exciting possibilities.

The 28 authors of these chapters are recognized experts in their areas of research and are major contributors to this field. They fully supported us by offering inputs and providing their respective materials in accordance with the main objectives of this project. For having accepted our invitation to participate in this buildup of this book, and also for the understanding they showed in the face of unexpected delays to reach to this point, we express our utmost gratitude. Our objective is to congregate competences of many experts to produce a text in optical sensing that would be useful for students, researchers, and users of these technologies. The output of this endeavor, we think, meets this key objective and, if so, it is very rewarding to us.

Editors

José Luís Santos earned his *licenciatura* in physics from the University of Porto, Portugal, and PhD from the same university, benefiting from a collaboration with the University of Kent at Canterbury, United Kingdom. He is currently a professor of physics at the Physics and Astronomy Department of Faculty of Sciences of the University of Porto, Portugal. He is also researcher of INESC TEC-Centre for Applied Photonics (former INESC Porto—Optoelectronics and Electronic Systems Unit). His main area of research is optical fiber sensing, with a focus on interferometric and wavelength-encoded devices. He is author or coauthor of more than 200 scientific articles and coauthor of 5 patents.

Faramarz Farahi earned his BS in physics from Aryamehr University of Technology (Sharif) in Tehran, his MS in applied mathematics and theoretical physics from Southampton University, and PhD from the University of Kent at Canterbury, United Kingdom. He is currently a professor of physics and optical science at the University of North Carolina at Charlotte, where he is a member of the Center for Optoelectronics and Optical Communications and Center for Precision Metrology. He was chair of the Department of Physics and Optical Science from 2002 to 2010. Dr. Farahi has more than 25 years of experience in the field of optical fiber sensors and devices. His current research interests include integrated optical devices, hybrid integration of micro-optical systems, optical fiber sensors, and optical metrology. He also pursues research on the application of optical imaging and optical sensors in the medical field. He is the author or coauthor of over 200 scientific articles and texts and holds 10 patents.

Contributors

Mehrdad Abolbashari
Department of Physics and Optical Science
University of North Carolina at Charlotte
Charlotte, North Carolina

Francisco Araújo
INESC TEC
FiberSensing
Porto, Portugal

Xiaoyi Bao
Department of Physics
University of Ottawa
Ottawa, Ontario, Canada

José Manuel Baptista
Centre of Exact Sciences and Engineering
University of Madeira
Funchal, Portugal

Liang Chen
Department of Physics
University of Ottawa
Ottawa, Ontario, Canada

Shahab Chitchian
Center for Biomedical Engineering
University of Texas Medical Branch
Galveston, Texas

Geoffrey A. Cranch
Optical Sciences Division
US Naval Research Laboratories
Washington, DC

Angela Davies
Department of Physics and Optical Science
University of North Carolina at Charlotte
Charlotte, North Carolina

Kert Edward
Department of Physics
University of the West Indies at Mona
Mona, Jamaica

Faramarz Farahi
Department of Physics and Optical Science
University of North Carolina at Charlotte
Charlotte, North Carolina

Luís Ferreira
INESC TEC
FiberSensing
Porto, Portugal

Orlando Frazão
Centre of Applied Photonics
INESC TEC
Porto, Portugal

Nathaniel M. Fried
Department of Physics and Optical Science
University of North Carolina at Charlotte
Charlotte, North Carolina

Kenneth T.V. Grattan
School of Engineering and Mathematical Sciences
City University
London, United Kingdom

Banshi D. Gupta
Department of Physics
Indian Institute of Technology Delhi
New Delhi, India

Wolfgang R. Habel
BAM Federal Institute for Materials Research and Testing
Berlin, Germany

Rajan Jha
School of Basic Sciences
Indian Institute of Technology Bhubaneswar
Bhubaneswar, India

Pedro Alberto da Silva Jorge
INESC TEC
Porto, Portugal

Wenhai Li
Department of Physics
University of Ottawa
Ottawa, Ontario, Canada

William N. MacPherson
Institute of Photonics & Quantum Sciences
and
School of Engineering & Physical Sciences
Heriot-Watt University
Edinburgh, United Kingdom

T.H. Nguyen
School of Engineering and Mathematical Sciences
City University
London, United Kingdom

Bishnu P. Pal
Department of Physics
Indian Institute of Technology Delhi
New Delhi, India

Marco N. Petrovich
Optoelectronics Research Centre
University of Southampton
Southampton, United Kingdom

José Luís Santos
Department of Physics and Astromy
University of Porto
Porto, Portugal

Tong Sun
School of Engineering and Mathematical Sciences
City University
London, United Kingdom

Robert K. Tyson
Department of Physics and Optical Science
University of North Carolina at Charlotte
Charlotte, North Carolina

Carmen Vázquez
Electronics Technology Department
School of Engineering
Carlos III University of Madrid
Leganés, Spain

David Webb
Aston Institute of Photonic Technologies
Aston University
Birmingham, United Kingdom

Optical Sensing and Measurement

1. Overview of Optical Sensing

José Luís Santos
University of Porto

Faramarz Farahi
University of North Carolina at Charlotte

1.1 Introduction

The evolution of science and technology of optical sensors has reached to a point that we can almost measure all physical parameters of interest and a broad range of biological and chemical quantities. Optical measurements are made in a wide variety of methods with a particular selection determined by factors such as local or remote operation, characteristics of the environment, and the size of the field being measured. For example, the measurement done on a point could be very different from the measurement done on a large two-dimensional surface or a large three-dimensional object. The incorporation of optical waveguides in the sensing process adds an interesting twist to this field. For example, the formation of modes in optical waveguides could be exploited for sensing, or measurement could be made on remote locations and in a large sensing network using optical fibers.

Optical sensors draw from many different disciplines of electrical engineering, mechanical sciences and engineering, civil engineering, chemical engineering, material science, biology, chemistry, physics, and optics. In this book, we will see a variety of examples of optical sensors and their applications, which clearly show the multidisciplinary nature of this field. But all sensors, regardless of their target applications and their specific technology and design, have some common tread and are based on some basic principles, which will be discussed in the next section. Because of its multidisciplinary nature, the field of optical sensing sometimes seems to be a collection of disjointed topics; therefore, it is instructive to develop an integrated view within a historical context, which will be discussed in Section 1.3. Finally, an outline of the book contents is provided in Section 1.4.

Handbook of Optical Sensors. Edited by José Luís Santos and Faramarz Farahi © 2015 CRC Press/Taylor & Francis Group, LLC. ISBN: 9781439866856.

1.2 General Concepts

Sensing is the process of getting information from the environment, constituting a basic functionality of a complex system, either a machine or a biological entity, with the knowledge acquired from its surrounding, triggering and guiding its internal operation to a level compatible with the external conditions, or eventually initiating actuation steps toward their modification. The technological progress and the fast development of intelligent systems inevitably require an increasingly effective and reliable sensing technology. This process has three steps: in the first step, the sensing element, also called the sensing head, interacts with the environment and is being affected by its environment, which causes a change on some characteristics of the sensing element; the second step is the transmission of raw information about these changes to a processing unit; and in the final step, raw information will be analyzed depending on the sensing mechanism and application. Most notably, biological sensing systems well illustrate this three-step process, as is the case for our vision system. In this system, the eye is the sensing head with its adjustable focal length lens to form an image of the outside scene on the retina, which contains photosensitive elements (rods and cones) that convert light into electrical impulses; the transmission channel is the optic nerve, which is the route by which these impulses are sent to the brain to be processed.

Physical laws expressed as quantitative relations between variables are the design tool for a sensing system to measure one or more parameters. This means a specific sensing function can be achieved from a variety of phenomena if they involve interrelated variables, the reason why there is a plethora of approaches to detect and measure a particular quantity. Phenomena of the type mechanical, thermodynamic, electric, magnetic, and others demand distinct classes of sensing approaches, with relative advantages and disadvantages determined by factors such as adequacy to a specific need, performance, implementation cost, and, not the least, the scientific background of the designer/end user.

Independent of the physical principles behind sensing, two major classes can be identified. In one, it is the action of the measurand field that induces a signal in the sensing head that propagates down to the processing unit. The example of the vision fits within this class. The optics of the eye projects an image of the incident light wavefront on the retina, a *screen* where each pixel generates signal dependent on the intensity and wavelength of the light impinging on it. The other class refers to sensing systems where the targeted measurand changes specific physical properties of the sensing element, perturbing the conditions of interaction of a probe field or signal with this structure, and the measurand information recovered from the analysis of such perturbation. An electrical strain gauge is an example of a sensor of this type, where the sensing element is an element of an electric circuit, and when it is subject to strain, its resistance changes and consequently the circuit characteristics will also change. This classification helps with the organization of sensing systems, but it is sometimes possible that a sensing system possesses features associated with both classes.

Optical sensing is concerned with the detection of an optical field and the analysis of its intensity, phase, wavelength, polarization, spectral distribution, or wavefront to obtain information from the illumination background, or the source of the optical field and eventually from the objects and propagation medium that interacted with this field (systems of class one), or to obtain information from the measurand that, through the sensing element, impacted a probe optical field (systems of class two). In the second case, the optical field is generated with characteristics tailored for the sensing head and its mechanism of interaction with measurand, and transmission of such an optical field to and from the sensing head requires certain care. Figure 1.1 illustrates these classes of optical sensing systems.

The requirement to control propagation of light from the optical source to the sensing region and then to the detection/processing unit has been fulfilled following two approaches. In the classical one, lens, mirrors, apertures, and other bulk optic elements direct the light, as happens, for example, in an optical microscope to get a magnified image of a sample. Because of the complexity of aligning these optical elements and keeping them stable relative to each other, in most cases, such systems are compact and in one location, although the optical field they analyze could be from far distances, such as happens with telescopes. The second approach is based on guided wave optics, where the light field is guided from the source to the sensing region and to the detection/processing unit using optical waveguides. Waveguides could be divided into two categories: waveguides

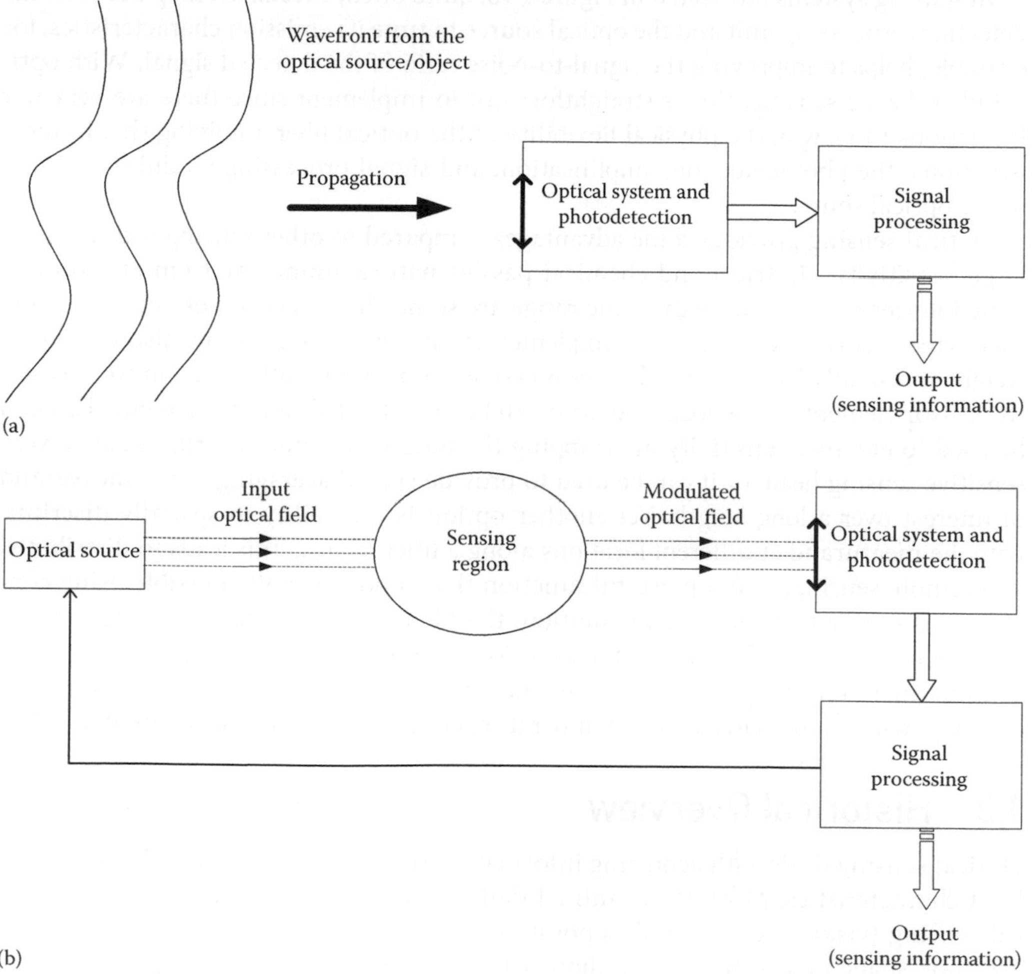

FIGURE 1.1 Sensing systems of (a) class one and (b) class two.

made on substrates such as planar waveguides and integrated channel waveguides, and optical fibers. The first one is, in general, a preferred choice where the proximity of optical source, sensing region, and photodetection is possible, a limitation that is overcome with the application of optical fibers, a development that provided a degree of freedom and consequently unforeseen opportunities in optical sensing.

The development of low-loss optical fibers allows light to be guided from the optical source to the sensing volume that can be remotely located, to interact with the measurand field, and to be guided back for detection by the same or a second fiber. The interaction with the measurand field could occur while light is still guided in the fiber or, alternatively, after leaving the fiber for free space propagation or coupled to a planar waveguide where the measurand-induced light modulation takes place. The situation where light does not leave the fiber in the sensing region (intrinsic sensing), meaning the optical fiber has the dual role of sensing element and communication channel, is unique to optical fiber sensing technology. This characteristic is very important for remote and distributed sensing.

In sensing systems illustrated in Figure 1.1b, quite often, a feedback loop between the detection/processing unit and the optical source to tune its emission characteristics, for example, helps to improving the signal-to-noise ratio of the detected signal. With optical fiber–based sensing, this is straightforward to implement since there are very few limitations in view of the physical flexibility of the optical fiber, implying that in most situations, the photodetection, amplification, and signal processing modules are close to the optical source.

Optical sensing provides some advantages compared to other sensing technologies. High sensitivity, electrical and chemical passive nature, immunity from electromagnetic interference, and wide dynamic range are some characteristics of optical sensors that make them attractive. In some implementations, optical sensors are also very lightweight and small-size devices. Fiber sensors offer some unique attributes; in some cases, the sensor element can be localized, or it can be an extended sensing element. This can be used to enhance sensitivity by wrapping the fiber in a compact form to create very sensitive sensing head, or it can be used to provide spatial averaging of the measurand of interest over a long length. Yet another option is the ability to spatially discriminate the measurand at different locations along a fiber length. This leads to distributed and remote sensing, a very powerful function that is not generally possible using conventional sensor technologies. In addition, the fiber transmits the optical field to the measurement region, eliminating the need for wiring for power delivery to the sensor, a requirement for many other sensor technologies. Finally, this type of sensors has the ability to withstand relatively high temperatures due to the high melting point of silica.

1.3 Historical Overview

Optical sensing deals with acquiring information from the surroundings by analyzing light characteristics. Most of the animal daily life on Earth depends largely on optical sensing (vision). Recent studies point out that all types of eyes, as varied as they are, took shape in less than 100 million years, evolving from a simple light sensor for circadian and seasonal rhythms around 600 million years ago to becoming an optically and neurologically sophisticated system 500 million years ago, with the majority of the advancements in early eyes believed to have taken only a few million years to develop, with multiple eye types and subtypes evolving in parallel (Halder et al. 1995, Lamb 2011).

The humanity has for long tried to correct deficiencies of human vision, and the utilization of lenses in glasses can be considered the first technological upgrade of an optical sensing system. The oldest lens artifact known is the Nimrud lens, which is over 3000 years old, dating back to ancient Assyria, and the earliest written records of lenses are from Ancient Greece, known as the burning glass in the Aristophanes's play *The Clouds* (424 BC), associated to the biconvex lens effect of focusing the Sun's light to produce fire (Sines and Sakellarakis 1987). There are several references indicating that burning glasses were well known to the Roman Empire, and clues of what can eventually be the first utilization of lens to correct the vision when it is said that Nero watched the gladiatorial games using an emerald, presumably with a concave shape to correct for myopia (Pliny).

Following excavations at the harbor of the town Fröjel in Gotland, Sweden, there is evidence that in the eleventh to twelfth centuries, the Vikings were able to make rock crystal lenses with remarkable quality (Tilton 2005). However, the widespread use of lens to correct the vision occurred in Europe by the end of the thirteenth century after the invention of spectacles in Italy at about 1286 (Ilardi 2007). It is interesting to note that this development occurred not long after the translation of Ibn al-Haytham's *Book of Optics* into Latin. Ibn al-Haytham (known as Alhacen or Alhazen in Western Europe: 965–1040) is often regarded as the *father of modern optics*. He formulated and worked out what is considered the first comprehensive and systematic alternative to Greek optical theories (Bala 2006). One of the topics that most attracted his attention was the understanding of the vision process and the revolutionary statement that *vision only occurred because of rays entering the eye* is credited to him, discarding the accepted view since the Ancient Greece that the rays proceed from the eye.

The telescope and the microscope are the first engineered optical sensing systems that human has developed, and they expanded the world that we could see. The earliest known working telescopes were refracting telescopes, a type that relies entirely on lenses for magnification. Their development in the Netherlands in 1608 is credited to three individuals, Hans Lippershey (1570–1619) and Zacharias Janssen (1580–1638), who were spectacle makers in Middelburg, the Netherlands, and Jacob Metius (1571–1630), a Dutch instrument maker and optician of Alkmaar. In Venice, in 1609, Galileo Galilei (1564–1642) heard rumors of these developments in the Netherlands and became determined to build his own instrument. First, he made a 3-power instrument, then a 10-power one, and in 1610 a 30-power telescope. It was with this instrument that Galileo discovered that satellites were rotating about Jupiter, implying the existence of celestial bodies that did not move around the Earth, contrary to what was believed that the Earth was the center of the Universe.

Isaac Newton (1643–1727) realized, from his studies on the dispersion of light, that refracting telescopes would suffer from the dispersion of light into colors and worked out to bypass that problem. In 1668, his efforts led to the invention of the reflecting telescope based on mirrors instead of lenses (today known as the Newtonian telescopic configuration), with a performance far better than the refracting telescope with the same aperture. It is interesting to mention that he did himself the grinding of the mirrors, judging the quality of the optics using an interference phenomenon later known as Newton's rings, perhaps the first time an optical metrological technique was consistently used to guide the fabrication of optical elements (Watson 2005).

The origins of the optical microscope are not easy to identify, but it is accepted that an early microscope was made in 1590 in Middelburg, the Netherlands, an invention credited to Hans Lippershey and Sacharias Jansen, referenced earlier in virtue of their contribution to the invention of the telescope. The name attributed to this instrument has been associated to Giovanni Faber (1574–1629), who gave the designation microscope to the apparatus developed by Galileo in 1625 (Gould 2000). What is certain is that from these early days, the progress in optical telescope and microscopy instrumentation fertilized reciprocally, generating up to the present notable advances in our ability to see the very small and the very large.

The early optical sensing systems were vision systems of some type. The broad meaning of the term *optical sensing* as understood today had to wait up to the twentieth century, particularly after the 1960s. This is confirmed by a search in the website of Optical Society of America, established in 1916 and one of the oldest and largest scientific associations in optics and photonics in the world, allowing a good insight into the time line of optical sensors' development (Viegas 2010). In 1961, for the first time, the word *sensor* in the abstract of a paper in an OSA journal appeared, when Astheimer and co-workers reported infrared radiometric instruments on TIROS II, a space satellite application of optics (Astheimer *et al* 1961). This was followed in the next two decades by a large number of papers regarding radiometric and positioning sensors for astronomy, satellite-based remote sensing, and atmospheric optics; examples are works of Palser (1964), Goss (1970), and Mende and O'Brien (1968).

The work in optical sensors was limited to free space optics and developments of photoelectric sensors for imaging, with the exception of a paper mixing biology and optical waveguides (Callahan 1968), reporting a high-frequency dielectric waveguide on the antennae of night-flying moths (Saturniidae). This is probably one of the first studies of biological specimens with analogy to optical devices, nowadays referred as bio-inspired photonics.

Guided-wave optical sensors started in the late 1960s as reported in a NATO work entitled *Optical Fiber Sensors* (Chester et al. 1967). In the same year, the celebrated *Fotonic Sensor* appears, which is considered the first effective demonstration of a fiber-optic sensor, where object displacement was measured based on the fraction of light that was coupled to fibers of a bundle after being emitted by other fibers of the bundle and reflected from the object surface (Menadier et al. 1967). However, at the time the fiber-optic technology was incipient, and it was only when its development accelerated, triggered by the optical fiber communications needs, exploration of the sensing potential of the optical fiber began. It is reasonable to state that this second phase emerged in 1976 when the operation of a fiber ring interferometer was demonstrated and its sensing applications were suggested (Vali and Shorthill 1976). In 1977, the fiber-optic-based acoustic wave sensor was reported (Bucaro et al. 1977), as well as the application of this technology to voltage and current sensing, exploring its unique advantages, namely, immunity to electromagnetic interference (Rogers 1977). In an important development, published in 1978, a fiber-optic strain gauge was developed, indicating an optical alternative to the electrical strain gauge (Butter and Hocker 1978). The progress of this sensing technology continues at an accelerated pace, leading to the measurement of physical, chemical, and biological parameters in close or remote locations with single point or distributed sensing. A historical overview of fiber-optic sensing can be found in Chapter 11.

Sensing based on integrated optical devices appeared later, probably through the work of Johnson and collaborators in 1982, where an integrated optical temperature sensor consisting of a parallel array of unequal arm-length waveguide Mach-Zehnder interferometers in $LiNbO_3$ with a projected range and resolution of ≥700°C and 2×10^{-2}°C was developed (Johnson et al. 1982). In the following year, other integrated optical sensors were reported, namely, an optical-waveguide pressure sensor at the *Second European Conference on Integrated Optics* (Izutsu et al. 1983), and the work on wavefront measurement based on waveguide Mach-Zehnder interferometers in $LiNbO_3$ permitting a resolution in the wavefront optical path difference of less than $\lambda/80$ (Rediker et al. 1983). In 1984, Tiefenthaler and Lukosz reported what was probably the first work on integrated optical sensors for chemical sensing (Tiefenthaler and Lukosz 1984). Humidity and gas sensing based on grating couplers and Bragg grating reflectors were reported. A multipass interferometer with resonant cavity sensor in integrated optics fabricated in $LiNbO_3$ was proposed by Syms (1985). In 1986, Boucouvalas and Georgiou presented the possibility of using the external refractive index influence on the transmitted intensity of a tapered coaxial waveguide coupler as a refractive index sensor (Boucouvalas and Georgiou 1986), while in the same year, Cozens and Vassilopoulos also reported a work on couplers based on the backward wave propagation (Cozens and Vassilopoulos 1986).

The following years saw a steady growth in integrated optics sensing, which has accelerated since the early 2000s. It is interesting to combine optical fiber and integrated optics technology for higher-performance optical sensing, with selectivity offered by integrated optics and remote operation and multipoint measurement offered by optical fibers.

1.4 Concluding Remarks and Book Overview

A historical review of the field of optical sensing, excluding microscope and telescopic systems, shows a new level of interests and activities in the second half of the last century. Given the diversity and complexity of optical sensing systems available today, it is fair to say that optical sensing has been a rapidly changing technology. Optical sensors benefit from other major technological development such as the laser technology (demonstrated in 1960, now playing a central role in a large number of technological systems) and the optical fiber (only 50 years ago was considered not viable for communications, now the backbone of our information society). The review of its recent past and the awareness of current important developments allow us to be better prepared for what are conceivably possible in the future.

This book intends to provide the reader a comprehensive perspective of the field of optical sensing, which also includes optical metrology and imaging. Organizing the material in this book with such a broad scope was a challenge. At the first level, the approach followed was to invite well-known experts to write their contributions following a general trend where, in the first part of the chapter, general concepts, principles, and technologies are presented keeping in mind didactic objective, complemented in the second part with novel developments of the subject and state-of-the-art review. At the second level, the integration of the chapters was done such that when viewed collectively, a coherent view into the field of optical sensors emerges.

In summary, Part I, which includes this chapter and the next one, details the basic aspects of optical sensors and the principles of optical metrology. A historical perspective of the subject is also included. Part II addresses optical measurement principles and techniques, where, after a chapter focused on the role of optical waveguides in sensing, appear chapters that detail how this functionality can be achieved via intensity, interferometric, fluorescence, and plasmonic approaches, followed by chapters focused on wavefront sensing and adaptive optics, multiphoton microscopy, and imaging based on optical coherence tomography. Part III explores the specific but increasingly important field of optical fiber sensing. A chapter dedicated to a historic overview of the subject is presented, followed by one dedicated to the waveguide optical fiber, where the different types of optical fiber available nowadays, light propagating principles, fabrication techniques, and main characteristics are presented. Two chapters detail fiber sensing based on light intensity and phase modulation, followed by other two dedicated to multipoint and distributed sensing. Due to the relevance of fiber Bragg gratings for sensing, a specific chapter is included on the subject, followed by the description of the methodology and techniques that have been developed to apply fiber sensors to the chemical domain. This part includes two chapters more oriented to the application issues of fiber-optic sensing technology, one dealing with the specific but important topic of strain measurement and the other focused on standardization and its impact on measurement reliability.

The final chapter intends to provide a broad perspective of the field and identifies trends and new possible paths for the future. The first part discusses some of the well-established sensing technologies that seem to experience further progress in the future because of the high level of interest and current worldwide research and development resources allocated to them. Hence, the discussion will focus on wavelength-encoded sensors, such as fiber Bragg gratings, DFB fiber laser sensors, and plasmonic-based sensors (including surface-enhanced Raman scattering), and then distributed sensing, which is based on unique attributes of optical fibers with no counterpart in any other sensing technology.

The second part describes some of the recent developments in optics that could potentially impact the field of optical sensors beyond what can be envisioned today. It has already been demonstrated that metamaterials have applications in optical sensors, permitting improved performance and reduced size. Finally, a discussion on entangled quantum state of light will address some potential impact of entangled states on optical sensors.

The multidisciplinary field of *optical sensing* benefits from discoveries, innovations, and technological progress in many branches of science and engineering. It is hardly possible to foresee how this field is going to evolve and what would be the next significant development, but it is safe to say that what we have witnessed so far is the beginning of a fascinating ride toward many discoveries to come.

References

Astheimer R. W., R. DeWaard, and E. A. Jackson. 1961. Infrared radiometric instruments on TIROS II. *Journal of Optical Society of America* 51: 1386–1393.

Bala A. 2006. *The Dialogue of Civilizations in the Birth of Modern Science*. Palgrave Macmillan, New York.

Boucouvalas A. C. and G. Georgiou. External refractive-index response of tapered coaxial couplers. *Optics Letters* 11: 257–259.
Bucaro J. A., H. D. Dardy, and E. F. Carome. 1977. Fiberoptic hydrophone. *Journal of the Acoustical Society of America* 62: 1302–1304.
Butter C. D. and G. B. Hocker. 1978. Fiber optics strain-gauge. *Applied Optics* 17: 2867–2869.
Callahan P. S. 1968. A high frequency dielectric waveguide on the antennae of night-flying moths (Saturnidae). *Applied Optics* 7: 1425–1430.
Chester A. N., S. Martellucci, and A. M. V. Scheggi. 1967. *Optical Fiber Sensors*. NATO. ASI Series E: Applied Sciences No. 132, Wiley & Sons, New York.
Cozens J. R. and C. Vassilopoulos. 1986. Combined directional and contradirectional coupling in a three-waveguide configuration. *International Journal of Optical Sensors* 3: 263–274.
Gould J. G. 2000. *The Lying Stones of Marrakech*. Harmony, New York.
Goss W. C. 1970. The Mariner spacecraft star sensors. *Applied Optics* 9: 1056–1067.
Halder G., P. Callaerts, and W. J. Gehring. 1995. New perspectives on eye evolution. *Current Opinion in Genetics & Development* 5(5): 1–10.
Ilardi V. 2007. *Renaissance Vision from Spectacles to Telescopes*. American Philosophical Society, Philadelphia, PA.
Izutsu M., A. Enokihara, N. Mekada, and T. Sueta. 1983. Optical-waveguide pressure sensor. *Technical Digest, Second European Conference on Integrated Optics*, Firenze, Italy, 1983, pp. 144–147.
Johnson L. M., F. J. Leonberger, and G. W. Pratt Jr. 1982. Integrated optical temperature sensor. *Applied Physics Letters* 41: 134–136.
Lakowicz J. R. 1999. *Principles of Fluorescence Spectroscopy*, 2nd Edition. Kluwer Academic/Plenum Publishers, New York, pp. 1–698.
Lamb T. D. 2011. Evolution of the eye. *Scientific American*, July, p. 36.
Menadier C., C. Kissinger, and H. Adkins. 1967. The fotonic sensor. *Instruments and Control Systems* 40: 114–120.
Mende S. B. and B. J. O'Brien. 1968. A high sensitivity satellite-borne television camera for the detection of auroras. *Applied Optics* 7: 1625–1634.
Michelson A. A. and F. G. Pease. 1921. *Measurement of the diameter of alpha Orionis with the interferometer. Astrophysical Journal* 53: 249–59.
Palser W. E. 1964. A wide-angle, linear-output horizon sensor optical system. *Applied Optics* 3: 63–69.
Pliny the Elder. *The Natural History* (trans. John Bostock).
Rediker R. H., T. A. Lind, and F. J. Leonberger. 1983. Integrated optics wave front measurement sensor. *Applied Physics Letters* 42: 647–649.
Rogers A. J. 1977. Optical methods for measurement of voltage and current on power-systems. *Optics and Laser Technology* 9: 273–283.
Saha S. K. 2011. *Aperture Synthesis, Methods and Applications to Optical Astronomy*, Springer.
Simmons M. 1984. *History of Mount Wilson Observatory: Building the 100-Inch Telescope*. Mount Wilson Observatory Association (http://www.mtwilson.edu/).
Sines G. and Y. A. Sakellarakis. 1987. Lenses in antiquity. *American Journal of Archaeology* 91(2): 191–196.
Syms R. 1985. Resonant cavity sensor for integrated optics. *IEEE Journal of Quantum Electronics* 21: 322–328.
Tiefenthaler K. and W. Lukosz. 1984. Integrated optical switches and gas sensors. *Optics Letters* 9: 137–139.
Tilton B. 2005. *The Complete Book of Fire: Building Campfires for Warmth, Light, Cooking, and Survival*. Menasha Ridge Press, Birmingham, AL.
Vali V. and R. W. Shorthill. 1976. Fiber ring interferometer. *Applied Optics* 15: 1099–1100.
Viegas J. P. R. 2010. *Integrated Optical Sensors*. University of Porto, Porto, Portugal.
Watson F. 2005. *Stargazer: The Life and Times of the Telescope*. Da Capo Press, Cambridge, MA.

2. Principles of Optical Metrology

Angela Davies
University of North Carolina at Charlotte

2.1 Introduction

This chapter provides an introduction to the notion of measuring physical properties with light. Measurements exploit the wave nature of light where the properties of wave amplitude, polarization, frequency, and phase are connected to the property to be measured. A measurement begins with the detection of the energy delivered by the light field. The measured energy is then connected to the measurand, that is, the physical property or parameter to be measured, by understanding the way in which the light has interacted with the environment and the way in which the environment changes the wave properties.

The measurement is specifically engineered to have the light interact with the environment with a desired sensitivity, meaning a property of the light such as polarization or phase will change in a known way with interaction in the environment. For example, the principle of superposition and interference are often exploited to establish a connection between measured energy and a phase relationship between two light waves. And with a properly engineered optical configuration, the phase relationship can be connected to the displacement of a precision stage, a temperature change in the environment, or a chemical change

Handbook of Optical Sensors. Edited by José Luís Santos and Faramarz Farahi © 2015 CRC Press/Taylor & Francis Group, LLC. ISBN: 9781439866856.

in the propagation path, as examples. The basic mathematical model behind interferometry is discussed later in the chapter.

The polarization state of the wave is exploited in the broad sensing field of ellipsometry. Ellipsometry can be used to determine properties such as the refractive index and thickness of films (Tompkins 2005). This is done by reflecting a light wave with a known polarization state from a surface and configuring the detection such that the measured energy indicates the amount of polarization change. A model of the polarization change connects the measured energy to the surface properties to be sensed.

Spectroscopy is the broad sensing field where the frequency content of the light contains information about the property to be sensed (Workman 1998, Ball 2011). This is often chemical or electronic information about a sample. This is because discrete electronic energy levels in matter overlap with photon energies of visible light, and the photon energy is connected to the light wave frequency. Thus, absorption of light at a specific frequency results when the material has a specific electronic energy level corresponding to this frequency, and that in turn indicates a specific chemical state. The frequency content can be determined in many ways, most of which separate the light in space based on frequency. This allows the energy at each frequency or in a frequency range to be directed to a detector and directly measured. The spreading of light based on frequency can be done as simply as directing a beam to a prism of glass, called a prism spectrometer. The refractive index in glass varies with frequency, thus the light bending (refraction) at the interface depends on frequency. Light exiting a prism is spread into a fan based on frequency. We are familiar with this phenomenon as the rainbow of color when sunlight shines through a faceted crystal. Topics of spectroscopy are discussed in detail in several chapters of this book, particularly in Chapters 6, 9, and 18.

This chapter describes underlying principles and configurations of basic optical measurements associated with the propagation of light. It begins with the mathematical foundation used to describe a light wave, followed by an overview of radiometry, which is the formalism used to describe the energy content of light. The chapter ends with an introductory discussion of basic interferometry, holography, and fringe projection.

2.2 Mathematical Description of Light

Any light wave can be described as a superposition of harmonic waves. A harmonic light wave of optical frequency ω can be written as

$$\vec{E}(\vec{r},t) = \vec{E}_o \sin(\vec{k}\cdot\vec{r} - \omega t + \varphi_o) \tag{2.1}$$

This describes the electric field part of the light wave, and Maxwell's equations tell us the electric field is always accompanied by a similar magnetic field wave. The vector direction of the magnetic field is such that the vector $\vec{E}_o \times \vec{B}_o$ points in the direction of the vector $\vec{k}$. The vector $\vec{k}$ is the wavevector, the magnitude of which describes the repeat length scale of the wave, namely, $|\vec{k}| = 2\pi/\lambda$, where λ is the wavelength or repeat length scale of the harmonic wave. The repeat time scale of the wave is set by the value of the angular frequency, ω, namely, $\omega = 2\pi/T$, where T is the period of the harmonic wave. Waves are functions of both time (t) and position ($\vec{r}$). We see that the argument of the sine wave cycles 2π

when the time advances by a period (holding the position fixed); thus, the value of the wave cycles when the time advances by a period. Holding the time fixed, similarly we see that the value of the wave cycles when the position along the direction of $\vec{k}$ advances by a wavelength. We can see this by writing $\vec{k}\cdot\vec{r}$ as $|\vec{k}||\vec{r}|\cos(\theta) = (2\pi/\lambda)|\vec{r}|\cos(\theta)$ where θ is the angle between $\vec{r}$ and $\vec{k}$. Thus, $|\vec{r}|\cos(\theta)$ is the component of $\vec{r}$ in the direction of $\vec{k}$. These relationships are shown in Figure 2.1a. The speed of the wave, v, is related to these two periodic values. The frequency of a wave is set by the oscillation frequency of the source generating the wave, and the speed of the wave is set by the propagation properties of the medium. The wavelength is a consequence of the two properties, that is, $\lambda = vT$. In vacuum, a light wave travels at speed $c = 3.0 \times 10^8$ m/s and in a nonmagnetic dielectric material with permittivity ε travels with speed $v = c/n$, where n is the index of refraction and is given by $n = \sqrt{\varepsilon/\varepsilon_o}$. Waves described by Equation 2.1 are called plane waves because the surfaces of constant phase (and therefore wave value), the wave fronts, are planes, planes that are perpendicular to the advancing direction of the wave. This is illustrated in Figure 2.1b.

The accompanying harmonic magnetic field wave is written similarly with the magnitude and vector directions defined in terms of the $\vec{E}_0$ field as described. For typical transparent optical materials like dielectric glass and plastics, the force exerted on charge in the material by the light field's electric field is many orders of magnitude larger than the force exerted by the magnetic field; thus, the mathematical description usually focuses on the electric field wave, knowing the magnetic field is present as well.

In some applications, the polarization state of the wave (defined by the vector direction of $\vec{E}_0$) does not significantly impact the measurement. In this case, the scalar wave description can be used. The equation for the electric wave is the same as Equation 2.1, but the vector notation above $\vec{E}_0$ is dropped, $\vec{E}_0 \to E_0$, and the electric field is treated as a scalar quantity. Some measurements explicitly leverage the full vector nature of the light field and use the polarization state as a property to be impacted by the environment and, thus, is the property that carries information about the measurand.

Equation 2.1 describes a harmonic plane wave, that is, an electric field wave for which the surfaces of constant electric field (wave fronts) are a plane whose normal is perpendicular to the wavevector $\vec{k}$. A spherical wave is another common wavefront geometry. A harmonic spherical wave is described by

$$\vec{E} = \frac{\vec{\varepsilon}_o}{r}\sin(kr \pm \omega t + \varphi_o) \qquad (2.2)$$

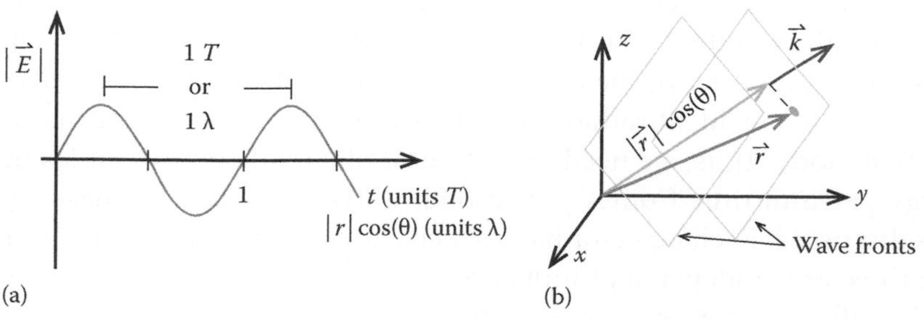

FIGURE 2.1 (a) Graph of harmonic electric field wave as a function of time or position. (b) Schematic of a harmonic plane wave. The wave fronts are planes perpendicular to the propagation direction.

where $\vec{\varepsilon}_0$ is the electric field times length, so $\vec{\varepsilon}_0/r$ has units of electric field strength. This equation describes waves whose surface of constant electric field are spheres with the parameter r representing the distance from the center of the sphere. The equation describes a spherical wave source or sink with the choice of the sign "+" or "−" for sink or source, respectively. The amplitude of a spherical wave must fall off with $1/r$ to conserve energy. The energy of any wave is proportional to the amplitude of the wave squared, and the surface area of a sphere is proportional to r^2. Thus, the energy is spread over the spherical wave fronts, and this area is larger for the wave front farther from the center. Notice this description is not physically meaningful at the center, $r = 0$. A different mathematical model would be needed for this region.

These basic harmonic descriptions for a light wave provide the starting equations for many sensing applications and are referred to in sections that follow. The discussion now turns to a description of radiometry, the measurement of energy delivered by electromagnetic (EM) radiation. Basic light-based techniques used to measure physical properties are then described such as interferometry, holography, fringe projection, and spectroscopy. Simple implementations and examples are presented to provide a foundational understanding of the technique. References are used to direct the reader to advanced discussions.

2.3 Radiometry

Sometimes, the absolute energy detected by a sensor is not important; rather, a sensor is used to measure relative changes in a signal. However, it is always advisable to go through a first-order radiometry analysis of the measurement to be sure the measured power or energy will be at an acceptable level compared to the saturation and noise floor levels of the detector.

Detailed discussions of the basic units of radiometry can be found in several introductory books on optics (Mouroulis 2002, Pedrotti 2012). There are four related quantities important for optical sensors. They are radiant energy Q [J], radiant flux ϕ [W], irradiance ε [W/m^2], and radiant intensity, I [W/sr]. The descriptors and symbols vary (e.g., irradiance is sometimes incorrectly called intensity); therefore, it is best to look to the units to confirm the physical meaning of a term. Radiometric quantities always correspond to the absolute energy, unlike photometry units that include the response of the human eye (Mouroulis 2002, Pedrotti 2012). Table 2.1 summarizes the quantities, the units, and important connections.

For sensors, we are concerned with the detection of EM energy. The output of a detector is usually a voltage that is proportional to the incident radiant energy such as a photodiode operating in the photovoltaic mode or the output is a current that is proportional to the incident radiant flux such as a photodiode operating in the photoconductive mode. Thus, we need units to describe total energy (joules in SI units) and energy per unit time (joules per second, watts, in SI units). These are the first two units discussed. Each detector has a specific area over which incident radiation is detected; thus, area-independent units are also useful, which can then be integrated over the specific detector area. This quantity is the radiant flux per unit area, called the irradiance (watts per unit area). Another common situation is to collect radiation in the far field from a localized source. This could be the collection of radiation

Table 2.1 Radiometry Quantities, Units, and Relationships

Radiometric Quantity	Units	Connection to Related Quantity
Radiant energy, Q	[J]	
Radiant flux, ϕ	[W]	$Q = \int_0^t \phi(t)dt$
Irradiance, ε (radiant flux density)	[W/m^2]	$\phi(t,\omega) = \int_{\text{detector effective area}} \varepsilon(t,\omega)dA$
Radiant intensity	[W/sr]	$\phi(t,\omega) = \int_{\text{detector solid angle}} I(t,\omega)d\omega$

at the entrance pupil of an optical system or collection at a detector. The detector or entrance pupil is well described in this case by the solid angle subtended in units of steridians. The radiant intensity (units of watts per steridian) is the best unit to use in this case.

2.3.1 Radiant Energy (Q) and the Radiant Flux (ϕ)

Radiant energy is the total energy delivered by an EM field. Light is always moving, so the total energy is the integration over time of the incident radiant flux,

$$Q = \int_0^t \phi(t)dt \tag{2.3}$$

where $\phi(t,\omega)$ is the radiant flux incident on the detector area. The radiant flux could vary with time and is a function of the frequency content, that is, the wavelength content, of the light. The wavelength content is important because detectors have unique wavelength dependencies, and this will impact the interpretation of the output.

The radiant flux is the radiant energy collected by the detector per unit time. When the radiation field is larger than the detector area, the flux collected depends on the area of the detector. As illustrated in Figure 2.2a and b, the radiant flux is best calculated as an integration of the incident radiation over the detector area. The incident irradiance and the incident radiant intensity are the two quantities commonly used to describe the incident radiation that is integrated.

2.3.2 Irradiance (ε)

The irradiance is the most common unit used to characterize the energy in an EM wave. It is often incorrectly referred to as the intensity. The irradiance is the radiant flux per unit area with units of W/m^2—it is the radiant energy per unit time per unit area

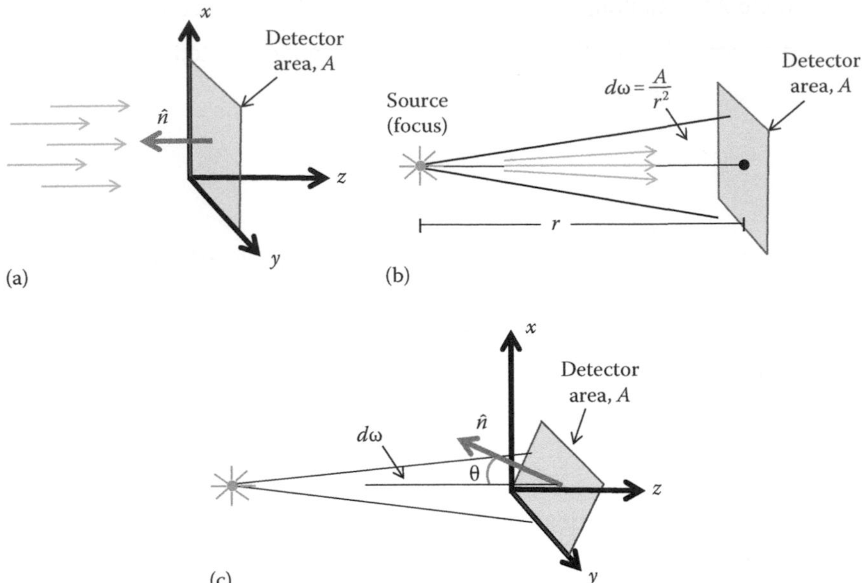

FIGURE 2.2 (a) Schematic of incident radiation on a detector of area A. (b) Schematic of the solid angle geometry from a point source to a detector of area A. (c) Schematic of solid angle dependence on detector orientation.

perpendicular to the flow of the radiation field. Thus, the radiant flux passing a region in space like a detector is the integration of the irradiance over the detector area,

$$\phi(t,\omega) = \int_{\text{detector area}} \varepsilon(t,\omega)dA. \tag{2.4}$$

The irradiance may vary with time and, in general, will be different for different optical frequencies, ω, in the light field. In many cases, the irradiance is constant over the area of the detector, and the radiant flux is easily calculated as

$$\phi(t,\omega) = \varepsilon(t,\omega)A \tag{2.5}$$

where A is the area of the detector. The total radiant energy can be calculated by subsequently integrating the radiant flux over time and optical frequency.

If the detector is at an angle to the incident field, the effective area in the direction of the incident field must be used in Equations 2.2 and 2.3. For a flat detector plane with area A, at an angle θ to the propagation direction of the light, the effective detector area is $A\cos(\theta)$ (see Figure 2.2c). For a field with a uniform irradiance and a detector at an angle θ then, Equation 2.3 becomes

$$\phi(t,\omega) = \varepsilon(t,\omega)A\cos(\theta). \tag{2.6}$$

Thus, a detector directly facing the incident radiation will correspond to $\theta = 0$, the effective detector area will be the full area A, and the maximum radiant flux will be collected. If the detector is 90° to the incident radiation, the effective area is zero, and the detector collects no radiation.

2.3.3 Radiant Intensity (*I*)

The radiant intensity is a convenient unit to use when the radiation field is a spherical wave diverging from a point, as shown in Figure 2.2b. Radiant intensity is similar to irradiance, but it is the radiant flux per unit solid angle rather than per unit area. The detector area is represented as a solid angle with this geometry. Thus, the radiant flux collected by a detector would be

$$\phi(t,\omega) = \int_{\text{detector solid angle}} I(t,\omega)d\omega. \tag{2.7}$$

where the integral is over the full solid angle of the detector, and $d\omega$ is the differential solid angle to the detector. The solid angle of a region on a sphere is defined as the area of the region on the surface divided by the radius squared, $d\omega = A/r^2$ (Figure 2.3). The detector is usually small compared to the distance between the detector and the source, and the radiant intensity is usually uniform over the solid angle of the detector. In this limit, the area used to calculate the solid angle is approximately flat (like the surface of a detector), the total solid angle of the detector Ω becomes A/r^2, and we have

$$\phi(t,\omega) = I(t)\Omega = \frac{I(t)A\cos(\theta)}{r^2} \tag{2.8}$$

where θ is the detector angle as shown in Figure 2.2c.

2.3.4 Connection between Radiometry and EM Wave Properties

Sensors measure the collected energy; thus, a fundamental understanding of the connection between energy and the parameters of the light wave (e.g., amplitude, polarization, and phase) is critical—after all, it is the wave parameters that are modified by the environment and therefore carry the information to be sensed. Fundamentally, the energy is related to the amplitude of the electric and magnetic fields. Sometimes, the energy of the field is directly modified by the environment and used to sense something of interest. More often, the principle of superposition is exploited where the energy of the *superposition* of waves is measured. The principle of superposition states that the total wave value

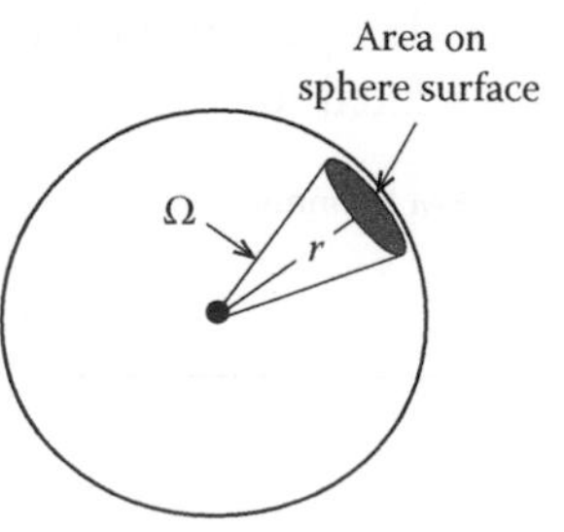

FIGURE 2.3 Geometry defining the solid angle.

at a region in space is the sum of the individual wave values present. The wave addition depends on detailed properties of each wave, like phase and polarization state; thus, the amplitude of the superposition carries information about changes in these other wave properties. This allows us to consider any of these wave properties as the parameter that will carry the information to be sensed. It is important to understand the connection between the amplitude of an EM wave and the radiometric quantities, where this amplitude may or may not be the result of superposition.

As discussed earlier, light is the oscillation of the coupled electric and magnetic fields propagating at the speed of light (in vacuum). There is an energy density associated with an electric field and a magnetic field, which is an energy per unit volume. We will look at the connection between this and irradiance for a light wave in vacuum. For an electric field, the energy density in vacuum is given by $u_E = \varepsilon_0 E^2/2$ in SI units, where ε_0 is the permittivity of free space and E is the amplitude of the electric field. Similarly, energy is present in a magnetic field, and the energy density in vacuum is given by $u_B = B^2/2\mu_0$, where μ_0 is the permeability of free space and B is the magnitude of the magnetic field. The total energy density in the fields of a light wave, then, is the sum of two contributions, $u_{tot} = \varepsilon_0 E^2/2 + B^2/2\mu_0$. The speed of light in vacuum is $c = 1/\sqrt{\varepsilon_0 \mu_0}$, and E and B are related by $E = cB$ so we can simplify the total energy density to $u_{tot} = \varepsilon_0 E^2$.

The energy density stored in the E and B fields in a light wave is always flowing, in vacuum traveling at the speed of light in the direction of propagation. So, the first simplification is to convert energy per unit volume into an energy flux (joules per second) per unit area. Energy density multiplied by a volume is a total energy, so we consider a small cross-sectional area A, oriented perpendicular to the flow, and moving along with the wave. A volume $dV = A\ell$ is swept out in an infinitesimal time interval dt where $\ell = cdt$. Thus, the total energy contained in this small volume is $Q = u_{tot}dV = u_{tot}Acdt$ (Figure 2.4). The flux is the energy per unit time, $\phi = Q/dt = u_{tot}Ac$.

The flux density is the flux per unit area perpendicular to the flow, $\phi/A = Q/Adt = u_{tot}c$. Substituting in our expression for u_{tot} in terms of the E and B amplitudes, we have

$$\frac{\phi}{A} = \varepsilon_o E^2 c = \varepsilon_o c E^2. \tag{2.9}$$

The flux density shown in Equation 2.3 is not quite one of our radiometric quantities. Notice that the electric field value in this expression is the magnitude of the electric field vector at any position and any time—the value of E oscillates between $\pm E_0$, where E_0 is the amplitude of the electric field rather than the time-varying value (see Equation 2.1). Light waves in the visible portion of the spectrum have optical frequencies on the order or 10^{15} rad/s. Thus, if we imagine a fixed location (e.g., the surface of a detector), the

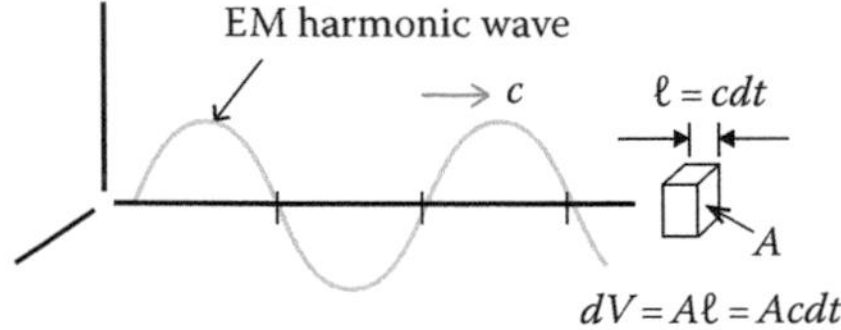

FIGURE 2.4 Determination of the flux density of an electromagnetic wave.

incident energy density is oscillating at a very high frequency as the light wave arrives and is detected. This frequency is far beyond the frequency response of the detector, so, in practice, the detector measures a time average of the incident flowing energy density. For a harmonic wave, the time average over one oscillation period is sufficient:

$$\langle \phi/A \rangle_T = \langle \varepsilon_o c E^2 \rangle_T = \frac{1}{T}\int_0^T \varepsilon_o c E^2 dt \tag{2.10}$$

Substituting in our expression for a harmonic electric field wave, Equation 2.1, we have

$$\langle \phi/A \rangle_T = \varepsilon_o c E_o^2 \frac{1}{T}\int_0^T \sin^2(\vec{k}\cdot\vec{r} - \omega t + \varphi_o)dt \tag{2.11}$$

where the time-independent factors, ε_0, c, and E_0 have been pulled out of the integral. The average of the $\sin^2(\theta)$ function over one full cycle is 1/2. Thus,

$$\langle \phi/A \rangle_T = \frac{1}{2}\varepsilon_o c E_o^2. \tag{2.12}$$

This is the practical quantity that captures what the detector can measure. It is much like Equation 2.9, but the time-varying value of the electric field, E, has been replaced with the amplitude of the electric field wave, E_0, with an additional factor of 1/2. This is what is called the irradiance:

$$\varepsilon = \langle \phi/A \rangle_T = \frac{1}{2}\varepsilon_o c E_o^2. \tag{2.13}$$

This is perhaps the most important equation of the chapter. The left side, the irradiance, is what we use to understand the response of the detector. The right side shows the connection to the fundamental property that we can measure, that is, the light field property, E_0. This is the magnitude of the resultant electric field wave at the detector. The value of E_0 may be directly impacted by the environment to provide a measure of the quantity to be sensed. Or more generally, E_0 may be the result of the superposition of two or more waves, whose properties like polarization, phase, and individual amplitudes impact the total value of E_0. It is these wave properties that then can be exploited and, with careful engineering, configured to vary with a quantity of interest in the environment. In this way, it provides a means of connecting the quantity measured, ε, to the quantity to be sensed, for example, displacement. The sections that follow provide basic examples of how this can be done.

2.4 Interferometry

Interferometry describes the general sensor category where the phase of an EM wave is used to sense a property of interest. The phase is connected to the measured irradiance by detecting the energy in the superposition of two waves. In the simplest implementation, a starting harmonic wave is divided into a test and a reference beam. The test beam interacts with a test part or the environment in a prescribed way so the phase of the test beam can be connected to the measurand. Both beams interfere at the detector, and the resulting irradiance of the superposition depends on the phase difference between the waves. This allows the phase or often a phase change of the test beam to be determined. The measurement is commonly implemented as a 0D or a 2D measurement. The 0D implementation is where the total irradiance of the combined beams is measured with a single detector. The 2D implementation is where the beams are expanded to have cross-sectional extent, and the interference between the beams vary over the cross-sectional region (Malacara 2007). The interference pattern across the beam is captured with a detector array where each pixel of the array measures the 0D total irradiance at this point.

Two basic interferometric configurations are shown in Figure 2.5, the Twyman–Green interferometer (a) and the Mach–Zehnder interferometer (b). The configurations start with a source that generates a light wave. For this introductory discussion, we assume the light is well described as a harmonic wave of optical frequency ω, such as light emitted by a laser. The harmonic wave propagates with speed v and can be described by

$$\vec{E}(s,t) = \vec{E}_o \sin(ks - \omega t + \varphi_o) \tag{2.14}$$

where all parameters have their meaning as described earlier, and we use the variable s to describe the optical path position along the propagation path. We define s to be 0 at the source (our origin), take the initial phase φ_0 to be zero, and use the variable s to represent the optical path length along the path, $s = n\ell$. The variable ℓ is the physical distance along the path, and n is the index of refraction of the medium in which the light travels. It may be air or an optical fiber, as examples. When light travels in a medium with an index of refraction n, the optical frequency remains the same, the speed is reduced by a factor of n, and as a consequence, the wavelength is reduced by a factor of n. Thus, the correct cycle behavior of the wave with position can be captured by using $k = 2\pi/\lambda_0$, where λ_0 is the wavelength in vacuum, and using an optical path length for the position variable, $s = n\ell$, so that $ks = 2\pi n\ell/\lambda_0$ captures the correct repeat length scale along the path.

The beam propagates through the system and reaches an initial amplitude-division beam splitter that divides the energy into two beams, a test beam and a reference beam. The path geometries for the beams differ between the two configurations in Figure 2.5, but ultimately, both beams encounter a second pass through a beam splitter that divides

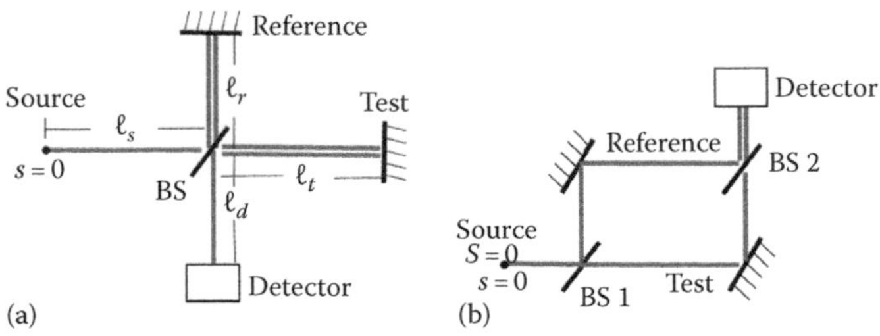

FIGURE 2.5 Basic Twyman–Green interferometer configuration (a). Basic Mach–Zehnder configuration (b).

each beam again and sends a portion of each toward a detector. The detector responds to the incident energy that now is the energy of the superposition of the test and reference beams. And as discussed earlier, the measured irradiance is the time-averaged energy flux density of the superposition of the two waves.

The resulting wave at the detector is written as

$$\vec{E}_{tot} = \vec{E}_r + \vec{E}_t = \vec{E}_{or}\sin(ks_r - \omega t) + \vec{E}_{ot}\sin(ks_t - \omega t) \tag{2.15}$$

where s_r and s_t are the optical path length coordinates at the detector for the reference and test beams, respectively. The time-averaged flux density, the irradiance ε, is given by

$$\varepsilon_{tot} = \langle \phi/A \rangle_T = \varepsilon_o c \left\langle \left| \vec{E}_{tot} \right|^2 \right\rangle_T = \varepsilon_o c \left\langle \vec{E}_{tot} \cdot \vec{E}_{tot} \right\rangle_T \tag{2.16}$$

This is the measured irradiance from the interference of the two beams. Expansion and simplification show the connection to the phase difference between the beams. First, consider the simplification of the magnitude squared of the resulting electric field term,

$$\left| \vec{E}_{tot} \right|^2 = \vec{E}_{tot} \cdot \vec{E}_{tot} = \left[\vec{E}_r + \vec{E}_t \right] \cdot \left[\vec{E}_r + \vec{E}_t \right] = \vec{E}_r \cdot \vec{E}_r + \vec{E}_t \cdot \vec{E}_t + 2\vec{E}_r \cdot \vec{E}_t \tag{2.17}$$

The first two terms are the magnitude squared of each of the two waves, and the contribution to the time average will be the sum of the irradiance for each, as we will see. The last term contains the information about the relationship between the term waves. This is called the interference term.

At this point, we make a simplifying assumption. Notice that the polarization (the electric field vector direction) information for each beam is expressed in the equation; that is, the electric field wave for each beam is written as $\vec{E}$ rather than E, the scalar approximation. In general, the electric field vector can point in any direction in the plane perpendicular to the propagation direction of the wave. Linear polarization is the state where the electric field vector oscillates along a line in the plane, and circular and elliptical polarizations correspond to the electric field vector sweeping out a circle or an ellipse in the plane, respectively (Hecht 2012). Circular or elliptical polarization states involve a phase difference between the two oscillating orthogonal components of the electric field in the plane, and the mathematical description becomes quite involved to carry out the next step with this most general treatment. We will keep the mathematics easier to follow, but allow the polarization role to be appreciated by assuming a simple polarization state. We will consider that each beam is linearly polarized; that is, the two components of the electric field are in phase. So the electric field in each wave oscillates along a line, but we allow the orientation of the line to be arbitrary.

We first focus on the last term in Equation 2.17, the interference term, and with this polarization assumption, we substitute in the expressions for the two waves and separate out the harmonic time-dependent factor,

$$2\vec{E}_r \cdot \vec{E}_t = 2\vec{E}_{or} \cdot \vec{E}_{ot}\sin(ks_r - \omega t)\sin(ks_t - \omega t) \tag{2.18}$$

Using a product-to-sum trigonometry identity, this can be written as

$$
\begin{aligned}
2\vec{E}_r \cdot \vec{E}_t &= 2\vec{E}_{or} \cdot \vec{E}_{ot}\frac{1}{2}\left[\cos\left((ks_r-\omega t)-(ks_t-\omega t)\right)-\cos\left((ks_r-\omega t)+(ks_t-\omega t)\right)\right] \\
&= \vec{E}_{or} \cdot \vec{E}_{ot}\left[\cos\left(k\Delta s\right)-\cos\left(ks_r+ks_t-2\omega t\right)\right]
\end{aligned}
\tag{2.19}
$$

where $\Delta s = s_r - s_t$. We are now ready to substitution back into the expression for the total irradiance, Equation 2.16, and we find

$$
\varepsilon_{tot} = \varepsilon_o c\left\langle \vec{E}_r \cdot \vec{E}_r + \vec{E}_t \cdot \vec{E}_t + 2\,\vec{E}_r \cdot \vec{E}_t \right\rangle_T = \varepsilon_o c\left\langle \vec{E}_r \cdot \vec{E}_r \right\rangle_T + \varepsilon_o c\left\langle \vec{E}_t \cdot \vec{E}_t \right\rangle_T + \varepsilon_o c 2\left\langle \vec{E}_r \cdot \vec{E}_t \right\rangle_T \tag{2.20}
$$

Substitution of Equation 2.19 for the last interference term leads to

$$
\varepsilon_{tot} = \varepsilon_o c\left\langle \vec{E}_r \cdot \vec{E}_r \right\rangle_T + \varepsilon_o c\left\langle \vec{E}_t \cdot \vec{E}_t \right\rangle_T + \varepsilon_o c\left\langle \vec{E}_{or} \cdot \vec{E}_{ot}\left[\cos(k\Delta s) - \cos(ks_r + ks_t - 2\omega t)\right]\right\rangle_T \tag{2.21}
$$

For the next simplification, we carefully consider possible time-varying dependencies in the interference term. The last cosine term shows that the only time dependence in the interference term is the last cosine function, so simplifying, we have

$$
\begin{aligned}
\varepsilon_{tot} &= \varepsilon_o c\langle \vec{E}_r \cdot \vec{E}_r \rangle_T + \varepsilon_o c\langle \vec{E}_t \cdot \vec{E}_t \rangle_T \\
&\quad + \varepsilon_o c\left[\langle \vec{E}_{or} \cdot \vec{E}_{ot} \rangle_T \cos(k\Delta s) - \left\langle \vec{E}_{or} \cdot \vec{E}_{ot} \cos(ks_r + ks_t - 2\omega t)\right\rangle_T\right]
\end{aligned}
\tag{2.22}
$$

We are now ready to evaluate the time average terms. The first two are the irradiance contribution from each of the two waves. The last time average term is zero—the time average of a cosine function oscillating at a frequency 2ω over a time interval of $T = 2\pi/\omega$ is zero. Therefore, Equation 2.22 becomes

$$
\varepsilon_{tot} = \varepsilon_r + \varepsilon_t + \varepsilon_o c\vec{E}_{or} \cdot \vec{E}_{ot} \cos\left(k\Delta s\right) \tag{2.23}
$$

This is a very important result. When the two beams are incident on a detector, the measured irradiance is that given by Equation 2.22, which is the sum of the irradiance of each wave plus an interference term. The interference term depends on two important aspects: the relative orientation of the linear polarization states of each wave, $\vec{E}_{0r} \cdot \vec{E}_{0t}$, and the optical path length *difference* between the two waves, $\Delta s = s_r - s_t$.

Consider the first aspect, the polarization state difference between the waves. If the two waves have the same polarization state, the vectors $\vec{E}_{0r}$ and $\vec{E}_{0t}$ are aligned, and the dot product becomes $E_{0r}E_{0t}$. In the other extreme, the polarization states are orthogonal, that is, $\vec{E}_{0r}$ and $\vec{E}_{0t}$ are always perpendicular, and $\vec{E}_{0r} \cdot \vec{E}_{0t} = 0$. In this limit, the entire interference term is zero, and the irradiance at the detector contains no information about a detailed difference between the waves.

The other aspect of the interference term is the optical path difference between the waves. This term will change if, for example, the test beam travels an extra distance or travels through a region where n changes. A sensor may exploit either one of these aspects. As an example, one could sense temperature with an interferometer by exploiting the temperature sensitivity to optical path length in an optical fiber. Imagine a fiber-based Mach–Zehnder configuration like Figure 2.5, where a segment of the fiber in the test path passes through a region for which the temperature is to be sensed. The path length difference in the interference term $k\Delta s$ contains a term $kn_s\ell_s$, which is the optical path length contribution from this sensing segment of the fiber. Both n_s and ℓ_s are sensitive to temperature, and the product $n_s\ell_s$ changes significantly on the length scale of λ_0 with changes in temperature, thus changing the total irradiance output of the sensor with temperature.

For the case where the polarization states of the two waves are equal, the dot product in Equation 2.23 becomes $E_{0r}E_{0t}$, and the interference part can be written in terms of the irradiance value of each beam, that is

$$\varepsilon_{tot} = \varepsilon_r + \varepsilon_t + 2\sqrt{\varepsilon_r\varepsilon_t}\cos(\delta) \tag{2.24}$$

The phase difference between the waves is often written as δ, where $\delta = k\Delta s$ from Equation 2.23. Figure 2.6 shows the behavior of the total irradiance as a function of the phase difference. The average value is the sum of the irradiance from each wave, $\varepsilon_r + \varepsilon_t$, and the interference term adds a modulation with an amplitude given by $2\sqrt{\varepsilon_r\varepsilon_t}$. The interference oscillations are often called *fringes*, and the visibility is quantified as the ratio, $F \equiv 2\sqrt{\varepsilon_r\varepsilon_t}/\varepsilon_r + \varepsilon_t$. It depends on the relationship between ε_r and ε_t. Maximum visibility is one when $\varepsilon_r = \varepsilon_t$, and this is graphed in Figure 2.6b. This condition potentially leads to the maximum variation of the total irradiance with phase difference between the waves.

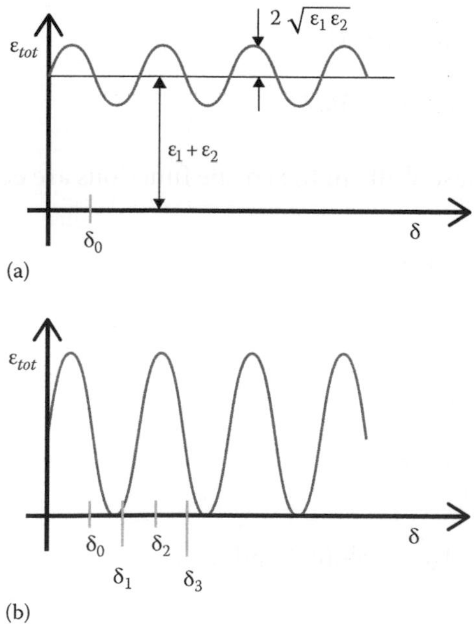

FIGURE 2.6 Irradiance dependence on phase difference of the superposition of two waves with unequal amplitudes (a). Irradiance dependence on phase difference with equal amplitudes (b).

This is the ideal limit when designing a sensor—design the configuration such that both beams have comparable irradiances when they reach the detector. If one beam carries significantly more energy, the interference effect is swamped by a large background. It is also important that the maximum irradiance is well above the noise threshold and below the saturation level of the detector.

With a single measurement, the behavior of Equation 2.24 would have to be known to extract the value of δ and therefore the value of the property to be sensed. In practice, the absolute irradiance is difficult to calculate or measure, and likely it will drift with time. This can be circumvented by taking a series of measurements with intentional changes in δ to effectively map out the behavior of Equation 2.24 and extract δ_0 for a particular configuration. Chapter 14 discusses modulation techniques in detail, and here we describe a simple means by which this can be done.

Consider engineering into the measurement a means of increasing the path length for the reference beam on the order of the wavelength of the light. This can be done by attaching the reference mirror in Figure 2.5a to a piezoelectric transducer that extends or contracts in the direction of the beam. A Twyman–Green configuration has free-space beam propagation here and a backward shift of the reference mirror by a distance $\lambda/8$ and introduces a path length increase of $s = \lambda/4$ (because the beam travels the additional distance once before it reflects and then again after reflection). This path length change results in a phase change between the test and reference beams at the detector of $\delta_1 = \delta_0 + \pi/2$, a 90° phase shift. Consider repeating this for additional 90° phase shifts so we have

$$\begin{aligned}
\varepsilon_{o,tot} &= \varepsilon_r + \varepsilon_t + 2\sqrt{\varepsilon_r\varepsilon_t}\cos(\delta_o) \\
\varepsilon_{1,tot} &= \varepsilon_r + \varepsilon_t + 2\sqrt{\varepsilon_r\varepsilon_t}\cos(\delta_o + \pi/2) \\
\varepsilon_{2,tot} &= \varepsilon_r + \varepsilon_t + 2\sqrt{\varepsilon_r\varepsilon_t}\cos(\delta_o + \pi) \\
\varepsilon_{3,tot} &= \varepsilon_r + \varepsilon_t + 2\sqrt{\varepsilon_r\varepsilon_t}\cos(\delta_o + 3\pi/2)
\end{aligned} \tag{2.25a}$$

These subsequent 90° phase shifts in the cosine functions are equivalent to

$$\begin{aligned}
\varepsilon_{o,tot} &= \varepsilon_r + \varepsilon_t + 2\sqrt{\varepsilon_r\varepsilon_t}\cos(\delta_o) \\
\varepsilon_{1,tot} &= \varepsilon_r + \varepsilon_t - 2\sqrt{\varepsilon_r\varepsilon_t}\sin(\delta_o) \\
\varepsilon_{2,tot} &= \varepsilon_r + \varepsilon_t - 2\sqrt{\varepsilon_r\varepsilon_t}\cos(\delta_o) \\
\varepsilon_{3,tot} &= \varepsilon_r + \varepsilon_t + 2\sqrt{\varepsilon_r\varepsilon_t}\sin(\delta_o)
\end{aligned} \tag{2.25b}$$

We see that $\varepsilon_{3,tot} - \varepsilon_{1,tot} = 4\sqrt{\varepsilon_r\varepsilon_t}\sin(\delta_0)$ and $\varepsilon_{0,tot} - \varepsilon_{2,tot} = 4\sqrt{\varepsilon_r\varepsilon_t}\cos(\delta_0)$, thus

$$\frac{\sin(\delta_o)}{\cos(\delta_o)} = \tan(\delta_o) = \frac{\varepsilon_{3,tot} - \varepsilon_{1,tot}}{\varepsilon_{0,tot} - \varepsilon_{2,tot}} \tag{2.26}$$

and therefore

$$\delta_o = \tan^{-1}\left[\frac{\varepsilon_{3,tot} - \varepsilon_{1,tot}}{\varepsilon_{0,tot} - \varepsilon_{2,tot}}\right] \tag{2.27}$$

Such a measurement procedure eliminates the need to know absolute irradiance values and arrives at δ_o by exploiting the harmonic nature of the interference equation. This modulation scheme is called the *4-bucket algorithm*. Many other algorithms and modulation techniques exist, each designed to minimize different errors. An overview of discrete bucket algorithms is in reference (Malacara 2007). In practice, continuous modulation techniques rather than discrete bucket algorithms are used with 0D interferometric measurements like the configuration discussed here. A 0D measurement is one in which the total irradiance of the two interfering beams is measured by a single detector. This is in contrast to a configuration in which an optical beam is expanded to cover a cross-sectional area to do pixel-by-pixel 2D interferometry over the cross-sectional area of the beam. Detector arrays measure the 0D interference irradiance at each pixel, and they have a much lower frequency response than a typical single 0D detector. A large array of irradiance measurements with a low-frequency response lead to a practical need to choose discrete phase step algorithms to extract δ_o at each pixel.

A more detailed discussion of interferometry is provided in Chapter 5.

2.5 Holography

Holography is a phenomenon based on interference and is used in many sensing applications. We are perhaps most familiar with it in an entertainment context where we observed the intriguing virtual 3D optical image of an object when a hologram is illuminated with a laser beam. A hologram is made by illuminating an object with a laser beam and exposing a recording media with the interference of the scattered light from the object and a coherent reference beam. The basic configuration is shown in Figure 2.7a. We can understand the scattered light from the object as a collection of spherical waves on the surface of the object. An example of one such spherical wave is

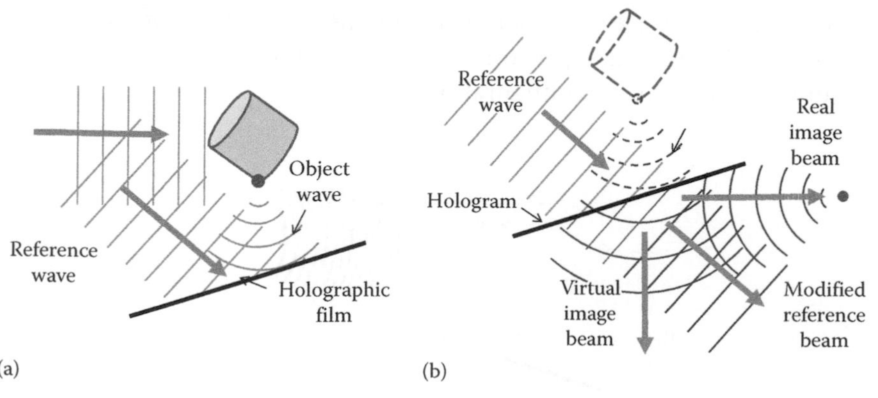

FIGURE 2.7 Schematic of the making of a hologram (a). Schematic of illuminating the hologram (b).

shown in the figure. When the reference and object beams are coherent, the superposition of the two creates a stable interference irradiance pattern, albeit complicated, on the holographic recording film. When developed, the film has a permanent signature of this interference pattern. When the developed film is illuminated with the same reference beam, the beam scatters and diffracts from the pattern (Figure 2.7b), resulting in specific scattered beams, one of which is a light wave identical to the scattered wave that left the object in the initial recording, a wave that leaves the back side of the film in the configuration shown as though simply continuing the propagation away from the object. This means that when this wave enters another optical system, like the eye, the same image will be formed as if the optical system were forming an image with light leaving the object itself.

Holographic sensors and holographic interferometry (Jones 1989) are two ways in which holograms are used for measurement. Holographic sensors take advantage of the sensitivity of diffraction to the detailed pattern of fine structure in the hologram causing the diffraction. The hologram recording medium is engineered to be sensitive to a chemical or substance to be sensed, meaning the hologram physically changes its optical properties if exposed to the substance. Even a slight change in the optical properties of the hologram causes the diffraction from this detailed recorded interference pattern to significantly change. The change in the diffraction pattern is then measured and connected back to the chemical or substance exposure that caused the change. The simplest implementation of holographic interferometry is double exposure of a hologram with an object in a reference state and the object in a new perturbed state to be measured. With double exposure, each of the recorded interference patterns on the film causes the corresponding object wave reconstruction when the hologram is played. Thus, for example, illumination of a double-exposure hologram with the reference wave will give rise to two virtual image beams (Figure 2.7b). One beam is the virtual image of the reference object, and another beam is the virtual image of the perturbed object. These two waves can be imaged, and the coherence between the two gives rise to an interference pattern superimposed on the image of the object. This additional interference pattern captures only the difference between the object in the reference state and the perturbed state. Thus, it provides a measure of the perturbation. It can be used to measure geometric deformation due to things like applied stress or temperature.

Scattered light from an object is well modeled as a collection of point source spherical waves leaving the surface of the object. The mathematical basis of holography can be developed by considering the superposition of one such spherical wave with a reference beam. Figure 2.8 shows such a configuration.

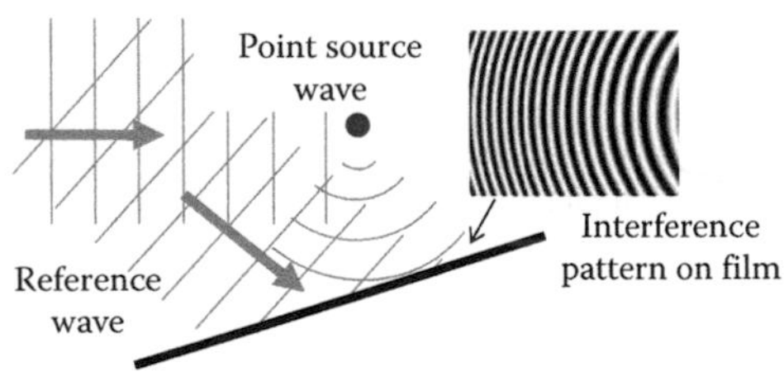

FIGURE 2.8 Model of hologram formed with reference wave and single point source.

The spherical wave in the figure is the object wave, E_o, and the collimated wave is the reference, E_r. The mathematical analysis is best understood using the complex description of each wave (Hecht 2012, Pedrotti 2012). The complex description for the electric field wave given by Equation 2.1 is written as

$$\tilde{E}(\vec{r},t) = E_o e^{i(\vec{k}\cdot\vec{r}-\omega t+\varphi_o)} \tag{2.28}$$

The amplitude of the wave is given by E_o and the phase by $\vec{k}\cdot\vec{r}-\omega t+\varphi_0$ The irradiance is still calculated as $\varepsilon = \varepsilon_0 c E_o^2/2$, where E_o is the amplitude of the wave, and the amplitude squared is calculated from the complex description by $E_o^2 = \tilde{E}\tilde{E}^*$. The complex description can be used in all analyses for the wave and is often more convenient, involving fewer trigonometric manipulations. Thus, we write a complex description for the object and reference waves for our holographic analysis as

$$\tilde{E}_o(x,y) = E_{oo}(x,y)e^{i(\varphi_o(x,y)-\omega t)} \tag{2.29}$$

$$\tilde{E}_r(x,y) = E_{or}e^{i(\varphi_r(x,y)-\omega t)} \tag{2.30}$$

These are the descriptions of the waves in the plane of the hologram. The superposition of the two is given by $\tilde{E}_{tot}(x,y) = \tilde{E}_o(x,y) + \tilde{E}_r(x,y)$ and the irradiance that exposes the holographic media as $\varepsilon_{tot} = C\tilde{E}_{tot}^2$ where $C = \varepsilon_o c/2$. That is,

$$\varepsilon_{tot} = C(\tilde{E}_o + \tilde{E}_r)(\tilde{E}_o + \tilde{E}_r)^* \tag{2.31}$$

which is

$$\varepsilon_{tot}(x,y) = C\tilde{E}_o\tilde{E}_o^* + C\tilde{E}_r\tilde{E}_r^* + C\tilde{E}_r\tilde{E}_o^* + C\tilde{E}_r^*\tilde{E}_o \tag{2.32}$$

This is the complex description of the irradiance resulting from the interference between two waves and is equivalent to Equation 2.24. The first two terms are the irradiance contribution from each wave independently, and the last two terms contain the interference contribution. The 2D irradiance pattern in the plane of the hologram that corresponds to the interference of the spherical wave and the plane wave in this example is a series of slightly curved and closely spaced lines, as shown in the inset. Exposure of the holographic film to this pattern and subsequent development lead to grayscale pattern that follows this irradiance pattern. For visible light holograms, the grayscale pattern is extremely fine and not visible to the eye.

The holographic image is generated by illuminating the hologram with the same reference beam. The transmittance of the hologram can be written as

$$T = D\varepsilon_{tot}(x,y) \tag{2.33}$$

where D is a proportionality constant, that is, the transmission properties of the hologram emulate the exposure irradiance pattern. The transmittance then is a function of the interference between the waves. The effect of the transmittance function is to modulate the reference beam to produce the reconstructed wave, E_f, mathematically described by

$$\tilde{E}_f = T\tilde{E}_r = D\left(C\tilde{E}_o\tilde{E}_o^* + C\tilde{E}_r\tilde{E}_r^* + C\tilde{E}_r\tilde{E}_o^* + C\tilde{E}_r^*\tilde{E}_o\right)\tilde{E}_r \tag{2.34}$$

Multiplying this out gives

$$E_f = (DC\tilde{E}_o\tilde{E}_o^*)\tilde{E}_r + (DC\tilde{E}_r\tilde{E}_r^*)\tilde{E}_r + DC\tilde{E}_r\tilde{E}_o^*\tilde{E}_r + (DC\tilde{E}_r^*\tilde{E}_r)\tilde{E}_o \tag{2.35}$$

The first two terms are amplitude-modified versions of the reference beam. They are the same reference wave, $\tilde{E}_r$, with amplitudes modified by factors $(DC\tilde{E}_o\tilde{E}_o^*)$ and $(DC\tilde{E}_r\tilde{E}_r^*)$. The last two terms are the holographic waves of interest.

The last term is the same as the object wave, $\tilde{E}_o$, differing only by an overall scale factor, $DC\tilde{E}_r\tilde{E}_r^*$. This means when the reference beam hits the hologram, one of the waves generated is identical to the scattered wave that left the object in the initial recording (aside from overall scale). This is the wave that appears as a simple continuation of the object beam beyond the hologram and can be imaged by a subsequent optical system like the eye, resulting in a final image that looks identical to the original object. This beam is a virtual image from the hologram, serving as a virtual object for the subsequent imaging system.

The second to last term is $DC\tilde{E}_r\tilde{E}_0^*\tilde{E}_r$. Substituting in the detailed description results in

$$DC\left(E_{or}e^{i(\varphi_r(x,y)-\omega t)}\right)\left(E_{or}e^{i(\varphi_r(x,y)-\omega t)}\right)\left(E_{oo}(x,y)e^{-i(\varphi_o(x,y)-\omega t)}\right) \tag{2.36}$$

This simplifies to $DCE_{0r}^2e^{i2\varphi_r(x,y)}E_{00}e^{i(-\varphi_0(x,y)-\omega t)}$. Notice the similarities and differences from the original object wave, $\tilde{E}_0(x,y) = E_{00}(x,y)e^{i(\varphi_0(x,y)-\omega t)}$. It is similar but differs in both amplitude and phase. The phase modification, $e^{i2\varphi_r(x,y)}$, is a term from the phase profile of the reference beam across the plane of the hologram. The value comes from the geometry of the plane wave fronts of the reference wave and the angle of the hologram. The effect of this term is an angular displacement of this wave. The other phase term is the same value but of the opposite sign of the original object wave, $e^{i(-\varphi_0(x,y))}$. This is a phase reversal of the original wave, meaning the original diverging object wave is now converging, focusing to a point and forming a real image. When the analysis is extended to a collection of point sources leaving an extended object for the hologram exposure, this last wave ends up leading to a real image of the object, but the object appears turned *inside out*, that is, points on the object that were farther away appear closer. The angular displacement of each point source also means the image displays what is termed *reverse parallax*, meaning moving the eye from left to right when looking at the real image is the same as moving the eye from right to left when looking at the virtual image (or the original object).

The recording of a hologram requires demanding experimental conditions. The interference pattern of the scattered object beam and the reference beam at the holographic film is a very fine detailed pattern, and it must be steady over the exposure time of the film. This means vibration must be low and the film must have sufficient high resolution to preserve the details of an object. A typical holographic film emulsion has a resolution of about (3000–5000) lines/mm (i.e., Agfa-Gevaert 8E75 HD plates, 5000 lines/mm, which is sensitive to red laser light). Photographic film emulsions are, however, not the only medium used to record holograms. Other photosensitive materials exist and are used in scientific experiments. They include photochromic thermoplastics, crystals such as lithium niobate and barium titanate, and other electro-optical media. The resolution of crystals is much lower than that of photographic emulsions, but the crystals operate in real time and do not have to be developed as do emulsions.

2.6 Fringe Projection

The term fringe projection represents the general category of measurements whereby an object's height profile can be determined by observing a distortion of an optical pattern when it is projected onto or reflected from an object (Structured-Light 3D Scanner 2013). It is a measurement that is based on irradiance patterns in a light wave and image formation, rather than on phase or frequency properties. It is a diverse and sophisticated field and is used to measure things from solder bump geometry in integrated circuit packaging (Li et al. 2013), tooth surface geometry in dentistry (Zhang and Alemzadeh 2007), component geometry for the automotive industry (Notni et al. 1999), and so on.

A simple demonstration is illustrated in Figure 2.9. A common optical pattern used is a sinusoidal grayscale series of lines, as shown in the figure. It might be generated with an LCD screen, for example. An optical system forms an image of this irradiance pattern on the surface of the object to be measured. The optical pattern on the object in turn serves as a new object to be imaged by a second optical system, as shown. Consider a simple implementation where the optical systems are telecentric, that is, their magnifications are independent of object and/or image plane locations. This means the pattern projected onto the object or imaged from the object is simply a straight line-of-light projection. This is shown in Figure 2.9b. The black array of dots represents the irradiance peaks in the periodic projected pattern in the projection arm, and the blue dots show

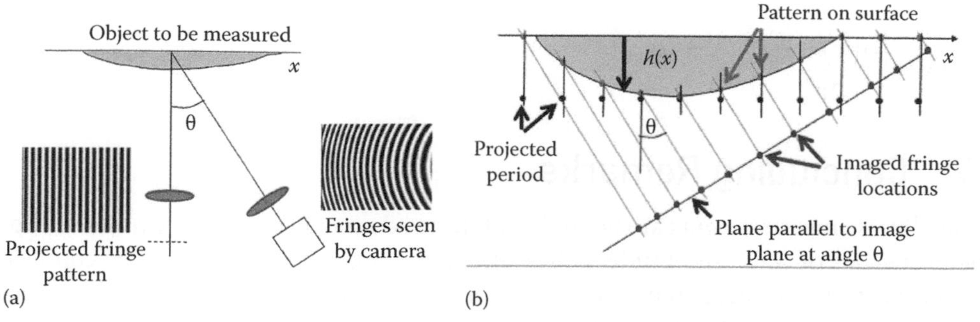

FIGURE 2.9 Diagram of basic fringe projection measurement configuration.

where these peaks intersect with the object's surface. These are the locations where the incident irradiance scatters and is therefore imaged in the observation direction. A telecentric optical system for observation means the irradiance peaks will be imaged as line-of-sight projections in the observation direction. This is shown as the red dots in Figure 2.9b. The observed spacing is no longer uniform. The peaks are shifted, and the amount of the shift is due to a geometric combination of the height profile of the object and the geometry of the setup.

The amount of pattern shift is measured, and this in turn is converted to the object's height profile. An analytic expression that connects the two is a function of the geometry of the configuration and is not trivial, even with telecentric optical systems. Systems are usually calibrated with an object of known height profile to empirically determine a conversion factor. This works well for many applications, but breaks down in the limit of precision measurements. Rigorously, the conversion factor depends on the object height and field of view location, thus calibration with the measurement of a known object at a single location will only yield an approximate conversion factor.

The amount of the line shift is determined with phase-shifting techniques like that used in interferometry. The projected pattern is a sinusoid, and this is the same irradiance profile as the interference of two waves. The shade of gray in the projected pattern at each location corresponds to a particular phase value of the sinusoid. With a series of projected sinusoidal lines, the phase varies linearly in the direction across the lines and is constant along the lines. Furthermore, sinusoidally varying the projected values in time results in a sinusoidal variation of the signal at each location on the object and then at each pixel in the camera. This is seen as the lines moving across the measurement field. The starting shade of gray corresponds to an initial phase offset at each location. The shift of a line due to object height means the phase no longer varies linearly over the measurement field. The same phase-shifting strategies discussed in the interferometry section can be used to extract this phase offset at each pixel. In interferometry, the phase shift is often done by changing the path length for one of the waves. In fringe projection, the irradiance pattern used for the projection is directly changed by, for example, sending a new array of irradiance values to each pixel on the projection LCD screen. For example, the 4-bucket algorithm with 90 phase shifts (Equation 2.25b) can be used and the phase offset, $\phi(x)$, calculated following the procedure that led to Equation 2.27. The calibration process yields the conversion factor between phase offset and height. The factor is usually expressed as an effective wavelength, λ_{eff}, and the height h, as seen at Figure 2.9b, is calculated from

$$h(x) = \frac{\phi(x)\lambda_{eff}}{2\pi} \tag{2.37}$$

2.7 Concluding Remarks

Many diverse measurements are possible by taking advantage of the wave nature of light. The properties of wave amplitude, polarization, frequency, and phase can be connected to a property to be measured in a variety of ways. Detectors fundamentally measure the energy in the light field, and the energy is connected to the property of interest by engineering the way in which the light interacts with the environment. Building from a foundational discussion of radiometry and the mathematical description of light, this chapter covers underlying principles and configurations of simple optical measurements. The ways in which light can be used is wide ranging, and introductory discussions of basic interferometry, holography, and fringe projection are used to illustrate this diversity.

References

Ball, D.W. 2011. *The Basics of Spectroscopy*. Bellingham, WA: SPIE—The International Society for Optical Engineering.

Geng, J. 2011. Structured-light 3D surface imaging: A tutorial. *Advances in Optics and Photonics* 3: 128–160. doi:10.1364/AOP.3.000128.

Hecht, E. 2012. *Optics*. New Delhi, India: Pearson.

Jones, R.W.C. 1989. *Holographic and Speckle Interferometry*. Cambridge, U.K.: Cambridge University Press.

Li, Z., Kang, Y.C., Moon, J. et al. 2013. The optimum projection angle of fringe projection for ball grid array inspection based on reflectance analysis. *International Journal Advanced Manufacturing Technology* 67: 1597–1607.

Malacara, D. 2007. *Optical Shop Testing*. Hoboken, NJ: John Wiley & Sons.

Mouroulis, P.M.J. 2002. *Geometrical Optics and Optical Design*. New York: Oxford University Press.

Notni, G., Schreiber, W., Heinze, M. et al. 1999. Flexible autocalibrating full-body 3D measurement system using digital light projection. *Proceedings of the SPIE - The International Society for Optical Engineering*, 3824: 79–88.

Pedrotti, F. 2012. *Introduction to Optics*. New Delhi, India: Pearson.

Structured-Light 3D Scanner. 2013. Available from: http://en.wikipedia.org/wiki/Structured-light_3D_scanner.

Tompkins, H.G.I.E.A. 2005. *Handbook of Ellipsometry*. William Andrew Pub, Norwich, New York; Springer, Heidelberg, Germany. Available from: http://site.ebrary.com/id/10305511.

Workman, J.S.A.W. 1998. *Applied Spectroscopy: A Compact Reference for Practitioners*. San Diego, CA: Academic Press.

Zhang, L. and Alemzadeh, K. 2007. 3-dimensional vision system for dental applications. Annual International Conference of the IEEE Engineering in Medicine and Biology, *29th Annual International Conference of IEEE-EMBS, Engineering in Medicine and Biology Society, EMBC'07*.

II

Optical Measurement Principles and Techniques

3. Optical Waveguide Sensors

Bishnu P. Pal
Indian Institute of Technology Delhi

Handbook of Optical Sensors. Edited by José Luís Santos and Faramarz Farahi © 2015 CRC Press/Taylor & Francis Group, LLC. ISBN: 9781439866856.

3.1 Introduction

3.1.1 History

Optical waveguide, by definition, is a central region called the core of certain refractive index, which is surrounded by either one medium of lower refractive index or two optical media, each of which has refractive index lower than that of the core. The former type is called a *symmetric* optical waveguide and the latter *asymmetric* optical waveguide. Due to the refractive index contrast, launched light into a waveguide could be confined within the core through total internal reflection. In its simplest form, a waveguide is planar in geometry. Optical fiber represents an optical waveguide in cylindrical geometry. These two variations are represented in Figure 3.1.

A planar waveguide is a 1D waveguide, in which refractive index is a function of only x, for example, and light is confined in one direction. On the other hand, 2D waveguides form the backbone of integrated optics, in which light is confined in two dimensions; few examples of 2D waveguides are shown in Figure 3.2.

A sensor typically converts a change in the magnitude of one physical parameter into a corresponding change in the magnitude of a more conveniently measurable second different parameter. For noncontact measurements and sensing, optical techniques have long played an important role in instrumentation and sensors. A substantial part of this

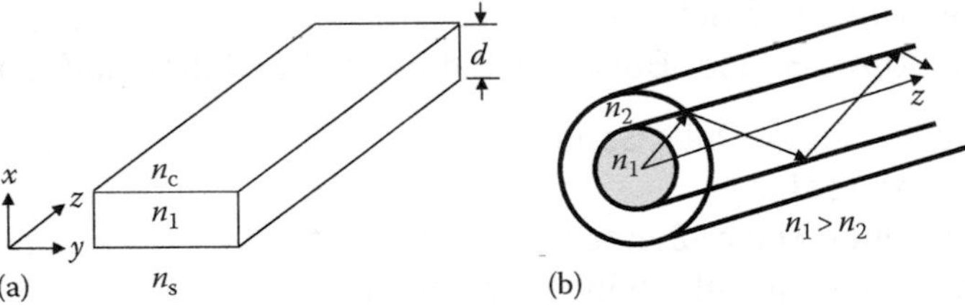

FIGURE 3.1 (a) The planar/slab waveguide geometry, here $n_1 > n_{c,s}$; for a symmetric waveguide, $n_c = n_s = n_2$; (b) optical fiber waveguide geometry.

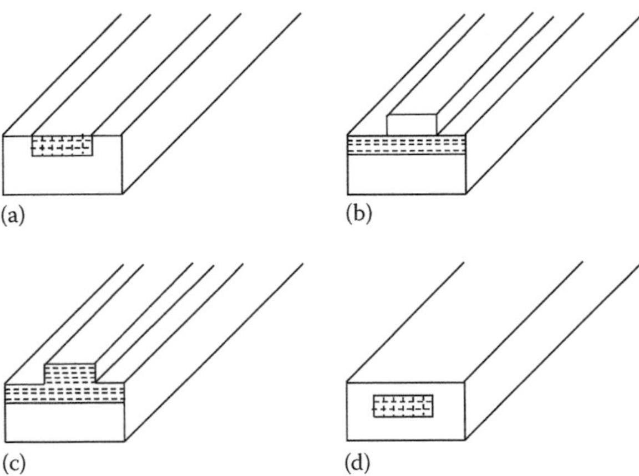

FIGURE 3.2 Examples of 2D optical waveguides: (a) embedded strip guide, (b) strip-loaded guide, (c) rib guide, and (d) immersed guide.

book addresses optical-fiber-based sensing with an emphasis on underlying principles, technologies, and applications, noting, however, that integrated optics platforms have also been extensively considered to perform the sensing functionality. In this chapter, we would essentially cover sensing principles and illustrative applications of fiber-optical waveguides, with details addressed in subsequent chapters.

In recent years, the scope of optical techniques in the area of instrumentation and sensors has made a quantum jump with the ready availability of low-loss optical fibers and associated optoelectronic components. The advantages and potentials offered by fiber-optical waveguides in sensing and instrumentation are many. Some of the key features of this new technology, which offer substantial benefits as compared to conventional electric sensors, are as follows: fiber being dielectric, sensed signal transported through it is immune to electromagnetic interference (EMI) and radio frequency interference, intrinsically safe in explosive environments, highly reliable and secure with no risk of fire/sparks, high-voltage insulation and absence of ground loops, and hence obviate any necessity of isolation devices like opto-couplers. Furthermore, a fiber has low volume and weight, for example, 1 km of 200 μm silica fiber would weigh only 70 g and occupies a volume of ~30 cm^2; as a point sensor, they can be used to sense normally inaccessible regions without perturbation of the transmitted signals. Fiber-optic sensors are potentially resistant to nuclear or ionizing radiations, and can be easily interfaced with low-loss telecom-grade optical fibers for remote sensing and measurements by locating the control electronics for LED/laser and detectors far away (could even be tens of kilometers away) from the sensor head. The inherent large bandwidth of a fiber offers the possibility of multiplexing a large number of individually addressed point sensors in a fiber network for distributed sensing that is, continuous sensing along the length of the sensing fiber at several localized points; these could be readily employed in chemical and process industries, health monitoring of civil structures, biomedical instrumentation, etc., due to their additional characteristics like small size, mechanical flexibility, and chemical inertness. These advantages were sufficient to attract intensive R&D effort around the world to develop fiber-optic sensors. This has eventually led to the emergence of a variety of fiber-optic sensors for accurate sensing and measurement of physical parameters and fields, for example, pressure, temperature, liquid level, liquid refractive index, liquid pH, antibodies, electric current, rotation, displacement, acceleration, acoustic, electric and magnetic fields, and so on. Initial developmental work had concentrated predominantly on military applications like fiber-optic hydrophones for submarine and undersea applications, and gyroscopes for applications in ships, missiles, and aircrafts. Gradually, a large number of civilian applications have also picked up. During the 1970s, the technology of optical fibers for telecommunication was evolving at a rapid pace after the reporting of the first low-loss (< 20 dB/km) high-silica optical fiber concomitantly with room temperature operation of semiconductor laser diodes for long hours without degradation as well as high-efficiency photodetectors (PDs). Researchers soon realized that the transmission characteristics of optical fibers exhibited strong sensitivity to certain external perturbation-like bends, microbends, and pressure. A great deal of effort was spent at that time to reduce the sensitivity of signal-carrying optical fibers to such external effects through suitable designs of fiber refractive index profiles and cabling geometries. An alternate school of thought took advantage of these observations and started to exploit this sensitivity of optical fibers to external effects, essentially representing a variety of measurands to construct a large variety of sensors and instruments. This offshoot of optical fiber telecommunication soon saw a flurry of R&D activities around the world to use optical fibers for sensing.

3.1.2 Classification

Today, fiber-optic sensors play a major role in industrial, medical, aerospace, and consumer applications. Broadly, a fiber-optic sensor may be classified as either *intrinsic* or *extrinsic* (Culshaw 2006). In the intrinsic sensor, the physical parameter/effect to be sensed modulates the transmission properties of the sensing fiber, whereas in an extrinsic sensor, the modulation takes place outside the fiber, as shown in Figure 3.3. In the former, one or more of the physical properties of the guided light, for example, intensity, phase, polarization, and wavelength/color, are modulated by the measurands, while in the latter case, the fiber merely acts as a conduit to transport the light signal and from the sensor head to a PD/optical power meter for detection and quantitative measurement.

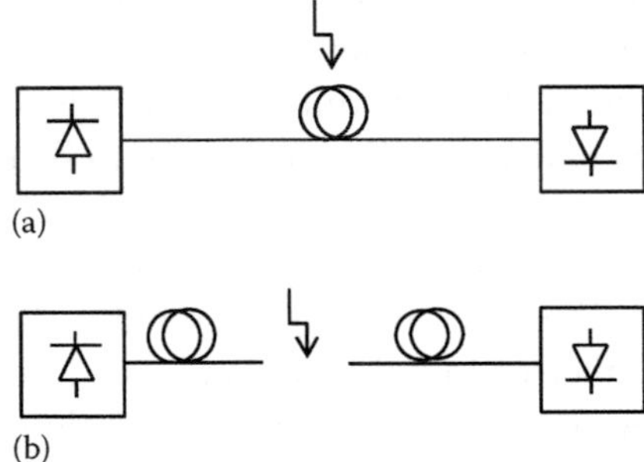

FIGURE 3.3 Examples of (a) intrinsic (b) extrinsic fiber sensors; in the intrinsic sensor, the measurands induce the modulation of one of the characteristics of the guided light, while in extrinsic sensors, the measurand-induced modulation takes place outside the fiber, for example, here in the gap between the two fibers. Measurand is debugged by the detector/power meter.

Table 3.1 Different Schemes for the Modulation of Light by a Measurand

Type of Information	Physical Mechanism	Detection Circuitry	Typical Example
Intensity	Modulation of light by emission, absorption, or refractive index change	Analog/digital	Pressure, displacement, refractive index, temperature, liquid level
Phase	Interference between measurand-induced signal and reference in an interferometer	Fringe counting, intra-fringe detection	Hydrophone, gyroscope, magnetometer, pressure
Polarization	Changes in gyratory optical tensor	Polarization analyzer and amplitude comparison	Magnetic field, large current measurement, e.g., in a busbar
Wavelength	Spectral-dependent variation of absorption and emission	Amplitude comparison at two fixed wavelengths	Temperature measurement

Four of the most commonly employed fiber-optic sensing techniques are based on either intensity modulation or interferometry or fluorescence, or spectral modulation of the light used in the sensing process (see Table 3.1). However, out of these four, the intensity and phase-modulated ones offer the widest spectrum of optical fiber sensors.

The advantage of intensity-modulated sensors lies in their simplicity of construction and they being compatible to the multimode fiber technology. The phase-modulated fiber-optic sensors necessarily require an interferometric measurement setup with associated complexity in construction, although as we shall see later in the chapter that they theoretically offer orders of magnitude higher sensitivity as compared to intensity-modulated sensors.

The rest of the chapter would describe functional principles and samples of a variety of fiber-optic sensors as a versatile sensor technology platform. Several other chapters in this book discuss further details of this optical sensing approach with guidelines for future developments.

3.2 Intensity-Modulated Fiber Sensors

3.2.1 General Features

Intensity-modulated fiber-optic sensors are the most widely studied fiber-optic sensors (Medlock 1987, Krohn 1988, Pal 1992a,b). The general configuration of an intensity-modulated sensor can be understood from Figure 3.4, in which the baseband signal (the measurand) modulates the intensity of the light propagating through the fiber that act as the sensor head. The resultant modulation envelope is reflected in the voltage output of the detector, which, upon calibration, can be used to retrieve measure of the measurand.

An alternative equivalent of the preceding Table 3.1 in an algebraic format is depicted in Figure 3.5. Intensity modulation can be achieved through a variety of schemes, for example, displacement of one fiber relative to the other, shutter type, that is, variable attenuation of light between two sets of aligned fibers, collection of modulated light reflected from a target exposed to the measurand, and loss modulation of light in the core or in the cladding through bending, microbending, or evanescent coupling to another fiber/medium.

A generic classification of intensity-modulated fiber-optic sensors is schematically depicted in Figure 3.6. In this figure, an example is shown of a measurand-induced variable attenuation of transmitted light across an input fiber and a receive fiber, and also a generic example of light from a source, which undergoes measurand-induced modulation either through reflection from a diaphragm subjected to a variable pressure environment or through reflection/scattering from a chemical environment, which could be a turbid solution or a solution that fluoresces on excitation with light at a suitable wavelength.

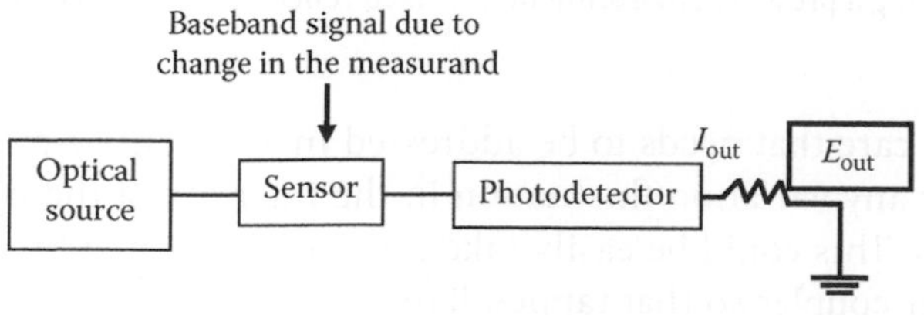

FIGURE 3.4 General principle of an intensity-modulated fiber-optic sensor, in which I_{out} represents modulated optical output in a guided wave optical circuit, and E_{out} is the modulation envelope in the voltage output.

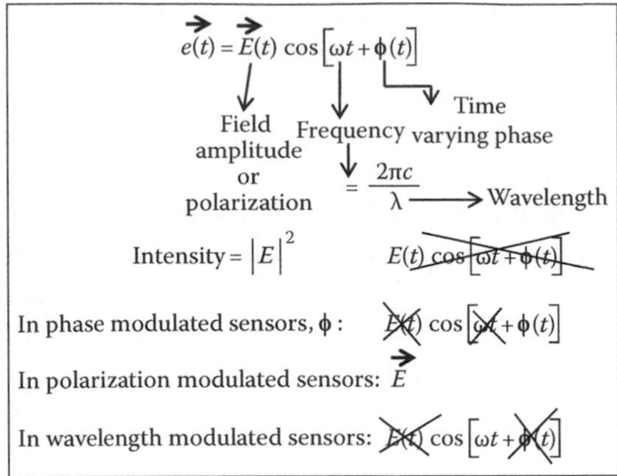

FIGURE 3.5 Algebraic interpretation of different modulation schemes.

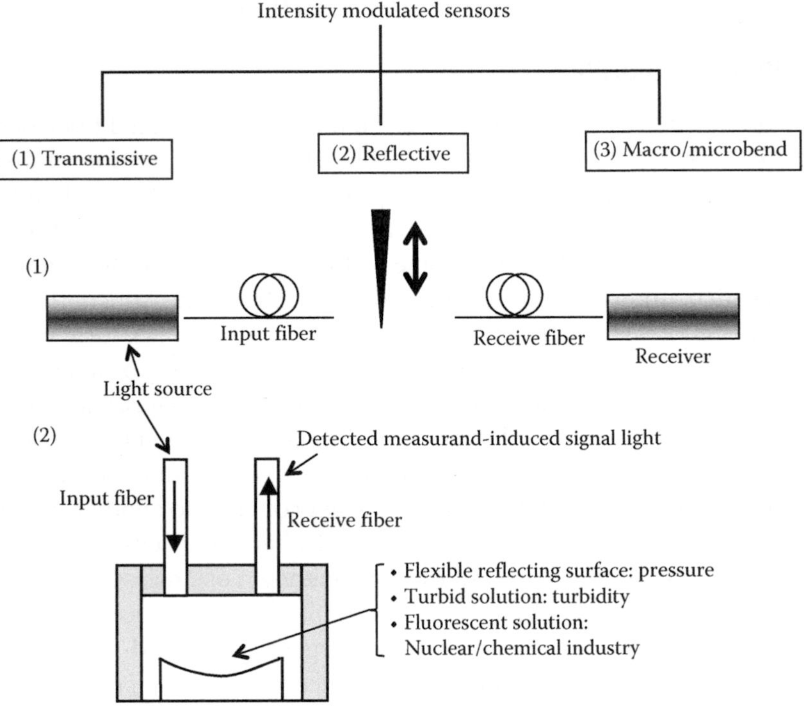

FIGURE 3.6 Classification of intensity-modulated sensors: (1) measurand-induced modulation of transmitted light across a gap between two fibers; (2) measurand-induced modulation of reflected light from a flexible reflector covering a pressure environment or a fixed reflector at the end of a turbid or a fluorescing solution.

One important care that needs to be addressed in all intensity-modulated sensors is to compensate for any variation/fluctuation in the intensity of the light source used in the measurements. This could be easily taken care of/accounted for by launching light through a tap fiber coupler so that tapped light intensity is continuously monitored by a PD/power meter as the *reference*; the ratio of measurand-induced variation in light intensity is always taken with respect to this *reference* intensity.

The principles and techniques associated with optical sensing based on intensity modulation are detailed in Chapter 4, while Chapter 13 further elaborates this sensing approach when supported by the optical fiber platform. In the following sections, some examples are presented illustrating the main characteristics of these sensors.

3.2.2 Intensity Modulation through Light Interruption between Two Multimode Fibers

Measurand-induced modulation of coupling of the light in a gap between two fibers can be exploited to detect threshold pressure, sound wave, threshold liquid level, etc. In its simplest form (see Figure 3.6, transmissive type), an interceptor in the form of a knife edge, for example, mounted on a diaphragm used as the lid of a pressure chamber could be used to detect/monitor if the pressure has exceeded a predetermined threshold value through suitable calibration. Optical fiber microswitches based on this basic principle are commercially offered to detect displacement (e.g., with pressure release valves) in hazardous environments (Pitt et al. 1985). Likewise, if the level of a liquid exceeds a predetermined threshold level/height, at which two fibers are kept perfectly aligned with a small air gap (approximately a few micrometers) in between from the wall of the liquid tank/container at a specific height, then an alarm signal could be triggered for taking corrective action as soon as the liquid level rises to that height. This configuration is particularly attractive if the liquid happens to be a potential source for fire hazard like gasoline, because there is no chance of electric spark in the measurement process. In an alternative version for detecting sound or displacement, one of these two fibers is mounted on an acoustically driven vibrating base/holder with a small length of the fiber extending from its holder as a cantilever and the other kept on a fixed base. Initially, without exposing to the measurand, the two fibers are kept perfectly aligned by maximizing the throughput power through the air gap in between. Then, as and when sound is switched on, the associated vibration would modulate the light transmitting through the gap between the two fibers, which could be detected and demodulated/processed to measure the frequency of the sound. The displacement between the fibers by one core diameter would result in approximately 100% light intensity modulation. Approximately, the first 20% of displacement yields a linear output (Krohn 1988). In the original experiment (Spillman and Mcmohan 1980), the device was found to detect deep-sea noise levels in the frequency range of 100 Hz to 1 kHz and transverse static displacements down to a few Angstroms. For higher accuracy measurements (Spillman and Gravel 1980, Pal 1992b), two opposed gratings were used in one such shutter mode actual device (as shown in Figure 3.7, in which acoustic waves were made to be incident on a flexible diaphragm made of rubber 1.5 mm thick and 2 cm diameter) to which one of the gratings was attached; the other was mounted on the rigid base plate of the housing. These gratings were made on two $9 \times 3 \times 0.7$ mm^3 cover strip glass substrates on which a 1.16 mm square grating pattern was produced from a 5 μm strip mask by means of photoresist lift-off technique through 1200 Å evaporation of chromium. An index matching liquid was inserted between the gratings, which were so aligned under a microscope that they were parallel and displaced relative to each other by one half strip width to ensure that the sensor works at the maximum and linear sensitivity region.

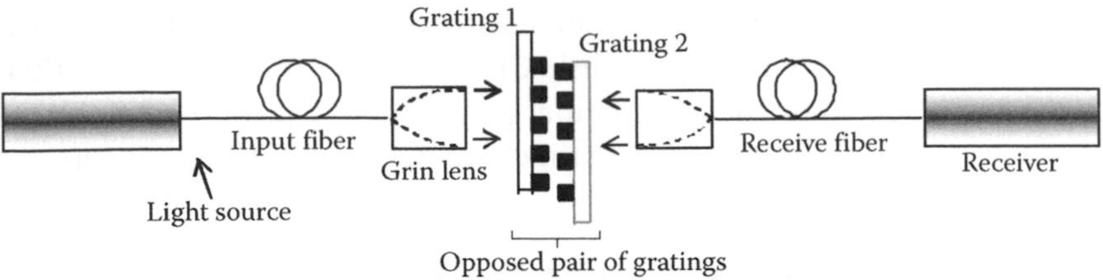

FIGURE 3.7 Schematic of an acoustic field/pressure sensor based on opposed gratings placed within the gap between a pair of multimode fibers: grating 1 (attached to a diaphragm) is movable by the measurand relative to the grating 2 (fixed to a base); a pair of GRIN microlenses were used to collimate light (from input fiber) and focus (into receiving fiber). (After Spillman, G.V. and Mcmohan, D.H., *Appl. Phys. Lett.*, 37, 145, 1980.)

The overlap area of two glass slides was sealed with a soft epoxy like RTV, which enabled the displacement of one grating relative to the other and also provided an elastic restoring force. The fibers consisted of two 200 μm plastic core, plastic clad fibers. The output He–Ne laser light from the input fiber was collimated by means of a GRIN Selfoc lens bonded to it. This collimated beam after transmission through the grating assembly was refocused by means of a second GRIN lens onto the input end of the receiving fiber; this procedure led to the isolation of the input and output coupling optics from the gratings. Such a device was shown to be sensitive to acoustic pressure less than 60 dB (relative to 1 μPa) over the frequency range of 100 Hz to 3 kHz, and it could resolve relative displacements as small as few Angstroms with a dynamic range of 125 dB. Before testing, the interior of the sensor assembly was filled with distilled water through the pressure relief hole. Further, it was relatively insensitive to static pressure head and responded well to variation in ac pressure for use as a hydrophone.

3.2.3 Reflective Fiber-Optic Sensors

In reflective sensors, the measurand induces modulation of the light reflected from a reflecting surface. In its simplest form, a Y-coupler fiber-optic probe consisting of two multimode fibers cemented/fused along some portion of their length (two bundles of fibers may also be substituted in their place) to form a power divider constitutes a reflective fiber-optic sensor. As an example, if light is injected through port 1 of the Y-power divider on to a light-reflecting diaphragm and the picked-up back-reflected light exits through port 3, its intensity would depend on the distance of the reflecting target from the fiber probe (see Figure 3.8). A dynamic range of such sensors can be enhanced by the use of a lens intermediate between the fiber probe and the reflective target. Such sensors can be used to detect displacement, pressure, or even the position of a float in a variable area flowmeter (Medlock 1987). The use of such reflective fiber-optic sensors have been demonstrated in determination of surface texture (Uena 1973), flow of pulp suspension in a tube in the range ~1–10 m/s (Oki et al. 1975), pressure over a range of ~100 psi (Tallman et al. 1975), in medical catheters as inter-cardiac pressure transducer with a sensitivity ~1 mm of Hg and linearity in the range of 0–200 mm of Hg (Lindstroem 1970, Matsmoto et al. 1978), vibrations (Uena et al. 1977, Parmigiani 1978), and also as a fiber laser Doppler anemometer (fiber LDA) (Kyuma et al. 1981).

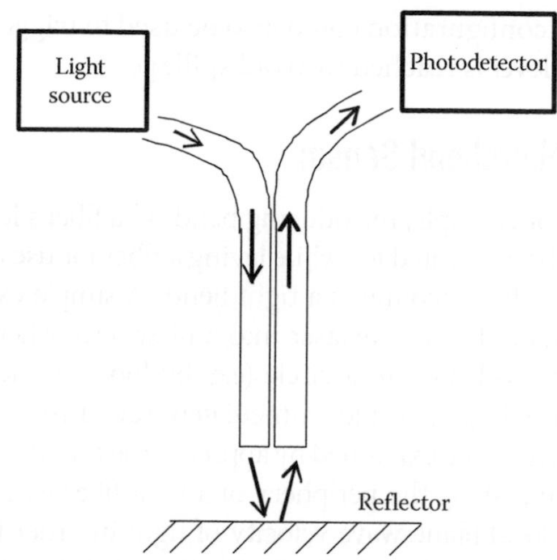

FIGURE 3.8 Schematic of a reflective sensor in which one fiber/a bundle of fibers transmits light from a source to a reflecting surface and another fiber/fiber bundle picks up the reflected light as a conduit for detection by a detector.

3.2.4 Fiber-Optic Liquid Level Sensing

Frustrated total internal reflection is exploited in a fiber-optic-based threshold liquid level sensor as shown in Figure 3.9. Light from a light source is coupled into a fiber, which is cemented with an optical adhesive at one end of the base of a 90° glass micro-prism, and a second fiber likewise is optimally fixed at the other end of the prism base to collect total internally reflected (TIR) light. As the liquid level rises and touches the prism, frustrated TIR takes place and the second fiber no longer receives any light and hence the power meter immediately detects a significant drop in the signal. This sensor configuration could be used as a threshold liquid level digital sensor, which could be used to stop, for example, an electric pump often used to fill an overhead water tank for

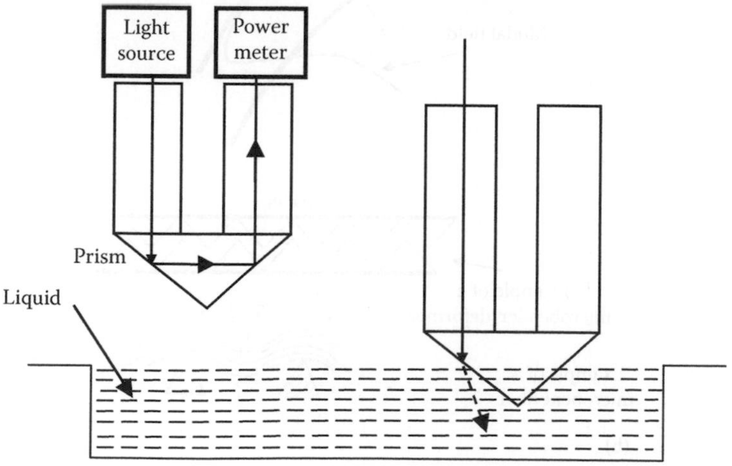

FIGURE 3.9 Schematic of a liquid level sensor, whose working principle is based on frustrated total internal reflection as described in the text.

domestic use. The same configuration could also be used to trigger an alarm signal once a predetermined liquid level is reached to avoid spillage.

3.2.5 Fiber-Optic Microbend Sensor

Extrinsic perturbation, for example, introducing bend on a fiber's lay leads to transmission loss. This effect needs to be accounted for while laying a fiber for use in a telecommunication link so that nowhere the fiber encounters a tight bend. A simple experiment that involves launching visible laser light like He–Ne laser into a fiber once when it is kept straight and once when the same fiber is bent into a circle (e.g. by looping a length of the same fiber around the experimenter's finger) would immediately reveal that a fiber suffers radiation loss at a bend. Physically, it can be explained by appreciating that the fractional modal power that travels in the cladding along the periphery of a bent fiber would be required to travel at a rate faster than the local plane wave velocity of light in order to maintain equi-phase fronts across radial planes. This being physically disallowed that part of the modal field radiates away (Marcuse 1982). In contrast to bend-induced transmission loss due to a constant curvature, if the fiber lay goes through a succession of very small bends (see Figure 3.10), the fiber exhibits transmission loss due to what is referred as microbend-induced loss in a fiber.

Physically, microbending leads to redistribution of power among the modes of the fiber and also transfer of some power from some high-order guided modes to radiation modes and hence loss in throughput power. Through a theoretical coupled mode analysis, it could be shown that strong coupling between the *p*th and *q*th modes of a fiber would occur if $\Delta\beta = |\beta_p - \beta_q|$ matches the spatial wave number ($=2\pi/\Lambda$) of the microbending deformer. This phenomenon could be exploited to detect a variety of

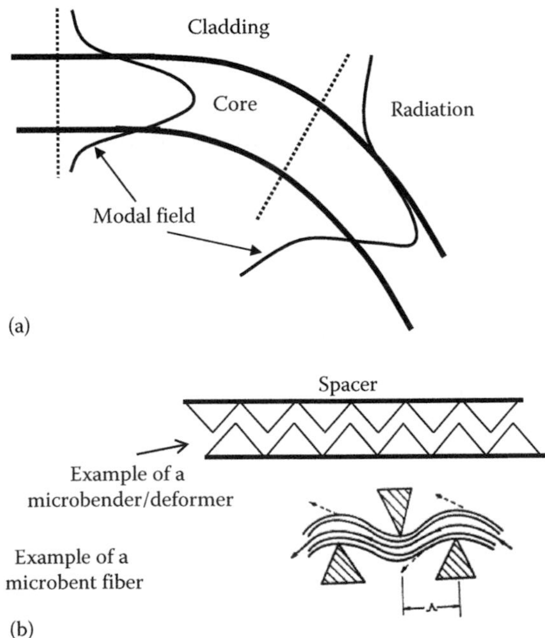

FIGURE 3.10 Schematic example of (a) macrobend and (b) microbend intensity-modulated fiber-optic sensor; Λ represents spatial wavelength of a microbender.

Table 3.2 Minimum Detectability of Environmental Changes

Environment	Minimum Detectability
Pressure	3×10^{-4} dyn/cm^2
Temperature	4×10^{-6} °C
Magnetic field	9.6×10^{-5} Oe
Electric field	1.7×10^{-1} V/m

fiber-optic sensors. Initially, these were used as hydrophones and displacement sensors (Fields and Cole 1980). Depending on the configuration of the microbender, same phenomenon could be exploited to construct a fiber-optic sensor, which could be used to determine several other environmental changes like temperature, acceleration, and electric and magnetic fields (Lagakos et al. 1987). For example, to function as a temperature sensor, the deformer should be made of a metal; for electric field, it should be a piezoelectric material and for magnetic field, it should be a magnetostrictive material. In Table 3.2, we tabulate minimum delectability of different environmental changes achievable in such fiber microbend sensors (Lagakos et al. 1987).

3.2.6 Fiber-Optic Chemical Sensing

Fiber-optic chemical sensors have been reported in a variety of forms. An interesting application of microbend sensors to chemical sensing exploits an offshoot of telecom industry, namely, the equipment optical time-domain reflectometer (OTDR), which is extensively used for monitoring the health of an optical telecommunication network. OTDR in combination with a localized measurand-induced microbend loss in a fiber has been used to detect localized water ingress in a civil structure or localized oil spillage, etc., in a distributed manner (Maclean et al. 2003). The basic functional principle of an OTDR is that it captures a snapshot of the backscatter signal from a fiber when an optical light pulse is launched into its input end. The signature of the backscatter signal in terms of detected power level as a function of length of the fiber is picked up at the input end itself through a fiber coupler as shown in Figure 3.11. Local signal at a point

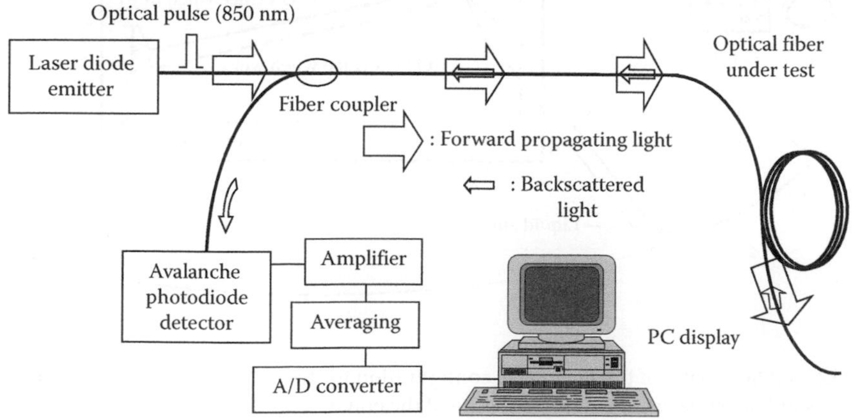

FIGURE 3.11 Schematic of an OTDR-based microbend intensity-modulated fiber-optic sensor. (Courtesy of B. Culshaw, EEE Department, University of Strathclyde, Glasgow, U.K.)

along the fiber's lay is distinguishable on the OTDR trace itself because the x-axis of the OTDR trace is calibrated in terms of fiber length while the y-axis represents the power level of the captured signal. Any local loss along the fiber length would be reflected as a drop in signal level at that location on the OTDR trace.

Figure 3.12a represents a schematic of the fiber cable configuration. These cables are tailored cables meant for distributed microbend sensing. A specific polymer material in the form of a gel, which would swell due to ingress of water, for example, surrounds a central GRP rod; this combination is held along its length within the cable along with a multimode optical fiber by helical wrapping with a Kevlar rope

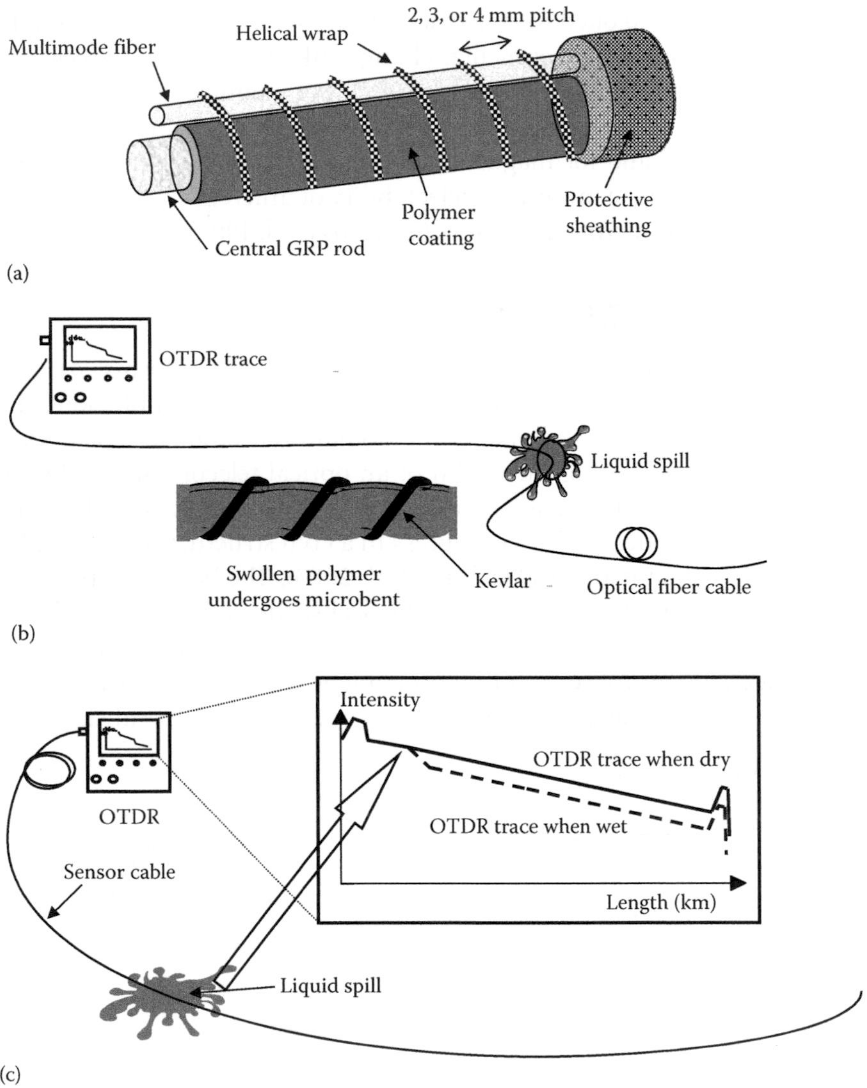

FIGURE 3.12 (a) Schematic of the fiber cable configuration for microbent fiber-optic sensor; (b) portion of the microbent fiber cross section due to swelling of the polymer due to water or liquid chemical ingress, and the resultant OTDR trace from the fiber; (c) schematic of the OTDR trace from the sensing fiber cable under dry and wet conditions. (Courtesy of B. Culshaw, EEE Department, University of Strathclyde, Glasgow, U.K.)

around the polymer. Overall combination is held within a protective sheath having micropores. If there is a localized water spillage anywhere along the cable lay, water would penetrate the cable through these pores and the chemically active hydrogel would swell due to absorption of water. Accordingly, the swelled gel would induce localized microbending of the fiber within the cable (see Figure 3.12b) and hence introduce a local loss there, which could be instantly captured in the OTDR trace as shown in Figure 3.12c. The same principle could be employed to detect spillage of other liquids like gasoline, if the hydrogel is substituted by another gel, which would swell due to absorption of gasoline, that is, another custom-designed cable with specific gel needs to be assembled for this purpose. In this fashion, custom-designed sensor fiber cables for sensing spillage of liquids like petrol, kerosene, diesel, crude oil, etc., could be designed and fabricated.

Another example of fiber-optic chemical sensing is shown in Figure 3.13, a schematic depicting measurement through excitation and collection of fluorescence emission. Since fluorescence is characteristic of a material, its spectroscopic measurement could yield information about the presence and its concentration in a solution. Fiber-optic probes of various designs could be used in such measurements. These probes are also referred to in the literature as *optrodes.*

A generic schematic of an optrode-based optical sensing is shown in Figure 3.14a, and few possible examples of optrode cross-sections are shown in Figure 3.14b. Such optrode-based sensing finds application in biomedical optics in oncology, for example, for the detection of cancerous cells, and diagnostics and monitoring of anatomical sites, while in chemical industries, these could be useful in hazardous and aggressive environments, for example, acidic, nuclear, and inflammable chemicals.

Similar optrodes could also be used as reflective sensors mentioned in Section 3.2.3 by using the central fiber as carrying the probe light and by collecting the reflected light from the reflecting surface through the peripheral fibers. Likewise, these could

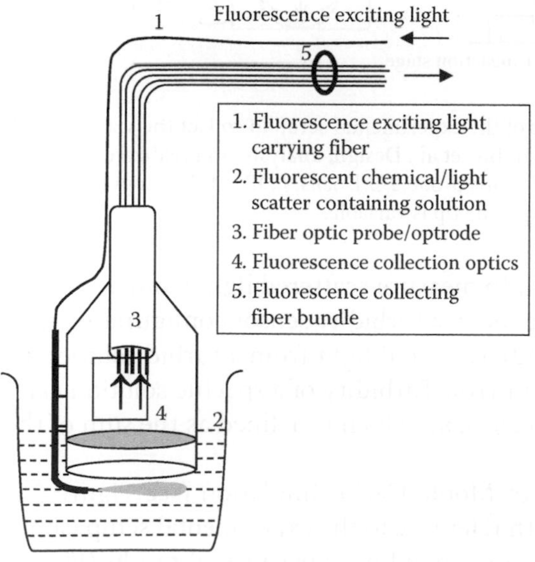

FIGURE 3.13 Schematic example of a fiber-optic fluorescence sensor.

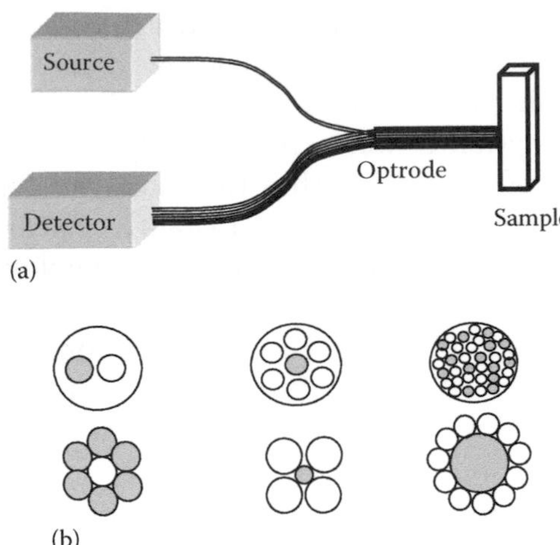

FIGURE 3.14 Schematic example of (a) a fiber-optic fluorescence sensor, in which fluorescence is excited and collected via an optrode; (b) examples of few possible variations in the optrode cross section; illuminating fiber is indicated as filled circles and light-collecting fibers are represented through open circles.

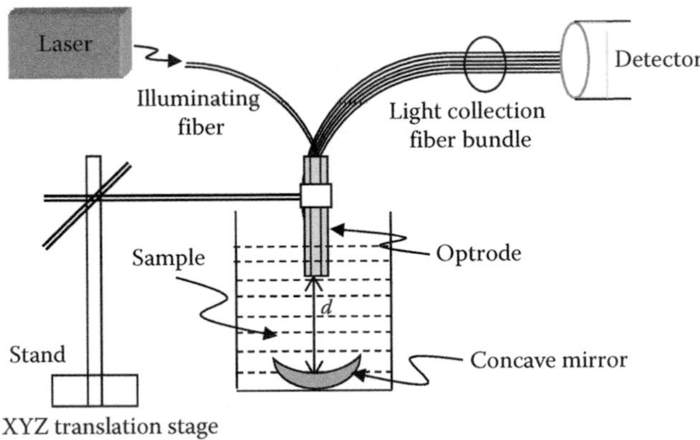

FIGURE 3.15 Schematic of the experimental setup to collect the scattered light from the samples. (After Prerana, Shenoy, M.R., Pal, B.P. et al., Design, analysis, and realization of a turbidity based on collection of scattered light by a fiber optic probe, *IEEE Sens. J.*, 12, 44–50, Copyright 2012 IEEE); distance *d* between concave mirror and the optrode tip is variable.

be gainfully deployed to measure scattered light from a solution also, for example, in turbidity measurements. In a turbidity sensor configuration (see Figure 3.15) reported in Prerana et al. (2012), scattered light from a turbid solution is collected after reflection from a concave mirror. Turbidity of a specific solution can be estimated in terms of *total interaction coefficient*, which is defined as the sum of the absorption and scattering coefficients.

Extensive results of Monte Carlo simulation (see Figure 3.16a) on optrode-based collection of light with reference to the experimental setup (shown in Figure 3.15) from a turbid solution and corresponding experimental results (Figure 3.16b) could be found

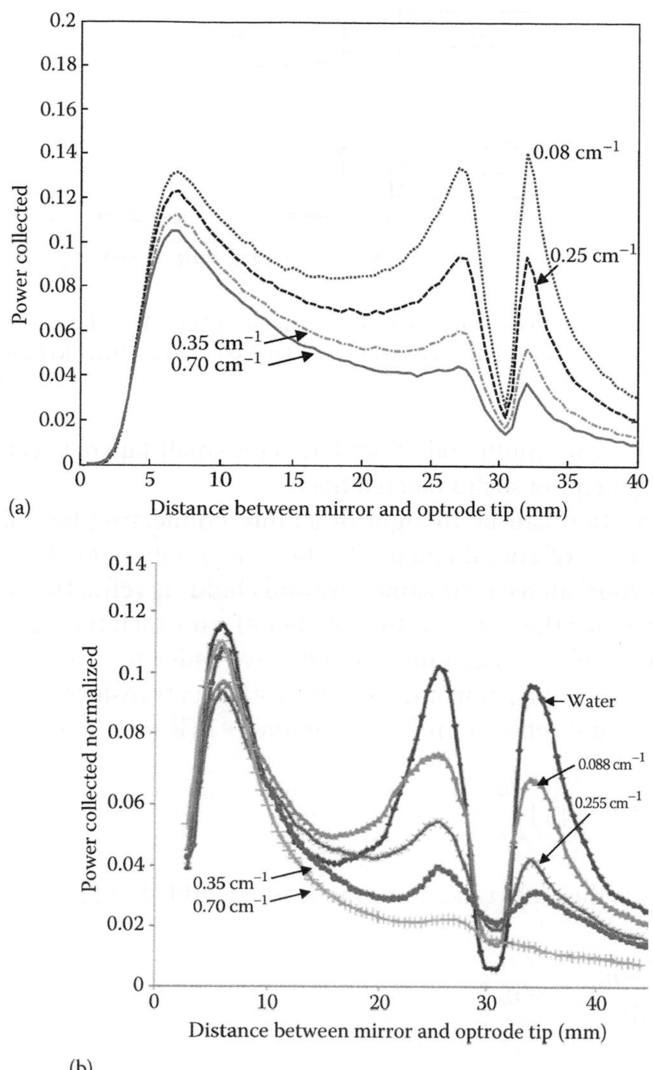

FIGURE 3.16 (a) Monte Carlo simulation results for collected power as a function of d for different turbid solution samples, each characterized by different interaction coefficient (μ_t), which labels the different curves; (b) experimental results for collected power as a function of d for different turbid solution samples, each characterized by different interaction coefficient (μ_t), which labels the different curves. (After Prerana, Shenoy, M.R., Pal, B.P. et al., Design, analysis, and realization of a turbidity based on collection of scattered light by a fiber optic probe, *IEEE Sens. J.*, 12, 44–50, Copyright 2012 IEEE.)

in Prerana et al. (2012). Turbidity is an important indicator for checking the quality of liquids like water and olive oil (Mignani et al. 2003).

3.2.7 Fiber-Optic Refractometer

Refractive index sensing is an important characteristic in many chemical industries. A multimode fiber-based refractometer reported in Kumar et al. (1984) is based on

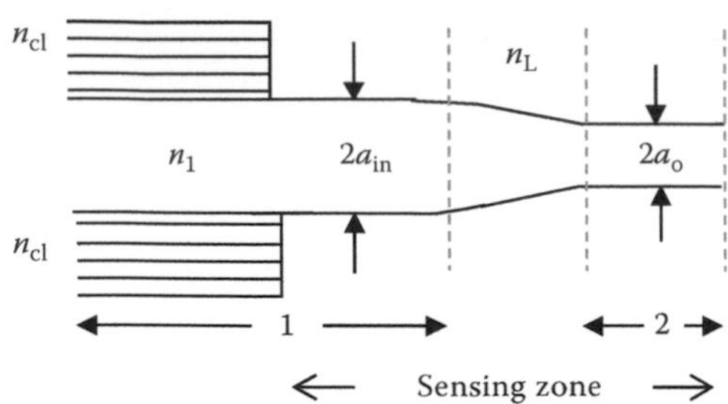

FIGURE 3.17 Refractometer based on a tapered multimode fiber. (After Kumar, A., Subrahmoniam, T.V.B., Sharma, A.D. et al., A novel refractometer using a tapered optical fiber, *Electron. Lett.*, 20, 534–535, Copyright 1984.)

a plastic clad silica core multimode fiber having a small tapered section. Figure 3.17 represents the geometry of such a tapered fiber.

The tapered portion can be thought of as interconnecting two fibers: #1 of core diameter $2a_{in}$ and #2 of core diameter $2a_o$ ($a_o < a_{in}$). Fibers #1, #2, and the tapered interconnecting zone, all have the same core and cladding refractive indices n_1 and n_2, respectively, except for the initial section of fiber #1, in which the cladding index is n_{cl}, that is, the plastic clad. A guided mode of effective index n_{e1} ($=n_1 \cos \theta_1$, θ_1 being the characteristic mode propagation angle) in fiber #1 gets transformed to a corresponding characteristic mode effective index n_{e2} in fiber #2 (Kumar et al. 1984) as

$$n_{e2} = \left[n_1^2 - R^2 \left(n_1^2 - n_{e1}^2 \right) \right]^{1/2} \tag{3.1}$$

where R ($=a_{in}/a_o$) is the taper ratio. For a mode to be guided in fiber #2, one requires

$$n_{e1} \geq \left[n_1^2 - \frac{n_1^2 - n_L^2}{R^2} \right]^{1/2} \equiv n_{e1}^{min} \tag{3.2}$$

If P_o represents the total power injected into the guided modes of fiber #1, then the power in the modes with $n_{e1} > n_{e1}^{min}$ would be given by

$$P_b = P_o \frac{n_1^2 - n_L^2}{R^2 \left(n_1^2 - n_{cl}^2 \right)} \tag{3.3}$$

Thus it is evident from Equation 3.3 that power coupled to fiber #2 through the taper increases linearly with proportional decrease in n_L^2. This result has been exploited to construct a fiber refractometer based on a plastic-clad silica core fiber, from a small portion of which plastic was removed and this bare portion of the fiber was then converted into a taper by heating and stretching in a flame burner. The tapered zone was immersed in a liquid of refractive index n_L ($<n_1$). By immersing the taper subsequently in a number of other liquids while taking care to clean tapered zone each time appropriately, and monitoring the corresponding power reaching fiber #2, one can generate a calibration curve for a given

fiber taper. Thus, a measure of the output power exiting fiber #2 for a liquid of unknown refractive index would yield a refractive index of that liquid. Kumar and coworkers have shown a linear decrease in P_b with n_L^2. In principle, the same technique could be exploited to construct a fiber-optic temperature sensor by encapsulating the taper with a metallic encapsulation filled with a thermo-optic liquid whose refractive index varies with temperature. In fact, plastic clad silica fibers provide an excellent platform to configure a range of intensity-modulated fiber-optic sensors. For example, a small bare section of such a fiber could be covered with a sol gel material to form a porous cladding, which could be used as a substrate to entrap a gas/liquid, which would form the local cladding there and hence would modulate the intensity of the propagating light in that region. Demodulation of this transmitted light could yield information about that gas/liquid.

3.2.8 Fiber-Optic Sensor Based on Side-Polished Fiber Half-Couplers

Side-polished single-mode fiber (SP-SMF) half-coupler offers a versatile sensor technology platform, in which phase resonant evanescent coupling of the SP-SMF with a multimode overlay waveguide (MMOW) could be exploited. A high-sensitive temperature sensor based on evanescent field coupling between a side-polished fiber half-coupler (SPFHC) and a thermo-optic MMOW was designed and realized in a paper by Nagaraju et al. (2008) (see Figure 3.18).

Such a structure essentially functions as an asymmetric directional coupler with a bandstop characteristic attributable to the wavelength-dependent resonant coupling between the mode of the SPFHC and one or more modes of the MMOW. The wavelength sensitivity of the device was ~5.3 nm/°C within the measurement range of 26°C–70°C; this sensitivity is more than five times higher compared to earlier reported temperature sensors of this kind. The SPFHC was fabricated by selective polishing of the cladding from one side of a bent telecommunication standard single-mode fiber, and the MMOW was formed on top of the SPFHC through spin coating. A seminumerical rigorous normal mode analysis was employed at the design stage by including the curvature effect of the fiber lay in the half-coupler block and the resultant z-dependent evanescent coupling mechanism. The agreement between theoretical and experimental results was excellent.

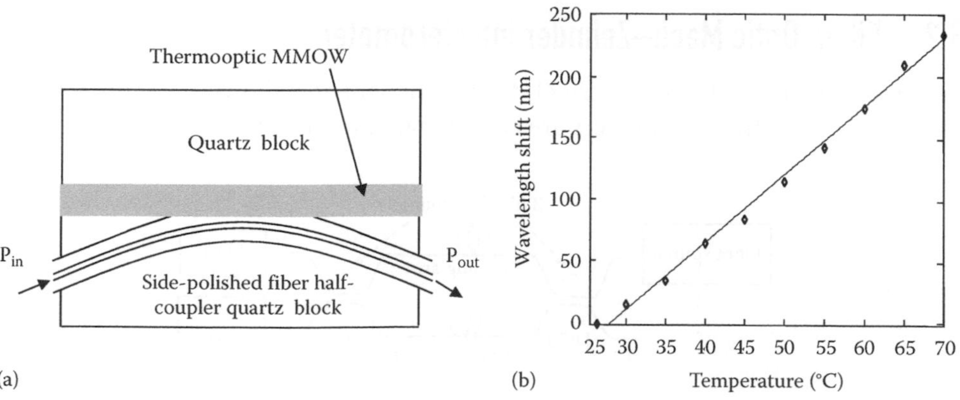

FIGURE 3.18 (a) Schematic diagram of the sensing device; (b) shift in the resonance wavelength with temperature (solid curve is a linear fit through experimental data points).

The temperature range measurable by this sensor was limited by the thermo-optic overlay material used; other suitable MMOW material of appropriate thickness could be chosen to adapt such a device for the measurement of temperature in other ranges. Further refinement in such a fiber refractometer was recently reported by Prerana et al. (2010), in which the role of a tapered MMOW was investigated. Such an SPF-MMOW configuration was earlier exploited by Johnstone et al. (1992) as a refractometer, whose performance was theoretically explained in Raizada and Pal (1996).

3.3 Interferometric Fiber-Optic Sensors

3.3.1 General Features

Interferometric fiber-optic sensors are ultra-high-sensitive sensors, which are based on interferometric measurements of measurand-induced change in the phase of the light propagating in a fiber (Giallorenzi et al. 1982, Dandridge 1991, Culshaw 1992). These sensors offer huge potential in a variety of applications especially for high-sensitive measurements of low-magnitude measurands. At times, their ultrasensitive nature poses a problem to accurately measure a measurand due to sensitivity to ambient conditions, and hence, signal processing of the measured parameter is an important issue. The sensitivity of an interferometer to a measurand involves two basic criteria: the efficiency of the interface between the measurand and the optical delay in the fiber and the ability to reject the interfering measurands at the same time (Culshaw 1992). Phase sensitivities vary with measurands, for example, in the case of strain, about 10 rad per microstrain per meter, for temperature about 100 rad per degree per meter, and for pressure about 10 rad per bar per meter. In view of the relatively higher sensitivity to strain, pressure or magnetic fields are measured through corresponding transformation to strain as in a fiber-optic hydrophone and magnetometer (Culshaw 1992).

The principles and techniques associated with optical sensing based on interferometric modulation are introduced in the following sections and detailed further in Chapter 14. Although most of the material presented is associated with the optical fiber, it is quite general and equally applicable to other optical sensing platforms.

3.3.2 Fiber-Optic Mach–Zehnder Interferometer

A Mach–Zehnder interferometer (MZI) fiber-optic sensor is shown in Figure 3.19. It can be seen that two 3 dB fiber couplers are concatenated to form the interferometer.

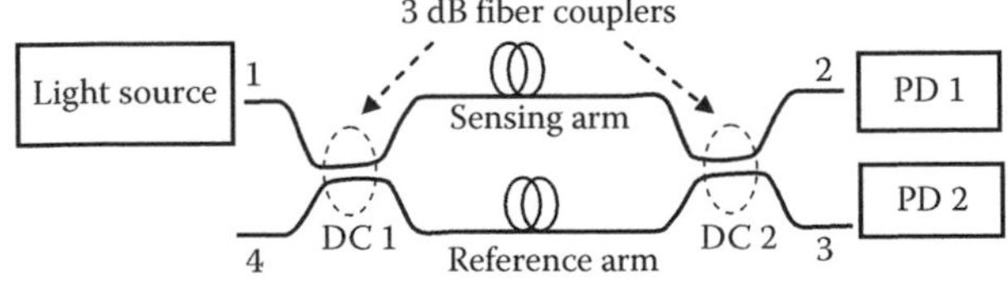

FIGURE 3.19 A fiber-optic Mach–Zehnder interferometer, in which two 3 dB fiber couplers (DC 1,2) are concatenated to form it; PD 1,2 are two photodetectors, either of which could be used to record the interference pattern.

If the amplitude of the light injected into port 1 is E_{in}, the corresponding intensity will be I_{in} ($=|E_{in}|^2$). If we consider the light from port 1 reaching port 2 via the upper and lower arms of the interferometer, its intensity detected by PD 1 would be

$$I_1 = E_1 \cdot E_1^* = \frac{1}{2} E_{in} \left[e^{i\phi_1} + e^{i(\pi+\phi_2)} \right] \cdot \frac{1}{2} E_{in} \left[e^{-i\phi_1} + e^{-i(\pi+\phi_2)} \right]$$

$$= \frac{I_{in}}{2} (1 - \cos\Delta\phi) = I_{in} \sin^2 \frac{\Delta\phi}{2} \tag{3.4}$$

where $\Delta\phi = \phi_1 - \phi_2$; $\phi_{1,2}$ correspond to respective phase accumulated along lengths of the upper and lower arms of the interferometer. In Equation 3.4, extra phase of $\pi/2$ accumulated by the coupled light at each of the two fiber couplers (Pal 2000) has been taken into account, which accounts for the additional phase of π accumulated by light taking path via the lower arm of the interferometer to reach PD 1 after crossing over twice at the two couplers. Likewise, the intensity of light reaching PD 2 would be given by

$$I_2 = I_{in} \cos^2 \frac{\Delta\phi}{2} \tag{3.5}$$

Normalized I_2 versus $\Delta\phi$ is plotted in Figure 3.20. For $\Delta\phi = 0$, $I_2 = I_{in}$ while for $\Delta\phi = \pi$, $I_2 = 0$.

The corresponding plot of I_1 with $\Delta\phi$ would be complementary to it. In practice, measurand-induced $\Delta\phi$ is small, and hence, variation in ϕ would be very small around the maximum and minimum of $I_{1,2}$ as a function of $\Delta\phi$. The most sensitive point of operation would correspond to the *quadrature point*, where $\Delta\phi = (2m + 1)\,\pi/2$; $m = 0, 1, 2, \ldots$ $\Rightarrow \Delta\phi$ is $\pi/2$, or $3\pi/2$, or $5\pi/2$, … . Around this point, $I_{2,1}$ varies linearly with phase. This phase bias can be introduced through stretching the fiber, for example, by wrapping the portion of the reference fiber arm on a piezoelectric drum driven by a signal generator; any small deviation in $\Delta\phi$ around the quadrature bias point could be actively controlled through the piezo driver. Thus, if the measurand-induced phase change is $\delta\phi$ around the quadrature point, $\Delta\phi = \pi/2 + \delta\phi$ then

$$I_1 = I_{in} \sin^2 \left(\frac{\pi}{4} + \frac{\delta\phi}{2} \right) \approx \frac{I_{in}}{2} (1 + \delta\phi) \tag{3.6}$$

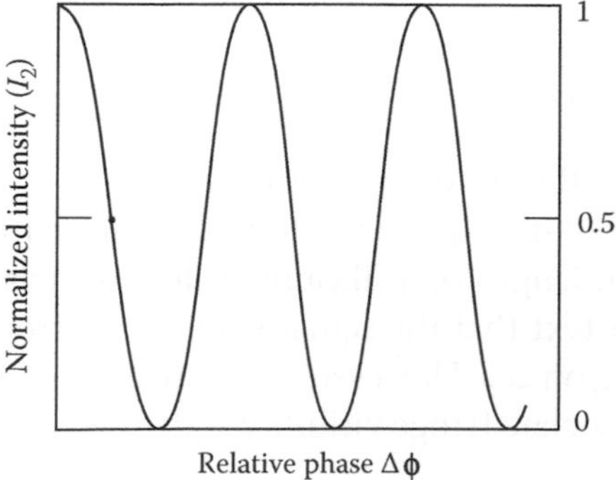

FIGURE 3.20 Mach–Zehnder interferometer output intensity I_2 with variation in phase; the quadrature point at which $\Delta\phi$ is $\pi/2$ is marked as a *dot* on the figure for which normalized I_2 is 0.5.

It shows that $I_1 \propto \delta\phi$, that is, the phase difference is converted into intensity, which can be measured by a square-law detector. For a shot-noise-limited detection system, assuming SNR = 1, it can be shown that minimum detectable phase change over a detection frequency bandwidth (Δv) is given by (Ghatak and Thyagarajan 1998)

$$\delta\phi_{\min} = 2\left(\frac{e\Delta v}{I_{in}\rho}\right)^{1/2} \tag{3.7}$$

where

ρ is the responsivity (A/W) of the detector
e is the electron charge = 1.6×10^{-19} C

If I_{in} = 1 mW, Δv = 1 Hz, and ρ = 0.5 A/W, $\delta\phi_{\min}$ from Equation 3.7 is $\cong 3.6 \times 10^{-8}$ rad, which is indeed very small. If this $\delta\phi_{\min}$ indeed occurs at a wavelength of λ_0 = 1550 nm due to a change in fiber length (δL), then

$$\delta L = \frac{1}{2\pi}\frac{\Delta\phi}{\lambda_0 n_{eff}} = \frac{3.6\times10^{-8}\times1.55\times10^{-6}}{2\times\pi\times1.449} \approx 6\times10^{-13}\ \text{m} \tag{3.8}$$

Thus, if a variation in fiber length induces earlier-mentioned phase change of $\delta\phi_{\min}$, and if that is measurable, then δL given earlier would amount to a miniscule change in the fiber length of approximately 100 fm (~10^{-13} m)! This is indeed extremely small. However, since a detector is usually not shot-noise-limited, measurable $\delta\phi_{\min}$ is about one to two orders of magnitude larger and hence δL correspondingly would be about a picometer.

If the phase of the propagating light in a fiber of length L varies due to a temperature change ΔT, then

$$\Delta\phi = k_0\left(L\cdot\Delta T\frac{dn_{eff}}{dT} + n_{eff}\cdot\Delta L\right) \tag{3.9}$$

where dn_{eff}/dT represents the thermo-optic coefficient. Thus,

$$\frac{\Delta\phi}{L\Delta T} = k_0\left(\frac{dn_{eff}}{dT} + n_{eff}\frac{\Delta L}{L\Delta T}\right) \tag{3.10}$$

In fibers, thermo-optic coefficient dominates over the linear expansion term; thus, the first term within the bracket in Equation 3.10 is the dominating term for determining the phase change per unit length of a silica fiber due to variation in temperature. It is apparent from the earlier text that the signal would vary from a maximum ($=I_{in}$) to a minimum (=0) depending on $\Delta\phi$. However, in practice, this complete modulation is not observed, and one can associate fringe visibility V defined as

$$V = \frac{I_{max} - I_{min}}{I_{max} + I_{min}} \tag{3.11}$$

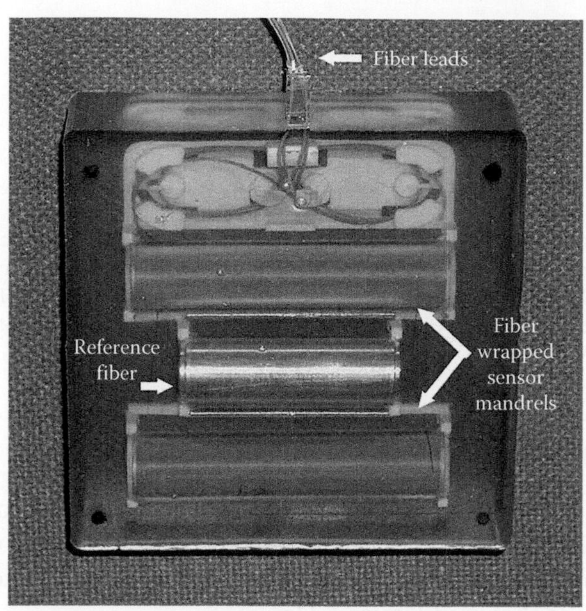

FIGURE 3.21 Photograph of a Virginia class fiber-optic hydrophone developed at the Naval Research Laboratory, Washington, DC. (Adapted from Cole, J.H. et al., *Wash. Acad. Sci. J.*, 90, 40, 2004. With the permission of the Washington Academy of Sciences.)

so that

$$I_{2,1} = \frac{I_{\text{in}}}{2}\left(1 \pm V\cos\Delta\phi\right) \tag{3.12}$$

In the early days of fiber-optic MZI development, detection of acoustic waves, which led to the development of fiber-optic hydrophones (see Figure 3.21) by scientists at Naval Research Laboratory in Washington, the United States, was the most widely pursued sensing scheme based on fiber-optic MZI (Cole et al. 2004).

In addition, fiber-optic MZI has been extensively used to detect temperature, linear strain, axial load, electric field, magnetic field, seismic signals, and vibration (Dandridge 1991). In several cases, the mechanical design of the sensing element is an extremely important issue such as the nature of the fiber coating. In more recent times, a pair of long-period gratings (LPGs) within a fiber was shown to effectively function as an MZI as illustrated in Figure 3.22 (Maier et al. 2007, Kim et al. 2008).

Since n_{eff} in the cladding is smaller than that at the core, the two different paths taken by the two distinct beams, namely, one as a core mode and the other as a core–cladding–core mode, lead to an optical path difference and hence form a two-beam interference pattern similar to that in an MZI described earlier. Such LPG pair-based MZI were shown to detect refractive index changes as low as $\sim 1.8 \times 10^{-6}$ for hydrogen detection (Kim et al. 2008, Lee et al. 2012). Hydrogen sensing/leak detection has become extremely important in the context of wide interest on fuel cells in recent times. Several other techniques have been reported for forming in-fiber MZIs including the use of photonic crystal fibers (Lee et al. 2012).

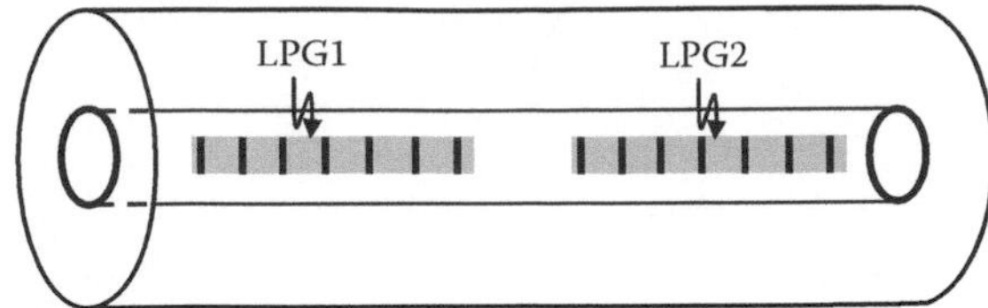

FIGURE 3.22 Schematic of an MZI formed with a pair of LPGs separated by a certain distance. LPG1 induces partial coupling of power from the core mode to a cladding mode, which after propagating as a cladding mode recouples back to the core mode after interacting with the second LPG. These two sets of beams taking different paths, when recombined after the second LPG, form an interference pattern that is characteristic of an MZI.

3.3.3 Fiber-Optic Michelson Interferometer

Fiber-optic Michelson interferometers (MIs) are another platform for two-beam interferometric sensors similar to MZI with the difference that it involves only one 3 dB fiber coupler to split one beam into two, both of which are reflected back by two mirrors placed at their ends as shown in Figure 3.23a.

The light from the source splits into two beams shown in the figure as full thick arrows at the 3 dB coupler. The reflected beams (full and dashed—both thin arrows for the two reflected beams from the mirrors M1 and M2 via arms *a* and *b*) reach the PD at port 4, where they interfere. These beams destructively interfere at port 1 as part of the reflected beams reaching back the source via arms *a* and *b* reach there with a phase difference of π. An algebraic analysis for the transmittance of an MZI presented in the

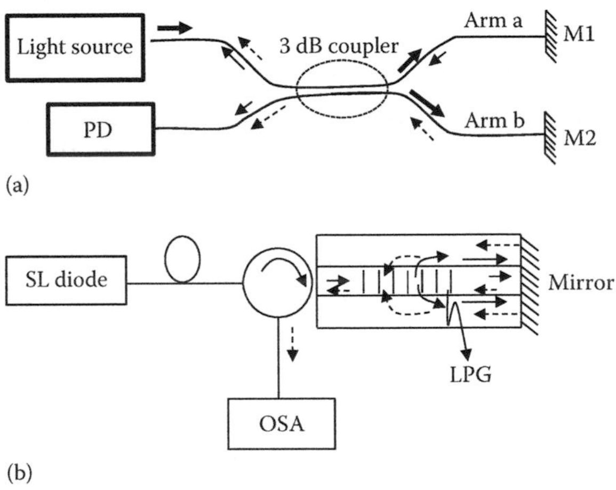

FIGURE 3.23 (a) A fiber-optic Michelson interferometer, in which one of the arms functions as a reference arm. The other arm as the signal arm is exposed to a measurand, which induces a change in the optical path length, and hence, the relative phase between the two light paths varies. As a result, an interference pattern is formed at the photodetector (PD), which varies with the measurand; (b) an alternative version of a fiber-optic Michelson interferometer for sensing in which an LPG is used to split and recombine reflected core and cladding mode lights to form interference and the same is captured on an OSA; a circulator is used in place of a fiber coupler. (After Swart, P.L., Long period grating Michelson refractometric sensor, *Meas. Sci. Technol.*, 15, 1576–1580, Copyright 2004 IOP.)

previous section could be extended to analyze transmittance of an MI, and it can be shown that at the port containing PD (Jones 1992),

$$I_4 = \frac{I_{in}}{2}\left(1 + V\cos\left(\Delta\phi\right)\right) \quad (3.13)$$

where $\Delta\phi = \phi_a - \phi_b$. Essentially, an MI is half of an MZI. Thus, all discussions made earlier for MZI would be valid here also. The fiber coupler, in principle, could be replaced with a circulator as shown in Figure 3.23b. An interesting compact version of a single fiber-based MI without involving a fused fiber coupler was proposed in the literature (Swart 2004, Kim et al. 2005), the basic concept of which is shown in Figure 3.23b. An LPG of about 50% coupling strength divides the propagating light in the core into two paths—one along the core via the LPG and the other along the cladding as a cladding mode—both of which are reflected by a common mirror fabricated directly at the end of the fiber. These two reflected beams form the two arms of the MI and form the interference fringes after being mixed through the same LPG at the input side, which is captured through a circulator on an optical spectrum analyzer. Such an MI has been used as a refractometer. Cross-sensitivity to temperature could be reduced in these measurements by using two types of fibers (Brakel and Swart 2005) or a photonic crystal fiber (Park et al. 2010).

3.3.4 Fiber-Optic Sagnac Interferometer

Rotation sensing is of significant interest in several areas, for example, in inertial navigation in aircraft/spacecrafts, surveying where accurate determination of geodesic latitude and azimuth is required, in determination of torsional oscillations in earth due to earthquakes, in determination of astronomical latitude, and in monitoring polar motion due to various geophysical effects (Ezekiel and Arditty 1982). Traditionally, rotation sensors have been mechanical gyroscopes based on spinning wheels and have relied on the conservation of angular momentum. However, fiber-based optical gyroscopes (FOGs) have attracted a much stronger interest in recent times due to the absence of moving parts, absence of warm-up time, and sensitivity to gravity (Ezekiel and Arditty 1982). FOGs have emerged as an offshoot of ring laser gyroscopes (RLGs), first reported by Rosenthal (1962). RLGs are now routinely used for inertial navigation in many passenger aircraft (Lee et al. 2006) whose navigational controls depend critically on accurate rotation sensors. Typical precision requirement in aircraft navigation lies in the range of 0.001–0.01°/h. In terms of rotation rate of earth (Ω_E = 15°/h), this amounts to 10^{-4}–10^{-3} times Ω_E. The first proposal for implementing an FOG in the form of a Sagnac interferometer was made in 1976 (Vali and Shorthill 1976). These sensors are produced commercially to support high-end automobile navigation systems, for the pointing and tracking of satellite antennas, inertial navigation for aircraft and missiles, and as backup guidance system for commercial aircraft such as Boeing 777 (Udd 2002). Other areas where fiber gyros find applications are mining, tunneling, radio-controlled attitude control of helicopters, guidance for unmanned trucks, and so on. The Sagnac interferometer in its classical form uses a beam splitter to split a light beam, which is directed to follow a square or triangular trajectory in opposite directions toward the input end, where these

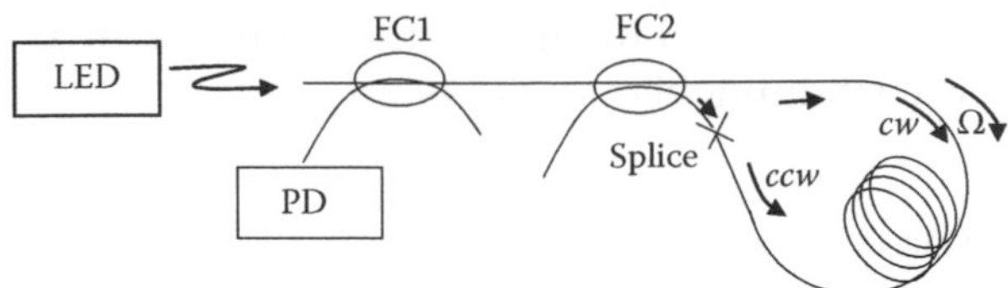

FIGURE 3.24 Fiber-optic Sagnac interferometer for rotation sensing; FC1 and FC2 represent two fiber couplers; PD is photodetector for measuring the interference between the counterpropagating cw and ccw beams; Ω represents angular velocity of the rotating frame on which the fiber loop is mounted.

are recombined to interfere. In an FOG, two oppositely directed light beams are divided into two counterpropagating beams through a few hundreds of meters long single-mode fiber coil/loop—one in clockwise (*cw*) and the other in counterclockwise (*ccw*) direction as shown in Figure 3.24.

These two counterpropagating *cw* and *ccw* beams exiting from the fiber loop are recombined to form the interference pattern at the PD for further signal processing. Two couplers FC1 and FC2 are required to make the *cw* and *ccw* propagating beams experience identical paths because an additional fixed phase retardation of $\pi/2$ (in the absence of measurand) is introduced by the coupler to the coupled beam. In order to understand the working principle, we outline below the simple analysis given in Ezekiel and Arditty (1982), which ignores relativistic mechanics. Let us consider a disk of radius R that is rotating clockwise with an angular velocity Ω about an axis perpendicular to the plane of the disk as shown in Figure 3.25.

If we assume that two identical photons are sent *cw* and *ccw* along the circumference starting at an arbitrary location 1, then, by the time *cw* propagating photon returns to its starting point, the disk would have rotated to location 2. Thus, this photon would be required to travel an extra linear distance (as compared to the case when the disk remains stationary, that is, for $\Omega = 0$, in which case physical path length $L = 2\pi R$ [i.e., the perimeter of the disk]) of

$$L_{cw} = 2\pi R + R\Omega t_{cw} \equiv c_{cw} t_{cw} \tag{3.14}$$

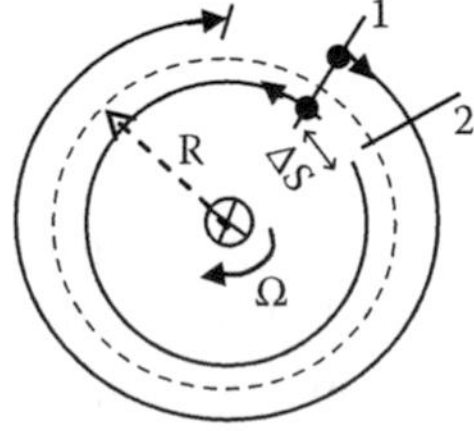

FIGURE 3.25 Functional principle of Sagnac interferometer for rotation sensing (see text for explanation). (After Ezekiel, S. and Arditty, H.J., Fiber optic rotation sensors: Tutorial review, in *Fiber-Optic Rotation Sensors and Related Technologies*, S. Ezekiel and H.J. Arditty, eds., Springer-Verlag, Berlin, Germany, 1982; Lee, B. et al., Principle and status of actively researched optical fiber sensors, in *Guided Wave Optical Components and Devices: Basics, Technology and Applications*, B. P. Pal, ed., Academic Press, Burlington, MA, 2006.)

and likewise the *ccw* propagating photon would take a shorter path length given by

$$L_{\mathrm{ccw}} = 2\pi R - R\Omega t_{\mathrm{ccw}} \equiv c_{\mathrm{ccw}} t_{\mathrm{ccw}} \tag{3.15}$$

where

$R\Omega$ (=ΔS)

$t_{\mathrm{cw,ccw}}$ represent the times taken to cover the distances

$L_{\mathrm{cw,ccw}}$ and $c_{\mathrm{cw,ccw}}$ correspond to the velocity of light

Thus, the net time difference between the two counterpropagating beams to cover $L_{\mathrm{cw,ccw}}$ would be

$$\Delta t = \frac{2\pi R}{c_{\mathrm{cw}} - R\Omega} - \frac{2\pi R}{c_{\mathrm{cw}} + R\Omega} \cong \frac{4\pi R^2}{c^2}\Omega = \frac{4A}{c^2}\Omega \tag{3.16}$$

where we have assumed that in vacuum light velocity, $c_{\mathrm{cw}} = c_{\mathrm{ccw}} = c$ and the product $R^2\Omega^2$ is negligible relative to c^2. Equivalently, this time difference amounts to a path length difference of

$$\Delta L = c\Delta t = \frac{4A}{c}\Omega \tag{3.17}$$

Even if the Sagnac effect is considered in a medium of refractive index n, for which the relativistic addition of the light velocity in that medium with the tangential velocity $R\Omega$ is required, Δt and ΔL would still be given by Equations 3.16 and 3.17, respectively (Ezekiel and Arditty 1982). If that medium is a single-mode fiber wound N number of turns in the form of a coil (see Figure 3.24), then

$$\Delta t = \frac{4AN}{c^2}\Omega \Rightarrow \Delta L = c\Delta t = \frac{4AN}{c}\Omega \tag{3.18}$$

If we consider phase difference $\Delta\phi$ between the *cw* and *ccw* propagating beams as they cover the total fiber length, $\Delta\phi$ would be given by

$$\Delta\phi = k_0 c\Delta t = \frac{8\pi AN}{\lambda_0 c}\Omega \Rightarrow \Delta L = \frac{\Delta\phi}{k_0} = \frac{4AN}{c}\Omega = \frac{LD}{c}\Omega \tag{3.19}$$

where

D is diameter of the fiber

$N = L/(\pi D)$

The output intensity from a Sagnac interferometer due to interference between the *cw* and *ccw* propagating beams can be shown by following the analysis identical to the one described earlier for MZI to be given by

$$I = \frac{I_{in}}{2}(1 + \cos\Delta\phi) \tag{3.20}$$

where $\Delta\phi = \phi_{cw} - \phi_{ccw}$. Following the same arguments as in the case of MZI before, the interferometer may be biased through the use of a PZT phase modulator at the quadrature point ($\Rightarrow \Delta\phi = \pi/2 + \delta$) so that Equation 3.20 becomes

$$I_{out} = \frac{I_{in}}{2}(1 - \sin\delta) \approx \frac{I_{in}}{2}(1 - \delta) \tag{3.21}$$

since δ in general would be small. If we assume shot-noise-limited detector as in the case of MZI described earlier (see Equation 3.7), then minimum measurable Ω would be given by [Ghatak and Thyagarajan (1998)]

$$\Omega_{min} = \frac{c\lambda_0}{4\pi AN}\left(\frac{e\Delta\nu}{\rho I_{in}}\right)^{1/2} \tag{3.22}$$

If we assume a fiber length to be 500 m spooled in a loop of 3.2 cm radius, $\rho = 0.5$ A/W, $I_{in} = 1$ mW, $\lambda_0 = 1.3$ μm, and $\Delta\nu = 1$ Hz, then $\Omega_{min} \sim 30 \times 10^{-8}$ rad/s. Several companies market application-specific FOGs, for example, for missile guidance systems, commercial aircraft's navigation systems, military helicopter as well as marine and submarine navigation systems, gas pipe mapping, compasses for tunnel construction, rockets, and even for automotive navigation systems (Lee et al. 2006). Birefringent fibers (including birefringent microstructured fibers) have been also used in fiber-optic Sagnac interferometers for the measurement of pressure (Fu et al. 2008), temperature (Moon et al. 2007), strain, and temperature (Frazao et al. 2006, Sun et al. 2007, Kim et al. 2009a,b), multiple-beam Sagnac topology (Baptista et al. 2000), twist sensor (Zu et al. 2011), cladding-mode resonance-based Sagnac interferometer (Dong et al. 2011), and curvature (Frazao et al. 2008).

3.3.5 Fiber-Optic Fabry–Perot Interferometer

In any of the sensors given earlier, the interaction of the measurand with the guided light in a fiber is most important and the same needs to be maximized for maximum sensitivity of the sensor. In order to enhance this interaction and hence the sensitivity of the sensor, one could design the sensor in such a way that the light is made to pass through the same length of the signal fiber several times, which could be possible through the use of multiple-beam interferometry like that in a Fabry–Perot interferometer (FPI). In general, FPI, also often referred to as FP etalon, is composed of two parallel reflecting surfaces with a small separation between them as shown in Figure 3.26a.

In intrinsic fiber form (referred to as FFPI in the literature), the reflecting mirrors are formed either within the fiber itself (see Figure 3.26b) through formation of Bragg gratings (Wan and Taylor 2002, Wang et al. 2007) or through micromachining (Ran et al. 2008, 2009) or by some other means like chemical etching (Machavaram et al. 2007) or

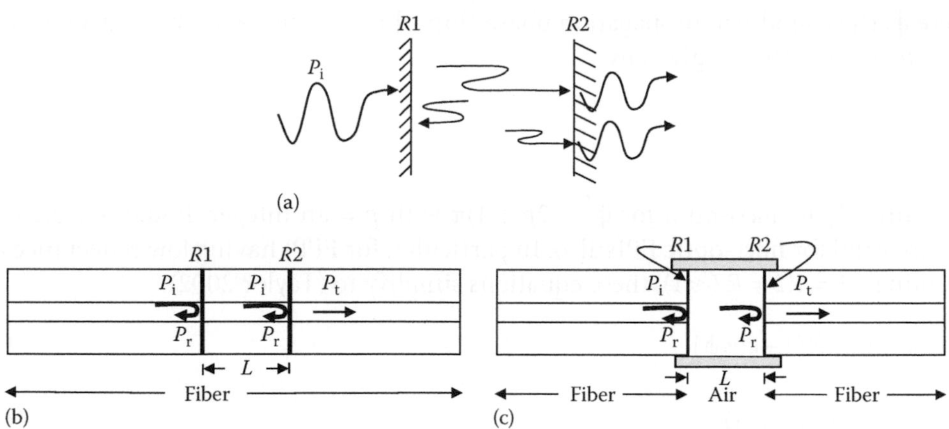

FIGURE 3.26 (a) Basic structure of Fabry–Perot interferometer (FPI); input power P_i is incident on a mirror having reflectivity $R1$. Part of this light is transmitted and partly reflected by the mirror, while the transmitted light gets again partially reflected and transmitted at the mirror $R2$. (b) Intrinsic and (c) extrinsic FPI-based fiber-optic sensor. (After Taylor, H.F., Fiber optic sensors based upon the Fabry–Perot interferometer, in *Fiber Optic Sensors*, F.T.S. Yu and S. Yin, eds., Marcel Dekker, New York, 2002; Lee, B.H. et al., *Sensors*, 12, 2467, 2012.)

by end cleaving followed by coating with titanium dioxide and re-splicing with a second piece of identical fiber (Udd 2002). On the other hand, the extrinsic version (referred to as EFPI) is shown in Figure 3.26c, in which one mirrored end of each fiber forms the two reflectors with an air gap in between; portions of the two mirrored fiber ends are kept inside a capillary for stability. In both cases, $R1$ and $R2$ together with a separation in between form a cavity of length L; light entering the cavity through partially reflecting $R1$ is partially reflected and partially transmitted through $R2$. The reflected wave from $R2$ undergoes further partial reflections at $R1$ and $R2$. Those wavelengths, for which L is an integral multiple of half the wavelength within the cavity (resulting one round-trip through the cavity is an integral multiple of the wavelength), would add in phase on transmission through $R2$. Any perturbation to the cavity in terms of its length or refractive index by a measurand through either of the mirrors would affect the optical path length of the cavity. In principle, the cavity could be extremely small mimicking a point sensor. It may be noted that a fiber-optic FPI sensor is less complex than fiber-optic MZI and MI sensors described earlier as the FPI sensor does not require any couplers; the earliest form of fiber-optic FPI sensors involved well-cleaved fiber end faces (Kersey et al. 1983) or dielectric mirrors as the mirrors (Petuchowski et al. 1981, Yoshino et al. 1982). If we neglect any potential loss due to scattering and absorption in the mirrors, classical expressions for transmittance, defined as the ratio of transmitted power to incident power and likewise for reflectance of the mirrors in a FPI are (Lee and Taylor 1995)

$$T_{FP} = \frac{T1T2}{1 + R1R2 + 2\sqrt{R1R2}\cos\phi} \tag{3.23}$$

$$R_{FP} = \frac{R1 + R2 + 2\sqrt{R1R2}\cos\phi}{1 + R1R2 + 2\sqrt{R1R2}\cos\phi} \tag{3.24}$$

where ϕ, the round-trip propagation phase shift, through the cavity having a refractive index n and length L, is given by

$$\phi = \frac{4\pi nL}{\lambda_0} \tag{3.25}$$

Naturally, T_{FP} is maximum for $\phi = (2p + 1)\pi$ with p = an integer. Equations 3.23 and 3.24 are valid for fiber-optic FPIs also. In particular, for FFPI having low reflectance and assuming $R1 = R2 = R$ ($\ll$1), these equations simplify to (Taylor 2002)

$$T_{FFP} \cong 1 - 2R(1 + \cos\phi) \tag{3.26}$$

$$R_{FFP} \cong 2R(1 + \cos\phi) \tag{3.27}$$

Equations 3.24 and 3.27 are plotted in Figure 3.27 as a function of round-trip phase ϕ. In order to test the validity of the approximate expression Equation 3.27 for low mirror reflectivity, R as a function of ϕ is shown in the same figure as a full curve.

The FFPI sensors could provide high sensitivity, large dynamic range, and fast response for the measurement of pressure, temperature, strain, displacement, magnetic field, flow rate, etc., and could be used also as embedded sensors in materials (Taylor 2002, Lee et al. 2012). The detection of acoustic noise burst produced by breaking a pencil lead on the surface of an aluminum sample by an EFPI has also been demonstrated as a pressure sensor (Tran et al. 1991). Other FFPI sensors have been reported for the measurement of humidity (Mitschke 1989), displacement (Li et al. 1995, Barrett et al. 1999), and magnetic field (Oh et al. 1997). FFPI sensors are suitable for implementing multiplexed sensing like space division multiplexing (Rao and Jackson 1995, Sadkowski et al. 1995), time division multiplexing (Lee and Taylor 1988), frequency division multiplexing (Farahi et al. 1988), and coherence multiplexing (Davis et al. 1988).

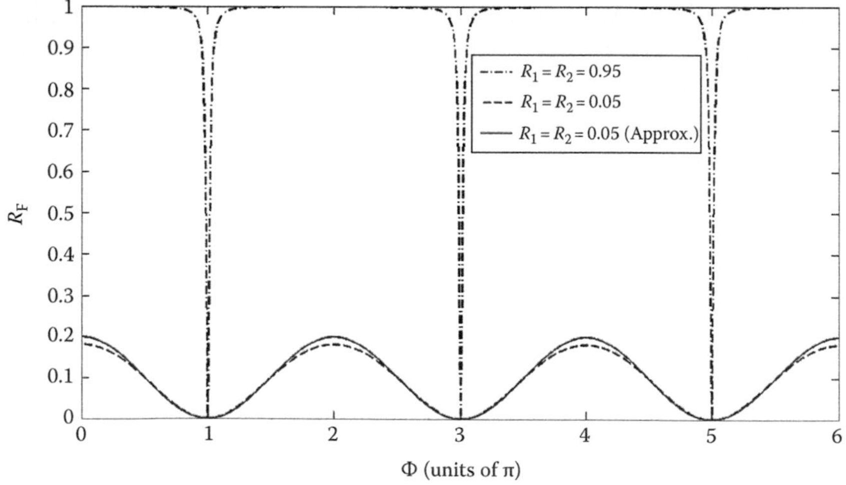

FIGURE 3.27 FPI reflectance as a function of phase ϕ as determined by Equation 3.21 for $R = 0.95$ and $R = 0.05$. For the latter case of low mirror reflectivity, the full curve is obtained from the approximate relation Equation 3.24, indicating that it is indeed a valid expression.

3.4 Fiber Grating–Based Sensors

In-fiber gratings are extensively used in dense wavelength division multiplexed optical communication links, in which fiber amplifiers are inevitably used. In-fiber gratings are also attractive for use as strain and temperature sensors in particular for distributed measurements (Jin et al. 2006). One important attribute of in-fiber grating sensors over typical fiber-optic intensity-modulated sensors is that the measurand information is wavelength encoded, which makes the sensor self-referencing independent of fluctuations in source intensity during a measurement (Orthonos and Kalli 1999, Measures 2001).

Within the types of fiber gratings, Bragg gratings are particularly attractive. Their operation, sensing characteristics, and applications as sensors are described in detail in Chapter 17. In this section, highlights of their functional properties and relevance as sensing elements are presented.

A schematic of a fiber Bragg grating (FBG) of spatial period Λ, which is formed/written through interference between two oblique UV beams made incident (through the silica cladding from one side) inside the photosensitive core made of Germania doped silica, is shown in Figure 3.28a. Due to photosensitivity of the core material, the interference pattern leads to a periodic variation in refractive index in the core region exposed to the interference pattern.

If light from a broadband source is launched into this fiber having an FBG in its core, a particular wavelength λ_B contained within the broadband source that satisfies the following phase matching Bragg condition

$$\lambda_B = 2n_{eff}\Lambda \tag{3.28}$$

would undergo strong reflection while the rest of the spectrum would get transmitted; here n_{eff} represents the fiber mode effective index. These features of an FBG are shown

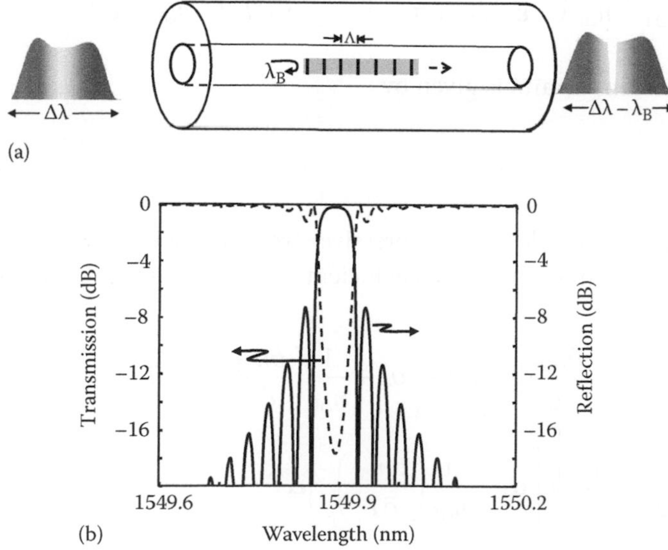

FIGURE 3.28 (a) Schematic of a fiber Bragg grating (FBG) inside the core, which is wavelength encoded with a specific Bragg wavelength λ_B, which gets reflected out of a broadband light launched into the fiber, and rest of the spectrum is transmitted; (b) typical reflection and transmission spectra of an FBG. (Courtesy of Parama Pal.)

in Figure 3.28b. Due to this wavelength-encoded response (for the reflected light), several fiber gratings of different spatial periods could be formed at different locations on the same fiber in a distributed manner in a network, which enable distributed measurements of strain and temperature. This feature is very attractive for local damage detection in civil structures like bridges, buildings (Kersey et al. 1997), and also aircraft (Cusano et al. 2006, Ben-Simon et al. 2007), as well as for internal strain mapping with high spatial resolution (Orthonos and Kalli 1999). The characteristic optical path length η in a fiber grating is given by the product of n_{eff} with spatial wavelength (Λ) of the grating. The parameter η would depend on both stress (σ) and temperature (T), and hence, a change $\Delta\eta$ in η with respect to a reference value would be given by (Pal 2003)

$$\Delta\eta(\Delta\sigma,\Delta T)=\eta(\sigma,T)-\eta(\sigma_r,T_r)=\left[\frac{\partial\eta}{\partial\sigma}\right]_T\Delta\sigma+\left[\frac{\partial\eta}{\partial T}\right]_\sigma\Delta T \tag{3.29}$$

where

The subscript r refers to the reference value

$\Delta\sigma$ and ΔT represent incremental changes in the local stress and temperature from their reference values

For an FBG, it is known that peak reflection, which occurs at the Bragg wavelength (λ_B), is given by (Pal 2000)

$$R_{\text{peak}}=\tanh^2(\kappa L) \tag{3.30}$$

Thus, Equation 3.29 could be rewritten as

$$\frac{\Delta\lambda_B}{\lambda_B}=\left[\left[\frac{\varepsilon}{\partial\sigma}\right]_T+\frac{1}{n_{\text{eff}}}\left(\frac{\partial n_{\text{eff}}}{\varepsilon}\right)_T\left(\frac{\varepsilon}{\partial\sigma}\right)_T\right]\Delta\sigma+\left[\left(\frac{\varepsilon}{\partial T}\right)+\frac{1}{n_{\text{eff}}}\left(\frac{\partial n_{\text{eff}}}{\partial T}\right)_\sigma\right]\Delta T \tag{3.31}$$

where ε, which means strain, is given by

$$\varepsilon=\pm\frac{\partial\Lambda}{\Lambda} \tag{3.32}$$

The signs $\pm$ represent tensile and compressive stresses, respectively. In terms of Young's modulus (Y_F) and thermal expansion coefficient (α_F), Equation 3.31 could be rewritten as

$$\begin{aligned}\frac{\Delta\lambda_B}{\lambda_B}&=\frac{1}{Y_F}\left[1+\frac{1}{n_{\text{eff}}}\left(\frac{\partial n_{\text{eff}}}{\varepsilon}\right)_T\right]\Delta\sigma+\left[\alpha_F+\frac{1}{n_{\text{eff}}}\left(\frac{\partial n_{\text{eff}}}{\partial T}\right)_\sigma\right]\Delta T\\&=\left[1-\frac{n_{\text{eff}}^2}{2}p_e\right]\varepsilon_z+\left[\alpha_F+\frac{1}{n_{\text{eff}}}\left(\frac{\partial n_{\text{eff}}}{\partial T}\right)_\sigma\right]\Delta T=S_\varepsilon\cdot\Delta\sigma+S_T\cdot\Delta T\end{aligned} \tag{3.33}$$

where

ε_z is the strain along fiber axis

p_e represents an effective strain-optic or photo-elastic coefficient defined through

$$p_e = \left[p_{12} - \nu \left(p_{11} + p_{12} \right) \right] \tag{3.34}$$

Here

p_{11} and p_{12} are Pockel's (piezo) coefficients, that is, components of the strain optic tensor

ν is Poisson's ratio

S_ε and S_T represent strain and temperature sensitivities, respectively

For a typical germano-silicate fiber, values of p_{11}, p_{12}, and ν are 0.113, 0.252, and 0.16, respectively. With n_{eff} as 1.482 at 1550 nm, strain sensitivity at this wavelength region is 1.2 pm/με (Lee et al. 2006) and shift in Bragg wavelength is typically almost linear with strain (ε).

The mirror characteristic of FBGs opens new possibilities for interferometric sensing. As an example, a fiber-optic MI based on a fiber coupler, in which an FBG is placed in one of its arms, as shown in Figure 3.29, could be used to measure strain-induced shift in Bragg wavelength and hence strain.

Figure 3.30 depicts sample results from measurements of small strain through a corresponding shift in Bragg wavelength carried out with in-house fabricated FBG by the Fiber Optics Group at the Central Glass and Research Institute (CGCRI, Kolkata, India); achieved strain sensitivity was 1.2 pm/με (Gangopadhyay et al. 2009a, Gangopadhyay 2012).

Results reported on fiber grating sensors indicate their capability to measure dynamic strain from DC to over 10 MHz with a strain resolution of 0.02 με (DC to 10 KHz) to 25 με (10 MHz). Many bridges around the world are now retro-fitted with FBGs as strain gauge for distributed measurements of strain. FBGs in general yield better and more reliable results than their electric counterpart due to smaller size and ruggedness and hence are suitable for embedding in civil structures, aircraft, ships, etc.; in fact, a large array of FBG strain sensors, each of which is uniquely wavelength encoded in terms of individual λ_B, could be sequentially interrogated by a broadband source (see Figure 3.31). Different topologies for the multiplexing of fiber-optic sensors have been discussed in detail in Bløtekjær (1992).

FBGs can be used to detect ultrasound propagating in structures and also for monitoring both static and dynamic strain fields (Culshaw 2004). Experiments (Perez et al. 2001) have indicated that dynamic strains up to 10^{-2} με or smaller and frequencies ~1 MHz

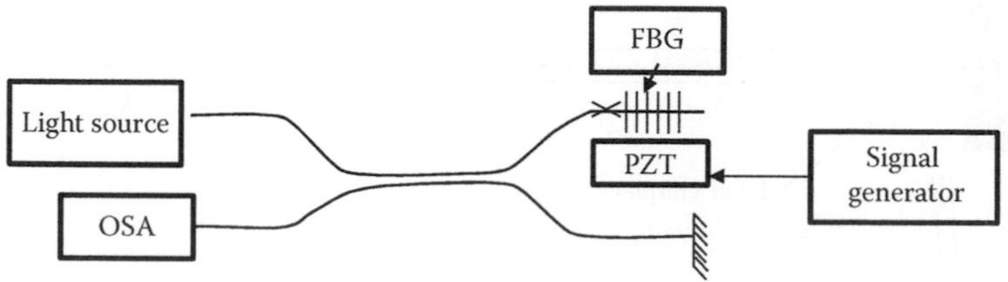

FIGURE 3.29 A fiber-optic Michelson interferometer containing an FBG spliced to one of its arms, which is mounted on a PZT ceramic resonator driven by a signal generator, which could induce stretching of the FBG containing arm. One of the wavelengths (λ_B) from a broadband light source is reflected by the FBG. The reflected signal in both strained and unstrained cases are detected and measured in an optical spectrum analyzer (OSA) placed at the second input port.

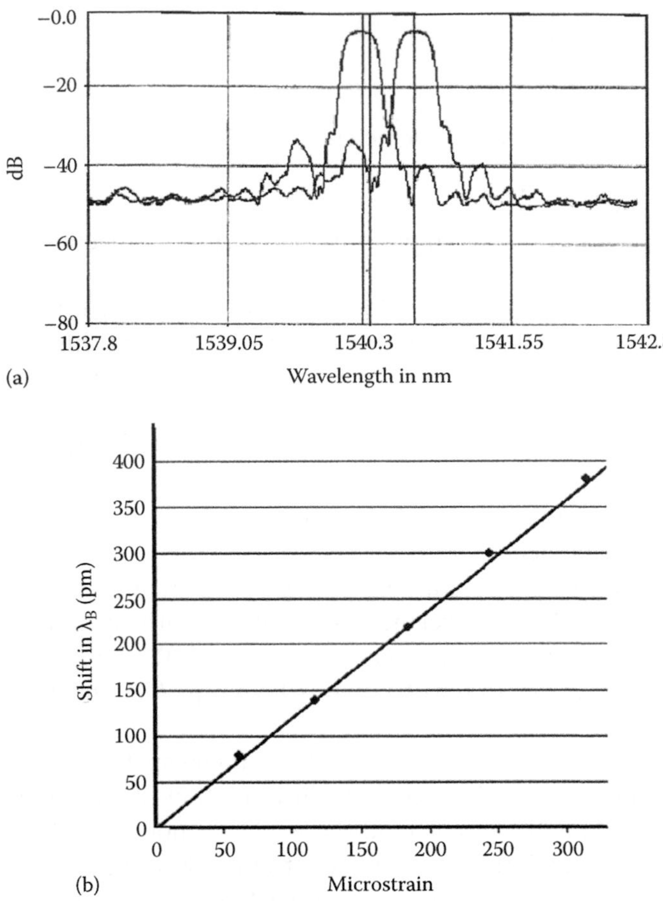

FIGURE 3.30 (a) Optical spectrum analyzer trace of reflected light from a fiber Bragg grating showing shift of the reflection spectrum due to strain and corresponding shift in the Bragg wavelength; (b) measured shift in Bragg wavelength as a function of strain (με). These measurements were carried out in the Fiber Optics Laboratory at CGCRI (Kolkata, India). (Courtesy of T.K. Gangopadhyay, CGCRI, Kolkata, India.)

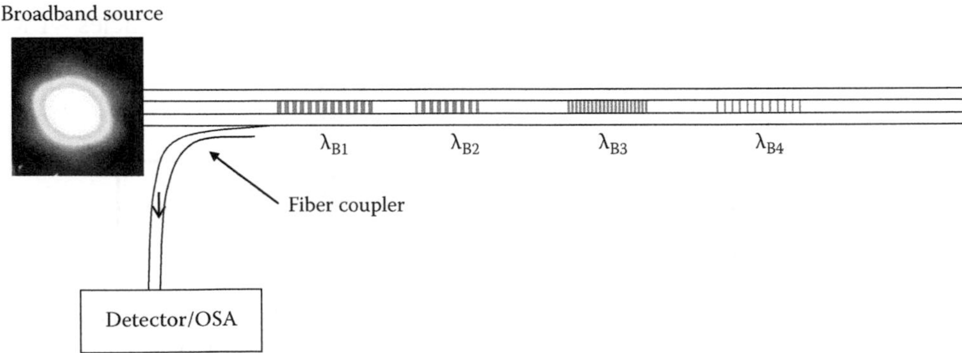

FIGURE 3.31 Distributed strain measurement with an array of FBGs of different pitch attached/embedded in, for example, a civil structure or an aircraft or a ship for monitoring its health. (Courtesy of Parama Pal.)

are detectable by FBGs. When cracks develop in a civil structure due to fatigue and loading, bursts of ultrasonic waves are generated that propagate through the structure. Detection of this acoustic emission (AE) could be exploited as alarm signal for structural failure. AE is due to transient elastic waves within a material, caused by the release of local stress energy, which implies the generation of sound waves when a material undergoes internal stress due to an external force.

If we consider temperature sensitivity S_T (Equation 3.33), the first term α_F representing thermal expansion coefficient of the fiber, which for silica is ~0.55×10^{-6}/°C, whereas the second term, namely, the thermo-optic coefficient term is ~8.6×10^{-6}/°C, which is naturally the more dominating term. At $\lambda_B = 1550$ nm, change $\Delta\lambda_B \sim 0.013$ nm/°C. Temperature change strongly impacts FBG signals, and hence, it is important to realize that precise strain measurements require proper temperature compensation. Since fractional change in λ_B depends on both strain and temperature, it is important to note that for absolute measurements of either of these parameters, one needs to de-convolve the second effect. An additional FBG could be deployed in parallel, which is isolated from the measurand, and through appropriate calculation, measurand signal could be corrected. Readers may find in Chapter 19 detailed information on strain sensing with FBGs (the white paper on strain measurements with fiber grating sensors from the company HBM by Kreuzer (2007)—http://www.hbm.com. optical—is also useful for further details). Figure 3.32 depicts sample results on temperature measurements with in-house fabricated FBG carried out by the Fiber Optics Group at CGCRI (Kolkata, India); λ_B at room temperature of this FBG was 1545 nm. CGCRI group in collaboration with SINTEF (Trondheim, Norway) had successfully installed FBGs to monitor real-time local temperature directly in real-world 400 kV electric power transmission lines of Power Grid Corporation (India) at a location near Kolkata (Gangopadhyay et al. 2009a,b); Figure 3.32b is a depiction of the experimental layout used for these measurements at that site (Gangopadhyay et al. 2009b, Gangopadhyay 2012).

LPGs (Vengsarkar et al. 1996) could also be used as fiber-optic sensor. Bhatia and Vengsarkar (1996) demonstrated LPG-based sensors written on standard telecommunication fibers with temperature, strain, and refractive index resolutions of 0.65°C, 65.8 µε, and 7.7×10^{-5}, respectively. The characteristic period Λ_{LPG} of such in-fiber gratings are typically few hundreds of micrometers in contrast to submicrometer period in FBGs. The transmission spectrum of the LPG exhibits a bandstop kind of filter characteristics as the LPG induces coupling of power from the core mode LP_{01} to cladding modes. In contrast to Equation 3.28, the corresponding phase-matching equation for an LPG leads to

$$\lambda_c = \Lambda \cdot (n_{e1} - n_{e2}) \tag{3.35}$$

where λ_c is the center wavelength at which the power coupling takes place, while $n_{e1,e2}$ represent mode effective indices of the core mode and the cladding mode involved in this power coupling process at this wavelength. Essentially, the LPG acts like a filter (Ghatak and Thyagarajan 1998, Pal 2000). The cladding modes interact with the surrounding environment as they are formed through total internal reflection at the cladding–air interface (see Figure 3.33), and hence any change in the ambient properties of the surrounding medium could be detected in terms of the transmittance of the LPG and measured.

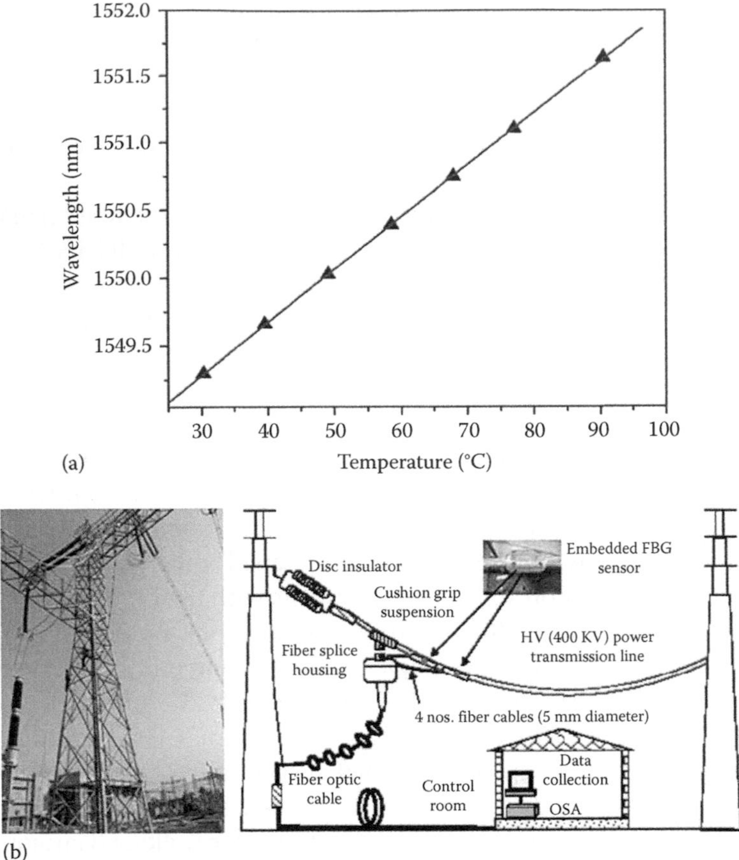

FIGURE 3.32 (a) Measured Bragg wavelength as a function of temperature of an FBG having Bragg wavelength of 1545 nm at room temperature; (b) schematic layout of the field experiment carried out for real-time monitoring of temperature with a pair of FBGs in a real-world 400 kV electric power transmission lines between power transmission towers. (Courtesy of T.K. Gangopadhyay, CGCRI, Kolkata, India.)

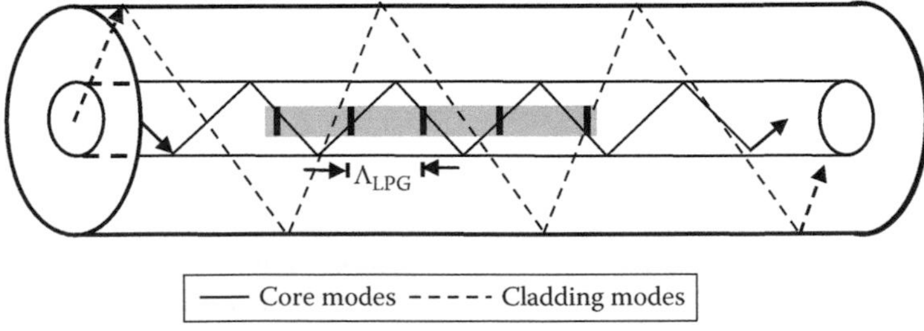

FIGURE 3.33 Schematic of a long-period grating (LPG) that induces power coupling from the core to the cladding modes at the phase matching wavelength λ_c given by Equation 4.8; core and cladding modes are represented schematically through ray diagrams showing core and cladding modes as being formed respectively through total internal reflections at the core–cladding interface and at the cladding–air interface.

A broadband light source, due to coupling of power from a guided to several of the cladding modes having little difference in their propagations constants, leaves a series of loss bands or resonance dips in the transmission spectrum of an LPG. Measurand-induced shift in these loss bands caused by the modified cladding mode spectrum could be exploited to get a measure of the measurand. Readers may find more details about use of in-fiber LPG as sensors in Bhatia (1999).

3.5 Fiber-Optic Current Sensor

Electric current measurement in electric power generation stations is a critical requirement because a sudden accidental surge in current needs to be detected, which should trigger isolation of the failed section from the power system network to minimize damage and system failure time for reliability and stability of the system. Current transformers (CTs) are most often used as current sensors in power protection relay systems (Sanders et al. 2002). In a CT, an iron core and suitable wire windings are used as secondary transformer to step down the high current flowing in the primary to a much lower current (typically 1–5 A). Unfortunately, these devices are subject to EMIs and could suffer distortions due to saturation and residual field in the magnetic cores besides chances of insulation failure due to large current in the system (approximately kiloamperes) (Lee et al. 2006). Due to these factors, optical sensors for large electric currents have assumed considerable importance in recent years (Sanders et al. 2002). With typical transmission voltages hovering around 400 kV requiring control and switching of currents approximately a few kiloamperes, dielectric optical fibers naturally are very attractive for use in communication and sensing in a power industry (Rogers 1992). The basic concept behind the optical detection of electrical current relies on the classical Faraday effect (Rogers 1988, 1992). The Faraday effect states that if a transparent solid or liquid is placed in a uniform magnetic field, and a beam of plane polarized light is passed through it in the direction parallel to the magnetic lines of force (e.g., through holes in the pole shoes of a strong electromagnet), its plane of polarization gets rotated (though remains plane polarized) by an angle proportional to the magnetic field intensity. The modes of light propagating in glass in the presence of a longitudinal magnetic field are right and left circularly polarized light waves, each of which propagates with different velocities. An optical fiber being made of silica exhibits Faraday effect. If a loop of one turn of a single-mode fiber encloses a current carrying conductor, for example, a busbar, the generated magnetic field around it due to the current influences the state of polarization (SOP) of the light propagating through the fiber through Faraday effect. The Faraday rotation of the SOP is given by

$$\theta = V\oint \vec{H}\cdot d\vec{l} \tag{3.36}$$

where

$\vec{H}$ is the applied magnetic field intensity

$\vec{l}$ is the length of the medium

V is the Verdet constant of silica

The Verdet constant for silica is ≅2.64 × 10^{-4}°/A (=4.6 × 10^{-6} rad/A). If there are N turns of fiber in the loop that surround the current-carrying conductor, then by Ampere's law,

$$\oint \vec{H} \cdot d\vec{l} = NI \tag{3.37}$$

By combining Equations 3.36 and 3.37, we get

$$\theta = V \cdot N \cdot I \tag{3.38}$$

where I is the current enclosed by a single fiber turn in the loop. Faraday rotation depends only on the magnitude of the electric current regardless of the shape or size of the loop and position of the conductor within the loop (Lee et al. 2006). A schematic of the Faraday effect-based fiber-optic electric current sensor is shown in Figure 3.34.

Light from a polarized laser after passing through a half-wave plate is focused onto a single-mode fiber (which is single-moded at the laser wavelength, e.g., on a lab-scale experiment, the laser could be a He–Ne laser, for which the fiber should be single-moded). The fiber is used in the form of a loop that surrounds the current-carrying conductor. Output light from the fiber is picked up by a microscope objective and passed through a Wollaston prism or a polarization beam splitter, which divides the Faraday rotated light into two mutually orthogonal linearly polarized components, which are picked up by two independent photodetectors PD 1 and PD 2. The difference between the two PD output intensities, $I_{1,2}$, is normalized with respect to their sum, which is proportional to θ (Lee et al. 2006), that is, for small θ, the ratio R is

$$R = \frac{I_1 - I_2}{I_1 + I_2} = \sin 2\theta \approx 2\theta \tag{3.39}$$

This procedure also makes the output independent of any laser power fluctuation during the measurement or laser drift. In real-world systems, some random birefringence could also be present; besides, bend-induced linear birefringence could also be present

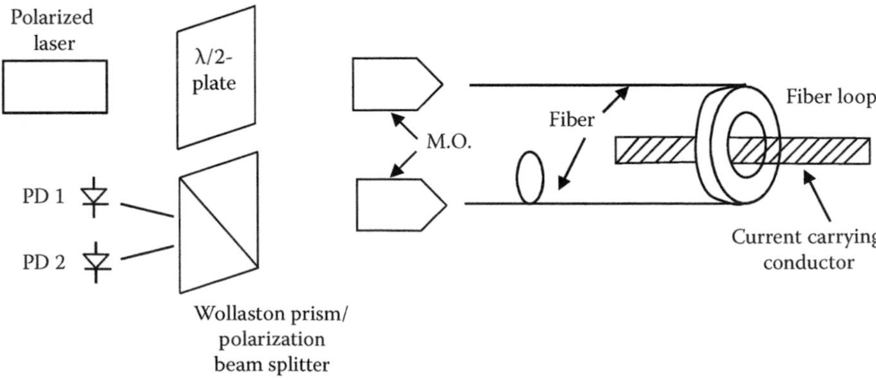

FIGURE 3.34 Schematic of fiber-optic Faraday sensor for the measurement of large electric current (MO stands for microscope objective; functional principle is described in the text).

in the fiber loop especially in case of small loop radii. In the presence of linear as well circular birefringence, R is given by (Ghatak and Thyagarajan 1998)

$$R = 2\theta\left(\frac{\sin\Delta}{\Delta}\right) \tag{3.40}$$

where

$$\Delta^2 = 4\theta^2 + \delta^2; \quad \delta = k_0 \Delta n_{\text{eff}} 2\pi R_1 N \tag{3.41}$$

with θ in rad, Δn_{eff} is the linear birefringence, and R_1 is the fiber loop radius. Linear birefringence Δn_{eff} is inversely proportional to the square of R_1 and directly proportional to the square of fiber cladding radius. For the case in which circular birefringence due to Faraday effect dominates over linear birefringence, R is given by Equation 3.40, whereas if linear birefringence dominates over circular birefringence, then

$$R \approx 2\theta\frac{\sin\delta}{\delta} \tag{3.42}$$

In the latter case, sensitivity is rather low (Ghatak and Thyagarajan 1998). Linear birefringence can be significantly reduced or compensated by the introduction of additional circular birefringence through twist in the fiber (Ulrich and Simon 1979). Fiber-optic Faraday current sensors have been successfully used to measure large currents, approximately a few kiloamperes. Faraday rotation could be also measured through fiber-optic Sagnac interferometer (Briffod et al. 2002).

3.6 Conclusions

In this chapter, we have attempted a unified description of the basic functional principles and applications of a variety of optical waveguide sensor platforms based on intensity, phase, polarization, and wavelength modulation of light, mostly supported by optical fibers. Wherever appropriated, applications of these sensors were also described and a large number of relevant references introduced. The chapter should be useful as an introduction to basics, technology, and applications of these types of sensors, further details are available in subsequent chapters of the book.

Acknowledgment

I wish to thank the editors of the book for their kind invite to write this chapter. I am particularly grateful to Professor Santos for his enormous patience, encouragement, and also for his concern about my recent health issues including my hospitalization for a surgery that I had to undergo, which resulted in long delay in the submission of my chapter. My graduate students Somnath Ghosh and Ajanta Barh are thanked for their help in the drawing of a few figures. My secretary Shashi is thanked for typing some portions of the manuscript. My daughter Parama and wife Subrata are thanked for their help in editing the manuscript and also for their encouragement. Parama is also thanked for pointing out technical corrections here and there and also helping me draw certain figures.

References

Baptista, J. M., J. L. Santos, and A. S. Lage. 2000. Self-referenced fiber optic intensity sensor based on a multiple beam Sagnac interferometer. *Optics Communication* 181: 287–294.

Barrett, M. D., E. H. Peterson, and J. W. Grant. 1999. Extrinsic Fabry–Perot interferometer for measuring the stiffness of ciliary bundles on hair cells. *IEEE Transaction on Biomedical Engineering* 46: 331–339.

Ben-Simon, U., I. Kressel, Y. Botsev et al. 2007. Residual strain measurement in bonded composite repairs for aging aircraft by embedded fiber Bragg grating sensors. Third European Workshop on Optical Fibre Sensors, July 2, 2007; *Proceedings of SPIE* 6619: 661944–661949. http://dx.doi.org/10.1117/12.738849.

Bhatia, V. 1999. Applications of long period gratings to single and multi-parameter sensing. *Optics Express* 4: 457–466.

Bhatia, V. and A. M. Vengsarkar. 1996. Optical fiber long period grating sensors. *Optics Letters* 21: 692–694.

Bløtekjær, K. 1992. Fiber optic sensor multiplexing. In *Fundamentals of Fiber Optics in Telecommunication and Sensor Systems*, B. P. Pal, ed. Wiley Eastern, New Delhi, India.

Brakel, A. V. and P. L. Swart. 2005. Temperature-compensated optical fiber Michelson interferometer. *Optical Engineering* 44: 1576–1580.

Briffod, F., D. Alasia, L. Thevenaz et al. 2002. Extreme current measurements using a fiber optic current sensor. *The 15th Optical Fiber Sensors Conference, Technical Digest*, Portland, OR, Post-deadline paper PD3.

Cole, J. H., C. Kirkendall, A. Dandridge et al. 2004. Twenty-five years of interferometric fiber optic acoustic sensors at the Naval Research Laboratory. *Washington Academy of Science Journal* 90: 40–57.

Culshaw, B. 1992. Interferometric optical fiber sensors. In *Fundamentals of Fiber Optics in Telecommunication and Sensor Systems*, B. P. Pal, ed. Wiley Eastern, New Delhi, India.

Culshaw, B. 2004. Optical fiber sensor technologies: Opportunities and pitfalls. *IEEE Journal of Lightwave Technology* 22: 39–50.

Culshaw, B. 2006. Principles of fiber optic sensors. In *Guided Wave Optical Components and Devices: Basics, Technology and Applications*, B. P. Pal, ed. Academic Press, Burlington, MA.

Cusano, A., P. Capoluongo, S. Campopiano et al. 2006. Experimental modal analysis of an aircraft model wing by embedded fiber grating sensors. *IEEE Sensors Journal* 6: 67–77.

Dandridge, A. 1991. Fiber optic sensors based on the Mach-Zehnder and Michelson interferometers. In *Fiber Optic Sensors: An Introduction for Engineers and Scientists*, E. Udd, ed. Wiley, New York.

Davis, C. M., C. J. Zarobila, and D. J. Rand. 1988. Fiber-optic temperature sensors for microwave environments. *Proceedings of SPIE* 904: 114–119.

Dong, B., J. Hao, C. Y. Liaw et al. 2011. Cladding-mode resonance in polarization-maintaining photonic-crystal-fiber-based Sagnac interferometer and its application for fiber sensor. *Journal of Lightwave Technology* 29: 1759–1763.

Ezekiel, S. and H. J. Arditty. 1982. Fiber optic rotation sensors: Tutorial review. In *Fiber-Optic Rotation Sensors and Related Technologies*, S. Ezekiel and H. J. Arditty, eds. Springer-Verlag, Berlin, Germany.

Farahi, F., T. P. Newson, P. A. Leilabady et al. 1988. A multiplexed remote fiber-optic Fabry-Perot sensing system. *International Journal of Optoelectronics* 3: 79–84.

Fields, J. N. and J. H. Cole. 1980. Fiber microbend acoustic sensor. *Applied Optics* 19: 3265–3267.

Frazao, O., J. M. Baptista, J. L. Santos et al. 2008. Curvature sensor using a highly birefringent photonic crystal fiber with two asymmetric hole regions in a Sagnac interferometer. *Applied Optics* 47: 2520–2523.

Frazao, O., L. M. Marques, J. L. Santos et al. 2006. Simultaneous measurement for strain and temperature based on long period grating combined with a high-birefringent fiber loop mirror. *Photonics Technology Letters* 18: 2407–2409.

Fu, H. Y., H. Y. Tam, L. Y. Shao et al. 2008. Pressure sensor realized with polarization-maintaining photonic crystal fiber-based Sagnac interferometer. *Applied Optics* 47: 2835–2839.

Gangopadhyay, T. K. 2012. Personal communication. CGCRI, Kolkata, India.

Gangopadhyay, T. K., M. Majumdar, A. K. Chakraborty et al. 2009a. Fiber Bragg grating strain sensor and study of its packaging material for use in critical analysis on steel structure. *Sensors and Actuators A* 150: 78–86.

Gangopadhyay, T. K., M. Paul, and L. Bjerkan. 2009b. Fiber-optic sensors for real-time monitoring of temperature on high voltage (400 kV) power transmission lines. *Proceedings of SPIE* 7503: 75034M01–75034M04.

Ghatak, A. and K. Thyagarajan. 1998. *Introduction to Fiber Optics.* Cambridge University Press, Cambridge, U.K.

Giallorenzi, T. G., J. A. Bucaro, A. Dandridge et al. 1982. Optical fiber sensor technology. *Journal of Quantum Electronics* QE-18: 626–664.

Jin, W., T. K. Y. Lee, S. L. Ho et al. 2006. Structural strain and temperature measurements using fiber Bragg grating sensors. In *Guided Wave Optical Components and Devices: Basics, Technology and Applications*, B. P. Pal, ed. Academic Press, Burlington, MA.

Johnstone, W., G. Thursby, D. Moodie et al. 1992. Fiber-optic refractometer that utilizes multimode overlay waveguide devices. *Optics Letters* 17: 1538–1540.

Jones, J. D. C. 1992. Signal processing in monomode fiber optic sensor systems. In *Fundamentals of Fiber Optics in Telecommunication and Sensor Systems*, B. P. Pal, ed. Wiley Eastern, New Delhi, India.

Kersey, A. D., M. A. Davies, J. P. Heather et al. 1997. Fiber grating sensors. *Journal of Lightwave Technology* 15: 1442–1463.

Kersey, A. D., D. Jackson, and M. Corke. 1983. A simple fiber Fabry–Perot sensor. *Optics Communication* 45: 71–74.

Kim, D. W., Y. Zhang, K. L. Cooper et al. 2005. In-fiber reflection mode interferometer based on a long-period grating for external refractive index measurement. *Applied Optics* 44: 5368–5373.

Kim, G., T. Cho, K. Hwang et al. 2009a. Strain and temperature sensitivities of an elliptic hollow-core photonic bandgap fiber based on Sagnac interferometer. *Optics Express* 17: 2481–2486.

Kim, H. I., H. Nam, D. S. Moon et al. 2009b. Simultaneous measurement of strain and temperature with high sensing accuracy. *Proceedings of the 14th Optoelectronics and Communication Conference* (*OECC2009*), Hong Kong, China, July 13–17, 2009. doi:10.1109/OECC.2009.5219750.

Kim, H. Y., M. J. Kim, M.-S. Park et al. 2008. Hydrogen sensor based on a palladium-coated long-period fiber grating pair. *Journal of Optical Society of Korea* 12: 221–225.

Kreuzer, M. 2007. Strain measurements with fiber grating sensors. White paper from the company: HBM Gmbh, Darmstadt, Germany. www.hbm.com.

Krohn, D. A. 1988. *Fiber Optic Sensors: Fundamentals and Applications*. Instrument Society of America, Research Triangle Park, NC.

Kumar, A., T. V. B. Subrahmoniam, A. D. Sharma et al. 1984. A novel refractometer using a tapered optical fiber. *Electronics Letters* 20: 534–535.

Kyuma, K., S. Tai, K. Hamanaka et al. 1981. Laser Doppler velocimeter with a novel fiber optic probe. *Applied Optics* 20: 2424–2428.

Lagakos, N., J. H. Cole, and J. A. Bucaro. 1987. Microbend fiber optic sensor. *Applied Optics* 26: 2171–2176.

Lee, B., Y. W. Lee, and M. Song. 2006. Principle and status of actively researched optical fiber sensors. In *Guided Wave Optical Components and Devices: Basics, Technology and Applications*, B. P. Pal, ed. Academic Press, Burlington, MA.

Lee, B. H., H. Y. Kim, K. S. Park et al. 2012. Interferometric fiber optic sensors. *Sensors* 12: 2467–2486.

Lee, C. E. and H. F. Taylor. 1988. Interferometric optical fiber sensors using internal mirrors. *Electronics Letters* 24: 193–195.

Lee, C. E. and H. F. Taylor. 1995. Sensors for smart structures based upon the Fabry–Perot interferometer. In *Fiber Optic Smart Structures*, E. Udd, ed. Wiley, New York, pp. 249–269.

Li, T., A. Wang, K. Murphy et al. 1995. White-light scanning fiber Michelson interferometer for absolute position-distance measurement. *Optics Letters* 20: 785–787.

Lindstroem, L. H. 1970. Miniature pressure transducer intended for intravascular use. *Transaction on Biomedical Engineering* BME-17: 207–219.

Machavaram, V. R., R. A. Badcok, and G. F. Fernando. 2007. Fabrication of intrinsic fiber Fabry–Perot sensors in silica fiber using hydrofluoric acid etching. *Sensors and Actuators A* 138: 248–260.

Maclean, A., C. Moran, W. Johnstone et al. 2003. Detection of hydrocarbon fuel spills using a distributed fiber optic sensor. *Sensors and Actuators A: Physical* 109: 60–67.

Maier, R. R. J., B. J. S. Jones, J. S. Barton et al. 2007. Fiber optics in palladium-based hydrogen sensing. *Journal of Optics Part A: Pure and Applied Optics* 9: S45–S59.

Marcuse, D. 1982. *Light Transmission Optics*. Van Nostrand Reinhold, New York.

Matsmoto, H., M. Saegusa, K. Saito et al. 1978. The development of a fiber optic catheter tip-pressure transducer. *Journal of Medical Engineering and Technology* 2: 239–242.

Measures M. 2001. *Structural Monitoring with Fiber Optic Technology*. Academic Press, London, U.K.

Medlock, R. S. 1987. Fiber optic intensity modulated sensors. In *Optical Fiber Sensors*, S. Martelucci and A. M. V. Scheggi, eds. Martinus Nijhoff, Dordrecht, the Netherlands.

Mignani, A., L. Ciaccheri, A. Cimato et al. 2003. Spectral nephelometry for making extra virgin olive oil fingerprints. *Sensors and Actuators B* 90: 157–162.

Mitschke, F. 1989. Fiber optic sensor for humidity. *Optics Letters* 14: 967–969.

Moon, D. S., B. H. Kim, A. Lin et al. 2007. The temperature sensitivity of Sagnac loop interferometer based on polarization maintaining side-hole fiber. *Optics Express* 15: 7962–7967.

Nagaraju B., R. K. Varshney, B. P. Pal et al. 2008. Design and realization of a side-polished single-mode fiber optic high-sensitive temperature sensor. *Proceedings of SPIE* 7138: 71381H1–71381H6.

Oki, K., T. Akehata, and T. Shirai. 1975. A new method of evaluating the size of moving particles with a fiber optic probe. *Powder Technology* 11: 51–54.

Oh, K. D., J. Ranade, V. Arya et al. 1997. Optical fiber Fabry–Perot interferometric sensor for magnetic field measurement. *Photonic Technology Letters* 9: 797–799.

Orthonos, A. and K. Kalli. 1999. *Fiber Bragg Gratings: Fundamentals and Applications in Telecommunication and Sensing*. Artech House, Boston, MA.

Pal, B. P. 1992a. Optical fiber sensors and devices. In *Fundamentals of Fiber Optics in Telecommunication and Sensor Systems*, B. P. Pal, ed. Wiley Eastern, New Delhi, India.

Pal, B. P. 1992b. Intensity modulated optical fiber sensors. In *Fundamentals of Fiber Optics in Telecommunication and Sensor Systems*, B. P. Pal, ed. Wiley Eastern, New Delhi, India.

Pal, B. P. 2000. All-fiber components and devices. In *Electromagnetic Fields in Unconventional Structures and Materials*, A. Lakhtakia and O. N. Singh, eds. John Wiley, New York.

Pal, B. P. 2003. In-fiber gratings: Evolution, optics and applications in sensing. *Asian Journal of Physics* 12: 263–274.

Park, K. S., H. Y. Choi, S. J. Park et al. 2010. Temperature robust refractive index sensor based on a photonic crystal fiber interferometer. *IEEE Sensors Journal* 10: 1147–1148.

Parmigiani, F. 1978. A high sensitive laser vibration meter using a fiber optic probe. *Optics and Quantum Electronics* 10: 533–537.

Perez, I., C. Hong-Liang, and E. Udd. 2001. Acoustic emission detection using fiber Bragg gratings. *Proceedings of SPIE* 2007: 209–215.

Petuchowski, S. J., T. G. Giallorenzi, and S. K. Sheem. 1981. A sensitive fiber-optic Fabry–Perot interferometer. *Journal of Quantum Electronics* 17: 2168–2170.

Pitt, G. D., P. Extance, R. C. Neat et al. 1985. Optical fiber sensors. *Proceedings of IEE* 132(Part J): 214–218.

Prerana, M. R. Shenoy, B. P. Pal et al. 2012. Design, analysis, and realization of a turbidity based on collection of scattered light by a fiber optic probe. *IEEE Sensors Journal* 12: 44–50.

Prerana, R. K. Varshney, B. P. Pal et al. 2010. High sensitivity fiber optic temperature sensor based on a side-polished single-mode fibercoupled to a tapered multimode overlay waveguide. *Journal of Optical Society of Korea* 14: 337–341.

Raizada, G. and B. P. Pal. 1996. Refractometers and tunable components based on side-polished fibers with multimode overlay waveguides: Role of superstrate. *Optics Letters* 21: 399–401.

Ran, J., Y. Rao, J. Zhang et al. 2009. A miniature fiber-optic refractive index sensor based on laser-machined Fabry-Perot interferometric tip. *Journal of Lightwave Technology* 27: 5426–5429.

Ran, Z. L., W. J. Rao, X. Liao et al. 2008. Laser-micromachined Fabry–Perot optical fiber tip sensor for high-resolution temperature-independent measurement of refractive index. *Optics Express* 16: 2252–2263.

Rao, Y. J. and D. A. Jackson. 1995. A prototype multiplexing system for use with a large number of fiber-optic-based extrinsic Fabry–Perot sensors exploiting coherence interrogation. *Proceedings of SPIE* 2507: 90–95.

Rogers, A. J. 1988. Optical fiber current measurement. *International Journal of Optoelectronics* 3: 391–407.

Rogers, A. L. 1992. Optical fibers for power systems. In *Fundamentals of Fiber Optics in Telecommunication and Sensor Systems*, B. P. Pal, ed. Wiley Eastern, New Delhi, India.

Rosenthal, A. H. 1962. Regenerative circulatory multiple-beam interferometry for the study of light-propagation effects. *Journal of Optical Society of America* 52: 1143–1148.

Sadkowski, R., C. E. Lee, and H. F. Taylor. 1995. Multiplexed fiber-optic sensors with digital signal processing. *Applied Optics* 34: 5861.

Sanders, G. A., J. L. Blake, A. H. Rose et al. 2002. Commercialization of fiber-optic current and voltage sensors at NxtPhase. *The 15th Optical Fibers Sensors Conference Technical Digest*, Portland, OR, pp. 31–34.

Spillman, G. V. and R. L. Gravel. 1980. Moving fiber optic hydrophone. *Optics Letters* 5: 30–33.

Spillman, G. V. and D. H. Mcmohan. 1980. Schlieren multimode fiber optic hydrophone. *Applied Physics Letters* 37: 145–147.

Sun, G., D. S. Moon, and Y. Chung. 2007. Simultaneous temperature and strain measurement using two types of high-birefringent fibers in Sagnac loop mirror. *IEEE Photonics Technology Letters* 19: 2027–2029.

Swart, P. L. 2004. Long period grating Michelson refractometric sensor. *Measurement Science and Technology* 15: 1576–1580.

Tallman, C. R., F. P. Wingate, and E. O. Ballard. 1975. Fiber optic coupled pressure sensor. *ISA Transactions* 19: 49–51.

Taylor, H. F. 2002. Fiber optic sensors based upon the Fabry–Perot interferometer. In *Fiber Optic Sensors*, F. T. S. Yu and S. Yin, eds. Marcel Dekker, New York.

Tran, T. A., W. V. Miller III, K. A. Murphy et al. 1991. Stabilized extrinsic fiber optic Fabry–Perot sensor for surface acoustic wave detection. *Proceeding of SPIE* 1584: 178–186.

Udd, E. 2002. Overview of fiber optic sensors. In *Fiber Optic Sensors*, F. T. S. Yu and S. Yin, eds. Marcel Dekker, New York.

Uena, S. 1973. New method of detecting surface textures by fiber optics. *Bulletin of the Japan Society of Precision Engineering* 7: 87–90.

Uena, S., N. Shibata, and J. Tsujiuchi. 1977. Flexible coherent optical probe for vibration measurements. *Optics Communication* 23: 407–410.

Ulrich, R. and A. Simon. 1979. Polarization optics of twisted single-mode fibers. *Applied Optics* 18: 2241–2251.

Vali, V. and R. W. Shorthill. 1976. Fiber ring interferometer. *Applied Optics* 15: 1099–1100.

Vengsarkar, A. M., P. J. Lemaire, and J. B. Judkins. 1996. Long period fiber gratings as band rejection filters. *Journal of Lightwave Technology* 14: 58–65.

Wan, X. and H. F. Taylor. 2002. Intrinsic Fabry-Perot temperature sensor with fiber Bragg grating mirrors. *Optics Letters* 27: 1388–1390.

Wang, Z., F. Shen, L. Song et al. 2007. A multiplexed fiber Fabry–Perot interferometer sensors based on ultra-short Bragg gratings. *Photonics Technology Letters* 19: 1388–1390.

Yoshino, T., K. Kurosawa, and T. Ose. 1982. Fiber-optic Fabry–Perot interferometer and its sensor applications. *IEEE Journal of Quantum Electronics* 18: 1624–1633.

Zu, P., C. C. Chan, Y. Jin et al. 2011. A temperature insensitive twist sensor by using low-birefringence photonic crystal fiber-based Sagnac interferometer. *Photonic Technology Letters* 23: 920–922.

4. Intensity Measurements

Principles and Techniques

Carmen Vázquez
Carlos III University of Madrid

4.1 Introduction

Most of the basic principles and techniques for photonic sensors have been known for more than 40 years but industrial applications are growing, fostered by increasing diffusion of low-cost telecommunication components and to the possibility of integrating many optical devices in a single chip. Intensity measurements are also benefited by both aspects; they keep being simple and low cost, even more than before, by using the advantages of electronic signal processing and high-scale integration. Also, novel reference intensity

Handbook of Optical Sensors. Edited by José Luís Santos and Faramarz Farahi © 2015 CRC Press/Taylor & Francis Group, LLC. ISBN: 9781439866856.

Chapter 4

techniques minimizing limited resolution can be developed. Intensity measurements show advantages and drawbacks, which must be considered depending on the specific application to be developed.

In this chapter, general concepts of intensity measurement are reported, such as which elements are present in any intensity measurement system, which are their main parameters, and how they are related to the performance of the sensing technique itself. Special attention is given to analysis of different noise sources responsible for the resolution limitation of the system, and how this is inherently related to the detection mechanisms, an essential background when addressing the task of minimizing the noise effects. Some examples are given providing general guidelines for estimating potential resolution, sensitivity and a dynamic range of intensity measurement techniques. Some strategies to help improve resolution on specific applications are also discussed. Sensing configurations using bulk optics, optical fibers, integrated optics, and hybrid systems are described. Undesirable intensity fluctuation is the main drawback in any intensity modulation–based measurement technique; therefore, a specific section is dedicated to describe different reported approaches to overcome this problem. Novel techniques for intensity measurement and future trends on this area aiming improved performance will also be discussed in this chapter.

4.2 Basics

Intensity measurements are based on modulating the loss of a light path to provide the measurement as an optical intensity modulation signal. It is important to be able to really associate the magnitude to be measured with that intensity modulation.

Any optical intensity measurement technique includes an optical source, a detector, and the sensor. Light can be guided through an optical fiber, a waveguide, or through the air (see Figure 4.1). Different bulk optical elements such as lens, mirrors, prisms, and integrated and fiber-optic devices can be used to direct light to the desired measuring point and to process the optical beam. If more than one sensing point is considered, different multiplexing techniques can be used to get the information from the different points (see Chapter 15). Input light modulation is not mandatory, but in

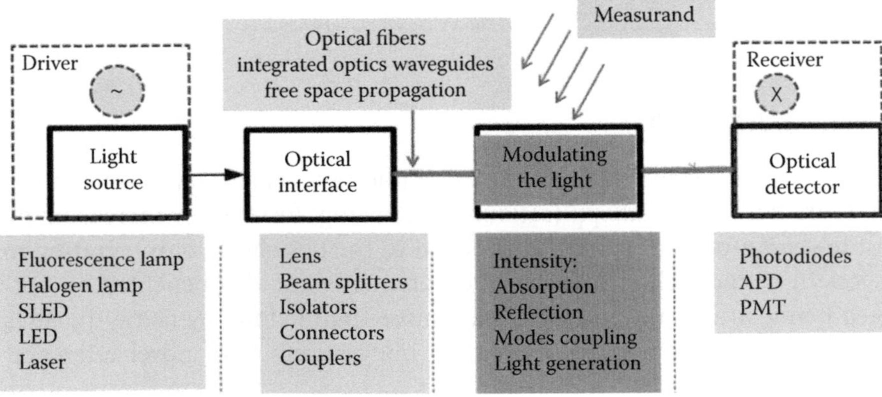

FIGURE 4.1 Schematic of optical intensity measurement system. APD, Avalanche photodiode; PMT, photomultiplier tube; SLED, super luminescence LED; LED, light emitting diode.

some cases, it can be used for improving resolution or for multiplexing purposes. The basic of a photonic sensor in intensity-modulated systems was reported a long time ago (Menadier et al. 1967).

Some simple measuring techniques are on/off modulation based of an optical signal, and in other cases, a technique for continuously monitoring the variation of a specific magnitude is required. Optical techniques can be used for providing either noncontact or contact measurements.

Intensity based-techniques can be classed according to the basic principle used by the optical sensor involved in the measurement (Yao and Asawa 1983, Busurin et al. 1985, Kersey and Dandridge 1990, Kersey 1996, Culshaw 2000). Sometimes a specific category is dedicated to spectroscopic sensors outside intensity category (Udd 1995), but in this section, it is given a general classification based on optical amplitude modulation irrespective of the presence or absence of wavelength dependence, but with potential to be interrogated as an intensity sensor (see Figure 4.2). From this perspective, intensity modulation can be performed by the following methods:

1. Direct attenuation of light in a medium by many different ways, including absorption changes
2. Change in the transverse cross section of an optical channel
3. Change in the reflectivity (absorptivity) due to a change in the refractive index, including the case of frustration of total internal reflection (TIR)
4. Controlled coupling of waveguides and mode interference by a change in the refractive index
5. Generation of additional radiation

Amplitude modulation techniques can be configured to detect a variety of fields using waveguide media and free space (Cole et al. 1981), either directly if optical intensity is dependent on the magnitude to measure or indirectly by using optical signals to be converted to measure the magnitude of interest. In the following, some examples of intensity measurement techniques based on previous classification are briefly described.

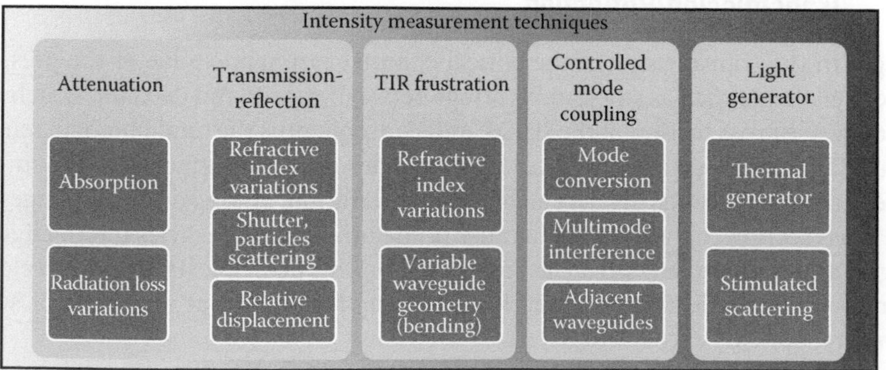

FIGURE 4.2 General classification of intensity measurement techniques.

4.2.1 Light Attenuation

The light propagating through a medium with an absorption coefficient κ, after a distance z, is attenuated by a factor $e^{-\kappa z}$. This also applies if attenuation coefficient, α, of a waveguide or fiber is considered. Control of the absorption coefficient of a specific substance filling the optical channel between source and detector can be used for measuring. Temperature sensing systems have been developed using a semiconductor absorber (Kazuo et al. 1982) and gas detection systems using gas species with well-known absorption lines. Chapters 6 and 18 also include the description of some of these spectroscopic techniques.

Control of the attenuation α of a section of a fiber or waveguide can be used for measuring high-energy radiation, as changes in α are proportional to radiation dose (Suter et al. 1992).

The evanescent field or wave is an effect experienced by light at boundaries with a refractive index change. Although the light can be totally reflected by the boundary, part of the electromagnetic field enters the other side, occupying the two involved media. This is the case for waveguides in which the light is guided by the inner medium of higher refractive index, but a small percentage of the field (i.e., the evanescent field) actually travels in the cladding. If the cladding is removed, or its properties can be modified by some external magnitude, the evanescent wave, and thus the guided light, is able to interact with the measurand, providing the basis for many sensing schemes. The main drawback of evanescent field sensors is the weak interaction with the measurand due to the small excursion of the field into the cladding. In fact, the field strength decreases in an exponential way outside the core (by a factor e^{-z/d_p}, z is the distance from the interface in the optically thinner medium and d_p is the penetration depth), regardless of the waveguide shape or the modal distribution. Coating the waveguide with a higher refractive index layer can be used for sensitivity enhancement while keeping the penetration deep (Hahn et al. 2008). This is important in optical monitoring of biological and chemical processes at boundary layers. About waveguide material selection, if low penetration depth is a key point, $LiNbO_3$ with a high refractive index is a good choice, providing low penetration depth of the evanescent field of the order of only a few tens of nanometers into the covering medium. With the appropriate selection of wavelength, many chemical species in liquid and gas forms can be directly detected through the absorption of the evanescent wave. A specific section with fiber-optic examples is reported in Chapter 13.

4.2.2 Transmission/Reflection

Changes in the transmission of the optical channel can happen due to the presence of different elements such as stops and gratings between source and detector, which reduce the transverse cross section of the signal and, consequently, the total photocurrent at the detector. Those elements can be fixed on both sides or can be attached to moving parts for measuring vibrations, displacement, and acceleration, among others. Changes in the beam cross section can also occur because of the presence of particles forcing light scattering. Many commercial optical barriers based on this principle have been developed in transmission and reflection configurations as part of automatic process control, security perimeters, optical barcode reader, etc.

Figure 4.3 shows an example of a noncontact intensity-measuring technique by a change in the transverse cross section of an optical channel in a reflection mode.

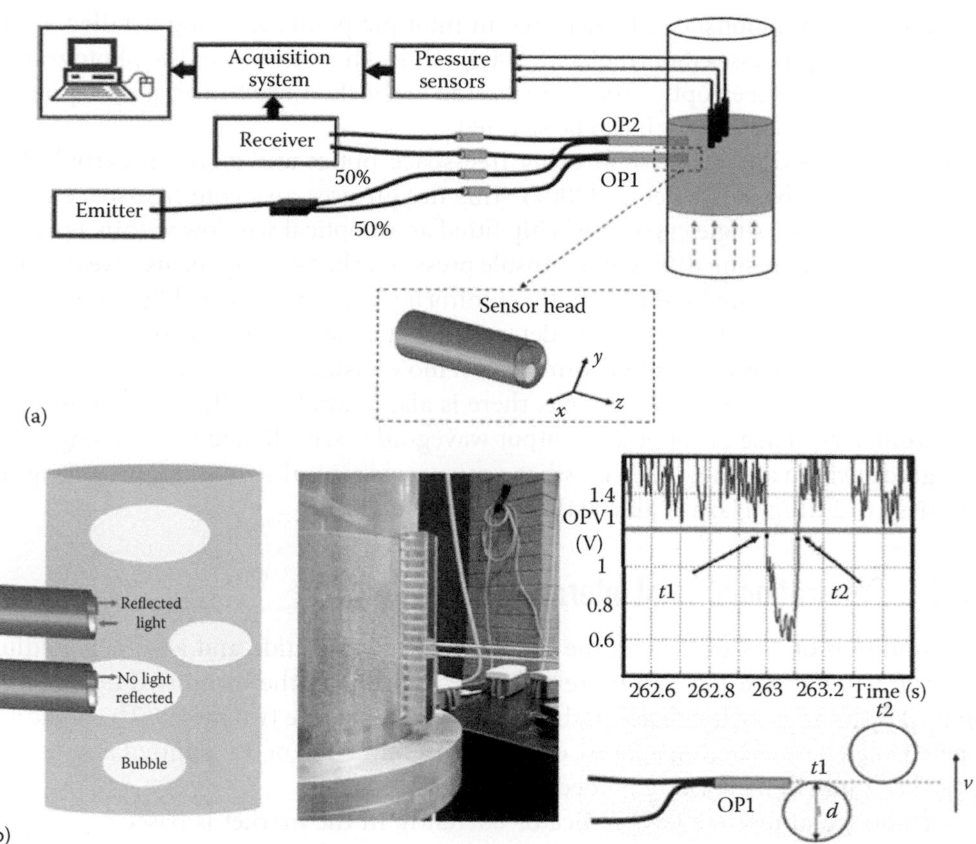

FIGURE 4.3 (a) Bubble optical measuring system; (b) measurement principle, photograph of a specific setup and output voltage measurement versus time.

In bubbling fluidized beds, a voidage (gas–solid) distribution optical measurement technique is used for optimizing combustion efficiency (Vázquez et al. 2007). The system has a red light-emitting diode (LED) as the optical source, a receiver with a Si photodiode connected to a data acquisition card within a PC for monitoring the measurement, and the sensor head (see Figure 4.3a). Light is guided to the measuring point through an optical fiber. When light exits the fiber, it emerges in a cone of radiation, which is partially scattered only in the presence of particles in the fluidized bed. Thus, if a second fiber, aligned in parallel with the first one, is allowed to intercept the reflected light, the fraction of light trapped by the second fiber can be converted to an electrical signal that increases in the absence of bubbles. A typical plot of the detected light versus time is shown in Figure 4.3b.

On the other hand, while a bubble is passing, light is not scattered, and no light is collected at the receiver fiber. Emitter and receiver fibers are part of each sensor head. At least two sensor heads are needed to measure not only bubble pierced length but also bubble velocity. Flow velocity of bubbles insensitive to system intensity fluctuations can also be measured.

In comparison with other conventional methods such as using commercial pressure sensors given an average behavior, more localized phenomena in the bed, in static and dynamic conditions, can be detected in multiple points allowing detailed studies of combustion processes (Sobrino et al. 2009). This can be shown in the photograph of Figure 4.3b, where fiber-optic probes are located in the fluidized bed at specific points in weight and penetration, which can be selected.

Another hybrid optical system using free-space optics and a silicon carbide (SiC) chip is reported by Nabeel et al. (2007). This design uses a remote free-space optical beam that targets a single crystal SiC chip fitted as an optical window within a pressure capsule. With increasing differential capsule pressure, the SiC chip forms a weak convex mirror with a changing focal length. By monitoring the chip reflected light beam magnification, pressure in the capsule is determined. It is capable of measuring from 0 to 600 psi (0–41 atm) differential pressures at a remote distance of 2.5 m.

Based on this modulation principle, there is also a whole family of integrated optic accelerometers, made of input and output waveguides self-aligned to a sensing (mass) waveguide, which are integrated on silicon micromechanical structures, providing sensitivities of ~2 dB/g (Plaza et al. 2004).

4.2.3 Frustration of Total Internal Reflection

The condition of TIR of light in the boundary of a waveguide and an outer medium, with refractive indexes n_1 and n_2, respectively, is given by the inequity $\theta \geq \sin^{-1} n_2/n_1$, where θ is the angle of incidence in the interface between the two media. This condition can be violated by changing n_1 or n_2, originating a reduction of transmitted power in the optical channel in the form of guided modes.

A commercial discrete level indicator currently in the market is based on a hybrid system made of two optical fibers and a prism sensor at the end. The refractive index of the outer medium can have a great variation, for instance, from $n_2 = 1$ (air) to $n_2 = 1.33$ (water), frustrating TIR in the last case, so no optical power reaches the detector.

Frustration of TIR also happens in waveguide bends by modifying the refractive index, n, in the boundaries. To this respect, bending effects on numerical aperture in a multimode integrated silica/silicon optical waveguide have been used for measuring temperature with excellent linearity response (1%) between 20°C and 200°C (Remouche et al. 2007). In a similar way, a temperature sensor based on bend plastic optical fibers (POFs) is shown in Figure 4.4a, where power fluctuations are associated with changes on the core and cladding refractive indexes. Sensitivities of 10^{-3} dB/°C and linearity below 1% are obtained (Tapetado et al. 2011). Sharp bends on fibers are also used to measure the level or type of a liquid, because of refractive index changes. Multiple points can be measured by adding additional bends in the same fiber as shown in Figure 4.4b, where a sensitivity of 1.2 dB/bend is obtained, fixed for the application to paramotoring and powered paragliding (Montero et al. 2010).

Microbending loss induced by pressure is used in systems for measuring human breathing (Kloppe et al. 2004) and acoustic waves (Giallorenzi et al. 1982) among others. More examples can be seen in Chapter 13.

4.2.4 Controlled Mode Coupling

In this technique, controlled coupling between waveguides or between different guided modes on multimode waveguides is used for developing intensity sensors. This can be

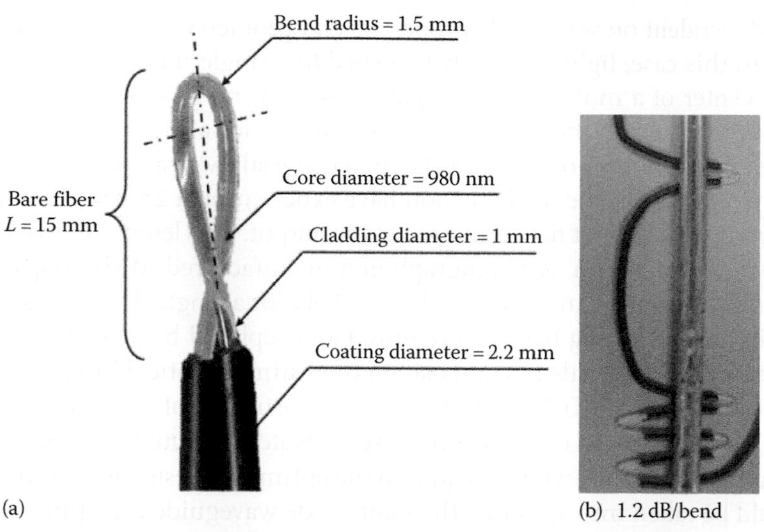

FIGURE 4.4 Photograph of macrobend fiber-optic sensors: (a) temperature sensor; (b) level sensor.

achieved by altering the phase properties of modes inducing multimode interference changes in the same waveguide, or transferring energy to a neighboring waveguide.

An example of an intensity measurement optical technique performed by controlling the coupling of waveguides by a change in n, and based on integrated multimode interference couplers (MMIs), is shown in Figure 4.5a. MMI couplers have been proposed as compact devices in integrated optic circuits (Soldano and Pennings 1995) with high tolerances to fabrication process, but if refractive index changes are induced, self-imaging

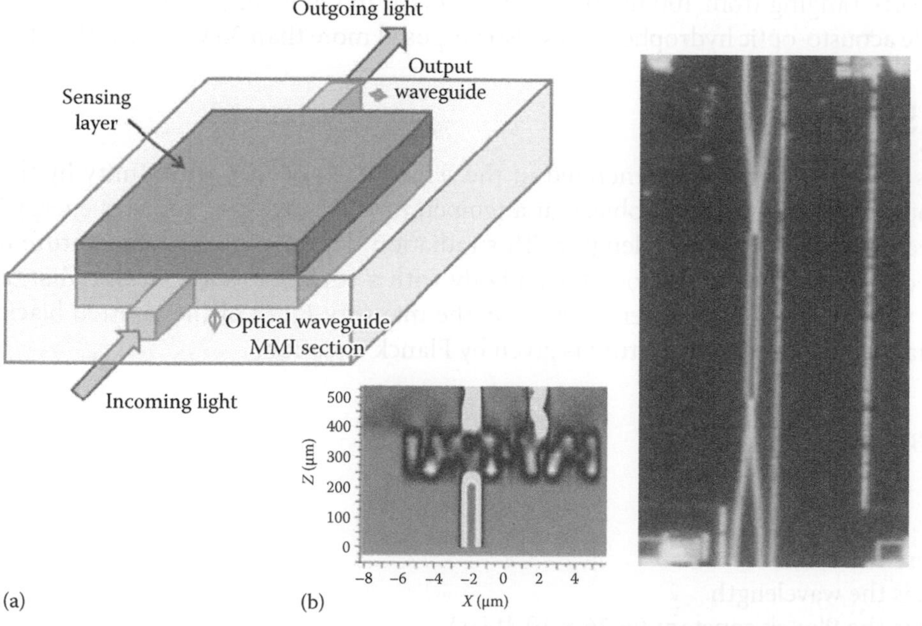

FIGURE 4.5 (a) Biosensor platform based on integrated optics MMI coupler; (b) photograph of a manufactured MMI coupler and mode propagation simulation.

properties dependent on waveguide properties and geometry (Vázquez et al. 1995a) can be altered. In this case, light is directly launched to a single-mode input waveguide that excited the center of a multimode waveguide (the sensing section) coupled to a single-mode output waveguide. The MMI section is designed for a specific width and nominal values of refractive indexes of core and cladding materials with a length L_z. At this length, all the excited modes in the MMI section have experienced a 2π phase change, and the intensity profile of the input field is imaged at the output. This length, L_z, is equal to $3L_\pi/4$, where L_π is the beat length. A photograph of a manufactured MMI coupler and mode propagation simulations can be seen in Figure 4.5b, for a length $L_z/2$. For sensing, at the MMI, section, the cladding has been removed and replaced by a layer material similar to the waveguiding materials in composition to minimize optical losses, but tunable by air humidity percentage (Kribich et al. 2005). The principle of detection is based on an increase in refractive index of the sensing layer as water molecules are absorbed.

Input/output waveguide width is 6 μm, so the optimized design to minimize coupling losses should be taken into account. The multimode waveguide is 50 μm wide; the cladding index is 1.49 and the core index is 1.5; the vacuum wavelength is 1310 nm. A great number of platforms using planar lightwave circuits (PLCs) of this type for biosensing applications are now under development. PLCs allow mass production for developing optical intensity technologies, which can compete with other existing technologies in key parameters such as cost, size, sensitivity, and flexibility.

As examples of controlling coupling between adjacent waveguides, fiber-optic couplers have been reported for measuring temperature and level (Montero et al. 2009a,b), among others. This principle has also been used in integrated optic devices by controlling longitudinal or vertical coupling between waveguides. As an example, a sensor made of flexible polymer waveguides suffering amplitude fluctuations of 20 dB for measuring pressure ranging from 100 to 500 kPa is reported in Oh et al. (2009). A coupled waveguide acousto-optic hydrophone was also proposed more than 30 years ago (Fields 1979).

4.2.5 Light Generator

This technique uses light generated in the sensor itself or in its proximity by thermal or optical interaction. Any object, at a temperature T above 0 K, radiates energy in all directions at different wavelengths. This radiation depends on the temperature of the object. For a perfect blackbody (i.e., a body with a surface absorbing all radiations, at any wavelength and at all temperatures), the intensity $B_\lambda(T)$ of the emitted blackbody radiation at a given temperature is given by Planck equation:

$$B_\lambda(T) = \frac{2hc^2}{\lambda^5 \cdot \left(e^{hc/k\lambda T} - 1\right)} \tag{4.1}$$

where

λ is the wavelength
h is the Planck constant (6626×10^{-34} J·s)
k is the Boltzmann constant ($13{,}806 \times 10^{-23}$ J·K)
c is the light velocity in vacuum ($29{,}979 \times 10^8$ m/s)

By integration over all wavelength range, it can be seen that total energy density of blackbody radiation increases at a rate of T^4.

Radiant heat transfer measurements are converted to temperature measurements in a pyrometer. Any pyrometer includes a radiation-receiving head, with a focusing optical system, and a radiation detector that produces an electrical output that, after conditioning and processing, is suitable for the display of temperature, data storage, or to perform a control function. The response of a pyrometer to incident radiation can be written as (Rugunanan and Brown 1991)

$$P(\lambda,T) = B_{\lambda}(T)\cdot r(\lambda)\cdot d\lambda \tag{4.2}$$

The function $r(\lambda)$ describes the spectral response characteristics of the detector.

There are basically three types of pyrometers: wideband, narrowband, and ratio (or two-color) pyrometers. Wideband pyrometers respond to a wide range of wavelengths and usually have thermistor bolometer or thermopile-type detectors. Narrowband pyrometers have filters that allow only a narrow range of wavelengths to pass through to a detector. This detector, usually a photodetector, has a maximum sensitivity in the wavelength band transparent to the filters. In ratio (two-color) pyrometry, filters allow two narrow wavelength bands to pass through into two independent photodetectors to avoid the influence on measurements of emittance wavelength dependence as well as intensity fluctuations. The cone of view of the detector depends on the optical system used; if fiber optics is put at the pyrometer head, the temperature of very localized areas can be measured (Müller and Renz 2003). Fiber-optic pyrometers are also attractive for measuring on hard environment conditions such as gas turbine engines (Mason 1989).

Additional radiation, in the form of scattered light, can be generated in an optical fiber by illumination with a high power source. In the framework of intensity-based sensors, Raman scattering should be considered. The Raman scattering process produces spectral components about the exciting (pump) wavelength comprising Stokes and anti-Stokes emission. The ratio of Stokes to anti-Stokes peak intensity is proportional to the factor $e^{(hc\Delta\nu/kT)}$ where $\Delta\nu$ is the frequency shift at which measurement is made, k is the Boltzmann constant, and T is the temperature in Kelvin. Monitoring this ratio at different points of the fiber allows to obtain distributed temperature information (further information will appear in Chapter 16).

4.3 Elements of Optical Intensity-Measuring Techniques

In the following, the different elements that are usually present on any optical intensity measurement are described. Some design rules and basic aspects to be addressed are also discussed.

4.3.1 Sources

Although the amplitude modulation systems make no special demands on the source of light compared to other optical techniques, there are several aspects to be taken into account when selecting the optical source such as the optical power, central wavelength, full width half-maximum, source noise, and, depending on the application, its long-term stability. In general, the majority of the amplitude modulation systems do not require the use of coherent radiation, although some of them can operate only when polarized light is used.

Optical techniques generally take advantage of the small size and low power consumption offered by solid-state optical sources, for example, LEDs and laser diodes (LDs). Any of them can be used in intensity measurement techniques, so we will focus on them in this section. Although when very stable sources are needed, such as in potentiometric probes bounded to the membrane of different cells for detecting their membrane potential, different halogen lamps can be used (Salzberg et al. 2005).

Optical power is critical because it determines the available power to be used in the whole system, and the level required depends on the sensing principle and the transmissive/reflective coefficients along the optical system, as well as light coupling at different elements (like the emitter, receiver, and the sensing head). Sometimes, this available optical power is referred to as the power budget of the optical measuring technique. This power budget affects the maximum distance to the measuring point in case of remote operation, the maximum number of points where the measurement can be made, and the resolution of the sensing technique.

Source wavelength is also a basic parameter because, depending on the measuring technique, the targeted measurand can interact with the sensing head only in a specific optical spectrum range, determining the needed spectral response of the receiver, which must be able to detect the optical power on that spectral window.

On the other hand, the selection of the optical source can also be determined by the way in which online calibration is developed, as it will be described in Section 4.6, devoted to self-reference techniques. The minimum measurable amplitude variation, or resolution of the system, is limited by system noise, and depending on the light source, source noise can have a major impact on the system.

There are several noise sources in optoelectronic systems such as thermal, shot, and $1/f$ noise, and some of them are dependent on the source type (see Table 4.1). Most of the

Table 4.1 Main Noise Sources and Detector Noise Characteristics

Noise Source	Optical Sources (LD, LED)	Photodetectors (PIN, APD)
Noise type	Amplitude noise[a] (A^2/Hz)	Shot noise
	$\left(\approx \frac{1}{f^{\gamma}}; \gamma = 1, \text{or } \gamma \in [1,2]\right)$	$\langle i_{shot}^2 \rangle = 2qI_{phd}\Delta f M^2$
		Thermal noise
		$\langle i_{th}^2 \rangle = \frac{4kT\Delta f}{R_d}$
		Dark noise
		$\langle i_{dark}^2 \rangle = 2qI_{dark}\Delta f M^2$

[a] Noise power spectral density; f = measurement frequency, T = absolute temperature (K), q = electron charge, k = Boltzmann constant, Δf = equivalent noise bandwidth, I_{phd} = average photodiode current, I_{dark} = photodiode dark current, R_d = parallel photodiode resistance, and M = avalanche factor (only for APD).

studies have been done on laser sources due to their impact on optical interferometric techniques, so amplitude and phase noise fluctuations need to be addressed. Anyhow, in this section for amplitude measurements, only amplitude noise terms are relevant. As some noise terms depend on frequency, spectral noise density S_I (in units of A^2/Hz) is considered. Shot noise and $1/f$ noises are dominant at low frequencies.

Low-frequency intensity fluctuations in LEDs are much smaller than those in superluminescent LEDs and LDs. Therefore, when coherent light sources are not required, LEDs are preferred for low-noise applications.

Photodetector shot noise (described in the next section) can be partially suppressed at high frequencies by driving the LED via a constant-current source (Lynam et al. 2003), but at low frequencies, that is not the case. A detailed characterization of different non-coherent sources is reported in Rumyantsev et al. (2004). At low frequency [$f < (10^2–10^4)$ Hz], the noise spectra of LEDs depends on frequency as $1/f^\gamma$ with γ in the interval [1–2]. Halogen lamp noise spectra have also been analyzed, showing a more unstable behavior, close to a $1/f^4$ law.

This $1/f$-like noise, dominant at low frequencies, may either increase or decrease with an increase in the LED current for different devices.

The LED area should also be taken into account for the comparison of the low-frequency noise properties of different LEDs. If the noise sources are uniformly distributed within the area, the overall noise can be reduced by connecting several LEDs in parallel and/or in series up to the level of the shot noise at a given frequency.

LDs have higher noise values, and the biasing point and modulation frequency are critical. LD intensity noise has a maximum relative noise value around lasing threshold current, I_{th}. Relative intensity noise (RIN) is defined as 20 log ($\Delta I_l/I_l$) with I_l the average laser emission intensity and ΔI_l its *rms* fluctuation. So, a 10^{-5} laser intensity fluctuation corresponds to a −100 dB noise level, normalized to a 1 Hz bandwidth. As the current is further increased, the relative noise decreases around 10–20 dB and appears to reach an asymptotic value at around $1.5I_{th}$. It should be noted that the rapid increase in the laser's output, as the current is increased beyond threshold, causes a large reduction in the quantum noise limit. Lasers show a frequency dependence of noise power approximately proportional to $1/f$. RIN values from −100 to −120 dB/Hz for different LDs at 850 nm, 1 kHz, 1 Hz bandwidth have been reported (Dandridge and Taylor 1982).

On the other hand, LDs are commonly provided with integral thermoelectric coolers for maintaining constant-temperature operation providing intensity output stabilization. The forward current of a typical LD is on the order of 50 mA. LDs are easily damaged by transient current. Thus, it is essential that a laser driver incorporating current regulation and transient suppression is used. Alternatively, battery power may be employed. Furthermore, LDs are sensitive to static electricity, and appropriate handling techniques must be employed (Davis and Zarobila 2000).

4.3.2 Optical Transmission Media and Coupling

There are different ways in which light modulation can be developed, as it has been described before where the amount of light at the receiver depends on the presence of particles on a fluidized bubbling bed, or on the coupling between modes in a multimode waveguide versus refractive index variations. In any of them, coupling losses from source to sensing point and to receptor should be taken into account, and if optical fibers are present, connectors should be used for making temporary junctions, and they may also be a source of drift and loss.

In waveguide sensors, aspect ratio between optical fiber and waveguide numerical aperture should be taken into account. In some cases, light from laser source is directly launched onto waveguide input by using appropriate optical systems (GRIN lenses, three-axis micropositioning platforms, and others) to collimate and focus the light. Different optical alignment fixtures are required when modulation takes place in free-space optics.

Either monomode or multimode optical fibers can be used as the transmission media in intensity measurements. Silica fibers or POFs can be used depending on the application. The POFs permit the realization of remote-sensing heads with superior light-collecting capabilities that translate into an overall ease of handling. The POFs typically have a 0.98 mm core (which is much larger than the 9 μm core of glass single-mode fibers used for high-performance optical communications and the 62.5 μm core of multimode glass fibers) that is made of poly-methyl-methacrylate and surrounded by a thin (about 20 μm) fluorinated polymer cladding. The large core size, together with the high numerical aperture (about 0.5), is the advantage of the POF over glass fibers. They also have easier connectivity, less attention to tolerances with a subsequent reduction in the connector requirements and costs, and possibility to use low-cost transceivers based on LED instead of LDs. The POFs also exhibit some disadvantages, such as much larger attenuation and dispersion, which are drawbacks that can have relevant impact depending on the sensing applications. On the other hand, more passive components are developed specifically for multimode and monomode silica fibers as they are most commonly used in optical networks all around the world. Recently, novel photonic crystal fibers (PCFs) are also proposed to be used in some intensity measurement techniques.

Different materials and geometries are used in integrated optic platforms for sensing purposes whose main target is implementing multifunctional miniaturized circuits, possibly of the size of few cm, if not mm (Righini et al. 2009). Integrated optic waveguides can be fabricated on dielectrics, polymers, liquid crystals, and semiconductors. Those based on Si/SiO_2 can benefit from using a manufacture process quite developed with an easier integration of optics and electronics. Active and passive elements can be integrated in a single chip in III–V materials such as InP/InGaAsP, and promising approaches in other technologies are under development. Large-core low-loss waveguides can be manufactured on polymer waveguides (Lee et al. 2007) for improving the measurements and being compatible with doping using organic dyes for specific applications. In other cases, it is interesting having light confined in material with high refractive indexes or with high electro-optic (EO) coefficients; in those cases, $LiNbO_3$ waveguides are used.

Special care must be taken to reduce back-reflections, either in connectors or in interfaces between different materials, mainly if single-mode lasers are employed as the optical source. Isolators can be used for eliminating the tendency of the single-longitudinal-mode lasers to modehop, which appears as noise in the system output. At common communication wavelengths of 1300 nm and C+L bands (1520–1600 nm), these isolators are integral to the laser package. At other wavelengths (e.g., 830 nm), bulk isolators may be employed.

In optical intensity techniques using fiber optics, an erbium-doped fiber amplifier can be used for optical amplification. This is important for sensitivity enhancement because the sensitivity of optical intensity–based techniques is proportional to the input light level. But additional cost and noise terms will be added if optical amplification is used (Souza and Newson 2004).

Optical switches for space optics, hybrid, or waveguide applications have been developed in different technologies, and they can be used for providing reference channels, to implement redundant paths (Vázquez et al. 2003), to address different sensors, or to allow external modulation of the optical sources in specific configurations.

4.3.3 Optical Detectors

There is no special demand on the detection of amplitude modulation systems since an amplitude-modulated (AM) signal can be measured directly using a conventional photodetector. For the purpose of optical detection, *PIN* photodiodes or avalanche photodiodes (APDs) are employed. Anyhow, in most applications, *PIN* photodiodes are used because they have the advantage of not requiring voltage back-biasing. For wavelengths between the visible and 1.1 μm, silicon detectors are generally used. Above 1.1 μm, germanium, GaAs, and InGaAsP are a common choice. In some cases, it is interesting to reproduce human eye response in the optical intensity measurements, or sometimes no high resolution is required. In such situations, photoconductive cells such as light detector resistance are often a choice, as in the system responsible to control switch-off and switch-on of streetlights.

Main optical receiver parameters to be considered in the design are spectral bandwidth, calibration curve showing output photocurrent versus input light intensity (identified as responsivity, $\mathfrak{R}$, in A/W units), response time, and noise figures. Those values determine the resolution and sensitivity of the sensing technique. Spectral bandwidth must be aligned with the optical source and the sensing technique, while the responsivity should be as high as possible at the central wavelength of the measuring system. Additional elements can be placed at the receiver to improve the resolution that depends on the threshold of optical power to be detected (there are different ways of reducing the threshold of optical power to be detected, which are based on noise level reduction).

There are three main types of noise in photodetectors: shot noise, dark noise, and thermal or Johnson noise (see Table 4.1). They are white noise terms proportional to the bandwidth of the detected signal. Shot noise is proportional to the generated photocurrent when light reaches the photodetector; dark noise is proportional to the inverse current on the photodetector when there is no light; and thermal noise increases with temperature. In APDs, an additional term related to the avalanche effect as a multiplying factor should be taken into account (Hamamatsu Photonics 1998). Anyhow, the manufacturer usually gives a parameter that takes into account all the noise effects, the noise-equivalent power (NEP) of the photodetector, which is defined as

$$\mathrm{NEP} = \sqrt{\frac{\langle i_{rt}^2 \rangle}{\mathfrak{R}}} \quad \mathrm{W}/\sqrt{\mathrm{Hz}} \tag{4.3}$$

Chapter 4

where $\left\langle i_{rt}^2 \right\rangle$ is the root mean square total photodiode noise, which can be written as

$$\sqrt{\left\langle i_{rt}^2 \right\rangle} = \sqrt{\left\langle i_{shot}^2 \right\rangle + \left\langle i_{dark}^2 \right\rangle + \left\langle i_{th}^2 \right\rangle} \quad \mathrm{W}/\sqrt{\mathrm{Hz}} \tag{4.4}$$

NEP is an important parameter that can be used in establishing system resolution. Typical values of NEP are in the range 10^{-11}–10^{-15} W/√Hz. Those values increase with photodiode area and biasing inverse voltage. As an example, in an optical intensity measurement system using a photodiode having an NEP value of 10^{-13} W/√Hz and biased in a receptor circuit with an equivalent noise bandwidth (ENB) of 1 kHz, the minimum optical intensity power that can be measured is ~100 pW.

The noise figure of the detector is directly proportional to the bandwidth of the signal to be measured, so sometimes this bandwidth is limited to reduce the noise increasing, therefore, improving the measurement resolution.

4.4 Resolution and Sensitivity

In any intensity measurement technique, it is important to have a characterization of system response to the magnitude to be measured, and this is done by its calibration curve. A general calibration curve for an intensity measurement shows variation of output optical power at the receiver, I, for any variation of the light modulating entity, x. It can be written as

$$I = I_i F\big(x(t)\big) \sim I_i \left[F(x_o) + \left(\frac{\partial F(x)}{\partial x} \right)_{x=x_0} \partial x + O'(x^2) \right] = I_o + \left(\frac{\partial I}{\partial x} \right)_{x=x_0} \partial x + O'(x^2) \tag{4.5}$$

Here, $F(t)$ is a nonlinear function characteristic of each intensity measurement system, I_i is the optical input intensity at the system, and I_o is the optical intensity reaching the detector when there is no optical modulation (at $x = x_o$), also named as output nominal value.

From this calibration curve, it can be quantified how much optical power changes by a specific change on the measurand. To do so, the normalized sensitivity of the optical instrumentation system is defined as

$$S_{n,o} = \frac{(\partial I / \partial x)_{x=x_o}}{I_o} \tag{4.6}$$

It can be seen that $S_{n,o}$ is the slope of the calibration curve at a specific value of the magnitude to be measured, divided by the output nominal value. Maximum sensitivity can be achieved by selecting the bias point or nominal operation value. This parameter is very attractive since the system response and the minimum detectable threshold are readily calculated from $S_{n,o}$ value as will be shown later.

As intensity changes are measured to determine how much the magnitude of interest has changed, those optical intensity techniques have a series of limitations imposed by variable losses in the system that are not related to the environmental effect to be measured.

In any instrumentation system, it is advantageous to have a linear response where the normalized sensitivity is a constant value:

$$I = I_0(1 + S_{n,o}\Delta x) = I_0 + \Delta I \tag{4.7}$$

Linear instrumentation systems are useful in giving the quantitative status of the parameter of interest. On the other hand, nonlinear sensors are useful as threshold sensors for the detection of the existence or proximity of objects or substances. But in any case, system response around the sensing point can always be described by $S_{n,o}$ defined in Equation 4.6.

System response in the electrical domain can be calculated from Equation 4.7 along with the system sensitivity, S. This sensitivity is defined as the incremental current at the photodiode for a unit change in the measurand, Δx, which is directly proportional to optical power and can be written as

$$S = \frac{\Delta I_{phd}}{\Delta x} = \Re I_o S_{n,o} \tag{4.8}$$

Sensitivity of the system can be improved by using optical amplification in fiber-optic-based systems and, in any case, by electrical amplification. Those amplifiers introduce noise terms that should be taken into account. The physical structure of the sensor can also help to improve system sensitivity.

Similarly, the noise-limited detection threshold, at $x_o = 0$, can be written as

$$\Delta x_{\min} = \frac{\sqrt{\langle i_{rt}^2 \rangle}}{\Re I_o S_{n,o}} \tag{4.9}$$

where $\sqrt{\langle i_{rt}^2 \rangle}$ is given by Equation 4.4, and $1/f$ source noise is not included.

In case shot noise limits detection threshold, from Equation 4.4 and Table 4.1, and substituting in Equation 4.9, detection threshold is now given by (Δf is the ENB)

$$\Delta x_{\min} = \frac{\sqrt{2qI_{phd}\Delta f}}{\Re I_o S_{n,o}} = \frac{1}{S_{n,o}}\sqrt{\frac{2q\Delta f}{\Re I_o}} \tag{4.10}$$

It is relevant to indicate that even lower detection threshold can be achieved if the system is designed using a specific optical bias that allows the system to only have dark current (the modulating zone is not illuminated) for $x_o = 0$ (Lagakos et al. 1981). In this new situation, detection threshold is now given by

$$\Delta x_{\min} = \frac{\sqrt{2qI_{dark}\Delta f}}{\Re I_o S_{n,o}} \tag{4.11}$$

A complete description of a hybrid fiber-optic and free-space optical system for level measurement, including resolution and sensitivity aspects, will be given in Section 4.6.2. Under this perspective, other sensor types are analyzed in Giallorenzi et al. (1982) and Carome and Koo (1980).

4.5 Advantages and Drawbacks

There are many advantages in using optical measuring techniques, which also hold to those based on intensity measurements. One of the most important is the elimination of electrical phenomena on the sensing point. As a result, optical techniques can provide extremely good electromagnetic interference immunity, which is crucial in automation, radioactive, harsh, high electric fields, or explosive environments. They can exhibit high sensitivity, are nondestructive, and permit noncontact operation. Additionally, they can have fast response, can be miniaturized and ruggedized, are versatile, and can be designed to sense many parameters, showing also compatibility to remote sensing and multiplexing.

In a general view, optical measurement techniques can be classified into two categories: there are optical intensity techniques and optical-phase techniques. Intensity measurement techniques are inherently less sensitive than their interferometric counterparts as they rely on amplitude instead of optical-phase modulation. Ordinary photodetectors do not respond to the phase variation in an optical waveform, unless an optical reference waveform is provided to establish optical interference. The stability of an optical interferometer demands special coherence properties of the optical wavefronts. These requirements do not appear on intensity measurements where modulated light intensity can be directly detected; so far more simple and cost-effective configurations are possible. It is also worthwhile to mention that sometimes the transducing mechanism used for converting physical or chemical measurands into a modulation of light intensity shows poor linearity.

As previously reported, a serious drawback of loss modulation happens when optical power fluctuations have a different origin than the interaction with the magnitude of interest, associated with unpredictable changes in losses of passive components such as fiber leads, beam splitters, optical couplers, or connectors, related to variable environmental conditions and aging. In free-space optical intensity measurements, environmental light can be treated as noise. In reflection sensors, changes on the reflectivity of the surface are also a noise source. Additionally, random power fluctuations of optical sources at the input of the network and detector responsivity changes can induce an intensity noise added to the desired signals. All these effects directly affect the accuracy of the measurements.

In order to neutralize them, a variety of self-referencing techniques for intensity sensors have been reported (Murtaza and Senior 1994, Vázquez et al. 2007), and some of them are described in Section 4.6. Those techniques should demonstrate long-term reliability of intensity sensors. As an example, if the accumulated system instability is 1 dB, then a system with 30 dB response range can resolve only 30 states of the sensor.

However, these limitations do not prevent optical intensity measurements from having a large potential. They still benefit from the aforementioned intrinsic advantages of optical detection such as remote sensing, electromagnetic immunity, and security in hazardous environment; they can use available standard fibers, waveguides, and sources; the transducers are simple and can easily be made selective to specific measurands; no sophisticated detection system is needed, absolute measurements with high bandwidth can be taken, and multiplexing of a number of sensors on the same fiber lead can be made. Therefore, there is a wide variety of intrinsically safe photonic sensors in which the external parameter of interest modulates the attenuation of an optical signal (Berthold 1995).

It is also important to remark that in some applications intensity techniques have achieved sensitivities and resolutions of the same order of magnitude as far more complex interferometric techniques. As an example, in a review of hydrogen measuring techniques (Silva et al. 2011), analyzing sensitivity and response time for interferometric, intensity-based, and Bragg grating techniques, quite a few intensity sensors show good performance, having higher sensitivity and lower response times. So, in each specific application, the different techniques should be considered before making a choice, having in mind all relevant parameters: sensitivity, measuring range, resolution, cost, portability, and long-term stability.

4.6 Detection Techniques

On intensity measurements, the magnitude to be measured modulates the light amplitude, so demodulation at the receiver is a key point to get the measurement. AM sensors make use of relatively simple detection techniques. It is important to stabilize the output amplitude of the optical source, since any variation appears as a signal. Appropriate signal conditioning is required to interface the demodulated output to the data collection equipment.

Essential to making this demodulation is an understanding of the influence of extraneous noise sources within the local environment and from associated electronic components, as well as the role of optimum photoreceiver design. The latter is especially significant when trying to make low-level noise measurements in the presence of high background signals, because measurement resolution (or minimum detectable measurand change) depends on the receiver noise.

The functional dependence of the different optical noise sources are listed in Table 4.1. There are white noise terms that dominate from a specific frequency depending on the different system elements, and there is a noise term dominant at low frequencies proportional to $1/f^{\gamma}$, with γ typically in the range between one and two. This low-frequency noise term is also present in electronics operational amplifiers (OAs) that are key elements of the receiver.

4.6.1 Simple Direct Detection

In direct detection of the modulated light, if sufficient power is available, photodetector shot-noise power, which is proportional to $I_{phd}^{1/2}$ (see Table 4.1), is not a limitation. However, in the case of multiplexing or high loss systems, the photodetector noise is a factor in determining the total number of sensors that may be powered by a single source or in the viability of the system.

Shot-noise limited *DR*, inherent in a continuous-wave (CW) laser beam that produces an average photocurrent $I_{phd} = \Re I_o$, equals the ratio of signal to shot-noise power, which can be written as

$$\mathrm{DR} = \frac{I_{phd}^2}{\langle i_{shot}^2 \rangle} = \frac{I_{phd}}{2q\Delta f} \tag{4.12}$$

which in dB normalized to 1 Hz bandwidth gives

$$\mathrm{DR}(\mathrm{dB}) = 155 + 10 \log_{10}\left[I_{phd}\,(\mathrm{mA})\right] \quad (\mathrm{dB/Hz}) \tag{4.13}$$

This is the maximum DR achievable in the measurement of a conventional (nonsqueezed) laser light source since any amplitude modulation, no matter how small, cannot be detected below this shot-noise level. DR is a complementary parameter to resolution and sensitivity, which were analyzed in Section 4.4. To preserve the maximum DR and the ultimate noise floor defined by this shot-noise limit, all electronic circuits following the photodetector must produce noise below the shot-noise floor (Scott et al. 2001).

Different receiver designs can be considered, but in terms of photodiode conditioning circuit, small noise levels are obtained by measuring photocurrent in a short-circuit photodiode in a transimpedance amplifier configuration (Hamamatsu Photonics 1998). A simplified version and noise model can be seen in Figure 4.6, where $1/f$ noise is not considered. Using superposition theorem in a linear circuit, the output mean-squared noise voltage spectral density of this simplified receiver is given by

$$\langle v_o^2 \rangle = \frac{2qI_{dark}R_d^2}{(R_s + R_d)^2} + \frac{4kT}{R_s + R_d} + i_n^2 + \frac{4kT}{Rf}R_f^2 + e_n^2\left(1 + \frac{Rf}{R_s + R_f}\right)^2 \quad \mathrm{V^2/Hz} \tag{4.14}$$

where

i_n is the OA input noise current spectral density
e_n is the OA input noise voltage spectral density
R_d is the photodiode dynamic resistance
R_s is the photodiode series resistance
R_f is the feedback resistance in the transimpedance amplifier (as shown in Figure 4.6)

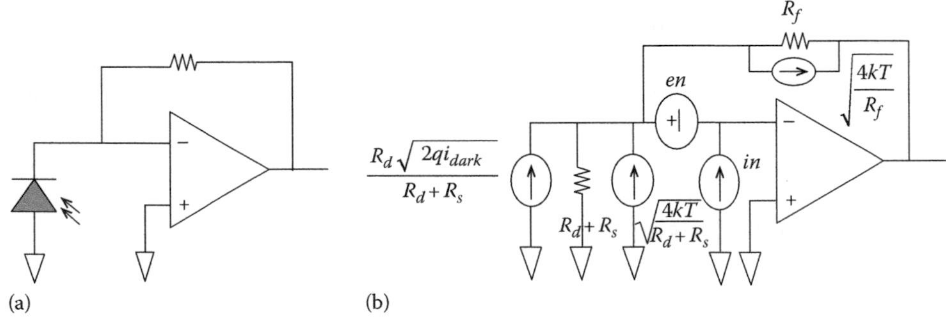

FIGURE 4.6 (a) Schematic of photodiode conditioning circuit based on a transimpedance amplifier; (b) equivalent circuit of a transimpedance amplifier driven by a photodiode including noise sources (except 1/*f* noise).

Output noise has two thermal noise terms associated with feedback and photodiode resistors, two OA noise terms and dark noise. For calculating output mean-squared noise voltage, an integral of output mean-squared noise voltage spectral density over receiver ENB should be calculated. This bandwidth depends on the filter order of the receiver (Pérez et al. 2004). Narrowband receiver bandwidth reduces the effects of measurement system noise contributions, thus improving the measurement resolution. Low-pass filter in the receiver will limit this bandwidth and time response of the system. For avoiding undesirable characteristics of the transimpedance amplifier related to source capacitance (not drawn in Figure 4.6 for the sake of simplicity), the system may benefit from integrating the amplifier into the detector, thereby eliminating interconnect capacitance.

In static measurements, $1/f$ noise should be considered, having in mind that also OAs suffer from its effect (flicker noise) (Texas Instruments 1999). On dynamic measurements, higher frequencies are detected, and this term can have no influence because receiver bandpass filters can be used.

Anyhow, if special care is taken in selecting the proper source and detection circuit, DC recording of slow optical changes as small as 8×10^{-5} without signal averaging can be achieved (Salzberg et al. 2005).

4.6.2 Homodyne Detection

As it has been shown, static sensors are generally intensity-modulated devices and, as such, suffer from lead sensitivity. In those measurements, low-frequency noises arising from temperature, shot noise, and $1/f$-type noise are dominant. To overcome this drawback, radio frequency (RF) modulation of the optical carrier and related demodulation techniques can be used. In this case, frequency selection has to be optimized, for instance, in measurements made above a few kilohertz, amplitude noise is generally not a problem.

About frequency selection, it is also important to develop a noise spectral analysis to check which frequencies should be avoided. Discrete spectral lines at multiples of 50 or 60 Hz are usually associated with power supplies. Cooling fan and other vibrationally or electromagnetically induced discrete spurious lines (spurs) may also be visible and may or may not affect the measurement. Spectrum analyzers are good tools for hunting down these sources of interference.

An AM optical carrier can be used as the input optical power in the intensity measurement technique. Internal modulation of the optical source is possible only with some specific sources such as LED and LD. External modulation using a Mach–Zehnder modulator or acousto-optic modulators for lower frequencies can be used with any source, but cost terms and power budget should be taken into account.

Output optical power in intensity measurement system using an AM optical carrier is given by

$$I_{AM} = I\left[1 + m\cos(2\pi f_m t)\right]\cos\left(2\pi f_o t\right) = (I_o + \Delta I)\left[1 + m\cos(2\pi f_m t)\right]\cos\left(2\pi f_o t\right) \quad (4.15)$$

where

m is the light source modulation index
f_o and f_m are the optical carrier and modulated frequency, respectively
I is the carrier output optical power, so modulated by the intensity sensor

The usual procedure for measuring amplitude changes on the carrier is to downconvert the whole spectrum about the carrier to baseband in a homodyne arrangement. The downconversion process takes place in a mixer, which produces an output proportional to the product of the two inputs. In this case, the input or *RF* port is fed with the signal to be analyzed, so $I_{RF} = I_{AM}$ shown at Equation 4.15, proportional to output optical power, while the local oscillator (LO) port is driven with signal I_{LO}, a pure sinusoid at frequency f_m, and amplitude P_{LO}. At the mixer output (the *intermediate frequency* or IF port), the output current signal is given by

$$I_{phd(IF)} = \Re I_{RF}(t)\Re I_{LO}(t) = \Re^2(I_o + \Delta I)P_{LO}\left[1 + m\cos(2\pi f_m t)\right]\cos(2\pi f_o t)\cos(2\pi f_m t) \tag{4.16}$$

After low-pass filtering, the output signal is

$$I_{phd(IF/LPF)} = \Re^2(I_o + \Delta I)\frac{mP_{LO}}{2} \tag{4.17}$$

The same result is obtained if the photodiode conditioning circuit used capacitors to filter DC and high-frequency noise, in order to reach the shot-noise limit.

This homodyne detection is used in the hybrid intensity level measurement technique (Vázquez et al. 2004) with bulk optics and optical fibers, shown in Figure 4.7. The measurement is based on a change in the transverse cross section of the optical channel.

Light is guided to the measuring point through an optical fiber. When light exits the fiber, it emerges in a cone of radiation that is collimated through a lens placed on the focal plane (focal length *f*), but with its center slightly shifted upward from the center of the emitter fiber. This lens is also used to focus the reflected beam. The cross section of the radiation cone after the lens is a function of the distance from the lens to the reflecting surface perpendicular to the axis of the optical fiber, *D*. Thus, if a second fiber, aligned in parallel with the first one (these two fibers are placed symmetrically about the optical axis of the lens), is allowed to intercept the light, the fraction of light trapped

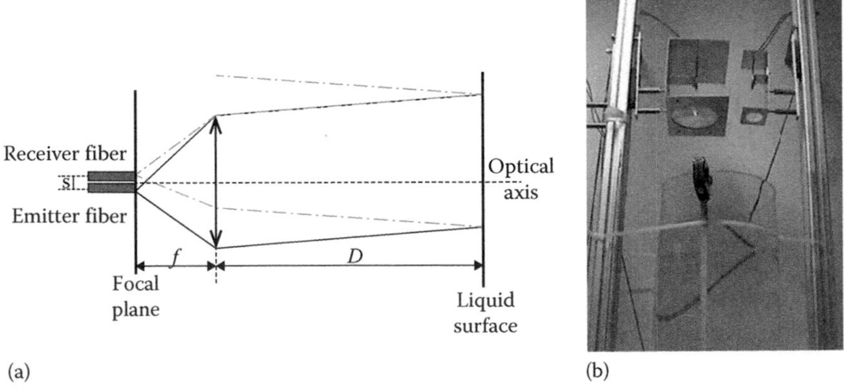

FIGURE 4.7 (a) A schematic showing the incident (–) and reflected (-–) beam paths depending on the location of the lens, *D*, and the emitter and receiver fibers separation, s, and location with respect to focal lens, *f*. (b) Photograph of the level measurement setup.

by the second fiber is a function of the liquid level. This amount of light is proportional to the rate of the common area between the reflected beam and the lens, to the total area of the reflected beam. In the setup shown in Figure 4.7, with an emitter-to-receiver fiber separation $s = 0.98$ mm, $f = 75$ mm, and $D > 200$ mm, the radius, a_1, of the lens used is smaller than the beam radius a_2 (slightly divergent beam) in the whole range of measurements. In this case, the function $f(k, \alpha)$ describing the behavior of the sensor is given by

$$f(k,\alpha) = k\frac{a_2^2}{a_1^2} \tag{4.18}$$

where $a_2 = a_1 + 2Dtg\alpha$, with α the beam divergence and k a constant taking into account effects such as the liquid reflectance and fiber attenuation. Typical plots of the detected light versus liquid level are shown in Figure 4.8.

For measuring the liquid level of different tanks with the same receiving fiber and detection circuit and to improve resolution, the homodyne approach was used. The system is made of a set of LDs modulated by sinusoidal signals at different frequencies (around 1 kHz to avoid $1/f$ noise of laser source) needed for measuring levels of 2 m in a system with more than 30 dB loss. Band-pass filtering at the receiver is included, prior to downconvert the detected signal (see Figure 4.8b). Passive optical components are used to combine signals. Selection of the sensing head is performed by using an optical switch based on bulk optics and liquid crystals.

If a laser source with a RIN of –100 dB/Hz at 1 Hz and –110 dB/Hz at 10 Hz is used, this homodyne technique with 100 kHz bandwidth for multiplexing at least eight sensors and DC filtering allows an improvement of noise in two orders of magnitude. The dominance of $1/f$ noise with photodiode currents below 1 μA in the whole measuring range was observed.

Homodyne detection can also be implemented for measuring low optical signals by means of an optical shutter (chopper) synchronized with the frequency of the modulated

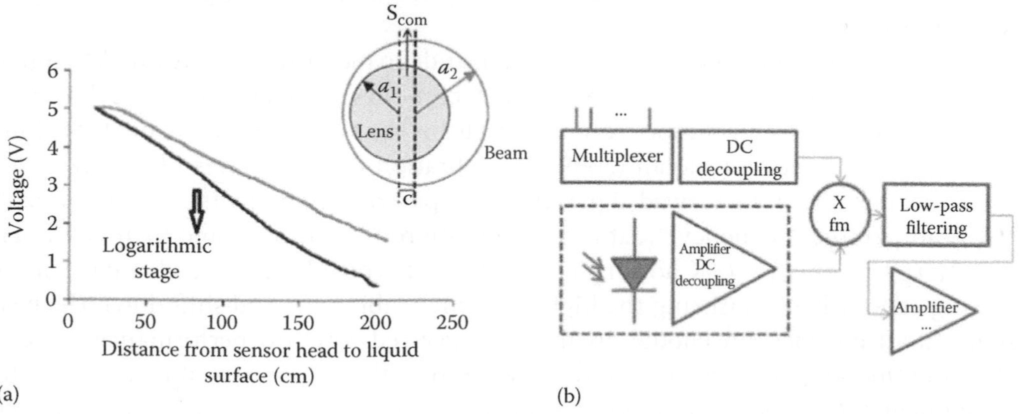

FIGURE 4.8 (a) Measured calibration curve before and after the logarithmic stage of the homodyne detector for a sensor head after the demodulation stage (a_1 and a_2 are the lens and reflected beam radius, respectively). (b) Schematic of the homodyne receptor circuit.

source through a lock-in amplifier with high resolution. The beams are recombined and sent to the same detector. In traditional analog lock-in amplifiers, the signal and reference are analog voltage signals. The signal and reference are multiplied in an analog multiplier, and the result is filtered with one or more stages of RC filters. In a digital lock-in amplifier, the signal and reference are represented by sequences of numbers. Multiplication and filtering are performed mathematically by a digital signal processing chip. As an example, this technique is used for measuring losses in waveguides and passive integrated optical circuits, as in the measuring technique described in Vázquez et al. (1995b).

4.6.3 Heterodyne Detection

As in the homodyne detection, an AM optical carrier is used in heterodyne detection as the input to the system to be modulated by the sensing magnitude. Intensity fluctuations on the carrier are measured by downconverting the whole spectrum about the carrier frequency, f_o, to an IF different from the AM optical carrier-modulated frequency, f_m, as in the homodyne scheme, and given by $f_{IF} = f_m - f_{LO}$.

In this technique, the output current signal is given by

$$I_{phd(IF)} = \Re^2 I_{LO}(t) I_{RF}(t) = \Re^2 (I_o + \Delta I) P_{LO} \left[1 + m\cos(2\pi f_m t)\right] \cos(2\pi f_o t)\cos(2\pi f_{LO} t) \tag{4.19}$$

The same result as in homodyne detection is obtained after low-pass filtering. This technique allows shifting the IF and sweeping a wider frequency range in the measurements.

A low-noise and high-frequency resolution EO technique for measuring RF near-fields such as cell phone's electric near-fields, using an external optical modulator in a heterodyning scheme, is reported in Sasagawa and Tsuchiya (2008). RF near-field signal is mixed with modulated light by an EO crystal and downconverted to the difference frequency. External modulation is done using a commercially available $LiNbO_3$ Mach–Zehnder modulator with operating bandwidth typically from DC to several tens of gigahertz, and its modulation frequency can be arbitrarily tuned by using an RF synthesizer.

In dynamic measurements, where the magnitude to measure generates an AM signal at a specific set of frequencies, heterodyning techniques are also used in the detection circuit for improving resolution. As an example, in the vibration measuring technique reported in Perrone and Vallan (2009), the vibration target behaves like a mixer for the optical beams. The received signal has five spectral components: DC component, at vibration target frequency, f_V, at LED modulation frequency, f_m, and their combinations at $f_V + f_m$, and $f_V - f_m$. Choosing f_m close to f_V, it is possible to take advantage of the mixing effect when measuring the high-frequency vibrations and shift down the beat signal at a frequency low enough to also be measured with low-performance devices. The vibration amplitude is measured as the ratio of the component at $f_V + f_m$ and the component at f_m. This system improves the resolution of the system and does not need additional calibration, as happens on those based only on intensity-modulated back reflected light from the target without source modulation (Dib et al. 2004).

4.7 Self-Referencing Techniques and New Trends on Intensity Techniques

As previously pointed out, a serious drawback in any intensity technique is that undesirable optical power fluctuations can occur, for example, in bends in fibers or waveguides, or because of slightly variable mechanical misalignments or background light fluctuations on free-space systems, as well as associated with component aging. In integrated optical systems, alignment tolerances can also induce error in the measurement. So, either the technique itself is self-calibrated (Perrone and Vallan 2009) or a reference signal is needed for calibration out of the sensor response. This reference signal should preferably undergo all losses in the system not related to the magnitude to measure, so it should follow the same optical path or being multiplexed in the same waveguide as the measurement signal.

In any case, at least source power fluctuations should be monitored by using a beam splitter (Riza et al. 2007) in free-space optics or a coupler on guided media. In semiconductor light sources, this monitoring can be done by electronic means through a monitoring photodiode integrated in the same chip with the light source (Vázquez et al. 2004).

Spatial referencing is the simplest to implement, with the reference and measurand signals being located within two separated optical beams (Murtaza and Senior 1994). This can be done with a beam splitter, an optical coupler, or an optical switch (Intelman et al. 2011), but sometimes, it is difficult to ensure similar common-mode variations in two separated beams.

In time division, one pulse is propagated through the sensor and another pulse follows the same path, spatially separated only in the sensing area (Spillman and Lord 1987). In time domain, optical time-domain reflectometry (OTDR) is also sometimes used for remotely interrogating intensity-based sensors, which shows self-reference properties (Yao and Asawa 1983).

In wavelength normalization (Kamiya et al. 2001), the system response at two wavelengths is used, as in a two-color pyrometer (Müller and Renz 2003). In case of using different broadband wavelength ranges in measuring, the technique is referred to as spectral splitting (Wang et al. 2001).

On the other hand, self-referencing can be based on RF signals modulating the optical carrier; only one is modulated at the sensing point and both are beating at the receiver. Those frequency-based self-referencing methods can measure the differential amplitude (MacDonald and Nychka 1991) at different RF frequencies, with an optical delay in the sensing point. Different configurations based on this approach have been proposed (Baptista et al. 2000, Vázquez et al. 2005, Vázquez et al. 2006). Others use amplitude-to-phase conversion (Abad et al. 2003) associated with optical delays, so the phase of the RF mixed signal is measured. Most recently, a single RF frequency combined with an electronic delay (Montalvo et al. 2008) was used. Other self-referencing techniques use counterpropagating signals (Pérez et al. 2004) or electronic means (Tsai et al. 1997).

Another relevant aspect of remote sensing is the possibility of multiplexing the response of multiple points to get a quasi-distributed sensing scheme. So, it is very important to be able to conjugate self-referencing and multiplexing (Abad et al. 2003, Baptista et al. 2004, Montalvo et al. 2006), which can be done using optical intensity measurement techniques. More information about multiplexing techniques can be found in Chapter 15.

Resolution improvement is another key aspect in any sensing technique; intelligent algorithms can be introduced in the system to reach this purpose as reported in Borecki (2007). The sensor resolution depends on the efficiency of some components such as an asymmetrical coupler, the precision of the optoelectronic signal conversion, and the learning accuracy of the neural network. Therefore, the number and quality of the points used for the learning process is very important. Being applied to a sensor intended for examining the concentration of liquid soap in water, the network has answered with an accuracy of 0.1%, and the construction of the detection block with the possibility of modulation enables properly detecting low-power signals of tens of nanowatts.

Another way to reach resolution improvement, by reducing impact noise, can be achieved using modern digital communication techniques such as code division multiple access or spread spectrum techniques (Aydin et al. 2005).

Recently, different goals such as improving resolution, self-referencing, and multiplexing are simultaneously achieved in the EO fiber-optic topology (Montero et al. 2010) shown in Figure 4.9. The system can remotely address *N* intensity-based optical sensors, and each sensor is placed between two fiber Bragg gratings (FBGs). The sensor loss modulation, H_i, which depends on the measurand, is only present in reflected signal centered at λ_{Si}. Coarse Wavelength Division Multiplexer allows addressing each sensor and recombining optical signals without typical losses of uncolored splitter or coupler configurations. RF modulation permits to avoid the effect of the $1/f$ noise at the reception stage, by using homodyne detection with low-noise lock-in amplifiers. Self-referencing is achieved by electronically processing the two signals reflected at both FBGs, as only one of them is dependent on the loss in the sensing head (Montalvo et al. 2008). Optical delays are also proposed as aforementioned, but less compact designs, greater than 10 km fiber-optic delay lines for a modulation frequency of 2 kHz, are

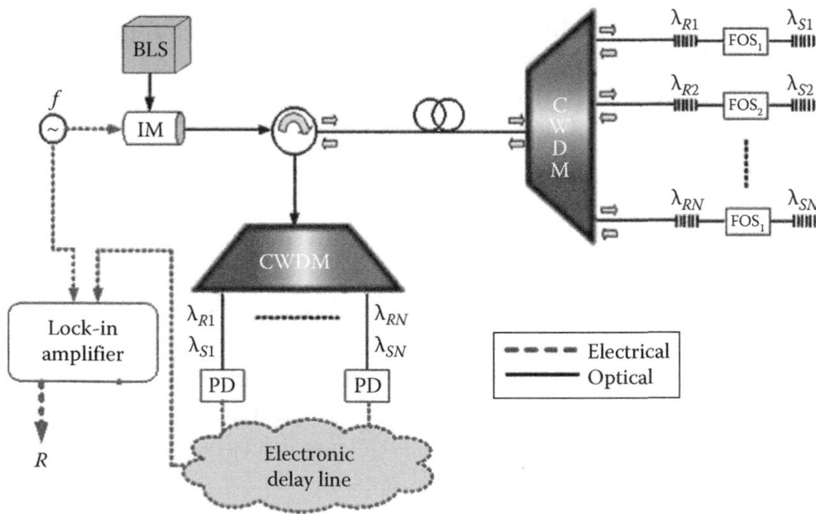

FIGURE 4.9 Schematic of the proposed electro-optic CWDM network for supporting *N* self-referenced optical fiber intensity sensors (FOS$_i$, $i = 1, \ldots, N$). BLS, broadband light source; IM, intensity modulator; PD, photodetector; FOS, fiber-optic sensor.

obtained. By including two electrical delay lines Ω_{1i} and Ω_{2i} at the reception stage, either amplitude or phase self-referencing parameters can be set. Electrical phase-shift reconfiguration can overcome tolerance errors permitting an easy-reconfigurable operation of the network (Montero et al. 2010, Montero and Vázquez 2013). This means that any operation point for each remote sensor can be selected by means of their associated electrical phase shifts at the reception stage, adding flexibility to tune the linearity response, sensitivity, resolution, or another system property demanded by specific requirements.

As an amplitude technique, a self-referencing parameter, R_i, for each generic remote sensing point i of the network is defined. R_i is defined as the ratio between output voltage values at the reception stage for different electrical phase shifts, being given by

$$R_i = \frac{V_o(f,\Omega_{2i})}{V_o(f,\Omega_{1i})} = \frac{M(f,\Omega_{2i})|_{\Omega_{1i}=0}}{M(f,\Omega_{1i})|_{\Omega_{2i}=0}} = \sqrt{\frac{\left[1+\left(\dfrac{2\beta_i}{1+\beta_i^2}\right)\cos\Omega_{2i}\right]}{\left[1+\left(\dfrac{2\beta_i}{1+\beta_i^2}\right)\cos\Omega_{1i}\right]}} \tag{4.20}$$

where

$$M(f,\Omega_1,\Omega_2) = \alpha_i\sqrt{(1+\beta_i^2)\left[1+\left(\frac{2\beta_i}{1+\beta_i^2}\right)\cos\Omega_i\right]} \tag{4.21}$$

with $\Omega_i = \Omega_1 - \Omega_2$. The sensor loss modulation, H_i, which depends on the measurand, is encoded in the transfer function of the self-referencing configuration by means of the parameter β_i, which can be written as

$$\beta_i = \frac{m_{Si}R(\lambda_{Si})d_{Si}}{m_{Ri}R(\lambda_{Ri})d_{Ri}}H_i^2 \tag{4.22}$$

where

m_{Ri}, $R(\lambda_{Ri})$, and d_{Ri} are the RF modulation index, the reflectivity of the FBG, and the photodetector response at the reference wavelength λ_{Ri} for the generic remote sensing point i, respectively

m_{Si}, $R(\lambda_{Si})$, and d_{Si} are the respective similar parameters for the sensor wavelength λ_{Si}

This modulation loss, $H_i(t)$, plays the role of $F(t)$ in Equation 4.5.

The results given in Figure 4.10 show that intensity fluctuations of 10 dB (90% optical power random fluctuations out of the sensing point) do not affect the measurements. This behavior has been tested for different values of the measurand (through β_i).

Around a specific measuring point, and considering small variations, the self-reference system resolution can be defined as the minimum loss change, $\Delta\beta_{min}$, which can be measured. Both reference and sensor wavelengths being close enough, it can be

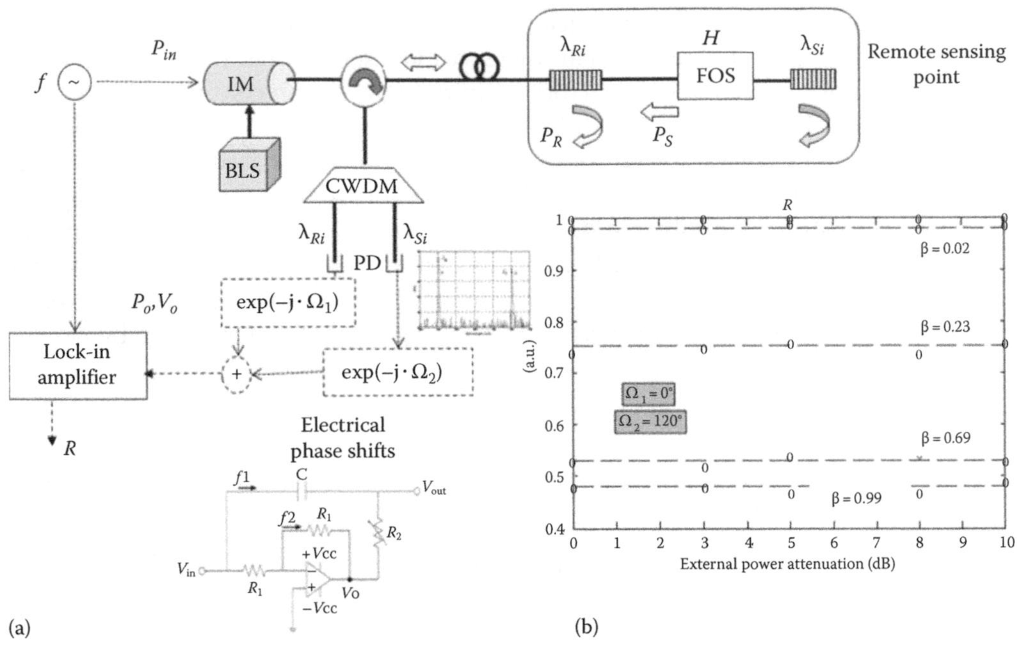

FIGURE 4.10 (a) Schematic of the proposed electro-optic configuration for sensor *i* with the schematic of electronic circuit for generating phase shifts. BLS, broadband light source; IM, intensity modulator; PD, photodetector; FOS, fiber-optic sensor. (b) System output signal (R parameter in arbitrary units) versus nondesirable intensity fluctuations.

considered from Equation 4.22 that $\Delta H_{\min} \sim \sqrt{\Delta\beta_{\min}}$, a typical behavior for a reflective configuration. This resolution is given by

$$\Delta\beta_{\min} = \frac{\Delta R_{\min}}{S_R} \Rightarrow \Delta x_{\min} = \frac{\Delta I_{\min}}{I_o S_{no}} = \frac{\Delta H_{\min}}{H_o S_{no}} = \frac{\sqrt{\Delta\beta_{\min}}}{\sqrt{\beta_o}\, S_{no}} \quad (4.23)$$

where

S_R is the slope of the calibration curve of self-reference parameter R versus β
β_o is the nominal output loss
S_{no} is the normalized sensitivity of the optical sensor

Using typical manufacturer specifications of amplitude resolution in lock-in amplifiers (~9 × 10^{-5}), resolution or $\Delta\beta_{\min}$ value can be calculated, as shown in Table 4.2. From these data, the influence on the final resolution of selecting each pair of electrical phase shifts at the reception stage can also be seen. The use of this two-phase-shift approach provides substantial flexibility on the application of the proposed self-referencing technique.

Further developments for new smart optical sensors that mix nanoelectronics and micro-/nano-optical devices on the same silicon chip are about to come using recent developments on nanophotonic integration in CMOS foundries (Orcutt et al. 2011).

Also, these integration capabilities, but in polymer substrate, including active and passive devices, and integrating self-referencing techniques are under development.

Table 4.2 Sensitivity (S_R) and Resolution ($\Delta\beta_{min}$) of the Two-Phase-Shift Self-Reference Configuration Shown in Figure 7.2

(Ω_1, Ω_2)	S_R	$\Delta\beta_{min}$
(135°, 60°)[a]	1.75	$\sim 5 \times 10^{-6}$
(30°, 150°)[b]	1.75	$\sim 5 \times 10^{-6}$
(160°, 10°)[c]	8	$\sim 1 \times 10^{-6}$
(104°, 180°)[d]	1	$\sim 9 \times 10^{-6}$

[a] Arbitrary electrical phase shifts.
[b] Best linearity condition for the output phase.
[c] Around highest *R* sensitivity.
[d] Best linearity condition for the *R* parameter.

A referenced waveguide chip package containing a pulsed central OLED light source, a chemistry sensor, and two opposite thin fill organic solar cell (OPV) photodetectors is reported in Ratcliff et al. (2010). Light is split into two paths that pass through the two identical chemistries, one exposed to the analyte, and the other not. Self-reference measurement of refractive index changes down to sensitivities of ~10^{-4} RI units are reported in this inexpensive, easily produced, and disseminated OLED/OPV sensor platform.

Another promising field is related to novel sensors using PCF infiltrated with high index materials, employed in intensity-based measurement schemes, as the electric field intensity sensor reported in Mathews et al. (2011). It is made of a 1 cm PCF infiltrated with a liquid crystal, and electric field is applied perpendicular to the fiber axis. A linear fit performed for this electric field range shows that the slope of the response is ~10.1 dB per kV_{rms}/mm with a 1 V_{rms}/mm resolution.

4.8 Conclusion

Intensity measurements benefit from the intrinsic advantages of optical measuring techniques such as remote sensing, electromagnetic immunity, and security in hazardous environments. They can use available standard fibers, waveguides, and sources; the transducers are simple and can easily be made selective to specific measurands; no sophisticated detection system is needed, absolute measurements with high bandwidth can be taken, and multiplexing of a number of sensors on the same fiber lead can be made. As the main drawback, undesirable intensity fluctuations should be overcome on the final application.

A detailed analysis of noise terms either in the optical and electrical domain is essential to delimit the resolution of the system. Shot-noise and $1/f$ noise are dominant at low frequencies. Low-frequency intensity fluctuations in LEDs are much smaller than those in superluminescent LEDs and LDs. Therefore, LEDs are preferred for low-noise applications, where coherent light sources are not required and, indeed, introduce excess noise. At low frequency [$f < (10^2–10^4)$ Hz], the noise spectra of LEDs depend on frequency as $1/f^{\gamma}$ with γ in the interval [1–2]. LD intensity noise has a maximum relative noise value around lasing threshold current. Lasers show a frequency dependence of noise power approximately proportional to $1/f$. The noise figure of the detector is directly proportional to the bandwidth of the signal to be measured, so sometimes this bandwidth is limited to reduce the noise, increasing therefore the resolution, which can also be improved by proper detector biasing.

There are different detection schemes, from direct detection to mixing signals, which can increase the complexity of the system, but providing improved sensitivity and resolution. In direct detection of the modulated light, if sufficient power is available, photodetector shot-noise power is not a limitation. However, in the case of multiplexing or high loss systems, the photodetector noise is a factor in determining the total number of sensors, which may be powered by a single source or in the viability of the system. Shot-noise limits DR. To preserve the maximum DR and the ultimate noise floor defined by the shot-noise limit, all electronic circuits following the photodetector must produce noise below the shot-noise floor. To overcome low-frequency noise effects, modulation of the optical carrier and related demodulation techniques can be used, either in a homodyne or in a heterodyne configuration.

Techniques for avoiding measurement errors because of unpredictable optical power fluctuations, for example, in bends in fibers or waveguides, or because of slightly variable mechanical misalignments or background light fluctuations on free-space systems, as well as associated with component aging, are very important. Spatial referencing systems with optical switches or beam splitters to feed a reference path can be used. In time domain, OTDR is also sometimes used for remotely interrogating intensity-based sensors with self-reference properties. In wavelength normalization, the system response at two wavelengths is used, as in a two-color pyrometer. On the other hand, self-referencing can be based on RF signals modulating the optical carrier; only one is modulated at the sensing point, and both are beating at the receiver. Those frequency-based self-referencing methods can measure the differential amplitude or phase at different RF frequencies, with an optical delay in the sensing point or electronic delays at the reception stage. Sometimes, it is interesting to be able to conjugate self-referencing, multiplexing, and improving resolution for remote measurement in multiple points. As an example, any intensity-based optical sensor can be placed between two FBGs, N sensors can be multiplexed in the same fiber lead, can be remotely address, and self-referencing can be achieved by electronically processing the two signals reflected at both FBGs, as only one of them is dependent on the loss in the sensing head. Electrical phase-shift reconfiguration can overcome tolerance errors permitting an easy-reconfigurable operation of the network. The system has flexibility to tune the linearity response, sensitivity, resolution, or another system property demanded by specific requirements. It has been tested that intensity fluctuations of 10 dB (90% optical power random fluctuations out of the sensing point) do not affect the measurements.

It is also important to remark that in some applications, some intensity techniques have achieved sensitivities and resolutions in the same order of magnitude of other interferometric or more complex techniques. Therefore, in each specific application the different techniques should be considered before making a choice, having in mind all relevant parameters: sensitivity, measuring range, resolution, cost, portability, and long-term stability.

First photonic sensors were proposed more than 40 years ago, some of them based on intensity measurements. Even now, intensity measurements keep playing an important role, being simple and with competitive cost due to increasing diffusion of low-cost telecommunication components and to the possibility of integrating many optical devices in a single chip.

New niches are appearing, discovering the potential of the optical techniques, in bioengineering, energy, and avionics sectors, and their simplicity puts intensity measurements in the first stage. Signal processing tools, novel self-referencing techniques, and integration are major aspects to be considered for improving resolution and global performance of intensity measurements. Further developments for new smart optical sensors, which mix nanoelectronics and micro-/nano-optical devices on the same silicon chip, are about to come using recent developments on nanophotonic integration in CMOS foundries. So this sensing approach, being the one that appeared first, benefits from the notoriety of photonics and will surely be around for years to come.

References

Abad, S., F. M. Araújo, L. A. Ferreira et al. 2003. Interrogation of wavelength multiplexed fiber Bragg gratings using spectral filtering and amplitude-to-phase optical conversion. *Journal of Lightwave Technology* 21: 127–131.

Aydin, N., T. Arslan, and D. R. S. Cumming. 2005. A direct-sequence spread-spectrum communication system for integrated sensor microsystems. *IEEE Transactions on Information Technology in Biomedicine* 9: 4–12.

Baptista, J. M., S. Abad, G. M. Rego et al. 2004. Wavelength multiplexing of frequency-based self-referenced fiber optic intensity sensors. *Optical Engineering* 43: 702–707.

Baptista, J. M., J. L. Santos, and A. L. Lage. 2000. Mach-Zehnder and Michelson topologies for self-referencing fiber optic intensity sensors. *Optical Engineering* 39: 1636–1644.

Berthold, J. W. 1995. Historical review of microbend fiber-optic sensors. *Journal of Lightwave Technology* 13: 1193–1199.

Borecki, M. 2007. Intelligent fiber optic sensor for estimating the concentration of a mixture-design and working principle. *Sensors* 7: 384–399.

Busuring, V. I., A. S. Semenov, and N. P. Udalov. 1985. Optical and fiber-optic sensors (Review). *Soviet Journal Quantum Electronics* 15: 595–621.

Carome, E. F. and K. P. Koo. 1980. Multimode coupled waveguide sensors. *Optics Letters* 5: 359–361.

Cole, J. H., T. G. Giallorenzi, and J. A. Bucaro. 1981. Advances in optical fiber sensors, integrated optics. *Proceedings of the Society of Photo-Optical Instrumentation Engineers* 269: 115–124.

Culshaw, B. 2000. Fiber optics in sensing and measurement. *IEEE Journal of Selected Topics in Quantum Electronics* 6: 1014–1021.

Dandridge, A. and H. E. Taylor. 1982. Correlation of low-frequency intensity and frequency fluctuations in GaAlAs lasers. *IEEE Journal of Quantum Electronics* QE-18: 1738–1750.

Davis, C. M. and C. J. Zarobila. 2000. Optical fibers. In *Electro-Optics Handbook*, 2nd edn., R.W. Waynant and M.N. Ediger (eds.). New York: McGraw-Hill, Chapter 21, pp. 12.1–12.8.

Dib, R., Y. Alayli, and P. Wagstaff. 2004. A broadband amplitude-modulated fibre optic vibrometer with nanometric accuracy. *Measurements* 35: 211–219.

Fields, J. N. 1979. Coupled waveguide acoustooptic hydrophone. *Applied Optics* 18: 3533–3534.

Giallorenzi, T., J. A. Bucaro, A. Dandridge et al. 1982. Optical fiber sensor technology. *IEEE Journal of Quantum Electronics* QE-18: 626–665.

Hahn, J., C. H. Rüter, F. Fecher et al. 2008. Measurement of the enhanced evanescent fields of integrated waveguides for optical near-field sensing. *Applied Optics* 47: 2357–2360.

Hamamatsu Photonics, KK. 1998. *Photodiodes*. http://www.hamamatsu.com/resources/pdf/ssd/e02-handbook_si_photodiode.pdf.

Intelman, S., H. Poisel, A. Bachmann et al. 2011. Switch for polymer optical fibers. *POF 2011 Bilbao Conference Proceedings*, Bilbao, Spain, pp. 321–332.

Kamiya, M., H. Ikeda, and S. Shinohara. 2001. Analog data transmission through plastic optical fiber in robot with compensation of errors caused by optical fiber bending loss. *IEEE Transactions on Industrial Electronics* 48: 1034–1037.

Kazuo, K., S. Tai, T. Sawada et al. 1982. Fiber optic instrument for temperature measurement. *IEEE Journal of Quantum Electronics* QE-18: 676–679.

Kersey, A. D. 1996. A review of recent development in fiber optic sensor technology. *Optical Fiber Technology* 2: 291–317.

Kersey, A. D. and A. Dandridge. 1990. Applications of fiber-optic sensors. *IEEE Transactions on Components, Hybrids and Manufacturing Technology* 13: 137–143.

Kloppe, A., K. Hoeland, S. Müller et al. 2004. Mechanical and optical characteristics of a new fiber optical system used for cardiac contraction measurement. *Medical Engineering & Physics* 26: 687–694.

Kribich, K. R., R. Copperwhite, H. Barry et al. 2005. Novel chemical sensor/biosensor platform based on optical multimode interference (MMI) couplers. *Sensors and Actuators B* 107: 188–192.

Lagakos, N., T. Litovitz, P. R. Macedo et al. 1981. Multimode optical fiber displacement sensor. *Applied Optics* 20: 167–168.

Lee, K. S., H. L. T. Lee, and R. J. Ram. 2007. Polymer waveguide backplanes for optical sensor interfaces in microfluidics. *Lab on a Chip* 7: 1539–1545.

Lynam, P., I. Mahboob, A. J. Parnell et al. 2003. Photon-number squeezing in visible spectrum light-emitting diodes. *Electronics Letters* 39: 110–112.

MacDonald, R. I. and R. Nychka. 1991. Differential measurement technique for fiber-optic sensors. *Electronics Letters* 27: 2194–2196.

Mason, R.A. 1989. Optical fibre radiation pyrometer, US4799787.

Mathews, S., G. Farrell, and Y. Semenova. 2011. Liquid crystal infiltrated photonic crystal fibers for electric field intensity measurements. *Applied Optics* 50: 2628–2635.

Menadier, C., C. Kissinger, and H. Adkins. 1967. The fotonic sensor. *Instruments and Control Systems* 40: 114–120.

Montalvo, J., F. M. Araujo, L. A. Ferreira et al. 2008. Electrical FIR filter with optical coefficients for self-referencing WDM intensity sensors. *IEEE Photonics Technology Letters* 20: 45–47.

Montero, D. S. and C. Vázquez. 2013. Remote interrogation of WDM fiber-optic intensity sensors deploying delay lines in the virtual domain, *Sensors* 13: 5870–5880.

Montalvo, J., C. Vázquez, and D. S. Montero. 2006. CWDM self-referencing sensor network based on ring resonators in reflective configuration. *Optics Express* 14: 4601–4610.

Montero, D. S., C. Vázquez, J. M. Baptista et al. 2010. Coarse WDM networking of self-referenced fiber-optic intensity sensors with reconfigurable characteristics. *Optics Express* 18: 4396–4410.

Montero, D. S., C. Vázquez, I. Möller et al. 2009b. A self-referencing intensity based polymer optical fiber sensor for liquid detection. *Sensors* 9: 6446–6455.

Montero, D. S., C. Vázquez, and J. Zubia. 2009a. Plastic optical fiber sensor for fuel level measurements applied to paramotoring and powered paragliding. *18th International Conference on Plastic Optical Fibers*, Sydney, Australia, Paper 47.

Müller, B. and U. Renz. 2003. Time resolved temperature measurements in manufacturing. *Measurement* 34: 363–370.

Murtaza, G. and J. M. Senior. 1994. Referenced intensity-based optical fiber sensors. *International Journal of Optoelectronics* 9: 339–348.

Nabeel, A., N. A. Riza, F. Ghauri, and F. Perez. 2007. Silicon carbide-based remote wireless optical pressure sensor. *IEEE Photonics Technology Letters* 19: 504–506.

Oh, M., J. Kim, K. Kim et al. 2009. Optical pressure sensors based on vertical directional coupling with flexible polymer waveguides. *IEEE Photonics Technology Letters* 21: 501–503.

Orcutt, J. S., A. A. Khilo, C. W. Holzwarth et al. 2011. Nanophotonic integration in state-of-the-art CMOS foundries. *Optics Express* 19: 2335–2346.

Pérez, C. S., A. G. Valenzuela, G. E. S. Romero et al. 2004. Technique for referencing of fiber-optic intensity modulated sensors by use of counterpropagating signals. *Optics Letters* 29: 1467–1469.

Perrone, G. and A. Vallan. 2009. A low-cost optical sensor for noncontact vibration measurements. *IEEE Transactions on Instrumentation and Measurements* 58: 1650–1656.

Plaza, J. A., A. Llobera, C. Dominguez et al. 2004. BESOI-based integrated optical silicon accelerometer. *Journal of Microelectromechanical Systems* 13: 355–364.

Ratcliff, E.L., P. A. Veneman, A. Simmonds et al. 2010. A Planar, chip-based, dual-beam refractometer using an integrated organic light emitting diode (OLED) light source and organic photovoltaic (OPV) detectors. *Analytical Chemistry* 82: 2734–2742.

Remouche, M., R. Mokdad, A. Chakari et al. 2007. Intrinsic integrated optical temperature sensor based on waveguide bend loss. *Optics & Laser Technology* 39: 1454–1460.

Righini, C., A. Tajani, and A. Cutolo. 2009. *An Introduction to Optoelectronic Sensors*. Series in Optics and Photonics, Vol. 7. Singapore: World Scientific.

Rugunanan, R. A. and M. E. Brown. 1991. The use of pyrometry in the study of fast thermal processes involving initially solid samples. *Journal of Thermal Analysis* 37: 2125–2141.

Rumyantsev, S. L., M. S. Shur, Y. Bilenko et al. 2004. Low frequency noise and long-term stability of noncoherent light sources. *Journal of Applied Physics* 96: 966–969.
Salzberg, B. M., P. V. Kosterin, M. Muschol et al. 2005. An ultra-stable non-coherent light source for optical measurements in neuroscience and cell physiology. *Journal of Neuroscience Methods* 141: 165–169.
Sasagawa, K. and M. Tsuchiya. 2008. A low-noise and high-frequency resolution electrooptic sensing of RF near-fields using an external optical modulator. *Journal of Lightwave Technology* 26: 1242–1248.
Scott, R. P., C. Langrock, and B. H. Kolner. 2001. High-dynamic-range laser amplitude and phase noise measurement techniques. *IEEE Journal of Selected Topics in Quantum Electronics* 7: 641–655.
Silva, S., L. Coelho, O. Frazão et al. 2011. A review of palladium-based fiberoptic sensors for molecular hydrogen detection. *IEEE Sensors Journal* 99.
Sobrino, C., J. A. Almendros-Ibáñez, D. Santana et al. 2009. Maximum entropy estimation of the bubble size distribution in fluidized beds. *Chemical Engineering Science* 64: 2307–2319.
Soldano, L. B. and E. C. M. Pennings. 1995. Optical multimode interference devices based on self-imaging: Principles and applications. *Journal of Lightwave Technology* 13: 615–627.
Souza, K. D. and T. P. Newson. 2004. Signal to noise and range enhancement of a Brillouin intensity based temperature sensor. *Optics Express* 12: 2656–2661.
Spillman, W. B. and J. R. Lord. 1987. Self-referencing multiplexing technique for fiber-optic intensity sensors. *Journal of Lightwave Technology* 5: 865–869.
Suter, J. J., J. C. Poret, and M. Rosen. 1992. Fiber optic ionizing radiation detector. *IEEE Transactions on Nuclear Science* 39: 674–679.
Tapetado, A., C. Vázquez, and J. Zubia. 2011. Temperature sensor based on polymer optical fiber macrobends. *20th International Conference on Plastic Optical Fibers*, Bilbao, Spain, pp. 207–212.
Texas Instruments. 1999. *Noise Analysis in Operational Amplifiers Circuits*. http://www.ti.com/lit/an/slva043b/slva043b.pdf
Tsai, C., H. Huang, S. Tsao et al. 1997. Error reduction of referenced intensity-based optical fiber sensor by adaptive noise canceller. *Electronics Letters* 33: 982–983.
Udd, E. 1995. An overview of fiber-optic sensors. *Review Scientific Instruments* 66: 4015–4030.
Vázquez, C., P. Baquero, and J. F. Hernández-Gil. 1995b. Fabry-Perot method for the characterization of integrated optical directional couplers. *Applied Optics* 34: 6874–6884.
Vázquez, C., A. B. Gonzalo, S. Vargas et al. 2004. Multi-sensor system using plastic optical fibers for intrinsically safe level measurements. *Sensors and Actuators A* 116: 22–32.
Vázquez, C., J. Montalvo, and P. C. Lallana. 2005. Radio-frequency ring resonators for self-referencing fiber-optic intensity sensors. *Optical Engineering Letters* 44: 1–2.
Vázquez, C., J. Montalvo, D. S. Montero et al. 2006. Self-referencing fiber-optic intensity sensors using Ring Resonators and fiber Bragg gratings. *Photonics Technology Letters* 18: 2374–2376.
Vázquez, C., J. Montalvo, P. C. Lallana et al. 2007. Self-referencing techniques in photonics sensors and multiplexing. SPIE's International Symposium on Microtechnologies for the New Millennium. *Photonics and Optoelectronics* 6593: 6593-63.
Vázquez, C., F. J. Mustieles, and J. F. Hernández-Gil. 1995a. Three-dimensional method for simulation of multimode interference couplers. *IEEE Journal of Lightwave Technology* 13: 2296–2299.
Vázquez, C., J. L. Nombela, C. Sobrino et al. 2007. Plastic fiber-optic probes for characterizing fluidized beds in bubbling regime. *Proceedings of the 16th International Conference on Plastic Optical Fibers*, Turin, Italy, pp. 202–205.
Vázquez, C., J. M. S. Pena, S. Vargas et al. 2003. Optical router for optical fiber sensor networks based on a liquid crystal cell. *IEEE Sensors Journal* 3: 513–518.
Wang, A., H. Xiao, J. Wang et al. 2001. Self-calibrated interferometric–intensity-based optical fiber sensors. *Journal of Lightwave Technology* 19: 1495–1501.
Werneck, M. M., R. C. Allil, D. M. C. Rodrigues et al. 2011. LPG and taper based fiber-optic sensor for index of refraction measurements in biosensor applications. *Proceedings of 20th International Conference on Plastic Optical Fibers*, Bilbao, Spain, pp. 545–550.
Yao, S. and C. K. Asawa. 1983. Fiber optical intensity sensors. *IEEE Journal on Selected Areas in Communications* 3: 562–575.

5. Interferometric Measurement

Principles and Techniques

Mehrdad Abolbashari
University of North Carolina at Charlotte

Faramarz Farahi
University of North Carolina at Charlotte

José Luís Santos
University of Porto

Handbook of Optical Sensors. Edited by José Luís Santos and Faramarz Farahi © 2015 CRC Press/Taylor & Francis Group, LLC. ISBN: 9781439866856.

5.1 Introduction

The wave nature of light is best demonstrated by the phenomenon of interference. As with other waves, such as sound and water waves, light obeys the principle of superposition, which states that the resultant disturbance at any point is the sum of the separate disturbances. Optical interference is therefore the interaction of two or more waves, with the generation of a pattern where the total intensity differs from what would be expected by the sum of the individual wave intensities. This is a fundamental physical concept that is behind a huge number of optical phenomena that can be found in the nature, inspiring, along the ages, novel technological applications following a trend that is becoming increasingly important. One of them is the optical interferometric measurement, associated with the highest performance levels that can be achieved nowadays.

This chapter addresses the principles and techniques associated with measurement by optical interferometry. Owing to the subject of this book, other chapters also more directly detail the concepts and applications associated with optical interferometry, particularly Chapters 2 and 14. In Section 5.2, the basis of optical interference is presented, and Section 5.3 describes the topic of optical coherence. The different types of optical interferometers are outlined in Section 15.4. Section 15.5 intends to illustrate the notable possibilities for imaging/measurement made possible with optical interferometry, in the case when applied to high-resolution astronomic measurement and imaging.

5.2 Optical Interference

One of the first scientists who experimentally demonstrated the interference of two waves was Thomas Young, as reported in 1804 at the *Philosophical Transactions of the Royal Society of London* (Young 1804), with his famous *Young's double-slit experiment*, shown in Figure 5.1. The light passes through a small aperture that acts as a point source for two other apertures in the middle plane. The two narrow slits in the middle

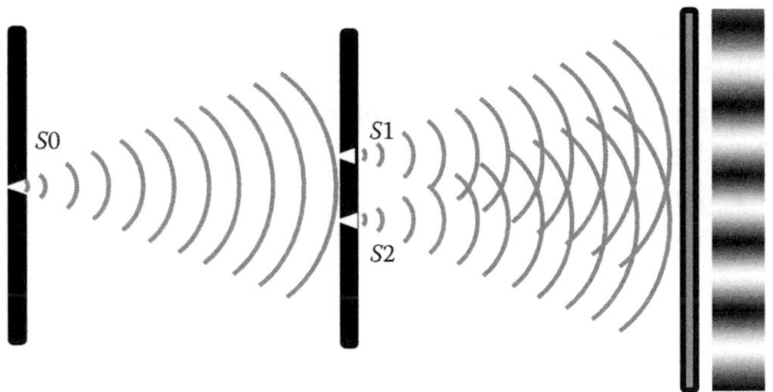

FIGURE 5.1 Young's experiment.

plane in turn are two light sources that interfere and form an interference pattern on the screen (the plane on the right).

The two slits in the middle plane act as two sources of electromagnetic waves with some degree of coherence. Ideally, the monochromatic plane wave electric field oscillation as a function of time can be described in a complex exponential form by a constant amplitude at a given frequency

$$\boldsymbol{E} = \boldsymbol{E}_0 \exp i(\boldsymbol{k} \cdot \boldsymbol{r} + wt + \theta) \tag{5.1}$$

where $\boldsymbol{k}$ is the wave vector. The magnitude of the wave vector, wave number, equals $\boldsymbol{k} = 2\pi/\lambda$, where λ is the wavelength of the monochromatic electric field, $\boldsymbol{r}$ is the position vector, w is the angular frequency of monochromatic electric field, t is the time, and θ is an arbitrary phase term. With these considerations, interference occurs when two or more waves combine to form a new wave, where its amplitude is the sum of the amplitudes of the waves being combined (Figure 5.2).

If we have two waves of the same frequency traveling in the z direction, then the two waves are given by

$$\boldsymbol{E}_1 = \boldsymbol{E}_{01} \exp i(\boldsymbol{k} \cdot z_1 + wt + \theta_1) \tag{5.2}$$

$$\boldsymbol{E}_2 = \boldsymbol{E}_{02} \exp i(\boldsymbol{k} \cdot z_2 + wt + \theta_2) \tag{5.3}$$

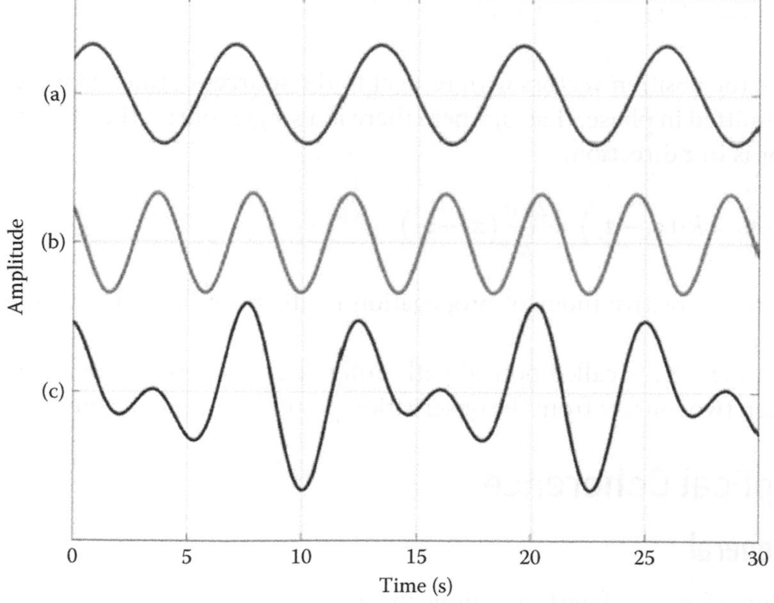

FIGURE 5.2 Interference between two sinusoidal waves with different frequencies. (a): Sinusoidal wave 1 with angular frequency $w_1 = 1$ and initial phase $\theta_1 = \pi/4$; (b): sinusoidal wave 2 with angular frequency $w_1 = 1.5$ and initial phase $\theta_1 = 3\pi/4$; (c): interference of wave 1 and wave 2.

And the superposition of the waves is

$$E_T = E_1 + E_2 = \{E_{01}\exp(i\varphi_1) + E_{02}\exp(i\varphi_2)\}\exp i(wt) \tag{5.4}$$

where

$\varphi_1 = \boldsymbol{k}\cdot \boldsymbol{z}_1 + \theta_1$

$\varphi_2 = \boldsymbol{k}\cdot \boldsymbol{z}_2 + \theta_2$

The intensity is given by

$$\boldsymbol{I} \propto \left\langle |\boldsymbol{E}_{01}|^2 \right\rangle_t = |\boldsymbol{E}_{01}|^2 + |\boldsymbol{E}_{02}|^2 + 2\boldsymbol{E}_{01}\cdot \boldsymbol{E}_{02}\cos(\varphi_1 - \varphi_2) \tag{5.5}$$

Assuming $\boldsymbol{E}_1$ and $\boldsymbol{E}_2$ have the same state of polarization,

$$I \propto \left\langle |\boldsymbol{E}_T|^2 \right\rangle_t = \left\langle |\boldsymbol{E}_1 + \boldsymbol{E}_2|^2 \right\rangle_t = \boldsymbol{E}_{01}^2 + \boldsymbol{E}_{02}^2 + 2\boldsymbol{E}_{01}\boldsymbol{E}_{02}\cos(\varphi_1 - \varphi_2) \tag{5.6}$$

The intensity is comprised of a uniform background $\boldsymbol{E}_{01}^2 + \boldsymbol{E}_{02}^2$ on which is superimposed a cosine variation $2\boldsymbol{E}_{01}\boldsymbol{E}_{02}\cos(\varphi_1-\varphi_2)$ known as the interferogram. The contrast of the fringes is given by the visibility defined as

$$V = \frac{I_{max} - I_{min}}{I_{max} + I_{min}} = \frac{2\boldsymbol{E}_{01}\boldsymbol{E}_{02}}{\boldsymbol{E}_{01}^2 + \boldsymbol{E}_{02}^2} \tag{5.7}$$

The interferogram varies as the cosine of $\delta\varphi = \varphi_1-\varphi_2$ where

$$\delta\varphi = \varphi_1 - \varphi_2 = \boldsymbol{k}\cdot(\boldsymbol{z}_1 - \boldsymbol{z}_2) + (\theta_1 - \theta_2) \tag{5.8}$$

$\boldsymbol{z}_1$ and $\boldsymbol{z}_2$ are the position vectors with respect to the sources of the two waves. If the two waves are emitted in phase, that is, where there is a single source, then $\theta_1 = \theta_2$; assuming wave vector is in $\boldsymbol{z}$ direction,

$$\varphi = \varphi_1 - \varphi_2 = \boldsymbol{k}\cdot(z_1 - z_2) = \frac{2\pi n}{\lambda}(z_1 - z_2) \tag{5.9}$$

where n is the refractive index of propagation medium and λ is the wavelength of the light emitted.

The term $n(z_1-z_2)$ is called optical path difference and is the equivalent path difference between two sources from the observation point, assuming the medium is vacuum.

5.3 Optical Coherence

5.3.1 General

So far, the interference of perfectly monochromatic light emanating from a point source has been considered. In reality, an optical source is not a single frequency, but the light emitted consists of a band of frequencies $\Delta\nu$ centered around a single frequency ν_0. Sources for which $\Delta\nu$ is small compared with ν are called quasi-monochromatic; such a source may be considered as emitting a series of randomly phased finite wave trains. The period of time over which we can predict the phase of wave, by knowing the phase at the beginning of the period, is known as the coherence time ($\Delta\tau$). Coherence time is related to the optical bandwidth of light source by $\Delta\tau = c/\Delta\nu$, where c is the speed of light in the medium. Corresponding to the coherence time, there is coherence length, which is the length at which the light wave is coherent and is given by $\Delta z = c \cdot \Delta\tau$. Since this form of coherence is related to the temporal property of the light source, it is called temporal coherence. Two extreme cases are zero and infinite bandwidths. The latter corresponds to complete temporal incoherence, zero coherence time, and zero coherence length, while the former is ideal temporal coherence and infinite coherence length.

A second form of coherence is related to the geometry of light source, which is known as spatial coherence. While temporal coherence relates to the finite bandwidth of the source, spatial coherence relates to the finite size of the source. In an ideal case of point source, the source is considered spatially coherent; with finite size, the source is considered as a combination of multiple sources. This means that at two different points, the phase of field is different due to the superposition of wavelets from different points of finite size source. A complete and comprehensive theory of coherence can be found elsewhere (Mandel and Wolf 1995, Wolf 2007). In the following, mathematical expressions for both temporal and spatial coherence are derived.

5.3.2 Temporal Coherence

A mathematical description of coherence may be derived in terms of correlation functions. This approach was first used in the development of the description of interference, the correlation of two beams of light, and has since been used to describe higher-order correlations. Consider a familiar Young's experiment that can be considered as a two-beam interferometer. Essentially, there is a source, S, a mean of splitting and recombining the beams (Figure 5.3), and a screen onto which the interference pattern is projected.

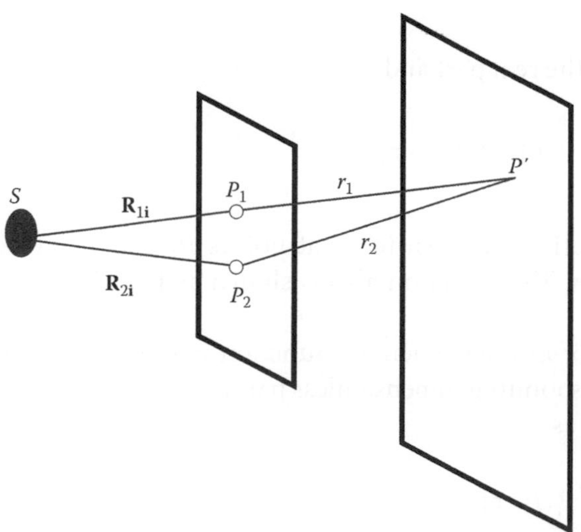

FIGURE 5.3 Young's experiment with an extended source.

Each beam of the interferometer will separately form an image at points P_1 and P_2 on the screen as seen from the source; the positions of P_1 and P_2 are denoted by the vectors $\mathbf{R}_{1i}$ and $\mathbf{R}_{2i}$ (Figure 5.3).

The time at which light travels from points P_1 and P_2 to point P' on the screen is $t_1 = r_1/c$, and $t_2 = r_2/c$, respectively. Considering the field as a function of time and space, the field at point P' is given by

$$E(P',t) = K_1E(P_1,t-t_1)+K_2E(P_2,t-t_2) = K_1E(P_1,t-t_1)+K_2E(P_2,t-(t_1+\tau)) \quad (5.10)$$

where $\tau = t_2 - t_1$. Coefficients K_1 and K_2 are factors that depend on geometry and the medium and are called amplitude transmission factors. The quantity that is measured in optics is usually intensity, which is proportional to the average amplitude squared of the electric field.

Ignoring the constants that relate intensity to the electric field,

$$\begin{aligned} I(P') &= \left\langle E(P',t)E^*(P',t)\right\rangle_t \\ &= \left\langle \{K_1E(P_1,t-t_1)+K_2E(P_2,t-(t_1+\tau))\}\{K_1E(P_1,t-t_1)+K_2E(P_2,t-(t_1+\tau))\}^*\right\rangle_t \end{aligned} \quad (5.11)$$

Assuming that the field is stationary, which means that the statistics of the field does not depend on the origin of time, the intensity at point P' is calculated as

$$\begin{aligned} I(P') &= \left\langle \{K_1E(P_1,t-t_1)+K_2E(P_2,t-(t_1+\tau))\}\{K_1E(P_1,t-t_1)+K_2E(P_2,t-(t_1+\tau))\}^*\right\rangle_t \\ &= |K_1|^2 I(P_1)+|K_2|^2 I(P_2)+2\,\mathrm{Re}\left\{K_1K_2^*\Gamma(P_1,P_2,\tau)\right\} \end{aligned} \quad (5.12)$$

where $\mathrm{Re}\{\cdot\}$ shows the real part and

$$\Gamma(P_1,P_2,\tau) = \left\langle E(P_1,t)E^*(P_2,t-\tau)\right\rangle_t \quad (5.13)$$

is the cross-correlation between fields at points P_1 and P_2, which is called mutual coherence function. This function also is shown as $\Gamma_{12}(\tau)$, and its real part is shown as $G_{12}(\tau)$.

The dimension of mutual coherence function is the same as the dimension of the intensity. The corresponding dimensionless parameter is defined as the degree of coherence and is defined as

$$\gamma(P_1,P_2,\tau) = \frac{\Gamma(P_1,P_2,\tau)}{\sqrt{I(P_1)I(P_2)}} \quad (5.14)$$

By absorbing the coefficients K_1 and K_2 into the intensities $I(P_1)$ and $I(P_2)$, we have

$$I(P') = I_1(P') + I_2(P') + 2\sqrt{I_1(P')I_2(P')}\,\mathrm{Re}\{\gamma(P_1,P_2,\tau)\} \tag{5.15}$$

where

$$I_1(P') = |K_1|^2 I(P_1) \tag{5.16}$$

$$I_2(P') = |K_2|^2 I(P_2) \tag{5.17}$$

The degree of coherence satisfies the following inequality:

$$|\gamma(P_1,P_2,\tau)| \le 1 \tag{5.18}$$

The radiation at points P_1 and P_2 are called fully coherent if $|\gamma(P_1,P_2,\tau)| = |\gamma_{12}(\tau)| = 1$; incoherent if $|\gamma_{12}(\tau)| = 0$; and partially coherent when $0 < |\gamma_{12}(\tau)| < 1$.

From Equations 5.7 and 5.15, the visibility of fringes produced by points P_1 and P_2 is calculated as

$$V = \frac{I_{\max} - I_{\min}}{I_{\max} + I_{\min}} = \frac{2\sqrt{I_1(P')I_2(P')}}{I_1(P') + I_2(P')}|\gamma_{12}(\tau)| \tag{5.19}$$

Assuming that intensities are the same $I_1(P') = I_2(P')$, the visibility is the same as absolute value of the degree of coherence

$$V = |\gamma_{12}(\tau)| \tag{5.20}$$

Therefore, the degree of coherence can be measured in terms of the visibility of fringes.

If points P_1 and P_2 coincide, the mutual coherence function reduces to the autocorrelation. For example, in an ideal amplitude-splitting interferometer, the beams are split into two beams from a single point, which means P_1 and P_2 coincide. In this case, the radiation is spatially fully coherent (considering P_1 and P_2 as points and not distributed apertures) and the only effect is the temporal effect, which results in different temporal coherence.

Since mutual coherence function reduces to the autocorrelation, the Fourier transform of that function is the spectrum

$$g_{11}(\nu) = g(P_1,P_1,\nu) = \int_{-\infty}^{+\infty} \Gamma(P_1,P_1,\tau)e^{j2\pi\nu\tau}d\tau = \int_{-\infty}^{+\infty} \Gamma_{11}(\tau)e^{j2\pi\nu\tau}d\tau\Gamma \tag{5.21}$$

Besides, mutual coherence function is an analytic function (i.e., it has no negative frequency component)

$$\Gamma_{11}(\tau) = \int_{0}^{+\infty} g_{11}(\nu) e^{-j2\pi\nu\tau} d\nu \tag{5.22}$$

where $g_{11}(\nu)$ is the spectrum of the interferogram. Therefore, interferogram and spectrum form a Fourier transform pair.

In addition to the power spectrum, which is the Fourier transform of autocorrelation function, we can define the Fourier transform of the mutual coherence function, which is a cross-correlation function

$$g_{12}(\nu) = g(P_1, P_2, \nu) = \int_{-\infty}^{+\infty} \Gamma(P_1, P_2, \tau) e^{j2\pi\nu\tau} d\tau = \int_{-\infty}^{+\infty} \Gamma_{12}(\tau) e^{j2\pi\nu\tau} d\tau \tag{5.23}$$

and since again mutual coherence function is an analytic function,

$$\Gamma_{12}(\tau) = \int_{0}^{+\infty} g_{12}(\nu) e^{-j2\pi\nu\tau} d\nu \tag{5.24}$$

Considering a quasi-monochromatic wave (i.e., bandwidth $\Delta\nu$ is very small compared to the central frequency ν_0), the spectrum of wave is nonzero only for frequencies around ν_0. Therefore, mutual coherence function can be written as

$$\Gamma_{12}(\tau) = e^{-j2\pi\nu_0\tau} \int_{-\nu_0}^{+\infty} j_{12}(\nu) e^{-j2\pi\upsilon\tau} d\nu = e^{-j2\pi\nu_0\tau} J_{12}(\tau) \tag{5.25}$$

where

$j_{12}(\nu) = g_{12}(\nu)$ and

$J_{12}(\tau)$ is the Fourier transform of $j_{12}(\nu)$ and is called mutual intensity

5.3.3 Spatial Coherence

So far, we have considered coherence effects due to the spectral properties of the source. Here coherence effects due to the spatial extent of the source are considered. The coherence between two points illuminated by a source of incoherent light is given by the *van Cittert–Zernike theorem* (Zernike 1934, 1938).

Consider two points P_1 and P_2 (Figure 5.3), illuminated by an extended quasi-monochromatic source of mean frequency ν_0. The source may be considered to be a collection of point source elements. If $S(\mathbf{R}_i,t)$ denotes the waves at P_i due to each element, then

$$S_1(t)=\sum_{n=1}^{N}S(R_{1n},t) \tag{5.26}$$

$$S_2(t)=\sum_{n=1}^{N}S(R_{2n},t) \tag{5.27}$$

where $S(R_{in},t)$ shows the effect of the nth element of source at point P_i. Therefore, the mutual coherence function can be calculated as

$$\Gamma(P_1,P_2,\tau)=\left\langle S_1(t)S_2^*(t-\tau)\right\rangle_t=$$
$$=\sum_{n=1}^{N}\langle S(R_{1n},t)S^*(R_{2n},t-\tau)\rangle_t+\sum_{n=1}^{N}\sum_{m=1,\neq n}^{N}\langle S(R_{1n},t)S^*(R_{2m},t-\tau)\rangle_t \tag{5.28}$$

Considering that different elements of source radiate independently, the second term on the right side of Equation 5.28 equals zero. Furthermore, if we assume that each element of source radiates spherical waves, then

$$S(R_{1n},t)=\frac{1}{R_{1n}}A_n\left(t-\frac{R_{1n}}{c}\right)e^{-j2\pi\nu_0(t-(R_{1n}/c))} \tag{5.29}$$

$$S(R_{2n},t)=\frac{1}{R_{2n}}A_n\left(t-\tau-\frac{R_{2n}}{c}\right)e^{-j2\pi\nu_0(t-\tau-(R_{2n}/c))} \tag{5.30}$$

and

$$\left\langle S(R_{1n},t)S^*(R_{2n},t-\tau)\right\rangle_t=\frac{1}{R_{1n}R_{2n}}\left\langle A_n\left(t-\frac{R_{1n}}{c}\right)A_n^*\left(t-\tau-\frac{R_{2n}}{c}\right)\right\rangle_t e^{-j2\pi\nu_0(\tau-(R_{1n}-R_{2n}/c))} \tag{5.31}$$

Considering zero time delay, which eliminates the temporal incoherence effects, we have

$$\left\langle S(R_{1n},t)S^*(R_{2n},t)\right\rangle_t=\frac{1}{R_{1n}R_{2n}}\left\langle A_n\left(t-\frac{R_{1n}}{c}\right)A_n^*\left(t-\frac{R_{2n}}{c}\right)\right\rangle_t e^{-j2\pi\nu_0((R_{2n}-R_{1n})/c)} \tag{5.32}$$

Quasi-monochromatic wave results in slow change in amplitude, and by assuming that $(R_{2n}-R_{1n})/c$ is much smaller than the inverse of bandwidth, we have

$$\left\langle S(R_{1n},t)S^*(R_{2n},t)\right\rangle_t \approx \frac{1}{R_{1n}R_{2n}}\left\langle A_n(t)A_n^*(t)\right\rangle_t e^{-j2\pi\nu_0((R_{2n}-R_{1n})/c)} = \frac{1}{R_{1n}R_{2n}}I_n e^{-j2\pi\nu_0((R_{2n}-R_{1n})/c)} \tag{5.33}$$

where I_n is the intensity of nth element of the source. From Equations 5.28 and 5.33, the mutual coherence function is calculated as

$$\Gamma(P_1,P_2,0)=\left\langle S_1(t)S_2^*(t)\right\rangle_t = \sum_{n=1}^{N}\frac{1}{R_{1n}R_{2n}}I_n e^{-j2\pi\nu_0((R_{2n}-R_{1n})/c)} \tag{5.34}$$

and the integral form of this equation will be

$$\Gamma(P_1,P_2,0)=\int_S \frac{1}{R_1(s)R_2(s)}I(s)e^{-j2\pi\nu_0((R_2(s)-R_1(s))/c)}ds \tag{5.35}$$

where

S is the source surface

$R_1(s)$ and $R_2(s)$ are distances of partial surface ds from points P_1 and P_2

$I(s)$ is the intensity distribution of the source

From Equation 5.14, the degree of coherence is defined as

$$\Gamma(P_1,P_2,0)=\frac{\Gamma(P_1,P_2,0)}{\sqrt{I(P_1)I(P_2)}}=\frac{\int_S (1/R_1(s)R_2(s))I(s)e^{-j2\pi\nu_0((R_2(s)-R_1(s))/c)}ds}{\sqrt{\int_S \left(I(s)/R_1^2(s)\right)ds \int_S \left(I(s)/R_2^2(s)\right)ds}} \tag{5.36}$$

A more generalized form of the mutual coherence function can be obtained by considering both the finite size and finite bandwidth of the source. In this general form, the mutual coherence function can be stated as

$$\Gamma(\tau)=\int_0^{+\infty} J(P_1,P_2,\nu)e^{-j2\pi\nu\tau}d\nu \tag{5.37}$$

where

$$J(P_1,P_2,\nu)=\int_S \frac{I(s,\nu)}{R_1(s)R_2(s)}e^{-j2\pi\nu((R_2(s)-R_1(s))/c)}ds \tag{5.38}$$

The effects of both spatial and temporal coherence upon the fringe visibility have been considered. A further factor that affects the fringe visibility of two interfering beams is their relative polarization states. According to Equation 5.5, if the polarization states of the two interacting beams are orthogonal ($\boldsymbol{E}_{01} \perp \boldsymbol{E}_{02}$), then the last term on the right side of equation, which is the interference term, equals zero ($\boldsymbol{E}_{01} \cdot \boldsymbol{E}_{02} = 0$). For example, vertically and horizontally polarized or left and right circularly polarized light are orthogonal states of polarization, and therefore, they do not interfere.

In general, a beam is neither perfectly monochromatic nor fully polarized; however, the coherence and polarization properties may be represented by a coherence matrix (Gori et al. 1998), and the fringe visibility may be determined via a complex degree of coherence.

5.4 Optical Interferometers

Interferometers are instruments that employ interference of two or more waves to measure a wide range of physical parameters (Malacara 2007). The first work on interferometry was done by Michelson and Morley (1887). They performed an experiment to show the existence of *aether drift* and implemented an interferometer. The null result of their experiment led to rejection of the concept of aether. The application of interferometry followed, after this experiment, mainly in metrology and spectroscopy. After the invention of laser, the interferometry experienced a rapid development. The mathematical description of interferometers has been discussed in detail elsewhere (Steel 1983, Hariharan 1986).

There are different configurations for interferometers including the Michelson interferometer, the Mach–Zehnder interferometer, and the Fizeau interferometer. In all interferometers, two or more waves interfere, and the resultant interference pattern is used to calculate a wavefront change that results from transmission through or reflection from a test part of interest. The wavefront change, in turn, corresponds to the parameters and/or features of interest concerning the object under measurement.

There are different ways to categorize the interferometers. One category is based on the number of light waves (beams) that interfere, which can be two or more than two beams. Michelson interferometer is an example of two-beam interferometer, and Fabry–Pérot interferometer is an example of multi-beam interferometer.

Another categorization method divides interferometers into interference by division of amplitude and interference by division of wavefront. The Rayleigh interferometer is an example of former interferometer, while the Mach–Zehnder interferometer is an example of the latter one. In the following, those categories of interferometers will be introduced.

5.4.1 Two-Beam Interferometers

It was mentioned earlier that a condition for a stable interferogram was that the interfering light should emanate from a single source and that the interferometer's path length difference should be within the coherence length of the source. Two-beam interferometers may be classified as to how a beam is split to form the two interfering beams.

The general form of two-beam interferometer was given in Equations 5.5 and 5.15 for an ideal monochromatic point source and finite size, finite bandwidth source respectively.

Two beams can be separated using a beam-splitter, a birefringent element (e.g., Wollaston prism), a grating, or two apertures. The two separate beams then are brought together to interfere and form an interference pattern.

There are different configurations that realize two-beam interferometer including Michelson, Rayleigh, Mach–Zehnder, and Sagnac interferometers that will be explained briefly.

5.4.1.1 Michelson Interferometer

Michelson interferometer is the most common interferometer that uses the division of amplitude. A Michelson interferometer configuration is illustrated in Figure 5.4a.

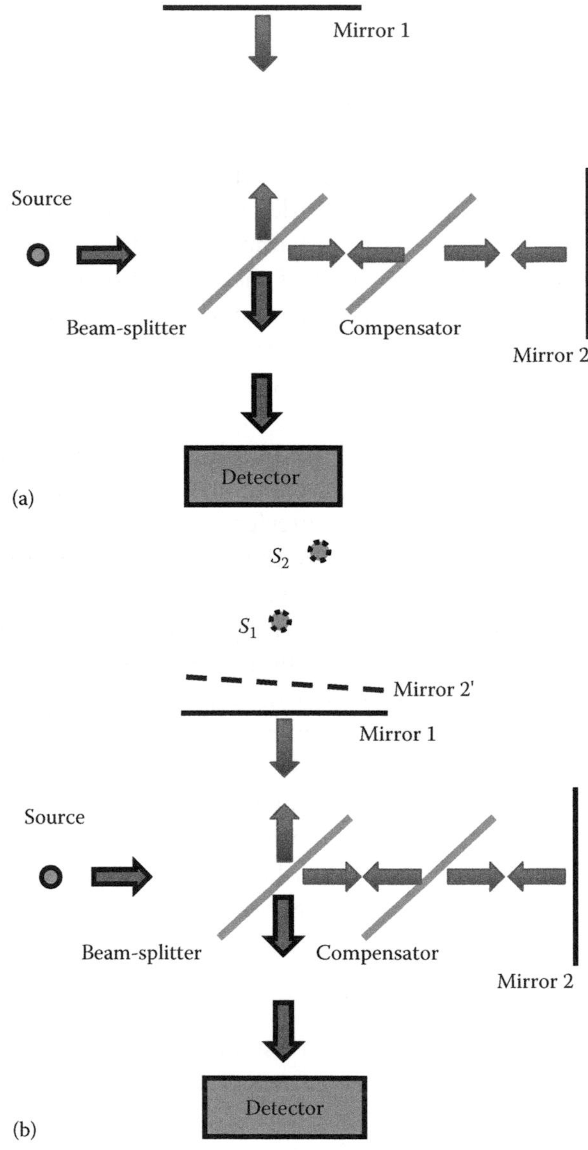

FIGURE 5.4 Schematic of Michelson interferometer with (a) mirrors with no tilt and (b) mirrors with a slight tilt with respect to one another.

Light wave from a source is divided into two beams, after passing through a beam-splitter. Both beams are reflected from mirrors and parts of each beam that go downward are combined to make an observable interference pattern, which is captured by the detector. A compensator is placed along the path of one beam to compensate for any additional unwanted phase differences due to the presence of optical components in the system and therefore any path difference comes from actual optical path difference that is of interest.

The image of mirror 2 is mirror 2′ (Figure 5.4b), which is the result of reflection from beam-splitter. The interference, which is acquired by the detector, is the interference between light from a source located at mirror 1 and a source located at virtual mirror 2′. Also the images of source in mirror 1 and virtual mirror 2′ are indicated by S_1 and S_2.

If mirror 1 and virtual mirror 2′ are parallel, then the line that connects images of the source (S_1S_2) is perpendicular to the mirror plane. Otherwise, the line S_1S_2 has an angle with the normal to the mirror 1 plane.

The straight-line fringes can be seen if the source is a monochromatic point source and line S_1S_2 is parallel to mirror 1. These parallel lines are actually hyperbolic fringes that are approximated as straight lines when the distance of line S_1S_2 from the detector is much larger than the separation between S_1 and S_2. The straight-line fringes can also be formed with an extended monochromatic source if there is a small angle between mirror 1 and virtual mirror 2′ and again the separation between S_1 and S_2 is very small compared to the distance of virtual sources from the detector.

The circular fringes are formed with a monochromatic point source, when mirror 1 and virtual mirror 2′ are parallel. With an extended monochromatic source, circular fringes are formed when mirror 1 and virtual mirror 2′ are parallel, but they have a finite distance from each other. If mirror 1 and virtual mirror 2′ coincide, then the fringe pattern will have a uniform intensity pattern.

Michelson interferometer can be used for very precise length measurements (Li et al. 1995), gravitational wave detection (McKenzie et al. 2002), surface acoustic wave imaging (Knuuttila et al. 2000), refractive index measurement (Becker et al. 1995, Tian et al. 2008a), and precise displacement measurement (Lawall and Kessler 2000).

5.4.1.2 Rayleigh Interferometer

The Rayleigh interferometer is a division by wavefront interferometer. The schematic of the Rayleigh interferometer is shown in Figure 5.5. The source wavefront is split into two beams: each beam is passed through the objects, and the beams interfere afterward. Lens 1 is used to collimate the light, and Lens 2 is used to bring the beams together to interfere. As it is seen, the Rayleigh interferometer is a simple interferometer and can

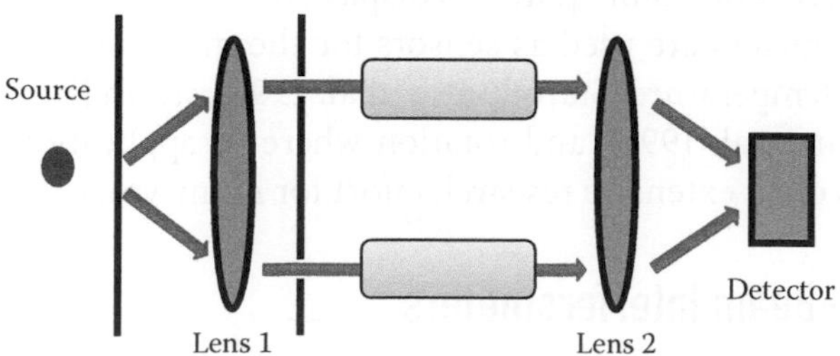

FIGURE 5.5 Schematic of Rayleigh interferometer.

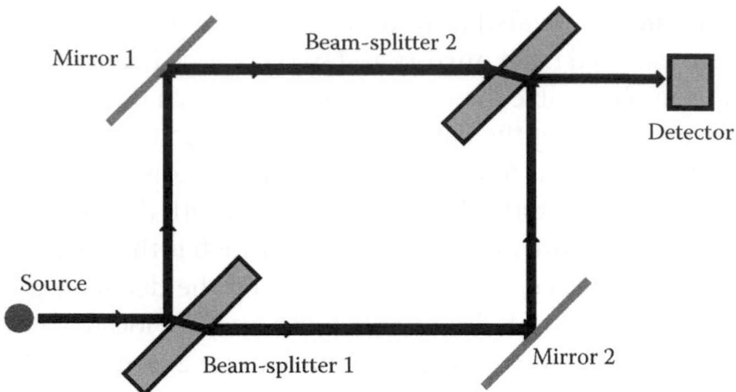

FIGURE 5.6 Schematic of Mach–Zehnder interferometer.

measure the optical path difference very precisely. One of the drawbacks of this interferometer is a high density of fringes that requires magnification in order to analyze the interferogram. Rayleigh interferometers are used in determination of diffusion coefficients (Rard and Miller 1979, Annunziata et al. 2005, Zhang et al. 2008, Annunziata and Vergara 2009).

5.4.1.3 Mach–Zehnder Interferometer

The schematic of a Mach–Zehnder interferometer is demonstrated in Figure 5.6. The beams are divided using beam-splitter 1. Each beam reflects from a mirror, and they are combined using beam-splitter 2. The optical difference between beams can be produced using a slight tilt in one of the mirrors or by inserting a wedge in the path of one of the beams. This interferometer has been used for label-free detection of liquids (Lapsley et al. 2011), biological and chemical properties of materials (Qi et al. 2002, Sepúlveda and del Río 2006), and refractive index sensor (Tian et al. 2008b).

5.4.1.4 Sagnac Interferometer

In Sagnac interferometer, both beams go through the same loop but in opposite directions and then recombine, which makes this interferometer very stable and easy to align. Figure 5.7 shows the schematic of the Sagnac interferometer with three mirrors. It is also possible to make a Sagnac interferometer with one beam-splitter and two mirrors.

Sagnac interferometer with three mirrors is insensitive to the displacement of mirrors and beam-splitter, while the Sagnac interferometer with two mirrors produces a shear when the mirrors or beam-splitter is displaced.

Sagnac interferometers are used as sensors for the measurement of electric current (Blake et al. 1996), temperature (Starodumov et al. 1997), strain (Dong et al. 2007), gravitational waves (Sun et al. 1996), and rotation where its application as fiber gyroscope has been the focus of an extensive research effort for many years (Lefevre 2012).

5.4.2 Multiple-Beam Interferometers

The principle of superposition is not restricted to two beams but may be applied to many beams. This also applies to interference. Again the beams must emanate from

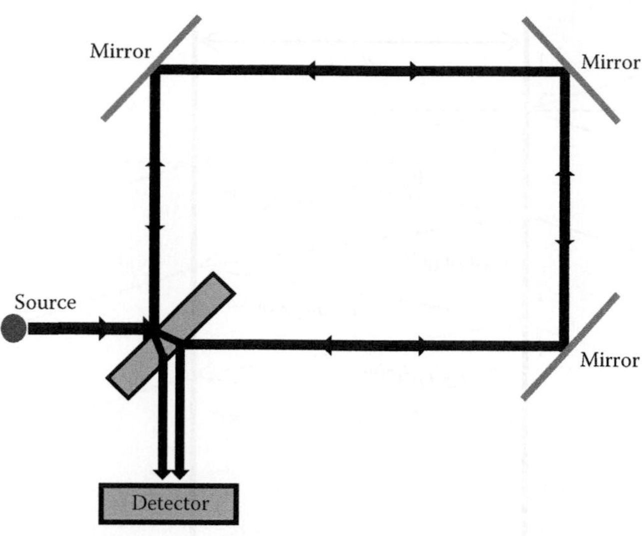

FIGURE 5.7 Schematic of Sagnac interferometer.

a single source in order that they may interfere. Like two-beam interferometers, beams of multiple-beam interferometer can be produced using the division of wavefront or the division of amplitude.

5.4.2.1 Fabry–Pérot

The most common form of multiple-beam interferometer with the division of amplitude is the Fabry–Pérot. In the Fabry–Pérot configuration, two highly reflective mirrors are placed such that when light is injected into the system, the light bounces back and forth many times. The transfer function of such an interferometer is, unlike the two-beam interferometer described earlier, the well-known Airy function. However, if the mirrors used are of a low reflectance, the transfer function becomes similar to the $(1 + k\cos\phi)$ as seen for two-beam interferometers.

Figure 5.8 shows the schematic of a Fabry–Pérot interferometer. The transmission from medium i to medium j is t_{ij}, and the reflection in medium i with the interface with the medium j is r_{ij}. Assuming the refractive index 1 for mediums 1 and 3, and refractive index of n for medium 2, and the thickness of d for medium 2, the phase change from point A to point B is

$$\delta = \frac{4\pi nd}{\lambda}\cos\theta_2 \tag{5.39}$$

The interference between transmitted fields equals the superposition of all transmitted fields:

$$E_t(\delta) = Et_{12}t_{23}e^{-j\delta/2} + Et_{12}t_{23}r_{23}r_{21}e^{-j3\delta/2} + Et_{12}t_{23}r_{23}^2r_{21}^2e^{-j5\delta/2} + \cdots = \frac{Et_{12}t_{23}e^{-j\delta/2}}{1-r_{23}r_{21}e^{-j\delta}} \tag{5.40}$$

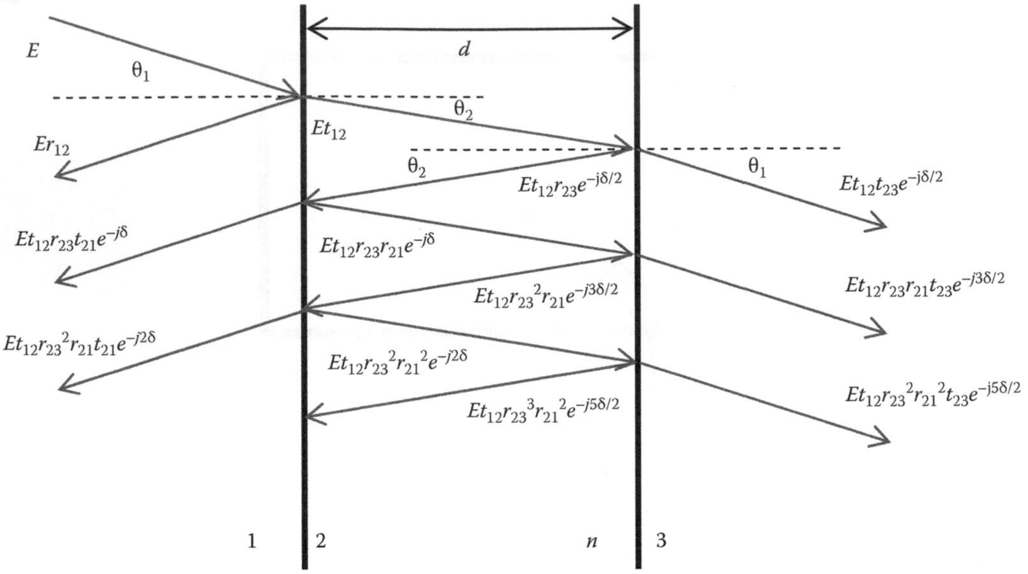

FIGURE 5.8 Schematic of the Fabry–Pérot interferometer.

The transmission intensity is calculated as

$$I_t(\delta) = I_{in}\left|\frac{t_{12}t_{23}e^{-j\delta/2}}{1-r_{23}r_{21}e^{-j\delta}}\right|^2 = I_{in}\frac{T_{12}T_{23}}{1+R_{23}R_{21}-2\sqrt{R_{23}R_{21}}\cos(\alpha+\delta)} \quad (5.41)$$

where

$T_{12} = |t_{12}|^2$
$T_{23} = |t_{23}|^2$
$R_{23} = |r_{23}|^2$
$R_{21} = |r_{21}|^2$
$\alpha = -(\arg(r_{23}) + \arg(r_{21}))$

Having the same mediums 1 and 3 results in

$$T_{12} = T_{23} = T, \quad R_{23} = R_{21} = R, \quad \text{and} \quad \alpha = 0, 2\pi$$

Therefore, the transmitted intensity would be written as

$$I_t(\delta) = I_{in}\frac{T^2}{1+R^2-2R\cos(\delta)} = I_{in}\frac{T^2}{(1-R)^2+4R\sin^2(\delta/2)} \quad (5.42)$$

which as mentioned before is an Airy function.

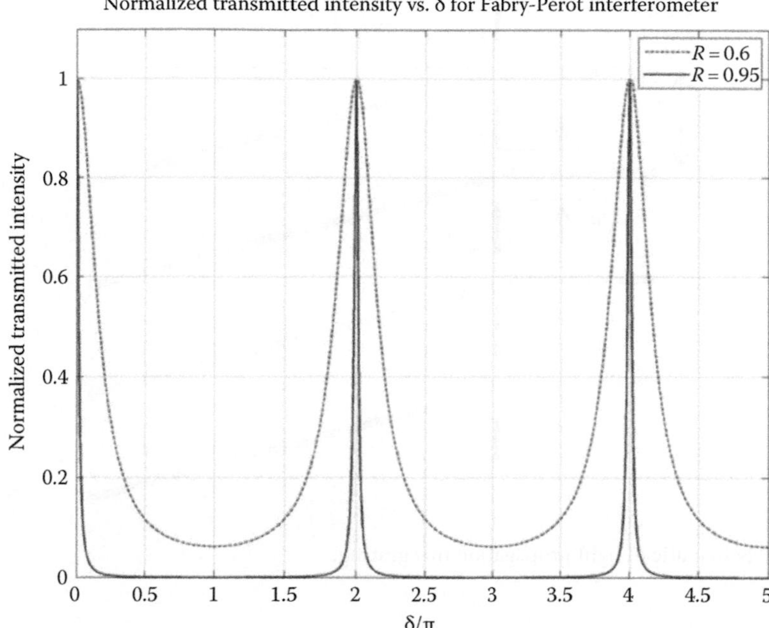

FIGURE 5.9 Normalized transmitted intensity $[I_t(\delta)/I_{in}]$ versus normalized phase (δ/π).

The intensity in reflection can be calculated by subtracting the incident intensity from the transmitted intensity.

$$I_r(\delta) = I_{in} - I_t(\delta) = I_{in}\frac{4R\sin^2(\delta/2)}{(1-R)^2 + 4R\sin^2(\delta/2)} \tag{5.43}$$

since $R + T = 1$ and $1 + R^2 - 2R - T^2 = (T - R)\{1 - (T + R)\} = 0$.

Figure 5.9 shows the normalized transmission intensity for the Fabry–Pérot interferometer. Fabry–Pérot interferometers are used both as sensors (Wei et al. 2008, Favero et al. 2012) and as optical filters (Atherton et al. 1981).

5.4.2.2 Grating

The most common multiple-beam interference with the division of the wavefront is the grating. Figure 5.10 shows the schematic of the grating.

A wave is incident on a grating with N apertures with the width of a and pitch of d. The phase change between successive apertures of grating is δ. The output of the grating at an angle θ is

$$E_o(\theta) = E + Ee^{-j\delta} + Ee^{-j2\delta} + \cdots + Ee^{-j(N-1)\delta}$$

$$= E\left(1 + e^{-j\delta} + e^{-j2\delta} + \cdots + e^{-j(N-1)\delta}\right) = E\frac{\left(1 - e^{-jN\delta}\right)}{1 - e^{-j\delta}} \tag{5.44}$$

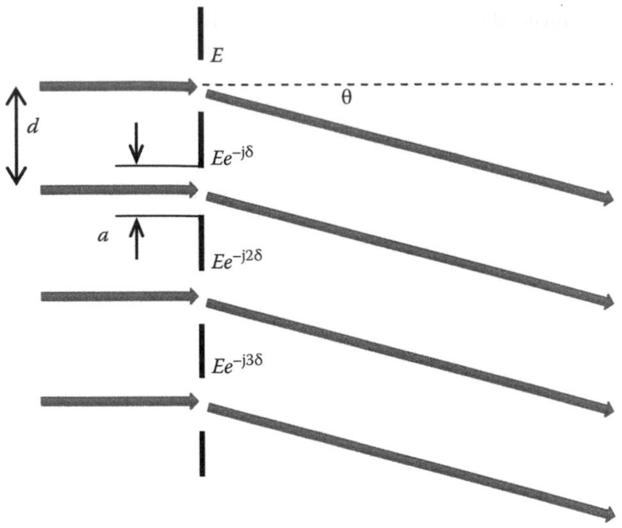

FIGURE 5.10 Schematic of light propagation in a grating.

where

$$\delta = \frac{2\pi nd}{\lambda}\sin(\theta) \tag{5.45}$$

n is the refractive index of the medium
λ is the wavelength of the wave

The output intensity then would be (by ignoring constant α where $I = \alpha|E|^2$)

$$I_o(\theta) = |E_o(\theta)|^2 = \left|E\frac{\left(1-e^{-jN\delta}\right)}{1-e^{-j\delta}}\right|^2 = |E|^2\frac{\sin^2(N\delta/2)}{\sin^2(\delta/2)} \tag{5.46}$$

The diffraction pattern for a single slit with the width of a equals

$$|E|^2 = I(0)\text{sinc}^2\left(\frac{na}{\lambda}\sin(\theta)\right) \tag{5.47}$$

where $I(0)$ is the intensity of diffraction at $\theta = 0$ for single slit. From Equations 5.46 and 5.47,

$$I_o(\theta) = I(0)\text{sinc}^2\left(\frac{na}{\lambda}\sin(\theta)\right)\frac{\sin^2(N\delta/2)}{\sin^2(\delta/2)} \tag{5.48}$$

Figure 5.11 shows the normalized transmitted intensity versus angle θ for $N = 2$, $N = 5$, $N = 20$, $d = 2a$, and $a = \lambda/4$.

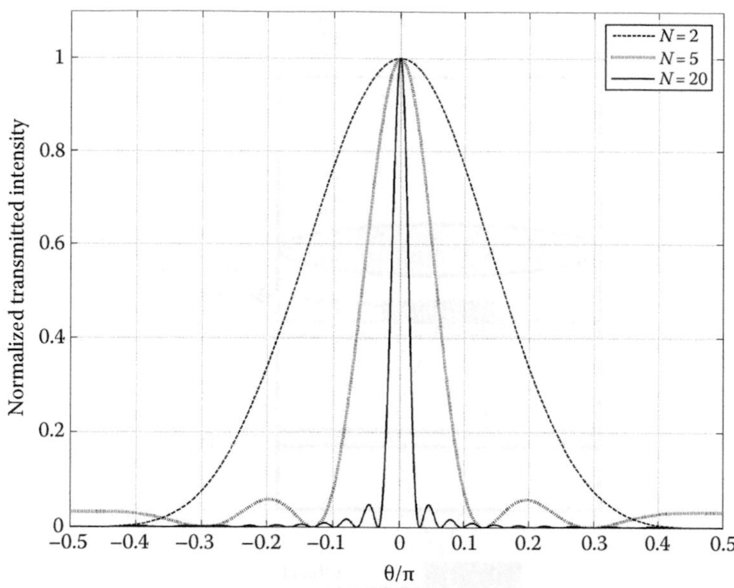

FIGURE 5.11 Normalized transmitted intensity versus θ for a grating.

Gratings have a wide range of applications in spectrometry (Lord and Mccubbin 1957, Mountain et al. 1990, Herder et al. 2001), sensors (Bhatia and Vengsarkar 1996, Kersey et al. 1997, Rao 1997), stabilization of diode lasers (Ricci et al. 1995), and as a modulator (Solgaard et al. 1992).

5.4.3 White-Light Interferometers

White-light interferometer is characterized by its wide spectral bandwidth and finite size source. These characteristics result in low temporal and spatial coherence of light source. Each wavelength from a wide bandwidth source produces its interferogram; these interferograms superimpose and are visible only if the interferometer is almost balanced; that is, the optical path difference is around zero. By moving away from the balance, the interferogram becomes blurry, and it vanishes if the imbalance becomes more than the coherence length, which is typically a couple of micrometers. The intensity pattern at each point of interferogram can be written as

$$I(z) = I_o\left\{1 + \nu(z)\cos\left(\frac{2\pi n}{\lambda_{avg}}z\right)\right\} \tag{5.49}$$

where

I_o is the background intensity
z is the scanning distance parameter
$\nu(z)$ is the visibility
λ_{avg} is the central wavelength of the light source

Chapter 5

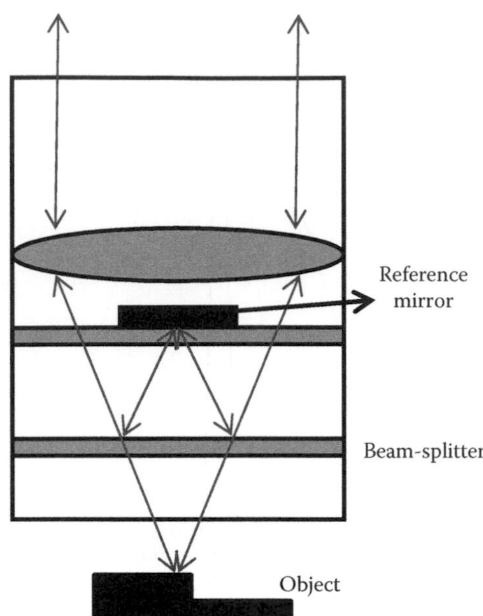

FIGURE 5.12 Mirau white-light interferometer.

White-light interferometer is a powerful tool to measure the profile of surfaces (Deck and Groot 1994, Groot and Deck 1995), thickness (Kim and Kim 1999), and absolute distance (Schnell et al. 1996). White-light interferometer can be implemented in several ways including Michelson interferometer and Mirau interferometer. Figure 5.12 shows the schematic of a Mirau white-light interferometer.

The light that passes through an objective lens partially reflects from the surface of the beam-splitter. The reflected light from the object and reference mirror interferes to form a fringe pattern. The interferometer is scanned vertically to measure the profile of the object. Figure 5.13 shows a typical fringe pattern for a single point when the interferometer is scanned vertically. The maximum intensity (peak of the envelope of the fringe pattern) corresponds to the zero optical path difference, which is a measure of the height of the object.

5.4.4 Fiber Interferometers

Fiber interferometers are the implementation of interferometers using optical fibers. The optical fibers that are used in fiber interferometers are usually single-mode fibers although birefringent fibers are also used. Figure 5.14 shows the schematic of a fiber Michelson interferometer. A laser source, usually a semiconductor laser, is coupled with the optical fiber. A coupler is used as the beam-splitter. Lights reflect back from the end faces of the fibers, which act as mirrors, interfere at the fiber coupler. The end of fibers can be just cleaved to reflect about 4% of the light or can be coated in order to form high-reflectivity fiber mirrors. The fiber interferometer can be used to measure relative phase difference between beams from the two arms, of which one acts as a reference arm (Jackson et al. 1980a). Piezoelectric fiber stretcher has been used in one arm of such fiber interferometers for signal processing of output signals (Jackson et al. 1980b).

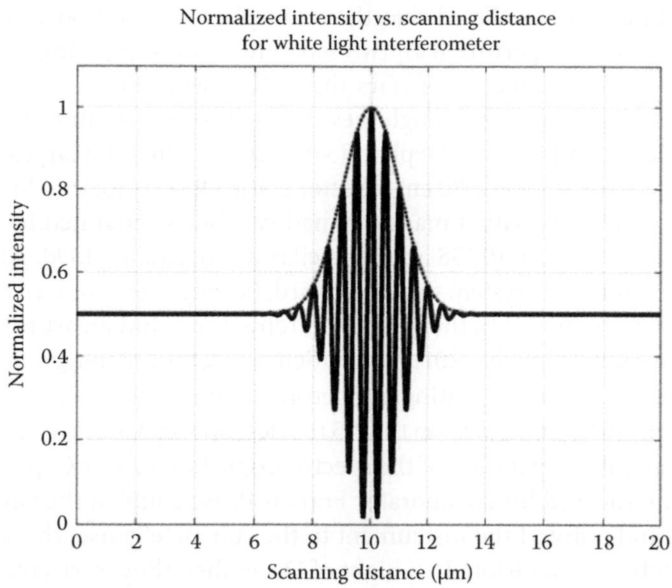

FIGURE 5.13 A typical output pattern of a while-light interferometer.

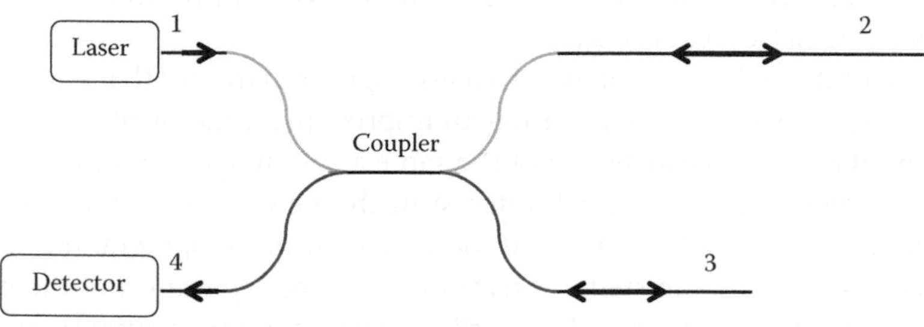

FIGURE 5.14 Fiber Michelson interferometer.

The fiber interferometers are widely used as sensors for the measurement of different physical, chemical, and biological parameters (Lee 2003) including displacement (Ruan 2000), temperature (Starodumov et al. 1997), strain (Zhang et al. 2008), humidity (Chen et al. 2012), and multiparameter sensors (Farahi et al. 1990, Udd 2007). The basic principle of operation of a fiber interferometer as sensor is the fact that the physical quantities can change the refractive index and the length of the optical fiber that results in change of optical path difference of fiber interferometer.

5.5 Astronomic Optical Interferometry

5.5.1 Introduction

Since the time of Galileo Galilei, astronomers believed that increasing the size of telescopes was the way to improve the quality of astronomical observations. It was by the time of William Herschel (1738–1822) when it became clear that the image sharpness is limited by atmospheric rather than telescope optics. An important breakthrough came with Fizeau in 1851, who predicted that the atmospheric limitation could be overcome by masking the telescope aperture with two separate holes, suggesting it may be possible to measure the angular diameters of stars through observations of interference fringes produced from starlight passing through these holes. The first attempts to apply this technique were carried out in 1874 by Stephan (Saha, 2011). With the largest reflecting telescope in existence at the time, the 80 cm reflector at the Observatory de Marseille, Stephan could not resolve any stars with a mask that had two holes separated by 65 cm, but was able to derive the upper limit (0.158 arcs) of stellar diameters. In 1890, Albert Michelson designed an interferometric system that, afterward, became the standard configuration in this field and, when coupled with the 1 m Yerkes refractor, could measure the diameters of Jupiter's Galilean satellites (Saha, 2011). Later, Schwarzschild managed in 1896 to resolve a number of double stars with a grating interferometer on the 25 cm Munich Observatory telescope, while in 1920, Anderson, on the 2.5 m telescope on Mount Wilson Observatory, determined the angular separation of the spectroscopic binary star Capella ($\alpha - Aur$).

In 1920, Michelson and his collaborator Francis Pease coupled the two mirrors of the 2.5 m telescope and pointed the instrument to the star Betelgeuse, the second brightest star in the constellation of Orion. In a night of December, they were able to measure the angular diameter of the star: $\theta_{Betelgeuse} = 0.047$ arcs. The distance of the star was known from parallax measurement (520 light years); therefore, the combination of these data permitted to determine its diameter as ~350 million km, approximately 280 times the Sun diameter (Michelson et al. 1921).

By the time of this achievement, the smallest angular diameter that could be resolved with a full aperture was ~1 arcs, meaning an improvement factor of ~21. Although the measurement of a stellar diameter is not the same as an image, the substantial increase in angular resolution attracted much interest in the new method (Glindemann 2011). Michelson and Pease moved further, and soon after this development, they constructed a separate-element Michelson stellar interferometer. The separation of the mirrors was equivalent to the slit separation in their earlier interferometers. Separations of over 6 m were possible (inset of Figure 5.15), enabling the measurement of the diameters of several large stars to be performed. An interferometer with 15 m of mirror separation was built in 1930 by Pease, with mirrors attached to 9 tons of steel girderwork on the front of a 1 m aperture optical telescope (Pease 1930). However, very few astronomical measurements were made by this instrument due to the difficulty of operating it. Atmospheric fluctuations produced phase variations, which caused the fringes to *shimmer*, making observation quite difficult. For these reasons, further improvements in this area, named astronomic interferometry,* needed five decades to appear, when the technical difficulties and mechanical instabilities could be overcome.

Meanwhile, much of the activity in interferometric astrometry was done by radio astronomers. Cosmic radio emission was discovered in the 1930s and radio interferometry developed after World War II. In 1946, Ryle and Vonberg constructed a radio analog of the Michelson interferometer and soon located a number of new cosmic radio sources (Ryle and Vonberg 1946). The signals from two radio antennas were added

* Astronomic interferometry is also known as *stellar interferometry*, designation derived from the fact these concepts were first applied to get an improved angular resolution of stellar observations, although nowadays, it has a more restricted meaning considering astronomic interferometry is gathering importance as a tool for sky imaging.

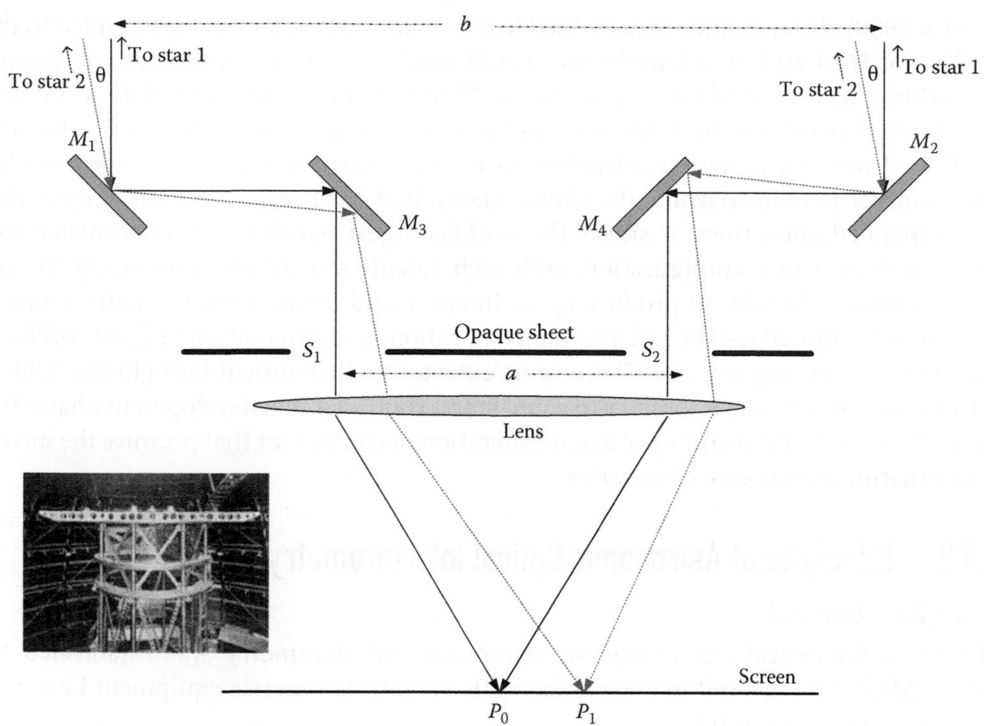

FIGURE 5.15 Layout of the Michelson stellar interferometer (inset image: 6 m Michelson interferometer for measuring star diameters, attached to upper end of the skeleton tube of the 2.5 m Hooker telescope of Mount Wilson Observatory). From Simons, M. *History of Mount Wilson Observatory: Building the 100-Inch Telescope*. Mount Wilson Observatory Association (http://www.mtwilson.edu/).

electronically to produce interference, with the rotation of Earth used to scan the sky in one dimension. This electronic processing facility, possible at radio wavelengths (and not in the optical window), enabled the substantial progress of radio-based interferometry in the following years, leading to the development of large instruments such as the emblematic *Very Large Array* in New Mexico, the United States (equipment with 27 independent antennas, each with a dish diameter of 25 m, formally inaugurated in 1980 and upgraded in 2011, which permitted to expand its technical capacities by factors as much as 8000), and the *ALMA-Atacama Large Millimeter Array* (an array of radio telescopes in the Atacama desert of northern Chile, consisting of sixty-six 12 and 7 m diameter radio telescopes observing at millimeter and submillimeter wavelengths, consequence of an international partnership between Europe, the United States, Canada, East Asia, and the Republic of Chile; this equipment was inaugurated in March 2013).

Astronomic optical interferometry reappeared in 1974 with a setup of two separate telescopes operating in the infrared (Johnson et al. 1974), followed in 1975 by a similar configuration but working in the visible region of the spectrum (Labeyrie 1975). In the late 1970s, improvements in computer processing permitted to design the first fringe-tracking interferometer operating fast enough to follow the blurring effects of astronomical seeing, leading to the development of the Mark I prototype interferometer, which used modern optical detectors and mechanical control systems to measure and track phase variations induced by fluctuations in the Earth's atmosphere (Shao and

Staelin 1980). This equipment permitted to test concepts and technology that led to the buildup of *Mark III Stellar Interferometer* at Mount Wilson Observatory, an interferometer with a maximum mirror separation of 32 m working in the [450–800] nm wavelength window (Shao et al. 1988), which operated in the period of 1987–1992. This line continued with the *Navy Precision Optical Interferometer*, a major astronomic equipment built at 1992 in Arizona, the United States, that has been continuously upgraded and expanded since then. It shows the world's largest baselines (interferometer laid out in a three-arm *Y* configuration, with each equally spaced arm measuring 250 m) and operated primarily to produce space imagery and astrometry, the latter a major component required for the safe position, navigation, and orientation of Earth-orbiting satellites (Armstrong et al. 1997). To some extent, this instrument brought the field of astronomic optical interferometry to a new stage, routing it to a development phase that led to the design and assembly of a new generation of equipment that promise the unveil of fascinating astronomic discoveries.

5.5.2 Principles of Astronomic Optical Interferometry

5.5.2.1 General

The basic theoretical description of astronomic interferometry was established by Albert Michelson in 1890 in connection with the interferometric equipment he developed, illustrated in Figure 5.15.

Two movable mirrors, M_1 and M_2, separated by b, reflect the rays essentially parallel to a stellar object. The light is guided to the openings S_1 and S_2 (separated by a) of an opaque sheet by mirrors M_3 and M_4, with the transmitted light collected by the telescope objective. The optical paths $M_1M_3S_1$ and $M_2M_4S_2$ are equal by construction. On the screen, located in the focal plane of the objective, appears a fringe pattern typical of the Young layout shown in Figure 5.1. The stellar object is supposed to be a binary star system (star 1, star 2), with the interferometer pointed to one of its components (it needs to rotate the angle θ to point to the other component). To simplify, let us assume the light is quasi-monochromatic centered in λ. When the interferometer points to star 1, the interference pattern in the screen is centered in P_0. When it points to star 2, there appears a dephasing of the optical fields in S_1 and S_2, $\delta\phi_{S_1S_2|star2}$, given by

$$\delta\phi_{S_1S_2|star2} = \frac{2\pi}{\lambda} b\theta \tag{5.50}$$

with the consequence that the fringe pattern in the screen becomes now centered in P_1 (Figure 5.15). The two stars of the binary can be considered as incoherent point optical sources; therefore, their interference patterns in the screen just add. The separation of the maximums in each of these patterns depends only on a, but the visibility is a function of b. When b is close to zero, the intensity maximums of the two patterns coincide in the screen; therefore, the visibility is maximum. When b starts to increase, the interference pattern associated with star 2 begins to move relative to the one of star 1, and when the maximums of one pattern are aligned with the minimums of the other, no intensity modulation appears in the screen, and the visibility becomes zero.

This happens for $b = b_{V=0}$ where $\delta\phi_{S_1S_2|\text{star2}} = \pi$. From Equation 5.50, the angular separation, seen from Earth, of the two stars of the binary system is given by

$$\theta_{binary\ system} = \frac{\lambda}{2b_{V=0}} \tag{5.51}$$

This technique can also be applied to measure the angular diameter of stars. The visibility of the fringe pattern associated with light from a circular star, considered as a distribution of incoherent point sources inside a circle of uniform intensity, is given by a Bessel function of first order (Mandel and Wolf 1995):

$$V = 2\left|\frac{J_1(\pi b\theta/\lambda)}{\pi b\theta/\lambda}\right| \tag{5.52}$$

where

now θ is the angular diameter of the star

$J_1(x)$ is the Bessel function of first kind and first order

The first zero of this function occurs for $x = 3.83$, meaning the fringe visibility becomes zero when $b = b_{V=0} = 1.22(\lambda/\theta)$, resulting for the angular diameter of the star

$$\theta_{star} = 1.22\frac{\lambda}{b_{V=0}} \tag{5.53}$$

As already stated in Section 5.5.1, in 1920, Michelson and Pease used the interferometric configuration shown in Figure 5.15 to measure the angular dimension of the star Betelgeuse; they found that the visibility of the fringe pattern vanished for $b_{V=0} = 3.025$ m. From Equation 5.53, it results $\theta_{Betelgeuse} = 22.6 \times 10^{-8}$ rad = 0.047 arcs.

5.5.2.2 Angular Resolution

Astronomic optical interferometry is a technique associated with the spatial sampling of the optical wavefront coming from an astronomic object and interference of the sampled light elements, permitting to generate an intensity pattern with additional structure from where improved angular resolutions can be obtained, opening the possibility to detect small features of the optical source and to generate images with improved quality. It is illustrated in Figure 5.16 for the situation of monochromatic light arriving from a well-defined direction (vertical in the case).

Figure 5.16a shows a diagram of a telescope with aperture diameter D_T, whose theoretical angular resolution is approximately λ/D_T, as indicated by the intensity pattern drawn in the second row of the figure. Figure 5.16b represents a masked aperture, a configuration equivalent to the Michelson interferometer. The intensity pattern displays a channeled structure with fringe spacing determined by the separation of the mask holes (B), and an envelope fixed by the size of these holes (D). Therefore, if the angular resolution of the instrument is defined as the fringe spacing (of the order of λ/B), it is no more limited by the diameter of the telescope but by the separation of the mask holes. Figure 5.16c depicts a similar structure as the middle figure, but this time, the two

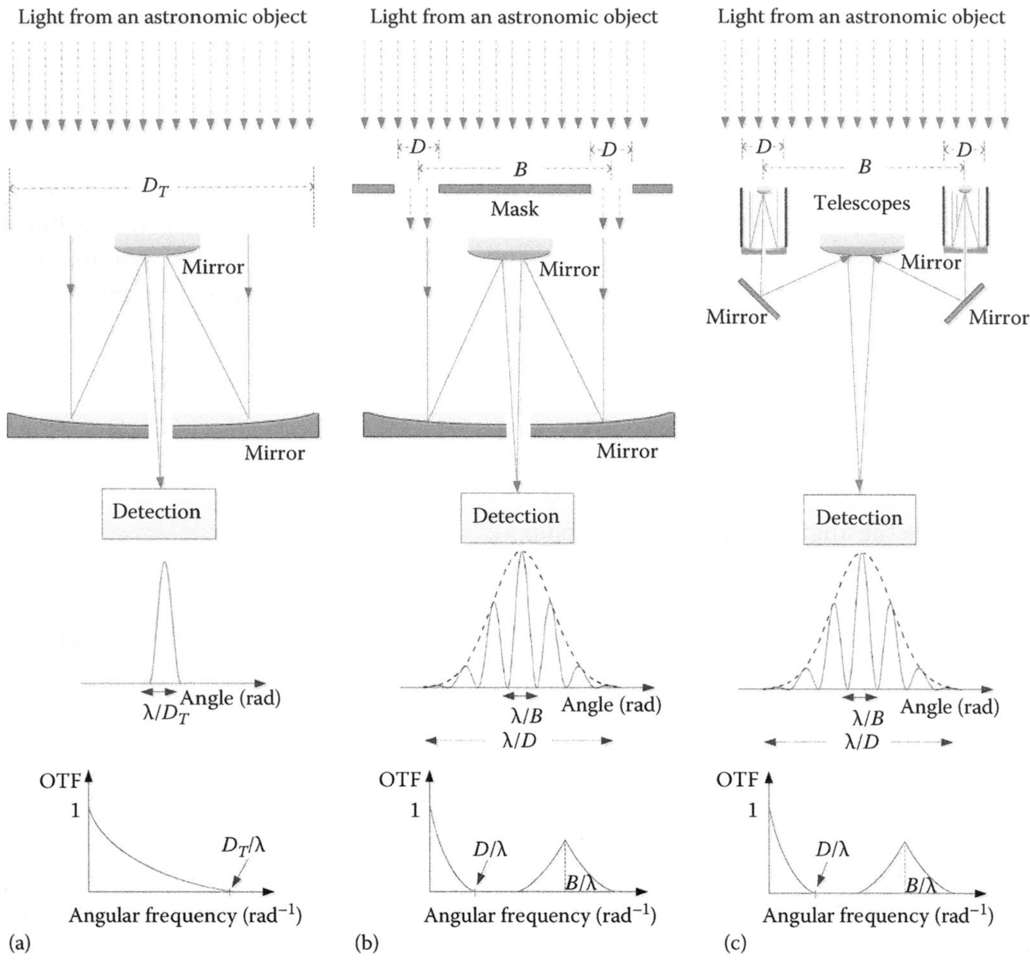

FIGURE 5.16 Illustration of the principle of optical astronomic interferometry. Imaging with a monolithic aperture (a), with a masked aperture (b), and with a combination of light from two telescopes (c). In each case, the imaging signal in the spatial (angular) and spectral domains (optical transfer function—OFT) is shown.

apertures are two telescopes with diameters D, separated by a distance B (known as *baseline*). The angular resolution of this arrangement is also approximately λ/B, which can be improved by increasing the separation of the two telescopes.

The observation in the spectral domain of the image formed in each case clearly illustrates the possibility of resolution improvement associated with astronomic interferometry. In the third row of Figure 5.16, the optical transfer function for each case is shown. For the standard telescope arrangement, the highest angular frequency that can be resolved is of the order of D_T/λ. Concerning the masked structure in Figure 5.16b, the highest angular frequency in the baseband is close to D/λ, substantially smaller than the one that can be achieved without mask. The wavefront sampling generates spectral components around B/λ (and $2B/\lambda$, $3B/\lambda$, …), but with increasingly reduced amplitude, indicating that when the configuration of Figure 5.16c is considered, smaller angular features of astronomic objects can be revealed just by increasing the separation between the telescopes (increasing B), coupled with the application of specific signal processing methodologies (characteristic that has enhanced importance when the limiting effects associated with the atmosphere are taken into account).

In the layouts of Figure 5.16b and c, the fringe visibility is unitary, a consequence of the assumptions of monochromatic and unidirectional radiation. When the radiation comes from an extended source seen from the Earth with an angular size 2θ, the component that arrives at the instrument at angle θ (with origin defined by the vertical direction) generates a shifted fringe pattern, which is incoherent relative to the one associated with the vertical direction. Therefore, the intensities of these patterns add to give a pattern of reduced visibility. If $\theta = \lambda/B$, the fringes add to give zero visibility (actually, the value of the visibility is determined by the sum of the patterns defined by the full angular range of the astronomic object seen by the instrument). When $\theta \ll \lambda/B$, the fringe visibility approaches 1, corresponding to the situation in which the angular source size is substantially smaller than the angular resolution of the interferometer, indicating that only an upper limit of λ/B for the angular source size can be obtained. When the baseline B decreases, for the same size of the source, there is a smaller reduction of the fringe visibility. Therefore, the smaller the separation between the two apertures (telescopes), the larger the source size that can be probed using interferometry.

This behavior is summarized in Figure 5.17, which illustrates the relation between source brightness as a function of angular distance and visibility of interference fringes as a function of B.

It can be observed for sources that subtend a small angle on Earth, the visibility remains high even for large values of B (in the limit of point sources, to infinite apertures separation), whereas for sources associated with larger values of θ, the visibility patterns fall off quickly as the aperture separation increases. This relation between $I(\theta)$ and $V(B)$ is one that maps a large Gaussian function into a small Gaussian function and

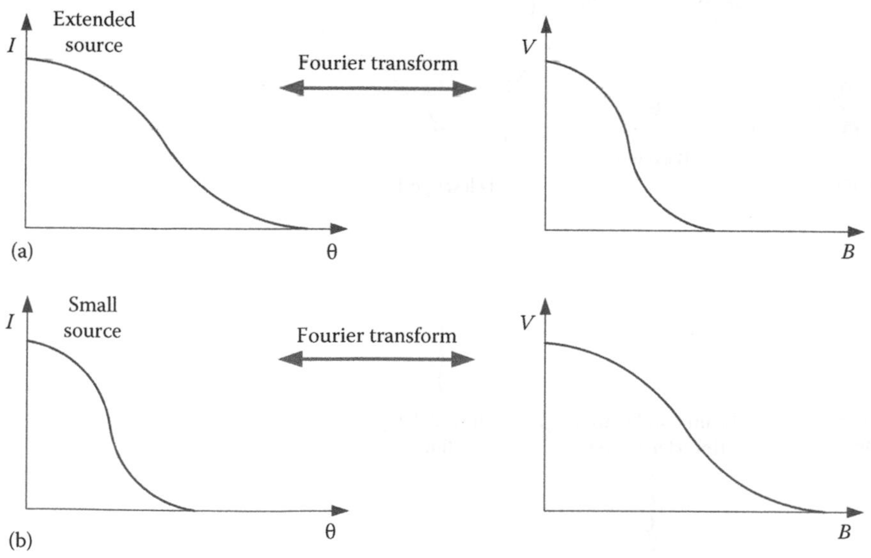

FIGURE 5.17 Dependence between sky source brightness versus angular view from the instrument (a) and fringe visibility versus interferometer slit separation B (b). (Adapted from Jackson, N., *Lect. Notes Phys.*, 742, 193, 2008.)

vice versa, mathematically connected with a Fourier transform as expressed by the van Cittert–Zernike theorem detailed in Section 5.3.3.

The technique of astronomic optical interferometry permits not only to determine the angular size of astronomic objects, but also to obtain a mapping of its light intensity emission, that is, an image of the object, as described in the next section.

5.5.3 Imaging with Astronomic Optical Interferometry

Let us consider the configuration shown in Figure 5.18, which shows two telescopes separated by a distance B, associated with the baseline vector $\vec{\mathbf{B}}$. Its projection into a direction perpendicular to the unit vector $\hat{\mathbf{s}}$ that points to the astronomic source is $\vec{\mathbf{b}}$.

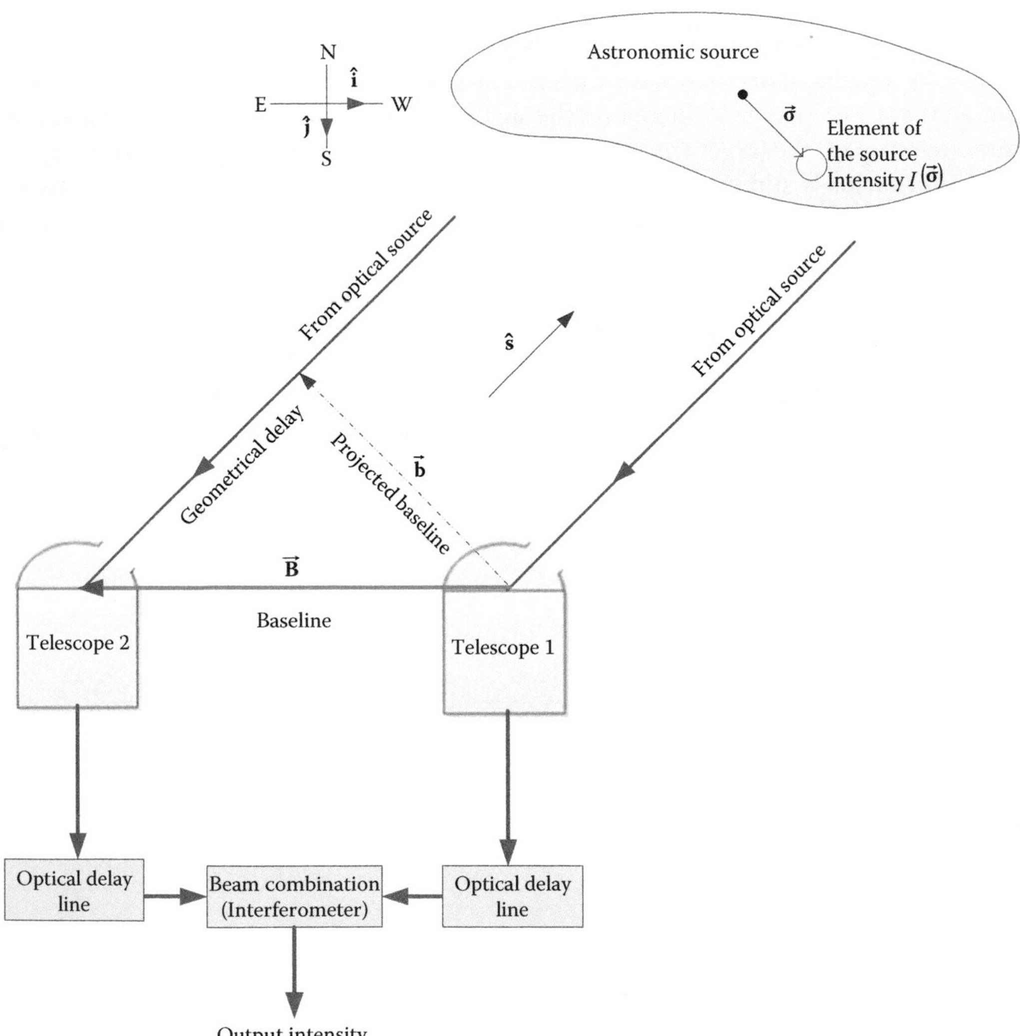

FIGURE 5.18 Parameter definition for an imaging astronomic interferometric layout with two telescopes observing an optical sky source along the unit vector $\hat{\mathbf{s}}$.

The geometric delay of the optical wave arriving at telescope 2 relative to the wave reaching telescope 1 is $-\vec{\mathbf{B}}\cdot\hat{\mathbf{s}}$, and the phase delay is $-k\vec{\mathbf{B}}\cdot\hat{\mathbf{s}}$, where $k = 2\pi/\lambda$ (it is assumed the source emits quasi-monochromatic radiation centered at λ). If the optical field (electric component) received by the first telescope is $\vec{\mathbf{E}}$, that received by the second is $\vec{\mathbf{E}}e^{-ik\vec{\mathbf{B}}\cdot\hat{\mathbf{s}}}$. Combining these two fields in the interferometer results in an intensity value associated with the light field arriving into the system along the direction defined by $\hat{\mathbf{s}}$. An elementary region of the optical source $d\sigma$, identified by the vector in the sky plane $\vec{\sigma}$ (for convenience, with origin in the central region of the source) and emitting the intensity $I(\vec{\sigma})$, illuminates the telescope system from the direction $\hat{s}+\vec{\sigma}$, resulting into another intensity for the interference of the optical fields detected by the two telescopes, which adds to the one arriving from $\hat{\mathbf{s}}$. Integrating for all elementary parts of the astronomic source, it comes out for the interferometric output that

$$I_{int} = \int I(\vec{\sigma})e^{-ik\vec{\mathbf{B}}\cdot(\hat{s}+\vec{\sigma})}d\sigma \tag{5.54}$$

Having in mind that $\vec{\sigma}$ and $\vec{\mathbf{b}}$ are parallel, $\vec{\mathbf{B}}\cdot\vec{\sigma} = \vec{\mathbf{b}}\cdot\vec{\sigma}$. Also, $\vec{\mathbf{B}}\cdot\hat{\mathbf{s}}$ does not depend on the characteristics of the optical source; therefore,

$$I_{int_{|\vec{b}}} = e^{-ik\vec{\mathbf{B}}\cdot\hat{\mathbf{s}}}\int I(\vec{\sigma})e^{-ik\vec{\mathbf{b}}\cdot\vec{\sigma}}d\sigma \tag{5.55}$$

The constant factor $e^{-ik\vec{\mathbf{B}}\cdot\hat{\mathbf{s}}}$ can be removed generating four values for I_{int}, each one associated with interference beams with an extra phase shit of $\pi/2$ relatively to the previous one. These phase shifts can be introduced by the delay lines shown in Figure 5.18. If $I_1 = I_{int_{|\vec{b}}}$, I_2, the intensity that results by adding an extra phase of $\pi/2$ to one of the interfering beams that originates I_1, I_3 with such extra phase equal π and I_4 when it is $3\pi/2$, then defining V as (Glindemann 2011)

$$V \equiv 2\frac{(I_1 - I_3) + i(I_2 - I_4)}{I_1 + I_2 + I_3 + I_4} \tag{5.56}$$

it comes out

$$V(\vec{\mathbf{b}}) = \int I(\vec{\sigma})e^{-ik\vec{\mathbf{b}}\cdot\vec{\sigma}}d\sigma \tag{5.57}$$

This equation shows that $V(\vec{\mathbf{b}})$, the complex visibility (with amplitude and phase information) of the interferometric fringes that result from the combination of light of the two telescopes, is the Fourier transform of the source intensity distribution evaluated for the projected baseline $\vec{\mathbf{b}}$.

In Equation 5.57, the scalar product $\vec{\mathbf{b}}\cdot\vec{\sigma}$ is better expressed by its form in Cartesian coordinates. For that, the sky plane is defined as the plane where two axes of the astronomic source are situated, one identified with the unit vector $\hat{\mathbf{i}}$ that points in the east–west direction and another with the unit vector $\hat{\mathbf{j}}$ that points in the north–south direction (Figure 5.18). Therefore,

$$\vec{\sigma} = x\hat{\mathbf{i}} + y\hat{\mathbf{j}}$$
$$\frac{\vec{\mathbf{b}}}{\lambda} = u\hat{\mathbf{i}} + v\hat{\mathbf{j}} \tag{5.58}$$

Resulting from Equation 5.57,

$$V(u,v) = \int I(x,y)e^{-i2\pi(ux+vy)}dxdy \tag{5.59}$$

a more familiar form of a 2-D Fourier transform. In view of definition (5.58), (u,v) are expressed in units of wavelength.

Due to the rotation of the Earth, the baseline vector $\vec{\mathbf{B}}$ changes its orientation relative to the astronomic source, and, therefore, its projection $\vec{\mathbf{b}}$ also changes, tracing an ellipse in the (u,v) plane. In this way, the rotation of the Earth permits to determine V for the points (u,v) situated on the ellipse, that is, for just two telescopes, it is possible to obtain a set of values for the function $V(u,v)$. This is relevant because what is looked for is the source brightness distribution, $I(x,y)$, which is analytically expressed as the inverse Fourier transform of $V(u,v)$, given from Equation 5.59 as

$$I(x,y) = \int V(u,v)e^{i2\pi(ux+vy)}dudv \tag{5.60}$$

This relation indicates a major problem with using this technique to get an image of the optical source, since it is impossible to measure the visibility function $V(u,v)$ over the whole (u,v) plane. To do this would require many telescopes such that, when combined with the Earth rotation, a number of projected baselines compatible with the sampling theorem for image reconstruction would become available. In general, this is not feasible, meaning the existence of a number (large) of different images consistent with the available data.

Stating in another way, what we get is not the $I(x,y)$ provided by Equation 5.60, but instead the distribution

$$I_S(x,y) = \int V(u,v)S(u,v)e^{i2\pi(ux+vy)}dudv \tag{5.61}$$

which results from the incorporation of the sampling function $S(u,v)$, which is one in parts of the (u,v) plane where there are measured values for the visibility and zero otherwise (Jackson 2008). This equation indicates that what is involved is the Fourier pair $I_S(x,y) \leftrightarrow V(u,v)S(u,v)$; therefore, the use of the convolution theorem allows to write

$$I_S(x,y) = I(x,y)^{*}B(x,y) \tag{5.62}$$

where $B(x,y)$ is the inverse Fourier transform of the sampling function

$$B(x,y) = \int S(u,v)e^{i2\pi(ux+vy)}dudv \tag{5.63}$$

Since the location of the telescopes is known as well as the baseline variation associated with the Earth rotation, the function $S(u,v)$ is well determined and so $B(x,y)$. Recovering the image $I(x,y)$ is therefore a classical deconvolution problem in which additional information or constraints need to be supplied in order to perform such operation. Examples of these constraints are (Monnier 2003) as follows: *limited field of view*, always present in real observation systems, introduces correlations in the complex visibility in the (u,v) plane; *positive-definite*, that is, source brightness distributions cannot be negative; *smoothness*, meaning the criterion of selecting the smoothest image consistent with the data available; *a priori information*, providing previously known information to constrain the possible image reconstructions.

Originally, it was thought necessary to make measurements at essentially every baseline length and orientation out to some maximum to obtain the image of the astronomic object. Such a fully sampled Fourier transform formally contains the information exactly equivalent to the image from a conventional telescope with an aperture diameter equal to the maximum baseline, hence the name *aperture synthesis*. Later it was discovered that by introducing the type of constraints just mentioned, useful images could be made with a relatively sparse and irregular set of baselines. The alternative name *synthesis imaging* acknowledges the shift in emphasis from trying to synthesize the complete aperture (allowing image reconstruction by Fourier transform) to trying to synthesize the image from the combination of available data, a set of reasonable constraints, and the application of powerful computational algorithms such as the *CLEAN* algorithm introduced by Högbom (1974) and the *maximum entropy method*, which appeared in 1977 for image restoration (Wernecke and d'Addario 1977), both in the context of radio astronomy. This was the field where aperture synthesis imaging was first developed, by Martin Ryle and coworkers from the Radio Astronomy Group of Cambridge University (Martin Ryde and Tony Hewish jointly received the Nobel Prize in Physics in 1974).

As a final note in this section, it is relevant to point out that interferometric (Fourier) imaging has important differences from direct imaging as it can be understood from the light patterns generated by the Young's slit setup; long baselines record effectively small-scale structures of the optical source, which are insensitive to large-scale structures. This is so because when the angular view of the source becomes larger than λ/B, the fringes wash out and do not return as the source size increases. Therefore, interferometer arrays should be chosen to match resolution to the spatial scales required by any astrophysical problem (Jackson 2008).

5.6 Final Remarks

Interference is an omnipresent phenomenon in the world with countless appearances in the physical and biological domains, some relatively straightforward to understand, others situated far deep into the realms of nature, as are those associated with certain types of quantum interference. The technologies based on interference effects derived from the fundamental understanding of waves have been around for many years, with continuous growing importance in virtually all application domains. For obvious reasons, technological developments connected with electromagnetic interference in the optical spectrum have been a most targeted field, with particular focus on precision measurements.

Interferometric measurement has been in each epoch associated with leading-edge performance, a feature that is confirmed nowadays and most probably will stand in the future. This chapter addressed its main principles and techniques. It was also pointed out that guided-wave optics allows interferometry to move beyond controlled laboratory environments. The huge impact and potential of interferometric measurement were illustrated in the final section of the chapter by presenting the area of optical astronomic interferometry. Several following chapters of this book will detail other applications where interferometry plays a key role in optical sensing and measurement.

References

Annunziata, O., D. Buzatu, and J. Albright. 2005. Protein diffusion coefficients determined by macroscopic-gradient Rayleigh interferometry and dynamic light scattering. *Langmuir* 21: 12085–12089.

Annunziata, O. and A. Vergara. 2009. Quaternary diffusion coefficients in a protein–polymer–salt–water system determined by Rayleigh interferometry. *The Journal of Physical Chemistry B* 113: 13446–13453.

Armstrong, J. T., D. Mozurkewich, T. A. Pauls et al. 1997. The Navy Prototype Optical Interferometer (NPOI) is operational. *American Astronomical Society* 191: 1234–1240.

Atherton, H. T., N. K. Reay, and J. J. Ring. 1981. Tunable Fabry-Perot filters. *Optical Engineering* 20: 206805.

Becker, A., W. Kohler and B. Muller. 1995. A scanning Michelson interferometer for the measurement of the concentration and temperature derivative of the refractive index of liquids. *Berichte der Bunsengesellschaft für Physikalische Chemie* 99: 600–608.

Bhatia, V. and A. M. Vengsarkar. 1996. Optical fiber long-period grating sensors. *Optics Letters* 21: 692–694.

Blake, J., P. Tantaswadi, and R. T. de Carvalho. 1996. In-line Sagnac interferometer current sensor. *IEEE Transactions on Power Delivery* 11: 116–121.

Chen, L. H., T. Li, C. C. Chan et al. 2012. Chitosan based fiber-optic Fabry–Perot humidity sensor. *Sensors and Actuators B—Chemical* 169: 167–172.

Deck, L. and P. Groot. 1994. High-speed noncontact profiler based on scanning white-light interferometry. *Applied Optics* 33: 7334–7338.

Dong, X., H. Y. Tam, and P. Shum. 2007. Temperature-insensitive strain sensor with polarization-maintaining photonic crystal fiber based Sagnac interferometer. *Applied Physics Letters* 90: 151113.

Farahi, F., D. J. Webb, J. D. C. Jones et al. 1990. Simultaneous measurement of temperature and strain: Cross-sensitivity considerations. *Journal of Lightwave Technology* 8: 138–142.

Favero, F. C., L. Araújo, G. Bouwmans et al. 2012. Spheroidal Fabry–Perot microcavities in optical fibers for high-sensitivity sensing. *Optics Express* 20: 7112–7118.

Glindemann, A. 2011. *Principles of Stellar Interferometry*. Springer, Berlin, Germany.

Gori, F., M. Santarsiero, S. Vicalvi et al. 1998. Beam coherence-polarization matrix. *Pure and Applied Optics A* 7: 941–951.

Groot, P. and L. Deck. 1995. Surface profiling by analysis of white-light interferograms in the spatial frequency domain. *Journal of Modern Optics* 42: 389–401.

Hariharan, P. 1986. *Optical Interferometry*. Academic Press, Boston, MA.

Herder, J. D., A. C. Brinkman, S. Kahn et al. 2001. The reflection grating spectrometer on board XMM-Newton. *Astronomy and Astrophysics* 365: L7–L17.

Högbom, J. A. 1974. Aperture synthesis with a non-regular distribution of interferometer baselines. *Astronomy and Astrophysics Supplement* 15: 417–426.

Jackson, D. A., A. Dandridge, and S. K. Sheem. 1980a. Measurement of small phase shifts using a single-mode optical-fiber interferometer. *Optics Letters* 5: 139–141.

Jackson, D. A., R. Priest, A. Dandridge et al. 1980b. Elimination of drift in a single-mode optical fiber interferometer using a piezoelectrically stretched coiled fiber. *Applied Optics* 19: 2926–2929.
Jackson, N. 2008. Principles of interferometry. *Lecturer Notes in Physics* 742: 193–218.
Johnson, M. A., A. L. Betz, and C. H. Towes. 1974. 10-μm heterodyne stellar interferometer. *Physics Review Letters* 33: 1617–1620.
Kersey, A. D., M. A. Davis, H. J. Patrick et al. 1997. Fiber grating sensors. *Journal of Lightwave Technology* 15: 1442–1463.
Kim, S. W. and G. H. Kim. 1999. Thickness-profile measurement of transparent thin-film layers by white-light scanning interferometry. *Applied Optics* 38: 5968–5973.
Knuuttila, J. V., P. T. Tikka, and M. M. Salomaa. 2000. Scanning Michelson interferometer for imaging surface acoustic wave fields. *Optics Letters* 25: 613–615.
Labeyrie, A. 1975. Interference fringes obtained on VEGA with two optical telescopes. *Astrophysical Journal* 196: L71–L75.
Lapsley, M. I., I.-K. Chiang, Y. B. Zheng et al. 2011. A single-layer, planar, optofluidic Mach–Zehnder interferometer for label-free detection. *Lab on a Chip* 11: 1795–1800.
Lawall, J. and E. Kessler. 2000. Michelson interferometry with 10 pm accuracy. *Review of Scientific Instruments* 71: 2669–2676.
Lee, B. 2003. Review of the present status of optical fiber sensors. *Optical Fiber Technology* 9: 57–79.
Lefevre, H. C. 2012. The fiber-optic gyroscope: Actually better than the ring-laser gyroscope? *Proceedings of the 22nd International Conference on Optical Fiber Sensors*, October 15–19, Beijing, China, 8421, 575.
Li, T., A. Wang, K. Murphy et al. 1995. White-light scanning fiber Michelson interferometer for absolute position-distance measurement. *Optics Letters* 20: 785–787.
Lord, R. C. and J. Mccubbin. 1957. Infrared spectroscopy from 5 to 200 microns with a small grating spectrometer. *Journal of the Optical Society of America* 47: 689–697.
Malacara, D. 2007. *Optical Shop Testing*. John Wiley & Sons, New York.
Mandel, L. and E. Wolf. 1995. *Optical Coherence and Quantum Optics*. Cambridge University Press, New York.
McKenzie, K., D. Shaddock, D. McClelland et al. 2002. Experimental demonstration of a squeezing-enhanced power-recycled Michelson interferometer for gravitational wave detection. *Physical Review Letters* 88: 231102.
Michelson, A. A. and E. W. Morley. 1887. On the relative motion of the Earth and the luminiferous ether. *American Journal of Science* 34: 333–345.
Monnier, J. D. 2003. Optical interferometry in astronomy. *Reports on Progresses in Physics* 66: 789–857.
Mountain, C. M., D. J. Robertson, T. J. Lee et al. 1990. An advanced cooled grating spectrometer for UKIRT. *Proceedings of the Instrumentation in Astronomy VII*, Bellingham, WA, pp. 25–33.
Pease, F. G. 1930. The 50ft stellar interferometer at Mount Wilson. *Scientific American* 143: 290–295.
Qi, Z., N. Matsuda, and K. Itoh. 2002. A design for improving the sensitivity of a Mach–Zehnder interferometer to chemical and biological measurands. *Sensors and Actuators B* 81: 254–258.
Rao, Y. J. 1997. In-fibre Bragg grating sensors. *Measurement Science and Technology* 8: 355–375.
Rard, J. and D. Miller. 1979. The mutual diffusion coefficients of Na_2SO_4–H_2O and $MgSO_4$–H_2O at 25°C from Rayleigh interferometry. *Journal of Solution Chemistry* 8: 755–766.
Ricci, L., M. Weidemuller, T. Esslinger et al. 1995. A compact grating-stabilized diode laser system for atomic physics. *Optics Communications* 117: 541–549.
Ruan, F. 2000. A precision fiber optic displacement sensor based on reciprocal interferometry. *Optics Communications* 176: 105–112.
Ryle, M. and D. Vonberg. 1946. Solar radiation on 175Mc/s. *Nature* 158: 339–340.
Schnell, U., R. Dändliker, and S. Gray. 1996. Dispersive white-light interferometry for absolute distance measurement with dielectric multilayer systems on the target. *Optics Letters* 21: 528–530.
Sepúlveda, B. and J. del Río. 2006. Optical biosensor microsystems based on the integration of highly sensitive Mach–Zehnder interferometer devices. *Journal of optics A* 8: S561–S566.
Shao, M., M. M. Colavita, B. E. Hines et al. 1988. The Mark III stellar interferometer. *Astronomy and Astrophysics* 193: 357–371.
Shao, M. and D. H. Staelin. 1980. First fringe measurements with a phase-tracking stellar interferometer. *Applied Optics* 19: 1519–1522.
Solgaard, O., F. S. A. Sandejas, and D. M. Bloom. 1992. Deformable grating optical modulator. *Optics Letters* 17: 688–690.
Starodumov, A. N., L. A. Zenteno, D. Monzon et al. 1997. Fiber Sagnac interferometer temperature sensor. *Applied Physics Letters* 70: 19–21.

Steel, W. H. 1983. *Interferometry*. Cambridge University Press, London, U.K.
Sun, K.-X., M. Fejer, E. Gustafson et al. 1996. Sagnac interferometer for gravitational-wave detection. *Physical Review Letters* 76: 3053–3056.
Tian, Z., S. S.-H. Yam, and J. Barnes. 2008a. Refractive index sensing with Mach–Zehnder interferometer based on concatenating two single-mode fiber tapers. *Photonics Technology Letters* 20: 626–628.
Tian, Z., S. S.-H. Yam, and H. P. Loock. 2008b. Refractive index sensor based on an abrupt taper Michelson interferometer in a single-mode fiber. *Optics Letters* 33: 1105–1107.
Udd, E. 2007. Review of multi-parameter fiber grating sensors. *Proceedings of SPIE* 6770: 677002.
Wei, T., Y. Han, Y. Li et al. 2008. Temperature-insensitive miniaturized fiber inline Fabry–Perot interferometer for highly sensitive refractive index measurement. *Optics Express* 16: 5764–5769.
Wernecke, J. and L. R. d'Addario. 1977. Maximum entropy image reconstruction. *IEEE Transactions on Computers* C-26: 351–364.
Wolf, E. 2007. *Introduction to the Theory of Coherence and Polarization of Light*. Cambridge University Press, Cambridge, U.K.
Young, T. 1804. Experiments and calculations relative to physical optics. *Philosophical Transactions of the Royal Society of London* 94: 1–16.
Zernike, F. 1938. The concept of degree of coherence and its application to optical problems. *Physica* 5: 785–795.
Zernike, F. and P. van Cittert. 1934. Die wahrscheinliche schwingungsverteilung in einer von einer lichtquelle direkt oder mittels einer linse beleuchteten ebene. *Physica* 1: 201–210.
Zhang, H. and O. Annunziata. 2008. Effect of macromolecular polydispersity on diffusion coefficients measured by Rayleigh interferometry. *The Journal of Physical Chemistry B* 112: 3633–3643.
Zhang, J., J. Yang, W. Sun et al. 2008. Composite cavity based fiber optic Fabry–Perot strain sensors demodulated by an unbalanced fiber optic Michelson interferometer with an electrical scanning mirror. *Measurement Science and Technology* 19: 085305.

6. Fluorescence Measurement
Principles and Techniques

Pedro Alberto da Silva Jorge
INESC TEC

Handbook of Optical Sensors. Edited by José Luís Santos and Faramarz Farahi © 2015 CRC Press/Taylor & Francis Group, LLC. ISBN: 9781439866856.

6.1 Introduction

Luminescence phenomena happen whenever an electronically excited species is deactivated by emission of ultraviolet (UV), visible, or infrared radiation. Depending on how the material is excited in the first place, the luminescent event can fit into different categories. If high-energy photons, from UV to blue, are the source of excitation, the emission is called photoluminescence; conversely, chemiluminescence happens when the excitation mechanism is a chemical reaction. A source of curiosity and fascination when observed in nature in earlier times (radioluminescence of aurora borealis, bioluminescence of fireflies), presently, luminescent phenomena are well understood and the basis for some of the most powerful analytical methods, providing the highest sensitivity. In this regard, photoluminescence is, by far, the most widely studied technique routinely applied in many well-established imaging and sensing methods (Lakowicz 1999, Kuhn and Forsterling 2000, Valeur 2002, Orellana 2004).

Traditionally, fluorescence-based analytical methods were associated with high-cost bulky laboratory equipment. However, a diversity of technological advances taking place over the last 20 years concurred to change this picture. In the field of optoelectronics, the advent of low-cost blue and UV lasers and LEDs, together with new miniature CCD spectrometers and high-sensitivity avalanche photodiodes (APDs), is enabling a new generation of spectroscopy tools to be developed and deployed in field applications. On the other hand, chemistry and material sciences are delivering new types of fluorescent materials with enhanced sensing properties, such as long-lived metallo-organic complexes or highly photostable quantum dots (QDs). Such developments are expanding the uses of fluorescence and enabling new applications such as long-term multiwavelength imaging, flow cytometry analysis in microfluidic chips, high-throughput DNA sequencing, and fiber-optic-based diagnostic probes.

In this chapter, the fundamental principles of photoluminescence spectroscopy and associated detection techniques will be presented. Standard methods and future trends of fluorescence-based sensing technologies and its applications will be discussed.

6.2 Principles of Photoluminescence

Photoluminescence phenomena take place in a spectral band ranging roughly from 200 to 1000 nm. Following the absorption of a photon, a molecule in an electronic excited state arises having available a diversity of decay mechanisms by which it can return to its fundamental state. In addition to photon emission, there are nonradiative decay routes in which energy is released in the form of heat. Alternatively, interaction with other molecules can result in nonemissive quenching processes. In some cases, the excited molecule can undergo a chemical reaction losing its ability to fluoresce (photodegradation). Typically, following the excitation of a population of fluorescent molecules, all these processes can take place simultaneously, and whatever is observed depends on the relative probabilities and time scales of each phenomena. These processes depend on environmental conditions (temperature, relative concentrations, type of molecules, intensity of excitation, etc.). From the observed behavior, a wealth of information can be retrieved about a molecular system making photoluminescence a powerful analytical tool.

6.2.1 Fluorescence and Phosphorescence

Depending on the spin multiplicities of the ground and excited states involved, photoluminescence phenomena can be classified as fluorescence or phosphorescence. Fluorescent transitions usually take place between the excited singlet state of lowest energy and the ground singlet state; therefore, the electron gets to keep its spin in the transition. These are spin-allowed transitions characterized by high transition rates and lifetimes in the nanosecond range. Phosphorescence, on the other hand, results from transitions involving levels with electrons having the same spin orientation, called triplet states. Electrons have to change their spin during the transition. These are quantum mechanically forbidden transitions, with very small but nonzero probability of occurrence, resulting in low transition rates and much slower time scales (ranging from few microseconds to several milliseconds or even longer). The border, in the time domain, between these two processes is ill-defined but can be roughly set around 1 μs.

6.2.2 Molecular Orbitals

The absorption and emission of radiation observed in most luminescent dyes result from transitions between different energy levels, involving molecular orbitals. Depending on the shape of the initial atomic orbitals and on the degree of overlap, different kinds of molecular orbitals arise. Sigma (σ) bonds are the strongest type of covalent bonds resulting from strongly overlapping *s* or *p* atomic orbitals. Weaker overlap of *p* orbitals results in weaker bonding orbitals. These bonds are formed by π orbitals and usually arise in molecules having multiple bonds. In formaldehyde (H_2CO), for example, the double bond between carbon and oxygen is formed by a π and a σ orbital. The structural formula of formaldehyde is depicted in Figure 6.1 where the bonds are indicated. Due to the

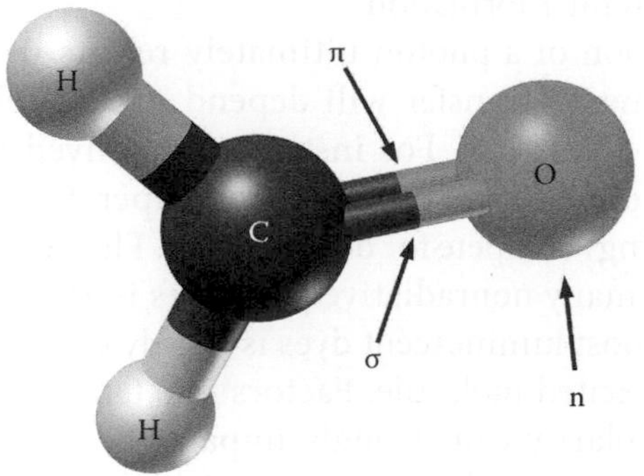

FIGURE 6.1 Structural formula of formaldehyde indicating different types of molecular orbitals.

presence of heteroatoms like oxygen and nitrogen, dye molecules can also have electrons that do not contribute to bonding. The corresponding molecular orbitals are called n orbitals and are also represented in the figure.

The promotion of electrons in π orbitals to antibonding orbitals, designated π^*, can be achieved by absorption of photons with appropriate energy. The same is true for σ orbitals. Transitions between nonbonding n orbitals and antibonding orbitals are also possible. The transitions $\pi \rightarrow \pi^*$ and $n \rightarrow \pi^*$ occur by absorption of UV and visible photons (in the 200–700 nm region) and are the most relevant to fluorescence phenomena.

Considering a molecule in the fundamental state, the highest occupied molecular orbital (HOMO) and the lowest unoccupied molecular orbital (LUMO) are particularly important for spectroscopy applications. The difference in energy between these two orbital corresponds to the minimum necessary to optically excite the molecule and establishes the onset of absorption. In formaldehyde, the HOMO is an n orbital and the LUMO is a π^* orbital.

In molecules like formaldehyde with a single double bound, the orbitals are localized between pairs of atoms, confining the electronic wave function. In longer molecules, however, where double and single carbon–carbon bonds alternate, the overlap of different π orbitals allows for electrons to spread over the whole molecule. The reduced confinement of the electronic wave function results in lower energies for the $\pi \rightarrow \pi^*$ transitions and in a corresponding shift of the absorption to longer wavelengths. This way, the absorption and emission properties of molecular dyes are strongly dependent on the molecular structure and will be defined by the orbitals arrangements. This provides a mechanism for chemical engineers to control the luminescent properties of dyes, which can be tailored for a specific application (Kuhn and Forsterling 2000).

6.2.3 Nonradiative Processes

Besides photon emission, there are many possible mechanisms through which an excited molecule can return to the fundamental state. Intrinsic processes include internal conversion (nonemissive vibrational relaxation), intersystem crossing (conversion from singlet to triplet states, with higher probability of nonradiative deactivation), intramolecular charge transfer, and conformational changes. Extrinsic processes, on the other hand, involve interaction of the excited luminophores with other molecules. Through these interactions, the excited states can be quenched by electron transfer, proton transfer, energy transfer, or excimer formation.

Whether the absorption of a photon ultimately results in the emission of a photon or in some other energy transfer will depend on the relative time scales of all the processes involved (Table 6.1). For instance, long-lived phosphorescence is seldom observed in solutions because, at ambient temperature, many faster processes (relaxation and quenching) compete for deactivation. Fluorescence, on the other hand, being much faster than many nonradiative processes is easily observed. This way, the emission properties of most luminescent dyes is highly sensitive to the microenvironment surrounding the excited molecule. Factors like temperature, pH level, pressure, presence of ions, and polarity can strongly impact the luminescence output. Many of the luminescence parameters, like quantum yield, lifetime, and spectral distribution, can thus be used to probe the environment with extreme sensitivity, providing

Table 6.1 Characteristic Time Scales of the Different Radiative and Nonradiative Transitions Involved in Luminescence Processes

Transition	Characteristic Times (s)
Absorption	10^{-15}
Vibrational relaxation	10^{-12}–10^{-10}
S_1 excited state lifetime	10^{-10}–10^{-7} fluorescence
Intersystem crossing	10^{-10}–10^{-8}
Internal conversion	10^{-11}–10^{-9}
T_1 excited state lifetime	10^{-6}–1 phosphorescence

Source: Data from Valeur, B., *Molecular Fluorescence: Principles and Applications*, Wiley VCH, Weinheim, Germany, 2002.

both spatial and temporal information. Depending on the lifetime of the molecular probes, dynamic processes with different time scales can be monitored (e.g., diffusion, chemical reactions, biological process). The high sensitivity and specificity (different molecular probes have different susceptibilities) of luminescence to environment parameters explain its success as an analytical tool and provide the basis for its application in optical sensing (Wolfbeis 1997).

Usually, the spectroscopic properties of a molecule are determined by the intramolecular processes. The effect of the external intermolecular processes over these properties can then be used as a probe of the molecular environment.

6.2.4 Absorption and Emission Spectra

The absorption and emission spectra of a molecular dye are primarily established by the possible electronic transitions. In addition, a set of vibrational sublevels, of much lower energy, is associated to each electronic state, multiplying the number of possible transitions.

The study of the processes that determine the spectral distribution of absorption and luminescent emission is usually performed using Perrin–Jablonski diagrams (Lakowicz 1999). In Figure 6.2a, an example of such diagram is represented, showing the radiative and nonradiative transitions involved in luminescent processes. External mechanisms, like quenching by other molecules, are not shown. The singlet (S) and triplet states (T) are represented by thick horizontal lines. The thin horizontal lines represent the vibrational levels associated with each electronic state.

At room temperature, most electrons are in the lowest vibrational level of the fundamental state following a Boltzmann distribution. Higher energy electronic levels can be populated by absorption of visible or UV radiation. Depending on the photons energy, electrons in the ground level can be promoted to different vibrational levels of the excited singlet states (S_1 or S_2). In the timescale of the absorption process (10^{-15} s), nuclear displacement can be neglected, so electronic transitions tend to occur between vibrational levels having identical nuclear configuration in both states (Franck–Condon principle). As a consequence, electrons are usually promoted to higher vibrational levels of the electronic excited state. As the nuclear configuration accommodates and releases the excess energy, electrons rapidly relax to the lowest vibrational level of S_1.

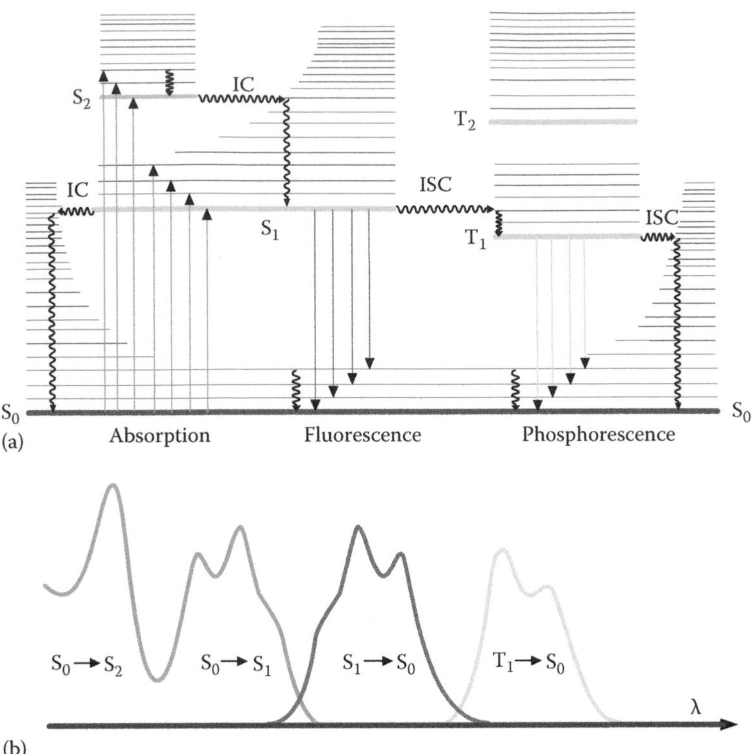

FIGURE 6.2 (a) The Perrin–Jablonski diagram illustrating the different transitions involved in luminescent processes. IC, internal conversion; ISC, intersystem crossing. Straight arrows represent radiative transitions and curly arrows nonradiative processes. (b) Corresponding absorption and emission spectra. (Adapted from Valeur, B., *Molecular Fluorescence: Principles and Applications*, Wiley VCH, Weinheim, Germany, 2002.)

The structure of the molecular absorption spectrum will therefore reflect the structure of the excited vibrational levels. The relative strength of each vibrational absorption line will depend on the molecular configuration and is proportional to the overlap of the wave functions of the two states involved in the transition. In Figure 6.2b, the absorption spectrum that results from the transitions represented in the respective Jablonski diagram is shown. In the band corresponding to the $S_0 \rightarrow S_1$ transition, different vibrational peaks can be identified. Due to inhomogeneous broadening, the vibrational structure is smoothed and can sometimes become unnoticeable.

Because internal conversion processes are usually faster than the fluorescence lifetimes, emission tends to occur from the lowest vibrational level of the excited state. This way, the energy of the emitted photons is always smaller than the energy of the absorbed photons. The wavelength difference between the absorption and emission peaks is called the Stokes shift and enables the spectral discrimination of the excitation and emission radiations. In practice, its value will determine the filtering and signal processing requirements of the instrumentation.

The return to the ground state usually occurs to excited vibrational levels. Further vibrational relaxation will then restore the system in thermal equilibrium placing almost all the electrons in the fundamental state. This way, the vibrational level structure, which depends mostly on the nuclear configuration, can also be seen in the emission spectrum. In fact, the emission spectrum is usually a mirror image of the absorption spectrum (see Figure 6.2b and corresponding transitions in the diagram of Figure 6.2a). The same homogeneous and inhomogeneous broadening processes that affect absorption will also impact the resulting emissions spectrum. This way, molecular dyes usually present relatively broad emission spectra.

Through intersystem crossing, an electron in the short-lived excited singlet level, S_1, can have its spin inverted being transferred to a long-lived excited triplet state, T_1. Radiative transition rates of molecules in excited triplet states correspond to phosphorescence and are orders of magnitude smaller than those of fluorescence. Because triplet states are of lower energy, phosphorescence occurs at longer wavelengths. Their longer radiative lifetimes favor nonradiative return to the ground state, through vibrational relaxation. These processes are particularly favored in solutions and at ambient temperature due to collisions with solvent molecules. In these conditions, radiative quantum yields are usually very low and phosphorescence is seldom observed. Conversely, at low temperatures or in solid state, phosphorescence can be detected, showing greatly increased lifetimes (ms to min). In addition, in molecules containing heavy atoms, there is higher spin–orbit coupling and intersystem crossing is favored increasing phosphorescence quantum yields.

The luminescence emission spectrum observed is usually independent of the wavelength of the excitation radiation. This happens because, even when the absorbed energy exceeds that of the HOMO/LUMO gap, due to the fast relaxation processes, electrons always end up in the lowest vibrational levels of S_1 or T_1. This way, the intensity of the luminescent signal observed, at a fixed wavelength, is proportional to the excitation power and to the absorbance at that particular wavelength. The plot of the luminescent intensity as a function of the incident wavelength is called excitation spectrum. If corrections are made to compensate for variation of the excitation source spectral distribution, the excitation spectrum will coincide with the absorption spectrum.

6.2.5 Excited State Lifetime

After excitation, not all molecules emit precisely at the same instant. Spontaneous emission is a random process with all molecules having identical emission probability in a certain interval. The average time a molecule stays in the excited state before returning to the fundamental state is called the fluorescence lifetime, τ_0, and it depends on the rate constants associated with all the radiative and nonradiative mechanisms involved:

$$\tau_0 = \frac{1}{k_r^S + k_{nr}^S} \tag{6.1}$$

In this relation, k_r^S stands for the radiative rate constant of fluorescence, associated with the $S_1 \rightarrow S_0$ transition. k_{nr}^S, on the other hand, results from adding the nonradiative rates associated with the processes of internal conversion, k_{IC}^S, and intersystem crossing, k_{ISC}^S.

This way, when a fluorescent sample is excited by a δ-like pulse of light, a given concentration of excited molecules is initially obtained that decreases at a rate of $1/\tau_0$. Because the intensity of the luminescent signal, *I*, is proportional to the concentration of excited molecules, following a pulsed excitation, the output luminescence intensity decreases in an exponential fashion with a time constant given by τ_0:

$$I(t) = I_0 e^{-t/\tau_0} \tag{6.2}$$

where I_0 is the luminescence intensity, immediately after excitation. This way, in practice, τ_0 can be defined as the time it takes for the fluorescence intensity to decrease by $1/e$ of its initial value. The function $I(t)$ is the impulse response of the system. The response of the luminescence output to complex excitation patterns can be obtained by convolution of this function with the exciting impulse. In the case where intersystem crossing is favored, phosphorescence will be obtained instead. Similar expressions can be obtained using the corresponding rate constants.

6.2.6 Quantum Yield

The quantum yield of a luminescent dye is a measure of its efficiency and can also be estimated from its rate constants. It is defined as the ratio between the number of emitted photons by the number of absorbed photons and it can be calculated from

$$\Phi = \frac{k_r}{k_r + k_{nr}} \tag{6.3}$$

The quantum yield depends on the relative magnitudes of the radiative and nonradiative rate constants, and it will always be less than unit, on account of the Stokes shift. In practice, the value of Φ results from the summation of all the photons emitted during the decay interval divided by the number of absorbed photons. Considering the same excitation optical power, the higher the quantum yield, the easier it will be to detect the luminescent signal of a given sample.

The lifetime and the quantum yield are two fundamental parameters of luminescent emission. They determine the intensity of the luminescent signal and its temporal behavior. In addition to the intrinsic deactivation mechanisms described earlier, the interaction with other molecular species introduces further nonradiative decay routes that can have a strong impact in both these parameters. The monitoring of the lifetime and/or quantum yield, under the influence of such mechanisms, provides the basis for very sensitive sensing techniques.

6.3 Fluorescence Measurement Techniques

In order to implement a practical fluorescence spectroscopy system, measurements of the luminescent properties must be performed. Luminescence detection techniques can be broadly classified into two main types of measurements, steady-state and time-resolved measurements (Valeur 2002).

6.3.1 Intensity-Based Spectroscopy

In steady-state spectroscopy, the luminescence spectra or the luminescence intensity are recorded while submitting the sample to continuous illumination with an adequate excitation source. Although simple in principle, the measurement of luminescence intensity can be a complex task as it depends on a great deal of intrinsic properties as well as experimental parameters.

6.3.1.1 Absorbance

An efficient excitation depends on the spectral overlap of the exciting radiation with the dye absorption spectrum. The absorption efficiency depends on the intrinsic properties of the fluorescent dye as well as on practical experimental parameters. Macroscopically, the absorption efficiency can be evaluated by the absorbance of a sample:

$$A(\lambda) = \log \frac{I_0(\lambda)}{I_T(\lambda)} \tag{6.4}$$

where $I_0(\lambda)$ and $I_T(\lambda)$ are the excitation light intensities entering and leaving the absorbing medium, respectively (excluding the luminescent emission).

The absorbance is proportional to the dye concentration, c, the sample thickness, l, and the molar absorptivity, ε, following the Beer–Lambert law:

$$A(\lambda) = \varepsilon(\lambda) lc \tag{6.5}$$

The molar absorptivity is an intrinsic parameter determined by quantum mechanical factors defining the strength of absorption at each particular wavelength.

According to the Beer–Lambert law, the sample absorbance increases linearly with increasing dye concentration. This way, the more concentrated or thicker the sample, the more efficient will be the excitation process. However, for very high concentrations, deviations from linearity are often observed. This may be due to saturation effects leading to the formation of dye aggregates, which scatter light or may have modified absorption. Sample turbidity and light scattering can also impact the measured absorbance. The accurate application of the Beer–Lambert law, in an experimental situation, must also take into account the reflections at the sample interfaces, detection solid angle, etc. In practice, it is not straightforward to determine the optical power actually reaching the sample or the luminescent power actually leaving the sample. Absorbance also increases with the sample thickness; however, for practical reasons, l cannot be made too large. All these factors should be carefully considered when preparing a luminescent sample.

6.3.1.2 Steady-State Luminescence

Considering a luminescent sample under continuous illumination, equilibrium is usually achieved between the excitation and the deactivation mechanisms, and the concentration of excited luminophores remains constant. This means that the excited state population attains a steady state. It can be shown that in such situation, in addition to the dependence on the excitation efficiency, the luminescent output will depend also on the quantum yield:

$$I_L = I_0\left(1-10^{-\varepsilon cl}\right)\Phi \tag{6.6}$$

The quantum yield is an intrinsic parameter of the luminescent material, proportional to the relative magnitudes of the radiative and nonradiative decay rates. This way, for obtaining a strong luminescence output, materials with high quantum yields should be chosen. Although a smaller quantum yield could be compensated by increasing the excitation intensity, this is not a desirable situation as it will also increase photodegradation processes.

6.3.1.3 Emission Spectrum

In practice, the photons emitted follow a certain energy distribution, $F(\lambda_L)$, which defines the characteristic emission spectrum of a given luminophore. Experimentally, such distribution together with the associated Stokes shift will determine the characteristics of the necessary detection instrumentation (filtering, detector spectral range, etc.).

Ultimately, the detected luminescent intensity, I_L, will depend on the incident excitation intensity $I_0(\lambda_{Ex})$, the absorption and emission spectral features of the luminophore, and the particular experimental arrangement:

$$I_L\left(\lambda_{Ex},\lambda_L\right) = 2.3kF\left(\lambda_L\right)I_0\left(\lambda_{Ex}\right)\varepsilon\left(\lambda_{Ex}\right)lc \tag{6.7}$$

The parameter k accounts for a diversity of experimental factors affecting the detected signal (detection geometry, wavelength dependence of detector sensitivity, system spectral resolution, polarization effects, filtering, electronic gain, etc.), some of which can hardly be quantified. This way, the numerical value of a luminescence intensity measurement has no direct physical meaning and is usually expressed in arbitrary units or simply normalized to a reference value.

This equation also demonstrates that the detected luminescent signal depends on a great number of intrinsic and experimental parameters, which should be considered thoroughly for an accurate interpretation of the obtained data. Conversely, it also provides a good insight into the many ways the signal detection, and thus a luminescent sensor sensitivity, can be optimized.

Equation 6.7 is an approximation for low luminophore concentrations and deviations from linearity can be observed with increasing absorbance. Inner filter effects, due to self-absorption of attenuation of excitation in thick samples, may take place that reduce the observed signal. The presence of impurities providing scattering centers or background luminescence can further complicate the observed behavior.

When I_L is measured as a function of λ_L, for a fixed excitation wavelength, λ_{Ex}, the emission spectrum is obtained (i.e., a signal proportional to $F(\lambda_L)$). On the other hand, if the luminescent intensity is measured at a fixed wavelength, λ_L, while changing the excitation wavelength, λ_{Ex}, the excitation spectrum can be obtained instead (i.e., a signal proportional to $A(\lambda_{Ex})$).

6.3.1.4 Instrumentation

Spectral measurements of fluorescent intensity are usually performed using a spectrofluorometer. These instruments usually consist in a broadband optical source, followed by a monochromator to select the excitation wavelength. The selected radiation is then shined upon the luminescent sample and detection is usually made at 90° with the incident beam. In order to perform wavelength discrimination of the luminescence emission, a second monochromator is placed before the detector, usually a photomultiplier. An ideal spectrofluorometer would have an optical source with a constant photon output at all wavelengths ($I_0(\lambda_{Ex})$ = const.), monochromators with polarization and wavelength independent efficiency, and a detector with equal sensitivity at all wavelengths. Unfortunately, such ideal components are not available and, in practice, corrections must be performed in order to account for these imperfections.

Instrumentation used in practice introduces a series of wavelength dependencies that cause distortion of the measured spectra if no corrective measures are taken (Lakowicz 1999). High-pressure xenon arc lamps are widely used as excitation sources. Except for a few sharp lines near 450 and 800 nm, these lamps provide a relatively continuous light output in a broad wavelength range (250–700 nm), which can be used to excite most luminophores. Selection of the excitation wavelength using a monochromator results in low optical power reaching the sample. It is possible to use other lamps that provide higher intensities in the UV region, like mercury lamps or mercury–xenon lamps. In these cases, however, the optical power is further concentrated in sharp spectral lines.

The monochromators also have a wavelength-dependent efficiency. This dependence arises mostly from their dispersive element, typically a diffraction grating, which usually has maximum diffraction efficiency at a given wavelength. Also, scattering and higher-order diffraction can increase the levels of stray light reaching the detector and further distort the measured spectra. On top of this, the efficiency of monochromators is typically dependent on the polarization of the incident radiation. This can be a problem because the luminescent emission is often anisotropic and can be partially polarized (Valeur 2002). In order to avoid polarization dependence, usually the excitation radiation is linearly polarized, and detection is performed through a second polarizer at 54.7° with the input polarization (called magic angle). In these conditions, it can be demonstrated that the signal detected is always proportional to the total luminescent intensity.

Typically, photomultiplier tubes are used for detection in steady-state measurements due to their high sensitivity. More recently, linear and 2D CCD detectors are also being used in many devices and practical applications. Unfortunately, all these detectors have a wavelength-dependent sensitivity that must be accounted for.

Due to the wavelength dependence of the optical source, most of the time, the only way to obtain accurate spectral measurements is to provide for a reference channel. For this purpose, in spectrofluorometers, a few percent of the excitation radiation is usually deviated toward a second detection channel. Generally, this light is detected by a photodiode with an approximately spectrally flat response. In alternative, a quantum counter is used. A quantum counter usually consists in a concentrated solution of a dye with a quantum yield that does not depend on λ_E. This way, all the incident photons are absorbed and the emitted luminescent signal detected by the reference detector is proportional to the photon flux incident in the sample under study. In principle, ratiometric detection of the signal from the main detector with the signal from the reference detector will allow compensation of the wavelength dependences of the excitation optical source, excitation monochromator efficiency, and detector sensitivity. In addition, fluctuations of the excitation source output intensity over time will be compensated as well.

Most of the time, however, because the optical path of both channels is not exactly the same, full compensation cannot be achieved. In addition, with this method, the detection monochromator is not accounted for. In cases where very accurate measurements are needed, usually a standard dye is used with known emission and excitation spectra. The spectra of this dye are then measured and compared with the expected spectra, allowing the calculation of adequate correction factors. Alternatively, instrument correction factors can also be obtained by measuring the spectrum of a calibrated light source transmitted through a transparent scattering solution. The determination of quantum yields is an example of a measurement only possible using accurately corrected spectra. For this particular application, the use of standard dyes, with known quantum yield, must be considered for accurate calibration of the system.

Whenever it is adequate to measure fluorescence intensity at a single excitation and emission wavelength, filter-based spectrometers can be used instead. The use of filters to select wavelength allows much higher excitation and luminescence collection efficiencies. Therefore, filter-based instruments are often used in ultratrace analysis, where it is crucial to maximize the fluorescence signal at the expense of selectivity. This approach is often used in low-cost portable analytical instruments and in a variety of luminescence-based sensors targeting specific analytes.

In summary, the main difficulty associated with intensity-based measurements arises from the fact that the detected intensity is not an intrinsic parameter of the luminescent dye. It depends on intrinsic properties of the luminophore, like the quantum yield, but also on a great diversity of external parameters. The control of all this variables is critical when aiming to obtain accurate information on the dye intrinsic properties like quantum yields and emission and absorption spectra. Nevertheless, whenever the quantification of an analyte is at stake, simpler configurations can be explored for low-cost high-sensitivity detection.

6.3.2 Time-Domain Spectroscopy

Time-resolved measurements of the luminescence intensity are a powerful alternative to steady-state measurements (Lakowicz 1999, Valeur 2002). Lifetime is an intrinsic characteristic of the luminophore; therefore, measurements are not so susceptible to system changes such as optical alignment, dye concentration, or even photodegradation.

The observation of luminescent decay, which occurs in very short time scales, provides the opportunity to monitor many transient behaviors of molecular systems, which may become unnoticeable under steady-state excitation. In general, the study of the dynamics of excited states provides much more information about the system being studied than is possible to achieve with steady-state data. These techniques therefore have a fundamental role in the study of the kinetics of many photophysical, photochemical, and photobiological processes.

Time-resolved measurements can be made in the time domain or in the frequency domain. The two methods are widely used nowadays both in laboratorial environment and in sensing applications. Time-domain methods will be described first.

6.3.2.1 Pulse Fluorometry

The basic principle of time-domain techniques is the excitation of the sample with a short pulse of light and the observation of the subsequent luminescent decay. The pulse width should be made as short as possible, preferably much shorter than the lifetime of the dye under study.

In the simplest case, the luminescent sample response, $I(t)$, to a δ-pulse excitation, is single exponential with a time constant equal to the excited state lifetime (Equation 6.2). A simulation of the data obtained in this ideal situation is represented in the diagrams of Figure 6.3. If the intensity is measured as a function of time, the decay time can then be estimated from the slope of the linear plot of $\log I(t)$ (Figure 6.3b) or from the time at which the intensity drops to $1/e$ of its initial maximum value at $t = 0$ (Figure 6.3a).

In practice, however, the excitation pulse has a short but finite time width, not negligible in comparison with the lifetime to be measured. In addition, the response time of the detection system can introduce further broadening in the measured signals. In such cases, the luminescent response will be given by the convolution of the ideal δ-pulse response, $I(t)$, with a function, $E(t)$, describing the real excitation impulse as perceived by the detection system:

$$R(t) = E(t) \otimes I(t) = \int_{-\infty}^{t} E(t')I(t - t')dt' \tag{6.8}$$

In order to obtain the lifetime of the luminophore from the measured data, the impulse response of the excitation/detection system ($E(t)$) must be known very accurately, and a deconvolution must be performed. Much of the complexity of lifetime data analysis comes from the difficulty in extracting the true luminophore δ-impulse response

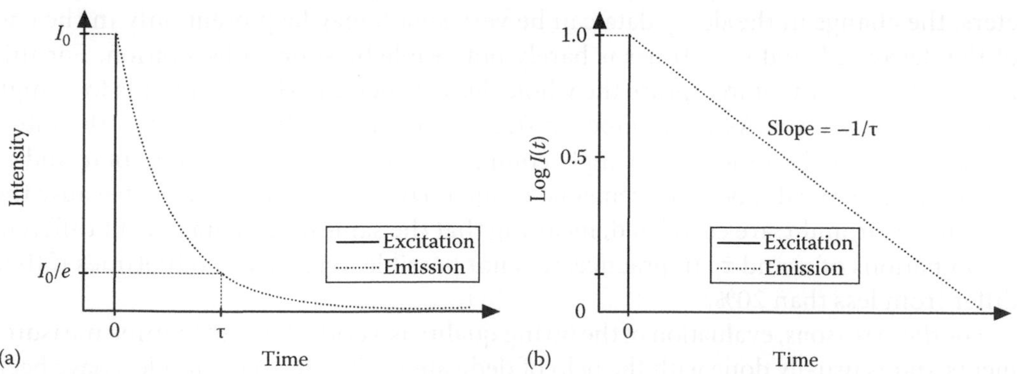

FIGURE 6.3 Principles of time-resolved measurements. (a) After an impulse excitation, luminescent intensity decays exponentially with a time constant given by the lifetime. (b) The decay time can be calculated from the slope of $\log I(t)$. (Adapted from Lakowicz, J.R., in *Principles of Fluorescence Spectroscopy*, 2nd Edition, Kluwer Academic/Plenum Publishers, New York, pp. 96, 1999.)

from the measured signals. For this reason, a great effort is placed in the optimization of the deconvolution process. Nevertheless, interpretation of data arising from single-exponential decays is relatively straightforward. The real challenge comes from the fact that most samples display more than one decay time.

6.3.2.2 Multiple-Exponential Decays

Multiple lifetimes can arise from the presence of different luminescent species in the same sample or from a single dye in a heterogeneous environment. In such case, different lifetimes may be assigned to each population. In general, complex decay functions can be fitted by a multiexponential model described by

$$I(t) = \sum_i \alpha_i e^{(-t/\tau_i)} \quad (6.9)$$

where

α_i are called the pre-exponential factors and determine the amplitudes of each component at $t = 0$

τ_i are the respective lifetimes

Depending on the system being studied, the meaning of these parameters may differ. For a single dye existing in different environments, usually both populations have the same radiative decay rate, in such case α_i represents the relative fraction of each luminophore. For a mixture of different luminescent species, however, the relative values of α_i depend on many parameters, like the quantum yield and relative concentration. In these cases, assigning a molecular meaning to each α_i is not straightforward. Nevertheless, it is always possible to estimate the fractional contribution of each species.

The simplest case to consider is the presence of two different populations. Usually, the presence of a second lifetime will introduce a curvature in the plot of log$I(t)$. Fitting the obtained data with Equation 6.9 will then allow recovering the decay parameters. However, depending on the relative magnitudes of the different parameters, the change in the decay data can be very small, may be present only in the end of the decay tail, and sometimes is barely noticeable by simple observation. For this reason, it is important to capture the whole decay function, which demands for a high signal-to-noise ratio. As the number of fractional contributions increases, the interpretation of data becomes increasingly complex, and to extract the values of α_i and τ_i, from the observed decay measurements, is not a trivial task. This happens because the parameters α_i and τ_i are correlated, meaning that the same decay data may fit different combinations of α_i and τ_i. In practice, it is not possible to resolve two lifetimes if they differ from less than 20%.

For these reasons, evaluation of the fitting quality is a critical step of lifetime measurements and is usually done with the help of dedicated software. Many models have been explored to evaluate the quality of fitting: methods of least squares, moments, Fourier transforms, Laplace transforms, and others. The most widely used method is based in nonlinear least squares where the basic principle is to minimize a correlation parameter, χ^2, that represents the mismatch between the theoretical model and the experimental data.

Although mathematical tools provide valuable help, careful analysis and a lot of experience is necessary for accurate interpretation of lifetime data.

Regardless of the number of exponential terms considered, it is always possible to define an average or apparent decay time, $\langle\tau\rangle$. Such average parameters are of particular interest for sensing applications where, for instance, the analyte concentration can be derived from the apparent lifetime. In this context, analysis of lifetime data can be greatly simplified.

6.3.2.3 Techniques and Instrumentation

There are many ways to record lifetime data, such as streak cameras and boxcar integrators. Nevertheless, most instruments used for the determination of lifetimes are based in the time-correlated single-photon counting (TCSPC) method, also called as single-photon timing (SPT). This technique is based on the fact that the probability of detecting a single photon at the time t after excitation is proportional to the luminescence intensity at that time. This way, measuring the time between the excitation pulse and the first arriving photon, at different times and for a large number of excitation impulses, it is possible to reconstruct the intensity decay curve, based on the obtained histogram of photon counts per time channel. Because only a single photon is detected by measurement cycle, this technique requires a large number of measurements in order to completely reconstruct the decay function. SPT can be very sensitive but, although conceptually simple, it is associated with complex instrumentation, which, ultimately, will determine the sensitivity and time resolution that can be achieved.

The excitation source is of primary importance and it should provide narrow pulses with a relatively high repetition rate. Initially, excitation was provided mainly by flash lamps, which delivered low-intensity *ns* pulses with repetition rates in the 10–100 kHz range in a limited spectral range (200–400 nm). Flash lamps are relatively inexpensive, and by using deconvolution, it is possible to measure lifetimes in the hundreds of picoseconds range. However, low repetition rates demand for long acquisition times where lamp drift can become a problem. Nowadays, although more expensive, lasers are used as the preferred excitation source. A mode-locked laser associated with a dye laser or Ti–sapphire laser can easily provide picoseconds pulses, at high repetition rates (MHz) and in a wide range of wavelengths. More recently, *ps* systems based in laser diodes are becoming available at a variety of wavelengths.

The time resolution of the instrument is also critically determined by the sampling and detection electronics. While standard dynode chain photomultiplier tubes (PMT) can provide response times in the *ns* range, nowadays, microchannel plate (MCT) PMTs are available capable of operating 10–20 times faster. Nevertheless, this new generation of PMTs is considerably more expensive. Although photodiodes are much cheaper and can respond faster than MCT PMTs, they lack sensitivity. APDs have adequate gain, but their fast response is dependent on a very small size active area (10×10 μm). This way, APDs do not have the necessary sensitivity for most measurements. With mode-locked lasers and MCT photomultipliers, widths of the instrument pulse response in the range 30–40 ps can be achieved. This allows the measurement of decay times as short as 10–20 ps.

Although source and detector are the key elements in lifetime measurements, these techniques rely also on a variety of sophisticated electronic devices like synchronization electronics, time-to-amplitude converters, filtering, and delays lines, which add to the cost and complexity of the measurement system. In addition, several problems in data collection can arise that must be considered like polarization effects, scattering in turbid solutions, and wavelength dependence of the measurement system. Solving these problems often demands for additional data processing, use of filtering, or even using a reference channel.

Generally, the shorter the lifetime, the more expensive and complex the measurement system becomes. This way, the use of standard organic fluorophores, with lifetimes in the nanosecond range, can be a serious obstacle to the implementation of low-cost sensing applications using time-resolved detection techniques. In this regard, the advent of long-lived luminescent probes, like organometallic complexes, can provide interesting solutions. Because these sensing dyes present lifetimes in the microsecond range, the instrumentation can be greatly simplified. In these time scales, system impulse responses of tens of nanoseconds wide are acceptable, which can be achieved by diode lasers or even low-cost LEDs in combination with photodiodes.

6.3.3 Frequency-Domain Spectroscopy

The measurement of excited state lifetimes can also be performed in the frequency domain, as an alternative to the time-domain techniques just described (Zhang et al. 1993, Lakowicz 1999).

6.3.3.1 Frequency-Domain or Phase Modulation Fluorometry

In frequency-domain spectroscopy, the excitation optical source is modulated in amplitude at high frequency. Typically, sinusoidal modulation is applied at frequencies that should be in the range of the reciprocal of the decay time. In these circumstances, the obtained luminescent emission is also modulated in amplitude at the same frequency. Due to the time lag between excitation and emission, however, the luminescent signal has a phase delay relative to the excitation. This way, the phase delay, ϕ, can be used to calculate the lifetime. The scheme in Figure 6.4 shows the typical

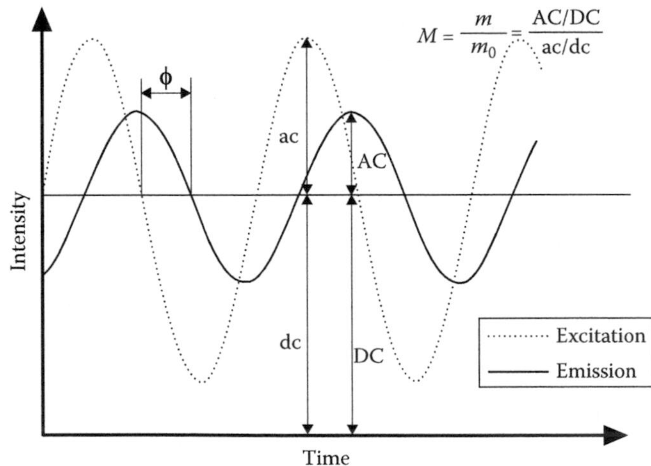

FIGURE 6.4 Response of luminescence emission to intensity-modulated excitation, showing the relative phase delay, ϕ, and modulation ratio, M.

response of a luminophore submitted to amplitude modulated excitation. This scheme also shows that, in addition to the phase delay, the resulting signal can present a decreased modulation depth, m = AC/DC, in comparison to the excitation signal, m_0 = ac/dc. The decrease in modulation depth is also called demodulation and it happens because the luminescent response is given by the convolution of the impulse response with the sinusoidal modulation. This way, for a fixed modulation frequency, the longer the decay tail of the luminophore impulse response, the greater the extent of the demodulation will be. Evaluation of the modulation ratio, $M = m/m_0$, can also be used to calculate the lifetime.

Typically, the relative phase, ϕ, and modulation ratio, M, are evaluated in a range of frequencies and the curves obtained characterize the harmonic response of the system. It is possible to demonstrate that the harmonic response corresponds to the Fourier transform of the δ-pulse response. This way, the two techniques are theoretically equivalent, in the sense that they analyze the same physical phenomenon although operating in different domains (time and frequency). However, they differ in the measurement methods and necessary instrumentation. When the data are evaluated directly in the frequency domain, no deconvolution is necessary.

6.3.3.2 Calculation of the Lifetime

In order to extract the decay information, from the frequency-domain data, it is necessary to relate ϕ and M with the lifetime τ. Assuming a single-exponential decay, the δ-impulse response of the luminophore can be described by Equation 6.2. In such case, the differential equation describing the time-dependent luminescent intensity is given by

$$\frac{dI(t)}{dt} = -\frac{1}{\tau}I(t) + E(t) \tag{6.10}$$

where $E(t)$ is the function describing the excitation radiation. In the particular case of sinusoidally modulated excitation, this function can be described by

$$E(t) = a + b\sin\omega t \tag{6.11}$$

where

ω is the angular modulation frequency

a and b correspond to the dc and ac components, respectively, of the incident intensity

This way, the modulation depth of the excitation function is given by $m_0 = a/b$. The luminescent dye will respond to this excitation with a luminescent output at the same frequency but shifted in phase and presenting a different modulation depth. Then, it can be assumed that the time-dependent luminescent intensity, $I_L(t)$, is described by a function like

$$I_L(t) = A + B\sin(\omega t - \phi) \tag{6.12}$$

When the sample is submitted to the sinusoidal excitation described by Equation 6.11, Equation 6.12 should be a solution to the differential equation (6.10). From the substitution of $I_L(t)$ and $E(t)$ into Equation 6.10, we can easily derive the following relations between the lifetime and the phase delay:

$$\tan\phi = \omega\tau = 2\pi f\tau \tag{6.13}$$

and the lifetime and the modulation ratio

$$M = \frac{B/A}{b/a} = \frac{1}{\sqrt{1+\omega^2\tau^2}} \tag{6.14}$$

This way, experimental measurements of the phase angle, ϕ, and the modulation ratio, M, can be used to calculate the lifetime.

6.3.3.3 Frequency-Domain Data

In practice, these parameters are usually evaluated in a range of frequencies, and the corresponding theoretical equations are then fitted to the measured data. A typical set of frequency-domain data was simulated using Equations 6.13 and 6.14, assuming a single-exponential decay with τ = 1.0 μs. The results obtained are shown in Figure 6.5.

For lower modulation frequencies, the decay tail is negligible in comparison to the modulation period; this way, the emission closely follows the excitation, the phase delay is very small, and the relative modulation depth is nearly unit. However, as the modulation frequency approaches values near the reciprocal of the lifetime, the phase delay rapidly increases, and the modulation starts to fade. This is the frequency range where the rates $\partial\phi/\partial f$ and $\partial M/\partial f$ are higher and which contains more information about the sample parameters. The calculation of the optimal modulation frequency, where the phase has maximum sensitivity to lifetime changes, yields a frequency given by the reciprocal of

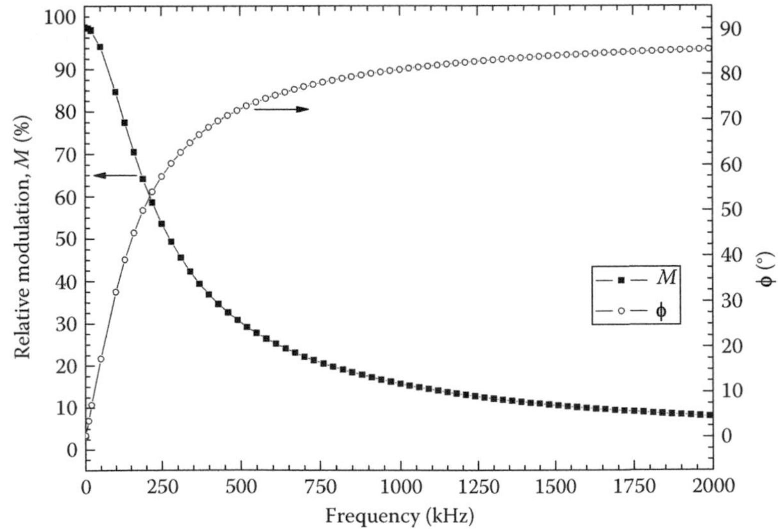

FIGURE 6.5 Simulated data showing relative phase, ϕ, and modulation ratio, M, as a function of the modulation frequency, for a single-exponential decay with τ = 1.0 μs.

the luminophore lifetime ($f_{opt} = 1/2\pi\tau$). For higher frequencies, as the decay tail starts to overlap several modulation periods, the modulation depth decreases toward zero and the luminescent emission becomes continuous. Simultaneously, the phase delay grows very fast, and it converges asymptotically to 90° as the modulation vanishes. For samples with shorter lifetimes, the frequency range of interest (where the two curves cross) is shifted for higher frequencies.

This way, frequency-domain data provide two independent measures of the lifetime, which can be evaluated by measuring phase, τ_ϕ, or assessing the relative modulation, τ_M. If the decay is a single exponential, the values obtained with the two measurements should be the same, identical to the sample lifetime, and independent of the modulation frequency. In such cases, measurement can be performed at a single frequency and lifetime determination can be very fast.

6.3.3.4 Multiple-Exponential Decays

Most of the time, however, multiple-exponential decays are observed. In such cases, the two independent measures of the lifetime, τ_ϕ and τ_M, yield different values that are dependent of the modulation frequency, decreasing for higher frequencies. The measured values are just apparent and result from a complex weighting of various decay parameters. Although characteristic of the system under study, these are not true molecular parameters. Nevertheless, it can be useful to monitor the apparent lifetime, at a fixed frequency, using one of these techniques. Many sensing applications were reported, where the apparent lifetime was monitored in order to determine changes in the quencher concentration, using frequency-domain techniques.

In general, however, to fully characterize a complex decay, measurements should be performed in the widest possible frequency range, centered at a frequency near the reciprocal of the lifetime. The frequency-domain data for a multiexponential decay can be calculated from the sine and cosine transforms of the δ-impulse response $I(t)$. The experimental data obtained must be fitted to a theoretical curve. Judgment of the fitting quality is a critical step in frequency-domain measurements. Evaluation is usually done by nonlinear least squares analysis, seeking to obtain a correlation parameter, χ^2, near unit. Mathematical tools are available providing valuable help, but careful analysis and a lot of experience is necessary for full insight of the data.

6.3.3.5 Techniques and Instrumentation

In addition to all the instrumentation associated to steady-state measurements, frequency-domain spectrofluorometers require a modulated light source and the associated RF electronics. The light source can be an arc lamp followed by a monochromator, in which case the modulation can be achieved by an electro-optic (EO) modulator, usually a Pockels cell. Typically, high driving voltage is required to operate these devices. In addition, careful optimization is needed in these setups because EO modulators have small active areas and work best with collimated light. This way, the excitation optical power obtained with this configuration is usually low. Nevertheless, commercial systems are available using this arrangement, in which arc lamps can be modulated up to 200 MHz. In applications where higher excitation powers are needed, the EO modulators can be used with a variety of *CW* laser sources. These sources, however, are more expensive and cover only limited wavelength ranges. Intrinsic modulated light sources, like diode lasers and LEDs, can overcome most of these problems. Presently, low-cost LED sources are widely available in the blue-violet range of the spectrum. Although more expensive, some UV LEDs are becoming available. Many frequency-domain sensing applications for chemical species have been reported where low-cost LEDs were used as the light source (McDonagh et al. 2001).

In order to perform measurements of phase angles and modulation at such high frequencies, special techniques must be used. Presently, most frequency-domain instruments are based on a multifrequency cross-correlation detection technique (MFCC). This technique requires that the gain of the PMT should be modulated at the same frequency as the excitation beam, f_m, plus a small frequency offset Δf, that is, at a frequency $f_m + \Delta f$. With adequate electronic filtering, this system yields a signal at the difference frequency, Δf, usually in the 100–1000 Hz range, that contains the same phase and modulation information as the original high-frequency signal. This way, all the measurements can be performed in a low frequency range, using a zero-crossing detector and a ratio digital voltmeter, independently of the modulation frequency. With this method, high-frequency harmonics and other sources of noise are rejected, allowing measurements at lower frequencies to be performed with high accuracy. For instruments using a *CW* laser, the standard deviations of 0.1–0.2° for the phase shift and 0.002–0.004 for the modulation ratio can be obtained.

Similar to time-domain techniques, detectors with short impulse responses are also needed in frequency-domain measurements, in order to avoid signal demodulation due to the instrumentation. The necessary bandwidth will depend on the lifetimes to be measured. For a given luminophore, the frequency range of interest is approximately given by $\omega = 1/\tau$. Typically, for luminophores with lifetimes around 10 ns, modulation frequencies in the 2–200 MHz are adequate. Picosecond lifetimes, on the other hand, will require modulations in the GHz range. Due to their fast responses, PMTs and MCT PMTs are also widely used in frequency-domain spectrofluorometers. The gain of regular PMTs can be easily modulated by injection of the RF signal in one of its dynodes. Typical bandwidths are in the 200–300 MHz range. With MCT PMTs, much higher bandwidths can be achieved (2–5 GHz) with these devices; however, cross correlation must be performed using an external circuit. Although they are seldom used in laboratorial instrumentation because of limited gain, photodiodes, in combination with lock-in amplifiers, are being increasingly used for frequency-domain measurements particularly in sensing application based in long-lived luminophores.

Frequency-domain spectrofluorometers usually have two detectors, one for the sample and one to serve as a reference. The reference PMT usually measures part of reflected or scattered excitation light. On the other hand, the sample detector is alternatively exposed to luminescence from the sample or to excitation light from a scattering solution. This way, performing a relative measurement the arbitrary phase angles due to delays in cables and electronic circuits can be eliminated and an absolute phase shift and modulation ratio, accounting only for the luminophore lifetime, can be retrieved. Although this referencing technique allows compensating for the influence of many external parameters, some problems common to time-domain devices may still be observed, such has wavelength dependence of detector sensitivity and polarization effects.

Presently, from the instrumental point of view, laboratorial devices for time-domain and frequency-domain measurements are very similar in cost and performance. Nevertheless, differences exist that may be important in specific applications. For the particular case of low-cost sensing applications, frequency-domain measurements present some considerable advantages. Frequency-domain measurements are usually faster; the sinusoidal signals are easier to measure than the very short decay signals characteristic of time-domain measurements; no deconvolution is necessary, saving computation time and money in complex analyses software. Overall, considering specific applications, frequency-domain systems can usually be implemented with simpler electronics and at lower cost than time-domain systems. Nevertheless, luminescent emission can sometimes be too weak for frequency-domain detection, where signal must be strong enough to make zero-crossing detection. In these cases, time-domain SPT with longer acquisition times are extremely sensitive and can still be used.

6.4 Fluorescent Sensing and Imaging

Sensing and imaging applications based in luminescence can be extremely sensitive. Unlike absorption where measurements are performed comparing two light intensities that, sometimes, differ less than 1%, luminescence measurements are performed against a dark background. This way, signal-to-noise ratios in luminescence can be very high, even when faint signals are being measured. In certain conditions, even the luminescence of a single molecule can be measured. Luminescence, on the other hand, can be very sensitive to the external medium. This way, it is the perfect probe to perform a wide variety of measurements.

One of the most attractive features of fluorescence-based techniques lie in the possibility of using fluorescent molecules, called probes or indicators, to study nonfluorescent species. While a great deal of molecules can display intrinsic fluorescent emission, more often than not, a fluorescent probe must be used to label the target analyte. Indeed, fluorescent labeling allows targeting particular components with exquisite sensitivity and selectivity enabling trace detection of analytes or imaging of complex biomolecular assemblies, such as living cells.

6.4.1 Fluorescent Indicators

The luminescent probe or indicator is one of the most critical components in any luminescence-based sensing or imaging system. Ultimately, the photophysical and photochemical properties of the sensing dye will determine the characteristics of the whole system. The luminophore absorption and emission spectra will establish the spectral range of the optical source and detector. The modulation frequencies or pulse widths, on the other hand, will be determined by the probe lifetime. In addition, the properties of the dye will determine available sensing mechanisms, with which analytes can be studied. For this reason, a great deal of research has been devoted to the development of luminescent probes with adequate characteristics for a range of applications. This way, in addition to many substances that are naturally luminescent, many other synthesized organic and inorganic compounds are available nowadays.

6.4.1.1 Intrinsic Luminophores and Organic Dyes

A distinction can be made between intrinsic and extrinsic luminescent probes. Intrinsic luminophores are those occurring naturally and include aromatic amino acids, NADH, and chlorophyll, among others. These probes are ideal in the sense that they are already part of the studied environment, and measurements can be performed with minimum interference. For instance, the presence of tryptophan and tyrosine residues (luminescent amino acids) in proteins is often used to study the shape or the biological function of enzymes. However, such examples are limited, and most of the time, the molecule of interest is nonluminescent and an extrinsic luminescent probe must be introduced.

Most of the naturally occurring luminescent bio-probes have absorption and emission spectra in the UV/blue region of the spectrum. In the particular case of amino acids, absorption usually occurs in the 260–300 nm range and emission spans from 280 to 350 nm. Such energetic radiation can have a harmful effect when studying living systems. In addition, these wavelengths are out of the range of cheaper solid-state sources and detectors and require the use of quartz optics. Moreover, the fast lifetimes usually associated (3–5 ns) require more expensive instrumentation. A particular interesting example of a naturally occurring luminescent probe comes from GFP or green fluorescent protein. Originally isolated from marine organisms, it has been reengineered in different ways to improve its spectral properties (e.g., absorption at 488 nm, coinciding with Argon laser excitation, and emission at 509 nm). As a protein, it could be harmlessly incorporated in different processes of living cells (including genetic expression) revolutionizing the bio-applications of fluorescence microscopy.

Thousands of synthesized luminophores, mostly organic, are now available for the labeling of biological systems or for sensing applications (Wolfbeis 1997, Lakowicz 1999). Usually, they are either covalently bound to the molecule of interest or simply associated to it (by polarity or some other chemical affinity). Most of these dyes have some kind of sensitivity to the environment, through quenching or some other mechanism, enabling their use as sensor of some analyte (pH, ions, oxygen). It is usually desirable that these luminophores have longer absorption and emission wavelength. Primarily, the aim is to avoid interference with or from the intrinsic biological luminescence that can act as background luminescent noise. In addition, it is desirable from the instrumental point of view (sources, detectors, optical fibers transmission properties) to work with visible and near-infrared wavelengths instead. Although this has been accomplished to some extent, nowadays, luminophores are available covering the whole visible spectrum, and some options are even available in the near infrared. Nevertheless still a majority of dyes available require UV or blue excitation.

Although there are exceptions, these properties are common to most organic luminophores. In addition, it should be considered that each of these dyes usually presents different chemical properties. This further complicates their combined use and demands for the development of dedicated protocols for synthesis, functionalization, and immobilization. Also, most organic dyes present high degree of photodegradation hindering their use in long-term sensing applications. In this context, the advent of a family of long-lived and photostable dyes, with more homogeneous spectral and chemical properties and large Stokes shifts, suitable for the sensing of chemical and biological species is highly desirable. Presently, this gap is being filled by hybrid metallo-organic luminescent complexes.

6.4.1.2 Long-Lived Transition Metal Complexes

Originally developed for solar-energy conversion applications, transition metal complexes (TMCs) are now being extensively explored in a wide range of sensing applications. These complexes are characterized by a transition metal containing one or more organic ligands (typically diimines). Special attention is being devoted to those complexes with platinum metals (Ru, Os, Re, Rh, and Ir), from which ruthenium is by far the most studied case (Watts and Crosby 1971, Juris et al. 1988, Carraway et al. 1991, Demas and DeGraff 1991, Mills and Williams 1997).

Ruthenium complexes have very desirable features for sensing applications. They are long lived, with lifetimes ranging from 100 ns to 7 μs, making them suitable for simpler and less expensive time-domain and frequency-domain measurements. They present relatively high quantum yields, independent of the excitation wavelength and that can go up to 0.5. Most ruthenium complexes display strong absorption bands in the visible range (around 470 nm), which overlap perfectly with the emission of low-cost blue LEDs. In addition, they present a large Stokes shifts with emission typically occurring around 600 nm. In Figure 6.6, the absorption and emission spectra of the ruthenium dye $Ru(dpp)_3$ having diphenyl-phenanthroline ligands can be seen, alongside with its molecular structure.

In comparison to traditional organic dyes, TMC are photochemically and thermally very stable. Chemically, they are extremely versatile due to the organic ligands. One or more of these ligands can be chemically modified to adjust the net electrical charge of the complex, to provide functional groups in order to attach the dye to a polymer support, to neutralize or enhance the interaction of the dye with a given analyte, or to adjust its spectroscopic properties.

Many of the interesting photophysical and photochemical properties of these complexes derive mostly from their unique electronic states. In ruthenium complexes of interest, the ligands are easily reduced and electrons are promoted from the metal to the ligand. These states are also called metal-to-ligand charge-transfer (MLCT) states and are the most remarkable feature of TMC, being responsible for their absorption bands in the visible range (shown in Figure 6.6).

After a fast and efficient intersystem crossing process, electrons rapidly relax to the MLCT triplet states. Emission from these triplet states is responsible for the characteristic orange-red luminescence associated with ruthenium dyes and is formally classified

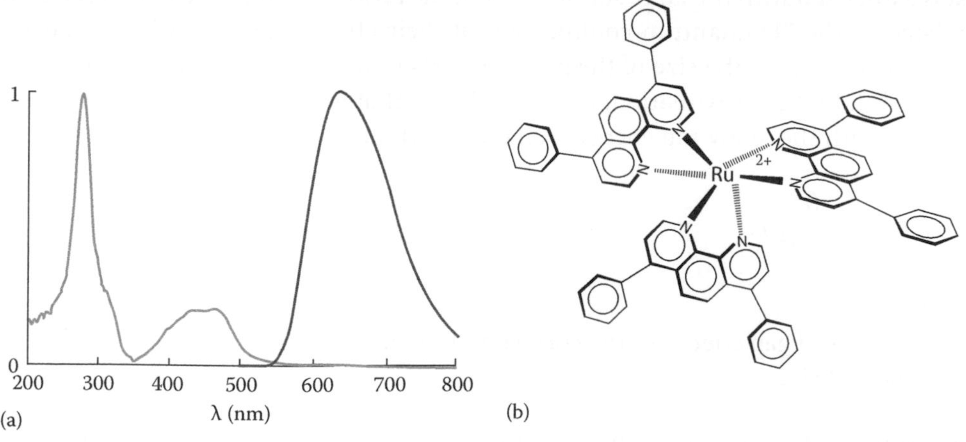

FIGURE 6.6 (a) Absorption (gray trace) and emission (black trace) spectra of the ruthenium complex $Ru(dpp)_3^{2+}$ dissolved in methylene chloride. (b) Dye molecular structure.

as phosphorescence. Nevertheless, the presence of the heavy metallic atom favors spin–orbit coupling, thus resulting in shorter lifetimes (μs) than typical phosphorescent compounds (ms or longer). This way, emission can occur prior to total quenching and these dyes usually display strong luminescence.

Due to their long lifetimes, most ruthenium dyes are very sensitive to collisional quenching by oxygen. The ruthenium complex represented in Figure 6.6 is one the most widely used for oxygen sensing. A long list of applications of these dyes as oxygen sensors has been reported (Carraway et al. 1991, MacCraith et al. 1993, 1994a, McEvoy et al. 1996), including in commercially available products.

In addition to oxygen sensing, the use of complexes with modified ligands has also been reported where the dye luminescence was made sensitive to humidity, pH, or CO_2. Some other examples include applications as quantum counters, DNA probes, study of membranes and proteins, immunoassays, and ion sensing.

These applications demonstrate the possibility of developing a homogeneous set of sensing dyes, with identical photophysical and photochemical properties, based on a single family of coordination compounds (Orellana and Fresnadillo 2004). Although this versatility may raise some concern about cross sensitivities, particularly to oxygen, the combination of the dye with adequate solid membranes can greatly enhance sensor selectivity. This way, conditions are created to the development of a family of chemical sensors, suitable for a wide variety of analytes and with similar enough characteristics to be operated with the same basic instrumentation.

6.4.1.3 Luminescent Quantum Dots

With the advent of nanotechnology, new types of luminescent materials are becoming available having unique properties. Such is the case of QDs or semiconductor nanocrystals. QDs are extremely small particles of semiconductor material, consisting of a few hundreds to a few thousands of atoms. Their small size, ranging from 1 to 20 nm, is mostly responsible for their unique optical, electrical, and chemical properties (Gaponenko 1998). The main differences between the macrocrystalline material and the corresponding nanocrystal arise from two fundamental factors that are size related. The first is associated with the large surface-to-volume ratio of nanoparticles and the second is related to the 3D quantum confinement of their charge carriers. When reduced to the nanometer scale, the size of the particle starts to interfere with its electronic distribution. To a first approximation, it can be shown that a QD behaves as an atomic-like structure with the following energy level distribution:

$$E_{nl} = E_g + \frac{\hbar^2}{2\mu a^2}\chi_{nl}^2 \tag{6.15}$$

where

E_g is the band gap energy of the corresponding bulk semiconductor
a is the QD size

This shows that, in a QD, the band gap is increased by an amount inversely proportional to the size of the nanoparticle. Therefore, the smaller the particle, the higher is its emission energy. This way, it becomes possible to tune the emission wavelength of nanoparticles by simple control of its size. Bright and relatively narrow emission spectrum, wide range absorption, and extremely high photostability are some of the other features offered by QDs.

Like bulk semiconductor materials, nanocrystal QDs will absorb any wavelength to the blue of their emission peak, that is, any photon with energy higher than that of the lowest transition is absorbed. Moreover, the probability of absorption grows with increasing photon energy, as more and more transitions with higher energy become possible. The very broad absorption spectrum observed in real QDs is a major feature when compared with organic dyes whose absorptions are relatively narrow and in the vicinity of their emission (small Stokes shift). This means that QDs can be excited by any wavelength lower than its emission peak and, unlike dyes, several different kinds of QDs can be simultaneously excited by the same optical source.

Akin to semiconductor, QDs display a relatively narrow emission spectrum. The major source of emission broadening in QDs, however, comes from their size distribution. In a colloidal dispersion, the solid particles have approximately, but not exactly, the same size resulting in slight variations in the emission wavelength. As a consequence, the emission spectrum of a certain ensemble of nanocrystals will be much broader than the individual QDs spectra. Presently, size distributions with less than 5% variation are achievable. This translates into an FWHM of approximately 25 nm, which is quite narrow in comparison to many luminescent dyes. This way, in contrast with traditional dyes, which have broad emission spectra with a characteristic long red tail, nanocrystals present a symmetrical and relatively narrow emission.

The picture in Figure 6.7 was taken with no filters and shows the very distinct emissions of CdTe QDs of different sizes, simultaneously excited using a single blue laser (475 nm).

Most popular methods to produce semiconductor nanoparticles are based in colloidal chemistry where, by a series of engineered reactions using adequate precursors,

FIGURE 6.7 CdTe QDs with different sizes excited by a single blue laser (475 nm) showing emission at 620, 580, and 540 nm.

Table 6.2 Luminescence Emission Range of QDs Made with Different Semiconductor Materials and of Different Sizes

Semiconductor Material	Emission Range (nm)	QD Size Range (nm)
CdSe	465–640	1.9–6.7
CdSe/ZnS	490–620	2.9–6.1
CdTe/CdS	620–680	3.7–4.8
PbS	850–950	2.3–2.9
PbSe	1200–2340	4.5–9

nanoparticles arise by precipitation. Size can be controlled by timely arresting the reaction with passivating agents (Trindade et al. 2001).

There are practical limitations to the size of the particles that can be achieved with a certain material. When the nanocrystals are made progressively bigger, the quantum confinement is eventually lost. On the other hand, they cannot be made infinitely small. This way, any given material provides tunability over a limited wavelength range. In order to cover a wider range of wavelengths, a variety of materials should be used. Table 6.2 shows the semiconductor materials most commonly used to fabricate nanocrystals by chemical methods. The wavelength range and the corresponding particle size covered by each material are also shown. Roughly a wavelength range of 2000 nm is covered with four different materials.

While most dyes present severe photodegradation when illuminated by energetic radiation, QDs have demonstrated to be extremely photostable in most situations. Although photobleaching has been reported in bare dots, nanocrystals with an adequate protective shell of a different semiconductor material, also called core–shell QD, are known to remain extremely bright even after several hour of moderate to high levels of UV radiation exposure (Hines and Sionnest 1996). On the other hand, the luminescence emission of common dyes can vanish completely after few minutes.

The unique properties of QD have been explored in a wide diversity of imaging and sensing applications (Jorge et al. 2007). Besides outperforming dyes in traditional bioassays and imaging applications, QDs also introduced new possibilities. Their unique photostability and potentially low cytotoxicity allow for long-term in vivo imaging (Dubertret et al. 2002, Jain and Stroh 2004, Jaiswal and Simon 2004, Kirchner et al. 2005) and monitoring of dynamic cell processes. Although cadmium ions alone can be highly toxic to cells, coating the core–shell dots with lipid layers minimizes the risk of contamination. Nevertheless, the toxic nature of the materials used in most QD materials is a concern (Hardman 2006). Presently, heavy metal-free QDs are making its appearance along with alternative nontoxic nanomaterials like carbon dots (Silva 2011). Indeed, a diversity of nanoscale materials and devices is being explored, adding a new dimension of control over luminescent properties and enabling new applications of luminescent spectroscopy.

6.4.2 Fluorescence Quenching

In most sensing applications, the luminophore is used as a probe, and the goal is to measure the concentration of some analyte, through the influence it has on the probe. Depending on the combination luminophore–analyte, different detection schemes can be used. The analyte concentration can impact the luminescent intensity, the lifetime, the spectral characteristics of emission, etc. Conversely, this influence can take place through a variety of interaction mechanisms, namely, resonant energy transfer, photo-induced electron transfer, and proton transfer. Many of these phenomena have been used successfully as sensing mechanisms, each of them with advantages and disadvantages depending on the specific application. The detection of the concentration of an analyte through collisional quenching of luminescence, in which the analyte is usually the quencher, is a particularly successful approach that has been used for the sensing of many chemical and biological analytes (Geddes 2001).

This technique is often compatible with either intensity or lifetime detection. Furthermore, several models are available, describing different quenching mechanisms that provide simple calibration functions. The term quenching refers, broadly, to any decrease in the luminescence intensity due to the presence of an external agent. Depending on the deactivation mechanism, the excited state lifetime of the luminophore is often decreased as well. Quenching is a bimolecular process involving the interaction between an excited luminescent molecule and the external quencher molecule, Q. The result of this interaction is the nonradiative deactivation of the luminophore. This can be achieved by a variety of mechanisms such as electron transfer, proton transfer, energy transfer, molecular rearrangements, or even formation of new molecular compounds (Eftink 1991, Valeur 2002).

6.4.2.1 Dynamic Quenching

A great deal of quenching mechanisms are photophysical processes meaning that, after all the de-excitation steps, the luminophore gets back into the ground state unaltered. These are reversible dynamic quenching processes in which a reduction in the quencher concentration results in an increase of the luminescent signal. The presence of a quenching agent, in a given concentration, introduces an additional external nonradiative decay rate in the luminescence process. As a consequence of the competition between intermolecular processes and the intrinsic deactivation mechanisms, both the quantum yield and the lifetime of luminescence can be decreased. Dynamic quenching processes usually follow a Stern–Volmer (SV) kinetics (Lakowicz 1999):

$$\frac{\tau_0}{\tau} = 1 + K_{SV}[Q] \tag{6.16}$$

where $K_{SV} = k_q\tau_0$ is the SV constant, which depends on the intrinsic lifetime, τ_0, and on the rate constant of the luminophore–quencher interaction k_q. This relation provides a way to monitor the concentration of a given quencher by comparing the quenched lifetime, τ, with the lifetime in the absence of the quencher, τ_0. A similar SV relation can be obtained for the luminescence intensities as well:

$$\frac{I_0}{I} = 1 + K_{SV}[Q] \tag{6.17}$$

where I and I_0 represent the steady-state luminescence intensities (obtained for a given pair of excitation and emission wavelengths). Comparing Equations 6.16 and 6.17, it is found that

$$\frac{I_0}{I} = \frac{\tau_0}{\tau} \tag{6.18}$$

This is a very distinct characteristic of purely dynamic quenching processes where, in the presence of the quencher, both the lifetime and the luminescence intensity are reduced. The decrease in the lifetime occurs because the quencher introduces an additional decay rate that depopulates the excited state. Because the deactivation process is nonradiative, the quantum yield is decreased as well. That is not the case of static quenching where usually chemical reactions are involved that change the ground state population instead. Therefore, only the luminescence intensity is decreased, whereas the lifetime remains constant.

Quenching data are usually obtained in the form of an SV plot where I_0/I or τ_0/τ is represented as a function of quencher concentration, [Q]. If only purely dynamic quenching is present, a linear behavior will be observed as described by Equations 6.17 and 6.18. Overall, the observation of a linear variation in the SV plot is a good indication that a single class of luminophore, homogeneously distributed and with equal access to the quencher, is present in the solution. In rigid media, however, it is often the case that different environments are created in which the quencher has different accessibility. In such situation, the observed SV plot departs from linearity presenting a downward curvature. Transient phenomena arising at high concentration, temperature, viscosity, and other sample conditions can also distort the SV behavior. For many of these cases, however, modified SV models can be used that account for these modification and allow for correct data interpretation.

For the particular case of sensing applications, however, data analysis can be greatly simplified. Most of the time, the quencher is the analyte under study and the goal is simply to determine its concentration. In this context, the simplest quenching models can still be used, and the choice should be made based on the fitting quality. The sensing of chemical species by luminescence quenching techniques has been used successfully in many applications, where very simple models were used, but still taking advantage of the very high sensitivity that is intrinsic to luminescence techniques (MacCraith et al. 1993, Chuang and Arnold 1998, Geddes 2001, Zhoua et al. 2004).

Examples of collisional quenchers include oxygen, amines, halogens, and acrylamide. The actual quenching mechanism can vary with luminophore–quencher pair. For instance, while acrylamide quenches indole by electron transfer processes, quenching by halogens is usually due to spin–orbit coupling inducing intersystem crossing to triplet states. Oxygen is an efficient collisional quencher of almost all known luminophores and it has often to be removed from solutions in order for other quenching processes to be studied. On the other hand, the quenching of luminescence is a preferred method for the optical detection of oxygen and is in the base of some pioneer application of solid-state optrodes.

6.4.3 Fluorescent Optrodes

Luminescence-based techniques are very powerful tools frequently used in a range of applications, from material characterization to the study of molecular dynamics. Although very sensitive and versatile, luminescent methods are usually associated with very complex and expensive instrumentation and, therefore, its use was for long restricted to research laboratories. In the last 20 years, however, revolutionary developments in many different areas have concurred to change this picture (Kulmala and Suomi 2003). The developments in optoelectronics, with the advent of solid-state optical sources including the blue range of the spectrum, together with contributions from organic chemistry, with the creation of long-lived, photostable, metallo-organic luminescent complexes, have made possible the implementation of highly sensitive frequency-domain techniques in low-cost sensing applications (Demas and DeGraff 1997, Juris 1988, Wolfbeis et al. 1998). Great advances in material sciences contributed to the development of sophisticated immobilization techniques, with polymer and solgel chemistry, allowing the design of solid-state luminescent sensors (MacCraith et al. 1995, Mohr 2004, Podbielska et al. 2004). All these advances have made possible to associate the potentialities of luminescence-based techniques with the intrinsic capability of optical waveguides like fibers or integrated optics platforms. Presently, luminescence-based optrodes are being widely explored in the design of chemical sensing probes and lab-on-a-chip platforms, with commercial systems already in the market (Mignani and Baldini 1996, Holst and Mizaikoff 2002).

6.4.3.1 Membranes for Optical Chemical Sensing

One of the most critical steps for the implementation of a solid-state luminescent optrode is the immobilization of the luminescent indicator in a solid matrix. In order to fabricate a sensing probe capable of continuous measurement, the indicator chemistry must be immobilized over the waveguide. This is usually done by encapsulating the luminophores in a solid host. The resulting sensing membrane can either be directly bound to the fiber surface, in an intrinsic approach, or attached to it by some physical support in an extrinsic configuration. Ultimately, the properties of the sensing membranes will determine the sensor performance (Draxler et al. 1995, Demas and DeGraff 1999, Rowe et al. 2002). An ideal immobilization membrane would effectively entrap the indicator and preserve its optical and chemical properties, avoiding leaching and photobleaching. Simultaneously, it should allow the establishment of a fast and reversible equilibrium with the aqueous environment or gaseous atmosphere being probed. While permeability to the analyte is a highly desirable feature, penetration of potentially interfering chemical species into the sensing membrane should be avoided. On top of these requirements, a good adhesion to the waveguide surface, together with mechanical and chemical stability, is a particularly important feature for long-term practical applications.

The simultaneous fulfillment of all these requirements is extremely difficult to obtain, especially considering that every analyte–luminophore combination may have specific chemical properties. Traditionally, polymers are materials of choice for this task due to their chemical versatility and availability in a great variety (Mohr 2004). Although many successful sensing applications using polymers doped with luminescent indicators have been reported, the sensing membranes are far from ideal and some problems subsist. Major issues are the leaching of the sensor from the solid host, poor chemical resistance, and modification of the properties of the sensing dye.

In this context, solgel glasses are a particular kind of polymeric material having very suitable properties for sensing applications (Podbielska et al. 2004). With this technique, porous silica matrices can be formed that are highly compatible with silica fibers and waveguides. In addition, the sensing dye is physically encapsulated in the microporous structure, avoiding leaching while preserving the indicator chemical properties.

Solgel-derived materials, particularly those based in silica alkoxide precursors, have been attracting the attention of many researchers working in the immobilization of sensing luminophores in solid supports. In comparison to some polymers, these materials can be mechanically robust and chemically very resistant. In addition, by choosing the right ingredients and adjusting the process parameters, properties like polarity and porosity can be tailored to the specific needs of an application (Hench and West 1990, McDonagh et al. 1998). Furthermore, they are usually optically transparent, and their refractive index can be tuned within a wide range. The simplicity and low temperatures associated to the fabrication process, along with the compatibility with a variety of coating processes (spin coating, dip coating, casting, spraying), make them suitable for use in either very specific laboratorial applications or in large-scale industrial processes. High-quality thin-film layers where the luminescent dye preserves its sensing properties can be obtained using the solgel process. Shown in Figure 6.8 is a circular glass substrate spin coated with a thin solgel film (thickness around 1 μm) doped with an oxygen-sensitive luminescent ruthenium complex. The thin-film layers are excited by a LED source (475 nm) through a large-diameter fiber bundle at the center of the substrate, where an intense luminescent spot can be observed.

In addition, high brightness luminescence is also observed coming out of the substrate edges. This happens because part of the isotropic fluorescent emission is trapped inside the glass substrate being guided toward its edged by a waveguide effect. Taking advantage of such effects, it is possible to design highly efficient luminescent optrodes

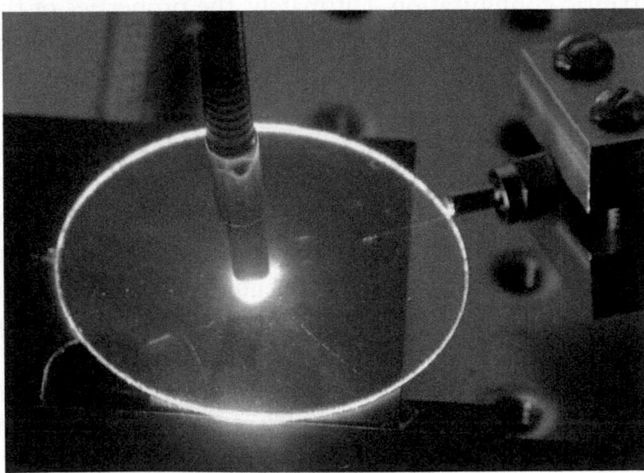

FIGURE 6.8 Thin film doped with 10 mM Ru(bpy) displaying strong luminescence when excited with a blue LED.

using planar waveguides or optical fibers. Indeed, solgel layers and ruthenium-based complexes are in the base of some of the most successful applications of luminescent-based optrodes for chemical sensing (Jorge et al. 2004).

6.4.3.2 Fiber-Optic Luminescent Optrodes

A wide variety of luminescent optrodes where a luminescent indicator is immobilized in a solid substrate and used as sensing probe, either in planar or in fiber-optic platforms, have been reported to date. Optical fiber-based optrodes, in particular, have made possible the use of conventional spectroscopy techniques in sites otherwise inaccessible. In addition, their unique characteristics have provided new possibilities like multiplexing capability and evanescent wave sensing. These features along with the technological advances in the field of optoelectronics and material science have allowed for the establishment a mature optical fiber biochemical sensing technology. Presently, with many fundamental proof-of-principle sensing schemes already demonstrated, a growing number of practical applications are being explored in a wide range of situations (Wolfbeis 2002, 2004, Narayanaswamy and Wolfbeis 2004). Major areas currently explored include medical and chemical analysis, molecular biotechnology, marine and environmental analysis, and bioprocess control.

One of the most popular and successful applications of the use of fiber-optic luminescent optrodes is the optical detection of oxygen by luminescence quenching. Its history is representative of the state of the art of luminescent optrode technology.

Although fiber optics has been used in very early luminescent oxygen sensing applications, the fiber was typically used simply as a light-guiding device through which excitation and detection of radiation were performed. Only after the development of polymer and solgel immobilization membranes could the first true optical fiber sensing probes be implemented. One of the first reports of a truly intrinsic oxygen fiber probe was made by MacCraith et al. (1994b). The authors used an evanescent wave configuration, where a solgel thin film, doped with Ru(dpp), was deposited over a fiber section with previously removed cladding.

Improvement of the encapsulation chemistry of solgel, using hybrid organic–inorganic materials, enabled to tailor the sensor properties improving sensitivity and photostability. Figure 6.9 shows a representative example fiber-optic luminescent optrode displaying two tapered multimode optical fibers coated with a hybrid solgel material, doped with an oxygen-sensitive ruthenium complex. Tapering is a commonly used technique to improve luminescent coupling efficiency and also to improve the probe spatial resolution (Jorge et al. 2004). In a pioneer work, tapered fibers were used to implement micro-optrodes with fiber tip diameters ranging from 10 to 50 μm (Klimant et al. 1999). The small dimensions of the sensing tip allowed measurements with very high spatial resolution and very fast response times (250 ms). Oxygen gradients with a spatial resolution of up to 20 μm were measured in freshwater sediments covered with a 2 mm layer of green algae.

The luminescence emission of most sensing dyes is generally temperature dependent. This way, most luminescent sensors (and also optical oxygen sensors) need to be provided with a temperature reference. Techniques explored to simultaneously measure oxygen and temperature included the combination of the oxygen-sensitive dye and a phosphorescent crystal (Liao et al. 1997). Alternatively, the use of QDs encapsulated

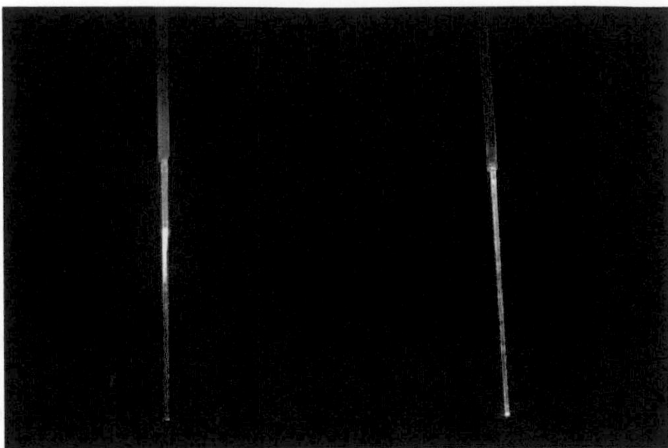

FIGURE 6.9 Picture of tapered fibers coated with solgel thin film doped with an oxygen-sensitive ruthenium complex and excited by a blue LED.

in polymer with low oxygen permeability and therefore sensitive only to temperature was combined with oxygen-sensitive ruthenium complex in a highly permeable solgel matrix (Jorge et al. 2008).

A diversity of practical applications made use of fiber-optic oxygen-sensitive optrodes. A prototype of a multiparameter sensing probe (pH, CO_2, and O_2) using a different fiber for each indicator dye and employed in intravascular monitoring was described (Mignani and Baldini 1996). A high-performance dissolved oxygen optical sensor based on phase detection using an external membrane configuration was developed for application in wastewater monitoring. A detection limit of <10 ppb was achieved and the long-term stability reached several months. In addition, the disposable sensing membrane could be easily substituted (McDonagh et al. 2001).

Many other similar applications were described with sensing optrodes being developed for special conditions. Oxygen measurements could be performed at high temperature, in hyperbaric chambers; sensing heads withstanding sterilization were developed; measurements could be performed in high-pressure fermentation reactors; and several in vitro and in vivo tests were demonstrated successfully (Wolfbeis et al. 1998, Remillard et al. 1999, Stokes and Somero 1999, Zhao et al. 1999).

Presently, a few companies offer a diversity of oxygen sensing probes. Ocean Optics offers a modular system where different kinds of fiber probes, a pulsed blue LED optical source, and a preconfigured CCD spectrometer with dedicated software can be acquired separately. PreSense is a German company offering similar oxygen sensing probes where time-resolved measurements are used instead. In addition to standard probes, micro-optrodes with 30 μm fiber tips are also available, which are suitable for performing high-spatial-resolution measurements of biological samples. A support unit capable of simultaneous interrogation of several sensors allows multipoint measurements to be implemented.

Other companies can be found that offer probes optimized for applications in wastewater treatment or oceanographic applications. Although not advertised, several companies selling oxygen sensors based on conventional technologies are making strong efforts to introduce optical technologies in their products and maintain close contact with the scientific community.

6.5 Conclusions and Future Outlook

A source of curiosity and fascination in earlier times, photoluminescence is presently the base for some of the most sensitive analytical techniques available and a reference method in many fields of science and industry.

Instrumentation for steady-state and time-domain spectroscopy has experienced dramatic evolution in the last decades, progressing from classical benchtop instruments to a diversity of analytical tools: powerful fluorescence microscopes enable visualization of the internal structure of cells unraveling the mysteries of life; unprecedented sensitivity supports single molecule analysis; fluorescent arrays facilitate fast DNA sequencing; fluorescent immunoassays are the base for some of the most selective biosensors.

Miniaturization, in particular, made possible by the fast progress of optoelectronics (sources and detectors) is taking fluorescence out of the lab into the world of field applications enabling in situ environmental monitoring. The combination of luminescent probes, immobilized in solid-state membranes, into guided wave platforms like optical fibers is the base for a new family of (bio)chemical sensing optrodes with remarkable features and potential of application (environmental monitoring, bioprocess control, single cell analysis, operation in chemically and electromagnetically hazard environments).

Presently, a new revolution is happening with the advent of nanotechnology. Control at the nanometer scale, combining different materials in different geometries, allows tailoring the optical, chemical, and mechanical properties of materials. New luminescent probes like QDs, with very high photostability, are enabling unprecedented applications like long-term in vivo cellular imaging that are literally shedding new light into biochemistry.

In the near future, combination with metamaterials made from the merging of different nanostructures like metallic nanopillars or graphene will allow unparalleled control over the properties of light pushing forward the possibilities of photoluminescence.

References

Carraway, E.R., Demas, J.N., DeGraff, B.A., and Bacon, J.R., Photophysics and photochemistry of oxygen sensors based on luminescent transition-metal complexes. *Analytical Chemistry*, 1991. 63: 337–342.

Chuang, H. and Arnold, M.A., Linear calibration function for optical oxygen sensors based on quenching of ruthenium fluorescence. *Analytica Chimica Acta*, 1998. 368: 83–89.

Demas, J.N., DeGraff, B.A., and Coleman, P.B., Oxygen sensors based on luminescence quenching. *Analytical Chemistry* 1999. A793–A800.

Demas, J.N. and DeGraff, B.A., Design and applications of highly luminescent transition metal complexes. *Analytical Chemistry*, 1991. 63: 829A–837A.

Demas, J.N. and DeGraff, B.A., Applications of luminescent transition metal complexes to sensor technology and molecular probes. *Journal of Chemical Education*, 1997. 74: 690–695.

Draxler, S., Lippitsch, M.E., Klimant, I., Kraus, H., and Wolfbeis, O.S., Effects of polymer matrices on the time-resolved luminescence of a ruthenium complex quenched by oxygen. *Journal of Physical Chemistry*, 1995. 195: 3162–3167.

Dubertret, B., Skourides, P., Norris, D.J., Noireaux, V., Brivanlou, A.H., and Libchaber, A., In vivo imaging of quantum dots encapsulated in phospholipid micelles. *Science*, 2002. 298: 1759–1762.

Eftink, M.R., Fluorescence quenching: Theory and applications, in *Topics in Fluorescence Spectroscopy: Principles*, J.R. Lakowicz, ed. 1991, New York: Plenum Press. pp. 53–120.

Gaponenko, S.V., *Optical Properties of Semiconductor Nanocrystals*. Cambridge Studies in Modern Optics, P.L. Knight and A. Miller, eds., Vol. 23. 1998, Cambridge, U.K.: Cambridge University Press.

Geddes, C.D., Optical halide sensing using fluorescence quenching: Theory, simulations and applications—A review. *Measurement Science and Technology*, 2001. 12: R33–R88.

Hardman, R.A., Toxicologic review of quantum dots: Toxicity depends on physicochemical and environmental factors. *Environmental Health Perspectives*, February 2006. 114(2): 165–172.

Hench, L.L. and West, J.K., The sol-gel process. *Chemical Reviews*, 1990. 90: 33–72.

Hines, M.A. and Sionnest, P.G., Synthesis and characterization of strongly luminescing ZnS-capped CdSe nanocrystals. *Journal of Physical Chemistry*, 1996. 100: 468–471.

Holst, G. and Mizaikoff, B., Fiber optic sensors for environmental applications, in *Handbook of Optical Fibre Sensing Technology*. 2002, Lôpez-Higuera, J.M. (Ed.), New York: John Wiley & Sons, Ltd. pp. 729–755.

Jain, R.K. and Stroh, M., Zooming in and out with quantum dots. *Nature Biotechnology*, 2004. 22(8): 959–960.

Jaiswal, J.K. and Simon, S.M., Potentials and pitfalls of fluorescent quantum dots for biological imaging. *Trends in Cell Biology*, 2004. 14(9): 497–504.

Jorge, P.A.S., Caldas, P., Rosa, C.C., Oliva, A.G., and Santos, J.L., Optical fiber probes for fluorescence based oxygen sensing. *Sensors and Actuators B—Chemical*, 2004. 103(1–2): 290–299.

Jorge, P.A.S., Martins, M.A., Trindade, T., Santos, J.L., and Farahi, F. Optical fiber sensing using quantum dots. *Sensors*, 2007. 7(12): 3489–3534.

Jorge, P.A.S., Maule, C., Silva, A.J., Benrashid, R., Santos, J.L., and Farahi, F., Dual sensing of oxygen and temperature using quantum dots and a ruthenium complex. *Analytica Chimica Acta*, 2008. 606(2): 223–229.

Juris, A., Balzani, V., Barigalletti, F., Campagna, S., Belser, P., and Zelewsky A., Ru(II) polypyridine complexes: Photophysics, photochemistry, electrochemistry, and chemiluminescence. *Coordination Chemistry Reviews*, 1988. 84: 85–277.

Kirchner, C., Liedl, T., Kudera, S., Pellegrino, T., Javier, A.M., Gaub, H.E., Stolzle, S., Fertig, N., and Parak, W.J., Cytotoxicity of colloidal CdSe and CdSe/ZnS nanoparticles. *Nano Letters*, 2005. 5(2): 331–338.

Klimant, I., Ruckruh, F., Liebsch, G., Stangelmayer, A., and Wolfbeis, O.S., Fast response oxygen micro-optrodes based on novel soluble ormosil glasses. *Mikrochimica Acta*, 1999. 131: 35–46.

Kuhn, H. and Forsterling, H.-D., *Principles of Physical Chemistry*. 2000, New York: John Wiley & Sons.

Kulmala, S. and Suomi, J., Current status of modern analytical luminescence methods. *Analytica Chimica Acta*, 2003. 500: 21–69.

Lakowicz, J.R., Fluorophores, in *Principles of Fluorescence Spectroscopy*. 1999, New York: Kluwer.

Liao, S.-C., Xu, Z., Izatt, J.A., and Alcala, J.R., Real-time frequency domain temperature and oxygen sensor with a single optical fiber. *IEEE Transactions on Biomedical Engineering*, 1997. 44(11): 1114–1121.

MacCraith, B.D., McDonagh, C.M., O'Keeffe, G., Keyes, E.T., Vos, J.V., O'Kelly, B., and McGilp, J.F., Fibre optic oxygen sensor based on fluorescence quenching of evanescent-wave excited ruthenium complexes in sol-gel derived porous coatings. *Analyst*, 1993. 118: 385–388.

MacCraith, B.D., McDonagh, C.M., O'Keeffe, G., McEvoy, A.K., Butler, T., and Sheridan, F.R., Sol-gel coatings for optical chemical sensors and biosensors. *Sensors and Actuators B*, 1995. 29: 51–57.

MacCraith, B.D., O'Keeffe, G., McDonagh, C.M., and McEvoy, A.K., LED-based fibre optic oxygen sensor using sol-gel coating. *Electronics Letters*, 1994a. 30: 888–889.

MacCraith, B.D., O'Keeffe, G., McEvoy, A.K., and McDonagh, C., Development of a LED-based fibre optic oxygen sensor using a sol-gel-derived coating. In Proc. SPIE 2293, Chemical, Biochemical, and Environmental Fiber Sensors VI, 110 (October 21, 1994b); doi:10.1117/12.190961.

McDonagh, C., Kolle, C., McEvoy, A.K., Dowlingm, D.L., Cafolla, A.A., Cullen, S.J., and MacCraith, B.D., Phase fluorometric dissolved oxygen sensor. *Sensors and Actuators B*, 2001. 74: 124–130.

McDonagh, C.M., MacCraith, B.D., and McEvoy, A.K., Tailoring of sol–gel films for optical sensing of oxygen in gas and aqueous phase. *Analytical Chemistry*, 1998. 70: 45–50.

McEvoy, A.K., McDonagh, C.M., and MacCraith, B.D., Dissolved oxygen sensor based on fluorescence quenching of oxygen-sensitive ruthenium complexes immobilized in sol-gel-derived porous silica coatings. *Analyst*, 1996. 121: 785–788.

Mignani, A.G. and Baldini, F., Biomedical sensors using optical fibres. *Reports on Progress in Physics*, 1996. 59(1): 1–28.

Mills, A. and Williams, F.C., Chemical influences on the luminescence of ruthenium diimine complexes and its response to oxygen. *Thin Solid Films*, 1997. 306: 163–170.
Mohr, G.J., Polymers for optical sensors, in *Optical Chemical Sensors*, F. Baldini et al., eds. 2004, Dordrecht, the Netherlands: Springer. pp. 297–321.
Narayanaswamy, R. and Wolfbeis, O.S., eds. *Optical Sensors—Industrial Environmental and Diagnostic Applications*. Springer Series on Chemical Sensors and Biosensors, O.S. Wolfbeis, ed. 2004, Berlin, Germany: Springer.
Orellana, G., Luminescent optical sensors. *Analytical and Bioanalytical Chemistry*, 2004. 379: 344–346.
Orellana, G. and Fresnadillo, D.G., Environmental and industrial optosensing with tailored luminescent Ru(II) polypyridyl complexes, in *Optical Sensors—Industrial Environmental and Diagnostic Applications*, R. Narayanaswamy and O.S. Wolfbeis, eds. 2004, Berlin, Germany: Springer. pp. 309–357.
Podbielska, H., Ulatowska-Jarza, A., Muler, G., and Eichler, H.J., Sol-gels for optical sensors, in *Optical Chemical Sensors*, F. Baldini et al., eds. 2004, Dordrecht, the Netherlands: Springer. pp. 353–385.
Remillard, J.T., Jones, J.R., Poindexter, B.D., Narula, C.K., and Weber, W.H., Demonstration of a high-temperature fiber-optic gas sensor made with a sol-gel process to incorporate a fluorescent indicator. *Applied Optics*, 1999. 38(25): 5306–5309.
Rowe, H.M., Xu, W., Demas, J.N., and DeGraff, B.A., Metal ion sensors based on a luminescent ruthenium(II) complex: The role of polymer support in sensing properties. *Applied Spectroscopy*, 2002. 56: 167–173.
Silva, J.G.G.E., Carbon and silicon fluorescent nanomaterials, in *Nanomaterials*, Prof. Mohammed Rahman, ed. 2011, Rijeka, Croatia: InTech. pp. 237–252. DOI: 10.5772/25809. Available from: http://www.intechopen.com/books/nanomaterials/carbon-and-silicon-fluorescent-nanomaterials.
Stokes, M.D. and Somero, G.N., An optical oxygen sensor and reaction vessel for high-pressure applications. *Limnology and Oceanography*, 1999. 44(1): 189–195.
Trindade, T., O'Brien, P., and Pickett, N. L., Nanocrystalline semiconductors: synthesis, properties, and perspectives. *Chemistry of Materials*, 2001. 13(11): 3843–3858.
Valeur, B., *Molecular Fluorescence: Principles and Applications*. 2002, Weinheim, Germany: Wiley VCH.
Watts, R.J. and Crosby, G.A., Spectroscopic characterization of complexes of ruthenium(II) and iridium(III) with 4,4'-diphenyl-2,2'-bipyridine and 4,7-diphenyl-1,10-phenanthroline. *Journal of the American Chemical Society*, 1971. 93: 3184–3188.
Wolfbeis, O.S., Chemical sensing using indicator dyes, in *Optical Fiber Sensors: Applications, Analysis and Future Trends*. Dakin, J. and Culshaw, B. (Eds.) 1997, Boston, MA: Artech House. pp. 53–107.
Wolfbeis, O.S., Fiber optic chemical sensors and biosensors. *Analytical Chemistry*, 2002. 72: 81R–89R.
Wolfbeis, O.S., Fiber optic chemical sensors and biosensors. *Analytical Chemistry*, 2004. 76: 3269–3284.
Wolfbeis, O.S., Klimant, I., Werner, T., Huber, C., Kosch, U., Krause, C., Neurauter, G., and Durkop, A., Set of luminescence decay time based chemical sensors for clinical applications. *Sensors and Actuators B*, 1998. 51: 17–24.
Zhang, Z., Grattan, K.T.V., and Palmer, A.W., Phase-locked detection of fluorescence lifetime. *Review of Scientific Instruments*, 1993. 64: 2531–2540.
Zhao, Y.D., Richman, A., Storey, C., Radford, N.B., and Pantano, P., In situ fiber-optic oxygen consumption measurements from a working mouse heart. *Analytical Chemistry*, 1999. 71(17): 3887–3893.
Zhoua, L.L., Suna, H., Zhang, X.H., and Wu, S.K., An effective fluorescent chemosensor for the detection of copper(II). *Spectrochimica Acta Part A*, 2004. 61: 61–65.

7. Surface Plasmon Measurement

Principles and Techniques

Banshi D. Gupta
Indian Institute of Technology Delhi

Rajan Jha
Indian Institute of Technology Bhubaneswar

Handbook of Optical Sensors. Edited by José Luís Santos and Faramarz Farahi © 2015 CRC Press/Taylor & Francis Group, LLC. ISBN: 9781439866856.

7.1 Introduction

Surface plasmon–based photonics or plasmonics is a part of the fascinating field of nanophotonics in which the confinement of electromagnetic field in the region smaller than the wavelength is explored. It is based on the interaction of electromagnetic radiation with the conduction electrons at the metal–dielectric interface, resulting in the enhancement of optical field in the region close to the wavelength of the radiation. The enhancement of the field depends on the dielectric properties of the metal and dielectric in addition to the properties of interacting electromagnetic radiation. First named in 2001, the field of *plasmonics* has become popular among physicists and engineers only recently, as scientists have developed tools to create nanosized structures that can guide and shape these light and electron waves for different applications. Surface plasmons have enormous potential in the fields of novel optical devices, optical computing, subwavelength optical imaging, and sensing. In this chapter, we shall confine ourselves to the application of surface plasmons to the field of sensing of various chemical and biological parameters of interest. Attention will also be focused on its principle and technique.

7.2 Fundamentals of Surface Plasmons

Surface plasmons are the guided waves at the metal–dielectric interface. These waves are produced because the free electrons in metal respond collectively to an electromagnetic disturbance induced by incident light or electrons. Due to this, surface plasmons possess the character of electromagnetic wave as well as of surface charge as shown in Figure 7.1. The electric field of the surface plasmon is perpendicular to the interface and decays exponentially into metal and dielectric medium. The penetration depth of the electric field in metal is much smaller than that in the dielectric medium. The surface plasmon can be excited by an incident light propagating along the interface.

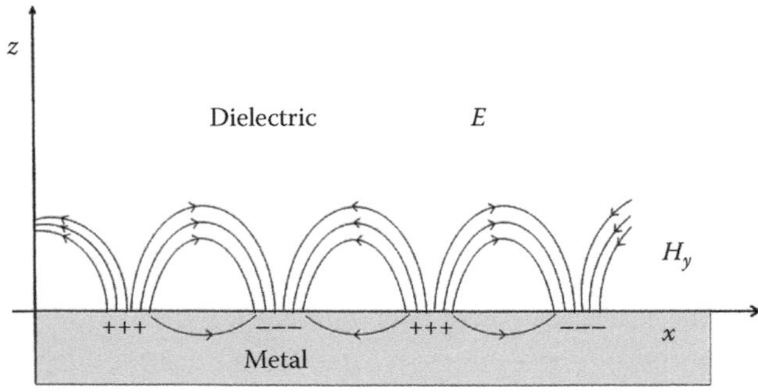

FIGURE 7.1 Electric field lines and charge distributions associated with a surface plasmon traveling on a metal–dielectric interface.

An electromagnetic plane wave propagating in a homogeneous and isotropic medium can, mathematically, be described by an electric field E:

$$\vec{E}(\vec{r},t) = \vec{E}_0 e^{i(\vec{k}\cdot\vec{r}-\omega t)} \tag{7.1}$$

where
- $\vec{E}_0$ is the amplitude of the electric field
- ω is the angular frequency
- $\vec{k}$ is the wave vector
- $\vec{r} = (x, y, z)$ is the position vector

The magnitudes of $\vec{k}$ and ω are related by the following relation known as dispersion relation:

$$k = \frac{\omega}{c}\sqrt{\varepsilon_d} = k_0 n_d \tag{7.2}$$

where
- c is the speed of light in vacuum
- ε_d is the relative permittivity (or the dielectric function) of the medium
- n_d is its refractive index of the medium
- $k_0 = \omega/c$ is the magnitude of the free space wave vector

The momentum of the associated photon is $\hbar k_0$.

As mentioned earlier, the electric field of the surface plasmon is normal to the interface and it propagates along the interface. The electric fields inside the dielectric ($z > 0$), E_z^d, and inside the metal ($z < 0$), E_z^m, can be written as

$$E_z^d = E_0^d e^{ik_z^d z} e^{i(k_{SP}x-\omega t)} \tag{7.3}$$

$$E_z^m = E_0^m e^{-ik_z^m z} e^{i(k_{SP}x-\omega t)} \tag{7.4}$$

where k_{SP} is the propagation constant of the surface plasmon wave along the interface and is given by the following dispersion relation:

$$k_{SP} = \frac{\omega}{c}\sqrt{\frac{\varepsilon_m \varepsilon_d}{\varepsilon_m + \varepsilon_d}} \tag{7.5}$$

where
- ε_m and ε_d represent the relative permittivity of the metal and dielectric, respectively
- k_z is a complex wave vector that accounts the extension of the electric field to both sides of the interface

The wave vectors, k_z, in dielectric and metal are given as follows:

$$\left(k_z^d\right)^2 = \varepsilon_d \left(\frac{2\pi}{\lambda}\right)^2 - \left(k_{SP}\right)^2 \tag{7.6}$$

$$\left(k_z^m\right)^2 = \varepsilon_m \left(\frac{2\pi}{\lambda}\right)^2 - \left(k_{SP}\right)^2 \tag{7.7}$$

where λ is the wavelength of light in free space. Since the relative permittivity or the dielectric constant of the dielectric medium is weakly dispersive as compared to metal, the dispersive nature of metal plays an important role in determining the characteristics of the surface plasmon. Figure 7.2 shows the dispersion curves of light in free space and the surface plasmon. It may be noted that at low frequencies, the surface plasmon mode and light line are close. As the frequency increases, the separation between two curves increases and the surface plasmon curve approaches an asymptotic frequency limit called plasmon frequency. This occurs when the dielectric constants of metal and dielectric media are equal in magnitude but opposite in sign. The dielectric constant of metal, ε_m, is a complex quantity and can be written as

$$\varepsilon_m(\lambda) = \varepsilon_m' + i\varepsilon_m'' = 1 - \frac{\lambda^2 \lambda_c}{\lambda_p^2 \left(\lambda_c + i\lambda\right)} \tag{7.8}$$

where

ε_m' and ε_m'' are the real and imaginary parts of the dielectric constant of metal, respectively
λ_p and λ_c denote the plasma and collision wavelengths, respectively

In Equation 7.8, we have used the Drude model for the dielectric constant. The surface plasmon propagation constant, k_{SP}, is also a complex quantity due to the absorbing property of the metal and can be written as

$$k_{SP} = k_{SP}' + ik_{SP}'' = \frac{\omega}{c}\sqrt{\frac{\varepsilon_m' \varepsilon_d}{\varepsilon_m' + \varepsilon_d}} + i\frac{\omega}{c}\left(\frac{\varepsilon_m' \varepsilon_d}{\varepsilon_m' + \varepsilon_d}\right)^{3/2}\left(\frac{\varepsilon_m''}{2\left(\varepsilon_m'\right)^2}\right) \tag{7.9}$$

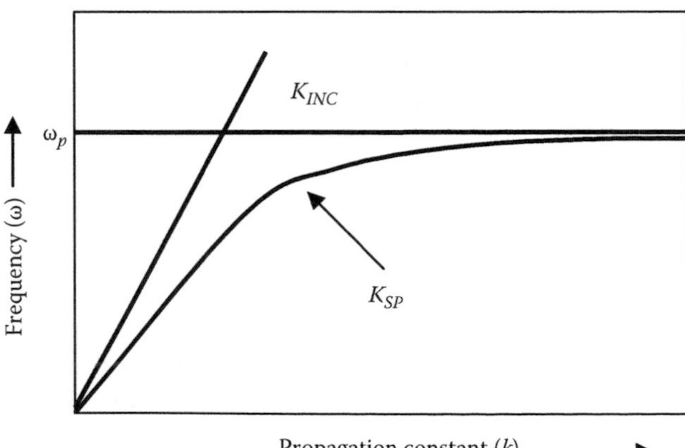

FIGURE 7.2 Dispersion curves of light in free space and the surface plasmon.

The surface plasmons propagate a short distance and get absorbed. Due to this, they have short lifetime.

Some of the characteristics of surface plasmon wave that play crucial role in determining the performance of the surface plasmon-based devices are its wavelength, its propagation length, and the penetration depth of its field in metal and dielectric medium. In the following, we discuss these characteristics.

7.2.1 Surface Plasmon Wavelength

The periodic variation of surface charge density is associated with the wavelength of the surface plasmon. The surface plasmon wavelength can be found from the real part of surface plasmon propagation constant defined in Equation 7.9 and given as:

$$k'_{SP} = \frac{\omega}{c}\sqrt{\frac{\varepsilon'_m \varepsilon_d}{\varepsilon'_m + \varepsilon_d}} \tag{7.10}$$

As $\lambda_{SP} = 2\pi / k'_{SP}$, the surface plasmon wavelength is given by

$$\lambda_{SP} = \lambda_0 \sqrt{\frac{\varepsilon'_m + \varepsilon_d}{\varepsilon'_m \varepsilon_d}} \tag{7.11}$$

From Equation 7.11, one can find that the surface plasmon wavelength is smaller than the free space wavelength (λ_0).

7.2.2 Propagation Length

Since the metal is absorbing, the energy of the surface plasmon decreases as it propagates. The propagation length (δ_{SP}) is defined as the distance in the direction of propagation at which the energy of surface plasmon becomes $1/e$ of the initial value. It is equal to $\delta_{SP} = 1/2k''_{SP}$. Using Equation 7.9 gives

$$\delta_{SP} = \frac{(\varepsilon'_m)^2}{2\pi\varepsilon''_m}\left(\frac{\varepsilon'_m + \varepsilon_d}{\varepsilon'_m \varepsilon_d}\right)^{3/2} \lambda_0 \tag{7.12}$$

The metal having large (negative) value of the real part of the dielectric constant and a small value of the imaginary part will have long propagation length. The propagation length calculated using Equation 7.12 for water–gold interface for wavelength 600 nm is 5512 nm. The propagation length suggests the limiting size of the photonic structure one can have. The achieved propagation length seems to be sufficiently high when thinking of highly miniaturized sensors and integrated optic devices for different applications. Nevertheless, by changing a metal–dielectric interface to a symmetrical structure of a thin metal embedded in a dielectric (Sarid 1981), one can significantly enhance the propagation length to few centimeters (Berini 2000) but that is at the cost of the subwavelength character of the surface plasmon mode.

7.2.3 Penetration Depth into Dielectric and Metal

The electromagnetic field of surface plasmon reaches its maximum at metal–dielectric interface and decays into both media. The field decay in the direction perpendicular to metal–dielectric interface is characterized by the penetration depth, which is defined as the distance from the interface at which the amplitude of the field becomes $1/e$ of the value at the interface. Assuming $|\varepsilon'_m| \gg |\varepsilon''_m|$, the penetration depths into dielectric (δ_d) and into metal (δ_m) can be written as follows:

$$\delta_d = \frac{\lambda_0}{2\pi}\left[\frac{\varepsilon'_m + \varepsilon_d}{\varepsilon_d^2}\right]^{1/2} \tag{7.13}$$

$$\delta_m = \frac{\lambda_0}{2\pi}\left[\frac{\varepsilon'_m + \varepsilon_d}{\varepsilon'^2_m}\right]^{1/2} \tag{7.14}$$

The values of penetration depths, in the case of water–gold interface, into water and gold region for 600 nm wavelength are 285 and 31 nm, respectively. The knowledge of the penetration depth of a surface plasmon helps in designing surface plasmon–based sensors and devices, which is a topic of upcoming research in the area of nanophotonics. The penetration depth in the dielectric gives us a measure of the length over which surface plasmon is sensitive to the changes in the refractive index of the dielectric medium, while the penetration depth into metal gives us an idea of the thickness of the metal film required for the coupling of light incident from the other interface of the metal film.

7.3 Excitation of Surface Plasmons

Surface plasmons are TM polarized and hence can be excited by p-polarized light. For excitation, the wave vector of excitation light along the metal–dielectric interface should be equal to the wave vector of the surface plasmon wave, a condition called the resonance condition. Mathematically, the resonance condition is $k_{INC} = k_{SP}$. In Figure 7.2, we have plotted both the wave vectors as a function of frequency. It can be seen from the figure that the two curves do not cross at any frequency and hence these cannot be equal. This implies that the surface plasmons cannot be excited by a direct light. To excite, the wave vector or the momentum of the excitation light has to be increased. Prism, waveguide, and grating have been used for this purpose. We discuss all the three methods in the following.

7.3.1 Prism-Based Method

In the prism-based configuration, the base of the prism is coated with metal. Light is incident on the prism–metal interface through one of the faces of the high-index prism as shown in Figure 7.3. When the angle of incidence of the light is greater than the critical angle, it is totally reflected from the base of the prism generating evanescent wave that

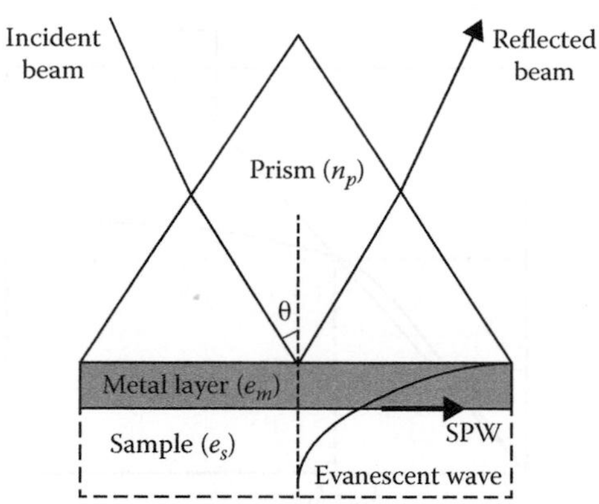

FIGURE 7.3 Excitation of a surface plasmon using a high-index prism and the Kretschmann configuration.

propagates along the prism–metal layer interface. The field of the evanescent wave penetrates in the metal layer. The propagation constant of the evanescent wave is given as

$$k_{ev} = \frac{\omega}{c}\sqrt{\varepsilon_p}\sin\theta = k_0 n_p \sin\theta \tag{7.15}$$

where

ε_p and n_p are the dielectric constant and the refractive index of the material of the prism, respectively

θ is the angle of incidence of light

Thus, the wave vector of the evanescent wave that will excite the surface plasmon can be increased by increasing the refractive index of the material of the prism.

The resonance condition, when the excitation of the surface plasmon at the metal–dielectric interface is carried out by evanescent wave, can be written as

$$\sqrt{\varepsilon_p}\sin\theta = \sqrt{\frac{\varepsilon'_m \varepsilon_d}{\varepsilon'_m + \varepsilon_d}} \tag{7.16}$$

In the preceding equation, $\varepsilon_p > \varepsilon_d$ and ε'_m is negative; hence, for some value of angle greater than the critical angle, Equation 7.16 or the resonance condition will be satisfied. In that case, the dispersion curves of surface plasmon and evanescent wave will cross each other as shown in Figure 7.4.

7.3.2 Waveguide-Based Method

The method of excitation of surface plasmon using waveguide is the same as that of the prism. In the case of a waveguide, a small length of the waveguiding layer in the middle is coated with metal layer. The light propagates in the waveguide by means of total internal reflection phenomenon, and hence there will be an evanescent wave at the waveguide–metal interface propagating along the interface. This will excite the surface plasmons at the metal–dielectric interface as shown in Figure 7.5. The requirement is that the refractive index of the core of the waveguide should be greater than the refractive index of the dielectric.

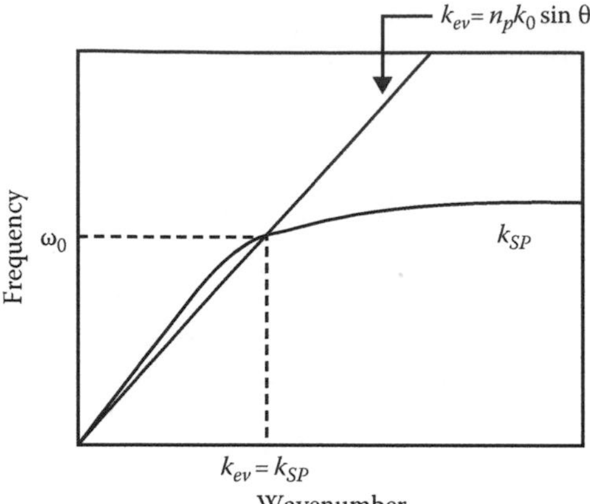

FIGURE 7.4 Dispersion curves, evanescent wave generated using high-index prism, and the surface plasmon.

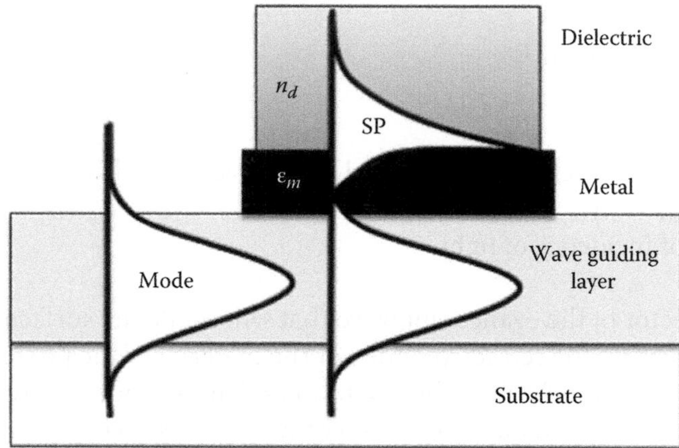

FIGURE 7.5 Excitation of a surface plasmon by a guided mode of an optical waveguide.

7.3.3 Grating-Based Method

Gratings are generally used to determine the spectrum of a light source; however, it can also be used for the excitation of a surface plasmon. For excitation, the grating is written on the metal layer shown in Figure 7.6. The light is incident through the dielectric medium on the grating that diffracts it in different directions. The component of the wave vector of the diffracted waves along the interface is given by

$$k_x = k_0 \pm m\left(\frac{2\pi}{\Lambda}\right) = \frac{2\pi}{\lambda} n_d \sin\theta \pm m\left(\frac{2\pi}{\Lambda}\right) \tag{7.17}$$

where

m is the order of diffraction

Λ is the grating element

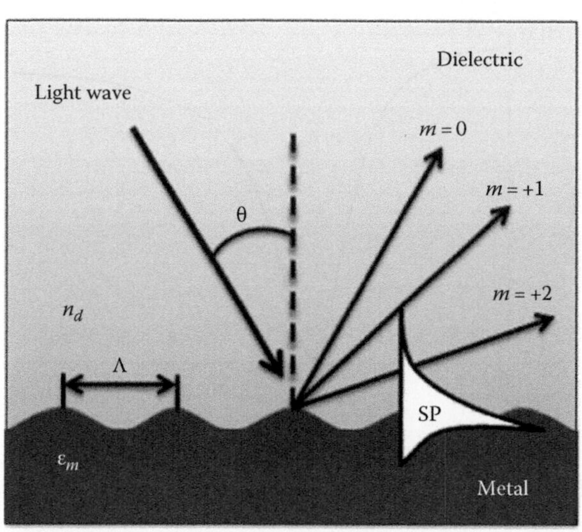

FIGURE 7.6 Excitation of a surface plasmon using diffraction order of the grating written on metal.

For positive values of m, the resonance condition can be written as

$$k_{SP} = \frac{2\pi}{\lambda} n_d \sin\theta + m\left(\frac{2\pi}{\Lambda}\right) \tag{7.18}$$

The aforementioned condition can be satisfied for a particular value of the angle of incidence. The disadvantage with this method is that the light is incident through the dielectric medium, and hence the dielectric medium and the flow cell, if it is a liquid, have to be optically transparent.

7.4 Interrogation Techniques

The interaction of light wave with surface plasmon wave can change the characteristics of the light in terms of amplitude, spectral distribution, phase, and polarization. Any change in the wave vector of surface plasmon will change any of these characteristics. Change in the dielectric constant or the refractive index of the dielectric medium resulting in the change in wave vector of the surface plasmon can be determined by knowing the changes in one of these characteristics. This is the principle of surface plasmon resonance (SPR)-based sensors. On the basis of change in light characteristics, the measurement techniques can be classified as angular, wavelength, intensity, and phase interrogations. In the following, we describe these interrogation methods in detail.

7.4.1 Angular Interrogation

For the excitation of surface plasmons, the resonance condition given by Equation 7.16 should be satisfied. For a prism of fixed refractive index and fixed wavelength of the light source, the resonance condition is satisfied at a particular angle of incidence called resonance angle (θ_{res}). At resonance, the transfer of energy of incident light to surface plasmons takes place resulting in the reduction in the intensity of the reflected light. In a sensing

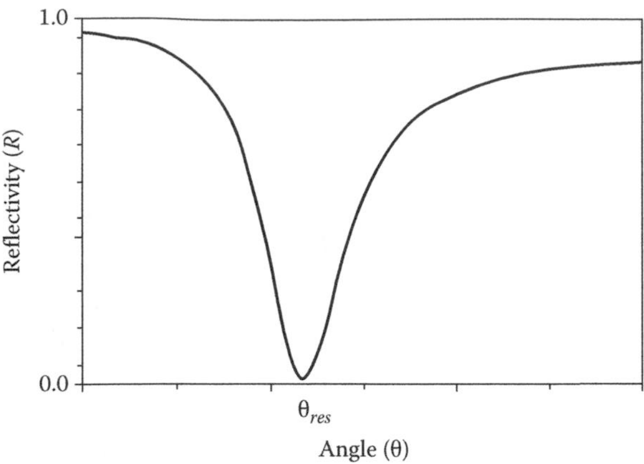

FIGURE 7.7 A typical SPR curve.

device based on the prism and angular interrogation method, the intensity of the reflected light is measured as a function of angle of incidence θ for fixed values of frequency, metal layer thickness, and dielectric constant of the sensing medium. At resonance angle, a sharp dip is observed due to an efficient transfer of energy to surface plasmons as shown in Figure 7.7. The plot can also be obtained with the help of Fresnel's equations for the three-layer system. For a given frequency of the light source and the dielectric constant of the metal film, one can determine the dielectric constant (ε_d) of the dielectric medium by using Equation 7.16 if the value of the resonance angle (θ_{res}) determined experimentally is substituted. The resonance angle is very sensitive to the variation in the refractive index of the dielectric medium. If the refractive index of the dielectric medium is changed, the resonance angle will change accordingly. A graph between the resonance angle and the refractive index of the dielectric medium serves as the calibration curve of the sensor. Any increase in the refractive index of the dielectric medium increases the resonance angle.

7.4.2 Wavelength Interrogation

In the aforementioned method, the wavelength of the excitation light is kept constant, while the angle of incidence is varied. If the angle of incidence is kept constant and the excitation light source is polychromatic, then Equation 7.16 will be satisfied at one particular wavelength of the light source where the maximum transfer of energy from the source to the surface plasmon wave will take place. This particular wavelength is called the resonance wavelength (λ_{res}) and the method is called wavelength interrogation. An SPR curve similar to Figure 7.7 is obtained in this method. Wavelength interrogation is mostly used in optical fiber–based SPR sensors where all the guided rays are launched and the light source used is polychromatic.

7.4.3 Intensity Interrogation

In this method, the wavelength of the light source and the angle of incidence of the light are kept constant. If the prism is used, then the intensity of the reflected light at a constant angle is measured as a function of the refractive index of the dielectric medium. Increase in refractive index increases the resonance angle, and hence the intensity of

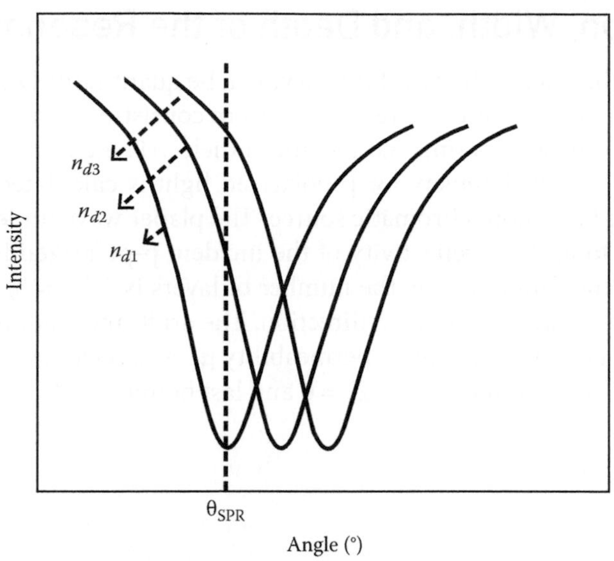

FIGURE 7.8 Intensity modulation. SPR curves for three different refractive indices of the dielectric.

the reflected light increases because of the shift in the SPR curve, as shown in Figure 7.8. Knowing the intensity, the refractive index of the dielectric medium can be determined. Sometimes, two light emitting diodes (LEDs) of different wavelengths are used and intensities are measured at these wavelengths. The knowledge of change in differential intensities is used to determine the change in dielectric constant of the medium.

7.4.4 Phase Interrogation

In this method, the phase difference between p-polarized and s-polarized light in the reflection spectrum for a monochromatic light source is measured as a function of angle of incidence. Around the resonance angle, the variation in intensity of the reflected light is small and hence the sensitivity is poor. If the difference in phase between p- and s-polarized light is plotted, then one can see from Figure 7.9 that the variation is very sharp around the resonance angle. Thus, the phase interrogation appears to be more sensitive.

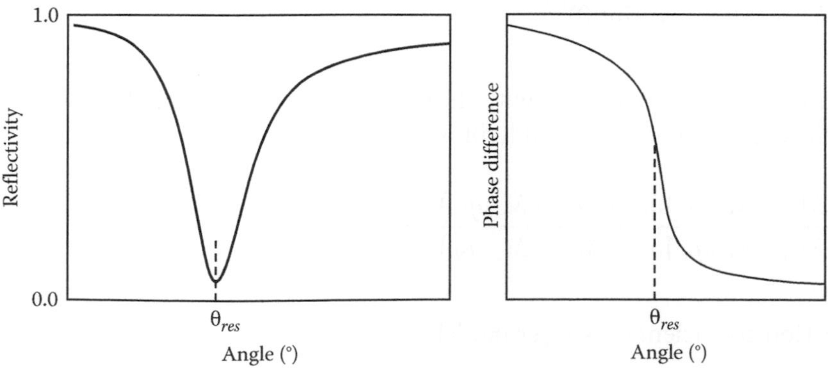

FIGURE 7.9 Phase interrogation. SPR curve based on angular interrogation and the corresponding plot of phase difference between p- and s-polarized light with angle of incidence.

7.5 Location, Width, and Depth of the Resonance Curve

The location, width, and depth of the SPR curve can be quantitatively described with the help of Fresnel's equations for the three-layer system consisting of high-refractive-index material prism (p), metal film (m), and low-index dielectric medium (d) such as air and water. The reflected light intensity for p-polarized light is calculated as a function of angle of incidence for a monochromatic source. The planer waveguide approach is generally considered to find the reflectivity of the incident p-polarized light (Sharma and Gupta 2005a). In the simplest case, the number of layers is 3 ($N = 3$) and all the layers are assumed to be stacked along the z-direction. The arbitrary medium layer is defined by thickness d_k, dielectric constant ε_k, permeability μ_k, and refractive index n_k. The tangential fields at the first boundary $Z = Z_1 = 0$ and last boundary $Z = Z_{N-1}$ are related as

$$\begin{bmatrix} U_1 \\ V_1 \end{bmatrix} = M \begin{bmatrix} U_{N-1} \\ V_{N-1} \end{bmatrix} \quad (7.19)$$

where

U_1 and V_1 are components of electric and magnetic fields, respectively, at the boundary of the first layer of structure

U_{N-1} and V_{N-1} are fields at the boundary of the $N-1$ layer

M is the characteristic matrix of this structure that is given by

$$M = \prod_{k=2}^{N-1} M_k \quad (7.20)$$

where

$$M_k = \begin{pmatrix} \cos\beta_k & -i\sin\beta_k/q_k \\ -iq_k\sin\beta_k & \cos\beta_k \end{pmatrix}$$

$$q_k = (\varepsilon_k - n^2\sin^2\theta)^{1/2}$$

$$\beta_k = d_k(2\pi/\lambda)(\varepsilon_k - n^2\sin^2\theta)^{1/2}$$

θ is the angle of incidence of light with the normal to the interface. The amplitude reflection coefficient (r_p) for p-polarized light is

$$r_p = \frac{(M_{11} + M_{12}q_N)q_1 - (M_{21} + M_{22}q_N)}{(M_{11} + M_{12}q_N)q_1 + (M_{21} + M_{22}q_N)} \quad (7.21)$$

The reflection coefficient of N layer model for p-polarized light is

$$R = |r_p|^2 \quad (7.22)$$

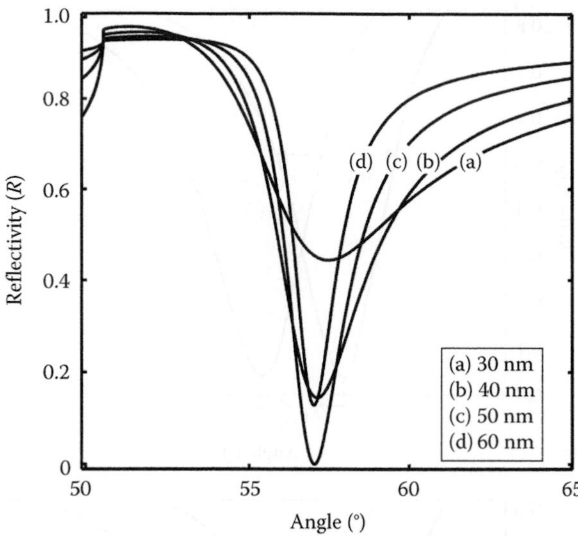

FIGURE 7.10 SPR spectra for different thicknesses of gold film over a prism-based sensor. The refractive index of the prism is 1.726, the sensing medium is water, and the wavelength is 600 nm.

To know the resonance angle, reflection coefficient R is plotted as a function of angle of incidence θ. The angle corresponding to minimum reflection coefficient (R_{min}), called resonance angle (θ_{res}), is determined from the plot. There is always a certain thickness of metal layer at a certain frequency for which R becomes zero. Further, the resonance point is different for different value of metal layer thickness, as shown in Figure 7.10.

7.6 Sensing Principle and Performance Parameters

Any optical sensor system requires a source, a transducer, and an electronic system for data processing. Sensors based on SPR technique consist of a source of optical radiation, an optical SPR configuration in which surface plasmon can be excited as a transducer and an interrogation system for analyzing and processing of the data. Here, the transducer transforms the changes in the refractive index into the changes in the quantity of interest that may be determined by the optical interrogation method of SPR. The optical system and the transducer determine the overall performance of the sensor in terms of sensitivity, detection accuracy or signal-to-noise ratio (SNR), stability, operating range, selectivity, and response time. Some of the detection approaches that have been commonly used in SPR sensor have been discussed earlier. The sensing principle of SPR sensors is based on Equation 7.16. For a given frequency of the light source and the metal film, one can determine the dielectric constant (ε_d) of the sensing layer adjacent to metal layer by knowing the value of the resonance angle. The dielectric constant (ε_d) of the sensing layer can also be determined if the angle of incidence is fixed and the resonance wavelength is determined. The resonance angle (or the wavelength) is very sensitive to the variation in the refractive index of the sensing layer. Increase in refractive index of the dielectric sensing layer increases the resonance angle or the wavelength.

The performance of an SPR sensor is analyzed with the help of two parameters: sensitivity and detection accuracy or SNR. For the best performance of the sensor, both the parameters

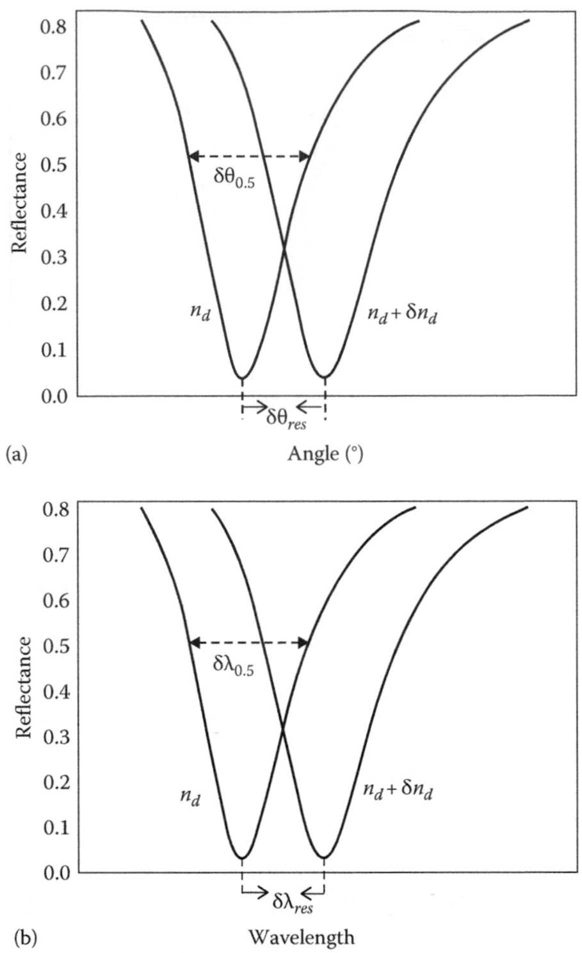

FIGURE 7.11 SPR curves for two different refractive indices of the sensing medium in the cases of (a) angular interrogation and (b) spectral interrogation for SPR system.

should be as high as possible to attain a perfect sensing procedure. Sensitivity of an SPR sensor depends on how much the resonance angle or the resonance wavelength shifts with a change in the refractive index of the sensing layer. If the shift is large, the sensitivity is large. Figure 7.11 shows a plot of reflectance as a function of angle of incidence of the light beam for sensing layer with refractive indices n_d and $n_d + \delta n_d$. Increase in the refractive index by δn_d shifts the resonance curve by $\delta\theta_{res}$ angle (Figure 7.11a) or $\delta\lambda_{res}$ wavelength (Figure 7.11b). The sensitivity of an SPR sensor with angular interrogation is defined as

$$S_n = \frac{\delta\theta_{res}}{\delta n_d} \tag{7.23}$$

The detection accuracy or the SNR of an SPR sensor depends on how accurately and precisely the sensor can detect the resonance angle and hence the refractive index of the sensing layer.

Apart from the resolution of the angle-measuring device, the accuracy of the detection of resonance angle depends on the width of the SPR curve. The narrower the SPR curve, the higher is the detection accuracy. Therefore, if $\delta\theta_{0.5}$ is the angular width of the SPR response curve corresponding to 50% reflectance, the detection accuracy of the sensor can be assumed to be inversely proportional to $\delta\theta_{0.5}$. A quantity that takes into account the sensitivity and the detection accuracy of the sensor is called the SNR of the SPR sensor and is defined as

$$SNR = \frac{\delta\theta_{res}}{\delta\theta_{0.5}} \tag{7.24}$$

The SNR should be high for the best performance of the SPR sensor. Another important parameter of an SPR sensor is its operating range. The operating range of intensity measurement-based SPR sensors is limited due to the limited width of the SPR dip, while the operating range of angular and wavelength interrogation-based SPR sensors may be made much longer. In principle, the operating range of these sensors is determined by the detection system, more specifically by the angular or spectral range covered by the optical system-angular position detector array or spectrum analyzer, respectively. In other words, it is the range of the sensing layer refractive index (n_d) that a sensor can detect for a given angular or wavelength range.

7.7 SPR as a Sensing Platform

We have discussed earlier how the SPR technique can be used for sensing the change in the refractive index of the dielectric medium around the metal layer. The technique is very versatile and accurate and has secured a very important place among the several sensing techniques due to its better performance and reliable procedure. In the following, we briefly discuss few chemical and biological SPR sensors based on prism, waveguide, and optical fiber.

7.7.1 Prism-Based SPR Sensors

In the simplest prism-based SPR sensor, the base of the glass prism is coated with a thin metal film (typically around 50 nm thick) and is kept in direct contact with the dielectric medium of the lower refractive index (such as air or some other dielectric sample). A p-polarized light beam is incident through one of the faces of the prism on the prism–metal interface at an angle greater than or equal to the critical angle. The intensity of the reflected light exiting from the other face of the prism is measured. For determining the refractive index of the dielectric, the angular interrogation method, discussed earlier, is used and the resonance angle is determined (Matsubara et al. 1988). Later on, a number of modifications were carried out in the probe to improve the performance of the sensor. Instead of using single metal layer, bimetallic layers were used (Zynio et al. 2002, Yuan et al. 2006). The purpose of using bimetallic layer (gold and silver) is to enhance the sensitivity and protect the silver layer kept in contact of the prism base from oxidation. Various configurations of films on the base of the prism have been used to improve the performance of the sensor. For example, the sensitivity of the sensor can be increased by an order of magnitude using a dielectric film of high value of the real part of the dielectric constant and thickness around 10–15 nm between the metal and the sensing dielectric medium (Lahav et al. 2008). In another configuration, the resolution of SPR curves is enhanced using a dielectric film between two metal layers (prism–Ag–dielectric–gold–surrounding medium). The inner silver layer over prism base couples incident light to a guided wave and makes more fields effectively concentrated on the outer gold layer. For ZnS–SiO_2 waveguide layer, a substantial enhancement in resolution is obtained (Lee et al. 2010). In another configuration, an alternating dielectric multilayer over the metal film in an SPR sensor is used (Lin et al. 2006). The thickness, number of layers, and other design parameters of the used material were optimized. The design resulted in higher sensitivity, better resolution, and wider dynamic range.

Prism-based SPR sensors have been reported for various applications. It can monitor photobleaching procedure with ultraviolet light (Feng et al. 2008). The probe configuration having an additional layer of a nonlinear polymer film over the metal film is reported to be very sensitive to the changes in the refractive index and nonlinear optical coefficient of the polymer. Hence, it can be used to determine the nonlinear coefficient. The SPR along with the optical waveguide mode coupling can also be used for the measurement of nanometric Ag_2O film thickness (Santillan et al. 2010). Recently, an SPR biosensor using electro-optically modulated ATR geometry has been reported (Wang et al. 2011). The variation of attenuated light power with applied voltage is utilized to determine the analyte concentration. Prism-based SPR sensors utilizing wavelength interrogation method have also been reported (Bolduc et al. 2009). Apart from these applications, prism-based SPR sensors have been reported for the detection of refractive index, various chemicals, and gases (Martinez-Lopez et al. 2011, Saiful Islam et al. 2011). A refractive index resolution of the order of 10^{-7} refractive index unit (RIU) can be achieved using these sensors. The use of high-quality bulk optics allows the development of SPR sensors with limited optical noise resulting in a high-resolution measurement possible. Generally, SPR measurements with conventional prism do not allow the detection in the infrared (IR) wavelength region, which may find application in environmental, medical, and security. Moreover, the SPR-based structures in IR have substantially different parameters of excitation and have advantages for sensing applications in terms of high probe depth (Nelson et al. 1999) and more accurate determination of SPR dip (Johansen et al. 2000). SPR sensing in the near IR can be carried out using silicon prism or chip (Patskovsky et al. 2004, Jha and Sharma 2009a). Apart from silicon, chalcogenide glasses are the potential candidate for designing SPR sensors to operate in the IR region owing to their transparent behavior in the near- to mid-IR region (Jha and Sharma 2011). The Al-based chalcogenide glass sensor has been found to have enhanced performance (Jha and Sharma 2009b). Prism-based SPR sensors utilizing phase interrogation for the detection of glucose and hazardous gases have also been reported (Chiang et al. 2006, Nooke et al. 2010). The phase interrogation improves the sensitivity by an order of magnitude compared to angular interrogation method.

7.7.2 Optical Waveguide-Based SPR Sensors

The process of exciting SPR using a waveguide is similar to that in the Kretschmann configuration utilizing prism. A light wave is guided by either a single- or multilayer (slab or channel) waveguide. When it enters the region with a thin metal overlayer, it evanescently penetrates through the metal layer as shown in Figure 7.5. When the surface plasmon mode and the guided mode of the waveguide are phase matched, the incident light excites the surface plasmon at the outer interface of the metal layer (Lambeck 1992). It has been reported that the sensitivity of the waveguide-coupled SPR sensor is approximately the same as the prism-coupled SPR sensor. The use of optical waveguides in exciting surface plasmons has some advantages such as the control of the optical path in the sensor system, ruggedness, and compactness due to the availability of the latest microfabrication techniques. In the waveguide-based SPR sensors, intensity modulation (Harris and Wilkinson 1995, Homola et al. 1997) and wavelength interrogation method (Lavers and Wilkinson 1994) have been used. In intensity modulation, a monochromatic light is used to excite the surface plasmon, and the intensity of the transmitted light is measured for the change in the refractive index of the medium around the metal layer. In some cases, two LEDs have been used. The use of the combination of two monochromatic light sources enhances the sensitivity. An immunoreaction sensor has been reported using two LEDs (Suzuki et al. 2005). The other options that have been used for the waveguide-based SPR sensors include single-mode waveguide (Matsushita et al. 2008) and the Otto configuration (Akowuah et al. 2009). Various possibilities for tuning the operating range of the sensor have also been explored, such as high-refractive-index overlayer (Ctyroky et al. 1997) and complex multilayer structures (Weiss et al. 1996a,b). However, all the approaches that introduce additional layers are found to yield less-sensitive SPR-sensing devices because of a relatively lower concentration of electromagnetic field in the analyte. Recently, detailed design principle and propagation characteristics of channel Bragg-plasmon-coupled waveguide (Srivastava et al. 2010) and highly accurate and sensitive SPR sensor based on channel photonic crystal waveguides (PCWs) (Srivastava et al. 2011a) have been reported. The sensitivity of the proposed design has been reported to be as high as 7500 nm/RIU.

7.7.3 Optical Fiber–Based SPR Sensors

Miniaturization of SPR probe is a key factor that can be achieved by the use of optical fiber in place of prism. There are many advantages of using optical fiber such as the small diameter of its core, which allows it to be used for a very small quantity of the sample, simplified optical design, and the capability of online monitoring and remote sensing. A schematic diagram of a fiber-optic SPR probe is shown in Figure 7.12. To fabricate

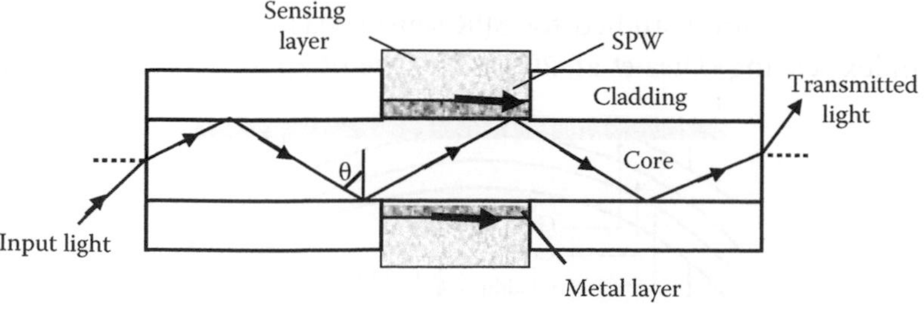

FIGURE 7.12 Illustration of a typical fiber-optic SPR sensor. (Reprinted from Sharma, A.K., Jha, R., and Gupta, B.D., Fiber-optic sensors based on surface plasmon resonance: A comprehensive review, *IEEE Sens. J.*, 7, 1118–1129, Copyright 2007 IEEE. With permission.)

the probe, cladding is removed from a small portion of the fiber and the unclad core is coated with a thin metal layer. The sensing medium is kept around the metallic layer. The concept of SPR in conjunction with optical fiber technology has been widely used for analyzing bimolecular interactions, interaction between low-molecular-weight molecules, conformational change of proteins, sensing of pesticides, and other biological entities. In all these cases, the local refractive index of the medium around the metal layer changes, and hence these changes can be detected by using the SPR phenomenon. The use of optical fibers for SPR sensing was first proposed by Jorgenson and Yee (1993). They used the wavelength interrogation technique and formed an SPR-sensing structure by using a conventional polymer-clad silica fiber with partly removed cladding and an SPR active metal layer deposited symmetrically around the exposed section of fiber core. The fiber-optic SPR sensor reported was capable of detecting variations in the refractive index within the operating range of 1.33–1.4 RIU. Further, the calibration curve was found to be nonlinear, giving an increase in the sensitivity with the increase in the refractive index of the sample.

Around the same time, an *in-line* optical fiber SPR sensor configuration was proposed for a large range of sensing applications (Alonso et al. 1993). In another significant application, a fiber-optic SPR sensor was used for the first time to monitor the deposition of a multilayered cadmium arachidate Langmuir–Blodgett film (Mar et al. 1993). Experiments showed constant shifts in resonance wavelength as the number of monolayers was increased. This provided the method of calculating the film thickness by measuring the changes in SPR spectra. Later, a fiber-optic SPR remote sensor for the detection of tetrachloroethane was proposed by using a gas-sensitive polymer film on the metal layer (Niggemann et al. 1995). The sensor showed good response time and reproducibility apart from a long-term stability. Around the same time, a fiber-optic SPR sensor with monochromatic excitation and angle of incidence was reported for the detection of refractive index (Ronot-Trioli et al. 1995). The sensor showed a resolution of the order of 10^{-4} RIU with a dynamic range between 1.33 and 1.40. Later, the modeling of sensing signal was reported by the same group (Trouillet et al. 1996). Homola and Slavik (1996) reported an SPR sensor using side-polished single-mode optical fiber and a thin metal overlayer as shown in Figure 7.13. The configuration of the probe is different from that shown in Figure 7.12. The study of self-assembled monolayers (SAMs) such as *n*-alkanethiol for the protection of silver film due to oxidation on fiber core has also been carried out (Abdelghani et al. 1996). Such thiol monolayers were found to be capable of protecting silver layer from oxidation.

The other configuration studied for SPR sensor was metal-coated tapered fiber for refractive index sensing (Diez et al. 2001). The use of zirconium acetate layer on silver

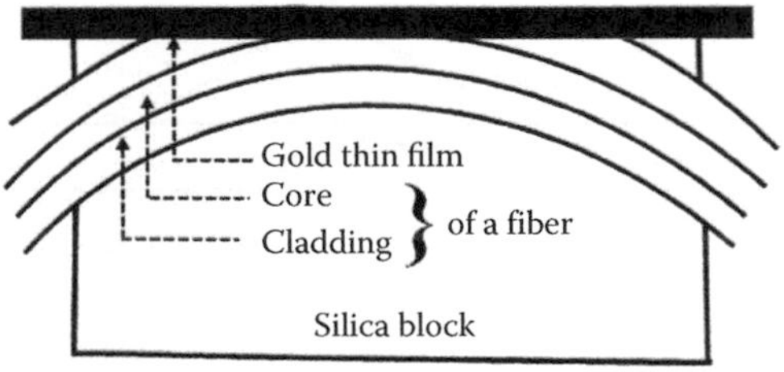

FIGURE 7.13 Schematic of an SPR sensor using side-polished single-mode optical fiber.

shifts the range of measurable refractive indices toward lower values. This has been used in a fiber-optic SPR sensor where silver, SAM, and zirconium acetate coatings were carried out for chemical and biological applications (Lin et al. 2001). Fiber-optic SPR sensor can also detect hydrogen (Bévenot et al. 2002). Measurement of concentrations, as low as 0.8%, of hydrogen (H_2) in pure nitrogen (N_2) has been reported. The response time varied between 3 s for pure H_2 and 300 s for the lowest H_2 concentration in N_2. Fiber-optic SPR sensor with asymmetric metal coating on a uniform-waist single-mode tapered fiber has also been tried (Monzón-Hernández et al. 2004). The variation of film thickness exhibits multiple resonance dips in the transmission spectrum. The multiple dips increase the dynamic range of the sensor. Apart from refractive index–based SPR sensors, absorption-based fiber-optic SPR sensor for the detection of concentration of chemicals has also been proposed (Sharma and Gupta 2004). Fast-responding fiber-optic SPR biosensor (Micheletto et al. 2005), analysis of a fiber-optic SPR sensor based on a crossing point of the two SPR spectra obtained from the sample fluid and the deionized water (Tsai et al. 2005), the application of single-crystal sapphire fiber-optic SPR sensor in the extreme environment (Kim et al. 2005a), and the use of tapered fiber-optic SPR sensor for vapor and liquid phase detection (Kim et al. 2005b) are some of the other advancements in this area. A detailed sensitivity analysis along with performance optimization for multilayered fiber-optic SPR concentration sensor has been reported (Gupta and Sharma 2005). The sensitivity and SNR analysis of a fiber-optic SPR refractive index sensor has been carried out for different conditions related to metal layer, optical fiber, and light launching conditions along with an extension to remote sensing (Sharma and Gupta 2005b). The presence of skew rays in the fiber reduces the sensitivity of the sensor, and hence the light launching is very important (Dwivedi et al. 2007). The performance of the probe can be improved if dopants are added in the fiber core (Sharma et al. 2007b, Jha and Badenes 2009). Many chemicals like naringin (Jha et al. 2006) and pesticide (Jha et al. 2007) and even temperature (Sharma and Gupta 2006c) can be sensed using fiber-optic SPR sensor with modification of the probe. However, the change in ambient temperature can influence the performance of the sensor (Sharma and Gupta 2006b).

A typical experimental setup used to study the response characteristics of an SPR-based fiber-optic sensor for the detection of a small percentage of water in ethanol is shown in Figure 7.14 (Srivastava et al. 2011b). The SPR curves were recorded for different percentages of water in ethanol, and from these curves, the resonance wavelengths were determined. The calibrated curve is shown in Figure 7.15. The SPR wavelength increases with a change in the percentage of water in ethanol due to the change in the refractive index of the liquid medium surrounding the metal film. The change increases up to 10% of water; after that, it starts decreasing due to the reasons reported in the study. The resonance wavelength can also be influenced by the presence of ions in the sample (Srivastava and Gupta 2011a). The shift in resonance wavelength is reported to be greater for ionic media than that for nonionic ones of the same refractive index. The shift in the resonance wavelength is proportional to the ion concentration and is attributed to the interaction of ions with the surface electrons of the metal film. Many attempts have been made to improve the sensitivity of the fiber-optic SPR sensors by changing the shape of the probe. For example, the tapering of the optical fiber enhances the sensitivity of the SPR sensor (Verma et al. 2007, 2008). However, the sensor becomes

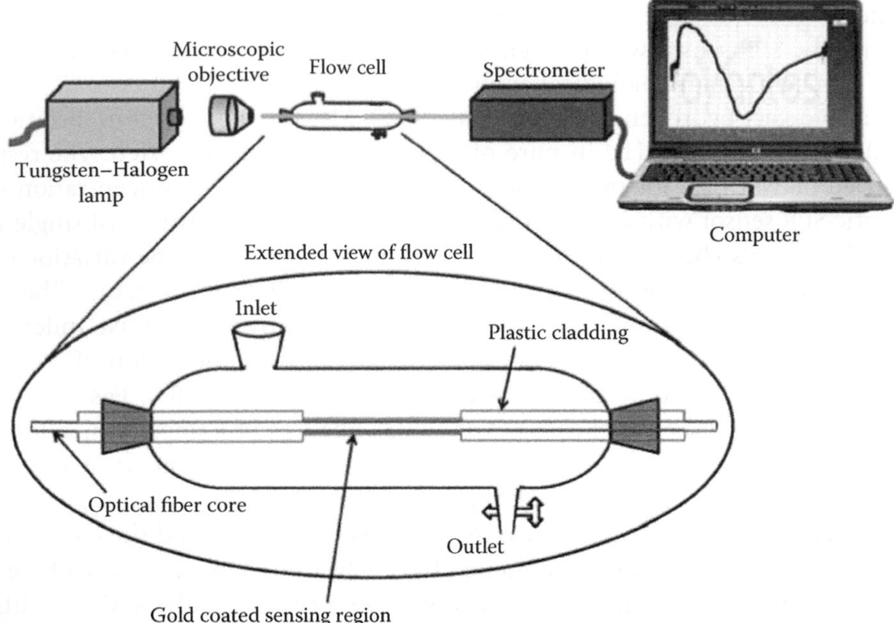

FIGURE 7.14 A typical experimental setup used for SPR-based fiber-optic sensor utilizing wavelength interrogation. (Reprinted from Srivastava, S.K., Verma, R., and Gupta, B.D., Surface plasmon resonance based fiber optic sensor for the detection of low water content in ethanol, *Sens. Actuat. B*, 153, 194–198, Copyright 2011, with permission from Elsevier.)

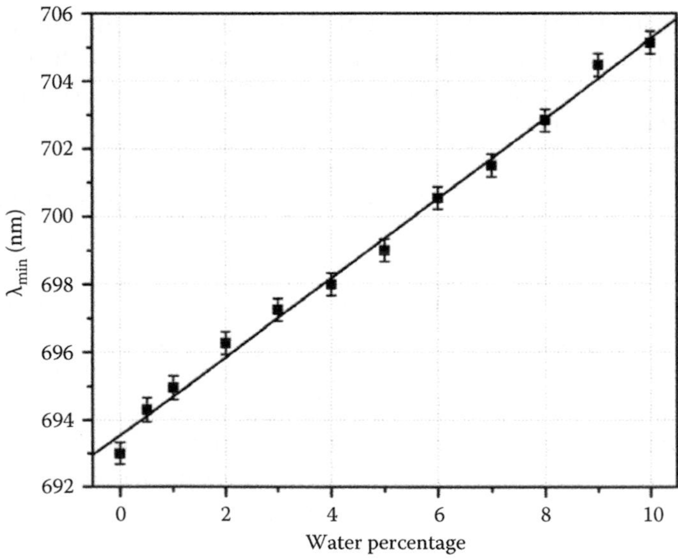

FIGURE 7.15 Variation of resonance wavelength with water percentage in ethanol. (Reprinted from Srivastava, S.K., Verma, R., and Gupta, B.D., Surface plasmon resonance based fiber optic sensor for the detection of low water content in ethanol, *Sens. Actuat. B*, 153, 194–198, Copyright 2011, with permission from Elsevier.)

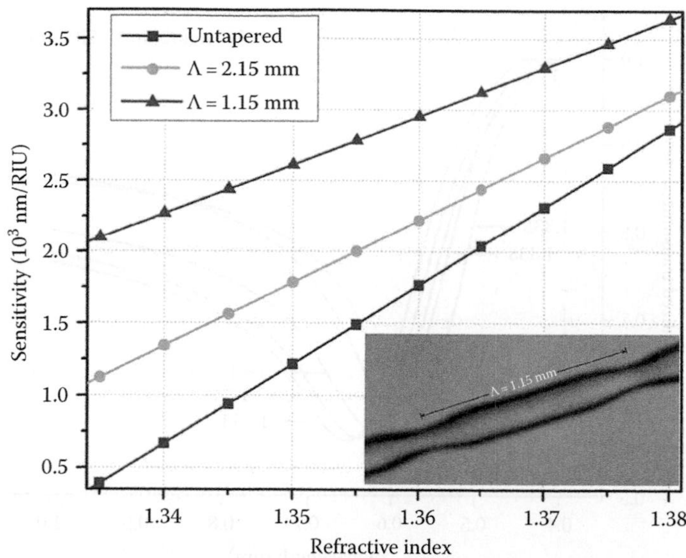

FIGURE 7.16 Variation of sensitivity with refractive index for different taper periods (Λ). Inset shows the microscopic image of a multitapered optical fiber with taper period of 1.15 mm. (Reprinted from Srivastava, S.K. and Gupta, B.D., A multitapered fiber-optic SPR sensor with enhanced sensitivity, *IEEE Photon. Technol. Lett.*, 23, 923–925, Copyright 2011 IEEE. With permission.)

very fragile with increase in the taper ratio. If instead of a single taper of large taper ratio a multitapered probe of small taper ratio is used, then the sensitivity can also be increased and the probe will be less fragile. Figure 7.16 shows the variation of sensitivity with refractive index as a function of taper period (Λ) for a multitapered fiber-optic SPR sensor that improves the sensitivity and makes the sensor less fragile (Srivastava and Gupta 2011b).

The sensitivity can also be increased by adding a high-index layer between the analyte and metal layer as mentioned in Section 7.7.1 for the prism-based SPR sensor. This has also been utilized in an optical fiber–based SPR sensor by adding a few nanometers thick film of silicon (Bhatia and Gupta 2011). The SPR and sensitivity curves are shown in Figures 7.17 and 7.18, respectively, for this kind of probe. The addition of 10 nm thick film of silicon doubles the sensitivity of the sensor.

Multimode fiber-optic SPR-sensing devices may suffer from rather low stability. The modal distribution of light in the fiber is very sensitive to mechanical disturbances, and the disturbances occurring close to the sensing area of the fiber may cause intermodal coupling and modal noise. Because of the cylindrical shape of the sensing area, fabrication of homogeneous SPR coatings and functionalization of the sensors' surface pose technological challenges. In order to overcome these drawbacks and to allow for further reduction of the sensing area, SPR sensors based on single-mode optical fiber have been proposed (Dessy and Bender 1994, Homola 1995). These SPR-sensing devices use a side-polished single-mode optical fiber with a thin metal layer supporting a surface plasma wave. The operating range of the single-mode fiber SPR sensors is very limited. The operating range of the sensor can be effectively tuned by a high-refractive-index dielectric overlayer; however, the presence of the dielectric overlayer may result in a drop in the sensitivity of the sensor. A major drawback of optical fiber SPR sensors

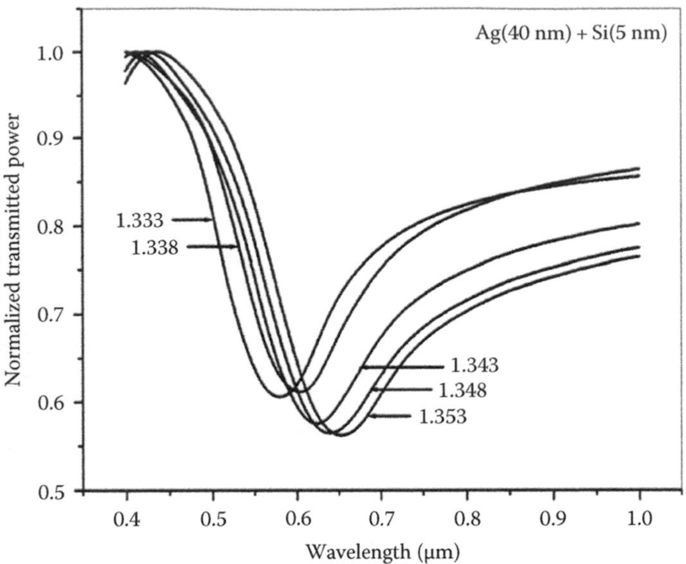

FIGURE 7.17 SPR spectra of fiber-optic SPR probe for a silicon layer of 5 nm thickness for refractive index of sensing region ranging from 1.333 to 1.353. (Reprinted from Bhatia, P. and Gupta, B.D., Surface-plasmon-resonance-based fiber-optic refractive index sensor: Sensitivity enhancement, *Appl. Opt.*, 50, 2032–2036, 2011. With permission of Optical Society of America.)

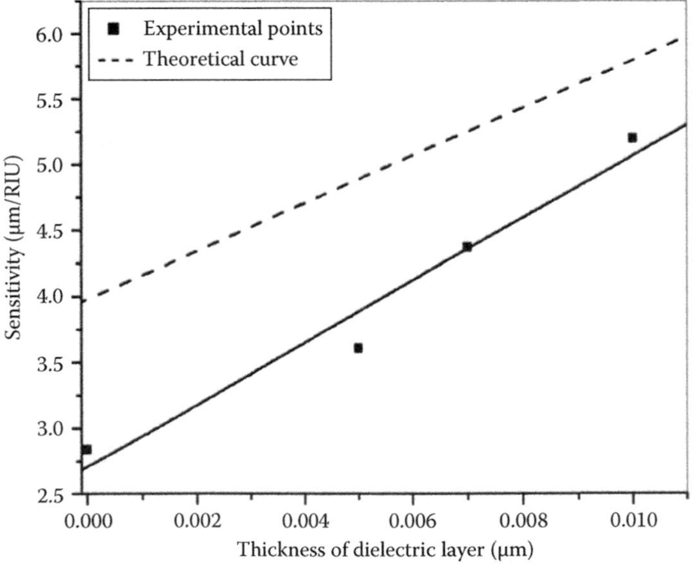

FIGURE 7.18 Variation of sensitivity of the sensor with the thickness of the silicon layer. (Reprinted from Bhatia, P. and Gupta, B.D., Surface-plasmon-resonance-based fiber-optic refractive index sensor: Sensitivity enhancement, *Appl. Opt.*, 50, 2032–2036, 2011. With permission of Optical Society of America.)

of this type is that they require reliable control of the polarization state of the optical wave propagating in the fiber (e.g., by using polarization-maintaining optical fibers). Among the most recent developments is the application of nanoparticle films in fiber-optic SPR sensors. The analyses of fiber-optic SPR sensor with metal-host nanoparticle layers (Sharma and Gupta 2005c) and silver–gold alloy layers (Sharma and Gupta 2006a) have been reported. Furthermore, the application of localized surface plasmon resonance (LSPR) in optical fiber sensors has been studied (Chau et al. 2006). The preceding sensor with long-period fiber grating has shown a detection limit for anti-DNP (dinitrophenyl compound) of 9.5×10^{-9} M.

7.8 SPR around Telecommunication Wavelength

Most of the sensors based on SPR have been fabricated in the visible region owing to practical applications, transparency window of the materials, and the easy availability of the sources and detectors. The surface plasmon–based model for the modes supported at the interface of a metal and periodically stratified medium (Bragg structure) has been reported around the telecommunication wavelength for various applications (Das and Jha 2009). A highly sensitive SPR sensor based on channel PCW has also been studied around the telecommunication window. The PCW is based on widely used lithographic and nanofabrication compatible materials like TiO_2 and SiO_2. By optimizing the different waveguide parameters, a significant amount of modal power can be transferred around phase matching or resonance wavelength that can be utilized to design a compact and highly sensitive sensor for lab on chip (Srivastava et al. 2011a). Recent experiments have demonstrated that multiple surface plasma waves can be excited in optical fibers at the 1.5 μm optical communications region for the range of refractive indices of aqueous media using doubly deposited tapered fiber structures, well known to exhibit small or zero sensitivity to polarization. Well-defined plasmon dips have been obtained with high sensitivity to the surrounding refractive index (Díaz-Herrera et al. 2010).

7.9 Terahertz Plasmonics and Its Applications

In the visible and ultraviolet range, the collective oscillation of free charge carriers at a metal–dielectric interface yields a surface plasmon wave propagating along the surface of the metal. The sensitivity of plasmon excitation to changes in the refractive index of the dielectric medium has been widely exploited for sensing applications. The plasma frequency imposes a lower frequency limit for the existence of these plasmons. At frequencies significantly below the plasma frequency such as the terahertz (THz) range, large negative permittivity strongly prohibits electromagnetic fields from penetrating inside a metal, and plasmon excitation on the metal–dielectric interface becomes challenging. Therefore, efficient plasmonic excitation at lower frequencies requires materials with lower plasma frequencies. Studies have recently confirmed that artificial subwavelength-sized metal structures can tune the plasma frequency of metals to the gigahertz (GHz) or THz frequency range and allow metals to support plasmon-like surface waves at frequencies much lower than the visible range, thereby opening a floodgate for sensing application using plasmonics (Van Zandt and Saxena 1989). Due to its nonionizing nature and the distinctive optical response of many molecules in the THz frequency range, interest has been emerged in using THz radiation for chemical and biological analysis (Maier et al. 2006). However, with THz wavelengths ranging from 100 μm to 1000 μm, the diffraction limit for this long wavelength light means that it is difficult to measure small-volume samples. It has been demonstrated that, it is possible to detect the presence of thin dielectric layers on the metal surface by allowing the THz surface wave to traverse a greater length of analyte thereby opening a new avenue for the development of a THz biosensor.

The ability of optical sensing systems to give rapid, direct information on key parameters facilitates a better understanding and will lead to the development of new biotechnological products. Perhaps the new THz technique will lead to new bioanalytical tools for bioprocess monitoring. It has tremendous potential for applications in chemical and biological identification (Zhang 2002). Pulsed terahertz spectroscopy (PTS) has also become a preferable method for performing low-frequency spectroscopy and imaging. PTS should enable the determination of the sequence of intermediate tertiary structures. Thus, it could become a very useful tool for bioprocess monitoring, since the correct folding of recombinant proteins is one of the most important factors in the pharmaceutical industry.

The significant advantage of working in THz regime is that the target sample required for reliable detection is reduced dramatically compared to other conventional approaches. This fact results in a much higher analytic sensitivity for the proposed device. THz technique leads to label-free, namely, no additional label molecules are needed to be attached to the target molecules for recognition. Labeling schemes not only impose additional cost- and time-consuming preparatory steps, but also they may introduce modifications in the target molecule structure, for example, DNA strand conformation and lowering the precision of the recognition. THz regime has been widely used for gaseous sensing. The results suggest that the laser THz emission system can detect a work function difference in the samples, and a fully noncontact, nondestructive evaluation of the hydrogen sensors can be realized (Kiwa et al. 2005).

7.10 Future of SPR-Based Sensors

The future of SPR devices, particularly biosensors, will be driven by the need of the consumers, and hence it is important that sensors should be made more consumer friendly. The area in SPR biosensor that really needs to be addressed presently and that throws a challenge to the researchers working in this field is its specificity, that is, how can sensors be used to detect specific molecules from a groups of molecules. The unwanted molecules will react with the biosensors and will alter the refractive index, hence affecting the different important parameters such as sensitivity, detection accuracy, and reproducibility of the biosensor. The other challenge lies in the nonspecific interactions between sensor surface with unwanted molecules and background refractive index variations. These variations can be because of temperature, humidity, and compositional fluctuations. The important issue, which eagerly waits the commercialization of SPR biosensor, is its use in out-of-laboratory environment. To use the biosensors in field, mobile analytical systems need to be developed, which may enable rapid detection of given biological entity. Future development of these systems requires significant advances in miniaturization of biosensing platform, high specificity, robustness, and user friendliness.

On a similar note, nanoparticle-based SPR sensors, also called LSPR, are also due to get attention in the future. The technique of LSPR with metal nanoparticle layers has shown a lot of promises. Further optimization of the crucial factors and parameters to improve the sensor's performance is required. The added phenomena like surface-enhanced Raman spectroscopy (SERS) are also there to have a look into. However, the future of the SPR technology will benefit from the use of optical waveguide technology compatible with the latest fabrication techniques that offer the potential for the development of miniaturized, compact, and rugged sensing elements with the prospect of fabrication of multiple sensors on one chip for various applications in the visible and around the telecommunication window.

Acknowledgments

The authors are thankful to Sachin Kumar Srivastava, Sarika Singh, Priya Bhatia, Roli Verma, and Satyendra Kumar Mishra for their help during the preparation of this chapter.

References

Abdelghani, A., J.M. Chovelon, J.M. Krafft et al. 1996. Study of self-assembled monolayers of n-alkanethiol on a surface plasmon resonance fibre optic sensor. *Thin Solid Films* 284–285: 157–161.

Akowuah, E.K., T. Gorman, and S. Haxha. 2009. Design and optimization of a novel surface plasmon resonance biosensor based on Otto configuration. *Optics Express* 17: 23511–23521.

Alonso, A., F. Villuendas, J. Tornos et al. 1993. New 'in-line' optical-fibre sensor based on surface plasmon excitation. *Sensors and Actuators A* 37: 187–192.

Berini, P. 2000. Plasmon-polariton waves guided by thin lossy metal films of finite width: Bound modes of symmetric structures. *Physical Review B* 61: 10484–10503.

Bévenot, X., A. Trouillet, C. Veillas et al. 2002. Surface plasmon resonance hydrogen sensor using an optical fibre. *Measurement Science and Technology* 13: 118–124.

Bhatia, P. and B.D. Gupta. 2011. Surface-plasmon-resonance-based fiber-optic refractive index sensor: Sensitivity enhancement. *Applied Optics* 50: 2032–2036.

Bolduc, O.R., L.S. Live, and J.F. Masson. 2009. High-resolution surface plasmon resonance sensors based on dove prism. *Talanta* 77: 1680–1687.

Chau, L.K., Y.F. Lin, S.F. Cheng et al. 2006. Fiber-optic chemical and biochemical probes based on localized surface plasmon resonance. *Sensors and Actuators B* 113: 100–105.

Chiang, H.P., J.L. Lin, and Z.W. Chen. 2006. High sensitivity surface plasmon resonance sensor based on phase interrogation at optimal incident wavelengths. *Applied Physics Letters* 88: 141105.

Ctyroky, J., J. Homola, and M. Skalsky. 1997. Tuning of spectral operation range of a waveguide surface plasmon resonance sensor. *Electronics Letters* 33: 1246–1248.

Das, R. and R. Jha. 2009. On the modal characteristic of surface plasmon polaritons at a metal Bragg interface at optical frequencies. *Applied Optics* 48: 4904–4908.

Dessy, R.E. and W.J. Bender. 1994. Feasibility of a chemical microsensor based on surface plasmon resonance on fiber optics modified by multilayer vapor deposition. *Analytical Chemistry* 66: 963–970.

Díaz-Herrera, N., A. González-Cano, D. Viegas et al. 2010. Refractive index sensing of aqueous media based on plasmonic resonance in tapered optical fibres operating in the 1.5 μm region. *Sensors and Actuators B* 146: 195–198.

Diez, A., M.V. Andres, and J.L. Cruz. 2001. In-line fiber-optic sensors based on the excitation of surface plasma modes in metal-coated tapered fibers. *Sensors and Actuators B* 73: 95–99.

Dwivedi, Y.S., A.K. Sharma, and B.D. Gupta. 2007. Influence of skew rays on the sensitivity and signal to noise ratio of a fiber-optic surface-plasmon-resonance sensor: A theoretical study. *Applied Optics* 46: 4563–4569.

Feng, W., L. Shenye, P. Xiaoshi et al. 2008. Reflective-type configuration for monitoring the photobleaching procedure based on surface plasmon resonance. *Journal of Optics A* 10: 095102.

Gupta, B.D. and A.K. Sharma. 2005. Sensitivity evaluation of a multi-layered surface plasmon resonance-based fiber optic sensor: A theoretical study. *Sensors and Actuators B* 107: 40–46.

Harris, R.D. and J.S. Wilkinson. 1995. Waveguide surface plasmon resonance sensors. *Sensors and Actuators B* 29: 261–267.

Homola, J. 1995. Optical fiber sensor based on surface plasmon excitation. *Sensors and Actuators B* 29: 401–405.

Homola, J., J. Ctyroky, M. Skalsky et al. 1997. A surface plasmon resonance based integrated optical sensor. *Sensors and Actuators B* 38–39: 286–290.

Homola, J. and R. Slavik.1996. Fibre-optic sensor based on surface plasmon resonance. *Electronics Letters* 32: 480–482.

Jha, R. and G. Badenes. 2009. Effect of fiber core dopant concentration on the performance of surface plasmon resonance based fiber optic sensor. *Sensors and Actuators A* 150: 212–217.

Jha, R. and A.K. Sharma. 2009a. SPR based infrared detection of aqueous and gaseous media with silicon substrate. *Europhysics Letters* 87: 10007.

Jha, R. and A.K. Sharma. 2009b. High performance sensor based on surface plasmon resonance with chalcogenide prism and aluminum for detection in infrared. *Optics Letters* 34: 749–751.

Jha, R. and A.K. Sharma. 2011. Design considerations for plasmonic-excitation based optical detection of liquid and gas media in infrared. *Sensors and Actuators A* 165: 271–275.

Jha, R., S. Chand, and B.D. Gupta. 2006. Fabrication and characterization of a surface plasmon resonance based fiber-optic sensor for bittering component-naringin. *Sensors and Actuators B* 115: 344–348.

Jha, R., S. Chand, and B.D. Gupta. 2007. Surface plasmon resonance based fiber-optic sensor for detection of pesticide. *Sensors and Actuators B* 123: 661–666.

Johansen, K., H. Arwin, I. Lundstrom et al. 2000. Imaging surface plasmon resonance sensor based on multiple wavelengths: Sensitivity considerations. *Review of Scientific Instruments* 71: 3530–3538.

Jorgenson, R.C. and S.S. Yee. 1993. A fiber-optic chemical sensor based on surface plasmon resonance. *Sensors and Actuators B* 12: 213–220.

Kim, Y.C., J.F. Masson, and K.S. Booksh. 2005a. Single-crystal sapphire-fiber optic sensors based on surface plasmon resonance spectroscopy for in situ monitoring. *Talanta* 67: 908–917.

Kim, Y.C., W. Peng, S. Banerji et al. 2005b. Tapered fiber optic surface plasmon resonance sensor for analyses of vapor and liquid phases. *Optics Letters* 30: 2218–2220.

Kiwa, T., K. Tsukada, M. Suzuki et al. 2005. Laser terahertz emission system to investigate hydrogen gas sensors. *Applied Physics Letters* 86: 261102.

Lahav, A., M. Auslender, and I. Abdulhalim. 2008. Sensitivity enhancement of guided wave surface plasmon resonance sensors. *Optics Letters* 33: 2539–2541.

Lambeck, P.V. 1992. Integrated opto-chemical sensors. *Sensors and Actuators B* 8: 103–116.

Lavers, C.R. and J.S. Wilkinson. 1994. A waveguide-coupled surface-plasmon sensor for an aqueous environment. *Sensors and Actuators B* 22: 75–81.

Lee, K.S., J.M. Son, D.Y. Jeong et al. 2010. Resolution enhancement in surface plasmon resonance sensor based on waveguide coupled mode by combining a bimetallic approach. *Sensors* 10: 11390–11399.

Lin, C.W., K.P. Chen, C.N. Hsiao et al. 2006. Design and fabrication of an alternating dielectric multi-layer device for surface plasmon resonance sensor. *Sensors and Actuators B* 113: 169–176.

Lin, W.B., M. Lacroix, J.M. Chovelon et al. 2001. Development of a fiber-optic sensor based on surface plasmon resonance on silver film for monitoring aqueous media. *Sensors and Actuators B* 75: 203–209.

Maier, S.A., S.R. Andrews, L. Martin-Moreno, and F.J. Garcia-Vidal. 2006. Terahertz surface plasmon-polariton propagation and focusing on periodically corrugated metal wires. *Physical Review Letters* 97: 176805.

Mar, M., R.C. Jorgenson, S. Letellier et al. 1993. In-situ characterization of multilayered Langmuir–Blodgett films using a surface plasmon resonance fiber optic sensor. *Proceedings of the 15th Annual Conference of IEEE Engineering Med. Biol. Soc.* San Diego, 28–31 October, pp. 1551–1552.

Martinez-Lopez, G., D. Luna-Moreno, D. Monzon-Hernandez et al. 2011. Optical method to differentiate tequilas based on angular modulation surface plasmon resonance. *Optics and Lasers in Engineering* 49: 675–679.

Matsubara, K., S. Kawata, and S. Minami. 1988. Optical chemical sensor based on surface plasmon measurement. *Applied Optics* 27: 1160–1163.

Matsushita, T., T. Nishikawa, H. Yamashita et al. 2008. Development of a new single-mode waveguide surface plasmon resonance sensor using a polymer imprint process for high-throughput fabrication and improved design flexibility. *Sensors and Actuators B* 129: 881–887.

Micheletto, R., K. Hamamoto, S. Kawai et al. 2005. Modeling and test of fiber-optics fast SPR sensor for biological investigation. *Sensors and Actuators A* 119: 283–290.

Monzón-Hernández, D., J. Villatoro, D. Talavera et al. 2004. Optical-fiber surface-plasmon resonance sensor with multiple resonance peaks. *Applied Optics* 43: 1216–1220.

Nelson, B.P., A.G. Frutos, J.M. Brockman et al. 1999. Near-infrared surface plasmon resonance measurements of ultrathin films. *Analytical Chemistry* 71: 3928–3934.

Niggemann, M., A. Katerkamp, M. Pellmann et al. 1995. Remote sensing of tetrachloroethene with a microfibre optical gas sensor based on surface plasmon resonance spectroscopy. *The Eighth International Conference on Solid-State Sensors and Actuators, and Eurosensors IX*, Stockholm, Sweden, Vol. 2, pp. 797–800.

Nooke, A., U. Beck, A. Hertwig et al. 2010. On the application of gold based SPR sensors for the detection of hazardous gases. *Sensors and Actuators B* 149: 194–198.

Patskovsky, S., A.V. Kabashin, M. Meunier et al. 2004. Near-infrared surface plasmon resonance sensing on a silicon platform. *Sensors and Actuators B* 97: 409–414.

Ronot-Trioli, C., A. Trouillet, C. Veillas et al. 1995. A monochromatic excitation of a surface plasmon resonance in an optical fibre refractive index sensor. *The Eighth International Conference on Solid-State Sensors and Actuators, and Eurosensors IX*, Stockholm, Sweden, Vol. 2, pp. 793–796.

Saiful Islam, M., A.Z. Kouzani, X.J. Dai et al. 2011. Investigation of the effects of design parameters on sensitivity of surface plasmon resonance biosensors. *Biomedical Signal Processing Control* 6: 147–156.

Santillan, J.M.J., L.B. Scaffardi, D.C. Schinca et al. 2010. Determination of nanometric Ag_2O film thickness by surface plasmon resonance and optical waveguide mode coupling techniques. *Journal of Optics* 12: 045002.

Sarid, D. 1981. Long-range surface-plasma waves on very thin metal-films. *Physical Review Letters* 47: 1927–1930.

Sharma, A.K. and B.D. Gupta. 2004. Absorption-based fiber optic surface plasmon resonance sensor: A theoretical evaluation. *Sensors and Actuators B* 100: 423–431.

Sharma, A.K. and B.D. Gupta. 2005a. Sensitivity evaluation of a multi-layered surface plasmon resonance-based fiber optic sensor: A theoretical study. *Sensors and Actuators B* 107: 40–46.

Sharma, A.K. and B.D. Gupta. 2005b. On the sensitivity and signal-to-noise ratio of a step-index fiber optic surface plasmon resonance sensor with bimetallic layers. *Optics Communications* 45: 159–169.

Sharma, A.K. and B.D. Gupta. 2005c. Fiber optic sensor based on surface plasmon resonance with nanoparticle films. *Photonics and Nanostructures: Fundamentals and Applications* 3: 30–37.

Sharma, A.K. and B.D. Gupta. 2006a. Fiber-optic sensor based on surface plasmon resonance with Ag-Au alloy nanoparticle films. *Nanotechnology* 17: 124–131.

Sharma, A.K. and B.D. Gupta. 2006b. Influence of temperature on the sensitivity and signal-to-noise ratio of a fiber optic surface plasmon resonance sensor. *Applied Optics* 45: 151–161.

Sharma, A.K. and B.D. Gupta. 2006c. Theoretical model of a fiber optic remote sensor based on surface plasmon resonance for temperature detection. *Optical Fiber Technology* 12: 87–100.

Sharma, A.K., R. Jha, and B.D. Gupta. 2007a. Fiber-optic sensors based on surface plasmon resonance: A comprehensive review. *IEEE Sensors Journal* 7: 1118–1129.

Sharma, A.K., Rajan, and B.D. Gupta. 2007b. Influence of different dopants on the performance of a fiber optic SPR sensor. *Optics Communications* 274: 320–326.

Srivastava, S.K. and B.D. Gupta. 2011a. Effect of ions on surface plasmon resonance based fiber optic sensor. *Sensors and Actuators B* 156: 559–562.

Srivastava, S.K. and B.D. Gupta. 2011b. A multitapered fiber-optic SPR sensor with enhanced sensitivity. *IEEE Photonics Technology Letters* 23: 923–925.

Srivastava, S.K., R. Verma, and B.D. Gupta. 2011b. Surface plasmon resonance based fiber optic sensor for the detection of low water content in ethanol. *Sensors and Actuators B* 153: 194–198.

Srivastava, T., R. Das, and R. Jha. 2010. Design considerations and propagation characteristics of channel Bragg-plasmon-coupled-waveguides. *Applied Physics Letters* 97: 213104.

Srivastava, T., R. Das, and R. Jha. 2011a. Highly accurate and sensitive surface plasmon resonance sensor based on channel photonic crystal waveguides. *Sensors and Actuators B* 157: 246–252.

Suzuki, A., J. Kondoh, Y. Matsui et al. 2005. Development of novel optical waveguide surface plasmon resonance (SPR) sensor with dual light emitting diodes. *Sensors and Actuators B* 106: 383–387.

Trouillet, A., C. Ronot-Trioli, C. Veillas, and H. Gagnaire. 1996. Chemical sensing by surface plasmon resonance in a multimode optical fibre. *Pure and Applied Optics* 5: 227–237.

Tsai, W., Y. Tsao, H. Lin, and B. Sheu. 2005. Cross-point analysis for a multimode fiber sensor based on surface plasmon resonance. *Optics Letters* 30: 2209–2211.

Van Zandt, L.L. and V.K. Saxena. 1989. Millimeter-microwave spectrum of DNA: Six predictions for spectroscopy. *Physical Review A* 39: 2672.

Verma, R.K., A.K. Sharma, and B.D. Gupta. 2007. Modeling of tapered fiber-optic surface plasmon resonance sensor with enhanced sensitivity. *IEEE Photonics Technology Letters* 19: 1786–1788.

Verma, R.K., A.K. Sharma, and B.D. Gupta. 2008. Surface plasmon resonance based tapered fiber optic sensor with different taper profiles. *Optics Communications* 281: 1486–1491.

Wang, T.J., C.C. Cheng, and S.C. Yang. 2011. Surface plasmon resonance biosensing by electro-optically modulated attenuated total reflection. *Applied Physics B* 103: 701–706.

Weiss, M.N., R. Srivastava, H. Groger et al. 1996a. A theoretical investigation of environmental monitoring using surface plasmon resonance waveguide sensors. *Sensors and Actuators A* 51: 211–217.

Weiss, M.N., R. Srivastava, H. Groger et al. 1996b. Experimental investigation of a surface plasmon-based integrated-optic humidity sensor. *Electronics Letters* 32: 842–843.

Yuan, X.C., B.H. Ong, Y.G. Tan et al. 2006. Sensitivity-stability-optimized surface plasmon resonance sensing with double metal layers. *Journal of Optics A* 8: 959–963.

Zhang, X.C. 2002. Terahertz wave imaging: Horizons and hurdles. *Physics in Medicine Biology* 47: 3667–3677.

Zynio, S.A., A.V. Samoylov, E.R. Surovtseva et al. 2002. Bimetallic layers increase sensitivity of affinity sensors based on surface plasmon resonance. *Sensors* 2: 62–70.

8. Adaptive Optics and Wavefront Sensing

Robert K. Tyson
University of North Carolina at Charlotte

8.1 Basic Principles

Adaptive optics is a technique whereby a combination of electro-optical technologies provides real-time wavefront control of a beam of light. The light can be incoming from a source far away, like a distant star or galaxy, or it can be outgoing light such as a laser used for free-space communications or a weapon. The reason that the wavefront must be controlled is because there is some unknown dynamic disturbance to the beam. Conventional adaptive optics (AO) systems consist of three principal components: (1) a wavefront sensor

Handbook of Optical Sensors. Edited by José Luís Santos and Faramarz Farahi © 2015 CRC Press/Taylor & Francis Group, LLC. ISBN: 9781439866856.

(WFS) to measure the disturbance, (2) a wavefront corrector such as a deformable mirror (DM) to provide compensation for the disturbance, and (3) a control system to decode the WFS information and provide proper control signals to the DM.

8.2 Conventional Inertial Adaptive Optics

Conventional inertial AO refers to the correction process where something is macroscopically moved, such as the DM. This leaves out the class of correction techniques such as nonlinear phase conjugation. A system operates typically with a linear controller; however, the control schemes can be quite complicated for the massively parallel operations that are needed.

As an example of the operation of an AO system, the astronomy application will be used. Referring to Figure 8.1, light from an object of interest has essentially a plane optical wavefront when it reaches the upper part of the atmosphere. For the approximately 40 km atmospheric path to the telescope (which takes about 1 ms), the wavefront is distorted by the various turbulent eddies along the path. If the telescope were to focus the light from this wavefront onto the detector, it would result in a blurred and speckled image. Most of the information would be lost. With AO, some of the light with the distorted wavefront is split from the main beam and sent to a WFS. Using one of many techniques, the WFS determines the shape of the wavefront or enough information about the wavefront to compensate for the distortion. An electronic control system converts the information about the wavefront to suitable drive signals for the DM which then takes on a shape opposite to the incoming wavefront. The opposite is the phase conjugate of the distortion to the extent that the DM can deform into the proper shape. The wavefront that was not split off for the WFS reflects from the DM with a restored plane wavefront, which is then focused onto the detector, either an imaging sensor or a spectrometer.

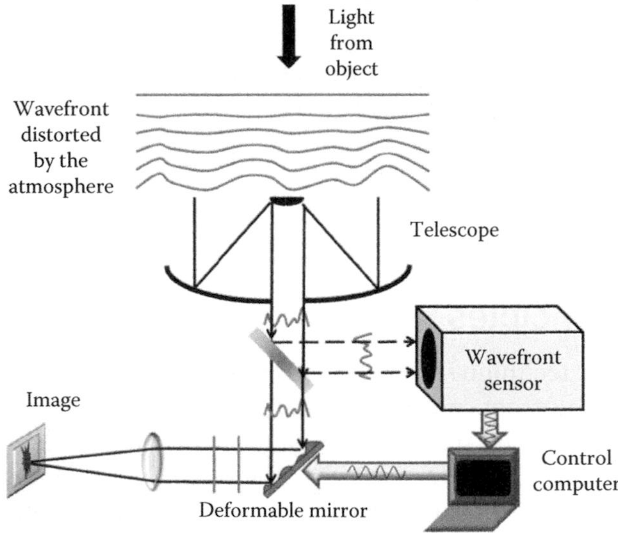

FIGURE 8.1 A conventional inertial AO system. Light from an object enters the telescope after being distorted by the atmosphere. A portion of the light is sent to a WFS. The sensor signals are relayed to a control computer that prepares drive signals for a DM. The DM counters the wavefront distortion, and the image is improved.

In most cases for astronomy, the light from the science object is not nearly bright enough to remove any light for the WFS. In that case, a nearby bright star can be used as a guide star, a source for the WFS. However, there can be a problem with that approach. The atmospheric turbulence for the light along the science path must be practically the same as the path from the guide star. But the turbulence is not isoplanatic, which would be the same in any direction up into the sky. It is anisoplanatic such that the turbulence differs in different directions. The limiting angular difference between the paths is known as the isoplanatic angle θ_0. It is calculated with the expression (Tyson 2011)

$$\theta_0 = \left[2.91 \left(\frac{2\pi}{\lambda} \right)^2 \int_{\text{path}}^{L} C_n^2(z) z^{5/3} dz \right]^{-3/5} \tag{8.1}$$

where λ is the wavelength of light and the integration is along the beam path to a distance L, where z is altitude and the structure constant C_n^2 is altitude dependent. If the science object and the guide star are separated by more than this angle, which is usually less than a few microradians, the wavefronts are too different for a useful measurement.

8.2.1 Laser Guide Stars

Anisoplanatic effects translate into restrictions on the field of view. For a pathlength L, the field of view is roughly $2L\theta_0$. A beacon's angular subtense should be as small as possible, ~1.22 λ/r_0 where r_0 is Fried's coherence length defined later by Equation 8.8.

Few natural visible guide stars fall within the isoplanatic angle of a science object. For infrared objects, the isoplanatic angle is larger and the system is less restricted. Since the early 1980s, researchers within the U.S. Air Force (Duffner 2009) and the astronomy community (Foy and Labeyrie 1985) have been investigating the use of artificial guide star techniques. The technique employs lasers to stimulate a portion of the upper atmosphere to create an artificial guide star for sampling the atmospheric turbulence. Figure 8.2 shows a configuration for an AO laser guide star system.

Guide stars can be created by Rayleigh scattering in the stratosphere around 20 km altitude. Since the atmosphere sampled and detected by the WFS is only the portion below the guide star in the cone between the guide star and the WFS aperture, there is the *cone effect* or focal anisoplanatism. The wavefront variance due to focal anisoplanatism was derived by Fried (1995):

$$\sigma_{\text{cone}}^2 = \left(\frac{D}{d_0} \right)^{5/3} \tag{8.2}$$

where D is the full telescope aperture diameter. The distance d_0 is found from Equation 8.3, where β is the zenith angle, z_{LGS} is the altitude of the laser guide star, and the C_n^2 profile is given.

$$d_0 = \lambda^{6/5} \cos^{3/5} \beta \left[19.77 \int \left(\frac{z}{z_{\text{LGS}}} \right)^{5/3} C_n^2(z) dz \right]^{-3/5} \tag{8.3}$$

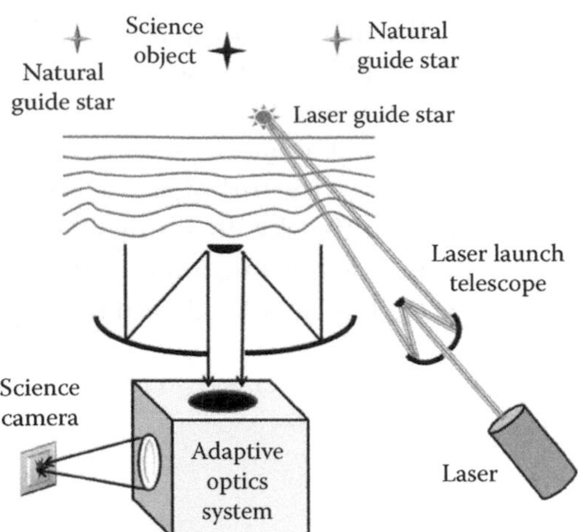

FIGURE 8.2 When the science object is not bright enough to discard any of the light for the WFS, other stars in the field of view can be used. When the natural guide stars are not within the isoplanatic angle, an artificial laser guide star can be generated by Rayleigh scattering in the lower atmosphere or by resonant sodium backscatter in the mesosphere.

Since the beacon cannot be placed directly in front of the object, angular anisoplanatism effects still must be considered. The guide star must be sufficiently bright to be detected and be within the anisoplanatic angle of the object.

8.2.2 Rayleigh Guide Stars

For a given laser energy, the brightness of the guide star is proportional to the density of the air within the laser volume. The detected Rayleigh photon flux is derived from the lidar equation (Tyson 2011):

$$F_{\text{Rayleigh}} = \eta T_A^2 \frac{\sigma_R n_R}{4\pi z_0^2} \frac{\Delta z \lambda_{LGS} E}{hc} \tag{8.4}$$

where

- η is the telescope and detector efficiency
- T_A is the one-way transmission of the atmosphere from the telescope to the guide star
- σ_R is the Rayleigh backscatter cross section
- n_R is the atmospheric density
- Δz is the thickness of the atmospheric layer that is observed
- λ_{LGS} is the wavelength of the laser guide star
- E is the laser energy per pulse
- z_0 is the height of the focused star above the telescope entrance
- h is Planck's constant (6.63×10^{-34} J s)
- c is the velocity of light (3.0×10^8 m/s)

The cross section-density product $\sigma_R n_R$ is a function of the altitude in the atmosphere where the guide star is formed and approximated by (Tyson 2011)

$$\sigma_R n_R \approx 2.0\times10^{-4}\exp\left[-\frac{(z_0+z_t)}{6}\right] \tag{8.5}$$

where z_t (in km) is the height of the laser transmitter above sea level. For Equation 8.5, a laser wavelength of 351 nm was assumed. The thickness of the layer, Δz, is derived from geometric considerations and the spreading of the laser energy up to the guide star altitude from a laser projector with aperture D_{proj}:

$$\Delta z = \frac{4.88\lambda z_0^2}{D_{proj} r_0} \tag{8.6}$$

For projector optics of 1 m, a site altitude of 3 km, and a guide star altitude of 20 km (above mean sea level), $\Delta z = 33$ m. Using the basic geometry shown in Figure 8.2, the flux can be determined.

There are numerous engineering concerns dealing with Rayleigh laser guide stars. There is Rayleigh backscatter from the surface up to the focus of the laser guide star. To eliminate the WFS from detecting the backscatter in the region below the guide star, range gating is employed (Fugate et al. 1991). The pulse should be short to stay within the Δz layer. About 66 μs after transmission, the pulse reaches the guide star altitude of 20 km. The return energy, the guide star pulse, reaches the WFS at 132 μs after transmission. Thus, to eliminate receiving energy continuously backscattered all along the path, an electronic range gate equal to the length of Δz, centered at 132 μs after each pulse, triggers the opening of the WFS detector. At 5 kHz, the next pulse is transmitted 200 μs after transmission and will not interfere or cause confusion from multiple pulses. If the guide star is pulsed at a slower rate, 300 Hz, for example, the pulse transmission and return in 132 μs is much quicker than the 3333 μs between pulses. It has been shown that varying the range gate depth during closed-loop operation can help to optimize performance (Morris and Wilson 2007).

Other engineering concerns include effects due to the direction of the laser guide star. Atmospheric tilt cannot be sampled from a single laser guide star (Rigaut and Gendron 1992). Because the propagation path up from the laser to the artificial star position follows the same reciprocal path back down to the WFS, the absolute position of the guide star, and therefore its relative position with respect to an object or fixed reference frame, is unknown.

With multiple guide stars, the effect of focal anisoplanatism can be reduced. However, because only one beacon can be on the optic axis, a *conic tilt error* is introduced along with a beacon position error due to turbulence on the upward propagation (Hardy 1998). The oblique path to the guide star and finite laser guide star size must be considered (Kibblewhite 1992). In many cases, the laser output aperture is not shared with the telescope receiving aperture, avoiding unwanted backscatter from the optics, possible fluorescence in the optics, and the complexity of high power optical coatings on the receiving optics. For off-axis laser guide stars, each one will have an oblique path and be elongated (Thomas et al. 2008). When an off-axis guide star is used, both angular anisoplanatism and focal anisoplanatism must be considered.

8.2.3 Sodium Guide Stars

Above 20 km, the atmosphere is too thin to provide enough backscatter to produce a sufficiently bright guide star. It was suggested by Will Happer that the use of resonant backscatter in the mesospheric sodium layer will produce a guide star near 92 km altitude well above most of the turbulent atmosphere. The flux of a resonant sodium guide star can be calculated in a manner similar to the Rayleigh guide star, recognizing that the density of sodium in the upper atmospheric layer and the resonant scattering limitations are considered (Papen et al. 1996). The atomic sodium, a result of meteoric disintegration, remains at an altitude between 89 and 92 km, because the conditions at this altitude prevent molecular reactions. The detected sodium photon flux, at a wavelength of 589.1583 nm, is given by (Happer et al. 1994)

$$F_{\text{Sodium}} = \eta T_A^2 \frac{\sigma_{\text{Na}} \rho_{\text{col}}}{4\pi z_0^2} \frac{\lambda_{\text{LGS}} E}{hc} \tag{8.7}$$

where the altitude z_0 is about 92 km, σ_{Na} is the resonant backscatter cross section, and ρ_{col} is the column abundance, usually between 3×10^9 atoms/cm^2 and 1×10^{10} atoms/cm^2. The product of the cross section and the column abundance is roughly 0.02 (Gavel et al. 1993). Because of the finite amount of sodium (less than 600 kg for the entire Earth), and the decay rate for the backscatter, a sodium guide star is limited to about 1.9×10^8 photons/s (Friedman et al. 1992).

It is not enough just to shine a sodium wavelength beam into the mesosphere. The sodium abundance (roughly 10^3–10^4 atoms/cm^3) represented by the column density can vary as much as 2½ magnitudes on a seasonal or diurnal scale or as short as 10 min (Davis et al. 2006). The effective height can vary between 89 and 92 km. The WFS must be focused correctly or a false defocus term results (Van Dam et al. 2006).

Actually, very few of the sodium atoms even in the beam will be excited. The return flux should be proportional to beam intensity; however, at the saturation intensity, all the atoms are excited and an increase in beam power will not increase flux. Continuous wave lasers often produce more return energy than pulsed lasers because they have a lower peak power.

For a sodium atom, the radiative lifetime is 16 ns (10 MHz). The scattering rate is inversely proportional to the radiative lifetime. Also, the atoms are moving because of thermal motion. This inhomogeneous Doppler broadening of 1.2 GHz enables the 589.2 nm beam to interact with only those in one *velocity group* or only about 1% of the sodium atoms. It is amazing that the sodium beacon will work.

8.3 Applications

The techniques of AO are important in a variety of applications, such as in astronomy, beam propagation systems, laser weapon systems, free-space laser communications, and retinal imaging (Sandler and Landis 1992), as detailed in the following sections.

8.3.1 Astronomy

Virtually every telescope larger than 4 m in diameter must employ AO. Failing to do so makes the telescope nothing more than large light bucket. It has spatial resolution no better than a telescope of a few centimeters in diameter. The atmosphere is a rapidly changing medium due primarily to mixing of air at various temperatures, which produces turbulent eddies throughout its 40 km thickness. By the 1960s, it was recognized that the size of the turbulent eddies was the limiting size for transmitting coherent light. Fried's *coherence length* (Fried 1966) or the *seeing cell size* can be calculated based upon the wavelength λ of the light and the atmospheric conditions as represented by the atmospheric refractive index structure constant, *C*. The effects are added along the path of the light and represented mathematically by the integral

$$r_0 = \left[0.423 \frac{(2\pi)^2}{\lambda^2} \sec\beta \int_{\text{path}}^{L} C_n^2(z)\,dz \right]^{-3/5} \tag{8.8}$$

The parameter β is the zenith angle (0° is directly overhead), and the integration is along the beam path where *z* is altitude and the structure constant is altitude-dependent.

Whereas the diffraction-limited resolution of the telescope is proportional to D/λ, the resolution in the presence of turbulence is D/r_0. For the optical (visible) band between 400 and 700 nm, the coherence length can be as small as a few centimeters, greatly limiting the utility of the telescope. Note Equation 8.8 shows that there is a 6/5 power wavelength dependence of r_0, which results in longer infrared bands being less impaired by the atmosphere. However, even increasing the coherence length to tens of centimeters still is an unacceptable limit for meter-class telescopes.

While the technology was being developed primarily for military high-energy laser propagation through the atmosphere, by 1991, the European Southern Observatory reported a functioning infrared AO system on the 3.6 m telescope at La Silla, Chile (Rigaut et al. 1991). As of this writing, dozens of telescopes around the world have operating AO systems, providing a wealth of new science obtained from ground-based telescopes. From the 0.7 m vacuum tower (solar) telescope to the twin 10 m Keck telescopes, new discoveries come with the aid of AO virtually every day, or night. Three gargantuan telescopes are currently being planned that incorporate complex AO in their design and operation. The Thirty Meter Telescope (TMT) in Hawaii, the 30 m Giant Magellan Telescope (GMT) in Chile, and the 42 m European Extremely Large Telescope (E-ELT) in Chile, all hope to be operational this decade.

AO systems on these astronomical telescopes have evolved from the single-conjugate systems using available light from the science object for one WFS and one DM optically conjugate to the strongest turbulence ground layer. For telescopes greater than 8 m, the systems are multiconjugate using light from multiple natural stars and additional light from artificial laser guide stars for multiple WFSs and DMs optically conjugated to various layers along the beam path (MAD 2011). These necessarily complex systems then exhibit very high spatial resolution in addition to large fields-of-view, just what is necessary for imaging and spectroscopy of distant early galaxies, discovery of extrasolar planets, and the search for dark matter and dark energy.

The concept of *multiconjugate adaptive optics* (MCAO) (Angel 1992) corrects for each *layer* of atmosphere independently. By placing correction devices (DMs) in series, and assigning each one a layer of atmosphere to correct, the isoplanatic patch is effectively increased. The terms MCAO, ground-layer adaptive optics (GLAO) (Travouillon et al. 2005), laser tomography (LTAO) (Gavel 2004), multiobject adaptive optics (MOAO), and extreme adaptive optics (XAO or ExAO) (Montagnier et al. 2007) are used to describe the various multiple guide star configurations, thus replacing what was once just an *AO system* with single conjugate adaptive optics (SCAO) (Babcock 1953, Tyson 2011).

In addition to conventional AO, many telescopes, like those used to search for extrasolar planets, have to use other novel optical techniques. For example, the light coming from an extrasolar planet is typically 7 or 8 orders-of-magnitude fainter than the parent star. The point spread function of the star often masks the planet. Long-exposure phase diversity (Mugnier et al. 2008), the use of coherent waveguides to remap pupil geometry (Kotani et al. 2009), or novel coronagraphs are often used to enhance the imaging. In one case, an optical vortex is imparted onto the phase of the incoming beam, effectively creating a zero-intensity hole in the center that can block the starlight (Foo et al. 2005).

8.3.2 Beam Propagation Systems

The same turbulent atmosphere that distorts an image by disturbing light passing down through the atmosphere will distort a beam passing up through the atmosphere. Even though propagation through the atmosphere (Yahel 1990) of weapons-class lasers for ballistic missile defense was the impetus for much of the early AO development, the technology and systems can be used for various peacetime applications. A system can be used for laser power beaming to recharge batteries in orbiting satellites (Neal et al. 1994), laser power beaming to supply energy for a lunar base (Landis), or ground-to-space laser communications (Tyson 1996, Wilson et al. 2003).

Often a dynamic disturbance or aberration occurs after propagation away from the optical system. In this instance, it is difficult to predict the disturbance and apply its phase conjugate without first sampling the aberration source. If a signal from the target, possibly a cooperative beacon, a laser guide star, or a reflection of sunlight, propagates back through the atmosphere, the AO system can sense the wavefront distortion of the beam returning from the target. By conjugating this field, remembering to scale for slightly different paths or wavelengths, the outgoing beam can be given a wavefront that allows radiation to reach the target near the diffraction limit.

8.3.3 Laser Weapon Systems

The early history of AO technology development was driven by the United States Department of Defense and its contractors who were building various laser weapon demonstrators and field experiments. The high-energy laser beams (tens of kilowatts to megawatts) were designed to propagate long distances through the atmosphere (tens to hundreds of kilometers) and still be lethal enough to engage and destroy enemy offensive weapons such as missiles. The beams encountered most of the same disturbances from the atmosphere as do astronomical telescopes. They also were powerful enough to have the atmosphere absorb significant amounts of radiation, which would heat up the local air and change its index of refraction, resulting in a phenomenon called thermal blooming. To overcome these difficulties, AO systems were developed. As the military missions changed throughout the 1980s and 1990s, the high energy lasers changed and the AO systems got more sophisticated.

The state-of-the-art for military high energy laser weapon systems with AO is the Airborne Laser Test Bed. This is a Boeing 747-400 freighter equipped with a megawatt-class Chemical Oxygen Iodide Laser (COIL) and a number of lower power lasers with AO in their beam trains that track, illuminate, and determine an aim point for the COIL. (Wolf 2010).

In addition to the airborne system tests, the U.S. Navy reported in April 2011 that a field test of the Maritime Laser Demonstrator was successful in destroying a small target boat by having the beam dwell on the engines of the target long enough for them to catch fire. The demonstrator is a ship-borne high energy solid state laser (reportedly 100 kW), which can operate continuously for as many as 35 firings.

8.3.4 Free-Space Laser Communications

There has been a recent growth in the need and use of AO for free-space optical communications. Using return-wave AO, compensation of the atmosphere by conjugating the phase of the outgoing transmitted beam can significantly improve performance. As the communication signal beam travels through the atmospheric turbulence, the phase changes result in scintillation, which cause fades and surges at the receiver (Andrews et al. 2001). The result of this is an increase in the bit-error rate (BER) that necessitates reducing the bandwidth. By compensating for the turbulence prior to transmission, much of the scintillation can be reduced (Tyson 2002). Various modulation schemes that have been studied include on-off-keying (OOK) (Tyson et al. 2005), pulse position modulation (PPM) (Wright et al. 2008), and optical vortex encoding (Gbur and Tyson 2008). Work on ground-to-space links (Wilson et al. 2003), space-to-ground links (Tyson 2002), air-to-space links (Nikulin 2005), air-to-ground links (Sergeyev et al. 2008), and ground-to-ground links (Mu et al. 2008) has been under investigation.

8.3.5 Retinal Imaging

Taking the conventional AO imaging technologies into the medical arena began in the late 1990s. The vitreous humor (fluid in the eyeball) is turbulent, much like the atmosphere. This turbulence prevented clear images of ganglion cells, retinal pigment epithelial cells (Gray et al. 2006), and photoreceptors (rods and cones) (Li and Roorda 2007). Laboratory systems have been built that use Shack-Hartmann WFSs and DMs to make clear images of these small features. In addition to the electro-optics needed for AO, keeping the eye steady and compensating for a dynamic tear film are engineering challenges met by a number of researchers. Results so far have advanced knowledge of perception, neural pathways, and color vision (Chui et al. 2008).

AO investigations have improved the use and manufacture of contact lenses and corneal surgery. The powerful imaging tools also have advanced the monitoring course of treatments for eye diseases and added to the potential for early diagnoses preventing blindness.

Many have built research and clinical systems using conventional Hartmann–Shack WFSs with low-power lasers acting as wavefront beacons reflecting off the retina (Prieto et al. 2000). Others have used different implementations such as a pyramid sensor (Chamot et al. 2006) or no sensor at all, but rather a simulated annealing algorithm applied to the images (Zommer et al. 2006). With multiple images, investigators have developed techniques to automatically identify cone photoreceptors (Li and Roorda 2007), or to evaluate changes in the lamina cribrosa, which is an early site of damage in glaucoma (Vilupuru et al. 2007).

In support of the imaging efforts, much work is being done to understand the scatter of light in the eye (Wanek et al. 2007) and the higher-order aberrations (Dalimier and Dainty 2008). The effect of compensating higher-order aberration is being investigated (Chen et al. 2006) in terms of visual acuity (Lundstrom et al. 2007).

8.4 Wavefront Sensors

An inertial AO system must have a way to sense the wavefront with enough spatial resolution and speed to apply a real-time correction (Geary 1995). The *direct* approach employs a step where there is an explicit determination of the phase or optical path difference (OPD) of the wave. After the wavefront is reconstructed, the information is used as feedback to correct for the unwanted components of the phase. The *indirect* approach never reduces the information to an explicit indication of wavefront; rather, it translates information related to the phase into signals that are used to compensate for the wavefront.

The requirements of wavefront sensing for AO differ from the requirements of phase or figure determination in optical testing in a number of ways (Hardy 1978). In optical testing, the phase can be measured and reconstructed very slowly (minutes, hours, even days). WFSs in an AO system are used for real-time operation. Since many disturbances can be on the order of hundreds of Hertz, the sensors must operate much faster. In some cases, the wavefront of individual pulses as short as 2 μs must be measured in real time (Rediker et al. 1989). The spatial resolution requirement of AO WFSs is usually very high. It is not uncommon to require resolution to 1/100 of an aperture diameter, which requires many channels of parallel sensing.

AO wavefront sensing requires a large dynamic range (many wavelengths) to account for the vast OPD over the pupil of interest.

The final, but fundamental, requirement of AO wavefront sensing is its ability to determine OPD independent of intensity. For many applications, imaging resolved objects through the atmosphere, for instance, the AO system will sense vast differences of intensity. Since the AO system is only normally capable of varying OPD over the aperture, the OPD must be determined without the confusing variation in the amplitude, manifested by the intensity nonuniformity. The use of broadband (white) light is required when the absolute OPD cannot be determined at a single wavelength.

8.4.1 Shack–Hartmann Wavefront Sensors

One method of testing a lens or mirror employs an opaque mask with holes placed behind the optical element under test. Each of the holes acts as an aperture, and since the light passing through the lens is converging, the image produced is an array of spots.

With proper calibration, the position of the spots is a direct indication of the local wavefront tilt at each hole and, thus, a description of the lens quality. This test is called the *Hartmann test* (Hartmann 1900).

Variations of this technique, especially for real-time wavefront measurement, are used in AO WFSs. Shack placed lenses in the holes, which increased the light-gathering efficiency of the mask and, with the spots focused, reduced the disturbing diffraction effects of the holes. A lens array for this purpose was first manufactured in 1971 (Shack and Platt 1971). Members of the astronomy community began to use this sensor in the late 1970s for testing of large telescope optics.

The Shack–Hartmann WFS is shown in Figure 8.3. The wavefront is divided by a mask, as in the classical test, an array of gratings, or an array of transmissive lenses. Each of the beams in the subapertures is focused onto a detector. To detect the position of the spot, various forms of modulation, detector geometry, and electro-optical processing are used. For atmospheric turbulence compensation, the local wavefront tilt must be measured accurately in each subaperture of size r_0. To do this, the subaperture must be large enough to resolve the isoplanatic patch. A Shack–Hartmann sensor is composed of an array of lenses for wavefront division and typically a charge-coupled device (CCD) array with multiple pixels used for spot position (wavefront tilt) determination. The physical construction of lens arrays is a major area for advanced engineering development. One must achieve high alignment accuracy of the array along with the required high optical quality.

If there is no expected channel-to-channel error, but only global misalignment, from bench vibration, for instance, one channel can be chosen as a reference and all measurements made in relation to the position of the spot in that channel.

One method of removing alignment errors is the introduction of a reference beam. This beam should be a plane wave. It does not have to be the same wavelength as the unknown wavefront beam, since no interference is needed or desired.

A number of other methods have been used to enhance specific aspects of Hartmann sensors. Optical binning (Basden et al. 2007) and optical amplification with an image intensifier (Witthoft 1990), or rotating the square quadcells with respect to square subapertures, eliminates the situation where zeros in the intensity pattern fall on the gaps between cells and introduce error. Each sensor must provide enough subapertures to

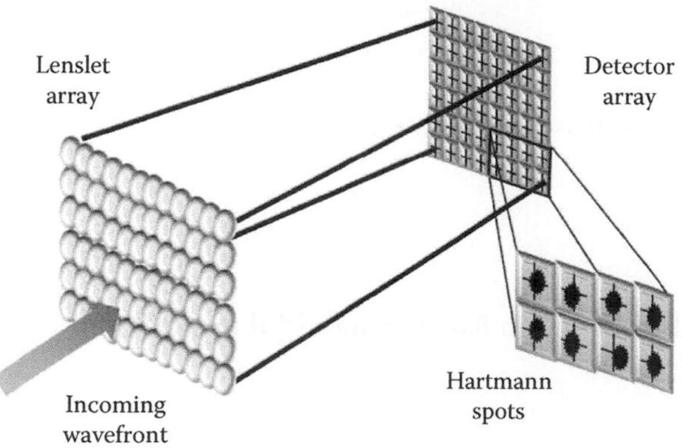

FIGURE 8.3 A Shack–Hartmann WFS uses a lenslet array to create a number of small beams that are focused onto a position-sensitive detector array. The positions of the spots are proportional to the local slope of the wavefront.

resolve the high-frequency aberrations. For turbulence compensation, the subapertures cannot be too small, since each subaperture must resolve the isoplanatic patch. If the subapertures are too large, however, they can possibly resolve the source of the wavefront. It has been shown that the subaperture size should be less than 0.16 λ/s, where s is the angular subtense of the object producing the unknown wavefront (Bikkannavar et al. 2008).

8.4.2 Curvature Wavefront Sensors

Roddier et al. have shown how focus sensing and the Hartmann subaperture division process can be combined (Roddier et al. 1988). The method, called *curvature sensing*, measures local wavefront curvature (the second derivative of the wavefront) in each subaperture (Erry et al. 2003). By comparing two near-simultaneous irradiance distributions at equally spaced points on either side of the subaperture focal plane, the local curvature is found (Fengjie et al. 2007). Two irradiance distributions $I_1(\vec{r})$ and $I_2(\vec{r})$ are detected a distance s from focus on either side of the focal plane. The relationship between the irradiance and the phase is given by

$$\frac{I_1(\vec{r})-I_2(-\vec{r})}{I_1(\vec{r})+I_2(-\vec{r})}=\frac{f(f-s)}{s}\left[\nabla^2\Phi\left(\frac{f}{s}\vec{r}\right)-\frac{\partial}{\partial\vec{n}}\Phi\left(\frac{f}{s}\vec{r}\right)\delta_c\right]=\frac{2f^2c_w}{s} \quad (8.9)$$

where

c_w is the local curvature at $\vec{r}$ expressed as $c_w = 1/r_w$
f is the focal length
r_w is the local radius of curvature

The Dirac delta δ_c represents the outward pointing normal derivatives on the edge of the signal. To reconstruct the wavefront from the known local curvatures and the edge derivatives, an iterative procedure can be followed (Roddier 1991) to solve Poisson's equation, or Zernike derivatives can be calculated (Han 1995). To use the approximations necessary to get the right side of Equation 8.9, the distance s should be

$$s \geq \frac{\lambda f^2}{\lambda f + r_0^2} \quad (8.10)$$

At optical wavelengths, this condition reduces to

$$s \geq \frac{\lambda f^2}{r_0^2} \quad (8.11)$$

where r_0 is defined in Equation 8.8. The standard deviation of a single curvature measurement is

$$\sigma_c = \frac{\lambda}{r_0^2 N_p^{1/2}} \quad (8.12)$$

where N_p is the photon count (Roddier 1988).

One appealing characteristic of this process is the possibility of applying a correction without the intermediate step of wavefront reconstruction. Since some correction devices, specifically membrane mirrors (Grosso and Yellin 1977) and bimorphs (Forbes 1989), can deform locally into near-spherical shapes, they can be driven directly by the output of a single subaperture that measures local wavefront curvature (Yang et al. 2006). When the mirror gain is considered and the WFS subapertures and corrector actuators are aligned, a very efficient process for wavefront control results (Graves 1996).

8.4.3 Pyramid Wavefront Sensors

Another pupil-plane WFS uses a pyramidal prism in the image plane to create four subbeams that are then optically relayed to a detector. The intensity at position $\vec{r}(x, y)$ in each of the subbeams in the detector plane ($I_{0,0} > I_{0,1} > I_{1,0} > I_{1,1}$) is used to find the x and y wavefront slopes at $\vec{r}(x, y)$:

$$S_x(\vec{r}) = \frac{I_{0,0}(\vec{r}) - I_{1,0}(\vec{r}) + I_{0,1}(\vec{r}) - I_{1,1}(\vec{r})}{I_t} \tag{8.13}$$

$$S_y(\vec{r}) = \frac{I_{0,0}(\vec{r}) + I_{1,0}(\vec{r}) - I_{0,1}(\vec{r}) - I_{1,1}(\vec{r})}{I_t} \tag{8.14}$$

I_t is the average intensity over the detector plane. One advantage of the pyramid technique over a Shack–Hartmann sensor is that the spatial resolution of the sensor is the size of the detector pixel in contrast to the larger lenslet subaperture size of the Shack–Hartmann (Verinaud et al. 2005).

8.4.4 Spinoffs: Laser Eye Surgery

Up until about a decade ago, defocus (myopia and hyperopia) and astigmatism were the only optical defects of the human eye that could be corrected. Using the WFS technologies spinoff from astronomy and beam propagation research, medical AO moved into the medical treatment arena rather than just as a diagnostic tool (Prieto et al. 2000).

By beaming a low power laser light into your eye and measuring the shape of the reflected light, a WFS can measure the lower order aberrations (nearsightedness, farsightedness, and astigmatism), as well as many different higher order aberrations, such as coma and spherical aberration (Miller et al. 1996). Then, with this detailed analysis of the ocular aberrations, laser ablation of the cornea by a method called laser-assisted in situ keratomileusis (LASIK) can be used to correct the defects.

8.5 Detectors Used in Wavefront Sensing

8.5.1 Commercial Off-the-Shelf Detector Arrays

Most WFSs such as the Shack–Hartmann WFS use detector arrays that are similar to the scientific image camera detectors. The characteristics and requirements of the detector arrays are quite similar. For wavefront sensing, one needs a detector with low read noise, low dark current and high quantum efficiency (for dim guide stars), high speed (to keep up with the changing atmosphere), a large field of view (to image hundreds or thousands of Hartmann spots over many pixels to get subpixel spot position measurement), and a large dynamic range (to avoid complicated automatic gain control as the wavefront source changes). There are a number of candidates for the wavefront sensing application. The most prominent in use today are interline CCD arrays, scientific complementary metal-oxide semiconductor (sCMOS) arrays, and electron-multiplying CCD (EMCCD) arrays. None of the devices is optimized for any one particular application of wavefront sensing; however, low-noise CCD arrays when cooled exhibit low dark current and EMCCD arrays have read noise of only a few electrons per pixel because the signal from each pixel is amplified before the readout electronics.

For very large arrays with many millions of pixels, the desired field of view can be achieved often with the limitations in speed. CCD arrays are serially read out and are frame-rate-limited to about 30 megapixels/s (30 Hz frame rate). Because of the well depth and desired readout rate, EMCCD arrays have dynamic range limited to 8000:1.

CMOS arrays, on the other hand, have internal circuitry for charge-to-voltage conversion and can be read out in parallel, greatly increasing the frame rate over CCDs. However, CMOS has inferior performance because of its higher dark current, read noise, and limited dynamic range. Recently, building upon the advances in field-programmable gate arrays (FPGA), sCMOS arrays have been shown to operate at 5500 megapixels/s (100 Hz frame rate) with a 30,000:1 dynamic range (Coates and Campillo 2011).

8.5.2 High-End Scientific Arrays

For many wavefront sensing applications, especially those needed for tracking, following, and imaging moving objects, the disturbances are moving much more rapidly than a static atmosphere. These are most apparent for imaging a low-Earth orbit object or a rapidly moving military target. For these systems, arrays with kilohertz frame rates or higher are needed. Most of these arrays are custom-made using cutting-edge technology. One example of the advanced technology is shown in the specification for the WFS detector array for the TMT (Adkins 2007). The WFS will operate with an array of laser guide stars that show as elongated spots on the detector array (the elongation is because the guide stars do not fall on the axis of the telescope and the WFS beam path). The prototype array is designed to sense one quadrant of the entire WFS focal plane. In that quadrant are 724 active subapertures with each one sensed by an array of 10 micron pixels with a configuration that varies because of the spot elongation from 6×6 to 6×15 pixels. The resultant WFS will have a 2 kHz readout rate, sufficient for the mission of the TMT.

8.6 Concluding Remarks

AO subsystems, the WFS, the DM, and the control computer, all have always been much like a three-legged stool. Each of the legs must be equal in length and sufficiently strong. Over the years, one of the legs often limited the overall system capability. For a period of time, the technology necessary to make the DMs limited the spatial extent of correction. Modern microfabrication technology has greatly reduced the limitations. At other times, the computing power limited the temporal response of the multichannel system. Moore's law applies here, and computing power is nearly sufficient for very high-speed operation. And for the longest time, the availability of high-speed, highly-sensitive detector arrays limited the available wavefront beacon sources. The current and future generations of visible and infrared arrays appear to be sufficient for most applications. Fortunately, we are now in an era where the subsystems do not necessarily limit the imaginations of the users and AO is fundamental to many practical applications.

References

Adkins, S. 2007. Next generation optical detectors for wavefront sensing. Polar coordinate detector final draft specification. Waimea, HI: W. M. Keck Observatory.

Andrews, L. C., R. L. Phillips, and C. Y. Hopen. 2001. *Laser Beam Scintillation with Applications.* Bellingham, WA: SPIE Press.

Angel, J. R. P. 1992. Use of natural stars with laser beacons for large telescope adaptive optics. *Proceedings of the Laser Guide Star Adaptive Optics Workshop*, U.S. Air Force Phillips Laboratory, Albuquerque, NM, Vol. 2, p. 494.

Babcock, H. W. 1953. The possibility of compensating astronomical seeing. *Publications of the Astronomical Society of the Pacific* 65: 229–235.

Basden, A., D. Geng, D. Guzman et al. 2007. Shack-Hartmann sensor improvement using optical binning. *Applied Optics* 46: 6136–6141.

Bikkannavar, S., C. Ohara, and M. Troy. 2008. Autonomous phase retrieval control for calibration of the Palomar adaptive optics system. *Proceedings of SPIE* 7015: 70155K.

Chamot, S. R., C. Dainty, and S. Esposito. 2006. Adaptive optics for ophthalmic applications using a pyramid wavefront sensor. *Optics Express* 14: 518–526.

Chen, L., P. B. Kruger, H. Hofer et al. 2006. Accommodation with higher- order monochromatic aberrations corrected with adaptive optics. *Journal Optical Society of America A* 23: 1–8.

Chui, T. Y. P., H. Song, and S. A. Burns. 2008. Adaptive-optics imaging of human cone photoreceptor distribution. *Journal Optical Society of America A* 25: 3021–3029.

Coates C. and C. Campillo. 2011. CCDs lose ground to new CMOS sensors. *Laser Focus World* March: 40–45.

Dalimier, E. and C. Dainty. 2008. Use of a customized vision model to analyze the effects of higher-order ocular aberrations and neural filtering on contrast threshold performance. *Journal Optical Society of America A* 25: 2078–2087.

Davis, D. S., P. Hickson, G. Herriot et al. 2006. Temporal variability of the telluric sodium layer. *Optics Letters* 31: 3369–3371.

Duffner, R. W. 2009. *The Adaptive Optics Revolution: A History.* Albuquerque, NM: University New Mexico Press.

Erry, G., P. Harrison, J. Burnett et al. 2003. Results of atmospheric compensation using a wavefront curvature based adaptive optics system. *Proceedings of SPIE* 4884: 245–250.

Fengjie, X., J. Zongfu, X. Xiaojun et al. 2007. High-diffractive-efficiency defocus grating for wavefront curvature sensing. *Journal Optical Society of America A* 24: 3444–3448.

Foo, G., D. M. Palacios, and G. A. Swartzlander. 2005. Optical vortex coronagraph. *Optics Letters* 30: 3308–3310.

Forbes, F. F. 1989. Bimorph PZT active mirror. *Proceedings of SPIE* 1114: 146–151.

Foy, R. and A. Labeyrie. 1985. Feasibility of adaptive optics telescope with laser probe. *Astronomic & Astrophysics* 152: L29–L31.

Fried, D. L. 1966. Limiting resolution looking down through the atmosphere. *Journal Optical Society of America* 56: 1380–1384.

Fried, D. L. 1995. Focus anisoplanatism in the limit of infinitely many artificial-guide-star reference spots. *Journal Optical Society of America A* 12: 939–949.

Friedman, H., J. Morris, and J. Horton. 1992. System design for a high power sodium laser beacon. *Proceedings Laser Guide Star Adaptive Optics Workshop*, U.S. Air Force Phillips Laboratory, Albuquerque, NM, Vol. 2, p. 639.

Fugate, R. Q., D. L. Fried, G. A. Ameer et al. 1991. Measurement of atmospheric wavefront distortion using scattered light from a laser guide star. *Nature* 353: 144–146.

Gavel, D. T. 2004. Tomography for multiconjugate adaptive optics systems using laser guide stars. *Proceedings of SPIE* 5490: 1356–1373.

Gavel, D. T., C. E. Max, K. Avicola et al. 1993. Design and early results of the sodium-layer laser guide star adaptive optics experiment at the Lawrence Livermore National Laboratory. *Proceedings ICO-16 Satellite Conference on Active and Adaptive Optics, ESO Conference and Workshop*, Munchen, Germany, Vol. 48, pp. 493–498.

Gbur, G. and R. K. Tyson. 2008. Vortex beam propagation through atmospheric turbulence and topological charge conservation. *Journal Optical Society of America A* 25: 225–230.

Geary, J. M. 1995. *Introduction to Wavefront Sensors*. Bellingham, WA: SPIE Optical Engineering Press.

Graves, J. E. 1996. Future directions of the University of Hawaii adaptive optics program. *Topical Meeting on Adaptive Optics*, Optical Society of America, Washington, DC, Digest Series 13, pp. 49–52.

Gray, D. C., W. Merigan, J. I. Wolfing et al. 2006. In vivo fluorescence imaging of primate retinal ganglion cells and retinal pigment epithelial cells. *Optics Express* 14: 7144–7158.

Grosso, R. P. and M. Yellin. 1977. The membrane mirror as an adaptive optical element. *Journal Optical Society of America* 67: 399–406.

Han, I. 1995. New method for estimating wave front from curvature signal by curve fitting. *Optical Engineering* 34: 1232–1237.

Happer, W., G. J. MacDonald, C. E. Max et al. 1994. Atmospheric-turbulence compensation by resonant optical backscattering from the sodium layer in the upper atmosphere. *Journal Optical Society of America A* 11: 263–276.

Hardy, J. W. 1978. Active optics: A new technology for the control of light. *Proceedings IEEE* 66: 651–697.

Hardy, J. W. 1998. *Adaptive Optics for Astronomical Telescopes*. New York: Oxford University Press.

Hartmann, J. 1900. Bemerkungen uber den Bau und die Justirung von Spektrographen. *Zeitschrift für Instrumentenkdunde* 20: 2–27.

Kibblewhite, E. 1992. Laser beacons for astronomy. *Proceedings Laser Guide Star Adaptive Optics Workshop*, U.S. Air Force Phillips Laboratory, Albuquerque, NM, Vol. 1, p. 24.

Kotani, T., S. Lacour, G. Perrin et al. 2009. Pupil remapping for high contrast astronomy: Results from an optical testbed. *Optics Express* 17: 1925–1934.

Li, K. Y. and A. Roorda. 2007. Automated identification of cone photoreceptors in adaptive optics retinal images. *Journal Optical Society of America A* 24: 1358–1363.

Lundstrom, L., S. Manzanera, P. M. Prieto et al. 2007. Effect of optical correction and remaining aberrations on peripheral resolution acuity in the human eye. *Optics Express* 15: 12654–12661.

Marchetti, E., N. Hubbin, E. Fedrigo et al. 2011. Mad the ESO multi-conjugate adaptive optics demonstrator. *Proceedings of SPIE* 4839: 317–328.

Miller, D. T., D. R. Williams, G. M. Morris et al. 1996. Images of cone photoreceptors in the living human eye. *Vision Research* 36: 1067–1079.

Montagnier, G., T. Fusco, J.-L. Beuzit et al. 2007. Pupil stabilization for SPHERE's extreme AO and high performance coronagraph system. *Optics Express* 15: 15293–13307.

Morris, T. J. and R. W. Wilson. 2007. Optimizing Rayleigh laser guide star range-gate depth during initial loop closing. *Optics Letters* 32: 2004–2006.

Mu, Q., Z. Cao, D. Li et al. 2008. Open-loop correction of horizontal turbulence: System design and result. *Applied Optics* 47: 4297–4301.

Mugnier, L. M., J-F. Sauvage, T. Fusco et al. 2008. On-line long-exposure phase diversity: A powerful tool for sensing quasi-static aberrations of extreme adaptive optics imaging systems. *Optics Express* 16: 18406–18416.

Neal, R. D., T. S. McKechnie, and D. R. Neal. 1994. System requirements for laser power beaming to geosynchronous satellites. *Proceedings of SPIE* 2121: 211–218.

Nikulin, V. V. 2005. Fusion of adaptive beam steering and optimization-based wavefront control for laser communications in atmosphere. *Optical Engineering* 44: 106001-1.

Papen, G. C., C. S. Gardner, and J. Yu. 1996. Characterization of the mesospheric sodium layer. *Topical Meeting on Adaptive Optics*, Optical Society America, Washington, DC, Technical Digest Series 13, p. 96.

Prieto, P. M., F. Vargas-Martin, S. Goelz et al. 2000. Analysis of the performance of the Hartmann-Shack sensor in the human eye. *Journal Optical Society of America A* 17: 1388–1398.

Rediker, R. H., B. G. Zollars, T. A. Lind et al. 1989. Measurement of the wave front of a pulsed dye laser using an integrated-optics sensor with 200-nsec temporal resolution. *Optics Letters* 14: 381–383.

Rigaut, F. and E. Gendron. 1992. Laser guide star in adaptive optics: The tilt determination problem. *Astronomy & Astrophysics* 261: 677–684.

Rigaut, F., G. Rousset, P. Kern et al. 1991. Adaptive optics on the 3.6-m telescope: Results and performance. *Astronomy & Astrophysics* 250: 280–290.
Roddier, F. 1988. Curvature sensing and compensation: A new concept in adaptive optics. *Applied Optics* 27: 1223–1225.
Roddier, F., C. Roddier, and N. Roddier. 1988. Curvature sensing: A new wave-front sensing method. *Proceedings of SPIE* 976: 203.
Roddier, N. 1991. Algorithms for wavefront reconstruction out of curvature sensing data. *Proceedings of SPIE* 1542: 120.
Sergeyev, A. V., P. Piatrou, and M. C. Roggemann. 2008. Bootstrap beacon creation for overcoming the effects of beacon anisoplanatism in a laser beam projection system. *Applied Optics* 47: 2399–2413.
Shack, R. B. and B. C. Platt. 1971. Production and use of a lenticular Hartmann screen. *Journal Optical Society of America* 61: 656.
Thomas, S. J., S. Adkins, D. Gavel et al. 2008. Study of optimal wavefront sensing with elongated laser guide stars. *Monthly Notices of the Royal Astronomical Society* 387: 173–187.
Travouillon, T., J. S. Lawrence, and L. Jolissaint. 2005. Ground layer adaptive optics performance in Antarctica. *Proceedings of SPIE* 5490: 934.
Tyson, R. K. 1996. Adaptive optics and ground-to-space laser communications. *Applied Optics* 35: 3640–3646.
Tyson, R. K. 2002. Bit error rate for free space adaptive optics laser communications. *Journal Optical Society of America A* 19: 753–758.
Tyson, R. K. 2011. *Principles of Adaptive Optics*, 3rd edn. Boca Raton, FL: CRC Press.
Tyson, R. K., J. S. Tharp, and D. E. Canning. 2005. Measurement of the bit error rate of an adaptive optics free space laser communications system. Part 2: Multichannel configuration, aberration characterization, and closed loop results. *Optics Engineering* 44: 096003.
Van Dam, M. A., A. H. Bouchez, D. Le Mignant et al. 2006. The W. M. Keck Observatory laser guide star adaptive optics system: Performance characterization. *The Publications of the Astronomical Society of the Pacific* 118: 310–318.
Verinaud, C., M. Le Louarn, V. Korkiakoski et al. 2005. Adaptive optics for high-contrast imaging: Pyramid sensor versus spatially filtered Shack-Hartmann sensor. *Monthly Notices of the Royal Astronomical Society* 357: L26–L30.
Vilupuru, A. S., N. V. Rangaswamy, L. J. Frishman et al. 2007. Adaptive optics scanning laser ophthalmoscopy for in vivo imaging of lamina cribrosa. *Journal Optical Society of America A* 24: 1417–1425.
Wanek, J. M., M. Mori, and M. Shahidi. 2007. Effect of aberrations and scatter on image resolution assessed by adaptive optics retinal section imaging. *Journal Optical Society of America A* 24: 1296–1304.
Wilson, K., M. Troy, M. Srinivasan et al. 2003. Daytime adaptive optics for deep space optical communications. *Proceedings 10th ISCOPS Conference*, Tokyo, Japan.
Witthoft, C. 1990. Wavefront sensor noise reduction and dynamic range expansion by means of optical image intensification. *Optical Engineering* 29: 1233–1241.
Wolf, J. 2010. US. successfully tests airborne laser on missile. www.reuters.com/article/2010/02/12/usa-arms-laser.
Wright, M. W., J. Roberts, W. Farr et al. 2008. Improved optical communications performance combining adaptive optics and pulse position modulation. *Optical Engineering* 47: 016003.
Yahel, R. Z. 1990. Turbulence effects on high energy laser beam propagation in the atmosphere. *Applied Optics* 29: 3088–3095.
Yang, Q., C. Ftaclas, and M. Chun 2006. Wavefront correction with high-order curvature adaptive optics systems. *Journal Optical Society of America A* 23: 1375–1381.
Zommer, S., E. N. Ribak, S. G. Lipson et al. 2006. Simulated annealing in ocular adaptive optics. *Optics Letters* 31: 939–941.

9. Multiphoton Microscopy

Kert Edward
University of the West Indies at Mona

9.1 Introduction and Background

Multiphoton microscopy (MPM) refers to a relatively new imaging technology whereby multiple photons are nearly simultaneously absorbed by a sample in the focal plane, resulting in the emission of a fluorescence signal. Since the probability of photon absorption is greatest in the excitation focal volume and negligibly small elsewhere, the procedure allows for inherent optical sectioning. This phenomenon was first predicted by Maria Göppert-Mayer in 1931 in her doctoral thesis (Göppert-Mayer 1931) but not verified until the 1960s by Franken et al. (1961) after the invention of the laser. This is due in part to the

Handbook of Optical Sensors. Edited by José Luís Santos and Faramarz Farahi © 2015 CRC Press/Taylor & Francis Group, LLC. ISBN: 9781439866856.

(a)

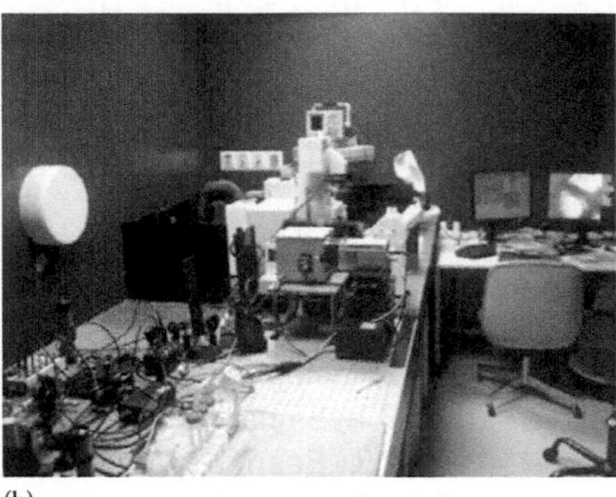

(b)

FIGURE 9.1 A modern laser scanning multiphoton microscope from Carl Zeis (LSM 7MP) is shown in (a) (from http://www.zeis.co.nz). The laser light source, viewing monitors, and computer control unit are not shown; (b) depicts a *home built* system that may be found in many research labs.

requirement of a very intense, temporally, and spatially confined coherent light source to maximize the likelihood of multiple photon absorption by a fluorophore (a molecule which fluoresces upon excitation). The most common example of a multiphoton system is the two-photon laser scanning fluorescence microscope that was first realized by Denk et al. (1990). Today, several commercial systems are available based on this initial prototype. In Figure 9.1, the LSM 7MP system from Zeiss is shown. In a typical system, a high power, high repetition rate (80–100 MHz) femtosecond laser is utilized in conjunction with a high numerical aperture (NA) objective, to achieve the required spatial and temporal photon confinement. This results in the near simultaneous absorption of two-photons in a subfemtoliter volume. Three-photon absorption is also possible; hence, the descriptive terminology of MPM is sometime used.

The technique has created a revolution in the biological sciences by facilitating unprecedented photochemical and morphological interrogation of samples at the cellular and subcellular level, at depths of up to 1 mm. In recent years, the technique has advanced to the point where the investigation of chemical interactions at the molecular level is routine (Xu and Webb 1996, Susana et al. 2005). In this chapter, the basic theory of MPM will be introduced, with an emphasis on two-photon microscopy. The requisite instrumentation and experimental setup will be reviewed in addition to a discussion of design considerations for optimal signal detection in thick samples (>600 nm). A separate section will be devoted to applications of MPM to the biological sciences and otherwise.

9.2 Laser Scanning Microscopy

Laser scanning microscopy (LSM) refers to a broad class of imaging instruments whereby images are generated by scanning a focused laser beam over a sample. Multiphoton microscopes are a subset of this group. In conventional optical microscopy, images are acquired by full-field irradiation of specific regions of the sample of interest. This is in direct contrast to the LSM, whereby two-dimensional raster scanning is typically utilized. Scanning is usually facilitated via two galvanometer scanning mirrors (galvos) that reflect incident light such that the region of interest on the sample is illuminated by a repeating pattern of line scans, which progress from top to bottom in each frame (raster scanning). The incident beam from a laser source is raster scanned by the galvos across the back aperture of an imaging objective lens. This results in a focused raster scanned beam at the focal plane of the objective. The motion of the galvos is computer controlled and corrected for the nonlinear angular displacement during each line scan (Callamaras and Parker 1999). Optimal positioning of the laser beam relative to the galvo mirrors, and objective is essential to avoid distorted images. Since only slight displacements of the mirrors are required, scan rates of several Hertz can be readily achieved. Three-dimensional imaging is possible by moving the focal plane into the sample while at the end of each 2D scan. Mechanical inertia of the galvos limits the imaging rate to a few frames per second. However, other modalities have been proposed for video frame rate imaging (30 frames or more per second), which include the spinning disk laser scanning confocal microscope (Graf et al. 2005) and programmable array microscope (PAM) (Vliet 1998). Endoscopic LSM systems rely on either MEMs to scan the laser beam onto a micro-objective (Yang et al. 2005), or miniature quartz tuning forks to dither a fiber probe relative to the sample (Polglase et al. 2005).

9.3 Fluorescence Microscopy

Most LSMs including multiphoton microscopes are based on the principle of fluorescence imaging. In this technique, a fluorophore (also interchangeably referred to as a fluorochrome) is excited by a suitable light source that may include a laser or laser diode, xenon arc lamp, or mercury-vapor lamp. In the case of the later two, excitation filters are required for source wavelength selection. This results in the transition of electrons from the ground state to an unstable excited state in the irradiated molecules. Electrons in the unstable state decay nonradiatively to an intermediate state before radiative decay to the ground state. An energy diagram summarizing this process is shown in Figure 9.2a. The emitted fluorescence signal is recorded by a photodetector.

This signal emission is always Stoke shifted, that is, emission photon has a lower energy than the excitation photon. Hence, the emission signal can be discriminated from the unwanted excitation signal by placing an appropriate emission filter in the detection path. Fluorophores can be endogenous biochemicals such as the metabolic coenzymes FAD (flavin adenine dinucleotide) and NAD+ (nicotine adenine dinucleotide), or commercial exogenous dyes such as DAPI. When native (endogenous) biochemicals are exploited for fluorescence imaging, this is referred to as autofluorescence imaging. A list of some common fluorophores is shown in Table 9.1. Judicious selection of two or more fluorescence dyes can allow for several components of a cell such as the nucleus and cell membrane to be selectively highlighted when excited at specific

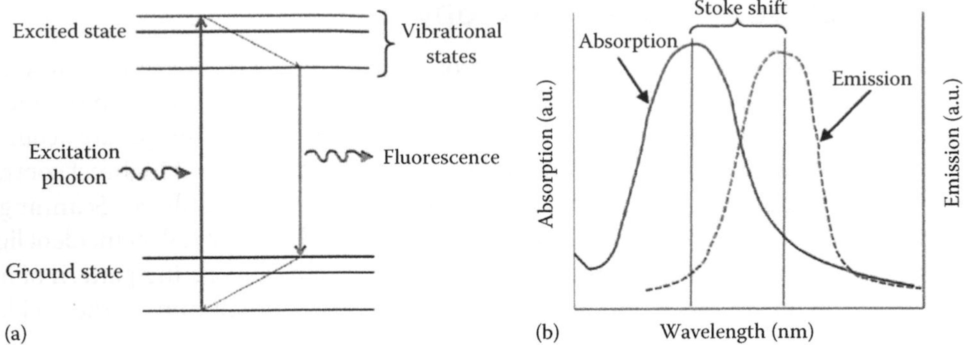

FIGURE 9.2 Diagram (a) is a Jablonski diagram of one-photon absorption, resulting in the emission of a fluorescence signal. The relationship between the absorption and emission spectra of a typical fluorophore is shown in (b). The emission spectrum is always red-shifted relative to the absorption curve.

Table 9.1 A List of Some Commonly Occurring Fluorophores in Biological Tissue Is Shown in the Table on the Left, While a List of Popular Exogenous Commercial Fluorophores Is Shown in the Table on the Right

Fluorophore	Absorption Wavelength	Emission Wavelength	Visible Color
Endogenous fluorophores			
Pyridoxamine	355	400	Blue
Elastin	290/325	340/400	Blue
Collagen	325/360	400/405	Blue
FAD	450	535	Green
NADH	290/351	440/460	Blue
Phospholipids	436	540/560	Yellow
Porphyrins	400–450	630/690	Red
Endogenous dyes			
Fluorescein FITC	495	518	Green (light)
Rhodamine Red X	560	580	Orange
Alexa Fluor 568	578	603	Red
Alexa Fluor 680	679	702	Red
DAPI	345	455	Blue
Hoechst 33258	345	478	Blue
Propidium Iodide	536	617	Red
TRITC	547	572	Yellow
CY3	550	570	Yellow
CY5	650	670	Red
CY5.5	675	694	Red
CY7	743	770	Red

wavelengths and with suitable filters. Postimaging, pseudocolors may be assigned to individual signals to create visually striking images.

9.4 Confocal Fluorescence Microscopy versus Two-Photon Microscopy

9.4.1 General

In fluorescence microscopy, a signal is elicited not only from the focal plane of the imaging objective, but also from an excitation volume, which extends from above to below this plane (see Figure 9.4b). Thus, it is impossible to avoid capturing extraneous out of focus signals, which results in blurry low resolution images. This problem is addressed by the confocal microscope by sacrificing signal intensity for spatial localization (Denk et al. 1990). As shown in Figure 9.3, a pinhole is placed in front of the detector, which acts as a spatial filter for out-of-focus light. This results in a significant reduction in image intensity, but the quantum efficiency of most exogenous fluorophores is such that this compromise is usually acceptable. Three-dimensional imaging is possible by moving the objective, and consequently the focal plane, into the sample after each 2D raster scan. In turbid media such as biological tissue, imaging depths of up to 100 μm is typical (Pawley 1995).

Since numerous fluorophores used in the biological sciences are excited in the UV to blue region of the electromagnetic spectrum, the risk of photodamage is high (Matsumoto 2003). Furthermore, the extended excitation volume of microscopy systems means that the damage is not restricted to the focal plane. Another disadvantage is that blue, shorter wavelength, light is highly scattered in biological tissue, which is detrimental to imaging. Many of these shortcomings are circumvented by the two-photon microscope.

In two-photon microscopy, favorable conditions for the otherwise rare event of near-simultaneous two photon absorption are created. A Jablonski energy diagram for the process is shown in Figure 9.4A. The idea of spatially localized excitation is exemplified

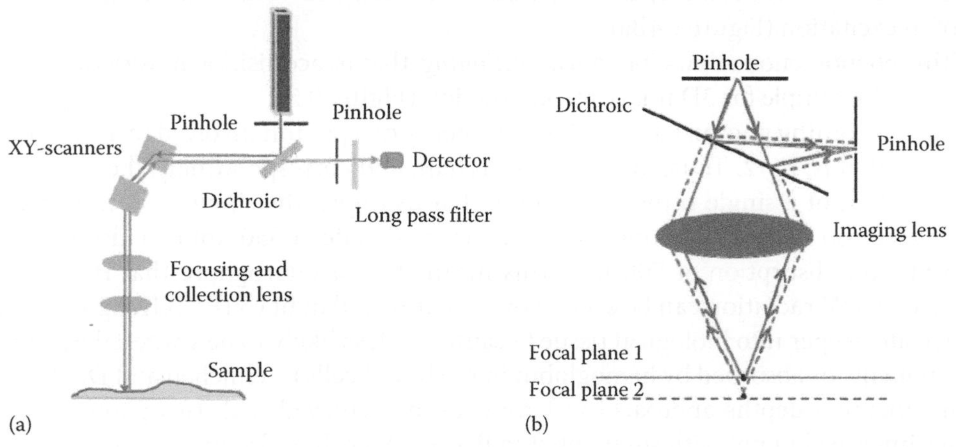

FIGURE 9.3 Schematic of experimental setup for a confocal microscopy system. Out-of-focus light is excluded by a pinhole position in front of a photodetector.

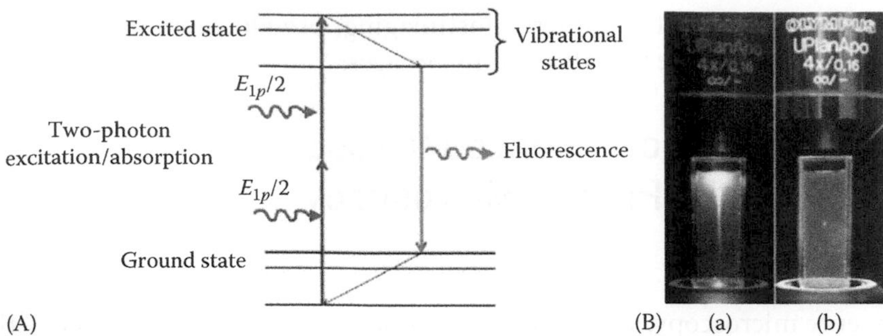

FIGURE 9.4 Jablonski diagram for two-photon absorption is shown in (A). The left and right images in (B) indicate the elicited signal from a fluorescence medium for one-photon (a) and two-photon (b) excitation.

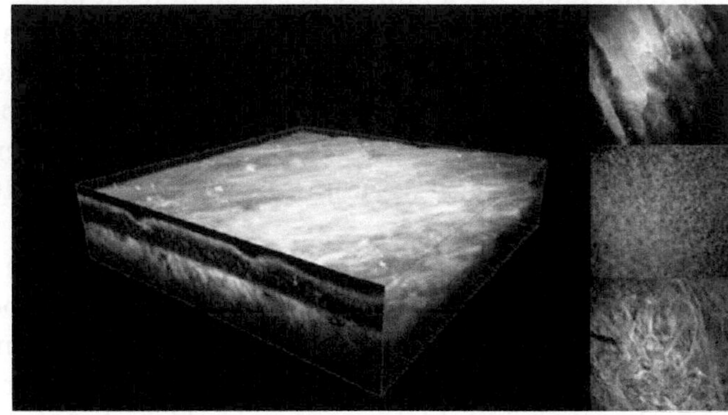

FIGURE 9.5 A 3D micrograph of in vivo normal mucosal tissue is shown. The three images on the right are 2D scans of the different layers in the tissue with increasing depth from top to bottom. Each slice in the image is 320 × 320 μm, and the imaging depth is approximately 150 μm.

in Figure 9.4B. For two-photon excitation (Figure 9.4Bb), fluorescence is elicited from a very narrow focal volume, as compared to an extended volume for confocal or one photon excitation (Figure 9.4Ba).

This phenomenon allows for *optical sectioning*, that is, acquisition of very thin optical slices of the sample for 3D image reconstruction (Figure 9.5).

Each incoming photon has half the total energy of a photon involved in a one-photon process, that is, $E_{1p}/2$. Thus, the frequency is half, and correspondingly, the wavelength is twice that of a single transition photon. For example, the single-photon transition of an electron from the ground state to an excited state at 350 nm can also occur via two-photon absorption at 700 nm. This means that a fluorophore that is otherwise excited by UV radiation can be excited by near-infrared light (NIR). NIR light is able to penetrate deeper into biological tissue because it is less likely to be scattered by cellular components or absorbed by hemoglobin in red blood cells (Helmchen and Denk 2005). Thus, imaging depths approaching 1 mm can be achieved with two-photon fluorescence imaging but not with single-photon fluorescence. It is also possible for three photons to be simultaneously absorbed, leading to three-photon fluorescence, although even higher photon densities are required compared to the two-photon process.

The term *mulitphoton microscopy* refers to the general phenomena whereby multiple photons are simultaneously absorbed by a fluorophore.

The interaction between a fluorophore and a photon can be modeled as the interaction between a molecule and an electromagnetic field. From quantum mechanics theory, can be described by a time-dependent Schrodinger equation where the Hamiltonian contains an electric dipole term. This term is given by $\vec{E}_\gamma \cdot \vec{r}$, where $\vec{E}_\gamma$ is the electric field vector of the photons and $\vec{r}$ position operator (So et al. 2000). The solution to this equation corresponds to photon transitions. For example, the first-order solution represents single-photon excitation, whereas multiple photon transition corresponds to higher order solutions. If we consider the case of two-photon transition from an initial state $|i\rangle$ to a final state $|f\rangle$, it can be shown that the probability of transition is given by Equation 4.1 (So et al. 2000).

$$P \sim \left| \sum_m \frac{\langle f | \vec{E}_\gamma \cdot \vec{r} | m \rangle \langle m | \vec{E}_\gamma \cdot \vec{r} | i \rangle}{\varepsilon_\gamma - \varepsilon_m} \right|^2 \tag{9.1}$$

where

ε_γ is the photonic energy associated with the electric field vector $\vec{E}_\gamma$

ε_m represents the energy difference between the ground and intermediate states m

The two-photon process is a nonlinear phenomenon in that the fluorescence intensity increases quadratically with increasing incident power. In fact, to prove that a particular process is indeed two-photon, a logarithmic plot of power versus intensity is obtained for which the gradient is expected to be approximately equal to 2. In their original paper, Denk et al. (1990) showed that the number of photon pairs absorbed per laser pulse per molecule (n_a) is given as

$$n_a \approx \frac{P_o \delta}{\tau_p f_p^2} \left(\frac{\pi A^2}{hc\lambda} \right)^2 \tag{9.2}$$

where

P_o is the average incident power of the laser source

δ is the two-photon cross section

τ_p is the pulse width

f_p is the repetition rate of the laser

A is the NA

h is the Plank's constant

c is the speed of light

λ is the wavelength of the incident light

Thus, it is seen that a high numerical and pulse repetition rate are critical to the two-photon process. As indicated earlier, multiphoton microscopes have found favor particularly in the biological sciences because of the numerous advantages that these instruments offer over conventional confocal microscopes, which can be summarized in the following points: (1) localized excitation that circumvents the need for a pinhole; (2) greater penetration depth due to the utilization of longer excitation wavelengths; (3) NIR excitation of UV fluorophores; (4) reduced likelihood of photo-damage and out-of-focus photo-damage due to NIR wavelengths; (5) the exploitation of second harmonic generation (SHG) signals as noninvasive image contrast modality in certain molecules such as collagen. This final aspect will be further discussed later in this chapter.

The main disadvantages associated with multiphoton systems are related to the laser source. In general, they are costly, bulky, and offer limited tenability (typically 700 nm to approximately 1000 nm). However, significant progress in the field of fiber-based pulsed lasers will inevitably lead to substantially reduced size and cost for these units. The ultimate goal is a portable unit for clinical applications. So far, noteworthy progress has been made (Myaing et al. 2006) and is likely that MPM systems will be as ubiquitous as conventional wide-field microscopes in the near future.

9.4.2 Two-Photon Experimental Setup

A MPM system consists of four main components: (1) a high power pulsed laser source; (2) optical components for power adjustment, focusing of excitation beam, discrimination of fluorescence signal from excitation source, etc.; (3) a high-speed 3D scanning system; (4) a low noise high sensitivity photodetector; and (5) electronics for scanning control and data acquisition. A schematic of a typical experimental setup is shown in Figure 9.6. The Tsumani laser (Spectra Physics) and Mai tai (Newport) are currently the industry standard for commercial systems. In the case of the Tsunami, a secondary laser acts as a pump in which a neodymium yttrium vanadate (Nd:YVO4) crystal is excited by two diode lasers and emits at 1064 nm. This light is frequency doubled using a SHG crystal to produce light at 532 nm. The output of this laser is used to excite a ti:sapphire crystal in the Tsunami laser. When mode-locked, this laser is capable of producing pulses down to 100 fs at a repetition rate of 82 MHz. The maximum output power of the Tsunami approaches 1 W,

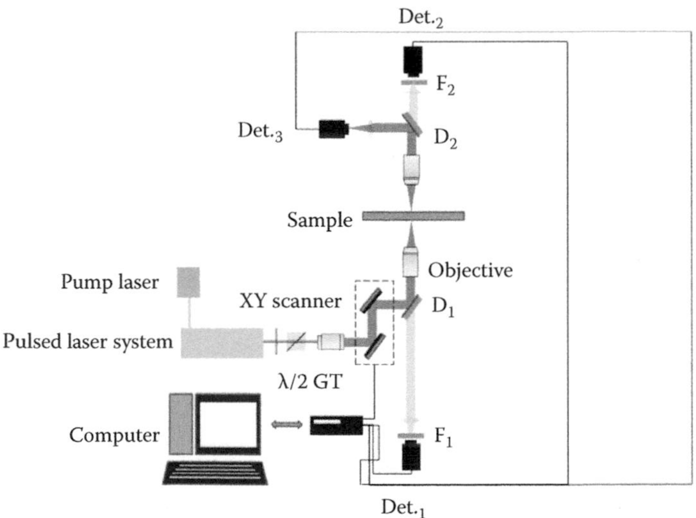

FIGURE 9.6 Schematic of experimental setup for two-photon system.

which is much more than is required for imaging, taking into account loses between the laser and the objective. Multiphoton imaging requires typical incident powers in the 5–50 mW range. The output power varies as the laser is tuned over the operation wavelength range and, as such, power control is an important aspect of this instrument.

9.4.2.1 Signal Attenuation Control

The incident power on the sample is controlled either via the use of neutral density filters or a combination of a half-wave plate (HWP) and Glan–Thompson prism. The HWP allows for the rotation of linearly polarized light from the laser. The GT prism consists of two birefringent crystals that allow for separation of the incoming signal into s- and p-polarized light. Only the s-polarized light is transmitted by the crystal (p-polarized is reflected), and the percentage of the original power in this component can be controlled by rotating the HWP.

9.4.2.2 Beam Expansion

The diameter of the laser beam from the femtosecond light source is typically less than 5 mm, whereas the back aperture of a high NA lens will accept a much larger beam. Thus, to utilize the full NA of the lens, a beam expander is required. The expander is often positioned before the galvos, which in turn establishes the mirror size for the instrument (mirror should be larger than the spot size). The dimensions of the mirror will determine the moment of inertia that directly impacts the scan rate. These considerations need to be taken into account in the design of the instrument. Underfilling the back aperture will result in a lower effective NA but may facilitate smaller scan mirrors.

9.4.2.3 Galvo Scanners

Two-dimensional galvo scanners are used to scan the beam over the samples. Scanning can also be achieved by moving the sample, but this is less common in most commercial systems. An acoustic-optic deflector (AOD) can be used to replace one of the mirrors (usually the horizontal line scanning mirror) for video frame rate imaging. This device utilizes sound waves to create a diffraction pattern, which deflects light.

9.4.2.4 High NA Lens

A high NA lens allows for a highly focused beam in the focal plane. Indeed, the NA is one of the most critical aspects for near-simultaneous two-photon absorption. The objectives in commercial systems are specially designed for transmission of NIR signals. Most microscopes utilize an epifluorescence geometry in which the objective and the detector are positioned above the sample. In such an arrangement, the back-reflected fluorescent signal is recorded. Considering a circular lens of numerical aperture NA = sin α, the spatial distribution at the focal plane can be expressed as

$$I(\mu,\nu) = \left| 2\int_0^0 J_o(\nu\rho)e^{-\frac{i}{2}\mu\rho^2}\,d\rho \right|^2 \qquad (9.3)$$

where

J_o is the zeroth-order Bessel function

μ and ν are the dimensionless axial and radial coordinates through normalization to the wave number $k = 2\pi/\lambda$ (Sheppard and Gu et al. 1990)

Chapter 9

For the two-photon emission of the photon flux, the point spread function is $I^2(\mu/2, \nu/2)$, as compared to $I(\mu,\nu)$ for one photon.

9.4.2.5 Detectors

A wide variety of detectors including photomultiplier tubes, GaAsP detector, and, occasionally, cooled CCD cameras are used for signal detection. Signal detection is achieved in either a descanned or a nondescanned mode. In the former, the fluorescent signal from the sample is sent back through the galvos, which results in a stationary, albeit weaker signal due to losses at the scan mirrors. In the latter, the signal is transmitted to the detector without being sent through the galvos a second time. Each channel of the system usually requires a separate detector and optics. For example, in a four channel system (three fluorescent dyes and SHG imaging), four detectors are required.

9.4.2.6 Filters

A dichroic (dichromatic) mirror is used to separate the excitation signal from the emission signal. In most instances, this takes the form of a 700 nm short pass filter that reflects the excitation signal and transmits the fluorescence and autofluorescence signals. A second short pass filter is placed in front of the detector to exclude any residual excitation signal. For SHG imaging (discussed later), a narrow bandpass filter centered at half the excitation wavelength is placed in front of the detector.

9.4.2.7 Computer Control

Scanning and image acquisition is achieved via computer control. In particular, the scan rate and the scan direction, for example, *xy* versus *xz*, the gain and offset of the detectors, and sometimes, the optical filters and laser power setting can usually be adjusted by a single computer unit in most systems.

9.4.2.8 Endoscopy Probe

As previously indicated, one of the ultimate goals of MPM is the development of a portable system for clinical application. This requires several major modifications of a conventional system including a portable light source, microscanners for raster scanning, and fiber-based excitation and signal collection. The main issue with respect to the latter is group velocity dispersion. However, it has been shown that this can be significantly reduced with hollow core photonic crystal fibers (Tai et al. 2005). Microscanners based on MEMs technology have proven to be particularly useful in two-photon fiber-based imaging systems (Fu et al. 2006). In such systems, a GRIN lens is the usual choice for both focusing the excitation beam and collection of the fluorescence signal. To date, several companies in Europe and Australia are actively involved in the development of these devices.

9.5 Applications of Multiphoton Microscopy

MPM has numerous applications ranging from noninvasive examination of structural components in normal tissue (So et al. 1998) to the early detection of various types of cancers (Skala et al. 2005, Durr et al. 2007, Paoli et al. 2007). This technique is also quite useful in gene expression imaging, tracking cell population and migration patterns, and quantifying tumor angiogenesis to give a few examples. MPM is often preferred over confocal microscopy primarily due to the improved imaging depth, although both

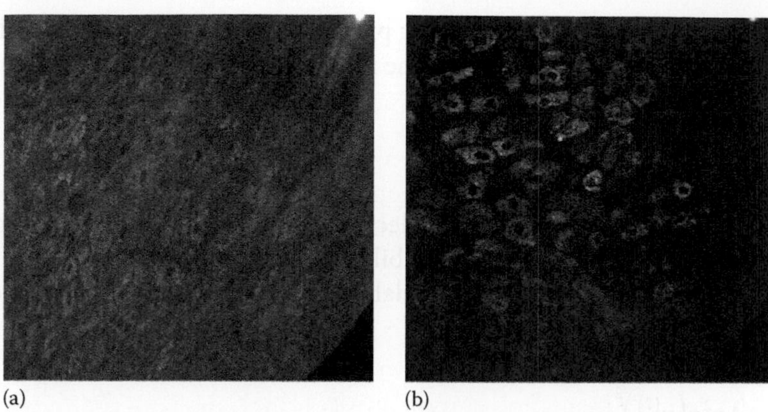

FIGURE 9.7 In vivo images of the superficial epithelial layer in the oral cavity of a normal hamster (a) and in a hamster classified as precancer subsequent to biopsy (b).

techniques have similar spatial and axial resolutions. In a recent publication, it was demonstrated that this technique could be used for the in vivo monitoring of blood flow. This is potentially important in transplant surgeries in evaluating the establishment of blood supply in new organs and tissue. A growing application of two-photon microscopy is as a tool for early cancer detection. Cancer is often preceded by dysplasia that is characterized by certain well-defined morphological and biochemical changes including increase in the nuclear to cytoplasmic ratio, increase in the size and variability of the nucleus, and increase in the metabolic ratio. Under normal circumstances, these parameters can only be measured subsequent to a biopsy, followed by sectioning and staining of the sample. The advantage of MPM is that it allows for all of these features to be determined in vivo and noninvasively. Examples of MPM images for normal and precancerous epithelium tissue from the oral cavity of a hamster are shown in Figure 9.7. Differences in the two sets of imaging are quite apparent. It is possible to image the full thickness of epithelium tissue where the earliest signs of precancer occur. Thus, it is clear that we have only touched the surface in terms of the potential of MPM for pathological diagnosis and physiological monitoring.

9.6 Second Harmonic Generation Imaging

It is well known that certain crystals and biological structures which are noncentrosymmetric can produce a SHG signal under similar conditions required for two-photon imaging. For a polarizable specimen, the nonlinear polarization is given by

$$\vec{P} = \chi^{(1)} \otimes \vec{E} + \chi^{(2)} \otimes \vec{E} \otimes \vec{E} + \chi^{(3)} \otimes \vec{E} \otimes \vec{E} \otimes \vec{E} + \quad (9.4)$$

where

$\vec{P}$ is the induced polarization vector
$\vec{E}$ is the electric field vector
$\chi^{(1)}$ the *i*th order nonlinear susceptibility tensor
$\otimes$ is a combination of a tensor product and integral over the frequencies

The $\chi^{(2)}$ term corresponds to second order processes such as SHG, sum and difference frequency generation, etc. It is related to the molecular hyperpolarizability β by

$$\chi^{(2)} = N_s < \beta > \tag{9.5}$$

where N_s represents the density of the molecules. Note that the expression includes the orientational average of the hyperpolarizability, which implies that a lack of a center of symmetry is required for $\chi^{(2)}$ to be appreciably large. The SHG signal can be shown to be given by

$$\text{SHG}_{sig} = Ap^2\tau\left(\chi^{(2)}\right)^2 \tag{9.6}$$

where
- p is the laser pulse energy
- τ is the pulse width
- A is the proportional constant

In the case of biological tissue, this signal is generated primarily by collagen. Since collagen plays an important biological role in terms of providing structural support in living tissue and is implicated in numerous degenerative diseases such as arthritis, SHG imaging has developed into a powerful investigative tool. This is so because the process is not invasive and the elicited signal is endogenous in its origin. In addition, SHG imaging can be performed with any two-photon system by simply placing an appropriate filter in front of the detector. As shown in Figure 9.8a, the SHG photon has exactly the same energy as the two incoming photons that were absorbed during the excitation process. Thus, the SHG signal is at exactly half the wavelength as the excitation signal. In order words, all that is required for SHG imaging is a narrow bandpass filter

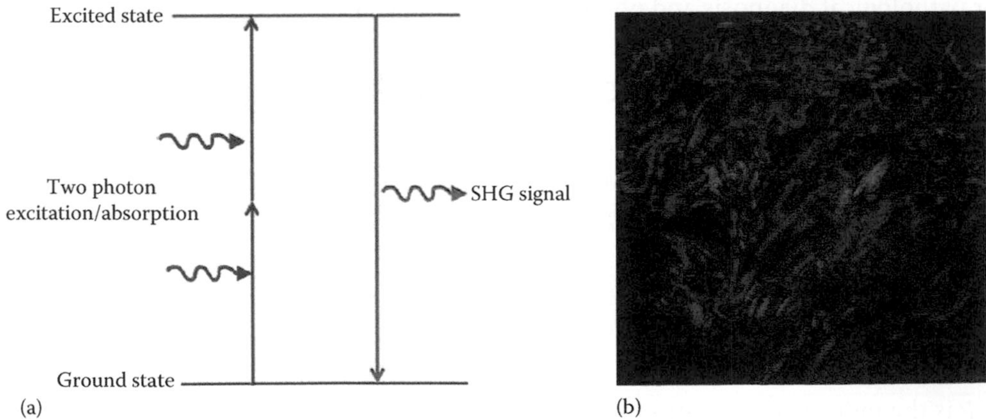

FIGURE 9.8 A Jablonski diagram for SHG is shown in (a). Image (b) is a two-photon pseudocolor image of pig lung tissue. The red structure highlights the autofluorescence from elastin, whereas the green structure represents the collagen distribution as determined from the SHG signal.

centered at half the excitation wavelength. This phenomenon is fundamentally different from two-photon fluorescence imaging in that the process is coherent. Thus, constructive and destructive interference plays a critical role in signal detection. In most situations, much of the SHG signal is directed in the forward direction and as such, forward detection is most efficient. This of course is not possible for thick samples. The image in Figure 9.8b illustrates how SHG imaging can be combined with two-photon autofluorescence imaging. The elastin structure of lung tissue (red) is highlighted via two-photon imaging at 780 nm excitation. A pseudocolor SHG image of the collagen present in the tissue was obtained by 840 nm excitation with a 40 nm bandpass filter centered at 420 nm. This technique has proven to be particularly useful for quantitative evaluation of collagen content such as in decellularized tissue for tissue engineering.

9.7 FLIM and CARS

There are several nonlinear techniques apart from multiphoton and SHG imaging for the investigation of biological processes. An emerging technique in this regard is fluorescence lifetime imaging (FLIM) (Lakowicz et al. 1992). When a fluorophore is excited, it decays exponentially as it returns to its normal nonfluorescent state. The fluorescence lifetime is the average time the molecule remains in its excited state before emitting a photon. This lifetime depends on the fluorophore in question. Therefore, it is possible to generate images based on the lifetimes of the constituent fluorescent components in a sample instead of the intensity values. Significant modification of the detection system of a two-photon system is required to allow for FLIM, since the lifetimes are in the nanosecond regime. Most detectors are not fast enough to capture these events. However, FLIM can often provide useful supplementary information to multiphoton imaging, particularly for different structures that fluoresce at the same wavelength. Coherent anti-Stokes Raman scattering (CARS) is yet another nonlinear technique with growing applications in numerous fields. This is a spectroscopic technique that is similar to Raman spectroscopy in that it is sensitive to the same vibrational signatures. In CARS however, multiple photons are exploited for assessing molecular vibrational characteristics in samples. Another difference is that the emitted signals are coherent in this technique. As such, the CARS signal is a significant stronger signal compared to spontaneous Raman emission. Numerous applications exist including the noninvasive imaging of lipids in biological tissue and the detection of roadside bombs.

9.8 Concluding Remarks

MPM is a nonlinear imaging modality, which facilitates high-resolution three-dimensional optical sectioning of thick tissue. This is of potentially great importance particularly for the noninvasive examination and diagnosis of pathological conditions such as cancer. The appeal of this technique resides in the inherent capability to image deep into tissue structures with cellular and subcellular resolution. This has in many instances allowed for in vivo visualization of cells in their natural environment for the first time. To date, MPM has found wide applications in the biological sciences including in vivo brain slice imaging (Oheim et al. 2001), early diagnosis of cancer (Durr et al. 2007), the study of protein dynamics (Schwille et al. 1999), and the in vivo monitoring of blood flow (Chaigneau et al. 2003). Current research is directed at new applications of the technology, technical improvement of the laser sources, and new fluorescence probes. Since the first publication by Denk et al. (1990), it is not hyperbole to state that MPM has revolutionized the biological sciences. It is now evident that as the cost of pulsed laser systems continues to fall, these instruments will make their way into most major research labs around the world.

References

Callamaras, N. and I. Parker. 1999. Construction of a confocal microscope for real-time x-y and x-z imaging. *Cell Calcium* 26: 271–279.

Chaigneau, E., M. Oheim, E. Audinat et al. 2003. Two-photon imaging of capillary blood flow in olfactory bulb glomeruli. *Proceedings of the National Academy of Sciences* 100: 13081–13086.

Denk, W., J. H. Strickler, and W. W. Webb. 1990. Two-photon laser scanning fluorescence microscopy. *Science* 248: 73–76.

Durr, N. J., T. Larson, D. K. Smith et al. 2007. Two-photon luminescence imaging of cancer cells using molecularly targeted gold nanorods. *Nano Letters* 7: 941–945.

Franken, P. A., A. E. Hill, C. W. Peters et al. 1961. Generation of optical harmonics. *Physics Review Letters* 7: 118–119.

Fu, L., A. Jain, H. Xie et al. 2006. Nonlinear optical endoscopy based on a double-clad photonic crystal fiber and a MEMS mirror. *Optics Express* 14: 1027–1032.

Göppert-Mayer M. 1931. Uber elementarakte mit zwei quantensprungen. *Annalen der Physik* (Leipzig) 5: 273–294.

Graf, R., J. Rietdorf, and T. Zimmermann. 2005. Live cell spinning disk microscopy. *Advances Biochemical Engineering Biotechnology* 95: 57–75.

Gu, M. and C. J. R. Sheppard. 1995. Comparison of three-dimensional imaging properties between two-photon and single-photon fluorescence microscopy. *Journal of Microscopy* 177: 128–137.

Helmchen, F. and W. Denk. 2005. Deep tissue two-photon microscopy. *Nature Methods* 2: 932–940.

Lakowicz, J. R., H. Szmacinski, K. Nowaczyk et al. 1992. Fluorescence lifetime imaging. *Analytical Biochemistry* 202: 316–330.

Matsumoto, B. (ed.) 2003. *Cell Biological Applications of Confocal Microscopy*. Academic Press, New York.

Myaing, M. T., D. J. MacDonald, and X. Li. 2006. Fiber-optic scanning two-photon fluorescence endoscope. *Optics Letters* 31: 1076–1078.

Oheim, M., E. Beaurepaire, E. Chaigneau et al. 2001. Two-photon microscopy in brain tissue: Parameters influencing the imaging depth. *Journal of Neuroscience Methods* 111: 29–37.

Paoli, J., M. Smedh, A. M. Wennberg et al. 2007. Multiphoton laser scanning microscopy on non-melanoma skin cancer: Morphologic features for future non-invasive diagnostics. *Journal of Investigative Dermatology* 128: 1248–1255.

Pawley, J. B. 1995. *Handbook of Biological Confocal Microscopy*. Kluwer Academic Publishers, New York.

Polglase, A. L., W. J. McLaren, S. A. Skinner et al. 2005. A fluorescence confocal endomicroscope for in vivo microscopy of the upper-and the lower-GI tract. *Gastrointestinal Endoscopy* 62: 686–695.

Sánchez, S. A. and E. Gratton. 2005. Lipid–protein interactions revealed by two-photon microscopy and fluorescence correlation spectroscopy. *Accounts of Chemical Research* 38: 469–477.

Schwille, P., U. Haupts, S. Maiti et al. 1999. Molecular dynamics in living cells observed by fluorescence correlation spectroscopy with one-and two-photon excitation. *Biophysical Journal* 77: 2251–2265.

Sheppard, C. J. R. and M. Gu. 1990. Image formation in two-photon fluorescence microscopy. *Optik* 86: 104–106.

Skala, M. C., J. M. Squirrell, K. M. Vrotsos et al. 2005. Multiphoton microscopy of endogenous fluorescence differentiates normal, precancerous, and cancerous squamous epithelial tissues. *Cancer Research* 65: 1180–1186.

So, P. T., C. Y. Dong, B. R. Masters et al. 2000. Two-photon excitation fluorescence microscopy. *Annual Review of Biomedical Engineering* 2: 399–429.

So, P. T., H. Kim and I. E. Kochevar. 1998. Two-photon deep tissue ex vivo imaging of mouse dermal and subcutaneous structures. *Optics Express* 3: 339–350.

Tai, S. P., M. C. Chan, T. H. Tsai et al. 2005. Two-photon fluorescence microscope with a hollow-core photonic crystal fiber. *Biomedical Optics* 2005: 146–153.

Vliet, V. 1998. Theory of confocal fluorescence imaging in the programmable array microscope (PAM). *Journal of Microscopy* 189: 192–198.

Xu, C. and W. Webb. 1996. Measurement of two-photon excitation cross sections of molecular fluorophores with data from 690 to 1050 nm. *Journal of Optical Society of America* 13: 481–491.

Yang, L., A. M. Raighne, E. M. McCabe et al. 2005. Confocal microscopy using variable-focal-length microlenses and an optical fiber bundle. *Applied Optics* 44: 5928–5936.

10. Imaging Based on Optical Coherence Tomography

Shahab Chitchian
University of Texas Medical Branch

Nathaniel M. Fried
University of North Carolina at Charlotte

10.1 Introduction

Various imaging modalities such as ultrasound (US), magnetic resonance imaging (MRI), computed tomography (CT), positron emission tomography (PET), and single-photon emission CT (SPECT) have been used for imaging of biological tissues (Manyak et al. 2006, 2007). The introduction of new imaging modalities may also provide enhanced accuracy for illness tissue localization and significant promise for improved diagnosis.

One of these newer imaging modalities is *optical coherence tomography* (OCT). OCT is a noninvasive optical imaging technique used to perform high-resolution, cross-sectional, in vivo, and in situ imaging of microstructure in biological tissues (Huang et al. 1991, Tearney et al. 1997a). OCT uses backscattered near-infrared light in an analogous way to US for biomedical imaging. OCT was originally developed in the early 1990s for imaging of the eye. Since the eye is relatively transparent with little attenuation of photons due to light scattering in tissue, OCT has proven to be indispensable for diagnosis of ophthalmic diseases and has since replaced US for many ophthalmic applications. However, imaging of opaque tissues, such as the prostate gland, has proven more challenging, due to the attenuation

Handbook of Optical Sensors. Edited by José Luís Santos and Faramarz Farahi © 2015 CRC Press/Taylor & Francis Group, LLC. ISBN: 9781439866856.

Table 10.1 Specifications for a Typical OCT System

Axial resolution (μm)	1–10
Transverse resolution (μm)	10–25
Imaging depth (mm)	1–2
Lateral imaging distance (mm)	1–10
Acquisition rate (fps)	1–30
3D reconstruction time (s)	10–60
Cost ($k)	50–150

Table 10.2 Comparison of OCT to Major Imaging Modalities

	Imaging Method					
Parameters	**OCT**	**US**	**MRI**	**CT**	**PET**	**SPECT**
Spatial resolution	Excellent	Good	Good	Good	Poor	Poor
Imaging depth	Poor	Good	Excellent	Excellent	Excellent	Excellent
Noncontact application	Yes	No	Yes	Yes	Yes	Yes
Soft tissue contrast	Excellent	Good	Excellent	Poor	Good	Good
Non-ionizing radiation	Yes	Yes	Yes	No	No	No
Functional imaging	Excellent	Good	Excellent	None	Excellent	Excellent
Data acquisition time	Fast	Fast	Slow	Slow	Slow	Slow
Size	Small	Small	Large	Large	Large	Large
Cost	Low	Low	High	High	High	High

of the OCT signal with depth in the tissue from multiple scattering. Table 10.1 shows a range of typical specifications for commercially available OCT systems.

OCT has a number of potential advantages over more established imaging modalities, such as US, MRI, CT, PET, and SPECT (Table 10.2). OCT is capable of providing cellular-level axial imaging resolution on the order of 1–20 μm, an order of magnitude better than high-frequency US and several orders of magnitude better than MRI, CT, PET, and SPECT. Unlike US, OCT is also capable of being operated in noncontact mode without the need of an index-matching medium. Since OCT utilizes near-infrared optical radiation for imaging tissue, it is also safe, unlike CT, PET, and SPECT that require the use of ionizing radiation. OCT also provides excellent tissue contrast and can be used for functional imaging as well. Real-time acquisition of 2D OCT images has been demonstrated at video rates, and soon-to-be-released next-generation commercial OCT systems will also offer 3D imaging in real time as well. Typical OCT systems are based on standard optical components first developed for the telecommunications industry, making them very compact, portable, and inexpensive, with a comparable cost to US systems. Furthermore, the all-optical-fiber OCT delivery probes can easily be integrated into existing surgical instrumentation such as catheters and endoscopes. For example, recent pilot studies have demonstrated integration of OCT fibers into laparoscopic and robotic surgical platforms for prostate cancer surgery. The only major limitation of OCT is its limited imaging depth in opaque tissues. OCT can be categorized as a superficial, high-resolution imaging technique providing anatomical imaging of tissue microstructure to a depth of approximately 1–2 mm. Thus, in general, OCT is not the imaging modality of choice for identification of tumors within bulk tissues.

However, there may be a role for OCT in selective applications involving tumor diseases. Focusing on prostate diseases to better illustrate the application of this imaging technique, over the past decade or so, about a dozen studies have been performed using OCT for imaging of the prostate gland. They generally fall into one of four categories: (1) image-guided ablation of prostate tissue for treatment of benign prostatic hyperplasia (BPH), (2) detection and staging of prostate cancer, (3) monitoring of prostate cancer therapy, and (4) identification and sparing of the cavernous nerves during prostate cancer surgery.

10.2 Theory of OCT

As previously mentioned, OCT can be considered in some ways to be the optical analog of US. However, back-reflected near-infrared radiation is measured instead of back-reflected sound waves. The optical index mismatch between tissue structures provides the contrast for optical imaging much as the acoustic impedance mismatch does for US. One major difference between sound waves and electromagnetic waves, however, is the wave velocity. Sound travels approximately 1500 m/s in soft tissues, while light travels approximately 300,000,000 m/s in tissue. This difference of roughly five orders of magnitude between sound and light waves raises some problems. Photons cannot be used for imaging by time gating their arrival from the tissue to the detector in the same method that sound waves are used for detection during US imaging. This would require a detector capable of measuring photon path times of about 1 fs (1×10^{-15}s).

Instead, OCT detection differs from US, in that OCT is a coherence-based imaging technique. In its simplest formulation, OCT consists of a Michelson interferometer (Tomlins and Wang 2005). The sensitivity of backscattered light as well as its selectivity from a specific backscattering site is achieved by using interference between backscattered light and a reference beam. Interference between propagating wave fronts of two light sources occurs when both wave fronts have a well-defined coherence (phase relation) within the overlapped region. This well-defined coherence of a wave front from a source is maintained within a distance called the coherence length. Therefore, if both the reference beam and beam back-reflected from a scattering site are derived from the same light source, a well-defined interference pattern will be produced only if their path-length difference is within the coherence length. A displacement of the reference beam produces a path-length difference between the light reflected from the reference mirror and backscattered photons from the scattering sample.

For a coherent light source, the interference between the reference beam and the backscattered beam can be maintained over a large path-length difference induced by the reference mirror displacement. Thus, selectivity to backscattering from a specific depth in the sample is impossible if a fully coherent light source is used. Therefore, in OCT, a short coherence length (or low-coherence source) is used so the interference pattern between the reference beam and backscattered beam is produced only when their path difference is within the coherence length. Thus, by scanning the reference beam, depth discrimination can be achieved in OCT.

The axial (depth) resolution is defined by the coherence length of the light source. A shorter coherence length provides better axial resolution. The axial resolution is equal to one-half the coherence length of the light source due to the round-trip propagation of the reference and sample beams. The equation for axial resolution is given by

$$\Delta z_R = 0.44 \frac{\lambda_0^2}{\Delta\lambda} \tag{10.1}$$

where
λ_0 is the center wavelength
$\Delta\lambda$ is the bandwidth of the light source, respectively

Typical specifications for current commercially available OCT systems include a wavelength of 1310 nm and a bandwidth of approximately 100 nm, providing an axial resolution of approximately 7.5 μm in air. The 1300 nm wavelength is generally the preferred wavelength for imaging of opaque tissues because attenuation due to both absorption of photons (from the water component in the tissue) and light scattering is relatively low, thus providing deep penetration of the near-infrared light into the tissue.

The OCT transverse resolution is defined as the focal diameter of the incident beam, which is dependent on the focusing optics used. The transverse resolution is given by

$$\Delta r_R = \frac{4\lambda_o f}{\pi D} \tag{10.2}$$

where
f is the focal length of the lens
D is the diameter of the beam at the lens

Typical transverse resolutions range from 10 to 25 μm.

The design of the common-path interferometer-based OCT system is shown in Figure 10.1 (Chitchian et al. 2009a). No separate arm for reference is required in the common-path-based OCT system, and this unique configuration provides the advantage of using any arbitrary length of detachable probe arm. The option to have detachable low-cost probes may simplify the instrumentation of endoscopic OCT (EOCT) and may also be very useful in cases where probes might need to be disposable to meet health safety standards. In our experimental setup, the probe is mounted on a high-speed translational stage (850G-HS, Newport, Irvine, CA), interfaced with a motor controller (ESP 300, Newport, Irvine, CA) with the sample scanned in the transverse plane. The reflection from the air–glass interface of the fiber tip provides the reference signal and the piezoelectric fiber stretchers implement the longitudinal scanning function. The scan range of the system is 7 mm (in air) at a scanning frequency of 30 Hz. The Doppler frequency of the interferometer is approximately 695 kHz. Faraday mirrors are used instead of simple reflectors in the scanning interferometer to ensure high interferometric visibility, as well as to reverse the effects of polarization or birefringence modulation caused by sweeping the fiber stretchers. This allows the entire design to be comprised of low-cost single-mode fiber.

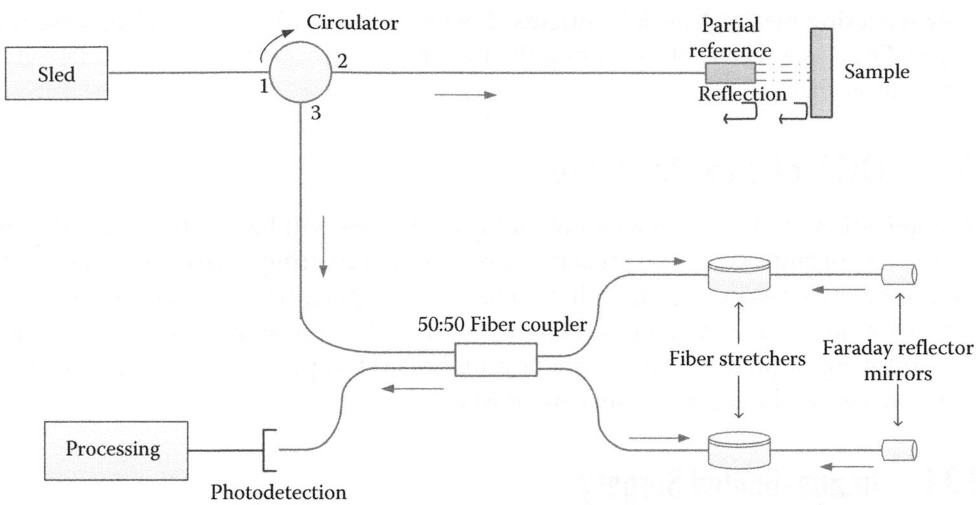

FIGURE 10.1 Common-path interferometer-based OCT system: this configuration was used in tandem with a translational stage to obtain 2D cross-sectional OCT images. SLED, C1, and D1 are the broadband light source, circulator, and detector, respectively.

A superluminescent light-emitting diode (DL-CS 3504, DenseLight Semiconductors, Singapore) with a center wavelength of 1310 and 35 nm optical bandwidth was used as the light source. Our optical receiver (LDPF0065, Tyco Elec PINFET, with 80 kV/W sensitivity and 80 kΩ transimpedance at 1300 nm) has a saturation limit of approximately 25 W and a responsivity of approximately 0.8 A/W. OCT imaging was performed with an output power of 5 mW. The average SNR of our system is measured to be approximately 35 dB when a rat prostate sample is illuminated by 5 mW incident power.

In the time domain (TD) OCT system described previously, if the reference path length is fixed and the detection system is replaced with a spectrometer, no moving parts will be required to acquire axial scans. The detected intensity spectrum is then Fourier transformed into the TD to reconstruct the depth-resolved sample optical structure. Fourier domain (FD) OCT has one main advantage that makes it particularly attractive for imaging applications: it obtains entire depth scans in one exposure. Therefore, FD-OCT has the potential to image a given biological sample much more rapidly than TD-OCT, and therefore FD-OCT is also the system of choice for rapid 3D image acquisition. FD-OCT can also be performed using a single detector by sweeping the source spectrum using a tunable laser. FD-OCT of this type is referred to as swept source (SS) OCT.

There are also a number of variants of OCT, such as polarization-sensitive (PS) OCT and Doppler OCT that may be useful for imaging. PS-OCT is similar to TD-OCT and FD-OCT, with the addition of a linear polarizer after the source and a polarizing beam splitter with an extra detector in the detection side. PS-OCT can resolve information about the birefringence of the tissue sample. Tissues that retain an orderly structure and exhibit strong birefringence, such as nerves in general and more specifically, potentially, the cavernous nerves on the prostate, may be used as a source of contrast during PS-OCT.

Doppler OCT detects the shift in frequency of waves reflected from a moving object, in an analogous way to Doppler US. This frequency shift can be used to determine the object's velocity. Doppler OCT is applied to quantitatively image and measure fluid flow in highly scattering media. As will be discussed below, Section 10.3.3, researchers have used Doppler OCT as a means of measuring the microvascular response in the prostate gland during cancer therapy.

10.3 OCT of the Prostate

The application of OCT to imaging of biological tissues will be illustrated for the specific but important case of the prostate. Several research groups have studied OCT for imaging of the prostate gland, with a wide range of potential applications, including (1) image-guided surgery, (2) optical biopsy for detection and staging of prostate cancer, (3) monitoring of prostate cancer therapy, and (4) intraoperative identification of the cavernous nerves during nerve-sparing prostatectomy.

10.3.1 Image-Guided Surgery

At least one preliminary study, by Boppart et al., has shown that OCT may have promise for image-guided therapy of urethral obstruction associated with BPH (Boppart et al. 2001). In this report, OCT was used to better understand differences in the mechanism of radio frequency (RF) and laser ablation of the prostate. An image-based comparison was made between the real-time dynamics of RF and laser ablation of in vitro prostate specimens. OCT provided high-speed, high-resolution subsurface imaging of the ablation process.

10.3.2 Optical Biopsy

A more recent study by Dangle and colleagues used OCT to evaluate prostate specimens immediately after removal during radical prostatectomy (Dangle et al. 2009). The objective of this study was to explore the feasibility of OCT for the real-time evaluation of positive surgical margins and extracapsular extension in robotic prostatectomy specimens and to compare these results to the gold standard of histology. The sensitivity, specificity, positive predictive value, and negative predictive value were reported as 70%, 84%, 33%, and 96%, respectively. The positive predictive value was reported to be low and may have to be significantly improved. However, the high negative predictive value may help to detect the true negative margins. These preliminary results show the promise of using OCT for assessing the microstructure of the prostate.

The introduction of new robot-assisted technology has led to declines of positive surgical margins. A study by Skarecky et al. has suggested that the integration of OCT into the da Vinci® robotic platform for identification of positive margin sites may provide promise for even further reduction of surgical margins (Skarecky et al. 2008).

Tearney et al. investigated the capability of OCT to differentiate the architectural morphology of various urologic tissues (Tearney et al. 1997b). OCT was used to image the microstructure of the urinary tract with the focus on a critical region consisting of the prostatic capsule, surrounding adipose tissue, and neurovascular bundles. This study suggested the potential of using OCT to noninvasively obtain information on tissue microstructure that could only previously be obtained with conventional biopsy.

OCT was also used by D'Amico and colleagues in the setting of the human prostate, ex vivo, to image tissue microstructure and compare the images with those obtained using standard histopathologic methods (D'Amico et al. 2000). Structures within the benign glandular epithelium, fibroadipose tissue, and malignant glandular epithelium were resolved on the order of 50–150 μm. Their study concluded that potential clinical applications of OCT may include assessing therapeutic response to neoadjuvant therapy, identifying neurovascular bundle invasion intraoperatively, and improving prostate cancer detection by decreasing the error associated with sampling. It should be noted that since this study was published, more recent OCT systems have been developed with significantly improved imaging resolution, as stated in Table 10.1.

10.3.3 Therapy Monitoring

Several recent studies have explored the use of interstitial Doppler OCT (IS-DOCT) to monitor the microvascular response of the rat prostate to photodynamic therapy (PDT) (Li et al. 2006, Standish et al. 2007, 2008). OCT has the ability to provide real-time localized blood flow measurements after PDT. OCT provided a direct correlation between the local PDT biological response and a corresponding local microvascular response. A strong correlation was found between the percent tumor necrosis 24 h after PDT treatment and the vascular shutdown rate. A slower shutdown rate corresponded to higher treatment efficacy (more necrosis). This technique may prove useful in monitoring critical boundary regions within and surrounding the prostate, such as the rectal wall.

10.3.4 Nerve-Sparing Prostate Surgery

Preservation of the cavernous nerves during prostate cancer surgery is critical for preserving sexual function after surgery. These nerves are at risk of injury during dissection and removal of a cancerous prostate gland because of the close proximity of the nerves to the prostate surface. Their microscopic nature also makes it difficult to predict the true course and location of these nerves from one patient to another. These observations may explain in part the wide variability in reported potency rates (9%–86%) following prostate cancer surgery (Burnett et al. 2007).

OCT imaging of the cavernous nerves in the human and rat prostates has recently been demonstrated (Aron et al. 2007, Fried et al. 2007a,b, Rais-Bahrami et al. 2008). Aron and coworkers performed preliminary clinical studies using OCT for tissue mapping and identifying the neurovascular bundles during laparoscopic and robotic radical prostatectomy. Fried and coworkers performed open surgical exposure of the rat prostate and OCT images of the prostate, cavernous nerve, pelvic plexus ganglion, seminal vesicle, blood vessels, and periprostatic fat were acquired. The prostate and cavernous nerves were also processed for histologic analysis. In these preclinical studies, the cavernous nerve and ganglion could be differentiated from the surrounding prostate gland, seminal vesicle, blood vessels, bladder, and fatty tissue.

10.4 Image Processing Algorithms for Improved OCT Images

Significant improvement in the quality of the OCT prostate images may be necessary before application in the clinic. Chitchian et al. have explored image processing algorithms for improving both OCT image contrast and imaging depth in prostate tissue

(Chitchian et al. 2009a,b, 2010). Three complementary image processing algorithms for improving identification and imaging of the cavernous nerves during OCT of the prostate gland were implemented: segmentation, denoising, and edge detection, as described in the following (Chitchian et al. 2010).

The input TD-OCT image of the rat prostate gland and cavernous nerve is shown in Figure 10.2a. A histologic section of the prostate gland and cavernous nerve is shown for comparison (Figure 10.2b). The OCT image was acquired in vivo using a clinical endoscopic TD-OCT system.

A 2D OCT image of the rat prostate was first segmented to differentiate the cavernous nerve from the prostate gland (Chitchian et al. 2009c). Three image features were

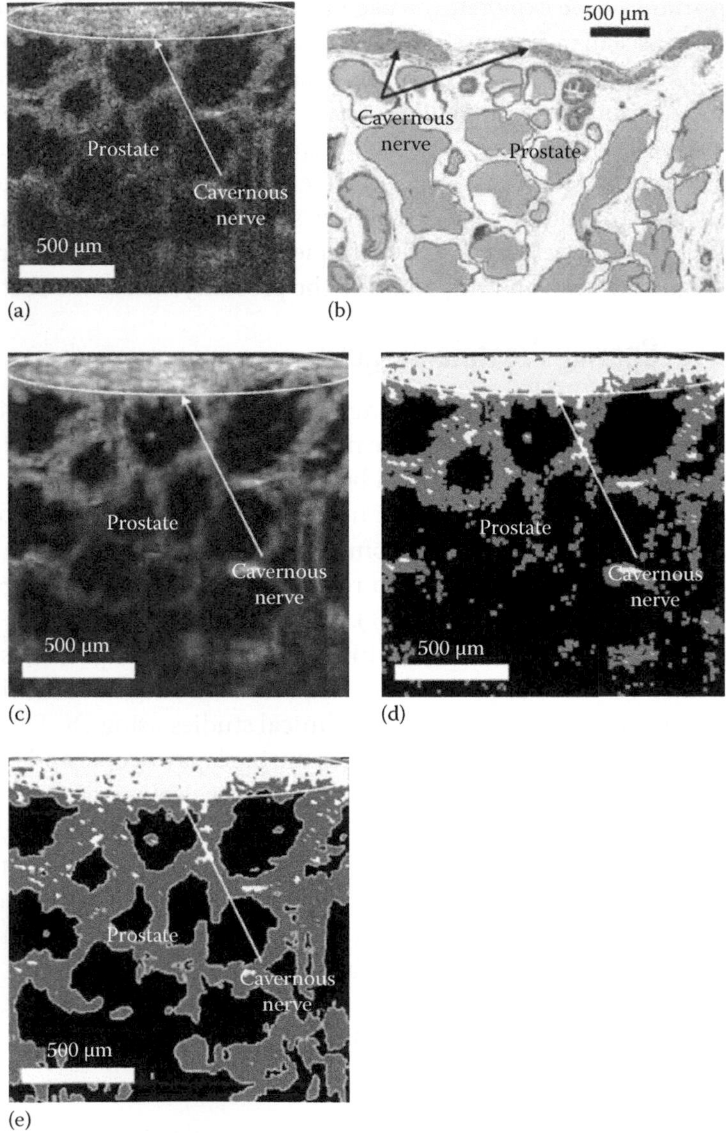

FIGURE 10.2 OCT image of the rat prostate gland and cavernous nerve: (a) original, (b) corresponding histology, (c) denoised, (d) segmented, and (e) edge detected.

employed: Gabor filter, Daubechies wavelet, and Laws filter. The Gabor feature was applied with different standard deviations in the x and y directions. In the Daubechies wavelet feature, an 8-tap Daubechies orthonormal wavelet was implemented, and the low pass sub-band was chosen as the filtered image. Finally, the Laws feature extraction was applied to the image. The features were segmented using a nearest-neighbor classifier. N-ary morphological postprocessing was used to remove small voids.

As a next step to improve OCT imaging of the prostate gland, wavelet denoising was applied. A locally adaptive denoising algorithm was applied before edge detection to reduce speckle noise in OCT image of the prostate (Chitchian et al. 2009b). The denoising algorithm was illustrated using the dual-tree complex wavelet transform. After wavelet denoising, an edge detection algorithm based on thresholding and spatial first-order differentiation was implemented to provide deeper imaging of the prostate gland (Chitchian et al. 2010). This algorithm addresses one of the main limitations in OCT imaging of the prostate tissue, which is the inability to image deep into the prostate. Currently, OCT is limited to an image depth of approximately 1 mm in most opaque soft tissues.

Figure 10.2c shows the image after denoising using CDWT. Figure 10.2d shows the same OCT image of Figure 10.2a after segmentation and Figure 10.2e combines edge detection of the denoised image and the segmentation result. The edge detection approach was successful in accentuating prostate structure deeper in the tissue, and the cavernous nerve could be differentiated from the prostate gland using the segmentation algorithm. The glandular structure of the prostate could be seen to a depth of approximately 1.6 mm in Figure 10.2e in comparison with an approximately 1 mm depth in the unprocessed OCT image in Figure 10.2a.

10.5 Future Directions

Superficial imaging depth (1–2 mm) remains one of the most important limitations during OCT imaging in many situations, particularly in tissue imaging, as shown when this technique is applied to imaging the prostate gland. Focusing on this specific situation, to overcome this limitation, OCT imaging of the prostate may benefit from replacement of commonly used 1310 nm broadband light sources with 1064 nm sources: this shorter wavelength provides a slightly improved optical penetration depth in prostate tissue (Chitichian and Fried 2010). US-enhanced OCT may be another effective solution for increasing image depth and resolution (Huang et al. 2008). The authors reported an improvement in resolution of up to 28% using the combination of OCT and US. The principle underlying US-enhanced OCT is that delivery of a US beam in parallel with the OCT beam reduces multiple light scattering, thus resulting in the improved penetration and image depth.

Typical OCT systems using compact, low-cost, broadband light sources, such as superluminescent diodes (SLDs), produce axial resolutions of 5–10 μm. However, Drexler and colleagues have demonstrated ultrahigh-resolution OCT with subcellular axial resolutions of 1–2 μm, using a Kerr-lens mode-locked Ti–sapphire laser, although these lasers are significantly larger and more expensive (Drexler et al. 1999).

Recent optical beam scanning designs based on microelectromechanical systems (MEMS) may be promising for integration into OCT endoscopic imaging systems and

Table 10.3 Commercially Available OCT Systems

Company	Location	System
Bioptigen	Durham, NC	SD-OCT
Imalux	Cleveland, OH	Endoscopic TD-OCT
LightLab	Westford, MA	Endoscopic SS-OCT
Michelson Diagnostics	United Kingdom	Dermatology SS-OCT
Santec	Japan	SS-OCT
Thorlabs	Newton, NJ	Laboratory SD-OCT
Zeiss Meditec	Dublin, CA	Ophthalmic OCT

OCT microscopes (Tran et al. 2004, Cobb et al. 2005, Gilchrist et al. 2009). Preliminary results of a novel radial-scanning MEMS probe have been presented by Tran et al. (2004). The OCT probe was designed and constructed with a 1.9 mm diameter MEMS motor. Cobb et al. (2005) introduced real-time OCT imaging with dynamic focus tracking. The reference arm length was scanned synchronously with the beam focus to maximize the OCT signal at the focus during focus tracking. Piezoelectric MEMS may additionally provide a solution for miniaturized scanning applications with high mechanical deflection in near static condition, good optical quality, and high resonance frequency (Gilchrist et al. 2009). MEMS devices may have direct application in clinical EOCT applications for laparoscopic and robotic nerve-sparing prostate cancer surgery.

Future studies using TD- and FD-OCT with faster data acquisition and processing power will result in 3D reconstruction and faster imaging for real-time applications. Recently, FD-OCT systems capable of acquiring data at up to 214 fps have been introduced, resulting in 3D imaging under a second.

Table 10.3 provides a summary list of some of the scientific and clinical OCT systems that are currently commercially available.

References

Aron, M., J. H. Kaouk, N. J. Hegarty et al. 2007. Second prize: Preliminary experience with the Niris optical coherence tomography system during laparoscopic and robotic prostatectomy. *Journal of Endourology* 21: 814–818.

Boppart, S. A., J. M. Herrmann, C. Pitris et al. 2001. Real-time optical coherence tomography for minimally invasive imaging of prostate ablation. *Computer Aided Surgery* 6: 94–103.

Burnett, A., G. Aus, E. Canby-Hagino et al. 2007. Function outcome reporting after clinically localized prostate cancer treatment. *Journal of Urology* 178: 579–601.

Chitchian, S., M. A. Fiddy, and N. M. Fried. 2009a. Speckle reduction during all-fiber common-path optical coherence tomography of the cavernous nerves. *Proceedings of SPIE* 7168: 71682N.

Chitchian, S., M. A. Fiddy, and N. M. Fried. 2009b. Denoising during optical coherence tomography of the prostate nerves via wavelet shrinkage using dual-tree complex wavelet transform. *Journal of Biomedical Optics* 14: 014031.

Chitichian, S. and N. M. Fried. 2010. Near-IR optical properties of canine prostate tissue using oblique-incidence reflectometry. *Proceedings of SPIE* 7548: 75480Z.

Chitchian, S., T. P. Weldon, and N. M. Fried. 2009c. Segmentation of optical coherence tomography images for differentiation of the cavernous nerves from the prostate gland. *Journal of Biomedical Optics* 14: 044033.

Chitchian, S., T. P. Weldon, M. A. Fiddy et al. 2010. Combined image-processing algorithms for improved optical coherence tomography of the prostate nerves. *Journal of Biomedical Optics* 15: 046014.

Cobb, M. J., X. Liu, and X. Li. 2005. Continuous focus tracking for real-time optical coherence tomography. *Optics Letters* 30: 1680–1682.

D'Amico, A. V., M. Weinstein, X. Li et al. 2000. Optical coherence tomography as a method for identifying benign and malignant microscopic structures in the prostate gland. *Urology* 55: 783–787.

Dangle, P. P., K. K. Shah, B. Kaffenberger et al. 2009. The use of high resolution optical coherence tomography to evaluate robotic radical prostatectomy specimens. *International Brazilian Journal of Urology* 35: 344–353.

Drexler, W., U. Morgner, F. X. Kärtner et al. 1999. In vivo ultrahigh-resolution optical coherence tomography. *Optics Letters* 24: 1221–1223.

Fried, N. M., S. Rais-Bahrami, G. A. Lagoda et al. 2007a. Imaging the cavernous nerves in the rat prostate using optical coherence tomography. *Lasers in Surgery and Medicine* 39: 36–41.

Fried, N. M., S. Rais-Bahrami, G. A. Lagoda et al. 2007b. Identification and imaging of the nerves responsible for erectile function in rat prostate, in vivo, using optical nerve stimulation and optical coherence tomography. *IEEE Journal Selected Topics in Quantum Electronics* 13: 1641–1645.

Gilchrist, K. H., R. P. McNabb, J. A. Izatt et al. 2009. Piezoelectric scanning mirrors for endoscopic optical coherence tomography. *Journal of Micromechanics and Microengineering* 19: 095012.

Huang, C., B. Liu, and M. E. Brezinski. 2008. Ultrasound-enhanced optical coherence tomography: Improved penetration and resolution. *Journal of Optical Society of America A* 25: 938–946.

Huang, D., E. Swanson, C. Lin et al. 1991. Optical coherence tomography. *Science* 254: 1178–1181.

Li, H., B. A. Standish, A. Mariampillai et al. 2006. Feasibility of interstitial Doppler optical coherence tomography for in vivo detection of microvascular changes during photodynamic therapy. *Lasers in Surgery and Medicine* 38: 754–761.

Manyak, M. J. 2007. The emergence of imaging technology in advanced prostate cancer. *Canadian Journal of Urology* 14: 32–138.

Manyak, M. J., M. Javitt, P. S. Kang et al. 2006. The evolution of imaging in advanced prostate cancer. *Urologic Clinics of North America* 33: 133–146.

Rais-Bahrami, S., A. W. Levinson, N. M. Fried et al. 2008. Optical coherence tomography of cavernous nerves: A step toward real-time intraoperative imaging during nerve-sparing radical prostatectomy. *Urology* 72: 198–204.

Skarecky, D. W., M. Brenner, S. Rajan et al. 2008. Zero positive surgical margins after radical prostatectomy: Is the end in sight. *Expert Review of Medical Devices* 5: 709–717.

Standish, B. A., X. Jin, J. Smolen et al. 2007. Interstitial Doppler optical coherence tomography monitors microvascular changes during photodynamic therapy in a Dunning prostate model under varying treatment conditions. *Journal of Biomedical Optics* 12: 034022.

Standish, B. A., K. K. Lee, X. Jin et al. 2008. Interstitial Doppler optical coherence tomography as a local tumor necrosis predictor in photodynamic therapy of prostatic carcinoma: An in vivo study. *Cancer Research* 68: 9987–9995.

Tearney, G. J., M. E. Brezinski, B. E. Bouma et al. 1997a. In vivo endoscopic optical biopsy with optical coherence tomography. *Science* 276: 2037–2039.

Tearney, G. J., M. E. Brezinski, J. F. Southern et al. 1997b. Optical biopsy in human urologic tissue using optical coherence tomography. *Journal of Urology* 157: 1915–1919.

Tomlins, P. and R. Wang. 2005. Theory, developments and applications of optical coherence tomography. *Applied Physics* 38: 2519–2535.

Tran, P. H., D. S. Mukai, M. Brenner et al. 2004. In vivo endoscopic optical coherence tomography by use of a rotational microelectromechanical system probe. *Optics Letters* 29: 1236–1238.

III

Fiber-Optic Sensors

11. Fiber Sensing
A Historic Overview

Orlando Frazão
INESC TEC

11.1 Introduction

The development of optical fiber sensors was first supported by the notable progresses in optical fiber technology associated with the need to improve the performance of optical fiber communication systems. Thus, as an introductory note to the description of the evolution of optical fiber sensors, the next paragraphs summarize some of those progresses according to their historical sequence.

In 1961, Elias Snitzer published an article in the *Journal of the Optical Society of America* entitled "Cylindrical dielectric waveguide modes" (Figure 11.1). The article included a theoretical description of a small core fiber, showing evidences that it was possible to obtain light propagation in a single guided mode (Snitzer 1961). In 1962, General Electric, IBM, and the Lincoln Laboratory at MIT demonstrated the first semiconductor laser using gallium arsenide (GaAs). In 1964, Charles Kao in a breakthrough development proposed the optical fiber as the core element of long-distance data transmission using the principle of optical communication (Kao and Hockham 1966).

In the summer of 1970, a team of researchers from Corning Glass performed experiments on light propagation using fused silica, an extremely pure material with a high melting point and a relatively low refractive index. Following that work, researchers Robert Maurer, Donald Keck, and Peter Schultz invented the optical fiber (United States, patent number 3.711.262). The same team had already developed a single-mode fiber with a titanium-doped core, registering a 17 dB/km loss at 633 nm. In June 1972, those researchers demonstrated the single-mode propagation in optical fibers with a germanium-doped

Handbook of Optical Sensors. Edited by José Luís Santos and Faramarz Farahi © 2015 CRC Press/Taylor & Francis Group, LLC. ISBN: 9781439866856.

Journal of the
OPTICAL SOCIETY
of AMERICA

VOLUME 51, NUMBER 5 MAY, 1961

Cylindrical Dielectric Waveguide Modes*

E. SNITZER
American Optical Company, Research Department, Southbridge, Massachusetts
(Received March 15, 1960)

The propagation of cylindrical dielectric waveguide modes near cutoff and far from cutoff are considered. The relative amounts of E_z and H_z, and the transverse components of the field are determined for both sets of hybrid modes. With the radial dependence of the z components of the field in the central dielectric given by $J_n(ur/a)$, the transverse components far from cutoff are given by $J_{n\pm 1}(ur/a)$, where u is a parameter found from the boundary conditions and which fixes the scale of the Bessel function relative to the boundary $r=a$. The two values $n+1$ and $n-1$ correspond to the two sets of modes. The designations of the hybrid modes are discussed. Field plots for the lower order modes are given.

FIGURE 11.1 Summary of the article by Snitzer published in the *Journal of the Optical Society of America.*

core, registering losses in the order of 4 dB/km. In 1973, John MacChesney, at the Bell Labs, changed the chemical vapor deposition process for the fabrication of optical fibers, thus reducing losses to 0.2 dB/km in the region of 1550 nm. Other major advance in the optical fiber domain was the possibility of fusion splicing two fibers with low losses using the electric arc technique. In that context, the study of the losses associated to the misalignment between fibers was performed by Marcuse (1977).

Other advance of optical fiber systems was the optical fiber coupler based on the techniques of fusion splicing and stretching of two twisted fibers (Abe et al. 1984). This development, which took place in 1981, made the fabrication of Michelson and Mach–Zehnder interferometers possible, as well as the reflectometric interrogation of the sensors and the fabrication of optical fiber rings. Later, the technical evolution of the fabrication led to the appearance of wavelength division multiplexers (WDMs), another fundamental component in the context of optical communications that is used to separate or group different wavelengths. In that same year, the operation of an optical circulator was demonstrated using polarization-sensitive optical prisms (Faraday rotator) (Shirasaki et al. 1981). In 1984, Stolen demonstrated the concept of the rocking filter (to couple polarization states) using high-birefringence fibers (Stolen et al. 1984). This type of filter uses a torsion periodical variation in the main axes of the fiber. The conversion of the polarization between the main axes depends on the wavelength, while the filter bandwidth is inversely proportional to the torsion period.

One of the first instruments developed to measure optical fiber losses was the *optical time-domain reflectometry* (OTDR), which consists of sending a light impulse and measuring the intensity of the Rayleigh scattering throughout the fiber. Another crucial development in the context of optical fiber systems took place in 1987 with the erbium-doped fiber amplifier (Desurvire et al. 1987). An equally important discovery was the possibility of using an ultraviolet interferometer to write optical fiber Bragg gratings (FBGs). However, these devices only started drawing attention of the scientific community in 1993 when it became possible to fabricate FBGs using the phase mask technique (Hill et al. 1993).

The fabrication of these devices directly in the core of the optical fiber—without affecting the physical integrity and the optical features of the fibers—proved to be, in the last decade, one of the most fertile research and development fields in optoelectronics. Equally important was the demonstration of long-period fiber gratings (LPGs) in 1996. These devices are periodical microstructures (with a period of the order of hundreds of micrometers) that couple the light of the fundamental guided mode in the fiber core to the copropagating modes of the cladding (Bhatia and Vengsarkar 1996). These modes decay rapidly as they propagate throughout the fiber due to losses in the air-cladding interface and due to possible curvatures in the fiber. These microstructures have multiple applications. For instance, it is possible to obtain a band rejection filter, as well as equalize gains in optical amplifiers and in optical fiber sensors.

The last structure progress up to date was probably the design and fabrication of a new type of optical fiber—the photonic crystal fiber—which took place in 1996 at the University of Southampton (Knight et al. 1996). This fiber has a pure silica core and a cladding with a periodical structure of channels through the fiber axis (in certain cases, the core is also a channel). This made way to new research and application areas in the context of optical fiber technology.

11.2 First Developments in Fiber-Optic Sensing

This exercise could have been made in several ways, but we chose to follow a chronological order of the publication records, as presented in Tables 11.1, 11.2, 11.3, and 11.4. Following those guidelines, some commentaries were also added to the mentioned works. Table 11.1 presents the historic chronology between 1967 and 1980.

It is historically accepted that the first research work involving optical fiber sensing was published in 1967 (Menadier 1967), as shown in Figure 11.2. This is a study on the efficiency of the light recoupling, which is reflected by a surface after abandoning the optical fiber. That efficiency depends on the surface distance to the tip of the fiber. Thus, this configuration is effectively a displacement sensor, emphatically designated as *Fotonic sensor* by the team that performed the study. In the subsequent years, progresses in this new scientific area were modest, a situation that changed in the second half of the 1970s.

Table 11.1 Progresses in Optical Fiber Sensing up to 1979

1967—The Fotonic sensor (Menadier 1967)
1974—A simple two-fiber optical displacement sensor (Powell 1974)
1976—Fiber ring interferometer (Vali and Shorthill 1976)
1977—Fiber-optic hydrophone (Bucaro et al. 1977); fiber sensor for voltage and current at high voltage (Rogers 1977); blackbody temperature sensor (Dakin and Kahn 1977); fiber-optic detection of sound (Cole et al. 1977)
1978—Mach–Zehnder fiber-optic strain gauge (Butter and Hocker 1978); Doppler anemometer (Dyott 1978)
1979—Displacement transducer (Cook and Hamm 1979), Mach–Zehnder pressure and temperature sensor (Hocker 1979); optical flowmeter (Culshaw and Hutchings 1979); multimode optical sensor (Hall 1979); air pollution sensor (Inaba et al. 1979); pressure sensitivity of a clad optical fiber (Budiansky et al. 1979); acoustic sensor utilizing mode–mode interference (Layton and Bucaro 1979)

Table 11.2 Progresses in Optical Fiber Sensing from 1980 to 1989

1980—Microbend sensor (Fields and Cole 1980); fiber-optic accelerometers (Tveten et al. 1980); fiber-optic rotation (Ulrich 1980); PH sensor (Goldstein et al. 1980); magnetic field fiber sensors (Dandridge et al. 1980); acoustic sensitivity using Brillouin effect (Lagakos et al. 1980)
1981—Extrinsic fiber Fabry–Pérot (Petuchowski et al. 1981); fiber spectrophone (Leslie et al. 1981); NO_2 sensor (Kobayasi et al. 1981); temperature sensor with Hi-Bi fiber (Eickhoff 1981); all-fiber gyroscope (Bergh et al. 1981); first review paper (Giallorenzi 1981)
1982—Confocal Fabry–Pérot sensor (Jackson et al. 1982a); interferometric signal processing (Jackson et al. 1982b); velocimeter (Erdmann and Soreide 1982); high-sensitivity accelerometer (Kersey et al. 1982)
1983—Ring-resonator sensor (Meyer et al. 1983); Michelson thermometer (Corke et al. 1983); methane sensor at 1.3 μm (Chan et al. 1983); fiber-optic sensor for surface acoustic wave (Bowers et al. 1983); multimode Michelson interferometer (Imai et al. 1983); temperature-distributed sensor with liquid-core fiber (Hartog 1983); optical fiber refractometer for infrared windows (Cooper 1983); coherence reading (Chalabi et al. 1983); simple fiber Fabry–Pérot sensor (Kersey et al. 1983); temperature sensor using a twin-core fiber (Meltz et al. 1983); *International Conference on Optical Fiber Sensors (OFS 1)*, London
1984—Monomode fiber-optic flowmeter (Leilabady et al. 1984); tapered sensor (Kumar et al. 1984); methane sensor at 1.6 μm (Chan et al. 1984); hydrogen sensor (Butler 1984); twisted fiber sensor (Abe et al. 1984); *edited the first book on optical fiber sensors* (Culshaw 1984)
1985—Raman distributed sensor (Dakin et al. 1985); optical fiber Faraday current sensor (Kersey and Dandridge 1985a); Fourier transform spectrometer (Kersey et al. 1985b); strain sensor with single-mode fiber (Leilabady et al. 1985); oxygen fiber-optic sensor (Wolfbeis et al. 1985); coherence multiplexing of interferometric sensors (Brooks et al. 1985); combined Michelson and polarimetric fiber-optic interferometric sensor (Corke et al. 1985)
1986—Side-hole fiber pressure sensor (Xie et al. 1986); refractometer using an optical coupler (Boucouvalas and Georgiou 1986); temperature-distributed sensor using Nd^{3+}-doped optical fiber (Farries et al. 1986); all-fiber polarimetric sensor (Mermelstein 1986)
1987—All-fiber hydrogen sensor (Farahi et al. 1987); optical seismometer (Gardner et al. 1987); liquid-level sensor (Morris and Pollock 1987)
1988—Intrinsic Fabry–Pérot (Lee and Taylor 1988); fiber-optic sensor for oxygen and carbon dioxide (Wolfbeis et al. 1988); polymer cure (Afromowitz 1988); fluorescence sensor (Grattan et al. 1988); hemispherical air cavity fiber Fabry–Pérot sensor (Gerges et al. 1988)
1989—Fabry–Pérot embedded in laminated composite (Lee et al. 1989); spun fiber sensor (Laming and Payne 1989); humidity sensor (Mitschke 1989); strain Brillouin sensor (Horiguchi et al. 1989); optical fiber rotary displacement sensor (West and Chen 1989); FBG sensor (Morey et al. 1989)

Powell (1974) described two different optical displacement sensors using only a pair of large-diameter fibers and a bundle of many small-diameter fibers. The operation mode is similar to the one described by Menadier (1967). The sensitivity of this two-fiber sensor is approximately the same as the most sensitive configuration of the many-fiber sensor.

Vali and Shorthill (1976) presented a communication demonstrating that the sensitivity of an optical fiber ring interferometer can be increased considerably when the number of fiber loops is higher. According to the authors, this effect can be used to design ultrasensitive gyroscopes.

Bucaro et al. (1977), in a famous publication, demonstrated the optical fiber hydrophone for the first time. The sensor was a Mach–Zehnder interferometer with optical fiber arms. The power dividers are conventional optical components using a dispatch

Table 11.3 Progresses in Optical Fiber Sensing from 1990 to 1999

1990—Temperature-distributed Brillouin sensor (Kurashima et al. 1990); simultaneous measurement (Farahi et al. 1990); fiber-optic based on electrochemiluminescence sensor (Egashira et al. 1990); water sensor (Tomita et al. 1990)
1991—Faraday rotator for sensors (Kersey et al. 1991); multiplexer of FBGs (Meltz 1991); white light interferometer (Chen et al. 1991); surface-mounted optical fiber strain sensor (Haslach and Sirkis 1991); fiber-optic pH sensor with solgel (Ding et al. 1991); miniaturized Michelson sensors (Murphy et al. 1991)
1992—Rare-earth-doped fiber sensor (Quoi et al. 1992); smoke sensor (Brenci et al. 1992); side-polished fiber sensors (Tseng and Chen 1992); liquid-level sensor (Iwamoto and Kamata 1992); FBG chemical sensor (Meltz et al. 1992); etched fibers as strain gauges (Vaziri and Chen 1992); evanescent wave methane detection using D-fiber (Culshaw et al. 1992); temperature and strain sensitivity in Hi-Bi elliptical fibers (Zhang and Lit 1992)
1993—FBG laser sensor (Kersey and Morey 1993); SPR in fiber-optic sensor (Jorgenson et al. 1993); Mach–Zehnder interferometer via concatenated biconical tapers (Brophy et al. 1993); fiber sensor for simultaneous measurement of current and voltage system (Rajkumar et al. 1993); D-fiber methane gas sensor with high-index overlay (Muhammad et al. 1993); temperature and strain sensitivity in Hi-Bi fibers (Zhang and Lit 1993); an FBG-tuned fiber laser strain sensor system (Melle et al. 1993); high-pressure grating sensors (Xu et al. 1993)
1994—Fluorescence-tapered sensor (Hale and Payne 1994); detection of O_2 in liquid (Singer et al. 1994); simultaneous measurement of strain and temperature using FBGs (Xu et al. 1994); oxygen sensor based on solgel (Craith et al. 1994)
1995—Chirped grating sensors (Xu et al. 1995); pH-distributed and O_2 optical sensor (Michie et al. 1995); optical fiber reflection refractometer (Suhadolnik et al. 1995)
1996—LPG sensor (Bhatia and Vengsarkar 1996); hybrid LPG/FBG sensor (Patrick et al. 1996); U-shape fiber sensor (Gupta et al. 1996a); FBG cryogenic temperature sensors (Gupta et al. 1996b); side-polished fiber sensor based on SPR (Homola and Slavik 1996); interferometric distributed optical ring fiber sensor (Spammer et al. 1996); FBG pressure sensor (Xu et al. 1996); distributed sensing using Raleigh and Brillouin (Wait and Newson 1996); simultaneous measurement of pressure and temperature (Ma et al. 1996)
1997—Hi-Bi fiber Sagnac sensor (Starodumov et al. 1997); acoustic FBG sensor (Takahashi et al. 1997); fiber sensor arrays using EDFAs (Wagener et al. 1997); fluorescence decay-time characteristics of erbium-doped optical fiber (Zhang et al. 1997); Fabry–Pérot for magnetic sensor (Oh et al. 1997); BOFDA (Garus et al. 1997); simultaneous measurement of strain and temperature using LPGs (Bhatia et al. 1997); FBG rosettes (Magne et al. 1997)
1998—Temperature Michelson LPG sensor (Lee and Nishii 1998a); Mach–Zehnder LPG sensor (Lee and Nishii 1998b); FBG refractometer (Asseh et al. 1998); LPG refractometer (Patrick et al. 1998); simple cantilever (Kalenik and Pajak 1998); thulium fiber sensor for high temperature (Zhang et al. 1998); quantum dots with fiber optic (Barmenkov et al. 1998); metal-coated FBG current sensor (Cavaleiro et al. 1998)
1999—FBG cavity sensors (Du et al. 1999); salinity sensor (Esteban et al. 1999); transmission Fabry–Pérot sensor (Kim et al. 1999); electric arc LPG (Hwang et al. 1999); DNA optical sensor (Liu and Tan 1999); nano-Fabry–Pérot sensor (Arregui et al. 1999); Hi-Bi LPG sensor (Zhang et al. 1999); LPG by CO_2 for high temperature (Davis et al. 1999); FBG in PCF (Eggleton et al. 1999); PCF for gas sensing (Monro et al. 1999); fiber distributed-feedback lasers as acoustic sensors (Lovseth et al. 1999)

Table 11.4 Progresses in Optical Fiber Sensing from 2000 up to the Present

2000—PCF sensor in composite (Blanchard et al. 2000); cure laminated composite with FBG (Murukeshan et al. 2000); FBG in textile composite (Tao et al. 2000); FBG sensor in multimode fiber (Zhao and Claus 2000)
2001—Fabry–Pérot sensor with hetero-core fiber (Tsai and Lin 2001); MEMS on top of the fiber (Abeysinghe et al. 2001); corrugated LPG sensor (Lin et al. 2001); intrinsic bend measurement with FBGs (Araujo et al. 2001); FBG refractometer (Schroeder et al. 2001); tilt FBG refractometer (Laffont and Ferdinand 2001)
2002—Whispering gallery modes for optical sensor (Vollmer et al. 2002); Fabry–Pérot FBG temperature sensor (Wan and Taylor 2002); LPG sensor in PCF (Kerbage et al. 2002); FBG in cryogenic applications (James et al. 2002)
2003—FBG ultrasonic sensor (Fomitchov and Krishnaswamy 2003); fluorescent sensing using biconical tapers (Wiejata et al. 2003); liquid crystal in PCF (Larsen et al. 2003); multimode interference fiber-optic sensor (Mehta et al. 2003); FBG sensor in side-hole fiber (Chmielewska et al. 2003); Fabry–Pérot sensors for detection of partial discharges in power transformers (Yu et al. 2003)
2004—Michelson LPG sensor (Swart 2004); fiber ringdown sensors (Wang and Scherrer 2004); thinned FBGs as refractive index sensor (Iadicicco et al. 2004); Sagnac interferometer sensor with PCF (Zhao et al. 2004); hollow core for gas sensing (Ritari et al. 2004); nanocarbon tubes for optical fiber sensors (Penza et al. 2004); nanotaper sensors (Brambilla et al. 2004); refractometer sensing using cladded multimode tapered optical fiber (Villatoro et al. 2004)
2005—Optical microfiber loop resonator (Sumetsky et al. 2005); SPR-based hetero-core-structured optical fiber sensor (Iga et al. 2005); biosensing using Fabry–Pérot (Zhang et al. 2005); Brillouin sensor in PCF (Zou et al. 2005); nanowire silica sensor (Lou et al. 2005); microspheres in the end of fiber (Hanumegowda et al. 2005); miniature fiber-optic pressure sensor (Zhu and Wang 2005)
2006—Michelson taper sensor (Yuan et al. 2006); liquid-core optical ring-resonator sensors (White et al. 2006); MMI reflection fiber sensor (Li et al. 2006); tapered holey optical fiber (Villatoro et al. 2006); fiber-top cantilever sensor (Iannuzzi et al. 2006); FBG-assisted SPR (Nemova and Kashyap 2006); all-fused-silica miniature optical fiber-tip pressure sensor (Wang et al. 2006)
2007—Suspended-core fiber sensor (Webb et al. 2007); Michelson with abrupt taper (Frazao et al. 2007); LPG with electrostatic self-assembly (Corres et al. 2007); carbon nanotube-based optical fiber sensors (Consales et al. 2007); FBG strain fiber laser sensor (Yang et al. 2007); elliptically spun birefringent PCF for current sensing (Michie et al. 2007)
2008—Fabry–Pérot by femtosecond (Wei et al. 2008); LPG in hollow-core fiber (Wang et al. 2008); sensor based on core-offset attenuators (Tian et al. 2008)
2009—Fiber taper interferometer with a subwavelength tip (Zhu et al. 2009); photonic bandgap fiber taper sensor (Ju et al. 2009); modal interferometer based on ARROW fiber (Aref et al. 2009)
2010—Sapphire FBG sensors (Mihailov et al. 2010); FBG in microfiber (Fang et al. 2010); Fabry–Pérot in fiber taper probe (Kou et al. 2010); hydrostatic pressure in Hi-Bi-microstructured fiber (Martynkien et al. 2010); optical microbubble resonator (Sumetsky et al. 2010); modal interferometry in Bragg fibers (Frazao et al. 2010); Fabry–Pérot sensor based on graded-index multimode fibers (Gong et al. 2010)
2011—Slow-light-based fiber-optic temperature sensor (Wang et al. 2011); micro-Michelson interferometer (Chen et al. 2011); long-active-length Fabry–Pérot strain sensor (Pevec and Donlagic 2011); fiber-optic hot-wire flowmeter (Caldas et al. 2011); refractometer based on large-core air-clad PCF (Silva et al. 2011); Fabry–Pérot with diaphragm-free (Ferreira et al. 2011)

Fiber Optics Techniques

THE FOTONIC SENSOR

C. MENADIER, C. KISSINGER, H. ADKINS
Instruments Division Mechanical Technology, Inc.

BY USING ADJACENT pairs of light sources and receivers, the distance to a reflecting surface can be measured by detecting the amount of reflected light (Fig. 1a). When sending and receiving elements are in contact with the surface, no light is reflected to the receiving element. As the distance between the elements and the surface increases, however, the cone of light from the transmitting element illuminates an increasingly larger area on the work surface. This area becomes, in effect, the source of a secondary cone of reflected light, which in turn increasingly illuminates the receiving element.

The relationship between surface displacement and receiver illumination remains essentially linear (Fig. 1b), until the entire surface of the receiving element is illuminated by the reflected light cone, at which point the curve reaches its peak. As the distance increases beyond this point, the illumination of the receiver decreases in approximately inverse proportion to the square of the distance.

Through the use of a large number of fiber optics fibers, the Fotonic sensor achieves the first claimed practical application of this principle. Although the principle is valid for even one pair of transmitting and receiving elements, the amount of returned light and slope of the resulting response curve is improved by the use of a large number of fibers.

The Fotonic fiber optics system permits, for the first time, the positioning of many pairs of elements in a very small space, giving the steep response curve necessary to provide extremely accurate displacement measurements.

Fiber Optics System

Remote illumination and viewing techniques employing fiber optics have been in use for several years, notably in medical and quality control applications. A number of firms, in this country and abroad, are marketing devices which use the phenomenon in one way or another. Such systems are in use in the process and information industries, in machine shops to illuminate specific work areas, in parts counting and inventory, in surgery and exploration, and in other diverse fields.

In the Fotonic sensor, the light-conducting properties of fiber optics permit positioning of the measuring components (light source and photocell) at a distance relatively remote from the measurement point. Also the physical flexibility permits measurements which would be impossible, because of space limitations or environmental conditions, if the optical path between measuring components and measuring point were rigid.

The sensitivity of a fiber optics measuring instrument is a function of the characteristics of the fiber bundle used. These characteristics can be carefully controlled, and include the quantity of fibers, their size and distribution, and the numerical aperture of the individual fibers (the angle at which light can enter or leave the end).

The probes used in the various Fotonic models employ high-quality fibers, each individually clad to prevent optical crosstalk. The core is flint glass with a refractive index of 1.62. The cladding—another type of glass—has a refractive index of 1.52. This difference in index causes total reflection of the light rays within the fiber core (Fig. 2) and prevents any of them from escaping and entering adjacent fibers.

This index difference also governs the maximum angle of internal reflection, thereby determining fiber aperture angle. In the standard Fotonic probe, 600 fibers, each 0.003 in. dia, are gathered in a cylindrical bundle. Near one end of the probe, the bundle is divided into two equal branches: one for light transmission, the other for reception. Synthetic or plastic fibers may also be used.

Distribution of fibers at the unbranched end of the bundle is a major factor controlling displacement range and slope sensitivity of the probe. For example, the optimum arrangement for the steepest response curve (and therefore the most sensitive response) involves surrounding each receiving fiber with four transmitting fibers (Fig. 3a). With this configuration, more and more receiving fibers are affected by the light which emanates from a single transmitting fiber as the distance between probe and work surface increases. Perfect realization of this configuration would imply careful deposition of alternate layers of transmitting and receiving fibers.

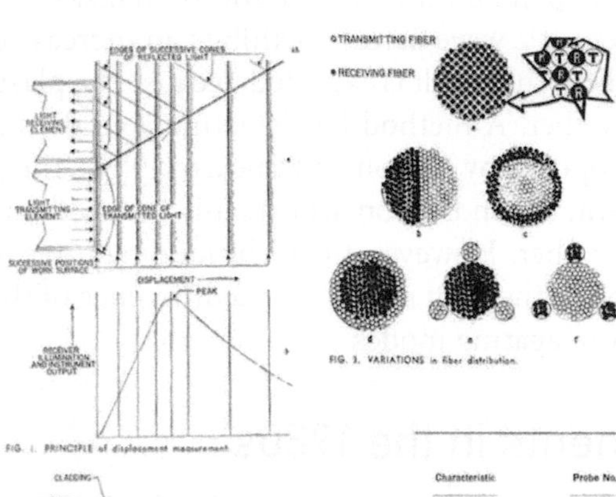

FIG. 1. PRINCIPLE of displacement measurement.

FIG. 3. VARIATIONS in fiber distribution.

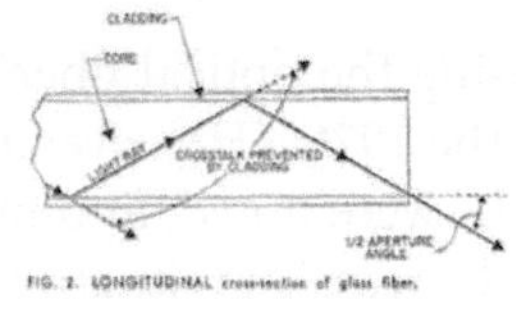

FIG. 2. LONGITUDINAL cross-section of glass fiber.

TABLE. 1. EFFECTS of varying fiber optics characteristics.

Characteristic	Probe No. 1	Probe No. 2	Probe No. 3	Probe No. 4
Probe Range	±0.0025"	±0.020"	±0.010"	±0.080"
Probe Outside Diameter	0.109"	0.109"	0.250"	0.250"
Fiber Size	0.003"	0.003"	0.008"	0.008"
Fiber Quantity	600	600	600	600
Fiber Distribution	random	hemispheric	random	hemispheric
Aperture Angle	60°	60°	60°	60°

FIGURE 11.2 First work on a photonic sensor published in *Instrument & Control Systems.*

demodulation technique that makes it possible for the phase variation induced by the acoustic wave to be read. In that same year, Rogers (1977) presented an article demonstrating the possibility of measuring electric currents using the magneto-optic effect on an optical fiber. Also, Dakin and Kahn (1977) presented a temperature sensor based on the blackbody radiation theory. The sensor structure consists of a step-index single mode fibers (SMF). A stainless steel tube was placed at the tip of the fiber. When heated to sufficiently high temperatures, the tube emitted radiation capable of being detected by a silica detector. The low-frequency acoustic wave response in optical fiber with water core was investigated by Cole et al. (1977).

Butter and Hocker (1978) demonstrated for the first time a strain sensor based on a Mach–Zehnder interferometer coupled to a cantilever. In that same year, Dyott (1978) presented an optical fiber anemometer based on the Doppler effect.

Cook and Hamm (1979) described a displacement sensor using a set of optical fibers. One of the fibers illuminated the flat target, while the remaining fibers captured the reflected signal, thus making it possible to mark power variations in the system that were not associated to the displacement of the target. In that same year, Hocker (1979) described an optical fiber Mach–Zehnder interferometric sensor to measure pressure and temperature. Also, Culshaw and Hutchings (1979) demonstrated a device to measure liquid flows using an optical fiber transducer. The flowmeter has no mobile mechanical parts, and the optical detection system, which is based on conventional optoelectronic components, is quite simple and robust. Furthermore, Hall (1979) implemented an acoustic–optic sensor for the first time using multimode fibers. The sensor system uses a heterodyne detection technique, showing a better performance when compared to equivalent systems based on single-mode fibers. Ikeda et al. (1979) presented an air pollution monitoring system based on the absorption differential method, using a multiple reflection configuration and recoupling to an optical fiber. The absorption differential method is based on the fact that the gases' absorption coefficients close to a resonance strongly depend on the wavelength. Budiansky et al. (1979) showed that the coated fiber in an acoustic wave detector exhibits an increase of the sensitivity compared to an uncoated fiber. The result is expected because the plastic is much more compressible than the glass fiber. A method for detecting sound using a single step-index multimode fiber was proposed by Layton and Bucaro (1979). The principle of operation, as in the two-arm system, depends upon acoustically induced phase modulation of the light propagating in the fiber. However, the technique requires a multimode fiber, for the intensity modulation of the light resulting in a difference of the induced phase shifts between two or more propagating modes.

11.3 Developments in the 1980s

Pioneering works based on the concept of using the optical fiber as a sensing element appeared in the late 1960s and mostly along the 1970s. However, serious limitations in the source–fiber and fiber–photodetector links were noted, and most of the configurations used bulk systems to connect multiple devices. The situation changed in the next decade with a notable burst of activity in fiber sensing, as described and summarized in Table 11.2.

In 1980, Fields et al. published results regarding single-mode microcurvature sensors to measure pressure and acoustic waves. These sensor structures, later significantly explored, enabled low-cost and easy-to-construct sensing solutions. Besides, Tveten et al. (1980) demonstrated an optical fiber accelerometer based on a Mach–Zehnder interferometer with one of the arms coupled to a seismic mass. With this sensor, it was possible to measure accelerations up to 400 Hz. This was the beginning of the application of optical fiber sensors as accelerometers, an area of intervention that has acquired a significant importance over the years. Ulrich (1980) carried out research on the optimization of rotation sensors using phase modulation and active stabilization of polarization techniques. It is also important to mention that, in the same year, Goldstein et al. (1980) presented, for the first time, an optical fiber-based pH sensor. Dandridge et al. (1980) proposed the first magnetic field fiber sensor. Magnetic sensors employ magnetostrictive jacketing materials in conjunction with conventional single-mode optical fibers. An all-fiber Mach–Zehnder interferometer was used to detect the magnetically induced changes in optical path length that arose because of strains transferred to the fiber from the magnetostrictive jacket. Lagakos et al. (1980) studied the elastic and elastooptic coefficients used to predict acoustic response sensitivity for two different single-mode optical fibers, determined from Brillouin scattering measurements.

Petuchowski et al. (1981) demonstrated a Fabry–Pérot external cavity created by the micrometric interval between two single-mode fibers with the ends coated with a partially reflective glass. Leslie et al. (1981) described an acoustic wave detector based on a Mach–Zehnder interferometer. In the context of NO_2 detection, Kobayasi et al. (1981) proposed a system similar to the one proposed by Inaba et al. (1979) yet with a better performance consequence of the optimization of the sensor head with three mirrors, two of them used to condition the light waves at the entry and exit of the system. In that same year, Eickhoff (1981) presented for the first time a temperature sensor using a high-birefringence fiber with an elliptical core. Bergh et al. (1981) also achieved an important progress by demonstrating the operation of a gyroscope totally implemented in optical fiber, which opened new possibilities for further developments with this type of sensor. In this year, Giallorenzi published the first review paper in fiber-optic sensors (Giallorenzi 1981).

In 1982, in the context of interferometric sensor demodulation, a conceptually simple technique was used to generate—from the phase modulation of one of the interferometer arms—an electric carrier with a phase that is proportional to the differential phase of the optical interferometer. This technique, included in the group of pseudoheterodyne techniques, has been frequently used since then, and its bibliographical reference is one of the most quoted in the optical fiber sensor domain (Jackson et al. 1982b). In that same year, Jackson et al. (1982a) published a work on a new configuration for a Fabry–Pérot cavity that uses confocal lenses coupled to the optical fibers. Erdmann and other collaborators proposed another sensor to measure velocity through the Doppler effect (Erdmann and Soreide 1982). A high-sensitivity accelerometer was described by Kersey et al. (1982) to detect the acceleration induced in a compliant cylinder based on a rubber.

The year 1983 was particularly productive as far as progresses in optical fiber sensors are concerned. Meyer et al. (1983) published a study on rotation sensors based on the Sagnac interferometric configuration. Corke et al. (1983) reported the features of an optical fiber Michelson interferometer operating at 800 nm. Chan et al. (1983) demonstrated a methane sensor using a semiconductor InGaAsP laser emitting at the absorption bandgap region ($\upsilon_2 + 2\upsilon_3$) of this gas at 1330 nm. Bowers et al. (1983) described an entire optical system that made it possible to analyze materials from the detection of acoustic waves propagating through the surface of those materials. Imai et al. (1983) studied the use of a Michelson interferometer implemented in a multimode fiber with homodyne processing to detect vibration. In a famous article, Hartog (1983) demonstrated, using the OTDR principle, the temperature-distributed measurement using a liquid-core optical fiber with a refractive index that was higher than the one of the cladding. Cooper (1983) used optical fiber refractometric configurations to measure the refractive index in several liquids. Chalabi et al. (1983) demonstrated a coherence interrogation concept for interferometric sensors. This was an important progress in a sense that it allowed a systematic use of low-coherence sources in the context of interferometric-based optical fiber sensors. Kersey et al. (1983) described a simple Fabry–Pérot fiber sensor based on a fiber section length, demonstrated as an accelerometer configuration. The first temperature sensor using a twin-core fiber was reported by Meltz et al. (1983). The first edition of the *International Conference on Optical Fiber Sensors (OFS 1)* in London also took place in April 1983. Ever since then, this international conference constitutes an important platform for the presentation of research on optical fiber sensors. The conference takes place every 18 months and the last edition (23rd edition) was held in June 2014.

Leilabady et al. (1984) demonstrated a monomode fiber-optic flowmeter. The configuration is based on an interferometric technique in which the operating principle is the flow-induced oscillation of the fiber caused by a vortex shedding. Kumar et al. (1984)

described a refractometer based on a biconical filter fabricated in a multimode fiber. On the other hand, Chan et al. (1984) presented their results on the detection of methane gas using its $2\upsilon_3$ absorption line at 1.6 μm. The illumination was provided by InGaAsP semiconductor lasers. The first hydrogen fiber sensor was demonstrated by Butler (1984) and consisted of a palladium coating that expanded upon exposure to hydrogen. Abe et al. (1984) reported an intensity sensor using twisted optical fibers to detect tensile-distributed strain. The sensor converts the tensile strain to optical loss. In that same year, Brian Culshaw published the first book dealing with optical fiber sensors and the associated signal processing.

Dakin et al. (1985) demonstrated a temperature-distributed sensor based on the Raman effect, achieving a 3 m spatial resolution in a 200 m extension. Kersey et al. (1985b) presented a configuration for close infrared spectroscopy based on a Michelson interferometer illuminated with a superluminescent diode, and a He–He laser for calibration, using Fourier processing techniques. On the other hand, Leilabady et al. (1985) reported a strain sensor using a high-birefringence single-mode fiber, thus changing the optical phase measurand and the polarization state. A Fabry–Pérot interferometer was also employed in this system in order to improve the resolution of the sensing structure. Furthermore, in that year, Wolfbeis et al. (1985) proposed a fluorescence-based optical fiber sensor to detect oxygen. It is also important to highlight that in 1985, the demonstration of interferometric sensor coherence multiplexing was also an important progress (Brooks et al. 1985). An optical fiber Faraday current sensor using a closed-loop configuration was described by Kersey and Dandridge (1985a). In the same year, Corke et al. (1985) combined a Michelson configuration with a polarimetric fiber-optic interferometer to achieve a temperature sensor.

In 1986, we witnessed, for the first time, the emergence of a fiber with special features, similar to a PANDA fiber. However, the tension region in the distinct material interface is replaced by two air holes (Xie et al. 1986). With the variation of air pressure in these holes, it is possible to tune the fibers' birefringence in a certain interval. An optical refractometer using a simple optical coupler was reported by Boucouvalas and Georgiou (1986). Farries et al. (1986) demonstrated a temperature-distributed sensor using Nd^{3+}-doped optical fiber interrogated by an OTDR. The first all-fiber polarimetric sensor was reported by Mermelstein (1986).

Farahi et al. (1987) presented an innovative configuration for a hydrogen sensor based on a Michelson interferometer where the fiber in one of the arms is fixed to a palladium wire. In the presence of hydrogen, this element expands proportionately to the gas concentration, which causes the fiber to strain and, consequently, the interferometer optical phase to change. In that same year, Gardner et al. (1987) created a seismograph, also associated to an optical fiber Michelson interferometer, where a 520 g seismic mass acted on a rubber mandrel to which one of the interferometer arms was rolled up. On the other hand, Morris and Pollock (1987) presented a sensor capable of measuring water levels, which was based on the deposition of a thin film on the optical fiber. Since the film is sensitive to the liquid, its mechanical properties are changed in the presence of liquids.

Lee and Taylor (1988) achieved an important progress when they demonstrated an intrinsically fabricated Fabry–Pérot cavity, where two fiber ends were previously coated, before the splicing of a reduced fiber extension between them (with a properly adapted fusion splicing technique). The device was demonstrated as a temperature sensor. Wolfbeis et al. (1988) published the configuration of an optical fiber sensor that was capable of simultaneously measuring oxygen and carbon dioxide. The two indicators were activated by a source with the same wavelength. Afromowitz (1988) also proposed the possibility of monitoring polymer curing processes with optical fiber sensors, taking into consideration the difference between these materials' refractive indexes when they are in their liquid and solid states. Grattan et al. (1988), on the other hand, presented a temperature sensor based on the neodymium fluorescence decay time on the silica matrix. A hemispherical air cavity Fabry–Pérot fiber sensor was used as temperature sensor by Gerges et al. (1988).

Bragg grating technology was probably the most important development that encouraged many researchers to guide their activity toward the design of sensing structures based on these devices. The pioneering reference in this area is the one by Morey et al. (1989), who demonstrated the use of an optical FBG to sense temperature and strain. In that same year, Lee et al. (1989) described a Fabry–Pérot cavity in a carbon fiber-based composite material. At the same time, Laming and Payne (1989) fabricated a spun optical fiber, which consisted of a high-birefringence fiber that was rotated at a constant speed while the preform was fabricated. As a consequence, the fiber got an elliptical birefringence whose eccentricity was reduced in the presence of an external magnetic field. The phenomenon was used on an electric current sensor configuration. Mitschke (1989) fabricated a Fabry–Pérot cavity based on TiO_2 thin films where the visibility was modulated by the degree of humidity. As a final reference to the year 1989, it is important to mention the work carried out by Horiguchi et al. (1989) where the distributed strain applied to a fiber was characterized using the Brillouin effect. West and Chen (1989) demonstrated an intensity fiber rotary displacement sensor whose basic structure consisted of a twist sensor based on a multimode fiber with a large core spliced between two single-mode fibers.

11.4 Developments in the 1990s

This decade was marked by a huge research effort on the sensing potentialities of FBGs and on a multitude of actions toward their widespread application in diverse technological domains. Also, other important developments took place in the fiber sensing field, benefiting from the outcome of new sensing concepts as well as of technological breakthroughs in fiber-optic technology, mostly induced within the optical fiber communication environment, with the optical amplification being a notable example. The most relevant progresses are summarized in Table 11.3 and described in the following.

Kurashima et al. (1990) demonstrated a temperature-distributed sensor using the Brillouin optical fiber time-domain analysis (BOTDA) concept. Farahi et al. (1990) analyzed the possibility of simultaneously measuring temperature and mechanic strain applied to a high-birefringence optical fiber. For that, the researchers considered the cross-sensitivity effects, as well as the presence of nonlinear terms associated with the high amplitude of these measurands. In that same year, Egashira et al. (1990) presented an optical fiber sensor in order to determine oxalates based on the electrochemical luminescence phenomenon. Furthermore, Tomita et al. (1990) studied a sensor configuration to detect water using attenuation induced by the presence of this liquid in the fiber mechanical splicing areas. An OTDR was used to obtain the spatial position of the splicing, as well as the level of attenuation.

Kersey et al. (1991) described the first configuration for Michelson interferometer that was insensitive to variations in the polarization state, achieved by Faraday rotation elements. This was an important evolution, consolidated over the subsequent years. Morey (1991) presented a multiplexing system for Bragg gratings used for sensing purposes. In that same year, Chen et al. (1991) used the white light interferometry concept to demonstrate strain and temperature sensors with an extensive dynamic range. Haslach and Sirkis (1991) studied different types of strain sensors applied (glued) to surfaces so as to be used as optical extensometers (strain gauges). Using the same concept, Ding et al. (1991) immobilized a pH indicator (bromocresol) as a pH sensor. A miniaturized Michelson sensor was described by Murphy et al. (1991). The geometrical configuration was based on a fused-biconical tapered coupler cleaved immediately after the coupled length and polished down to the region of the fused cladding.

Quoi et al. (1992) studied the use of optical fiber sensors in rare earths (Nd^{3+}, Pr^{3+}, Sm^{3+}, and Yb^{3+}) for temperature-distributed sensors. Brenci et al. (1992) proposed a smoke detecting sensor that had some advantages when compared to the conventional methods, especially in the presence of electromagnetic interference or explosive substances. The sensor used the triangulation technique in order to monitor particle scattering. Tseng and Chen (1992) presented a refractometer to measure the refractive index in liquids, which consisted of a side-polished (up to the core area) fiber. This caused the evanescent field to interact with the surrounding environment. Also, Iwamoto and Kamata (1992) reported an oil-level sensor based on the determination of the light reflection point on the surface of the liquid. The first FBG chemical sensor was described by Meltz et al. (1992). Vaziri and Chen (1992) demonstrated that the geometry of the features was defined photolithographically and the structures were formed by chemical etching. The etched fiber sensing elements were fabricated and tested as strain sensors. An evanescent wave methane detection using D-fiber was reported by Culshaw et al. (1992). Temperature and strain measurements in Hi-Bi elliptical fibers based on double-clad elliptical fiber were reported by Zhang and Lit (1992).

Kersey and Morey (1993) studied a laser structure using Bragg gratings as reflective elements. Therefore, the emission wavelength was a function of the measurand action over these elements. The cavity presented a ring geometry where the gain was achieved through an erbium-doped fiber amplifier. Jorgenson et al. (1993) demonstrated the possibility of using an optical fiber as a platform to the surface plasmon resonance (SPR) technique, removing the cladding by chemical etching and depositing a silver film. With this refractometer, it was possible to achieve a high sensitivity in the measurement of the external environment refractive index. A single-fiber Mach–Zehnder interferometer that combined core and cladding modes in a concatenated taper device has been described by Brophy et al. (1993). A single-mode fiber sensor for simultaneous measurement of current and voltage system was reported by Rajkumar et al. (1993). A D-fiber methane gas sensor with high-index overlay was demonstrated by Muhammad et al. (1993). Temperature and strain measurements in commercial Hi-Bi fibers were reported (Zhang and Lit 1993). A fiber laser that permitted the efficient interrogation of a sensing Bragg grating has been developed by Melle et al. (1993). A simple high-pressure Bragg grating sensor was demonstrated by Xu et al. (1993).

Hale and Payne (1994) studied a fluorescence sensor based on a biconical taper applied as a pH sensor. Furthermore, Singer et al. (1994) demonstrated the viability of using an optical fiber sensor to measure dissolved oxygen using a ruthenium complex. The first simultaneous measurement of strain and temperature performed by an FBG sensor using different wavelengths was reported by Xu et al. (1994). Craith et al. (1994) proposed an oxygen sensor based on a solgel deposition doped with ruthenium complex.

Xu et al. (1995) managed to demonstrate a temperature-insensitive strain sensor using an aperiodic Bragg grating written on a taper. In the same year, Michie et al. (1995) presented a pH-distributed and water detecting sensor using hydrogel optical fibers in the polymer coating. Furthermore, a reflective reading sensor was developed in order to measure the refractive index (Suhadolnik et al. 1995).

In 1996, an important progress was achieved with the introduction of the LPG used as a sensing element (Bhatia and Vengsarkar 1996). These structures were characterized for strain, temperature, and refractive index, considering several types of fibers and fabrication conditions. Patrick et al. (1996) presented a hybrid sensor based on an LPG and two Bragg gratings on the sides of the LPG attenuation bandgap. Gupta et al. (1996b) demonstrated the possibility of using Bragg gratings in cryogenic systems. Homola and Slavik (1996) proceeded with the development of the optical fiber supported SPR concept, now taking the lateral coating of the fiber into consideration. Using an interferometric ring, Spammer et al. (1996) managed to detect and locate mechanic or thermal perturbations throughout the fibers. Gupta et al. (1996a) presented a refractive index sensor based on the interaction of the measurand with the evanescent fields related to a U-shape multimode fiber from which the cladding was removed. Xu et al. (1996) demonstrated a pressure sensor based on a Bragg grating placed inside a hollow glass sphere. Finally, Wait and Newson (1996) studied the distributed measurement using the Landau–Placzek ratio, which consists of the intensity ratio of the Rayleigh and Brillouin scatterings. A simultaneous measurement of pressure and temperature based on highly elliptical core two-mode fiber was demonstrated by Ma et al. (1996).

Starodumov et al. (1997) demonstrated a Sagnac interferometer as a temperature sensor. The interferometer was based on a high-birefringence coupler where the two output gates were connected. However, both fibers suffered a 90° rotation, which made it possible to exchange radiation between the two polarization axes. In that same year, Takahashi et al. (1997) studied the characteristics of Bragg gratings when used as acoustic sensors in underwater environments. Wagener et al. (1997) designed erbium-doped fiber amplifiers in order to improve the efficiency of sensor multiplexing networks. Zhang et al. (1997) studied the fluorescence decay times of erbium-doped fibers for temperature sensing applications to be used in extreme conditions. Besides, Oh et al. (1997) presented a Fabry–Pérot cavity to measure magnetic fields where one of the reflectors was coupled to a magnetoresistive transducer. Finally, Garus et al. (1997) demonstrated strain and temperature measurements using the Brillouin optical fiber frequency-domain analysis (BOFDA) method. A simultaneous measurement of strain and temperature using different wavelength in the same LPG was reported by Bhatia et al. (1997). The first FBG rosettes were analyzed by Magne et al. (1997).

In 1998, Lee and Nishii (1998a) described self-interference fringes created in a single LPG. The same authors demonstrated a Mach–Zehnder interferometer based on two LPGs in series to measure curvature (Lee and Nishii 1998b). Asseh et al. (1998) proposed the use of Bragg gratings as refractometers, removing the cladding by chemical etching in the sensing area. Furthermore, Patrick et al. (1998) studied different types of fiber with LPGs that could be applied as refractive index sensors. Kalenik and Pajak (1998) demonstrated optical fiber cantilever configurations with the potential to sense several values. Also, Zhang et al. (1998) proposed, for the first time, the use of thulium fibers to measure temperatures up to 1250°C. For that same year, it is also important to mention the work carried out by Barmenkov et al. (1998). For temperature measurement processes, the authors demonstrated the potential of using semiconductor nanocrystals (quantum dots) inserted in a silica matrix illuminated by the optical fiber radiation. The new device was a Michelson interferometer and was tested as temperature sensor. Cavaleiro et al. (1998) reported a hybrid optical fiber current sensor combining a metal-coated FBG with a standard current transformer.

Du et al. (1999) proposed the simultaneous measurement of strain and temperature using a Fabry–Pérot cavity with Bragg gratings working as reflectors. The cavity was glued onto an aluminum tube and when subjected to temperature variations, the thermal expansion of the tube caused the interferometer phase to vary as well. Esteban et al. (1999) demonstrated the determination of water salinity using the SPR concept supported by an optical fiber taper. Also in that year, Kim et al. (1999) studied the features of Fabry–Pérot cavities as sensing elements in transmission operation and in the context of distributed measurements. Hwang et al. (1999) studied the fabrication of LPGs by electric arc, characterizing their properties as temperature sensors. Besides, Liu and Tan (1999) presented a DNA biosensor based on evanescent field interaction associated to a cladding mode. Additionally, Arregui et al. (1999) studied a humidity sensor based on the properties of the nanocavities fabricated with the ionic self-assembly technique. Zhang et al. (1999) studied LPGs written on high-birefringence fibers. Davis et al. (1999) characterized the stability of LPGs written with CO_2 laser for high temperatures. An important progress was reported by Eggleton et al. (1999), who demonstrated FBGs written in germanium-doped microstructured fibers (photonic crystal fibers). Finally, Monro et al. (1999) used these types of fibers to detect gases. New contributions to the acoustic signal sensitivity of fiber distributed-feedback DFB lasers in air were investigated both theoretically and experimentally by Lovseth et al. (1999).

11.5 Developments in the Last Decade

In optical fiber technology, this period will in the future be certainly connected with the confirmation of the breakthroughs offered by the concept of photonic crystal fibers, turning possible qualitatively new technological approaches in several domains. One of them is fiber-optic sensing, as shown from the most relevant progresses in this period summarized in Table 11.4.

Blanchard et al. (2000) presented a curvature interferometric sensor using two-core PCF fibers. Murukeshan et al. (2000) studied the curing mechanism for carbon fiber–reinforced plastic (CFRP) and glass fiber–reinforced plastic (GFRP) composites with embedded FBGs for sensing processes. In that same line, Tao et al. (2000) studied strain in laminated composites. Zhao and Claus (2000) wrote FBGs in multimode fibers and, as a result, they realized that there were several resonances associated to modal families. Thus, their properties were characterized as far as temperature and strain measurements were concerned.

Tsai and Lin (2001) studied Fabry–Pérot cavities based on fibers with different cores that were characterized as sensing structures for curvature and load. Abeysinghe et al. (2001) established an important evolution when they fabricated a microelectromechanical system (MEMS) device at the tip of an optical fiber, demonstrating it as a pressure sensor. Lin et al. (2001) designed and implemented corrugated LPG. These structures were characterized as sensing elements. In that same year, Araujo et al. (2001) described a configuration based on Bragg gratings written on D-type fibers, intrinsically sensitive to curvature. With this configuration, it is not necessary to couple the fiber to a mechanical structure in order to make these measurements. Schroeder et al. (2001) demonstrated an optochemical in-FBG sensor for refractive index measurement in liquids using fiber side-polishing technology. At a polished site where the fiber cladding has been partly removed, an FBG is exposed to a liquid analyte via evanescent field interaction of the guided fiber mode. Laffont and Ferdinand (2001) reported a tilt FBG sensor as optical refractometer.

Vollmer et al. (2002) presented a biosensor sensitive to predefined molecules, meaning that it gained functionalization. The device uses an optical microcavity with a particular set of resonances excited by the evanescent coupling to an optical fiber biconical filter. Wan and Taylor (2002) studied a Fabry–Pérot cavity setup from a uniform Bragg grating divided into two segments that were connected to each of the extremities of a single-mode fiber extension. This structure was characterized as a temperature sensor. Furthermore, it was also important to mention the work carried out by Kerbage et al. (2002) regarding the fabrication of LPGs in microstructured fibers. The objective was to study the dependence of these gratings' attenuation bandgap when some of the channels were filled with polymer materials. James et al. (2002) studied the temperature and strain response of FBG sensors in a cryogenic environment.

Fomitchov and Krishnaswamy (2003) studied thoroughly the use of Bragg gratings to detect sound waves. Wiejata et al. (2003) characterized taper-based fluorescent structures. Larsen et al. (2003) studied the effect of liquid crystals on microstructured fibers, particularly in temperature variation situations. Mehta et al. (2003) demonstrated a displacement sensor based on multimode interference using a step-index multimode fiber. Chmielewska et al. (2003) presented the simultaneous measurement of pressure and temperature using FBG imprinted in Hi-Bi side-hole fiber. Yu et al. (2003) used a diaphragm-based interferometric fiber-optic sensor for online detection of the acoustic waves generated by partial discharges inside high-voltage power transformers.

Swart (2004) demonstrated a Michelson modal interferometer associated to an LPG, with reflection at the tip of the fiber. The interferometer phase depended on the refractive index of the external environment, thus becoming a refractometer. Wang and Scherrer (2004) presented a ringdown-based pressure sensor. This concept uses an optical fiber ring as resonating cavity. The radiation is coupled to a closed-loop optical fiber, which causes the optical power to accumulate inside the ring. When the entry radiation is interrupted, the losses of power in the ring cause the detected radiation to decay. The decay time depends on the ring features, which are influenced by variations in pressure. Iadicicco et al. (2004) demonstrated the concept of the Bragg grating with phase shift that is originated by the removal of the cladding in the middle section of the grating and by the influence of an exterior liquid. The amplitude of the phase shift depends on the liquid refractive index. Zhao et al. (2004) studied a Sagnac interferometer where the ring included a microstructured fiber extension. The properties of this structure were then studied to sense temperature and strain. In addition, Ritari et al. (2004) studied the hollow-core microstructured fibers to be used as gas sensors. Besides, Penza et al. (2004) fabricated carbon nanotubes at the tip of the fiber in order to detect alcohol. Finally, Brambilla et al. (2004) developed nanotaper structures with low losses and high potential for sensing. Villatoro et al. (2004) proposed an optical refractometer based on a taper fabricated in a multimode fiber.

Sumetsky et al. (2005) presented a resonating cavity based on an optical fiber microring fabricated from a long taper. Iga et al. (2005) studied the use of the surface plasmonic resonance concept on fibers with different cores, combining single and multimode fibers. Zhang et al. (2005) proposed the use of micro-Fabry–Pérot cavities as biosensors. Zou et al. (2005) demonstrated the distributed measurement in a microstructured fiber using Brillouin scattering. Furthermore, Lou et al. (2005) studied silica nanowires, applying them as sensing structures. Hanumegowda et al. (2005) demonstrated a refractometer based on a resonating microsphere applied at the tip of an optical fiber. Finally, Zhu and Wang (2005) fabricated interferometric structures using fusion splicing between the single-mode/multimode fibers and between multimode/single-mode fibers, as well as chemical etching in the multimode fiber. These structures were characterized as pressure sensors.

Yuan et al. (2006) demonstrated a Michelson interferometer applying fusion splicing between the two fibers, one of them being a two-core fiber. A taper was fabricated in the splicing area so that light coming from the single-mode fiber could be coupled in the two cores of the following fiber. The reflection at this fiber extremity led to the configuration of a Michelson topology. White et al. (2006) demonstrated a new sensor based on a liquid-core optical ring resonator in a fused silica capillary. The architecture sensor was utilized to carry the aqueous sample and to act as the ring resonator. Li et al. (2006) demonstrated a temperature sensor using a step-index section fiber interrogated in reflection. In addition, Villatoro et al. (2006) studied taper structures fabricated in microstructured fibers with low sensitivity to temperature. Finally, Iannuzzi et al. (2006) proposed the fabrication of a cantilever structure on top of an optical fiber with optimized features for the measurement of temperature. Nemova and Kashyap (2006) studied a new surface plasmon–polariton (SPP) fiber sensor with an FBG imprinted into the fiber core for SPP excitation. An all-fused-silica pressure sensor fabricated directly onto a standard fiber tip was described by Wang et al. (2006). The simple fabrication steps only included the cleaving and the fusion process between a standard SMF and a hollow-core tube.

Webb et al. (2007) demonstrated the use of suspended-core optical fibers to gas sensing. Frazao et al. (2007) proposed the interferometric configuration that is probably the easiest to fabricate. It consists of a modal Michelson topology interferometer where the cladding mode is excited through a taper with certain features. Corres et al. (2007) studied the deposition of thin films on LPG using the electrostatic self-assembly process. As a result, it was possible to create several devices whose features were suitable for biochemical sensing, as demonstrated by the authors in the context of pH measurement. Consales et al. (2007) studied, for the first time, the excellent sensing properties of carbon nanotubes for detection of chemical pollutants in aqueous environments at room temperature using the tip of an optical fiber. Yang et al. (2007) demonstrated an FBG strain sensor based on erbium-doped fiber laser. A strain-sensing FBG element also acts as the lasing wavelength selecting component. Michie et al. (2007) preformed of a highly linearly birefringent PCF during the drawing process. The result was a spun Hi-Bi PCF with good sensitivity to magnetic fields for current measurements with greatly reduced temperature dependence.

Wei et al. (2008) presented a Fabry–Pérot cavity built inside the optical fiber. For that, they used femtosecond laser ablation techniques. Furthermore, Wang et al. (2008) demonstrated the viability of writing LPGs in hollow-core PCF fibers, applying an arc to the cladding area in order to periodically collapse the microstructured fiber channels. Tian et al. (2008) reported two configurations based on core-offset attenuators. These configurations were tested as optical refractometers.

Zhu et al. (2009) reported an ultraminiature fiber-optic sensor based on a fiber taper interferometer with a subwavelength tip. The interferometer was fabricated at the end face of a single-mode fiber by wet etching using buffered hydrofluoric acid. The interferometer was characterized in temperature and immersed in liquids of different refractive index. Ju et al. (2009) fabricated nonadiabatic tapers in hollow-core, air–silica photonic bandgap fibers. The interferometer was experimentally demonstrated for strain and temperature measurement. Aref et al. (2009) demonstrated interferometric sensors based on antiresonance reflecting optical waveguide (ARROW) fibers and characterized them in strain and temperature as well.

Mihailov et al. (2010) reported a dual-strain/temperature sapphire FBG sensor. The FBG was fabricated in the microfiber by the use of a femtosecond laser pulse irradiation by Fang et al. (2010). This structure was used as optical refractometer. Kou et al. (2010) presented an ultrasmall all-silica high-temperature sensor based on a reflective Fabry–Pérot modal interferometer. The sensing head was made of a microcavity (~4.4 μm) directly fabricated into a fiber taper probe. Martynkien et al. (2010) designed, manufactured, and characterized two Hi-Bi microstructured fibers that featured a fivefold increase in polarimetric sensitivity to hydrostatic pressure. Sumetsky et al. (2010) reported a new method for fabricating very small silica microbubbles having a micrometer-order wall thickness forming an optical microbubble resonator. Frazao et al. (2010) demonstrated strain and temperature discrimination using a modal interferometer based on two Bragg fibers. The special nature of this sensor is that the two Bragg fibers used present a different external cladding shape. Gong et al. (2010) reported the simultaneous measurement of the refractive index and temperature of aqueous solutions using a Fabry–Pérot cavity. This cavity was fabricated by cascading an etched micro–air gap and a short section of GI-MMF to a single-mode fiber. The periodic focusing effect of the graded index-multimode fiber (GI-MMF) was incorporated into the cavity sensors to improve its performance.

Wang et al. (2011) proposed and experimentally demonstrated a method for temperature sensing using stimulated Brillouin scattering-based slow light. Chen et al. (2011) reported a micro-Michelson interferometer using a sphered-end hollow fiber. Pevec and Donlagic (2011) fabricated a Fabry–Pérot sensor based on a new geometry. The sensor was composed of a lead-in fiber that also formed the first FP semireflective surface, an outer (semiconical) wall, a second FP semireflective surface, a gutter that surrounded the second FP semireflective surface, and a tail section of the sensor (which could be of an arbitrary length). Caldas et al. (2011) demonstrated an all-optical hot-wire flowmeter based on a silver-coated fiber combining an LPG and an FBG structure. Silva et al. (2011)

reported an optical refractometer using a large-core, air-clad PCF. Finally, a Fabry–Pérot based on a diaphragm-free hollow-core silica tube for refractive index measurement in gas was proposed by Ferreira et al. (2011).

11.6 Concluding Remarks

Seeking a global insight on the evolution of the concept of optical fiber sensing, it was possible to realize that the central motivation behind these studies has always been laid on the exploration of the unique features of this concept, in a sense that the fiber is simultaneously a sensing element and a communication channel. From the beginning, this feature has brought enormous competitive advantages when compared to other concepts and measurement technologies, namely, because, with this technology, it is possible to consider local or remote and point or multiplexed measurements. Based on this reality, at an initial stage, the optical fiber sensing field has developed by adapting well-known concepts, structures, signal processing, and technologies from other domains, most notably from the optical fiber communication area. After that, the greatest advances were related to technological developments—emergence of Bragg gratings and long-period gratings, optical amplification, and microstructured fibers. It is reasonable to foresee that in the near future, significant developments in the optical fiber sensor domain will be grounded on the exploration of the remarkable potential of nanoscience and nanotechnology, particularly when addressing sensing in the context of life sciences.

References

Abe, T., T. Mitsunaga, and H. Koga. 1984. Strain sensor using twisted optical fibers. *Optics Letters* 9: 373–374.

Abeysinghe, D. C., S. Dasgupta, J. T. Boyd et al. 2001. A novel mems pressure sensor fabricated on an optical fiber. *IEEE Photonics Technology Letters* 13: 993–995.

Afromowitz, M. A. 1988. Fiber optic polymer cure sensor. *Journal of Lightwave Technology* 6: 1591–1594.

Araujo, F. M., L. A. Ferreira, J. L. Santos et al. 2001. Temperature and strain insensitive bending measurements with D-type fibre Bragg gratings. *Measurement Science & Technology* 12: 829–833.

Aref, S. H., O. Frazao, P. Caldas et al. 2009. Modal interferometer based on ARROW fiber for strain and temperature measurement. *IEEE Photonics Technology Letters* 21: 1636–1638.

Arregui, F. J., Y. J. Liu, I. R. Matias et al. 1999. Optical fiber humidity sensor using a nano Fabry-Perot cavity formed by the ionic self-assembly method. *Sensors and Actuators B-Chemical* 59: 54–59.

Asseh, A., S. Sandgren, H. Ahlfeldt et al. 1998. Fiber optical Bragg grating refractometer. *Fiber and Integrated Optics* 17: 51–62.

Barmenkov, Y. O., A. N. Starodumov, and A. A. Lipovskii. 1998. Temperature fiber sensor based on semiconductor nanocrystallite-doped phosphate glasses. *Applied Physics Letters* 73: 541–543.

Bergh, R. A., H. C. Lefevre, and H. J. Shaw. 1981. All-single-mode fiber-optic gyroscope. *Optics Letters* 6: 198–200.

Bhatia, V., D. Campbell, R. O. Claus et al. 1997. Simultaneous strain and temperature measurement with long-period gratings. *Optics Letters* 22: 648–650.

Bhatia, V. and A. M. Vengsarkar. 1996. Optical fiber long-period grating sensors. *Optics Letters* 21: 692–694.

Blanchard, P. M., J. G. Burnett, G. R. G. Erry et al. 2000. Two-dimensional bend sensing with a single, multi-core optical fibre. *Smart Materials & Structures* 9: 132–140.

Boucouvalas, A. C. and G. Georgiou. 1986. External refractive-index response of tapered coaxial couplers. *Optics Letters* 11: 257–259.

Bowers, J., R. Jungerman, B. Khuri-Yakub et al. 1983. An all fiber-optic sensor for surface acoustic wave measurements. *Journal of Lightwave Technology* 1: 429–436.

Brambilla, G., V. Finazzi, and D. J. Richardson. 2004. Ultra-low-loss optical fiber nanotapers. *Optics Express* 12: 2258–2263.

Brenci, M., D. Guzzi, A. Mencaglia et al. 1992. Fiber-optic smoke sensor. *Sensors and Actuators B-Chemical* 7: 780–783.

Brooks, J. L., R. H. Wentworth, R. C. Youngquist et al. 1985. Coherence multiplexing of fiber-optic interferometric sensors. *Journal of Lightwave Technology* 3: 1062–1072.

Brophy, T. J., L. C. Bobb, and P. M. Shankar. 1993. In-line singlemode fiber interferometer via concatenated biconical tapers. *Electronics Letters* 29: 1276–1277.

Bucaro, J. A., E. F. Carome, and M. R. Layton. 1977. Optical fiber hydrophone. *Journal of the Acoustical Society of America* 62: S72–S72.

Budiansky, B., D. C. Drucker, G. S. Kino et al. 1979. Pressure sensitivity of a clad optical fiber. *Applied Optics* 18: 4085–4088.

Butler, M. A. 1984. Optical fiber hydrogen sensor. *Applied Physics Letters* 45: 1007–1009.

Butter, C. D. and G. B. Hocker. 1978. Fiber optics strain-gauge. *Applied Optics* 17: 2867–2869.

Caldas, P., P. A. S. Jorge, G. Rego et al. 2011. Fiber optic hot-wire flowmeter based on a metallic coated hybrid long period grating/fiber Bragg grating structure. *Applied Optics* 50: 2738–2743.

Cavaleiro, P. M., F. M. Araujo, and A. B. L. Ribeiro. 1998. Metal-coated fibre Bragg grating sensor for electric current metering. *Electronics Letters* 34: 1133–1135.

Chalabi, A., B. Culshaw, and D. Davies. 1983. Partially coherence sources in interferometric sensors. *Proceedings of the First International Optical Fiber Sensors* 1: 132–135.

Chan, K., H. Ito, and H. Inaba. 1983. Absorption measurement of Nu-2+2-Nu-3 band of Ch4 at 1.33 Mu-M using an ingaasp light-emitting diode. *Applied Optics* 22: 3802–3804.

Chan, K., H. Ito, and H. Inaba. 1984. All-optical remote monitoring of propane gas-using a 5-km-long, low-loss optical fiber Link and an Ingap light-emitting diode in the 1.68-Mu-M region. *Applied Physics Letters* 45: 220–222.

Chen, N. K., K. Y. Lu, J. T. Shy et al. 2011. Broadband micro-Michelson interferometer with multi-optical-path beating using a sphered-end hollow fiber. *Optics Letters* 36: 2074–2076.

Chen, S., A. J. Rogers, and B. T. Meggitt. 1991. Electronically scanned optical-fiber Young white-light interferometer. *Optics Letters* 16: 761–763.

Chmielewska, E., W. Urbanczyk, and W. J. Bock. 2003. Measurement of pressure and temperature sensitivities of a Bragg grating imprinted in a highly birefringent side-hole fiber. *Applied Optics* 42: 6284–6291.

Cole, J. H., R. L. Johnson, and P. G. Bhuta. 1977. Fiber-optic detection of sound. *Journal of the Acoustical Society of America* 62: 1136–1138.

Consales, M., A. Crescitelli, S. Campopiano et al. 2007. Chemical detection in water by single-walled carbon nanotubes-based optical fiber sensors. *IEEE Sensors Journal* 7: 1004–1005.

Cook, R. O. and C. W. Hamm. 1979. Fiber optic lever displacement transducer. *Applied Optics* 18: 3230–3241.

Cooper, P. R. 1983. Refractive-index measurements of liquids used in conjunction with optical fibers. *Applied Optics* 22: 3070–3072.

Corke, M., J. D. C. Jones, A. D. Kersey et al. 1985. Combined Michelson and polarimetric fibre-optic interferometric sensor. *Electronics Letters* 21: 148–149.

Corke, M., A. D. Kersey, D. A. Jackson et al. 1983. All-fiber Michelson thermometer. *Electronics Letters* 19: 471–473.

Corres, J. M., I. del Villar, I. R. Matias et al. 2007. Fiber-optic pH-sensors in long-period fiber gratings using electrostatic self-assembly. *Optics Letters* 32: 29–31.

Craith, B. D. M., G. O'Keeffe, C. McDonagh et al. 1994. LED-based fibre optic oxygen sensor using sol-gel coating. *Electronics Letters* 30: 888–889.

Culshaw, B.1984. *Optical Fiber Sensing and Signal Processing*. Stevenage, UK: Peregrinus.

Culshaw, B. and M. J. Hutchings. 1979. Optical-fibre flowmeter. *Electronics Letters* 15: 569–571.

Culshaw, B., F. Muhammad, G. Stewart et al. 1992. Evanescent wave methane detection using optical fibers. *Electronics Letters* 28: 2232–2234.

Dakin, J. P. and D. A. Kahn. 1977. Novel fiber-optic temperature probe. *Optical and Quantum Electronics* 9: 540–544.

Dakin, J. P., D. J. Pratt, G. W. Bibby et al. 1985. Distributed optical fiber Raman temperature sensor using a semiconductor light-source and detector. *Electronics Letters* 21: 569–570.

Dandridge, A., A. B. Tveten, G. H. Sigel et al. 1980. Optical fiber magnetic-field sensors. *Electronics Letters* 16: 408–409.

Davis, D. D., T. K. Gaylord, E. N. Glytsis et al. 1999. Very-high-temperature stable CO2-laser-induced long-period fibre gratings. *Electronics Letters* 35: 740–742.

Desurvire, E., J. R. Simpson, and P. C. Becker.1987. High-gain erbium-doped traveling-wave fiber amplifier. *Optics Letters* 12: 888–890.

Ding, J. Y., M. R. Shahriari, and G. H. Sigel. 1991. Fibre optic pH sensors prepared by sol-gel immobilisation technique. *Electronics Letters* 27: 1560–1562.

Du, W. C., X. M. Tao, and H. Y. Tam. 1999. Fiber Bragg grating cavity sensor for simultaneous measurement of strain and temperature. *IEEE Photonics Technology Letters* 11: 105–107.

Dyott, R. B. 1978. Fiber-optic Doppler anemometer. *Iee Journal on Microwaves Optics and Acoustics* 2: 13–18.

Egashira, N., H. Kumasako, and K. Ohga. 1990. Fabrication of a fiber-optic-based electrochemiluminescence sensor and its application to the determination of oxalate. *Analytical Sciences* 6: 903–904.

Eggleton, B. J., P. S. Westbrook, R. S. Windeler et al. 1999. Grating resonances in air-silica microstructured optical fibers. *Optics Letters* 24: 1460–1462.

Eickhoff, W. 1981. Temperature sensing by mode-mode interference in birefringent optical fibers. *Optics Letters* 6: 204–206.

Erdmann, J. C. and D. C. Soreide. 1982. Fiber-optic laser transit velocimeters. *Applied Optics* 21: 1876–1877.

Esteban, O., M. Cruz-Navarrete, A. Gonzalez-Cano et al. 1999. Measurement of the degree of salinity of water with a fiber-optic sensor. *Applied Optics* 38: 5267–5271.

Fang, X., C. R. Liao, and D. N. Wang. 2010. Femtosecond laser fabricated fiber Bragg grating in microfiber for refractive index sensing. *Optics Letters* 35: 1007–1009.

Farahi, F., P. A. Leilabady, J. D. C. Jones et al. 1987. Interferometric fiber-optic hydrogen sensor. *Journal of Physics E-Scientific Instruments* 20: 432–434.

Farahi, F., D. J. Webb, J. D. C. Jones et al. 1990. Simultaneous measurement of temperature and strain—cross-sensitivity considerations. *Journal of Lightwave Technology* 8: 138–142.

Farries, M. C., M. E. Fermann, R. I. Laming et al. 1986. Distributed temperature sensor using Nd-3+-doped optical fiber. *Electronics Letters* 22: 418–419.

Ferreira, M. S., L. Coelho, K. Schuster et al. 2011. Fabry-Perot cavity based on a diaphragm-free hollow-core silica tube. *Optics Letters* 36: 4029–4031.

Fields, J. N. and J. H. Cole. 1980. Fiber microbend acoustic sensor. *Applied Optics* 19: 3265–3267.

Fomitchov, P. and S. Krishnaswamy. 2003. Response of a fiber Bragg grating ultrasonic sensor. *Optical Engineering* 42: 956–963.

Frazao, O., L. M. N. Amaral, J. L. Santos et al. 2010. Simultaneous measurement of strain and temperature using modal interferometry in Bragg fibers. *Second Workshop on Specialty Optical Fibers and Their Applications* Wsof-2 7839.

Frazao, O., P. Caldas, F. M. Araujo et al. 2007. Optical flowmeter using a modal interferometer based on a single nonadiabatic fiber taper. *Optics Letters* 32: 1974–1976.

Gardner, D. L., T. Hofler, S. R. Baker et al. 1987. A Fiber-optic interferometric seismometer. *Journal of Lightwave Technology* 5: 953–960.

Garus, D., T. Gogolla, K. Krebber et al. 1997. Brillouin optical-fiber frequency-domain analysis for distributed temperature and strain measurements. *Journal of Lightwave Technology* 15: 654–662.

Gerges, A. S., T. P. Newson, F. Farahi et al. 1988. A hemispherical air cavity fiber Fabry-Perot sensor. *Optics Communications* 68: 157–160.

Giallorenzi, T. G. 1981. Fiber optic sensors. *Optics and Laser Technology* 13: 73–78.

Goldstein, S. R., J. I. Peterson, and R. V. Fitzgerald. 1980. A miniature fiber optic Ph sensor for physiological use. *Journal of Biomechanical Engineering-Transactions of the Asme* 102: 141–146.

Gong, Y. A., Y. Guo, Y. J. Rao et al. 2010. Fiber-optic Fabry-Perot sensor based on periodic focusing effect of graded-index multimode fibers. *IEEE Photonics Technology Letters* 22: 1708–1710.

Grattan, K. T. V., R. K. Selli, and A. W. Palmer. 1988. Fluorescence referencing for fiber-optic thermometers using visible wavelengths. *Review of Scientific Instruments* 59: 256–259.

Gupta, B. D., H. Dodeja, and A. K. Tomar. 1996a. Fibre-optic evanescent field absorption sensor based on a U-shaped probe. *Optical and Quantum Electronics* 28: 1629–1639.

Gupta, S., T. Mizunami, T. Yamao et al. 1996b. Fiber Bragg grating cryogenic temperature sensors. *Applied Optics* 35: 5202–5205.

Hale, Z. M. and F. P. Payne. 1994. Fluorescent sensors based on tapered single-mode optical fibers. *Sensors and Actuators B-Chemical* 17: 233–240.

Hall, T. J. 1979. High-linearity multimode optical fiber sensor. *Electronics Letters* 15: 405–406.

Hanumegowda, N. M., C. J. Stica, B. C. Patel et al. 2005. Refractometric sensors based on microsphere resonators. *Applied Physics Letters* 87: 201107.

Hartog, A. 1983. A distributed temperature sensor based on liquid-core optical fibers. *Journal of Lightwave Technology* 1: 498–509.

Haslach, H. W. and J. S. Sirkis.1991. Surface-mounted optical fiber strain sensor design. *Applied Optics* 30: 4069–4080.

Hill, K. O., B. Malo, F. Bilodeau et al. 1993. Bragg gratings fabricated in monomode photosensitive optical fiber by UV exposure through a phase mask. *Applied Physics Letters* 62: 1035–1037.

Hocker, G. B. 1979. Fiber-optic sensing of pressure and temperature. *Applied Optics* 18: 1445–1448.

Homola, J. and R. Slavik. 1996. Fibre-optic sensor based on surface plasmon resonance. *Electronics Letters* 32: 480.

Horiguchi, T., T. Kurashima, and M. Tateda. 1989. Tensile strain dependence of Brillouin frequency shift in silica optical fibers. *Photonics Technology Letters* 1: 107–108.

Hwang, I. K., S. H. Yun, and B. Y. Kim. 1999. Long-period fiber gratings based on periodic microbends. *Optics Letters* 24: 1263–1265.

Iadicicco, A., A. Cusano, A. Cutolo et al. 2004. Thinned fiber Bragg gratings as high sensitivity refractive index sensor. *Photonics Technology Letters* 16: 1149–1151.

Iannuzzi, D., S. Deladi, V. J. Gadgil et al. 2006. Monolithic fiber-top sensor for critical environments and standard applications. *Applied Physics Letters* 88: 053501.

Iga, M., A. Seki, and K. Watanabe. 2005. Gold thickness dependence of SPR-based hetero-core structured optical fiber sensor. *Sensors and Actuators B-Chemical* 106: 363–368.

Ikeda, H., M. Suzuki, and M. T. Hutchings. 1979. Neutron-scattering investigation of static critical phenomena in the 2-dimensional anti-ferromagnets—Rb2cocmg1-Cf4. *Journal of the Physical Society of Japan* 46: 1153–1160.

Imai, M., T. Ohashi, and Y. Ohtsuka. 1983. Multimode-optical-fiber Michelson interferometer. *Journal of Lightwave Technology* 1: 75–81.

Inaba, H., T. Kobayasi, M. Hirama et al. 1979. Optical-fibre network system for air-pollution monitoring over a wide area by optical absorption method. *Electronics Letters* 15: 749–751.

Iwamoto, K. and I. Kamata. 1992. Liquid-level sensor with optical fibers. *Applied Optics* 31: 51–54.

Jackson, D. A., A. D. Kersey, and M. Corke. 1982a. Confocal fabry-perot sensor. *Electronics Letters* 18: 227–229.

Jackson, D. A., A. D. Kersey, M. Corke et al. 1982b. Pseudoheterodyne detection scheme for optical interferometers. *Electronics Letters* 18: 1081–1083.

James, S. W., R. P. Tatam, A. Twin et al. 2002. Strain response of fibre Bragg grating sensors at cryogenic temperatures. *Measurement Science & Technology* 13: 1535–1539.

Jorgenson, R. C., S. S. Yee, K. S. Johnston et al. 1993. A novel surface-plasmon resonance based fiber optic sensor. Applied to Biochemical Sensing. *Proceedings of Fiber Optics Sensors in Medical Diagnostics* 1886: 35–48.

Ju, J., L. Ma, W. Jin et al. 2009. Photonic bandgap fiber tapers and in-fiber interferometric sensors. *Optics Letters* 34: 1861–1863.

Kalenik, J. and R. Pajak. 1998. A cantilever optical-fiber accelerometer. *Sensors and Actuators A-Physical* 68: 350–355.

Kao, K. C. and G. A. Hockham. 1966. Dielectric-fibre surface waveguides for optical frequencies. *Proceedings of the Institution of Electrical Engineers-London* 113: 1151–1158.

Kerbage, C., P. Steinvurzel, A. Hale et al. 2002. Microstructured optical fibre with tunable birefringence. *Electronics Letters* 38: 310–312.

Kersey, A. D. and A. Dandridge. 1985a. Optical fiber faraday-rotation current sensor with closed-loop operation. *Electronics Letters* 21: 464–466.

Kersey, A. D., A. Dandridge, A. B. Tveten et al. 1985b. Single-mode fiber Fourier-transform spectrometer. *Electronics Letters* 21: 463–464.

Kersey, A. D., D. A. Jackson, and M. Corke. 1982. High-sensitivity fiber-optic accelerometer. *Electronics Letters* 18: 559-561.

Kersey, A. D., D. A. Jackson, and M. Corke. 1983. A simple fiber Fabry-Perot sensor. *Optics Communications* 45: 71–74.

Kersey, A. D., M. J. Marrone, and M. A. Davis. 1991. Polarisation-insensitive fibre optic Michelson interferometer. *Electronics Letters* 27: 518–520.

Kersey, A. D. and W. W. Morey. 1993. Multi-element Bragg-grating based fibre-laser strain sensor. *Electronics Letters* 29: 964–966.

Kim, S. H., J. J. Lee, D. C. Lee et al. 1999. A study on the development of transmission-type extrinsic Fabry-Perot interferometric optical fiber sensor. *Journal of Lightwave Technology* 17: 1869–1874.

Knight, J. C., T. A. Birks, P. S. Russell et al. 1996. All-silica single-mode optical fiber with photonic crystal cladding. *Optics Letters* 21: 1547–1549.

Kobayasi, T., M. Hirama, and H. Inaba. 1981. Remote monitoring of No2 molecules by differential absorption using optical fiber link. *Applied Optics* 20: 3279–3280.

Kou, J. L., J. Feng, L. Ye, F. Xu and Y. Q. Lu. 2010. Miniaturized fiber taper reflective interferometer for high temperature measurement. *Optics Express* 18: 14245–14250.

Kumar, A., T. V. B. Subrahmanyam, A. D. Sharma et al. 1984. Novel refractometer using a tapered optical fiber. *Electronics Letters* 20: 534–535.

Kurashima, T., T. Horiguchi, and M. Tateda. 1990. Distributed-temperature sensing using stimulated Brillouin-scattering in optical silica fibers. *Optics Letters* 15: 1038–1040.

Laffont, G. and P. Ferdinand. 2001. Tilted short-period fibre-Bragg-grating-induced coupling to cladding modes for accurate refractometry. *Measurement Science & Technology* 12: 765–770.

Lagakos, N., J. A. Bucaro, and R. Hughes. 1980. Acoustic sensitivity predictions of single-mode optical fibers using Brillouin-scattering. *Applied Optics* 19: 3668–3670.

Laming, R. I. and D. N. Payne. 1989. Electric current sensors employing spun highly birefringent optical fibers. *Journal of Lightwave Technology* 7: 2084–2094.

Larsen, T. T., A. Bjarklev, D. S. Hermann et al. 2003. Optical devices based on liquid crystal photonic bandgap fibres. *Optics Express* 11: 2589–2596.

Layton, M. R. and J. A. Bucaro. 1979. Optical fiber acoustic sensor utilizing mode-mode interference. *Applied Optics* 18: 666–670.

Lee, B. H. and J. Nishii. 1998a. Self-interference of long-period fibre grating and its application as temperature sensor. *Electronics Letters* 34: 2059–2060.

Lee, B. H. and J. J. Nishii. 1998b. Bending sensitivity of in-series long-period fiber gratings. *Optics Letters* 23: 1624–1626.

Lee, C. E. and H. F. Taylor. 1988. Interferometric optical fibre sensors using internal mirrors. *Electronics Letters* 24: 193–194.

Lee, C. E., H. F. Taylor, A. M. Markus et al. 1989. Optical-fiber Fabry-Perot embedded sensor. *Optics Letters* 14: 1225–1227.

Leilabady, P. A., J. D. C. Jones, and D. A. Jackson. 1985. Monomode fiber-optic strain-gauge with simultaneous phase-state and polarization-state detection. *Optics Letters* 10: 576–578.

Leilabady, P. A., J. D. C. Jones, A. D. Kersey et al. 1984. Monomode fiber optic vortex shedding flowmeter. *Electronics Letters* 20: 664–665.

Leslie, D. H., G. L. Trusty, A. Dandridge et al. 1981. Fiber-optic spectrophone. *Electronics Letters* 17: 581–582.

Li, E. B., X. L. Wang, and C. Zhang. 2006. Fiber-optic temperature sensor based on interference of selective higher-order modes. *Applied Physics Letters* 89 (091119): 1–3.

Lin, C. Y., L. A. Wang, and G. W. Chern. 2001. Corrugated long-period fiber gratings as strain, torsion, and bending sensors. *Journal of Lightwave Technology* 19: 1159–1168.

Liu, X. J. and W. H. Tan. 1999. A fiber-optic evanescent wave DNA biosensor based on novel molecular beacons. *Analytical Chemistry* 71: 5054–5059.

Lou, J. Y., L. M. Tong, and Z. Z. Ye. 2005. Modeling of silica nanowires for optical sensing. *Optics Express* 13: 2135–2140.

Lovseth, S. W., J. T. Kringlebotn, E. Ronnekleiv et al. 1999. Fiber distributed-feedback lasers used as acoustic sensors in air. *Applied Optics* 38: 4821–4830.

Ma, J. J., W. Z. Tang, and W. Zhou. 1996. Optical-fiber sensor for simultaneous measurement of pressure and temperature: Analysis of cross sensitivity. *Applied Optics* 35: 5206–5210.

Magne, S., S. Rougeault, M. Vilela et al. 1997. State-of-strain evaluation with fiber Bragg grating rosettes: Application to discrimination between strain and temperature effects in fiber sensors. *Applied Optics* 36: 9437–9447.

Marcuse, D. 1977. Loss analysis of single-mode fiber splices. *Bell System Technical Journal* 56: 703–718.

Martynkien, T., G. Statkiewicz-Barabach, J. Olszewski et al. 2010. Highly birefringent microstructured fibers with enhanced sensitivity to hydrostatic pressure. *Optics Express* 18: 15113–15121.

Mehta, A., W. Mohammed, and E. G. Johnson. 2003. Multimode interference-based fiber-optic displacement sensor. *IEEE Photonics Technology Letters* 15: 1129–1131.

Melle, S. M., A. T. Alavie, S. Karr et al. 1993. A Bragg grating-tuned fiber laser strain sensor system. *IEEE Photonics Technology Letters* 5: 263–266.
Meltz, G., J. R. Dunphy, W. W. Morey et al. 1983. Cross-talk fiber-optic temperature sensor. *Applied Optics* 22: 464–477.
Meltz, G., W. W. Morey, and J. R. Dunphy. 1992. Fiber Bragg grating chemical sensor. *Chemical, Biochemical, and Environmental Fiber Sensors* 1587: 350–361.
Morey, W. W, J. R. Dunphy, and G. Meltz. 1991. Multiplexing fiber Bragg grating sensors. *Fiber and Integrated Optics* 10: 351–360.
Menadier, C., C. Kissinger, and H. Adkins. 1967. The fotonic sensor. *Instruments and Control Systems* 40: 114–120.
Mermelstein, M. D. 1986. All-fiber polarimetric sensor. *Applied Optics* 25: 1256–1258.
Meyer, R. E., S. Ezekiel, D. W. Stowe et al. 1983. Passive fiber-optic ring resonator for rotation sensing. *Optics Letters* 8: 644–646.
Michie, A., J. Canning, I. Bassett et al. 2007. Spun elliptically birefringent photonic crystal fibre for current sensing. *Measurement Science & Technology* 18: 3070–3074.
Michie, W. C., B. Culshaw, I. Mckenzie et al. 1995. Distributed sensor for water and Ph measurements using fiber optics and swellable polymeric systems. *Optics Letters* 20: 103–105.
Mihailov, S. J., D. Grobnic, and C. W. Smelser. 2010. High-temperature multiparameter sensor based on sapphire fiber Bragg gratings. *Optics Letters* 35: 2810–2812.
Mitschke, F. 1989. Fiber-optic sensor for humidity. *Optics Letters* 14: 967–969.
Monro, T. M., D. J. Richardson, and P. J. Bennett. 1999. Developing holey fibres for evanescent field devices. *Electronics Letters* 35: 1188–1189.
Morey, W. W., G. Meltz, and W. H. Glenn. 1989. Fibre optic Bragg grating sensors. *Proceedings of Fibre Optic and Laser Sensors VII*, Boston, MA, SPIE 1169, pp. 98–107.
Morris, J. and C. Pollock. 1987. A digital fiber-optic liquid level sensor. *Journal of Lightwave Technology* 5: 920–925.
Muhammad, F. A., G. Stewart, and W. Jin. 1993. Sensitivity enhancement of D-fiber methane gas sensor using high-index overlay. *Proceedings Journal Optoelectronics* 140: 115–118.
Murphy, K. A., W. V. Miller, T. A. Tran et al. 1991. Miniaturized fiber-optic Michelson-type interferometric sensors. *Applied Optics* 30: 5063–5067.
Murukeshan, V. M., P. Y. Chan, L. S. Ong et al. 2000. Cure monitoring of smart composites using fiber Bragg grating based embedded sensors. *Sensors and Actuators A-Physical* 79: 153–161.
Nemova, G. and R. Kashyap. 2006. Fiber-Bragg-grating-assisted surface plasmon-polariton sensor. *Optics Letters* 31: 2118–2120.
Oh, K. D., J. Ranade, V. Arya et al. 1997. Optical fiber Fabry-Perot interferometric sensor for magnetic field measurement. *Photonics Technology Letters* 9: 797–799.
Patrick, H. J., A. D. Kersey, and F. Bucholtz. 1998. Analysis of the response of long period fiber gratings to external index of refraction. *Journal of Lightwave Technology* 16: 1606–1612.
Patrick, H. J., G. M. Williams, A. D. Kersey et al. 1996. Hybrid fiber Bragg grating/long period fiber grating sensor for strain/temperature discrimination. *Photonics Technology Letters* 8: 1223–1225.
Penza, M., G. Cassano, P. Aversa et al. 2004. Alcohol detection using carbon nanotubes acoustic and optical sensors. *Applied Physics Letters* 85: 2379–2381.
Petuchowski, S. J., T. G. Giallorenzi, and S. K. Sheem. 1981. A sensitive fiber-optic Fabry-Perot-interferometer. *Journal of Quantum Electronics* 17: 2168–2170.
Pevec, S. and D. Donlagic. 2011. All-fiber, long-active-length Fabry-Perot strain sensor. *Optics Express* 19: 15641–15651.
Powell, J. A. 1974. Simple 2-fiber optical displacement sensor. *Review of Scientific Instruments* 45: 302–303.
Quoi, K. W., R. A. Lieberman, L. G. Cohen et al. 1992. Rare-earth doped optical fibers for temperature sensing. *Journal of Lightwave Technology* 10: 847–852.
Rajkumar, N., V. J. Kumar, and P. Sankaran. 1993. Fiber sensor for the simultaneous measurement of current and voltage in a high-voltage system. *Applied Optics* 32: 1225–1228.
Ritari, T., J. Tuominen, H. Ludvigsen et al. 2004. Gas sensing using air-guiding photonic bandgap fibers. *Optics Express* 12: 4080–4087.
Rogers, A. J. 1977. Optical methods for measurement of voltage and current on power-systems. *Optics and Laser Technology* 9: 273–283.
Schroeder, K., W. Ecke, R. Mueller et al. 2001. A fibre Bragg grating refractometer. *Measurement Science & Technology* 12: 757–764.

Shirasaki, M., H. Kuwahara, and T. Obokata. 1981. Compact polarization-independent optical circulator. *Applied Optics* 20: 2683–2687.

Silva, S., J. L. Santos, F. X. Malcata et al. 2011. Optical refractometer based on large-core air-clad photonic crystal fibers. *Optics Letters* 36: 852–854.

Singer, E., G. L. Duveneck, M. Ehrat et al. 1994. Fiber optic sensor for oxygen determination in liquids. *Sensors and Actuators a-Physical* 42: 542–546.

Snitzer, E. 1961. Cylindrical dielectric waveguide modes. *Journal of the Optical Society of America* 51: 491–498.

Spammer, S. J., P. L. Swart, and A. Booysen. 1996. Interferometric distributed optical-fiber sensor. *Applied Optics* 35: 4522–4525.

Starodumov, A. N., L. A. Zenteno, D. Monzon et al. 1997. Fiber Sagnac interferometer temperature sensor. *Applied Physics Letters* 70: 19–21.

Stolen, R. H., A. Ashkin, W. Pleibel et al. 1984. In-line fiber-polarization-rocking rotator and filter. *Optics Letters* 9: 300–302.

Suhadolnik, A., A. Babnik, and J. Mozina. 1995. Optical-fiber reflection refractometer. *Sensors and Actuators B-Chemical* 29: 428–432.

Sumetsky, M., Y. Dulashko, J. M. Fini, and A. Hale. 2005. Optical microfiber loop resonator. *Applied Physics Letters* 86: 161108.

Sumetsky, M., Y. Dulashko, and R. S. Windeler. 2010. Optical microbubble resonator. *Optics Letters* 35: 898–900.

Swart, P. L. 2004. Long-period grating Michelson refractometric sensor. *Measurement Science & Technology* 15: 1576–1580.

Takahashi, N., A. Hirose, and S. Takahashi. 1997. Underwater acoustic sensor with fiber Bragg grating. *Optical Review* 4: 691–694.

Tao, X. M., L. Q. Tang, W. C. Du et al. 2000. Internal strain measurement by fiber Bragg grating sensors in textile composites. *Composites Science and Technology* 60: 657–669.

Tian, Z. B., S. S. H. Yam, and H. P. Loock. 2008. Single-mode fiber refractive index sensor based on core-offset attenuators. *IEEE Photonics Technology Letters* 20: 1387–1389.

Tomita, S., H. Tachino, and N. Kasahara. 1990. Water sensor with optical fiber. *Journal of Lightwave Technology* 8: 1829–1832.

Tsai, W. H. and C. J. Lin. 2001. A novel structure for the intrinsic Fabry-Perot fiber-optic temperature sensor. *Journal of Lightwave Technology* 19: 682–686.

Tseng, S. M. and C. L. Chen. 1992. Side-polished fibers. *Applied Optics* 31: 3438–3447.

Tveten, A. B., A. Dandridge, C. M. Davis et al. 1980. Fiber optic accelerometer. *Electronics Letters* 16: 854–856.

Ulrich, R. 1980. Fiber-optic rotation sensing with low drift. *Optics Letters* 5: 173–175.

Vali, V. and R. W. Shorthill. 1976. Fiber ring interferometer. *Applied Optics* 15: 1099–1100.

Vaziri, M. and C. L. Chen. 1992. Etched fibers as strain-gauges. *Journal of Lightwave Technology* 10: 836–841.

Villatoro, J., V. P. Minkovich, and D. Monzon-Hernandez. 2006. Temperature-independent strain sensor made from tapered holey optical fiber. *Optics Letters* 31: 305–307.

Villatoro, J., D. Monzon-Hernandez, and D. Talavera. 2004. High resolution refractive index sensing with cladded multimode tapered optical fibre. *Electronics Letters* 40: 106–107.

Vollmer, F., D. Braun, A. Libchaber et al. 2002. Protein detection by optical shift of a resonant microcavity. *Applied Physics Letters* 80: 4057–4059.

Wagener, J. L., C. W. Hodgson, M. J. F. Digonnet et al. 1997. Novel fiber sensor arrays using erbium-doped fiber amplifiers. *Journal of Lightwave Technology* 15: 1681–1688.

Wait, P. C. and T. P. Newson. 1996. Landau Placzek ratio applied to distributed fibre sensing. *Optics Communications* 122: 141–146.

Wan, X. K. and H. F. Taylor. 2002. Intrinsic fiber Fabry-Perot temperature sensor with fiber Bragg grating mirrors. *Optics Letters* 27: 1388–1390.

Wang, C. J. and S. T. Scherrer. 2004. Fiber ringdown pressure sensors. *Optics Letters* 29: 352–354.

Wang, L. A., B. Zhou, C. Shu et al. 2011. Stimulated Brillouin scattering slow-light-based fiber-optic temperature sensor. *Optics Letters* 36: 427–429.

Wang, X. W., J. C. Xu, Y. Z. Zhu et al. 2006. All-fused-silica miniature optical fiber tip pressure sensor. *Optics Letters* 31: 885–887.

Wang, Y. P., W. Jin, J. Ju et al. 2008. Long period gratings in air-core photonic bandgap fibers. *Optics Express* 16: 2784–2790.

Webb, A. S., F. Poletti, D. J. Richardson et al. 2007. Suspended-core holey fiber for evanescent-field sensing. *Optical Engineering* 46: 010503.

Wei, T., Y. K. Han, H. L. Tsai et al. 2008. Miniaturized fiber inline Fabry-Perot interferometer fabricated with a femtosecond laser. *Optics Letters* 33: 536–538.

West, S. T. and C. L. Chen. 1989. Optical fiber rotary displacement sensor. *Applied Optics* 28: 4206–4209.

White, I. M., H. Oveys, and X. D. Fan. 2006. Liquid-core optical ring-resonator sensors. *Optics Letters* 31: 1319–1321.

Wiejata, P. J., P. M. Shankar, and R. Mutharasan. 2003. Fluorescent sensing using biconical tapers. *Sensors and Actuators B-Chemical* 96: 315–320.

Wolfbeis, O. S., H. E. Posch, and H. W. Kroneis. 1985. Fiber optical fluorosensor for determination of halothane and or oxygen. *Analytical Chemistry* 57: 2556–2561.

Wolfbeis, O. S., L. J. Weis, M. J. P. Leiner et al. 1988. Fiber-optic fluorosensor for oxygen and carbon-dioxide. *Analytical Chemistry* 60: 2028–2030.

Xie, H. M., P. Dabkiewicz, R. Ulrich et al. 1986. Side-hole fiber for fiber-optic pressure sensing. *Optics Letters* 11: 333–335.

Xu, M. G., J. L. Archambault, L. Reekie et al. 1994. Discrimination between strain and temperature effects using dual-wavelength fiber grating sensors. *Electronics Letters* 30: 1085–1087.

Xu, M. G., L. Dong, L. Reekie et al. 1995. Temperature-independent strain sensor using a chirped Bragg grating in a tapered optical fibre. *Electronics Letters* 31: 823–825.

Xu, M. G., H. Geiger, and J. P. Dakin. 1996. Fibre grating pressure sensor with enhanced sensitivity using a glass-bubble housing. *Electronics Letters* 32: 128–129.

Xu, M. G., L. Reekie, Y. T. Chow et al. 1993. Optical in-fibre grating high pressure sensor. *Electronics Letters* 29: 398–399.

Yang, X. F., S. J. Luo, Z. H. Chen et al. 2007. Fiber Bragg grating strain sensor based on fiber laser. *Optics Communications* 271: 203–206.

Yu, B., D. W. Kim, J. D. Deng et al. 2003. Fiber Fabry-Perot sensors for detection of partial discharges in power transformers. *Applied Optics* 42: 3241–3250.

Yuan, L. B., J. Yang, Z. H. Liu et al. 2006. In-fiber integrated Michelson interferometer. *Optics Letters* 31: 2692–2694.

Zhang, F. and J. W. Y. Lit. 1992. Temperature and strain sensitivities of high-birefringence elliptic fibers. *Applied Optics* 31: 1239–1243.

Zhang, F. and J. W. Lit. 1993. Temperature and strain sensitivity measurements of high-birefringent polarization-maintaining fibers. *Applied Optics* 32: 2213–2218.

Zhang, L., Y. Liu, L. Everall et al. 1999. Design and realization of long-period grating devices in conventional and high birefringence fibers and their novel applications as fiber-optic load sensors. *Journal of Selected Topics in Quantum Electronics* 5: 1373–1378.

Zhang, Y., H. Shibru, K. L. Cooper et al. 2005. Miniature fiber-optic multicavity Fabry-Perot interferometric biosensor. *Optics Letters* 30: 1021–1023.

Zhang, Z. Y., K. T. V. Grattan, A. W. Palmer et al. 1997. Fluorescence decay-time characteristics of erbium-doped optical fiber at elevated temperatures. *Review of Scientific Instruments* 68: 2764–2766.

Zhang, Z. Y., K. T. V. Grattan, A. W. Palmer et al. 1998. Thulium-doped intrinsic fiber optic sensor for high temperature measurements (> 1100 degrees C). *Review of Scientific Instruments* 69: 3210–3214.

Zhao, C. L., X. F. Yang, C. Lu et al. 2004. Temperature-insensitive interferometer using a highly birefringent photonic crystal fiber loop mirror. *Photonics Technology Letters* 16: 2535–2537.

Zhao, W. and R. O. Claus. 2000. Optical fiber grating sensors in multimode fibers. *Smart Materials & Structures* 9: 212–214.

Zhu, Y. Z., X. P. Chen, and A. B. Wang. 2009. Observation of interference in a fiber taper interferometer with a subwavelength tip and its sensing applications. *Optics Letters* 34: 2808–2810.

Zhu, Y. Z. and A. B. Wang. 2005. Miniature fiber-optic pressure sensor. *Photonics Technology Letters* 17: 447–449.

Zou, L. F., X. Y. Bao, and L. A. Chen. 2005. Distributed Brillouin temperature sensing in photonic crystal fiber. *Smart Materials & Structures* 14: S8–S11.

12. Optical Fibers

Marco N. Petrovich
University of Southampton

Handbook of Optical Sensors. Edited by José Luís Santos and Faramarz Farahi © 2015 CRC Press/Taylor & Francis Group, LLC. ISBN: 9781439866856.

12.1 Introduction

Optical fibers are the cornerstone of the global telecommunication revolution in the twentieth century leading to the advent of the digital age, as is testified by the Nobel Prize for Physics awarded to Charles Kao in 2009 (Kao 2010). The development of ultralow-loss optical fibers in the 1970s was arguably the most critical milestone among a series of groundbreaking developments, including the invention of the semiconductor laser and that of the erbium-doped optical amplifier (Mears et al. 1987), which have opened up the possibility for long-distance transmission of data at unprecedented rates and huge capacity. More recently, advances in rare-earth-doped optical fibers have enabled the development of multikilowatt ytterbium-doped fiber lasers that are revolutionizing the industrial materials processing sector (Richardson et al. 2010). An historical account on the development of optical fibers from the outset until recent activity can be found in Gambling (2000).

Much of the early progress in optical fiber technology was driven by the requirements of the telecom industry. However, as single- and multimode transmission fibers quickly established themselves, the need also grew for fibers with custom-tailored optical properties, beyond low transmission loss. The steady progress of fiber fabrication techniques, as well as enabling unprecedented levels of purity of silica and silica-based optical materials, also made it possible to realize more and more complex waveguide designs. This facilitated the development of *special* fibers, for instance, fibers with tailored chromatic dispersion, high or low birefringence, high or low nonlinearity, and photosensitive fibers. More recently, the invention of microstructured optical fibers (MOFs), that is, fibers with wavelength-scale features running along their full length, has dramatically increased the already considerable variety of fiber geometries and combination of optical properties available.

Special fibers have opened up entirely new possibilities in diverse application areas, and optical sensing is probably one of the most notable examples. In addition to low transmission loss and high bandwidth, optical fibers have other obvious features that are very attractive for sensing applications: They are relatively cheap and easy to manufacture, intrinsically safe, lightweight, and immune to EM interference; they have excellent mechanical properties, high chemical durability, and open up the possibility of passive sensor heads with extremely small footprint. But even more importantly, the ability to combine a waveguide with the sensing element in the same fiber has paved the way for distributed optical sensing. The possibility of building extremely long, common path fiber interferometers opened up the possibility of devices with unprecedented sensitivity. Nowadays, optical fiber sensors are not only a topic of acute scientific interest but have gained significant industrial relevance and are becoming well established in a growing number of niche areas, including, for instance, optical fiber gyroscopes (FOGs), structural monitoring, downhole pressure monitoring in oil and gas wells, and chemical and environmental sensing.

The purpose of this chapter is to introduce the fundamental concepts of light guidance in optical fibers, focusing on those transmission properties that are most relevant to sensing applications, as well as to provide a more in-depth description of the main types of *special fibers* with particular emphasis on those that have been most successfully used in optical fiber sensors.

12.2 Light Propagation in Optical Fibers

The simplest implementation of an optical fiber is composed of three axially concentric layers, that is, the inner region, or *core*, surrounded by the *cladding* and a further outer layer of polymer *coating*. The core element, about 8–10 μm in diameter in the case of standard single-mode fibers (SMFs) and 50 or 62.5 μm in the case of multimode fibers (MMFs), carries most of the optical power propagating along the fiber. The cladding layer is composed of a different glass with slightly lower (typically a few %) refractive index. The most common types include silica cladding/germanosilicate core fibers and fluorine-doped silica cladding/pure silica core fibers. As will be seen in the following section, the core size and refractive index difference with the cladding determine, to the first order, the propagation properties of the fiber. The outside diameter of standard fibers is 125 μm, but one or more additional layers of polymer coating are added in order to preserve the mechanical strength of the fiber, to reduce bend sensitivity and to provide additional chemical and environmental stability.

The following sections will provide a general overview of light guidance in optical fibers and a description of the basic transmission properties. This topic is covered by a number of classical journal papers and reference textbooks (for instance, Gloge (1971a,b), Okoshi (1982), Marcuse (1982), Snitzer (1961), and Snyder and Love (2010)), which the reader is encouraged to consult for a more comprehensive description and for further insight.

12.2.1 Guided Modes

In this section, we summarize the wave description of a uniform core optical fiber with a step-index profile under the so-called weak guidance approximation (Gloge 1971a). The general conclusions that can be drawn from this simplified picture in terms of mode classification and characteristics can be generalized to other fibers with more complex refractive index profiles, although analytical or semianalytical treatment is generally not possible.

The *step-index fiber* consists of two homogeneous core and cladding layers, as shown in Figure 12.1.

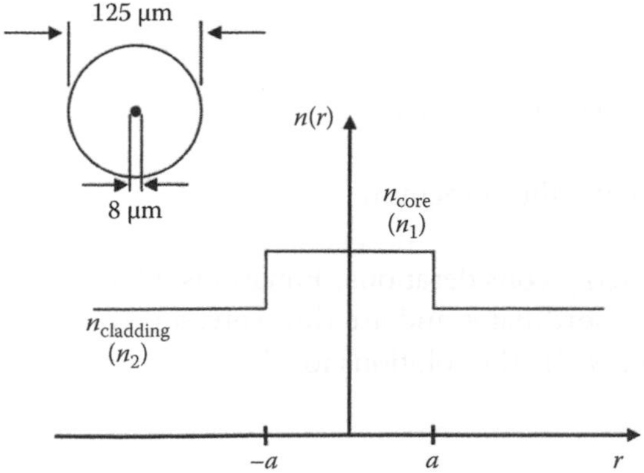

FIGURE 12.1 Refractive profile of the idealized step-index fiber and (top left) schematic of an SMF.

The core has radius a and refractive index n_1, while the cladding occupies the region $r \geq a$ and has refractive index n_2, with $n_1 > n_2$. The main design parameter (in addition to the core radius and refractive index) is the normalized index difference:

$$\Delta \equiv \frac{\left(n_1^2 - n_2^2\right)}{2n_1^2} \approx \frac{\left(n_1 - n_2\right)}{n_1} \tag{12.1}$$

which is related to the fiber's numerical aperture (NA):

$$NA \equiv \left(n_1^2 - n_2^2\right)^{1/2} = n_1\left(2\Delta\right)^{1/2} \tag{12.2}$$

The standard mathematical description for this waveguide consists in deriving the wave equations for the propagating electric and magnetic fields, $\vec{E}$ and $\vec{H}$, from Maxwell's equations for a uniform medium with permittivity ε and permeability μ (Okoshi 1982):

$$\nabla^2\vec{E} - \varepsilon\mu\partial_t^2\vec{E} = 0 \quad \text{and} \quad \nabla^2\vec{H} - \varepsilon\mu\partial_t^2\vec{H} = 0 \tag{12.3}$$

The earlier equation is a *vector* equation, equivalent to a set of three scalar equations for each of the electric and magnetic field components. In the step-index fiber, separate wave equations hold for the core and cladding, and suitable boundary conditions must be verified at their interface. By assuming the vector fields $\vec{E}$ and $\vec{H}$ as sinusoidal functions of time (with angular frequency ω), the wave equations can be rewritten as (Okoshi 1982)

$$\nabla^2\vec{E} + n_i^2k_0^2\vec{E} = 0 \tag{12.4a}$$

and

$$\nabla^2\vec{H} + n_i^2k_0^2\vec{H} = 0 \tag{12.4b}$$

where

$i = 1, 2$ for core and clad, respectively
$k_0 = 2\pi/\lambda$
λ is the wavelength of light in vacuum

Based on basic symmetry considerations, Equations 12.4 are expressed more conveniently in cylindrical coordinates and are thus solved through separation of variables, as detailed in Okoshi (1982). The solutions for the axial components of the electric field take the form

$$E_z(r,\vartheta,z) = A \cdot J_q\left(\frac{ur}{a}\right)\sin\left(q\vartheta\right)\exp\left(-j\beta z\right) \quad \text{for } r \leq a \text{ (core)} \tag{12.5}$$

$$E_z(r,\vartheta,z) = B \cdot K_q\left(\frac{wr}{a}\right)\sin(q\vartheta)\exp(-j\beta z) \quad \text{for } r \geq a \text{ (cladding)} \tag{12.6}$$

where

A and *B* are arbitrary constants
$q = 0, 1, 2, \ldots$ is the azimuthal mode number
$J_q(r)$ and $K_q(r)$ are *q*th-order Bessel functions of the first and second type, respectively
u and *w* are the normalized transverse propagation and decay constants, respectively, which are defined as

$$u \equiv a \cdot \left(n_1^2 k_0^2 - \beta^2\right)^{1/2} \tag{12.7a}$$

and

$$w \equiv a \cdot \left(\beta^2 - n_2^2 k_0^2\right)^{1/2} \tag{12.7b}$$

The wave equation for $\vec{H}$ (Equation 12.4b) can be solved following a similar procedure. The transverse field components for $\vec{E}$ and $\vec{H}$ can then be obtained from $E_z(r,\vartheta,z)$ and $H_z(r,\vartheta,z)$ through the use of Maxwell's equations. By imposing the continuity conditions of field components at the core/cladding interface ($r = a$), an *eigenvalue* equation is obtained; from the actual propagation constants, *u*, *w*, and β in Equations 12.5 and 12.6 are determined.

These solutions provide the *modes* of the step-index fiber waveguide. The modes are characterized by their azimuthal mode number, *q*, and by a radial mode number, or mode rank, *m* ($m = 1, 2, 3, \ldots$). The latter indicates that the propagation constant of the particular mode is calculated by the *m*th smallest root of *u*, which satisfies all the boundary conditions. The modes are first classified according to their azimuthal mode number. If $q = 0$, modes are classified into *transverse magnetic* (*TM*) modes for which H_z is zero ($H_z = H_r = E_\vartheta = 0$ and $E_z, E_r, H_\vartheta \neq 0$) and *transverse electric* (*TE*) modes for which E_z is zero ($E_z = E_r = H_\vartheta = 0$ and $H_z, H_r, E_\vartheta \neq 0$). If $q \geq 1$, modes are classified in EH_{qm} and HE_{qm} and are designated as *hybrid* modes.

The weak guidance approximation makes the assumption that the refractive index contrast between core and cladding is very small ($\Delta \ll 1$)—a condition that is satisfied by several practical fibers including standard telecom fibers. Under this approximation, a simplified expression for the eigenvalue equation can be formulated:

$$u \cdot \frac{J_{l-1}(u)}{J_l(u)} = -w \cdot \frac{K_{l-1}(w)}{K_l(w)} \tag{12.8}$$

where the new mode number *l* is defined as $l = 1$ for *TM* and *TE* modes, and $l = m + 1$, $l = m - 1$ for *EH* and *HE* modes, respectively. Within the weak guidance approximation, modes having the same values of the *q* and *l* mode numbers are degenerate, that is, have the same propagation constants. This allows defining a new set of *linearly polarized* modes (LP_{ql}) that are polarized in the transverse plane, thus of great practical usefulness.

We refer the reader to Tables 4.1 through 4.4 in Okoshi (1982) for a basic description of the lowest-order *LP* modes.

The modal properties of a fiber are frequently described using the *normalized frequency*, or *V-number*, which is defined as $V \equiv u^2 + w^2$ and can be calculated from the fiber design parameters via the following equation:

$$V = k_0 n_1 a \cdot (2\Delta)^{1/2} = \left(\frac{2\pi}{\lambda}\right) a \cdot NA \tag{12.9}$$

In general, it can be seen that single-mode propagation occurs for small values of *V*, while higher-order modes are supported at higher values of *V*. The *cutoff frequency* of a particular mode is the value of *V* for which it ceases to be confined to the core. For a generic LP mode, LP_{ql}, it can be seen that

$$V_c\left\{LP_{ql}\right\} = \rho_{(q-1),l} \tag{12.10}$$

where $\rho_{(q-1),l}$ is the *l*th root of the equation $J_{q-1}(r) = 0$. Values of *V* for a certain mode will fall within the range $V_c \leq V < \infty$. The *fundamental* or LP_{01} mode has no cutoff; the LP_{11} has cutoff at $V = 2.405$ (the first zero of the Bessel function $J_0(r)$); the LP_{21} and LP_{02} modes have cutoff at $V = 3.832$; and so forth. A table of cutoff frequencies of the first few lowest-order modes can be found in Table 3.2 of Buck (1995). A step-index optical fiber is thus single mode for $V \leq 2.405$. In the case of heavily MMFs, the *V*-number provides an estimate of the total number of supported modes:

$$\#\text{Modes} \approx \frac{4}{\pi^2} \cdot V^2 \tag{12.11}$$

The *V* parameter can also be employed to estimate the spatial extent of the mode within the fiber, that is, the mode field diameter (*MFD*) via Marcuse's equation (Marcuse 1982):

$$MFD \approx 2a \cdot \left[0.65 + \frac{1.619}{V^{3/2}} + \frac{2.879}{V^6}\right] \tag{12.12}$$

which is reasonably accurate for $0.8 \leq V < 2.5$. The *MFD* is intended within the Gaussian beam approximation as the twice the radius at which the field intensity is reduced by a factor of $1/e^2$ of the intensity at $r = 0$. In addition to the MFD, the quantity typically utilized to describe the spatial extent of an optical mode is the effective mode area, A_{eff}, defined as (Artiglia et al. 1989)

$$A^{eff} = 2\pi \cdot \frac{\left(\int_0^\infty |E(r)|^2\, r dr\right)^2}{\int_0^\infty |E(r)|^4\, r dr} \tag{12.13}$$

12.2.2 Fiber Attenuation

The principal function of passive optical fibers is to transmit light to a distance. Any optical signal traveling in an optical fiber will however undergo attenuation, or loss, which is an exponential decrease in its intensity with length, as a consequence of a variety of effects. In general, some sources of loss are fundamental to the glass material, and others are due to optically active impurities introduced, for example, during fabrication, and to waveguide design or imperfections. Loss mechanisms are thus broadly classified as *intrinsic* and *extrinsic*. Each mechanism is characterized by a wavelength-dependent intensity, and in general, different mechanisms prevail for different regions of the spectrum (see Figure 12.2). The optical attenuation is described through the imaginary part of the propagation constant, the *attenuation coefficient*, which is normally expressed in units of dB/km.

Intrinsic sources of loss in optical glasses include *electronic absorption* at UV/visible wavelengths, *Rayleigh scattering*, and *multiphonon vibrational absorption* at infrared (IR) wavelengths and are described with good approximation by the following relationship:

$$\alpha_{\text{intrinsic}} = \alpha_{UV} + \alpha_R + \alpha_{IR} = C_{UV} \cdot e^{c/\lambda} + A_R \cdot \lambda^{-4} + D_{IR} \cdot e^{-d/\lambda} \quad (12.14)$$

The electronic and multiphonon absorptions depend exponentially on the wavelength and define the edges of the *transmission window* of the particular glass material. For instance, the window of transmission of silica glass ranges from the UV to approximately 2.5 μm, while other glasses such as fluoride and chalcogenide glasses have

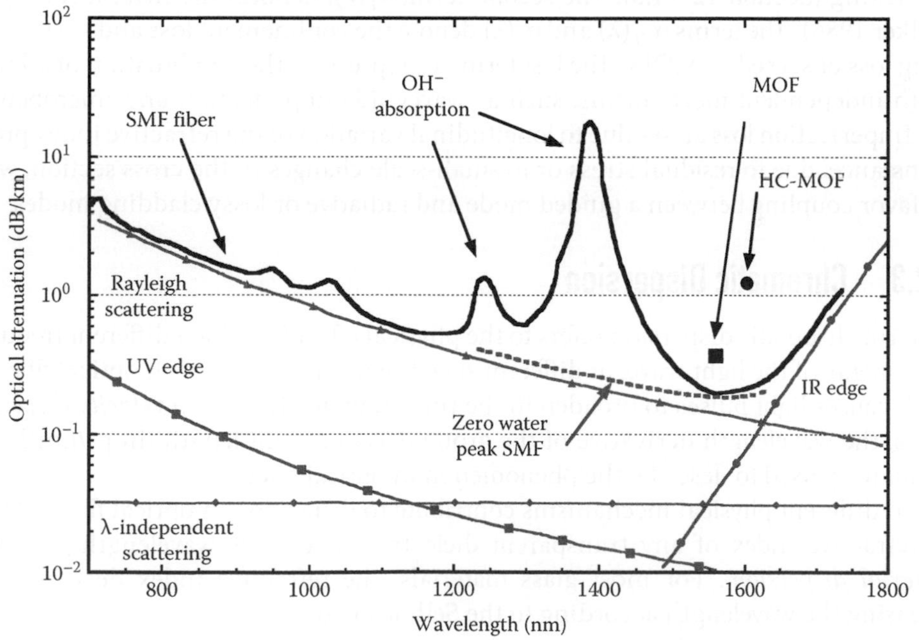

FIGURE 12.2 Spectral attenuation of a conventional SMF and a zero-water peak fiber, highlighting contributions due to individual loss mechanisms. The state-of-the-art losses of index-guiding MOFs and hollow-core MOFs (Section 12.4.5) are shown for comparison.

multiphonon edges extending further in the IR and conversely, their electronic edges are located at visible or near-IR wavelengths. The transmission window of IR glass materials will be discussed in Section 12.4.1. Within the transmission window and far from the absorption edges, the dominant intrinsic loss mechanism is Rayleigh scattering, which arises from nanometer-scale fluctuations in the density and composition, which are frozen in as the glass is formed. The Rayleigh scattering loss follows a typical $\propto \lambda^{-4}$ wavelength dependence law. The scattering coefficient, A_R, is approximately 0.74 dB/km/μm^4 for synthetic silica glass (Nagayama et al. 2002).

Non-intrinsic loss mechanisms are more difficult to capture in a single representation as they tend to be specific to the particular type of fiber waveguide and may also depend on the fabrication method. In the most general case, the following equation can be assumed:

$$\alpha_{\text{extrinsic}} = \alpha_I(\lambda) + \alpha_B(\lambda) + \alpha_{CL}(\lambda) + \alpha_S(\lambda) + \alpha_0 \tag{12.15}$$

The first term in Equation 12.15, $\alpha_I(\lambda)$, denotes the absorption from optically active impurities, located either in the bulk or, in the case of MOFs (Section 12.4.5), within the air holes or at the hole/glass interface. In silica fibers, the main impurity is the hydroxyl ion, which gives rise to absorption peaks at well-defined wavelengths in the near IR (Humbach et al. 1999), most notably at 1480 and 2215 nm. Other species that are known to have strong absorption at near-IR wavelengths are transition metals (such as Fe, Cu, V); however, since their concentration in synthetic silica is negligible, their loss contribution is typically only observed in some multi component glasses fabricated by powder melting (Section 12.3.1.3). The second term, $\alpha_B(\lambda)$, denotes the macrobending loss (Walker 1986). The terms $\alpha_{CL}(\lambda)$ and $\alpha_S(\lambda)$ denote the confinement loss and surface scattering loss observed in MOFs. The last term, α_0, represents the combination of all wavelength-independent mechanisms, such as waveguide imperfection and microbending loss. Imperfection loss arises due to longitudinal variation of the refractive index profile, for instance, due to residual stress or to small-scale changes in the cross section, which can favor coupling between a guided mode and radiative or lossy cladding modes.

12.2.3 Chromatic Dispersion

The term chromatic dispersion refers to the physical effect for which different frequency components of the light travel at different speeds within the core of an optical fiber and which causes light pulses to broaden in the time domain. The *group velocity dispersion*, that is, the wavelength derivative of the group delay per unit length (in ps/nm/km), is the quantity used to describe the phenomenon in optical fibers.

Two different physical mechanisms contribute to dispersion in optical fibers. Firstly, the refractive index of any transparent dielectric material is wavelength dependent (*material dispersion*). For most glass materials, the refractive index decreases with increasing the wavelength according to the Sellmeier equation:

$$n^2 - 1 = \sum_{j=1}^{N} \frac{\lambda^2 A_j}{\lambda^2 - \lambda_j^2} \tag{12.16}$$

The material-related dispersion of a medium of refractive index $n(\lambda)$ can then be expressed as

$$D_m = -\frac{\lambda}{c}\frac{d^2 n(\lambda)}{d\lambda^2} \tag{12.17}$$

Figure 12.3 shows D_m for a small selection of optical materials: a zero dispersion point is generally observed—1270 nm for silica glass, shifting progressively to longer wavelengths in IR-transmitting glasses such as fluoride and chalcogenide glasses.

In addition to the dispersion of the material, optical fibers also have a *waveguide dispersion* component, which arises from the modal field pattern of the guided light changing as a function of the wavelength. Intuitively, shorter wavelengths are more tightly confined in the core, while modes at longer wavelengths extend further in the cladding, resulting in a different propagation speed. A more complete formula for the *total* group velocity dispersion in optical fibers is then given by

$$D = -\frac{2\pi c}{\lambda^2}\frac{d^2\beta(\lambda)}{d\omega^2} \tag{12.18}$$

The dispersion of a few common SMF types is also shown in Figure 12.3. In conventional SMFs, the total dispersion is close to the material dispersion as the waveguide component is small, negative, and only weakly dependent on the wavelength. Dispersion-shifted, dispersion-flattened, and dispersion compensating fibers have comparatively larger waveguide dispersion, resulting in the profiles shown in Figure 12.3. The dispersion properties of microstructure fibers will be briefly discussed in Section 12.4.5.4.

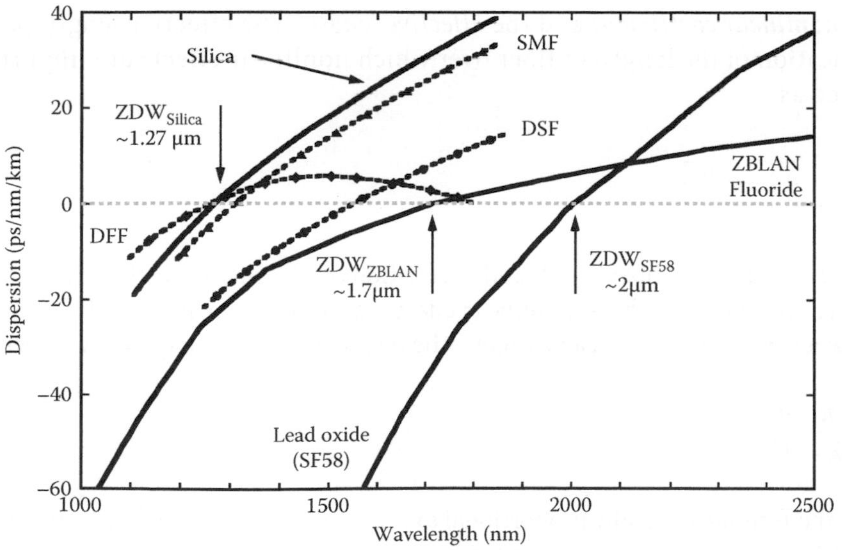

FIGURE 12.3 Dispersion of a few selected glass materials and fibers (including single-mode, dispersion-shifted, and dispersion-flattened fibers).

12.2.4 Optical Nonlinearities in Fibers

Optical nonlinearities in transparent dielectric materials arise when propagation of high-field-intensity light beams is considered. In this case, the simplifying assumption that light propagation can be described through a refractive index that is a function of the position and frequency *only* does not hold anymore. A new class of very diverse material effects is thus observed, which have all in common the fact that their magnitude is a *nonlinear* function of the incident electromagnetic field intensity.

The standard way of describing nonlinear effects is by rewriting the wave equation (Equation 12.3) via the nonlinear polarization:

$$P = P_0 + P_{NL} \approx \varepsilon_0 \left[\chi^{(1)}\mathbf{E} + \chi^{(2)}\mathbf{EE} + \chi^{(3)}\mathbf{EEE} \right] \tag{12.19}$$

where $\chi^{(i)}$ is the susceptibility—a tensor of rank i. The most important nonlinear phenomena involve either second or third order; however, since glass possesses inversion symmetry, only third-order nonlinearities are normally observed. Furthermore, it can be demonstrated that in a transparent isotropic medium, the third-order susceptibility tensor has only one independent component, and most nonlinear effects can thus be described by introducing an intensity-dependent refractive index:

$$n = n_0 + n_2 \left| \mathbf{E} \right|^2 \tag{12.20}$$

where

the *linear* refractive index n_0 is related to the linear susceptibility $\chi^{(1)}$

the nonlinear index n_2 is related to the third-order susceptibility $\chi^{(3)}$

Two quantities that are commonly introduced to describe nonlinear effects in fibers are the *nonlinear coefficient* and the *effective length*. The effective length provides a quantification of the length of fiber over which nonlinear effects are important and is defined as

$$L_{eff} = \frac{1}{\alpha} \cdot \left[1 - e^{-\alpha L} \right] \tag{12.21}$$

in which L and α are the fiber length and attenuation coefficient, respectively. Equation 12.21 shows that for high-loss fibers, nonlinear effects can only occur in a small fraction of the length, where intensity is sufficiently high. The *nonlinear coefficient* quantity is defined as

$$\gamma = \frac{2\pi}{\lambda} \cdot \frac{n_2}{A^{eff}} \tag{12.22}$$

Nonlinear effects are generally proportional to γ and thus to the ratio of n_2/A_{eff} (in W^{-1} km^{-1}). Table 12.1 shows the values of γ for a range of fiber types. Standard SMFs exhibit relatively small nonlinearity, which is a consequence of low intrinsic nonlinearity of silica glass. Substantially higher values of nonlinearity can be obtained by designing fibers to

Table 12.1 Values of Nonlinear Coefficient for Various Fiber Types

Fiber Type	γ (W^{-1} km^{-1})	Reference
SMF	2.7	Sugimoto et al. (2004)
Silica microstructured	60	Lee et al. (2001)
Lead silicate	860	Camerlingo et al. (2010)
Lead silicate microstructured	1860	Leong et al. (2006)
Bismuth oxide	1360	Sugimoto et al. (2004)
Chalcogenide microstructured	2750	El-Amraoui et al. (2010)

have a very small effective area. This is, for instance, the case of fiber tapers and of some MOFs. The highest nonlinearities are observed in fibers made of high-n_2 materials such as heavy metal oxide glasses and chalcogenide glasses.

A host of effects can be observed in highly nonlinear (HNL) optical fibers, depending on their composition and waveguide design and on the wavelength, temporal characteristics (e.g., continuous wave or pulsed), and peak intensity of the propagating light. Here, we briefly discuss stimulated scattering phenomena, due to their relevance to optical fiber sensor applications. The reader is referred to classical textbooks (Agrawal 2006, Boyd 2008) for a comprehensive description of nonlinear effects.

Stimulated scattering phenomena comprise *stimulated Raman scattering* (SRS) and *stimulated Brillouin scattering* (SBS). The well-known spontaneous Raman and Brillouin scattering effects arise as a consequence of the interaction between the propagating optical field and the vibrational modes in the host material, which causes the generation of additional spectral components at both upshifted (anti-Stokes) and downshifted (Stokes) frequencies. Spontaneous scattering effects in optical fibers have been investigated extensively for distributed sensing applications due to their dependence on temperature and/or strain (Dakin et al. 1985, Horiguchi et al. 1995). Stimulated Raman and Brillouin scatterings are observed at relatively high pump powers and in long fiber lengths, where the pump and scattered waves can consistently generate host vibrations thus amplifying the scattered signal.

SRS in optical fibers is typically observed as a forward-propagating, frequency downshifted Stokes wave. The threshold for SRS can be expressed as (Agrawal 2006)

$$P_0 = \frac{16A^{eff}}{g_r L^{eff}} \quad (12.23)$$

The Raman gain for standard SMF, shown in Figure 12.4, is relatively wideband and peaks at about 13.2 THz frequency difference from the pump wavelength; dispersion compensating fibers have ~2× higher Raman gain, owing to their smaller effective area. SRS is exploited in fiber laser technology and in Raman amplifiers for telecom applications (Islam 2002). In the context of optical fiber sensors, distributed Raman amplification is often utilized to amplify the weak backscattered signals, thereby extending the sensing range and response time of distributed temperature and strain sensors (Alahbabi et al. 2006).

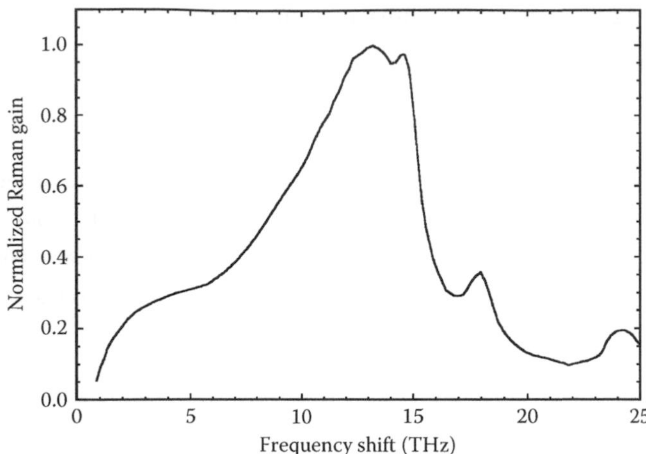

FIGURE 12.4 Raman gain of standard SMF. The gain peaks at about 13 THz from the pump frequency (1450 nm in this case). Pump and signal are copolarized.

SBS is generated by a pump signal interacting with an acoustic wave that it generates though electrostriction (Ippen and Stolen 1972). SBS gives rise to backward propagating, frequency downshifted radiation and is characterized by small frequency shifts (~11 GHz for at 1550 nm pump in SMFs), narrow gain bandwidth (~40 MHz), and the lowest power threshold among nonlinear effects. The Brillouin (peak) gain coefficient is given by

$$g_B = \frac{2\pi n^7 P_{12}^2}{c\lambda_p^2 \rho V_a \Delta\nu} \tag{12.24}$$

where

P_{12} is the longitudinal elasto-optic coefficient
ρ is the material density
λ_p is the pump wavelength
V_a is the acoustic velocity
$\Delta\nu$ is the Brillouin linewidth

The Brillouin gain spectrum of an SMF and a small-core, HNL fiber is shown in Figure 12.5. Differently from SRS, which essentially depends only on the core material, SBS gain also depends on the fiber structure, and a complex peak structure may be observed if multiple acoustic modes are excited (as is the case of the HNL fiber in Figure 12.5). SBS is also strongly dependent on the local parameters of the fiber, and thus it has been the subject of intense investigation for distributed sensing applications. The Brillouin frequency shift increases linearly with tensile strain and temperature (Garcus et al. 1997). Furthermore, Brillouin gain can be exploited in order to enhance the scattering cross section. A typical configuration (Brillouin optical fiber time-domain analysis [BOTDA]) involves a pair of counterpropagating beams (one cw and one modulated), injected from opposite ends of the fiber and having a small frequency separation matching the Brillouin frequency (Bao et al. 1993).

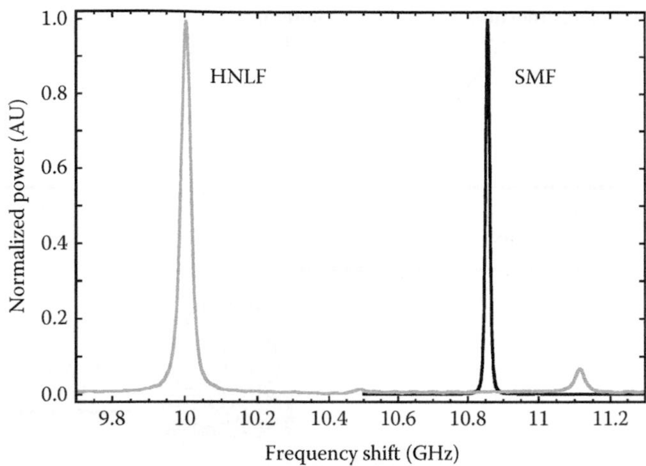

FIGURE 12.5 SBS spectrum for a standard SMF and a highly nonlinear (germanosilicate) fiber (HNLF). The multiple peaks for the HNLF correspond to the various longitudinal acoustic modes. Pump wavelength is 1550 nm. (Measurements courtesy of Dr. S. Dasgupta, ORC, University of Southampton, Southampton, U.K.)

12.3 Fabrication of Optical Fibers

In this section, the fabrication techniques most commonly used for most of the standard and special optical fibers are reviewed. The fabrication of MOFs is substantially different from conventional, all-solid fibers and is discussed in Section 12.4.5.2. Optical fiber fabrication is generally accomplished in two subsequent steps. The first step is the fabrication of a *preform*, that is, a rod of glass comprising several cylindrical layers of different composition and thus refractive index (core and one or more cladding layers), having the same relative aspect ratio of the final fiber but a much larger (macroscopic) cross section (10–100 mm in diameter). The second step is *fiber drawing*, in which the preform is elongated and thinned down to fiber size. The drawing process is accomplished at high temperature using suitably designed fiber draw towers and involves reducing the preform diameter by several orders of magnitude. Fiber fabrication techniques have been introduced in the early 1970s and have since undergone an extraordinary development, to the point that optical fibers with transmission approaching the theoretical limits are fabricated nowadays in several thousand kilometers' lengths in a single fiber draw.

12.3.1 Preform Fabrication Techniques

There are several established techniques to fabricate optical fibers, which provide different performance levels. The most common are shortly outlined in the following.

12.3.1.1 Modified Chemical Vapor Deposition

Modified chemical vapor deposition (MCVD) is the oldest and most established preform fabrication technique. In MCVD, suitable high-purity glass precursors are flowed and reacted inside of a glass tube. The process is schematized in Figure 12.6a. A commercially available synthetic silica tube (e.g., Suprasil F300 from Heraeus) is prepared via repeated

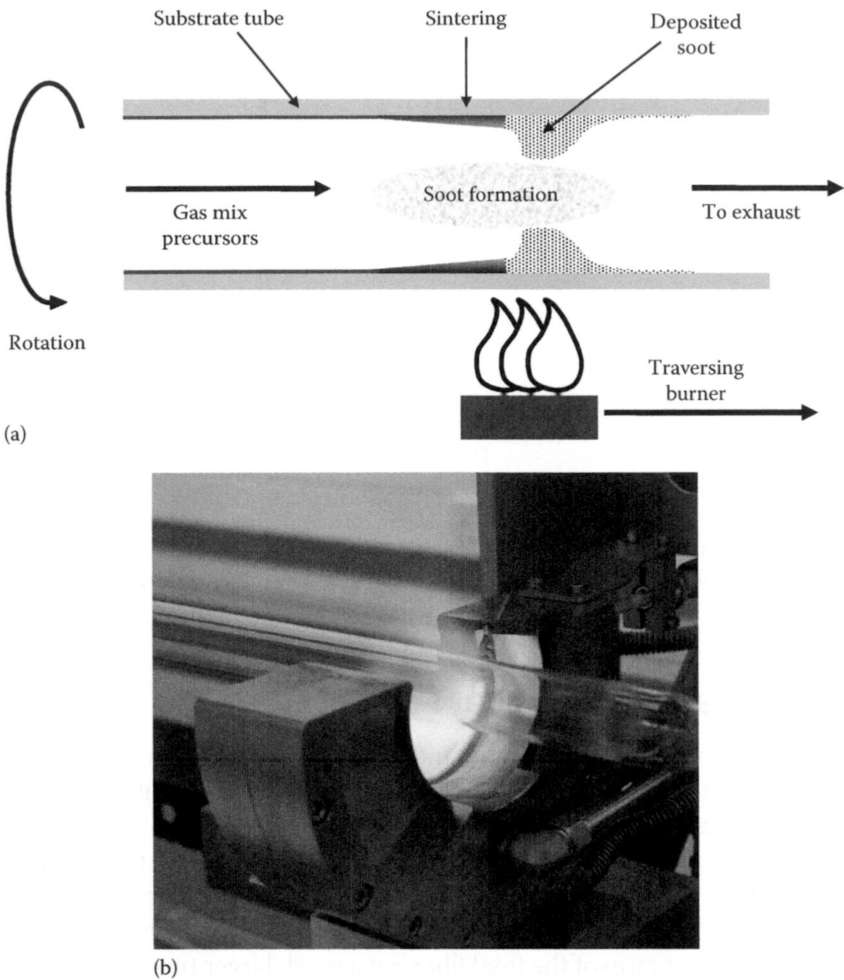

FIGURE 12.6 (a) Schematic of the MCVD process. (b) MCVD deposition lathe. (Photo courtesy of EPSRC Centre for Innovative Manufacturing in Photonics, University of Southampton, Southampton, U.K.)

cleaning and/or etching stages in order to make its inner surface chemically pristine. The tube is then mounted on a specially modified glass working lathe (*deposition lathe* shown in Figure 12.6b); one end of the tube is connected to a gas delivery system via a rotary seal, while the other end is connected to an exhaust. Subsequent heating stages (fire polishing) at temperature above 1850°C are applied in order to smooth out and eliminate any surface defects, which may prejudice the process or affect the mechanical strength of the final fiber. Next, the chemicals are fed into the tube in a stream of oxygen and an inert carrier gas (e.g., He or Ar). In the case of silica glass deposition, the precursor is silicon tetrachloride ($SiCl_4$), but a number of other chemical formulations (e.g., $GeCl_4$, BBr_3, $POCl_3$, and SF_6) can be added in order to either increase/decrease the refractive index, or decrease the processing temperature, or aid the subsequent incorporation of dopants via solution doping. The precursors of choice have orders of magnitude higher vapor pressures than the most common optically active contaminants (e.g., transition metals), which enable the process to produce extremely pure glass material. An oxyhydrogen burner is passed along the tube while the latter is rotated, which heats up the glass to ≥1850°C, causing the chemicals inside the tube to oxidize. A soot of glassy particles is nucleated and then deposited on the inside surface of the tube through thermophoresis. The silica soot is then sintered into a thin layer of glass by the flame itself. The process can be repeated tens of times in order to deposit thicker layers. When active dopant (e.g., rare-earth ions) incorporation is required, a pass is performed at suitably reduced temperature, which produces a layer of unsintered glass soot. The tube is then soaked in a liquid solution containing a suitable compound (e.g., $ErCl_3$ and $YbCl_3$), dried to remove the solvent, and then heated up to sinter the soot containing the dopant ions. After the deposition phase, the final step consists in collapsing the tube into a solid rod through a heating pass performed at temperatures in excess of 2100°C. The final preform consists of a number of cylindrical layers, the core being formed by the innermost deposited homogeneous layer; the geometry of the final preform is determined by the thickness of the glass layer during each deposition pass.

The MCVD technique is extremely flexible and versatile and relatively simple to implement and is still in use today particularly for the fabrication of active fibers, photosensitive fibers, polarization maintaining (PM) fibers, and some dispersion-tailored fibers. It should be however noted that the use of oxyhydrogen flames has some drawbacks (e.g., ingression of OH impurity in the glass), and thus alternative techniques based on plasma torches (plasma enhanced chemical vapor deposition or PECVD) have also been extensively investigated (Geittner et al. 1976). The main limitation of both MCVD and PECVD lies in their poor scalability, and thus it has been largely replaced by other vapor deposition techniques (namely, the vapor axial deposition [VAD] and outside vapor deposition [OVD] techniques discussed in the following) for the mass production of ultralow-loss transmission fibers.

12.3.1.2 Outside Vapor Deposition and Vapor Axial Deposition

OVD and VAD have been developed as a part of the effort to increase yields and thus reduce fiber manufacturing costs. These techniques can produce large amounts of glass leading to much bigger preform sizes than those possible using MCVD. On the other hand, these techniques are also significantly more complex and costly to implement. The OVD technique consists of three main steps: deposition, dehydration, and consolidation. The deposition (Figure 12.7a) is accomplished via flame hydrolysis: chloride precursors are injected in a methane/oxygen or hydrogen/oxygen flame where silica and/or doped silica particles are nucleated. The soot is deposited on a refractory *target rod*, which is continually rotated as the burner is traversed along its length. Preforms are obtained by depositing multiple layers, in which the soot composition can be varied as required. The following step involves removing the central rod and treating the soot preform under a chlorine atmosphere at temperatures around 1000°C in order to remove water and transition metal impurities. Last, the soot preform is sintered at 1400°C–1600°C in a consolidation furnace, where helium (which is fast diffusing in silica) is generally used in order to prevent the formation of bubbles.

Similar to OVD, the VAD technique (shown in Figure 12.7b) also involves the deposition of a cylindrical soot preform. However, the deposition is accomplished from one extremity rather than axially. In contrast to OVD, core and cladding can be deposited simultaneously by using two separate burners and vapor-phase precursor lines. The soot

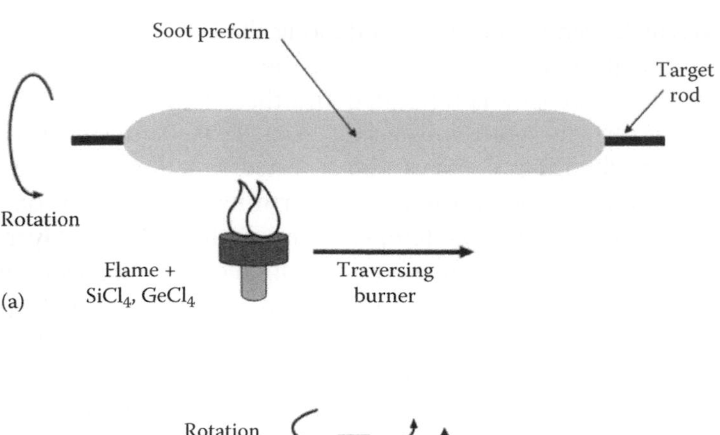

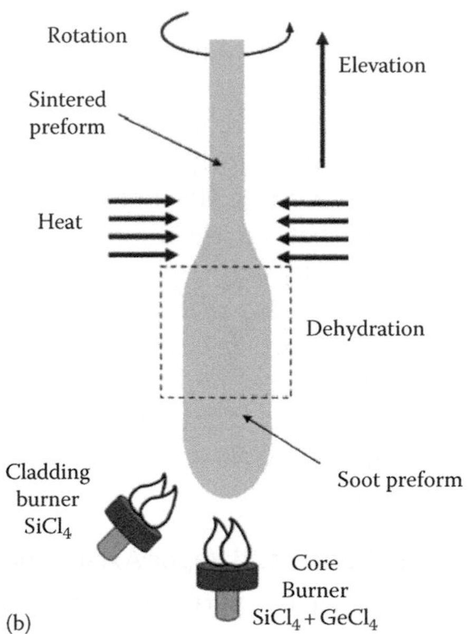

FIGURE 12.7 (a) Schematic of the OVD process and (b) schematic of the VAD process.

preform is continually rotated and pulled away as more soot is deposited at its tip by the burners. The composition profile of a VAD preform is determined to a large extent by the design of the burners; the burners used in VAD are therefore highly engineered. Once deposited, VAD preforms are dehydrated and consolidated using a similar procedure as described for OVD, but in this case, they can be implemented as a continuous process (Figure 12.7b). Currently, the VAD technique is the fabrication technique most widely used for the mass production of ultralow-loss transmission fibers.

12.3.1.3 Preform Fabrication Techniques for Multicomponent Glass

Chemical vapor deposition methods and their variants are suitable for silica glass and a rather restricted set of its modifications. In contrast, a large number of other glass-forming systems are obtained from mixtures of several components (*multicomponent* glasses), and as such, they are poorly suited to preparation via chemical deposition due to the lack of suitable precursors with similar vapor pressures. Multicomponent glasses, comprising *heavy metal oxide glasses*, *fluoride glasses*, and *chalcogenide glasses*, are of interest for a number of applications, most notably for IR-transmitting fibers (Section 12.4.1) and, in more recent years, for HNL fibers (particularly heavy metal oxides and chalcogenide glasses). Their preparation involves a completely different approach which is based on melt quenching: the glass components are individually purified, batched in the correct ratio, and melted in a crucible. Glass ingots are obtained by rapidly cooling down the melts (quenching) and subsequently annealed. The melt-quenching technique has two main limitations: Firstly, it cannot achieve the same level of purity as chemical vapor deposition methods. Secondly, glass ingots still need to be assembled or formed into preforms, and it is in general very challenging to obtain large and uniform preforms that are free of defects (e.g., bubbles, inclusions) at the interface between core and cladding layers.

The *rod-in-tube* technique is the simplest preform fabrication technique, which consists in elongating a core rod and inserting it in a suitable cladding glass tube; a tight fit and excellent surface quality are required in order to form a defect-free interface. Multiple stretching and sleeving steps may be required in order to obtain SMFs. While this technique is relatively simple to implement, it requires the glass to have excellent stability against devitrification in view of the repeated heating steps required.

Rotational casting (France et al. 1990) was applied with success to the fabrication of multimode preforms in fluorozirconate glass (Section 12.4.1). Molten glass is poured into a cylindrical mold, which is preheated at a temperature close to the glass transition temperature and immediately spun at about 5000 rpm. As the glass cools down, it forms a tube within the mold, which will constitute the cladding. Glass from a second melt (core composition) is then poured into the central hollow region and the preform obtained is then cooled down and annealed. This technique is well suited to fluoride glass due to the low melting point (~600°C) and steep viscosity/temperature curve (France et al. 1990).

More recently, *extrusion* has been employed in order to obtain solid and microstructured preforms in a number of glass compositions, including lead oxide (Petropoulos et al. 2003), bismuth oxide (Ebendorff-Heidepriem et al. 2004), and tellurium oxide glass (Feng et al. 2005a). The process is shown in Figure 12.8a. A cylindrical billet of glass with polished surfaces is heated up to the glass softening point and then pressure (20–40 MPa) is applied vertically through a ram. The glass is forced to flow through a suitably designed die, getting reshaped in the process. Typical extrusion speeds are slow (<1 mm/min). With this technique, it is possible to obtain core/clad structures (Feng et al. 2005b) and, by using suitably shaped dies, complex air/glass microstructured preforms (Figure 12.8b). One of the main advantages of extrusion lies in the low processing temperatures, which is very advantageous in the case of glasses with poor thermal stability as it reduces the possibility of crystal nucleation.

12.3.2 Fiber Drawing

Optical fibers are manufactured by heating and drawing preforms. The apparatus typically used, the fiber draw tower, is outlined in Figure 12.9. The preform is held vertically inside a furnace, where the lower end is heated up to above the glass softening point (~2000°C in the case of silica). The temperature profile inside the furnace is characterized

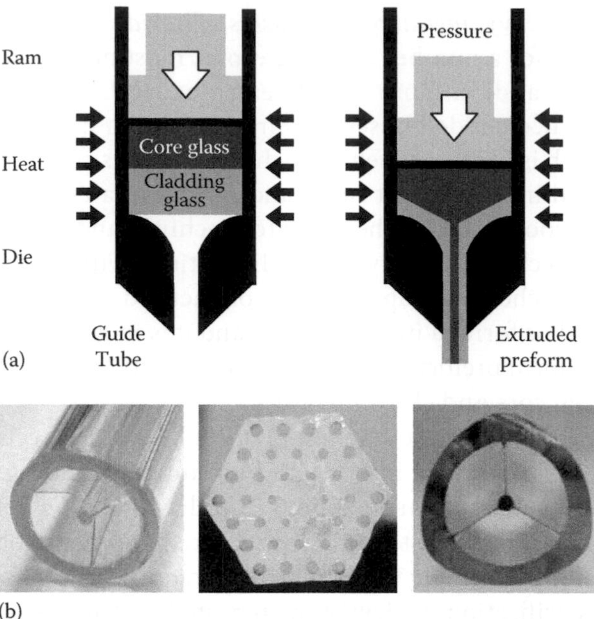

FIGURE 12.8 (a) Preform fabrication by extrusion and (b) extruded microstructured preforms obtained from various compound glasses. (Photo courtesy of Dr. X. Feng, ORC, University of Southampton, Southampton, U.K.)

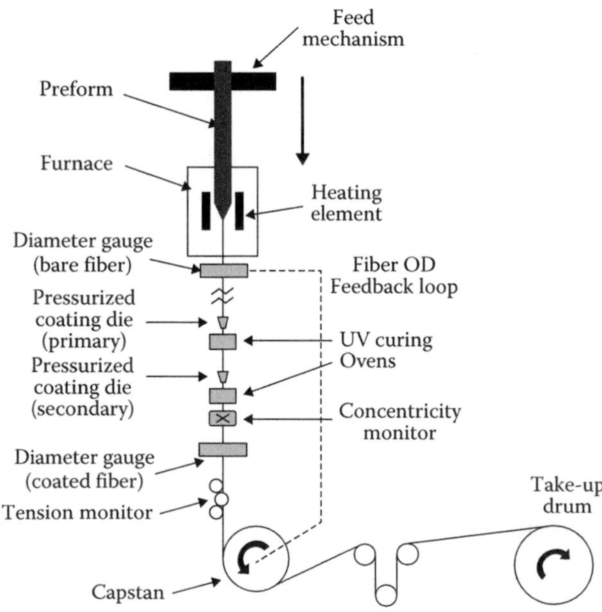

FIGURE 12.9 Layout of a fiber drawing tower showing the main functional blocks.

FIGURE 12.10 Fiber drawing tower equipment: furnace in operation. The preform is slowly fed into the furnace from the top while fiber is drawn from the bottom. (Photo courtesy of EPSRC Centre for Innovative Manufacturing in Photonics, University of Southampton, Southampton, U.K.)

by a narrow hot zone, which produces a transition or *neck-down* region in which the preform diameter is reduced to that of a thin thread of glass, typically 100–200 μm. Under stationary draw conditions, the fiber is pulled by a capstan motor located at the bottom of the tower, and the preform is slowly fed into the furnace (Figure 12.10). The diameter of the fiber is determined by the preform diameter and the ratio between feed rate and draw speed through mass conservation. As mechanical strength is a paramount requirement for optical fibers, one or several layers of polymer coating are applied in order to prevent direct contact between the bare fiber and the apparatus providing traction (wheels, pulleys, etc.). The coating applicator is a vessel with a pair of conical dies of appropriate diameter, where liquid polymer is fed at a constant pressure. The coated fiber is passed through a high-power UV lamp, which cures the polymer permanently on the fiber.

Longitudinal uniformity is a key fabrication requirement, as small diameter variations affect the performance of the optical waveguide and can induce excess loss. Diameter variations are classified according to their time constant. *Long-time constant* diameter variations may arise due to drifts in the furnace temperature, feed rate, and draw speed and from changes in the preform diameter. They can be minimized by monitoring the fiber diameter below the furnace and by adjusting the capstan speed to maintain a fixed diameter via a feedback loop (Figure 12.10).

In addition, *fast fluctuations* arise as a consequence of small perturbations such as temperature fluctuations at the neck-down region and mechanical vibrations and are minimized with a combination of equipment design and optimization of the process conditions. State-of-the-art preforms used for commercial production of telecom fibers have diameters up to 200 mm and drawing speeds in excess of 1000 m/min are common.

12.4 Overview of Key Special Optical Fibers

12.4.1 Infrared Fibers

Silica and silica-based materials are unsuitable for the transmission of wavelengths beyond the near IR (to about 2.5 μm). The optical attenuation of silica has a minimum at about 1.55 μm and increases rapidly beyond this wavelength due to vibrational absorption. On the other hand, several other glass-forming systems have transmission windows that extend to significantly longer wavelengths, as shown in Figure 12.11. In addition to glasses, a few IR-transmitting crystalline compounds exist, which are amenable to be drawn into fibers. These materials are often referred collectively to as low-phonon-energy materials. IR fibers have been investigated for over 40 years in an effort to match the properties of silica fibers. With very few exceptions however, the optical performance, mechanical properties, and chemical and environmental durability of these fibers remain inferior to those of telecom fibers. Table 12.2 summarizes the key properties of some of the most common IR fibers. As compared to silica, these fiber materials have lower softening point, higher refractive index and expansion coefficient, and higher optical attenuation.

Fluoride fibers based on the ZrF_4–BaF_2–LaF_3–AlF_3–NaF (ZBLAN) glass-forming system have a wide transmission window ranging from the visible to about 4 μm. Interest in

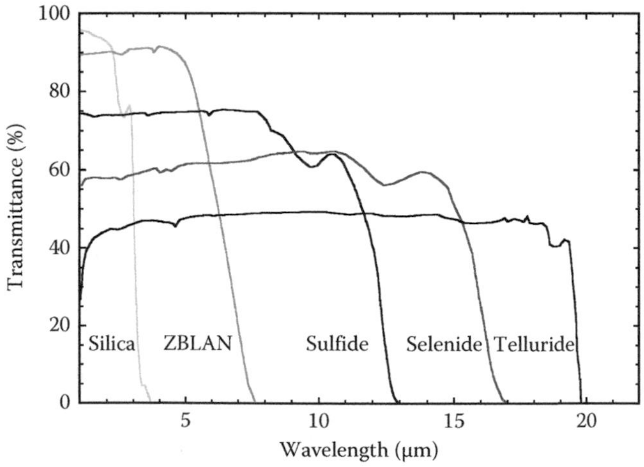

FIGURE 12.11 Transmission window of IR glasses (thickness 2–3 mm). (Adapted from *Comp. Rend. Chim.*, 5, Sanghera, J.S., Brandon Shaw, L., and Aggarwal, I.D., Applications of chalcogenide glass optical fibers, 873–883, Copyright 2002, with permission from Elsevier.)

Table 12.2 Properties of Selected IR Fibers Compared to Standard Silica Fibers

Property	ZBLAN	As_2S_3	As/Ge/Se/Te	Silver Halide	Silica
Glass transition (melting) temperature (°C)	265	203	203	412	1175
Refractive index	1.48	2.4	2.9	1.7	1.45
dn/dT (10^{-6}°C^{-1})	−1.5		10	−1.5	1.2
Transmission window (μm)	0.25–4.0	0.62–7	4–11	3–16	0.24–2.0
Lowest loss (dB/km)	0.45 (2.3 μm)	50 (2.2 μm)	800 (7–8 μm)	400 (10 μm)	0.16 (1.55 μm)

this material surged in the late 1980s when theoretical studies (Shibata et al. 1981) predicted their intrinsic loss to be one order of magnitude lower than silica. Several research establishments, most notably the British Telecom Labs in the United Kingdom, initiated substantial R&D programs aimed at developing a fiber technology that could rival standard silica telecom fibers. While these efforts largely failed to achieve their original target, fibers with very low loss were obtained (down to 0.45 dB/km at 2.3 μm) and a new fabrication technology, alternative to chemical vapor deposition methods, was developed (France et al. 1990). Preform fabrication is based on rotational casting (Section 12.3.1.3). A drawback of ZBLAN is its limited durability against attack from atmospheric moisture. Fibers are also significantly weaker than their silica counterparts. In more recent years, fluoride fibers have been investigated for applications in IR supercontinuum generation, sensing, imaging, and spectroscopy and for the delivery of Er^{3+}:YAG radiation at 2.9 μm for laser surgery (Harrington 2004). Rare-earth-doped fluoride fibers are of interest for fiber amplifiers, IR lasers, and visible upconversion lasers (Miyajima et al. 1994, Zhu and Peyghambarian 2010).

Chalcogenide glasses are a broad group of glass-forming systems comprising a *chalcogen* element (S, Se, Te) and other elements such as Ga, Ge, As, Sb, and P. Chalcogenide glasses are relatively easy to form, generally very stable, durable, and moisture resistant. Optically, chalcogenide glasses are characterized by a high refractive index ($n \geq 2$), high nonlinearity, high dn/dT, low transmission in the visible, and excellent transmission in the IR. The position of the transmission window varies substantially depending on the glass composition, as seen in Figure 12.11. Sulfide glasses generally transmit from the red end of the visible spectrum to about 5–6 μm; selenide and telluride glasses transmit over a broader range and further in the IR but are almost completely opaque in the visible. Several chalcogenide systems can be drawn into fibers and indeed some of them were investigated for optical transmission in fibers before silica (Kapany and Simms 1965). The most developed glass systems include AsS, AsSe, and AsGeSeTe. Glasses are generally obtained by mixing and melting appropriate amounts of the elemental components in sealed evacuated ampoules; long processing times and specially designed mixing (rocking) furnaces are commonly employed. Fibers have also been produced from the nontoxic gallium sulfide-based glasses such as GaNaS and GaLaS. Preforms can be obtained by rod in

FIGURE 12.12 Multimode chalcogenide gallium lanthanum sulfide (GLS) fiber fabricated by extrusion. The outer layer is graphite-based coating.

tube, rotational casting, or extrusion. Figure 12.12 shows a multimode GaLaS fiber fabricated via extrusion (Petrovich 2003). Mono- and multimode chalcogenide fibers can be fabricated directly from a melt using the double crucible method (Harrington 2004). Polymer clad multimode chalcogenide fibers are also common.

The transmission loss of chalcogenide fibers is believed to be dominated by absorption from impurities, particularly hydrogen groups (OH^-, SH^-, SeH^-) and oxygen. Substantial work from various groups worldwide was devoted to purifying the raw materials and reducing contamination during the glass fabrication. The typical loss of chalcogenide fibers is in the range 0.1–1 dB/m, although losses below 0.05 dB/m were reported for short lengths of AsS fiber (Sanghera et al. 2009). Step-index single and multimode AsS and AsSe are commercially available. More recently, a casting method has been developed, which has allowed the fabrication of chalcogenide MOFs with loss comparable to the best solid fibers (Coulombier et al. 2010). Chalcogenide glass fibers have been investigated for several applications, ranging from IR lasers and amplifiers, power delivery, IR imaging and thermometry, nonlinear applications, and IR sensing and spectroscopy (Sanghera et al. 2002). IR sensing is probably the most developed and successful application area. Chalcogenide glass fibers open up access to the wavelength region between 3 and 6 μm where the strongest vibrational fingerprints of most molecules are located. Probes based on various designs (including transmission, reflection, evanescent wave, and attenuated total reflection) for sensing of liquid and gas mixtures are commercially available.

In addition to IR glasses, some *crystalline materials* can also be manufactured in fiber form (Harrington 2004). Typical examples include single-crystal sapphire fibers and polycrystalline silver halide fibers. Fabrication of these fibers is a lot more challenging and can only be accomplished in meter-scale lengths. Sapphire fibers are exceptionally robust and transmit light to about 3.5 μm. Silver halide fibers can transmit to beyond 15 μm but are hygroscopic and photosensitive, requiring suitable packaging and protection from UV and visible light.

12.4.2 Polarization Maintaining Fibers

The PM fiber is a modified version of SMF specifically designed to preserve the polarization state of the light traveling through it. Standard SMFs support a degenerate pair of orthogonally polarized optical modes (LP_{01}). In real fibers, the two modes can exchange energy as they propagate due to of small-scale waveguide defects and/or perturbations such as temperature gradients and macro- and microbending. As a consequence, SMFs do not hold the polarization. This is undesirable for a number of applications ranging from interferometric sensors to fiber lasers and light delivery to/from polarization-sensitive components. In PM fibers, a constant and controlled difference in the propagation constant of the two orthogonally polarized modes is introduced, thus preventing energy transfer between the two (in absence of extreme perturbations). The *birefringence* is the difference of group index between the two modes:

$$B_G = N_G^{\text{slow}} - N_G^{\text{fast}} \tag{12.25}$$

The fiber birefringence is normally determined via a beat-length measurement (Figure 12.13). Light from a broadband source is coupled into the fiber at about 45° with respect to either axis, and a second polarizer is placed at the output. A pattern of fringes appears in the transmitted spectrum, and the beat length (at a wavelength λ_0) can be estimated as

$$L_B \approx L \cdot \frac{\Delta\lambda}{\lambda_0} \tag{12.26}$$

where

$\Delta\lambda$ is the wavelength spacing between two maxima
λ_0 is the central wavelength

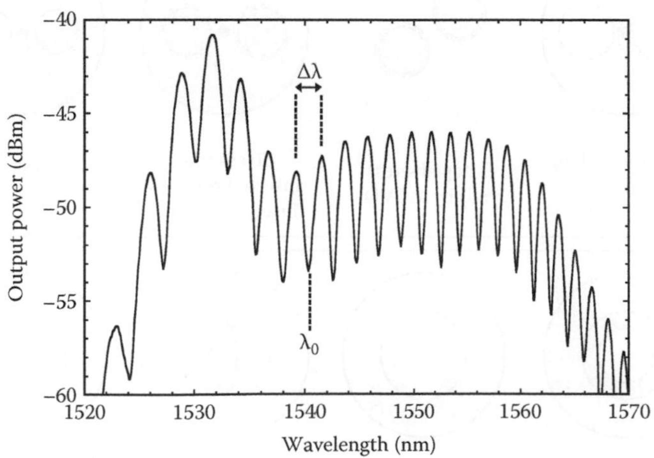

FIGURE 12.13 Beat-length measurement via wavelength scanning method. The variation of the fringe period indicates the birefringence is wavelength dependent. The fiber under test is a hollow core bandgap fiber (Section 12.4.5.6) HC-PBGF.

The beat length and birefringence are related by the following equation:

$$L_B = \frac{\lambda}{B} \tag{12.27}$$

Typical values of L_B in commercial PM fibers range from 1 to 2 mm (at 1550 nm).

Two different strategies can be followed in order to obtain highly birefringent fibers. In the first approach, the fiber geometry (e.g., the core shape) can be modified to make it noncircularly symmetric. This approach produces *form* birefringence. Typical examples include elliptical core fibers and some MOFs (Section 12.4.5.5) where spatial anisotropy can be obtained by elongating the core or by modifying the shape of the cladding holes (Ortigosa-Blanch et al. 2000, Chen et al. 2004). Form birefringence is in general wavelength dependent but less sensitive to temperature variations. The second approach relies on the *stress anisotropy*. In this case, uniaxial stress is introduced in an otherwise circular core; the stress anisotropy translates in refractive index anisotropy through the stress-optical coefficient. This is accomplished by introducing elements with high thermal expansion mismatch to silica (e.g., borosilicate glass) at diametrically opposite positions in the fiber cross section. To this second class belong the two most common types of PM fibers, the *PANDA fiber* and the *bow tie fiber*, which are shown in Figure 12.14, along with a simple representation of how they are manufactured. Stress birefringence is only weakly dependent on the wavelength, but it varies with temperature.

The PANDA fiber is obtained though a simple modification of the standard fabrication of telecom fibers. The starting point is a single-mode preform, in which two precision side bores are drilled at well-defined locations close to the core. Borosilicate stress rods are obtained via standard MCVD and milling or, more frequently, etching

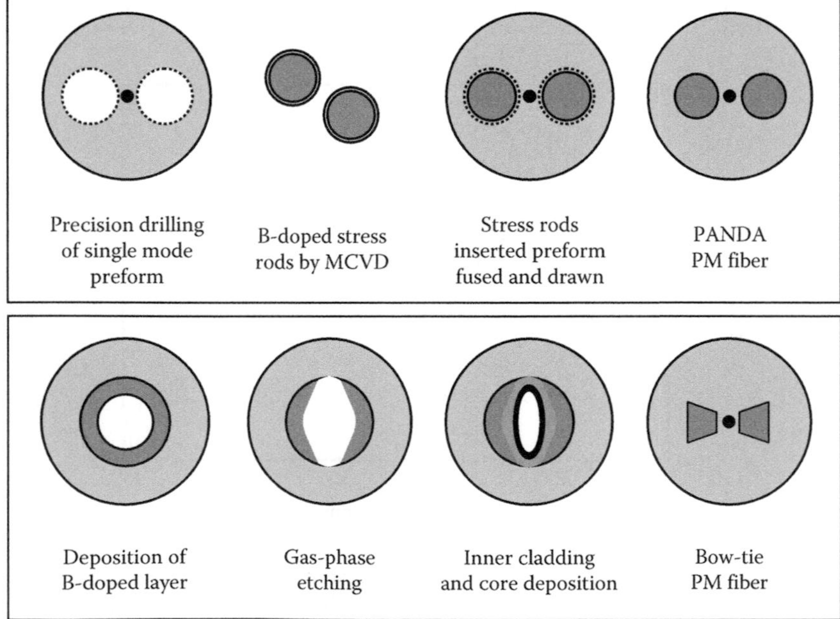

FIGURE 12.14 Schematic of PANDA and bow tie PM fiber types with their respective fabrication process.

in order to remove the substrate tube. The rods are then inserted in the two holes and the whole assembly is drawn into fiber (vacuum may be used to achieve a bubble-free interface). The fabrication process of the bow tie fiber is slightly more complex. First, a uniform thick layer of borosilicate glass is deposited in a tube using the MVCD technique. The tube is then gas-phase etched along longitudinal lines located at diametrically opposite positions. A further inner cladding and core layers are then deposited, and the whole structure is collapsed into a solid preform. The borosilicate layer collapses into two regions of trapezoidal shape, which generate the stress field. PM fibers are often drawn to a smaller form factor (80 μm) as compared to conventional telecom fibers in order to improve their resistance to static fatigue when tightly coiled. PM fibers are widely used for interferometric sensors such as the fiber-optic gyroscopes and current sensors.

12.4.3 Polymer Optical Fibers

Polymers are attractive as fiber materials due to their low cost, ease of manufacturing, and low processing temperatures as well as for their highly ductile properties. A comprehensive review on the development of polymer optical fibers (POFs) can be found in Zubia and Arrue (2001). Typical polymers used for fiber fabrication include polymethyl methacrylate (PMMA), polystyrene, and polycarbonate. Historically, POFs failed to make an impact in the data transmission application sector due to their higher loss and smaller bandwidth as compared to silica fibers. Intrinsic loss of polymer materials is dominated by Rayleigh scattering and by strong absorptions from C–H bond vibrations in the near IR. The region of minimum loss thus typically falls in the visible; for instance, the minimum loss of PMMA, one of the most common POF materials, is about 100 dB/km at ~600 nm (Figure 12.15). More recently, however, amorphous fluorinated polymers (CYTOP) have been developed with substantially reduced loss (~10 dB/km at 1100 nm), which has renewed the interest in these fibers especially for short-haul data transmission (Asai et al. 2011).

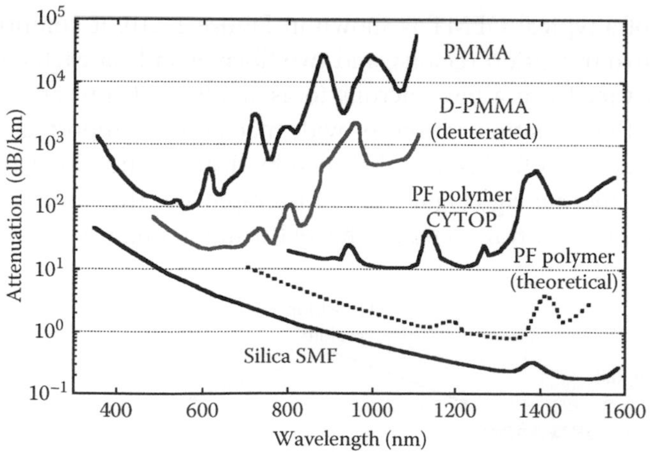

FIGURE 12.15 Attenuation of POFs. (Adapted from *Opt. Fiber Technol.*, 7, Zubia, J. and Arrue, J., Plastic optical fibers: An introduction to their technological processes and applications, 101–140, Copyright 2001, with permission from Elsevier.)

It should be noted that POFs are primarily available as step- and graded-index MMFs. Progress in fabrication techniques of single-mode POFs has led to a substantial loss reduction; single-mode POFs with losses of less than 1 dB/cm are commercially available. Single-mode POFs are however difficult to achieve at visible wavelengths (where material loss is minimum) through conventional core/clad design. A promising route to low-loss single-mode POFs is via MOF technology (Section 12.4.5). MOFs can be made of a single material, thus simplifying the fabrication, and their confinement and guidance properties can be accurately controlled via the fiber geometry to provide single-mode guidance at any wavelength. Single-mode PMMA can currently be fabricated with ~dB/m loss (Argyros 2009).

POFs are of great interest for sensing applications. As compared to conventional fibers, they have substantial advantages in terms of reduced weight and superior mechanical properties, such as high flexibility, high fracture toughness, and high elastic strain limits. The elastic deformation limit of PMMA is as high as 10%, as compared to a few percent for conventional fibers. POFs also have higher sensitivity to strain as compared to silica fibers (Silva-Lopez et al. 2005). POFs thus hold a clear advantage for strain measurements, for example, in high-load civil infrastructure monitoring where the strain can exceed the breaking strain of glass fibers. Recently, fiber Bragg gratings have been demonstrated in POFs, which greatly enhance their potential for high-sensitivity all-fiber strain gauges. In addition, materials used in POFs have excellent biocompatibility, opening up the possibility for in situ biochemical sensing.

12.4.4 Optical Fiber Microtapers

Optical fiber microtapers (OFMTs) are a novel and emerging class of waveguides that have attracted very strong interest in the last few years. OFMTs are obtained through extreme scaling down of conventional fibers and have remarkable optical and mechanical properties. Excellent reviews have recently appeared (Tong and Mazur 2007, Brambilla 2010) to which the reader is referred for a more comprehensive description of the subject.

A schematic of a typical OFMT is shown in Figure 12.16: it comprises a uniformly tapered region, two transition regions, and two fiber pigtails at each end. The diameter of the taper can vary from a few microns to as small as 20 nm and is most typically sub-wavelength. The transition regions provide mechanical support as well as input and output optical coupling. OFMTs are fabricated by heating and elongating conventional fibers. Usually, a tapering rig (Figure 12.17) is employed: the fiber is clamped on two translation stages, the central section is heated using a small flame or a miniaturized

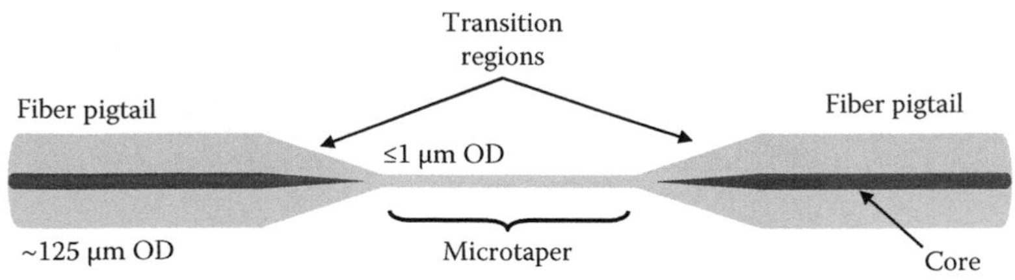

FIGURE 12.16 Schematic of an OMFT.

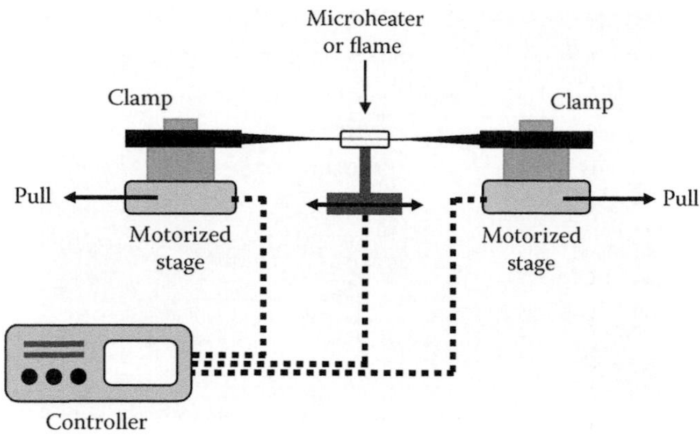

FIGURE 12.17 Schematic of fiber tapering rig.

furnace, and then the two extremities are pulled apart at controlled speed. Excellent uniformity and full control over the diameter profile of the various sections can be achieved in this fashion. OFMTs produced under optimized conditions have extremely smooth and defect-free surfaces, leading to high mechanical strength and low optical attenuation.

Guidance in OFMT is characterized by a transition from *core guided* (in the pigtails) to *cladding guided*—in the microtaper—where light is confined by the refractive index difference between glass and air or the surrounding medium. If the transition is sufficiently sharp, the modal content is preserved, that is, light couples from the fundamental mode of the pigtail to the fundamental mode of the OFMT. The OFMT is then termed as *adiabatic*. Two distinct guidance regimes can be obtained in OFMT depending on their diameter (relative to the wavelength of the guided light), and different classes of devices correspond to each of the two regimes. The two regimes can be easily described in terms of the V-number (Equation 12.9) of the OFMT. A *strong confinement* regime is observed for $V \approx 2$, which for silica glass equates to the diameter of the OFMTs being roughly equal to half the wavelength. Light is mainly confined to the glass, leading to very high optical field intensities. In this regime, OFMTs yield extremely efficient nonlinear devices. The chromatic dispersion can also be accurately controlled via the diameter, leading to the observation of a host of nonlinear effects, such as supercontinuum generation, high harmonic generation, and slow light. On the other hand, when $V < 2$ (i.e., $d < \lambda/2$ for silica), the OFMT operates in *high-evanescent field* regime, where most of the optical power travels in air. For instance, when $V = 1$, less than 10% of the optical field intensity is located in the glass (Brambilla 2010). The transmission properties of the OFMT are thus easily affected by any changes in the refractive index or absorption coefficient of the surrounding medium. This provides an alternative route to efficiently couple light to and from the OFMT and, especially, opens up the possibility for high-sensitivity refractometric or absorption sensors. The taper can be simply immersed in the medium to be probed or can be coated with a substance sensitive to a specific analyte.

One of the most intriguing applications of OFMTs are microresonators. These are straightforwardly obtained by forming one or more loops along the length of the OFMT. Light is resonantly coupled between different loops through the evanescent field and high-Q resonances appear in the transmitted spectra, which again can be

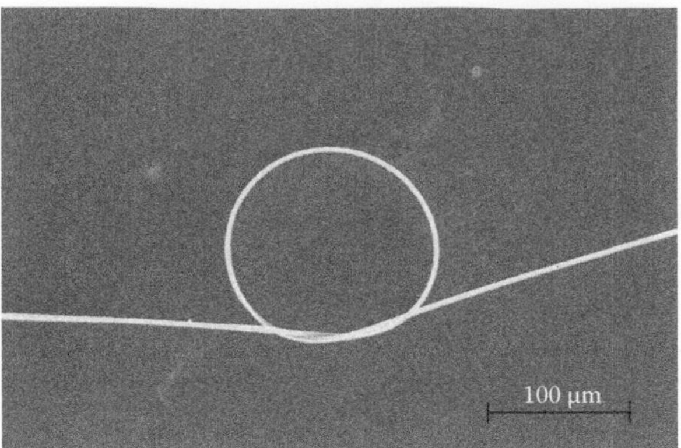

FIGURE 12.18 Single-loop knot microresonator obtained with an OFMT. (Photo courtesy of Dr. G. Brambilla, ORC, University of Southampton, Southampton, U.K.)

exploited for sensing purposes. Various designs have been demonstrated, including knot (Figure 12.18), single-, and multicoil resonators. These devices can be annealed at high temperature to remove stress (thus improving their long-term durability) and coated with polymers in order to preserve their optical performance. As compared to many other resonators, those based on OFMTs have the distinct advantage of an easier coupling via the fiber pigtails.

12.4.5 Microstructured Optical Fibers

12.4.5.1 Introduction

MOFs, also known as photonic crystal fibers, are among the most innovative developments in optical fiber technology in recent years (Russell 2006). These fibers have wavelength-scale features, often with high index contrast, in their transverse cross section, which define the waveguide properties. Optical confinement and guidance can be obtained through either modified total internal reflection (TIR) or photonic bandgap effects. Correspondingly, MOFs are classified into *index-guiding* MOFs (or equivalently holey fibers) and *photonic bandgap fibers* (PBGFs). In the most common type of MOFs, the regions of high index contrast are provided by arrays of tiny air holes that extend along the full fiber length. The vast majority of air/glass MOFs reported to date are fabricated either from undoped silica (Knight 2003, Monro and Richardson 2003) or compound glasses (including chalcogenide (Monro et al. 2000), lead silicate (Petropoulos et al. 2003), and bismuth silicate (Ebendorff-Heidepriem et al. 2004)); it should however be noted that MOFs made of polymers (van Eijkelenborg et al. 2001) and polycrystalline materials (Rave et al. 2004) have also been reported. Mode propagation in MOFs is strongly dependent on the wavelength and can be tailored through careful design of the fiber geometry. Since the latter can be specified with a wide degree of freedom (for instance, by controlling the number, size, shape, and arrangement of the air holes in air/glass MOFs), a substantially broader range of optical properties can be achieved in MOFs, than in the conventional fibers. These include, among others, broadband single-mode guidance, unique dispersion properties, extremely high (or extremely low) nonlinearities, highly birefringent fibers, and guidance in a low refractive index core.

12.4.5.2 Fabrication of MOFs

MOFs are typically fabricated using the *stack-and-draw* technique (Figure 12.19). Individual capillary and rod elements of appropriate size and aspect ratio are prepared from high-purity synthetic silica rods and tubes and then positioned to form the required structure. The most common geometry is based on a triangular lattice, although other lattices can also be obtained via this technique. The core is obtained by replacing one or more hollow capillaries at the center of the array with solid rods (or a suitable hollow element in the case of air-guiding fibers).

A preform, which contains the required structure on a ~mm scale, is obtained by inserting the stacked array into a jacketing tube (Figure 12.20). The preform is then drawn to fiber size in a conventional fiber drawing tower. This can be accomplished

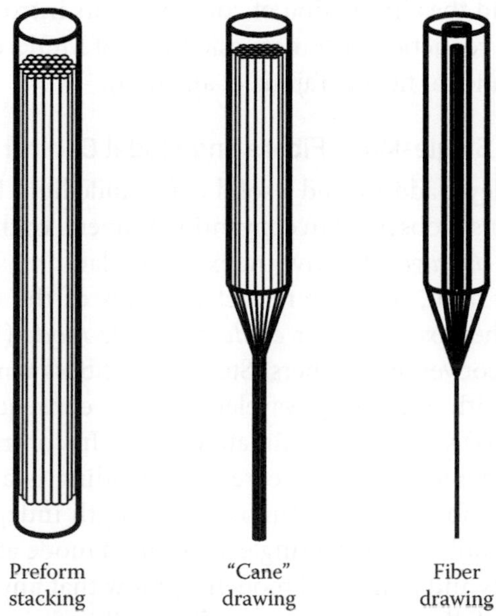

FIGURE 12.19 Fabrication of MOFs via the *stack-and-draw* technique (dual steps). Some types of MOF can be drawn directly from the stacked preform without the intermediate stage (single step).

FIGURE 12.20 Stacked microstructured preform. The target structure is a double-clad index-guiding MOF. The two different cladding layers are obtained using capillaries of different aspect ratios.

either in a single step or in two successive steps if a large scale-reduction factor is required (Figure 12.19). Positive and/or negative pressure differentials can be applied to different regions of the preform during the draw in order to maintain its geometry and help counteract the effect of surface tension at the drawing temperature. It should also be noted that microstructured preforms with simpler structure have also been obtained via other techniques, including drilling (Mukasa et al. 2006) and built-in casting of solgels (Bise and Trevor 2005). The structure of an MOF is characterized via the hole-to-hole spacing, Λ, and the relative hole size, d/Λ, where d is the diameter of the holes. Another important parameter is the air filling factor (*AFF*), which expresses the ratio between solid and hollow regions in the cross section and is related to d/Λ and to the hole shape (Mortensen and Nielsen 2004). The stack-and-draw method, though time intensive, is extremely flexible, and its refinement, particularly addressing the reduction of surface roughness and the elimination of contaminants introduced through manual handling, has led to the reduction of transmission loss of MOFs down to values that are close to those of conventional fibers (Tajima et al. 2004).

12.4.5.3 Endlessly Single-Mode Fibers and Modal Control

MOFs comprising a holey cladding and a solid core guide light by virtue of a modified form of the TIR mechanism observed in conventional fibers. In these fibers, the holes act to effectively reduce the *average* refractive index of the cladding, producing a difference, in the long-wavelength limit, with the refractive index of the core. A unique feature of MOFs, however, is the possibility for *endlessly single-mode* (ESM) guidance, which has no counterpart in conventional fibers. Step-index fibers tend to guide an increasing number of modes with decreasing wavelength, as a consequence of the *V*-number (Equations 12.9 and 12.11) scaling approximately as $1/\lambda$. In contrast, in MOFs, the effective refractive index difference between core and cladding is also wavelength dependent, and thus the *V*-number becomes almost wavelength independent, leading to the possibility of fibers that only support a single air-guided mode at all wavelengths (Birks et al. 1997). Experiments and numerical modeling show that any MOFs with $d/\Lambda < 0.43$ (independently from Λ) fall into this category. This striking property has two quite far-reaching consequences. Firstly, it is possible to design MOFs having very large scale factor while still retaining the ESM behavior. *Large mode area* (LMA) ESM MOFs can be obtained with values of the hole-to-hole spacing and thus core diameter, exceeding 20 μm. In order to achieve the same properties using a conventional design, extremely tight control over core/cladding refractive index profile would be required, which is beyond what can be afforded by conventional CVD preform manufacturing techniques. An example of LMA ESM MOF is shown in Figure 12.21a. The second consequence is that, by designing fibers with $d/\Lambda > 0.43$, it is possible to obtain fibers with an increasing number of modes, which again will be guided at all wavelengths. This property has been exploited to obtain modal interferometric sensors (Villatoro et al. 2007).

12.4.5.4 Properties of Small-Core Index-Guiding MOFs

The very large index contrast between air and glass makes it possible to design index-guiding MOFs with extremely small core sizes, of the order of 1–2 μm (shown in Figure 12.21b through d). These fibers are generally fabricated via a two-step stack-and-draw technique, due to the large scale-down factor required. When the d/Λ is made

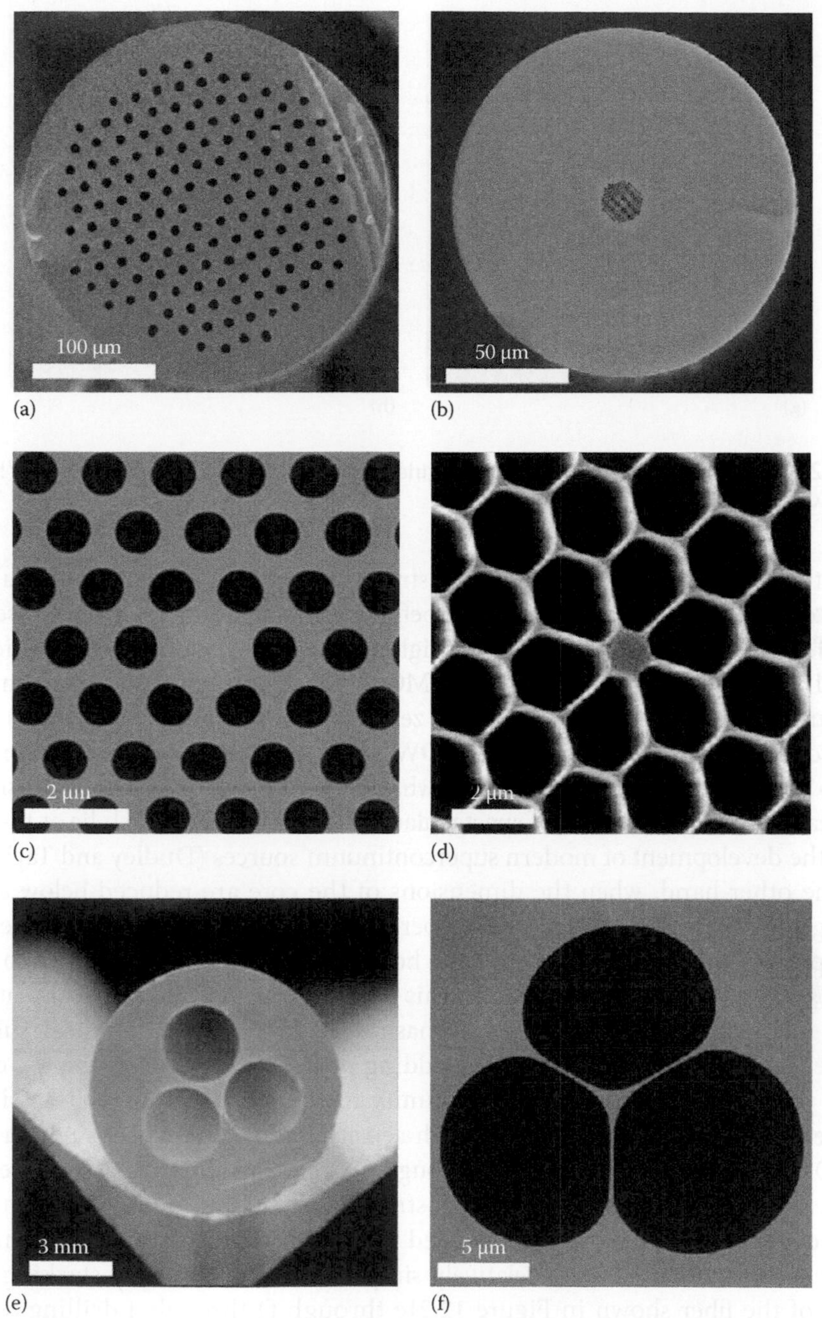

FIGURE 12.21 Examples of index-guiding MOFs: (a) large mode area endlessly single mode MOF; (b–d) small-core, highly nonlinear MOFs; suspended core MOF; (e) preform and (f) fiber. *(Continued)*

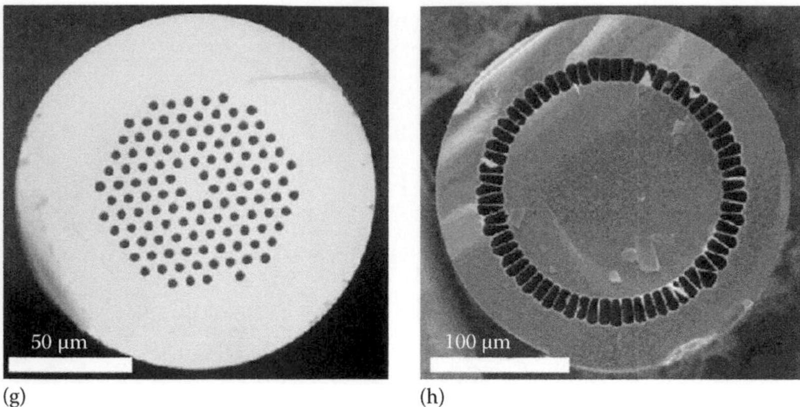

FIGURE 12.21 ***(Continued)*** Examples of index-guiding MOFs: (g) birefringent MOF and (h) jacketed air clad MOF.

sufficiently large, small-core MOFs show strong modal confinement and small MFDs, leading to high values of the nonlinear coefficient γ (Equation 12.22). As discussed in Section 12.2.4, MOFs achieve some of the highest nonlinearities observed in optical fibers. It should also be noted that in small-core MOFs, the group velocity dispersion acquires a very strong waveguide component. The zero dispersion wavelength (ZDW) can thus be shifted to shorter than the material's ZDW, as far down as to visible wavelengths. The ability to tailor the fiber dispersion profile while retaining high nonlinearity in small-core MOFs leads to the observation of spectacular nonlinear effects, which lie at the foundations of the development of modern supercontinuum sources (Dudley and Taylor 2009).

On the other hand, when the dimensions of the core are reduced below about the wavelength of the guided light, MOFs operate in a different regime. A large fraction of the optical field propagates within the holes located in close proximity to the core as evanescent field. MOFs operating in this regime thus provide an ideal platform for creating efficient interaction with gas-phase and liquid-phase chemical substances, since the arrays of air columns in the cladding provide natural microfluidic channels. The low propagation loss and close proximity to the core open up the possibility for long interaction lengths and thus for high-sensitivity absorption sensing. Suspended-core MOFs (SC-MOFs, Figure 12.21e through f) are a particularly useful type of high-evanescent field fiber, in which the microstructured region is composed of only three expanded holes, and the core is suspended on extremely thin struts. As compared to other structures, SC-MOFs are relatively simple to make, either by stacking or (as is the case of the fiber shown in Figure 12.21e through f) through a drilling technique. Furthermore, the core size can be chosen to be very small and the holes can be made comparatively very large. This has two advantages: First, it reduces the confinement loss, and second, from a microfluidic point of view, it greatly increases the ease and speed with which chemical formulations can be indiffused in the holes. The fiber shown in Figure 12.21f has a core of just 0.9 μm in diameter; the hole diameter is ~10 μm and the three struts supporting the core have just ~50 nm thick at their waist. SC-MOF has been demonstrated to achieve the highest fraction of evanescent field (>30%) among index-guiding holey fibers and has been used for absorption sensing (Petrovich et al. 2005).

12.4.5.5 Birefringent MOFs

The strong dependence of guidance properties of MOFs on the cladding structure yields an easy route to designing birefringent fibers. MOFs based on a triangular lattice and a standard center-omitted solid or hollow core formed by one, three, or seven substituted elements do not normally show birefringence. However, by simply introducing a controlled deviation from an ideal sixfold or threefold geometry in the core and/or cladding, a substantial amount of form birefringence can be introduced. Practical examples include fibers where an elongated core is obtained by replacing two or more capillaries with solid rods. An example of a birefringent MOF is shown in Figure 12.21g. Alternatively, the cladding can be modified by using capillaries of different aspect ratios. By introducing two larger or smaller holes at diametrically opposite positions in the ring surrounding the core, very large values of birefringence can be obtained (Ortigosa-Blanch et al. 2000).

Birefringent MOFs are of great interest for sensor applications due to the fact that these fibers are single material. In conventional Hi-Bi and PM fibers, stress birefringence is obtained through the combination of two glasses with different thermal expansion and is therefore temperature sensitive. In contrast, form birefringence in MOFs is largely insensitive to temperature (Michie et al. 2004). It should also be noted that, for applications in which temperature dependence of birefringence is not a concern, some MOF designs, particularly LMA-MOFs with moderate or low values of AFF, allow for the incorporation of stress rods (Folkenberg et al. 2004).

12.4.5.6 Photonic Bandgap and Hollow-Core MOFs

MOFs also enable a radically different type of light guidance, that is, *photonic bandgap* guidance. This is generally observed in fibers comprising a low-index core, in which propagation due to modified TIR cannot occur. PBGFs comprise regular arrays of high-/low-index elements (photonic crystal) in the cladding, which produce optical bandgaps, that is, regions in the (λ, β) phase space (β being the longitudinal component of the wave vector $\vec{\beta}$), in which no optical states can exist. The core is then obtained as a central *defect*, in which bound modes at wavelengths within the bandgap can propagate. Assuming the cladding to be infinite, such modes are confined to the core and exponentially decaying in the cladding. If, as in real fibers, the cladding is composed by a finite number of elements, a certain *confinement loss* will be observed, which represents the (weak) coupling to optical states in the solid region outside the microstructured cladding. Confinement loss is generally made negligible (as compared to other sources of loss) simply by adding a suitable number of rings. Bandgap guidance is a resonant guidance mechanism, that is, it can only occur at well-defined wavelength windows and requires a highly regular structure. In contrast, the perfectly symmetric cladding structure often observed in index-guiding MOFs is a consequence of fabrication, but not a prerequisite for guidance. A simple type of PBGFs is obtained by designing a fiber with arrays of high-refractive-index elements in the cladding and a low-index core. An example of such *all-solid PBGFs*, encompassing a silica core and circular germanosilicate rods in the cladding, is shown in Figure 12.22h. Such fibers can transmit light over multiple bandgaps (Argyros et al. 2005).

Arguably the most remarkable demonstration of bandgap guidance is provided by the hollow-core photonic bandgap fibers (HC-PBGFs). These fibers (Figure 12.22a

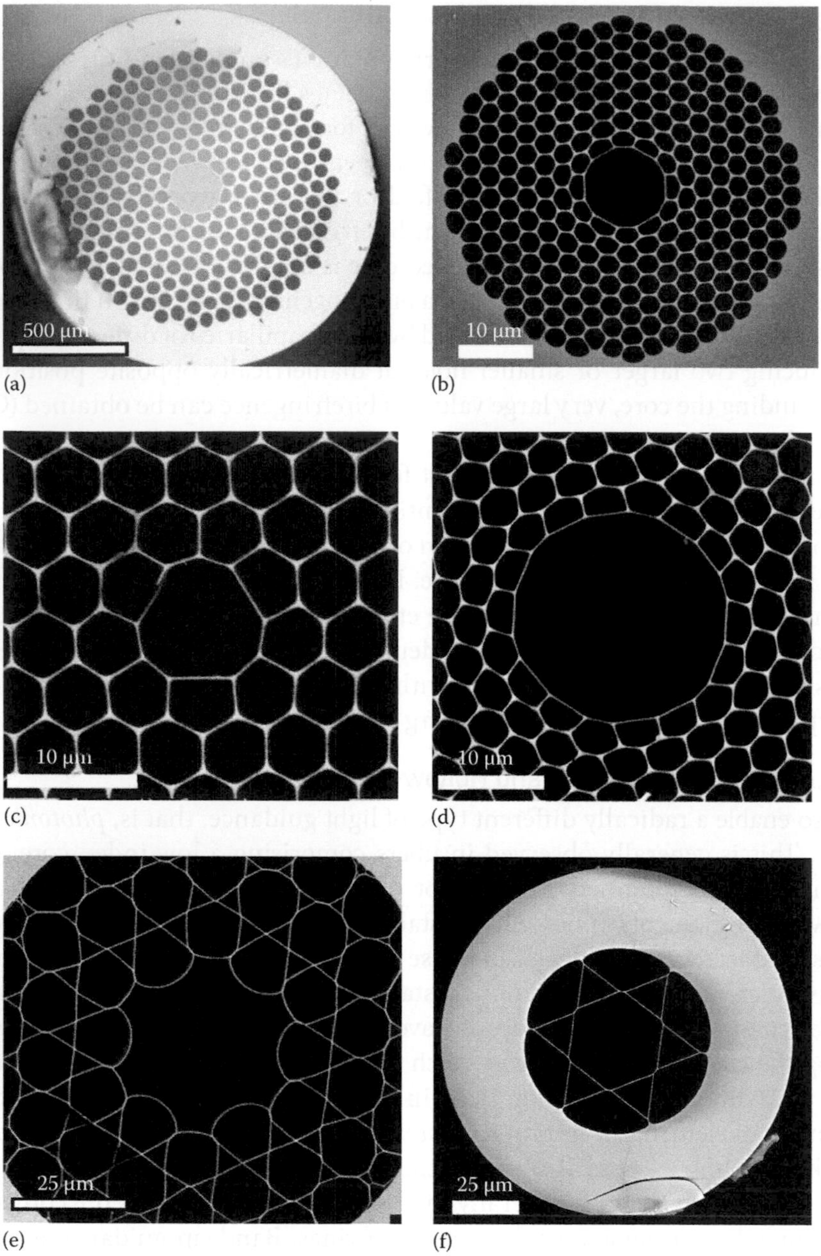

FIGURE 12.22 Examples of photonic bandgap and hollow-core MOFs: hollow-core photonic bandgap MOF with 7-cell core geometry, (a) cane and (b) fiber; HC-PBGFs with (c) 3-cell; (d) 19-cell core geometry; (e) Kagome MOF. (Photo courtesy of Dr. F. Benabid, CPPM Group, University of Bath, Bath, U.K.; Wang, Y.Y. et al., *Opt. Lett.*, 36, 669, 2011. With permission.); (f) Simplified Kagome-type MOF (Photo courtesy of J. Hayes, ORC, University of Southampton, Southampton, U.K.). *(Continued)*

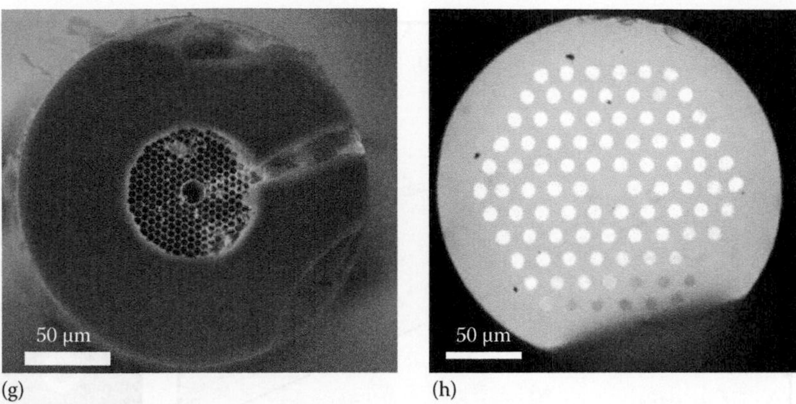

FIGURE 12.22 (*Continued*) Examples of photonic bandgap and hollow-core MOFs: hollow-core photonic bandgap MOF with 7-cell core geometry, (g) PBGF with microchannel fabricated by fs laser; (h) solid PBGF.

through d) comprise an array of high-aspect-ratio air holes and a core formed by replacing a certain number of elements at the center of the lattice with an oversized air hole. The most typical HC-PBGFs (including those available commercially) have cores composed of 7 (Figure 12.22b) or 19 (Figure 12.22d) missing unit cells, although, as will be discussed in the following, cores composed of fewer elements are also possible. A requisite for low-loss guidance in these fibers is the very high AFF: Bandgaps are only observed for *AFF* > 80% and their width increases exponentially with increasing *AFF* (Mortensen and Nielsen 2004). HC-PBGFs are fabricated using a two-step stack-and-draw technique (Section 12.4.5.2). The stacked preform and the intermediate stage (the *cane* shown in Figure 12.22a) have substantially lower *AFF* than is required in the final fiber in order to enable easier handling; the cane is expanded to the target *AFF* value during the second-stage fiber draw by use of suitable pressure differentials. As the bandgaps depend strongly on the structural parameters of the photonic crystal cladding, PBGFs are extremely sensitive to structural variations, and thus tight control over the fiber's structure and excellent longitudinal consistency are paramount, making the fabrication of PBGFs challenging.

Despite fabrication complexity and narrowband operation of HC-PBGFs, losses as low as 1.2 dB/km have been achieved (Roberts et al. 2005). The ultimate loss limit is believed to be due to scattering from nanometer-scale surface roughness at the air/silica interfaces. The origin of such roughness is thermodynamic in nature and determined by surface capillary waves (Jäckle and Kawasaki 1995), which exist on the glass surface at the fiber drawing temperature and are subsequently frozen in upon rapid cooling to below the glass transition temperature. Further loss reduction in these fibers will require reducing the roughness and/or identifying suitable fiber designs that minimize that optical field intensity at the air/glass interface.

An intriguing feature of HC-PBGFs is the complexity of their modal properties. The two most common types of HC-PBGFs (7 cell and 19 cell) are both multimoded. Figure 12.23 shows the modal properties of a seven-cell HC-PBGF. The shaded area corresponds to the bandgap, where light propagation in the cladding is prohibited. The core modes are represented through their dispersion curves and respective

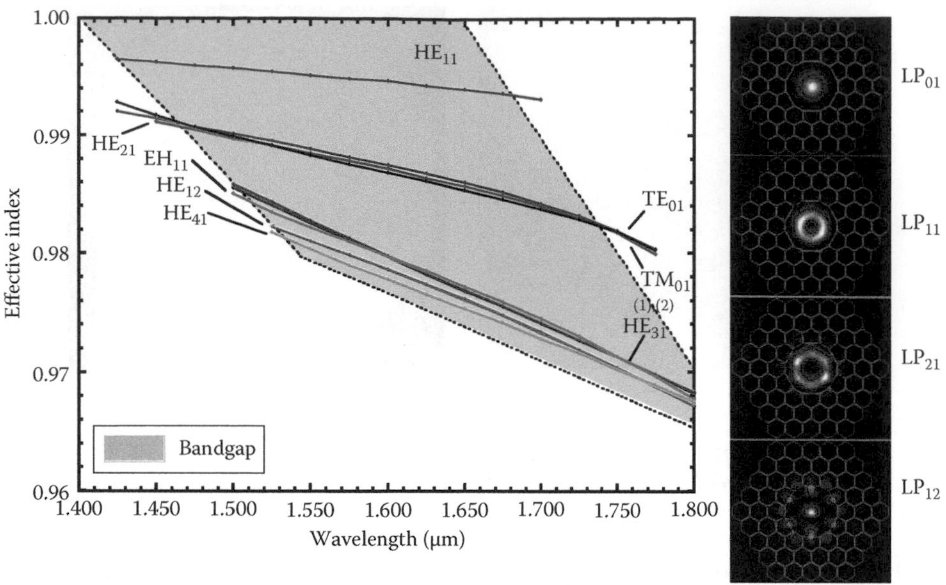

FIGURE 12.23 Modal properties of a seven-cell HC-PBGF. (Data courtesy of Dr. F. Poletti, ORC, University of Southampton, Southampton, U.K.)

modal profiles; at most wavelengths within the bandgap, multiple core modes are supported. These modes are characterized by very small overlap with the glass structure; in the case of the fundamental mode LP_{01}, this is typically less than 1%. In addition to core modes, *surface modes* may also be supported in HC-PBGFs at wavelengths within the bandgap. Such modes localize energy at the boundary between core and cladding. Energy transfer between core modes and surface modes occurs at specific anticrossing wavelengths due to small waveguide imperfections and microbends. Surface modes are deleterious as they introduce loss and large local variations of the modal dispersion.

The issue of controlling the modal properties of HC-PBGFs has attracted significant interest as these fibers evolved from simple lab novelty to the focus of practical applications. For some applications requiring a short device length, the beating between the various propagating modes in HC-PBGFs can cause severe impairment. Since, however, the modal properties depend strongly on the fiber structure, significant effort has been devoted to identifying those design regimes that reduce the number of undesirable optical modes. For instance, surface modes can be suppressed (or at least substantially reduced) by appropriate design of the core boundary and in particular by carefully selecting the boundary thickness to be half the thickness of the struts in the cladding (Amezcua-Correa et al. 2006). Furthermore, the number of core modes can be controlled—at least to some extent—by selecting the core size (Petrovich et al. 2008). It can be seen that the number of guided core modes roughly scales as the square of the core radii. While the 7- and 19-cell PBGFs are multimode, by reducing the core size to just 3 elements, a single-mode HC-PBGF can be obtained. A three-cell HC-PBGF is shown in Figure 12.22c.

Interest in sensing applications of HC-PBGFs has steadily grown over the past few years. A few potential application areas have emerged as promising, although still at the level of lab-based prototypes. The first and foremost area is *absorption sensing*, particularly of gas-phase formulations. With >99% of the optical power guided in the hollow microstructured region, losses as low as a few dB/km and negligible bend losses, HC-PBGFs open up the possibility for extremely long effective interaction lengths. More importantly, sensor heads based on PBGFs could be made extremely compact by coiling the fiber and would require minimal amounts of test gas. A drawback associated with the use of long fiber lengths is the very slow diffusion time of gas within the small channels of the HC-PBGF. Filling time is prohibitively long even for relatively short lengths of ~1 m, unless bulky pumping equipment is employed. The problem can be addressed by having periodic apertures in order to speed up the in- and outdiffusion; this can be accomplished either by segmenting the fiber (Lazaro et al. 2008) or, more elegantly, via femtosecond laser machining (Figure 12.22g). Despite the delicate structure, long microchannels of sufficient diameter and connecting the hollow core with the side surface of the HC-PBGF can be manufactured with very-low-loss penalty (van Brakel et al. 2007); the development of this technique to the fabrication of ~tens of channels per meter length would enable the use of these fibers for distributed chemical sensing.

The low overlap between guided optical field and glass may provide additional advantages for other sensor applications, for instance, in the context of novel fibers for FOGs (Kim et al. 2006). Though one of the most successful examples of fiber sensors, FOGs based on conventional fibers suffer from limitations due to nonlinear effects and noise due to thermal fluctuations (through dn/dT) and magnetic fields (via the magneto-optic effect). HC-PBGFs have extremely low optical nonlinearity (almost three orders of magnitude as compared to conventional fibers), which enables the use of narrowband laser sources (as opposed to the relatively broadband sources currently used in FOGs), and at the same time have much reduced sensitivity to thermal and magnetic field effects, which may prove crucial to developing a new generation of FOG sensors. It should however be observed that, in order for HC-PBGFs to achieve their full potential, a number of issues (including interconnections, low-loss splices, long-term stability against gas and moisture ingression, long-term mechanical stability) will need to be thoroughly investigated and appropriate technological solutions identified.

In addition to PBGFs, a different variety of hollow-core MOFs also exists, namely, the Kagome MOFs, which does not rely on photonic bandgap guidance (Couny et al. 2006). These fibers are obtained from conventional triangular stacks of thin-walled capillaries, but, differently from HC-PBGFs, the interstitial holes are kept open during fiber draw, resulting in a star-of-David, or Kagome, lattice in the cladding (Figure 12.22e). While Kagome MOFs do not possess a full photonic bandgap, they are able to support leaky core modes with losses in the region of 0.1–1 dB/m over broad wavelength intervals; further loss reduction is possible as these fibers are still under development. Kagome MOFs work due to a combination of antiresonance of the core surround and inhibited coupling to the cladding modes; quasi-single-mode propagation can be achieved by optimizing the launch conditions. Despite the higher loss, Kagome MOFs hold some advantages over HC-PBGFs for applications requiring short device lengths: Firstly, they operate over much wider wavelength intervals than HC-PBGFs (up to ~1000 nm), and secondly, they have large pitch and thus large core size, which correlates with lower nonlinearities and better ability to handle high peak powers. Kagome fibers are also easier to make as they comprise far less elements than typical HC-PBGFs: fibers with just seven elements were recently demonstrated (Hayes et al. 2011).

12.4.6 Multimaterial Microstructured Fibers

Similar to their air/silica counterparts, multimaterial microstructured optical fibers (M-MOFs) have wavelength-scale features in their cross section. In this case, however, the emphasis is toward the incorporation of materials with vastly different properties within the same fiber, aimed at substantially increasing the variety of its optical responses. Examples include combinations of one or more of the following materials: glasses, amorphous semiconductors, metals, and polymers. The availability of these fibers offers tremendous opportunities for sensing applications. This section—far from being exhaustive—aims to provide a snapshot of the most promising materials combinations and fabrication strategies, correlated with key references.

Activity in this emerging area was pioneered by Y. Fink's group (http://mit-pbg.mit.edu/), who first demonstrated a polymer/chalcogenide glass composite hollow-core Bragg fiber for IR transmission and power delivery (Temelkuran et al. 2002). A *co-drawing* fabrication technique was conceived, where suitable combination of materials with compatible thermal and mechanical properties is incorporated in the same preform and drawn into the fiber through standard techniques. Building on this early result, the same group went on to investigate a number of metal–semiconductor–insulator combinations providing some of the most compelling demonstrations to date of the integration of multiple devices in a single fiber (see Abouraddy et al. (2007) for a comprehensive review). Examples include high-power transmitting fibers with integrated heat sensing elements to monitor potential thermal failure; polymer/chalcogenide multilayers for optical guidance; narrowband filters and external reflection; fibers incorporating thermal and light-sensitive elements over their full length, which can be used for large-area 2-D detector arrays with arbitrary shape, or woven into fabrics; and the demonstration of a fiber incorporating a piezoelectric material for acoustic transduction (Egusa et al. 2010).

Other materials combinations and fabrication techniques are also being investigated. High-temperature codrawing was used to obtain glass fibers comprising a polycrystalline semiconductor core (such as silicon, germanium, and indium antimonide) (Ballato et al. 2010). In this approach, a preform composed of a semiconductor rod and glass jacketing tube is mechanically assembled and drawn. The materials are chosen so that the drawing temperature of the glass is higher than the melting point of the semiconductor: the glass then confines the molten semiconductor as it is drawn into fiber allowing for the fabrication of the multimaterial fiber. Current fibers have relatively simple geometry and relatively high loss (~dB/cm at 1.5 μm and ~dB/m in the mid-IR); however, as fabrication techniques improve, it is possible that loss will be reduced and it will become feasible to target more complex structures.

An alternative approach uses a high-pressure microfluidic chemical vapor deposition technique inside an air/silica MOF (Sazio et al. 2006). The hollow channels act like microscopic reaction chambers where suitable precursors are injected and amorphous semiconductors are deposited. Subsequent annealing is used to produce polycrystalline material. Since the silica MOF acts as a template, this technique allows achieving very complex structures. Deposition of Si, Ge, and more recently ZnSe (Healy et al. 2012) was demonstrated. Figure 12.24 shows a silicon-filled M-MOF based on a three-cell PBGF template. Typical deposition lengths are limited to below 1 m and losses are again in the

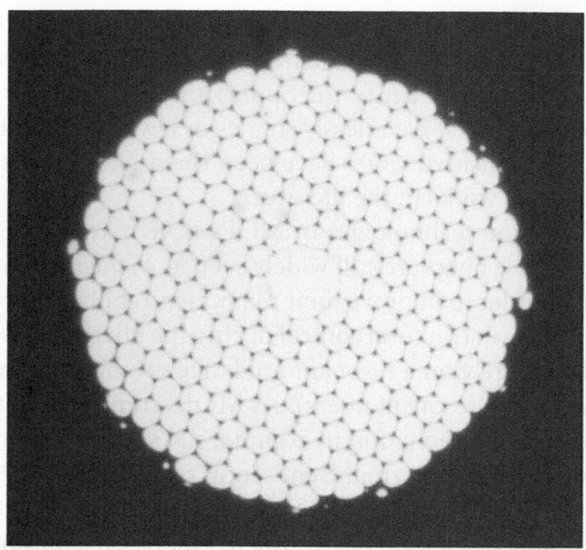

FIGURE 12.24 M-MOF comprising a three-cell PBGF filled with silicon through high-pressure microfluidic CVD method. The silicon-filled holes appear brighter as the image is taken in reflection. (Photo courtesy of Dr. A.C. Peacock, ORC, University of Southampton, Southampton, U.K.)

dB/cm range. The main interest of these fibers is currently for IR transmission and nonlinear optics. A simpler route to obtaining M-MOF is by pumping molten semiconductors and/or metals inside the hollow channels. Similar to the previous, this technique can only achieve lengths below 1 m. M-MOFs based on a silica template and filled with metals, semiconductor, and, more recently, chalcogenide glass have been obtained using this technique (Schmidt et al. 2008, Tyagi et al. 2008, Granzow et al. 2011).

12.5 Conclusions and Future Outlook

Optical fiber technology has advanced at a remarkable pace over the last 40 years and has revolutionized a number of application areas beyond optical telecommunications. Preform fabrication techniques based on chemical vapor deposition have allowed an unprecedented level of material purity to be achieved, and progress in waveguide design and fiber drawing has made it possible to attain optical properties close to the theoretical limits of silica fibers. Fibers with ultralow-loss and highly tailored transmission properties (e.g., dispersion, nonlinearity, and birefringence) are nowadays produced in very large volumes and thus are cheap and easy to source. Within the optical sensing application area, fibers lie at the heart of a substantial and ever-increasing number of practical sensor schemes. By virtue of their high mechanical resilience and extremely small footprint, and because they can act at the same time as a waveguide and as a sensing element, optical fibers offer great flexibility and adaptability to suit very diverse applications in this highly segmented field.

As fiber fabrication technologies matured, they stimulated an exponential increase of the variety of structures and thus of optical responses available. The advent of MOFs—arguably the most remarkable development in optical fiber technology in the last decade—has substantially contributed to this trend. This technology allows for single material fibers with a much greater variety of optical properties as compared to conventional, all-solid fibers based on a matched core/clad glass structure. The sensor community has been ready to identify the opportunities offered by this development. Most notably, hollow-core PBGFs, endlessly single- or few-mode MOFs, and highly birefringent, temperature-insensitive MOFs have been the subject of sustained interest as is testified by the number of works in the literature.

Although the fabrication technologies of MOFs have matured at an impressive rate over the last 15 years, these fibers are still widely seen as *challenging*, and some key issues are still outstanding in order to improve their manufacturability. One such issue revolves around methods to improve structural control during the fiber draw, which is particularly important for MOFs that rely on a resonant guidance mechanism (such as hollow-core PBGFs and Kagome or anti-resonant fibers). As MOFs complete the transition from lab use to industrial R&D and exploitation in commercial systems, the need for better and cheaper fibers is bound to grow. Current trends look set to address the issue of increasing fiber yields, improving longitudinal consistency and reducing fabrication cost.

Finally, exciting opportunities are beginning to emerge in a number of novel areas of optical fiber technology. Among the others, we single out two that appear particularly promising: optical fiber micro-tapers and multimaterial microstructured fibers. While both technologies are still very much in their infancy, recent research clearly demonstrates their potential for application in several areas relevant to optical sensing.

References

Abouraddy, A. F., M. Bayindir, G. Benoit et al. 2007. Towards multimaterial multifunctional fibres that see, hear, sense and communicate. *Nature Materials* 6: 336–347.

Agrawal, G. 2006. *Nonlinear Fiber Optics*. Academic Press, San Diego, CA.

Alahbabi, M. N., Y. T. Cho, and T. P. Newson. 2006. Long-range distributed temperature and strain optical fibre sensor based on the coherent detection of spontaneous Brillouin scattering with in-line Raman amplification. *Measurement Science and Technology* 17: 1082–1090.

Amezcua-Correa, R., N. G. Broderick, M. N. Petrovich, F. Poletti, and D. J. Richardson. 2006. Optimizing the usable bandwidth and loss through core design in realistic hollow-core photonic bandgap fibers. *Optics Express* 14: 7974–7985.

Argyros, A. 2009. Microstructured polymer optical fibers. *Journal of Lightwave Technology* 27: 1571–1579.

Argyros, A., T. Birks, S. Leon-Saval, C. M. B. Cordeiro, and P. St. J. Russell. 2005. Guidance properties of low-contrast photonic bandgap fibres. *Optics Express* 13: 2503–2511.

Artiglia, M., G. Coppa, P. Di Vita et al. 1989. Mode field diameter measurements in single-mode optical fibres. *Journal of Lightwave Technology* 7: 1139–1152.

Asai, M., Y. Inuzuka, K. Koike, S. Takahashi, and Y. Koike. 2011. High-bandwidth graded-index plastic optical fiber with low-attenuation, high-bending ability, and high-thermal stability for home-networks. *Journal of Lightwave Technology* 29: 1620–1626.

Ballato, J., T. Hawkins, P. Foy et al. 2010. Advancements in semiconductor core optical fiber. *Optical Fiber Technology* 16: 399–408.

Bao, X., D. J. Webb, and D. A. Jackson. 1993. 32-km distributed temperature sensor based on Brillouin loss in an optical fiber. *Optics Letters* 18: 1561–1563.

Birks, T. A., J. C. Knight, and P. St. J. Russell. 1997. Endlessly single-mode photonic crystal fiber. *Optics Letters* 22: 961–963.

Bise, R. and D. J. Trevor. 2005. Sol-gel derived microstructured fiber: Fabrication and characterization. *Proceeding of the Optical Fiber Communication Conference OFC 2005*, Anaheim, CA, Paper OWL6.

Boyd, R. W. 2008. *Nonlinear Optics*. Elsevier, Amsterdam, the Netherlands.

Brambilla, G. 2010. Optical fiber microwires and nanowires: A review. *Journal of Optics* 12: 043001.

Buck, J. A. 1995. *Fundamentals of Optical Fibers*. Wiley, New York.

Camerlingo, A., X. Feng, F. Poletti et al. 2010. Near-zero dispersion, highly nonlinear lead silicate W-type fiber for applications at 1.55 μm. *Optics Express* 18: 15747–15756.

Chen, X., M. J. Li, N. Venkataraman et al. 2004. Highly birefringent hollow-core photonic bandgap fiber. *Optics Express* 12: 3888–3893.

Coulombier, Q., L. Brilland, P. Houizot et al. 2010. Casting method for producing low-loss chalcogenide microstructured optical fibers. *Optics Express* 18: 9107–9112.

Couny, F., F. Benabid, and P. S. Light. 2006. Large-pitch kagome-structured hollow-core photonic crystal fiber. *Optics Letters* 31: 3574–3576.

Dakin, J.P., D.J. Pratt, G.W. Bibby, J.N. Ross. 1985. Distributed optical fibre Raman temperature sensor using a semiconductor light source and detector. *Electronics Letters* 21: 569–570.

Dudley, J. M. and J. R. Taylor. 2009. Ten years of nonlinear optics in photonic crystal fibre. *Nature Photonics* 3: 85–90.

Ebendorff-Heidepriem, H., P. Petropoulos, S. Asimakis et al. 2004. Bismuth glass holey fibers with high nonlinearity. *Optics Express* 12: 5082–5087.

Egusa, S., Z. Wang, N. Chocat et al. 2010. Multimaterial piezoelectric fibres. *Nature Materials* 9: 643–648.

El-Amraoui, M., J. Fatome, J. C. Jules et al. 2010. Strong infrared spectral broadening in low-loss As-S chalcogenide suspended core microstructured optical fibers. *Optics Express* 18: 4547–4556.

Feng, X., T. M. Monro, V. Finazzi et al. 2005a. Extruded single-mode, high-nonlinearity tellurite glass holey fiber. *Electronics Letters* 41: 835–837.

Feng, X., T. M. Monro, P. Petropoulos et al. 2005b. Extruded single-mode high-index-core one-dimensional microstructured optical fiber with high index-contrast for highly nonlinear optical devices. *Applied Physics Letters* 87: 081110–081112.

Folkenberg, J., M. Nielsen, N. Mortensen et al. 2004. Polarization maintaining large mode area photonic crystal fiber. *Optics Express* 12: 956–960.

France, P. W., M. G. Drexhage, J. M. Parker et al. 1990. *Fluoride Glass Optical Fibres*. Blackie, Glasgow, Scotland.

Gambling, W. A. 2000. The rise and rise of optical fibers. *IEEE Journal of Selected Topics in Quantum Electronics* 6: 1084–1093.

Garcus, D., T. Gogolla, K. Krebber, and F. Schliep. 1997. Brillouin optical-fiber frequency-domain analysis for distributed temperature and strain measurements. *Journal of Lightwave Technology* 15: 654–662.

Geittner, P., D. Kuppers, and H. Lydtin. 1976. Low loss optical fibers prepared by plasma-activated chemical vapour deposition (PCVD). *Applied Physics Letters* 28: 645–646.

Gloge, D. 1971a. Weakly guiding fibres. *Applied Optics* 10: 2252–2258.

Gloge, D. 1971b. Dispersion in weakly guiding fibres. *Applied Optics* 10: 2442–2445.

Granzow, N., P. Uebel, M. A. Schmidt et al. 2011. Bandgap guidance in hybrid chalcogenide–silica photonic crystal fibers. *Optics Letters* 36: 2432–2434.

Harrington, J. A. 2004. *Infrared Fibers and Their Applications*. SPIE Press, Bellingham, WA.

Hayes, J. R., F. Poletti, and D. J. Richardson. 2011. Reducing loss in practical single ring antiresonant hollow core fibres. *Proceedings of the Conference on Lasers and Electro-Optics Europe* (*CLEO EUROPE 2011*), Munich, Germany, Paper CJ2_2.

Healy, N., J. R. Sparks, R. He et al. 2012. Mid-infrared transmission properties of step index and large mode area ZnSe microstructured optical fibers. *Proceedings of the Advances in Optical Materials Topical Meeting* (*AIOM 2012*), San Diego, CA, Paper ITH3B.3.

Horiguchi, T., K. Shimizu, T. Kurashima et al. 1995. Development of a distributed sensing technique using Brillouin scattering. *Journal of Lightwave Technology* 13: 1296–1302.

Humbach, O., H. Fabian, U. Grzesik et al. 1999. Analysis of OH absorption bands in synthetic silica. *Journal of Non-Crystalline Solids* 203: 19–26.

Kao, C. K. 2010. Nobel lecture: Sand from centuries past: Send future voices fast. *Reviews of Modern Physics* 82: 2299–2303.

Kapany, N. S. and R. J. Simms. 1965. Recent developments of infrared fiber optics. *Infrared Physics* 5: 69–75.

Kim, H. K., M. J. F. Digonnet, and G. S. Kino. 2006. Air-core photonic-bandgap fiber-optic gyroscope. *Journal of Lightwave Technology* 24: 3169–3174.

Knight, J. C. 2003. Photonic crystal fibres. *Nature* 424: 847–851.

Ippen, E. P. and R. H. Stolen. 1972. Stimulated Brillouin scattering in optical fibers. *Applied Physics Letters* 21: 539–541.

Islam, M. 2002. Raman amplifiers for telecommunications. *IEEE Journal of Selected Topics in Quantum Electronics* 8: 548–559.

Jäckle, J. and K. Kawasaki. 1995. Intrinsic roughness of glass surfaces. *Journal of Physics: Condensed Matter* 7: 4351–4358.

Lazaro, J. M., A. M. Cubillas, M. Silva-Lopez et al. 2008. Methane sensing using multiple-coupling gaps in hollow-core photonic bandgap fibers. *Proceedings of the 19th Conference on Optical Fiber Sensors (OFS-19)*, Perth, Western Australia, Australia, Paper 7004-213.

Lee, J. H., Z. Yusoff, W. Belardi et al. 2001. A holey fibre Raman amplifier and all-optical modulator. *Proceedings of the European Conference on Optical Communications (ECOC 2001)*, Amsterdam, the Netherlands, Paper PDAI.1.

Leong, J. Y. Y., P. Petropoulos, J. H. V. Price et al. 2006. High-nonlinearity dispersion-shifted lead-silicate holey fibres for efficient 1 micron pumped supercontinuum generation. *Journal of Lightwave Technology* 24: 183–190.

Marcuse, D. 1982. *Light Transmission Optics*. Van Nostrand Reinhold, New York.

Mears, R. J., L. Reekie, M. Jauncey et al. 1987. Low-noise erbium-doped fiber amplifier operating at 1.54 μm. *Electronics Letters* 26: 1026–1028.

Michie, A., J. Canning, K. Lyytikäinen et al. 2004. Temperature independent highly birefringent photonic crystal fibre. *Optics Express* 12: 5160–5165.

Miyajima, Y., T. Komukai, T. Sugawa et al. 1994. Rare earth-doped fluoride fiber amplifiers and fiber lasers. *Optical Fiber Technology* 1: 35–47.

Monro, T. M. and D. J. Richardson. 2003. Holey optical fibres: Fundamental properties and device applications. *Comptes Rendus Physique* 4: 175–186.

Monro, T. M., Y. D. West, D. W. Hewak et al. 2000. Chalcogenide holey fibers. *Electronics Letters* 36: 1998–2000.

Mortensen, N. A. and M. D. Nielsen. 2004. Modeling of realistic cladding structures for air-core photonic bandgap fibers. *Optics Letters* 29: 349–351.

Mukasa, K., M. N. Petrovich, F. Poletti et al. 2006. Novel fabrication method of highly-nonlinear silica holey fibres. *Proceedings of CLEO/QELS 2006*, Long Beach, CA, Paper CMC5.

Nagayama, K., M. Kakui, M. Matsui et al. 2002. Ultra low loss (0.1488 dB/km) pure silica core fiber and extension of transmission distance. *Electronics Letters* 38: 1168–1169.

Okoshi, T. 1982. *Optical Fibers*. Academic Press, New York.

Ortigosa-Blanch, A., J. C. Knight, W. J. Wadsworth et al. 2000. Highly birefringent photonic crystal fibers. *Optics Letters* 25: 1325–1327.

Petropoulos, P., H. Ebendorff-Heidepriem, V. Finazzi et al. 2003. Highly nonlinear and anomalously dispersive lead silicate glass holey fibers. *Optics Express* 11: 3568–3573.

Petrovich, M. N. 2003. Gallium lanthanum sulphide glasses for near-infrared photonic applications. PhD thesis, University of Southampton, Southampton, UK.

Petrovich, M. N., F. Poletti, A. van Brakel et al. 2008. Robustly single mode hollow core photonic bandgap fiber. *Optics Express* 16: 4337–4346.

Petrovich, M. N., A. van Brakel, F. Poletti et al. 2005. Microstructured fibers for sensing applications. In: *Photonic Crystals and Photonic Crystal Fibers for Sensing Applications*, Du, H. H. (ed.), SPIE, Bellingham, WA; *Proceedings of the SPIE*, Vol. 6005, pp. 78–92.

Rave, E., P. Ephrat, M. Goldberg et al. 2004. Silver halide photonic crystal fibers for the middle infrared. *Applied Optics* 43: 2236–2241.

Richardson, D. J., J. Nilsson, and W. A. Clarkson. 2010. High power fiber lasers: Current status and future perspectives. *Journal of the Optical Society of America* B27: B63–B92.

Roberts, P. J., F. Couny, H. Sabert et al. 2005. Ultimate low loss of hollow-core photonic crystal fibres. *Optics Express* 13: 236–244.

Russell, P. St. J. 2006. Photonic-crystal fibers. *Journal Lightwave Technology* 24: 4729–4749.

Sanghera, J. S., L. Brandon Shaw, and I. D. Aggarwal. 2002. Applications of chalcogenide glass optical fibers. *Comptes Rendus Chimie* 5: 873–883.

Sanghera, J. S., L. Brandon Shaw, and I. D. Aggarwal. 2009. Chalcogenide glass-fiber-based mid-IR sources and applications. *IEEE Journal of Selected Topics in Quantum Electronics* 15: 114–119.

Sazio, P. J. A., A. Amezcua-Correa, C. E. Finlayson, et al. 2006. Microstructured optical fibers as high-pressure microfluidic reactors. *Science* 311: 1583–1586.

Schmidt, M. A., L. N. Prill Sempere, H. K. Tyagi et al. 2008. Wave guiding and plasmon resonances in two-dimensional photonic lattices of gold and silver nanowires. *Physical Review B* 77: 033417.

Shibata, S., M. Horiguchi, K. Jinguji et al. 1981. Prediction of loss minima in infra-red optical fibers. *Electronics Letters* 17: 775–776.

Silva-Lopez, M., A. Fender, W. N. MacPherson et al. 2005. Strain and temperature sensitivity of a single-mode polymer optical fiber. *Optics Letters* 30: 3129–3131.

Snitzer, E. 1961. Cylindrical dielectric waveguide modes. *Journal of the Optical Society of America* 51: 491–498.

Snyder, A. W. and J. Love. 2010. *Optical Waveguide Theory*, 2nd edn. Springer, New York.

Sugimoto, N., T. Nagashima, T. Hasegawa et al. 2004. Bismuth-based optical fiber with nonlinear coefficient of 1360 W^{-1}*km^{-1}. *Proceeding of the Optical Fiber Communication Conference, OFC 2004*, Los Angeles, CA, Paper PDP26.

Tajima, K., J. Zhou, K. Nakajima et al. 2004. Ultralow loss and long length photonic crystal fiber. *Journal of Lightwave Technology* 22: 7–10.

Temelkuran, B., S. D. Hart, G. Benoit et al. 2002. Wavelength-scalable hollow optical fibres with large photonic bandgaps for CO_2 laser transmission. *Nature* 420: 650–653.

Tong, L. and E. Mazur. 2007. Silica nanofibers and subwavelength diameter fibers. In: *Specialty Optical Fiber Handbook*, Mendez, A. and Morse, F. T. (eds.), Academic Press, Amsterdam, the Netherlands.

Tyagi, H. K., M. A. Schmidt, L. Prill Sempere et al. 2008. Optical properties of photonic crystal fiber with integral micron-sized Ge wire. *Optics Express* 16: 17227–17236.

van Brakel, A., C. Grivas, M. N. Petrovich et al. 2007. Micro-channels machined in microstructured optical fibers by femtosecond laser. *Optics Express* 15: 8731–8736.

van Eijkelenborg, M. A., M. C. J. Large, A. Argyros et al. 2001. Microstructured polymer optical fiber. *Optics Express* 9: 319–327.

Villatoro, J., V. P. Minkovich, V. Pruneri et al. 2007. Simple all-microstructured-optical-fiber interferometer built via fusion splicing. *Optics Express* 15: 1491–1497.

Walker, S. S. 1986. Rapid modeling and estimation of total spectral loss in optical fibers. *Journal of Lightwave Technology* LT-4: 1125–1131.

Zhu, X. and N. Peyghambarian. 2010. High-power ZBLAN glass fiber lasers: Review and prospect. *Advances in OptoElectronics* 2010: 501956.

Zubia, J. and J. Arrue. 2001. Plastic optical fibers: An introduction to their technological processes and applications. *Optical Fiber Technology* 7: 101–140.

13. Point Sensors

Intensity Sensors

José Manuel Baptista
University of Madeira

Handbook of Optical Sensors. Edited by José Luís Santos and Faramarz Farahi © 2015 CRC Press/Taylor & Francis Group, LLC. ISBN: 9781439866856.

13.1 Introduction

This chapter is focused on single-point measurement by intensity-based optical fiber sensors. General concepts presented in Chapter 4 are now applied to the situation where optical fiber is the sensing element. As mentioned there, these sensors rely on the variation of the received optical intensity to analytically measure the variation of the parameter under observation. They are simple in concept, and the cost of implementation is low when compared to other types, such as the ones relying on interferometric techniques. Nevertheless, they may require special referencing methods to guarantee that the change in the received optical intensity is only due to the measurement parameter and not from other undesired sources.

There are several approaches to classify sensors based on optical fibers. One of them is by the type of spatial measurement. If the sensor system operates by doing measurement in a single point, it may be called point sensor. On the other hand, if the sensor system comprehends multiple discrete measurements along an optical fiber line, it may be called multipoint sensor or quasi-distributed sensor. Another possibility is to have a measurement system that is able to measure continuously along an optical fiber line, and in this case, the system is named distributed sensor.

First, the focus will be on the working principle of point sensors: intensity sensors and then on the most relevant modulation techniques. Then, the most common referencing methods will be addressed, and finally an overview of the current state of the art of fiber-optic point intensity sensors will be presented.

13.2 Working Principle of Fiber-Optic Point Sensors: Intensity Sensors

The intensity sensors were the first generation of fiber-optic sensors to be implemented. For over a decade, the modulation of intensity remained the most used form of signal modulation in optical fiber sensor applications (Lagakos et al. 1981, Spillman and McMahon 1983, Krohn 1986, Berthold 1988, Jones et al. 1989). As the name itself indicates, the intensity of the radiation that propagates in the fiber is the modulating parameter. The simple reason for the extensive and diverse use of this modulation scheme is the inherent potential benefits, such as simplicity, reliability, ease of reproduction, reduced cost, and the fact that they can provide high sensitivities.

The fiber-optic intensity sensors require a small amount of components that may include multimode fiber and noncoherent light sources, which makes them very attractive in terms of cost. There are several possible configurations of sensing heads with varying degrees of complexity, and these sensors may be operated in transmission or in reflection (Figure 13.1).

The light emitted from an optical source is coupled to the optical fiber for transmission to the sensing head where the light is modulated according to the state of the measurand. The sensing head can be the fiber only or an external sensing device. When the measure is implemented in transmission mode, a second section of optical fiber is required for the transport of the modulated signal to the detector. On the other hand, when the measure is performed in reflection mode, the modulated optical signal is reinjected in the same fiber for signal transmission to the detector.

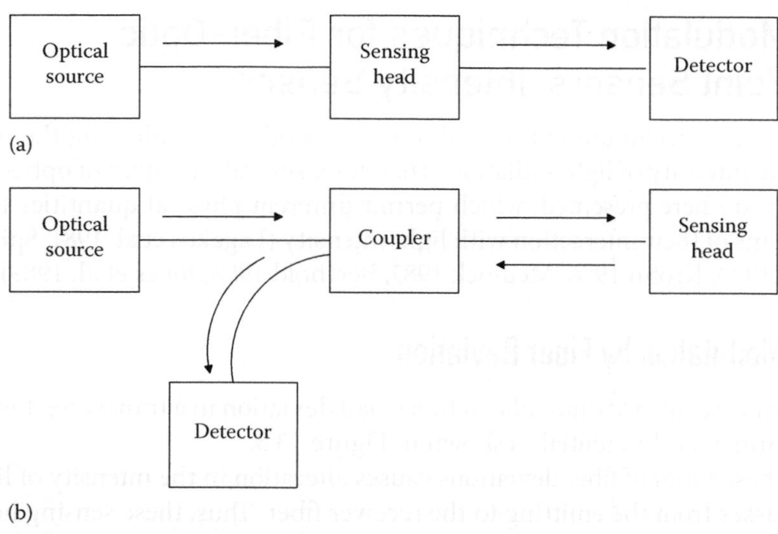

FIGURE 13.1 Optical fiber intensity sensors in transmission (a) and reflection (b) modes.

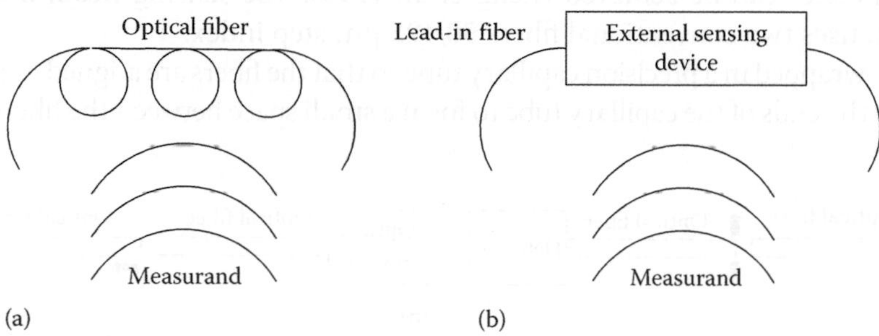

FIGURE 13.2 Intrinsic (a) and extrinsic (b) optical fiber point intensity sensors.

Another possible classification in terms of operation relates whether the measurand affects the light inside or outside the fiber. Thus, we may have an intrinsic point intensity sensor or an extrinsic point intensity sensor as illustrated in Figure 13.2.

In an intrinsic sensor, the measurand acts directly on the fiber, changing the desired optical property of the guided radiation, in this case, the intensity. Therefore, the light radiation remains inside the waveguide, that is, the fiber. For instance, in certain applications, the sensitivity of the optical fiber itself may be low, so you can cover the fiber glass with some adequate material to become sensitive in the presence of a specific measurand, thus enhancing the measurand effect on the intensity radiation that propagates in the fiber.

In an extrinsic sensor, the measurand is present in a region outside the fiber, and the optical fiber is only used as an optical channel transport of radiation to the monitoring site, where the optical intensity signal is modulated, that is, the fibers route the optical beam to an external sensing device that will modulate the optical intensity in response to an external measurand.

13.3 Modulation Techniques for Fiber-Optic Point Sensors: Intensity Sensors

There are various techniques of optical intensity modulation allowing the measurand to act on the intensity of light radiation. Therefore, several examples of optical intensity modulation are here presented, which permit different physical quantities to be measured in terms of their interaction with light intensity (Lagakos et al. 1981, Spillman and McMahon 1983, Krohn 1986, Medlock 1987, Berthold 1988, Jones et al. 1989).

13.3.1 Modulation by Fiber Deviation

Many sensors are intensity modulated by a small deviation in a transverse, longitudinal, angular manner, or differential as shown in Figure 13.3.

Any of these types of fiber deviations causes alteration in the intensity of light radiation that passes from the emitting to the receiver fiber. Thus, these sensing heads allow the measurement of transverse, longitudinal, differential, or angular displacement by analyzing the variation of the transmitted optical intensity.

Using this principle, an application to monitor a vibration signal in smart structures using composites can be achieved (Leng et al. 1999). The sensing head, illustrated in Figure 13.4, uses two conventional fibers 50/125 μm step index.

The set is wrapped in a precision capillary tube so that the fibers are aligned. Optical fibers are glued to the ends of the capillary tube to form a small space between the fibers inside the

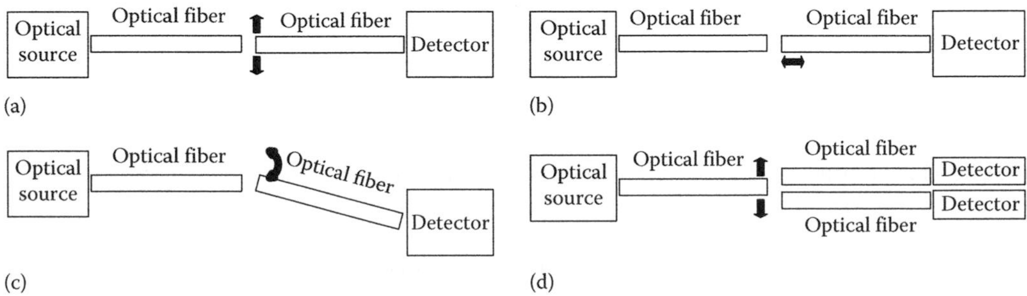

FIGURE 13.3 Types of sensing heads of intensity modulation by fiber deviation: (a) transversal deviation; (b) longitudinal deviation; (c) angular deviation, and (d) differential deviation.

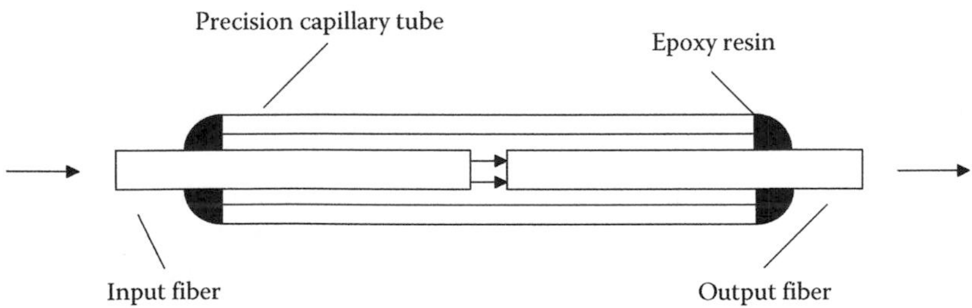

FIGURE 13.4 Example of sensing head of intensity modulation to measure vibration.

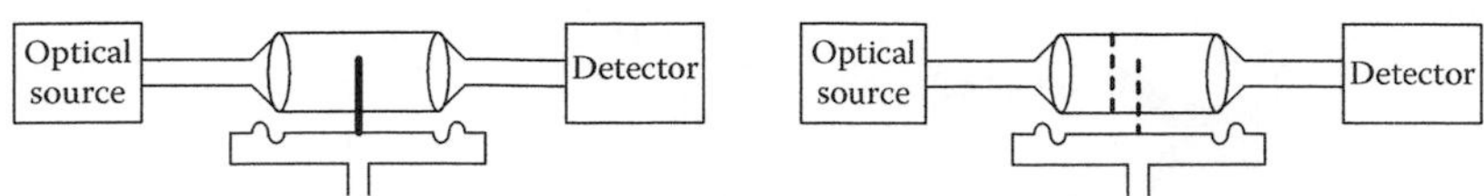

FIGURE 13.5 Examples of intensity sensing heads based on window modulation.

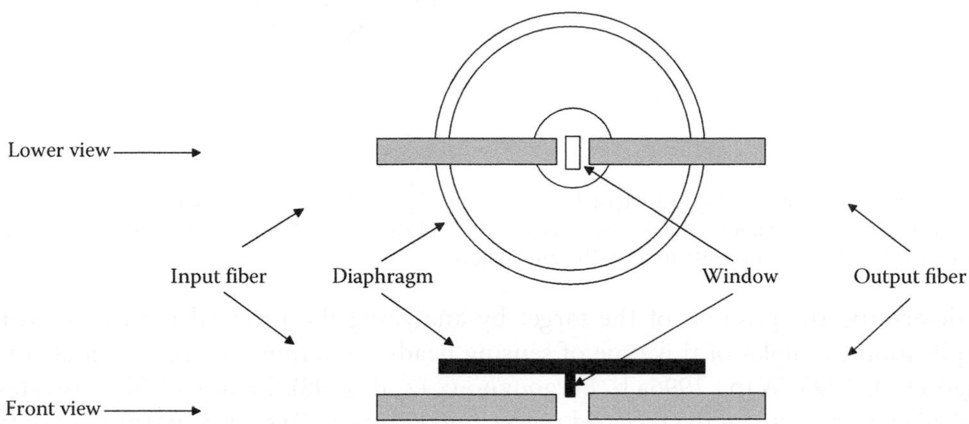

FIGURE 13.6 Intensity sensing head based on a window modulation by a sound diaphragm.

capillary tube. The vibrations, which are under monitoring, cause a transverse deviation between the two aligned fibers resulting in a change in the transmitted optical intensity.

13.3.2 Modulation by Window

Such modulators normally present the end of the fibers with collimating lens fixed to a rigid structure. Between the ends of the fibers then appears a window operated by a shift mechanism, such as a diaphragm, an element, or other means (Figure 13.5).

The windows need not be made of solid material. For example, liquids in pipes can also be used in a similar way. An application that uses window modulators allows the determination of sound spatial distribution in a two-dimensional array of sensing heads (Figure 13.6). The sensor head consists of a small window, placed between two fibers, which moves as a function of the diaphragm movement (Nakamura 2000).

This diaphragm is sensitive to sound pressure by changing the light intensity between the two coupled optical fibers.

13.3.3 Modulation by Reflection

This type of intensity sensing head was among the first to be built and may take several configurations: multiple fibers (also called bundle), dual fiber, or single fiber coupler (Figure 13.7a through c).

Its working principle is the emission of light through one or more fibers, which hits a target. The intensity of light reflected by the target is collected by one or more fibers and is dependent on the distance between the target and the fiber(s). So, it is possible

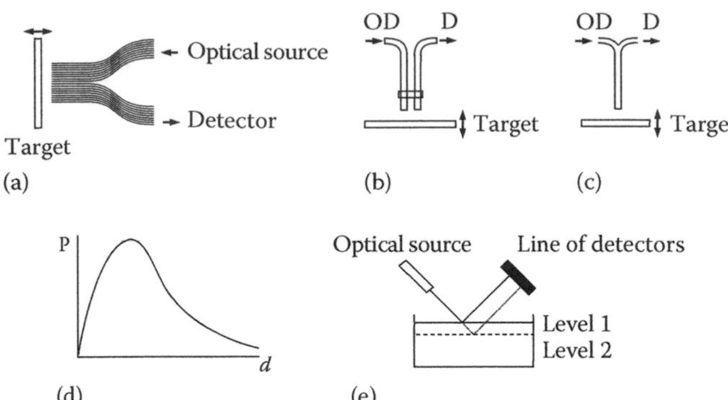

FIGURE 13.7 Intensity sensing heads based on modulation by a reflection (OS, optical source; D, detector). (a) Multiple, (b) double fiber, (c) single, (d) characteristic of the received power as function of the target distance (*d*) for the case of configuration (b), and (e) level.

to determine the position of the target by analyzing the received optical intensity. Application examples of this type of sensing heads are numerous (Brenci et al. 1988, Zhao et al. 1995, Wang 1996a,b, Bergougnoux et al. 1998). Figure 13.7d illustrates a typical characteristic of the received power, and Figure 13.7e shows another use of this type of sensing head to determine the level of a liquid. For a higher liquid level (Level 1), the light reflecting off the surface of the liquid is collected by a specific detector. For a lower liquid level (Level 2), the light reflecting off the surface of the liquid is collected by another detector.

13.3.4 Modulation by Fiber Loss

A fourth form of intensity modulation is based on the losses caused in the core or the cladding of the fiber, being this type of modulation very popular for fiber-optic point intensity sensors. A typical example is the microbend sensor, which is illustrated in Figure 13.8.

It can be shown that for a step index optical fiber, the loss is maximum when the applied curvature has a period (Λ) equal to

$$\Lambda = \frac{2\pi a n_o}{NA} \tag{13.1}$$

being a the radius of the core of the fiber, n_o the refractive index of the core, and NA the numerical aperture of the fiber (Lagakos et al. 1987, Berthold 1995).

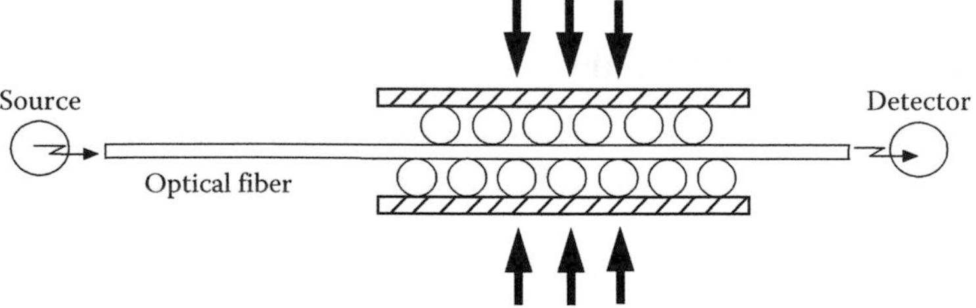

FIGURE 13.8 Intensity sensing head based on modulation by fiber loss (in this case microbend-induced loss).

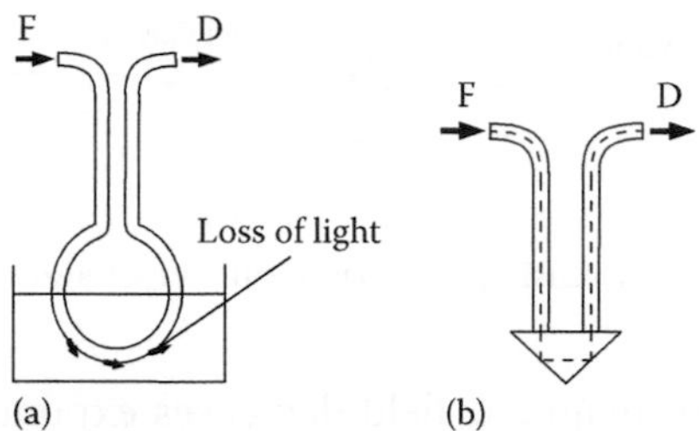

FIGURE 13.9 Intensity sensing heads based on modulation by fiber loss to detect liquid level. (a) Level detector and (b) two fibers level detector.

Another type of fiber-optic point sensor with intensity modulation by fiber loss, directed for the detection of liquid level, can be constructed from a fiber ring, as depicted in Figure 13.9a.

When the liquid begins to submerge the fiber ring, the higher-order modes of light that propagate in the fiber are lost from the core to the liquid. This device was designed to measure the level of acid in batteries (Spenner et al. 1983). A greater attenuation of light can be achieved when two fibers are bonded to a glass prism, as illustrated in Figure 13.9b. When the prism is in the air, the light inside the prism is reflected twice with 90° angles. When the prism is inserted in the liquid, the light exits the prism and is no longer detected. It is thus an on/off type sensing structure.

13.3.5 Modulation by Evanescent Field

This class of intensity modulators uses two phenomena associated with the change of the evanescent field: one is called frustrated total internal reflection (FTIR) and the other is called attenuated total reflection (ATR).

In the FTIR phenomenon, the amount of total internal reflection of light is dependent on the parameter under measurement (Rahnavardy et al. 1997). Other applications of the FTIR phenomenon can detect the presence or absence of liquid at a certain level (Ilev and Waynant 1999) or in the characterization of liquids (Ramos and Fordham 1999). In the ATR phenomenon, the evanescent field is guided by the fiber, and part of it interacts with an external medium under observation. Considering the propagation of light in a waveguide of refractive index n_1 surrounded by the sample to be analyzed with a refractive index n_2 ($<n_1$), the electromagnetic field is as shown in Figure 13.10a.

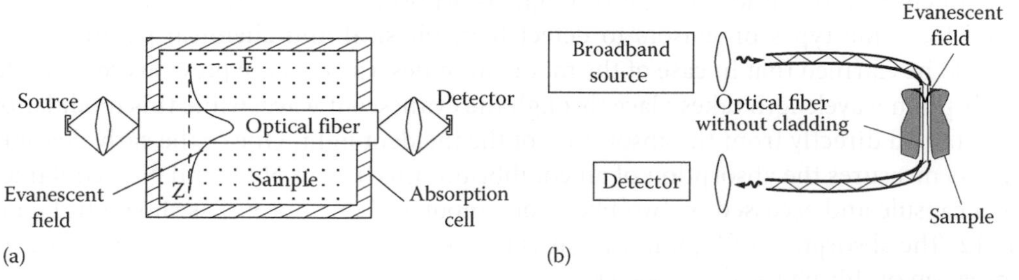

FIGURE 13.10 Example of intensity sensing head (a) and sensor (b) based on modulation by evanescent field via the ATR phenomenon.

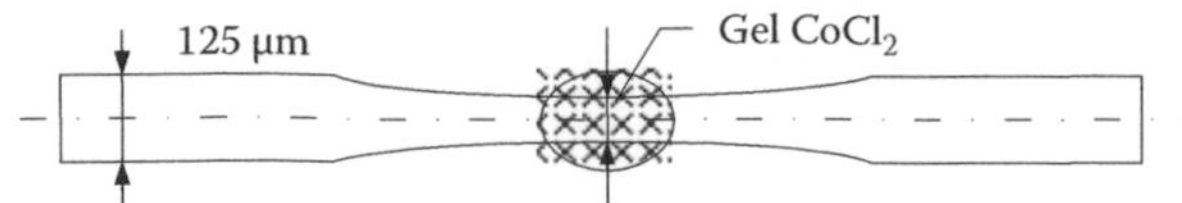

FIGURE 13.11 Example of intensity sensing head based on a taper and using modulation by evanescent field via the ATR phenomenon.

In the sample, the electromagnetic field decreases exponentially with distance z of the following form:

$$E(z) = E_o \exp\left(\frac{-z}{d_p}\right) \quad (13.2)$$

where d_p is the penetration depth, which is greater when the difference between n_1 and n_2 is small. In this situation, the evanescent wave is partially absorbed by the sample, particularly in specific wavelengths, and the total transmission through the waveguide changes at these wavelengths. The spectrum of transmission through the waveguide is then determined by the composition of the sample. Figure 13.10b presents the fiber-optic point intensity sensor.

Another example of modulation in the fiber by evanescent field using the ATR phenomenon is through the use of a fiber taper. A fiber taper is obtained by heating and stretching the fiber so as to achieve a thinner region. Heating the fiber is usually run through a flame or an electric arc. In the region thus formed, the taper, the core, and the cladding are fused into a single silica filament in the order of 10–50 μm of diameter. In this case, the filament only guides the light, without shielding it, and exposes the evanescent field to the external medium. Therefore, the conditions of propagation of light are very sensitive to the surrounding environment. Figure 13.11 presents a fiber-optic point intensity sensor based on a taper to measure the level of humidity.

The region of the taper was coated with a $CoCl_2$ sol-gel material whose refractive index varies in accordance with the existing level of humidity. This change in refractive index around the taper causes ATR in the intensity of transmitted light, thus allowing the detection of the present level of moisture (Bariárin et al. 1998, Arregui et al. 2000).

13.3.6 Modulation by Absorption

The intensity modulation of light through absorption at certain wavelengths has been used for some types of sensors to detect both physical and chemical parameters. It should be clarified that in case of the modulation described in the previous section, the absorption wavelength takes place through the evanescent wave, while this modulation is obtained directly from the absorption of the incident light. One of the early developments measures the absorption of an equilibrium mixture of N_2O_4 and NO_2 contained in a capsule and accessed by two fibers (one input and one output) as shown in Figure 13.12. The absorption of light in the gas is temperature dependent thus allowing determination of this physical parameter.

Another type, very common, of intensity modulation by absorption of wavelength uses a method called FIR—fluorescence intensity ratio—which allows self-referencing

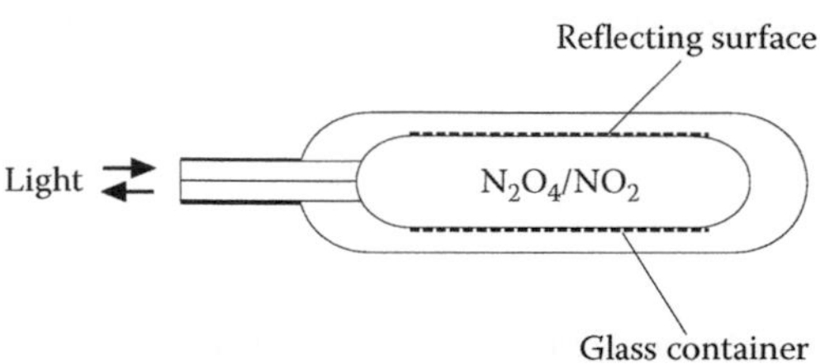

FIGURE 13.12 Sensing head based on the intensity modulation by absorption.

(Wade et al. 1999, Kharaz et al. 2000, Vergara et al. 2000). One example of this technique is based on thermalization of two adjacent energy levels of a rare earth ion, where the relative populations of the two levels vary according to temperature, regardless of the level of optical intensity with which they are illuminated. The optical fiber doped with such rare earths allows the measurement of a range of high temperatures (Zhang et al. 1998).

13.3.7 Modulation by Light Scattering

Turbulence and dispersion of light are two parameters that can be measured using fiber-optic point sensors using intensity modulation. The presence of oil in water can be detected by the method illustrated in Figure 13.13a, which is based on measuring the dispersion of light in drops of oil in the sample (Extance et al. 1983).

Another application for intensity modulation by light scattering incorporates a special fiber that contains particles that reflect or scatter light, which is propagated in the fiber in both directions to be collected by a detector. This technique allows the construction of a sensor for measuring the ambient light intensity (Figure 13.13b).

13.3.8 Modulation by Refractive Index Change

Changing the refractive index of the surroundings at the edge of a fiber determines the conditions of propagation of the light intensity of the fiber from the interior to its exterior, specifically the numerical aperture of the light that exits the fiber. If using a second fiber to receive light from the emitting fiber, the intensity of light collected is dependent on the opening of the cone of light emerging from the emitting fiber. Thus, it is possible to characterize the medium between the two fibers in terms of its refractive index (Pham et al. 1998). Figure 13.14 shows such a possible sensing head for the characterization of refractive index. The light cone angle is changed accordingly with the refractive index of the external

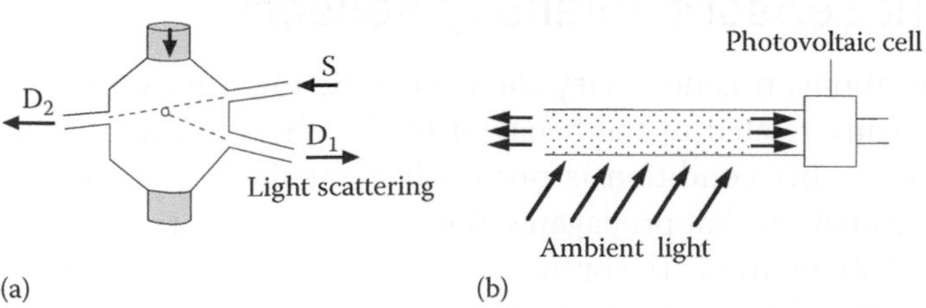

FIGURE 13.13 Intensity sensing heads based on modulation by light scattering. (a) Detector of oil in water and (b) light intensity detector.

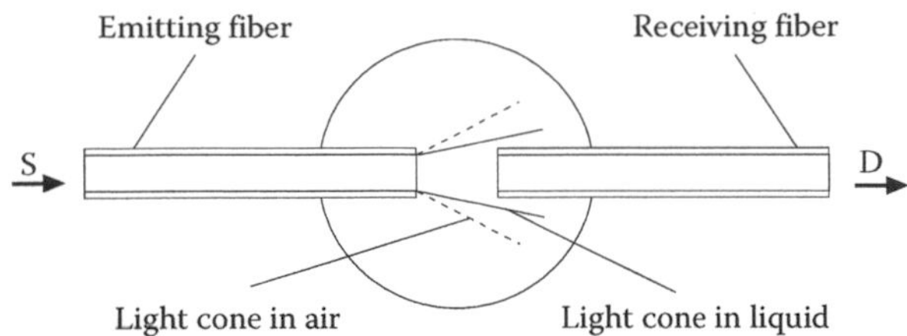

FIGURE 13.14 Intensity sensing head based on modulation by refractive index change.

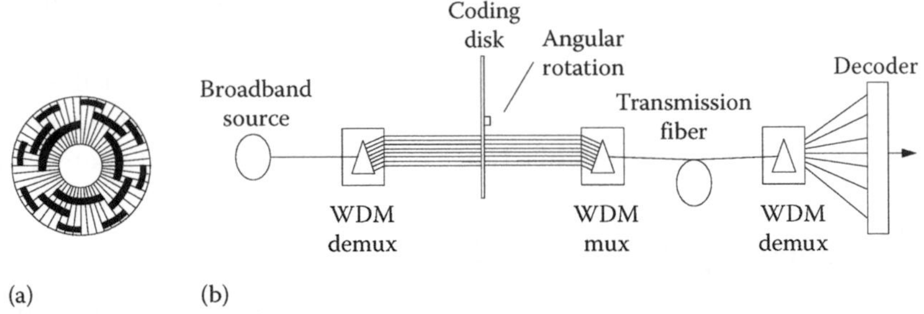

FIGURE 13.15 Intensity sensing heads based on modulation by digital coding. (a) Coding disk and (b) intensity sensor by digital coding.

medium, and the flow of light captured by the receiving fiber varies in accordance. So it is possible to characterize a particular environment in terms of its refractive index.

13.3.9 Modulation by Digital Coding

One way to measure the angular position of an object is to associate a coding disk, as presented in Figure 13.15a.

The coding disk associates a different eight-bit code for each position. If the bit is 0, there is no transmission of light associated with this bit. If the bit is 1, light transmission occurs. So, the coding disk incorporates transparent and opaque windows in order to form a different code for each position. An experimental setup is illustrated in Figure 13.15b, where a white light source is divided into eight spectral bands and allows for its independent decoding through the spectral separation.

13.4 Referencing Techniques for Fiber-Optic Point Sensors: Intensity Sensors

In analog applications, it is necessary the fiber-optic intensity sensor systems produce a result that is truly related to the behavior of the physical quantity to be measured. In practical terms, this condition is not easily verifiable due to various factors. The intensity of the radiation that propagates along the fiber is not constant as it undergoes various effects of attenuation. The optical components such as optical couplers and connectors have variable losses. Another type of additional uncertainty may come from the optoelectronic components, for example, the fluctuation of the intensity of optical sources and the sensitivity of the detector system. Finally, the performance degradation of optical and optoelectronic components, and variations in environmental conditions can produce errors in the final result (Adamovsky 1987).

Although it is impossible to eliminate these variations in any project of optical fiber systems, compensation can be applied through the monitoring of undesirable variations of the optical signal. The usual methods consist in generating, at least, one additional reference signal that, when applied together with the measurement signal, gives a result independent of the undesirable variations of the optical signal. Thus, the reference optical signal calibrates the sensor response, subject to the influence of the environment in the same way as the sensor optical signal (Murtaza and Senior 1994b).

Two important sources of common mode error can be identified due to the optical transmitter and optical transmission components. The first source of error is caused by thermal instability of the optical source that produces fluctuations of the intensity and of the wavelength of the optical signal emitted. This thermal instability is induced either by changes in temperature or by heat dissipation in the device, caused by the effect of Johnson. The second source of error is due to changes in the optical path, which also originate variations in the intensity of the optical signal. Both variations of intensity given earlier add to the intensity variations introduced by the physical quantity to be measured. On the other hand, the fluctuations of the wavelength of light can bring new sources of error associated with the spectral selectivity of the components of the measuring system. Moreover, in addition to errors caused by fluctuations of the optical source and variable losses in the connectors and couplers, the measurement of the parameter under test can also be affected by changes in detector sensitivity and signal processing unit in face of environmental perturbations.

The referentiation is usually achieved using techniques of spatial, wavelength, temporal, frequency separation or through a combination of these methods (Adamovsky 1987, Murtaza and Senior 1994b). General concepts have been presented in Chapter 4, now detailed for the important case of fiber-optic-based intensity sensors.

13.4.1 Spatial Referencing

The spatial separation technique can be used when the optical signals (measuring and reference) have the same spectral constitution, as is the case when they are generated in the same optical source in conjunction with an optical coupler. Referencing, although partial, is then possible by the existence of two physically separate channels, one being under the influence of the measurand and the other not (Xiao-Jun et al. 1992). An example of spatial referencing is given in Figure 13.16.

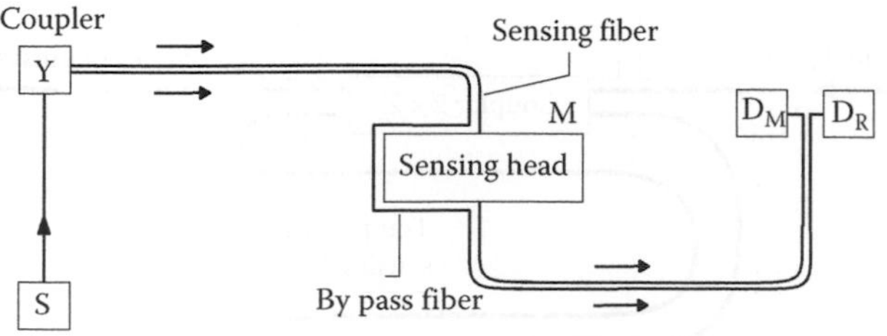

FIGURE 13.16 Example of a fiber-optic point intensity sensor spatially referenced (S, source; Y, coupler; M, measurand; D_M, measurand detector; D_R, reference detector).

Using two independent optical fibers, one for the measuring signal and the one for the reference signal, does not avoid the system to be sensitive to variable losses in the optical fibers and the corresponding variation of optical intensity in photodetectors. However, a careful implementation associated with a common coating for optical fibers will minimize any errors. It is important to note that doubling the optical fiber length, as well as the need to use two detectors, imply increased costs. It should be mentioned that it is advantageous the coupler be close to the sensor head, not only because it minimizes the extent of extra fiber required, but also, and mainly, because the common mode rejection factor between the signal and the reference is optimized.

13.4.2 Temporal Referencing

The temporal referencing method can be implemented in various ways, using, for example, a Fabry–Perot configuration, an optical ring recirculator or other schemes. In these topologies, a short optical pulse is periodically injected into the structure, unfolding a series of pulses.

To illustrate a practical case of a fiber-optic point intensity sensor temporally referenced, to measure temperature, a sensor configuration whose sensing head introduces intensity absorption of the light radiation as a function of temperature is illustrated. Figure 13.17 shows the implementation scheme of the sensor with temporal referencing. The sensor is based on a recirculating fiber ring that allows the introduction of delays between pulses (Adamovsky and Piltch 1986).

The optical source is modulated by pulses adequately separated. Each corresponding optical pulse, generated by the source, gives rise to a series of optical pulses in the detector, being taken the first two pulses of this series. The first optical pulse received serves as the optical intensity reference signal, while the second passes through the sensing head and is attenuated as a function of the temperature to be measured and acts as a measuring signal. Thus, when the optical pulse emitted by the source varies its amplitude due to fluctuations, the measurement is not affected since this variation is canceled by dividing the signal measured by the reference signal.

13.4.3 Spectral Referencing

When both signals (measurand and reference) are contained in separate spectral bands, they can be transmitted in the same fiber and separated afterward by using simple wavelength demultiplexing components. Once again, one has to ensure that the measurand

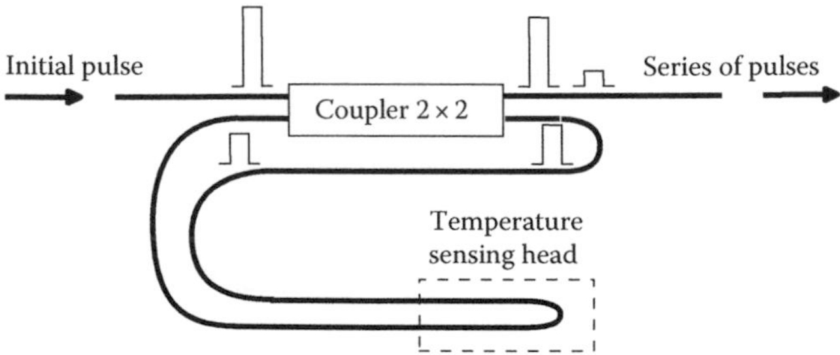

FIGURE 13.17 Example of a temperature fiber-optic point intensity sensor temporally referenced.

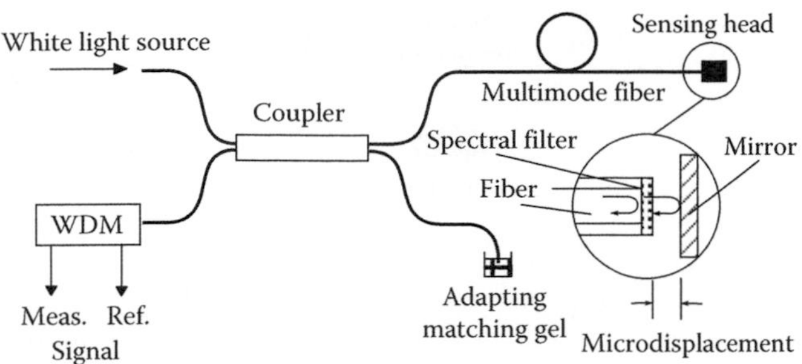

FIGURE 13.18 Example of a fiber-optic point intensity sensor spectrally referenced.

acts exclusively on the measurement signal that is associated to a specific spectral band, thus leaving the reference spectral band unaffected by the measurand. The wavelengths chosen for the two optical signals must be as close as possible to minimize unwanted variations resulting from different spectral content of signals. However, it is important to note that the decrease in separation between the wavelengths of the measurand and reference signals may increase the optical cross talk. There is usually a compromise between these two factors taking into account the desired accuracy for the measurement system. There are a few different configurations based on this type of referencing (Murtaza and Senior 1994a, 1995). A practical application of this type of referencing is given in Wang et al. (1996), where the cutting technique of spectral emission permits compensation for an intensity sensor for measuring fiber-optic microdisplacement, strain, and pressure. In the present case, the sensor can measure microdisplacement. The scheme is illustrated in Figure 13.18.

A white light source injects light into a multimode optical fiber that passes through a coupler until the sensing head. The use of the coupler allows the implementation of the sensor in reflection. The sensing head contains a spectral filter that is responsible for the distinction of the reference signal of the measurement signal. The intensity of light reflected by the filter acts as a spectral reference signal, while the light intensity, not affected by the filter, reaches the mirror, returns to the fiber partially as a function of distance from the mirror to the fiber, and acts as a measurement signal. Thus, using a wavelength division multiplexing (WDM) coupler in the receptor, it is possible to separate the reference signal and the measurement signal. The power ratio of these two signals allows obtaining a microdisplacement measurement that is immune to fluctuations of the optical source and losses outside of the sensing head, including the fiber and the coupler.

13.4.4 Frequency Referencing

One way of referencing fiber-optic point intensity sensors is in frequency and can be based on the degree of constructive interference between waves of sinusoidal modulation of an optical power source in a configuration where one of the paths uses an optical delay line. For a fixed length of the optical delay line, the degree of constructive interference is dependent only on the frequency of the sine wave. Thus, when taking the ratio of the amplitude of two optical signals that are generated at two different modulating frequencies, the result depends only on the losses induced by the sensing head, that is,

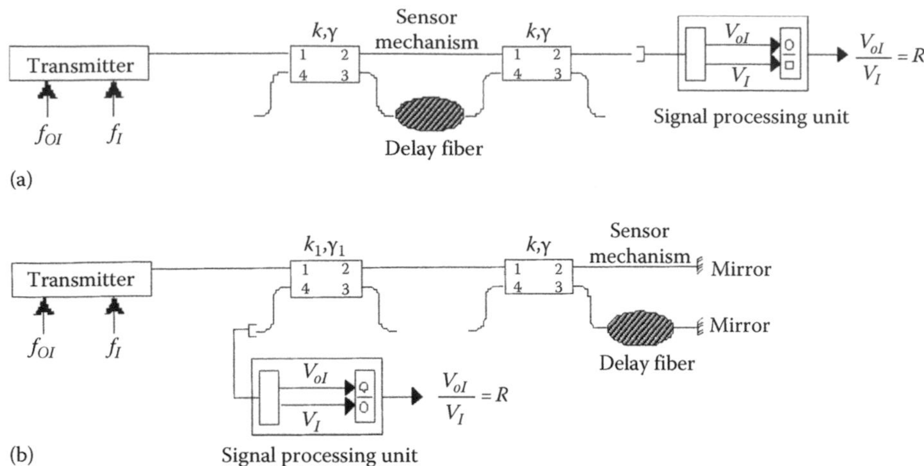

FIGURE 13.19 Two examples of fiber-optic point intensity sensors referenced in frequency using a (a) Mach–Zehnder topology and (b) Michelson topology. V_{OI} and V_I are voltages proportional to the amplitudes of the output optical power sinewave waveforms at an off-constructive interference frequency (f_{OI}) and at a constructive interference frequency (f_I), respectively.

inside the chosen configuration. This result is, therefore, independent of other losses that may occur during the remainder of the optical system.

Figure 13.19 shows a Mach–Zehnder (a) and a Michelson topology (b) as a fiber-optic point intensity sensor referenced in frequency (Baptista et al. 2000). The frequency response of a Mach–Zehnder or Michelson configuration when the optical power input comes from an optical source intensity modulated at a particular frequency shows that for some frequencies the amplitude of the output optical power waveform is maximum (constructive interfering frequencies—f_I), while for other frequencies, it results in a decreased output optical power (the off-constructive interference frequencies f_{OI}).

Therefore, the intensity modulation at different frequencies of the light injected into these fiber structures provides different amplitudes for the output optical power waveforms via the transfer functions. The ratio between two of these amplitudes, one obtained in a frequency where off-constructive interference occurs and another obtained in a situation of constructive interference, depends on the light modulation induced inside the fiber sensing interferometers and is free from light source fluctuations and variable transmission losses that can occur outside those structures. Figure 13.20 illustrates the frequency response of the Mach–Zehnder (Figure 13.20a) and Michelson (Figure 13.20b) configurations subject to losses. The parameter g indicates the attenuation factor externally induced in the fiber structures. For $g = 1$, there is no induced loss and the amplitude of the modulated output light is maximum, while for $g = 0$, the light in one arm of the interferometer is completely lost.

For both topologies, we can see that when the value of the induced losses rises, in other words when g gets closer to zero, the difference between the peaks and the valleys shortens until the frequency responses become a horizontal line. On the other hand, when there are no external losses induced in the fiber cavities ($g = 1$), the differences between the peaks and the valleys are at their maximum, being dependent only on the internal losses of those fiber structures. An example of this type of fiber-optic point

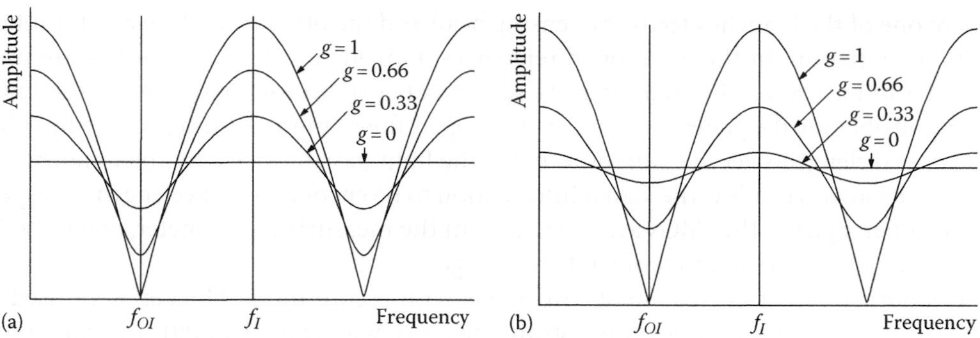

FIGURE 13.20 Transfer function of (a) Mach–Zehnder and (b) Michelson topology with the induced loss factor g as a parameter. In the vertical scale, the amplitude of the output optical power waveform at a particular frequency is indicated.

intensity sensor referenced in frequency for measuring microdisplacement or bending is presented in Baptista et al. (2006).

13.5 Current State of the Art for Point Sensors: Intensity Sensors

With the extensive use of fiber grating devices such as fiber Bragg gratings (FBGs) and long-period fiber gratings (LPGs), in the last decade or so, the fiber-optic point intensity sensors have been also incorporating those fiber structures to build different sensor types with better characteristics. Moreover, the advent of new optical sources and optical amplifiers with better operational characteristics at much lower cost permitted the development of fiber-optic point intensity sensors based, for instance, on nonlinear effects like Brillouin and Raman. Lately, due to the current availability of high-performance digital acquisition boards and personal computers, associated with accessible friendly user graphic programming languages, it has been possible to increase the control and compactness of the fiber-optic point intensity sensors, allowing also to transfer some of the required operations in the optical domain to the electrical or virtual domain. More recently, different intensity sensing heads based on hetero-core fiber for measuring microdisplacement (Koyama et al. 2009) and photonic crystal fibers filled with magnetic fluid for measuring temperature (Miao et al. 2011) have been presented. These are some examples of possible evolution trends for the fiber-optic point intensity sensors. Indeed, as the optical fiber communication industry progresses, so the optical fiber sensors can also advance as well as the fiber-optic point intensity sensors. Therefore, a continuous research effort is expected in this area. In the next subsections, some recent examples of fiber-optic point intensity sensors will be addressed.

13.5.1 Fiber-Optic Point Intensity Sensors Using Fiber Grating Devices

Fiber-optic point intensity sensors to measure temperature incorporating a laser generated from an erbium-doped fiber ring structure and an FBG have been presented (Passaro et al. 2009, Fernandez-Vallejo et al. 2010). The light is divided by a coupler, where one of the branches feeds the sensing head and the other branch serves for intensity referentiation. In the mentioned references, four identical pairs of FBGs are used in order to have a multiplexed intensity sensor network. In each pair, one of the FBGs is responsible for the generation of erbium-doped fiber ring laser line, while the other acts as wavelength-selective mirror to make the light passing twice by the temperature sensing head and to allow the signal information to be encoded in wavelength in respect to the other signals. The FBGs are isolated from the measuring parameter, and only the sensing head is subjected to temperature change.

In another recent development, two LPGs with a core mode blocker were used to have an intensity-based optical fiber strain sensor (Hwuang et al. 2009). The core mode blocker is fabricated by the arc discharge method between two LPGs forming a bandpass filter. The authors used an LED and an optical power meter to measure the transmittance near the resonance wavelength of the LPGs and obtained a nonreferenced intensity relation between the axial strain applied to one of the LPGs and the transmitted power. The measurement results indicate that the sensitivity of the power meter output voltage to the applied strain is 6.37 pV $\mu\varepsilon^{-1}$.

Other work using FBGs for implementing intensity sensors is based on a frequency-reflectometry technique (Pérez-Herrera et al. 2009). The experiment combines the concept of frequency-modulated continuous wave with the spectrally selective mirror properties of FBGs to interrogate with referencing properties of intensity-based sensors. Multiplexing two of these sensors using this technique in a parallel topology system was also experimentally demonstrated. Due to the use of different fiber lengths (delay lines), sensor and referenced signals from the FBGs located at different positions in an array are separated in the frequency domain.

Several recent works on fiber-optic point intensity sensors using fiber gratings based on the Fresnel reflection in a fiber tip for measuring refractive index have also been presented. For instance, a twin-grating interferometer for intensity reference uses signal processing in the Fourier domain to allow intensity-independent refractive index measurements for liquids and gases. A resolution of 5×10^{-5} was demonstrated experimentally (Shlyagin et al. 2009). Other work combines the use of an FBG and an LPG to obtain an intensity-referenced fiber-optic point sensing system for simultaneous measurement of refractive index and temperature. The intensity from the FBG and from the Fresnel reflection tip that passes twice by the LPG allows the intensity measurement operation of this referenced fiber-optic point intensity sensor to detect temperature and refractive index (Frazão et al. 2006).

Another work on fiber-optic point intensity sensor using fiber gratings for simultaneous measurement of refractive index and temperature, based on an LPG and two FBG arrangement, has been presented (Jesus et al. 2009). The gratings are arranged as depicted in Figure 13.21: the first grating is the LPG and next were the two FBGs where the relative spectral position of each grating was chosen in order to have one reflection peak on each side of the LPG resonance.

With the proposed configuration, the resonant peak of the LPG shifts in wavelength in accordance with the variations of the refractive index of the surrounding medium. This perturbation thus changes the intensity of light reflected by the two FBGs. The measurement can be obtained in reflection, by simple calculation of the ratio between the intensities reflected by the two FBGs. This ratio is proportional to the wavelength

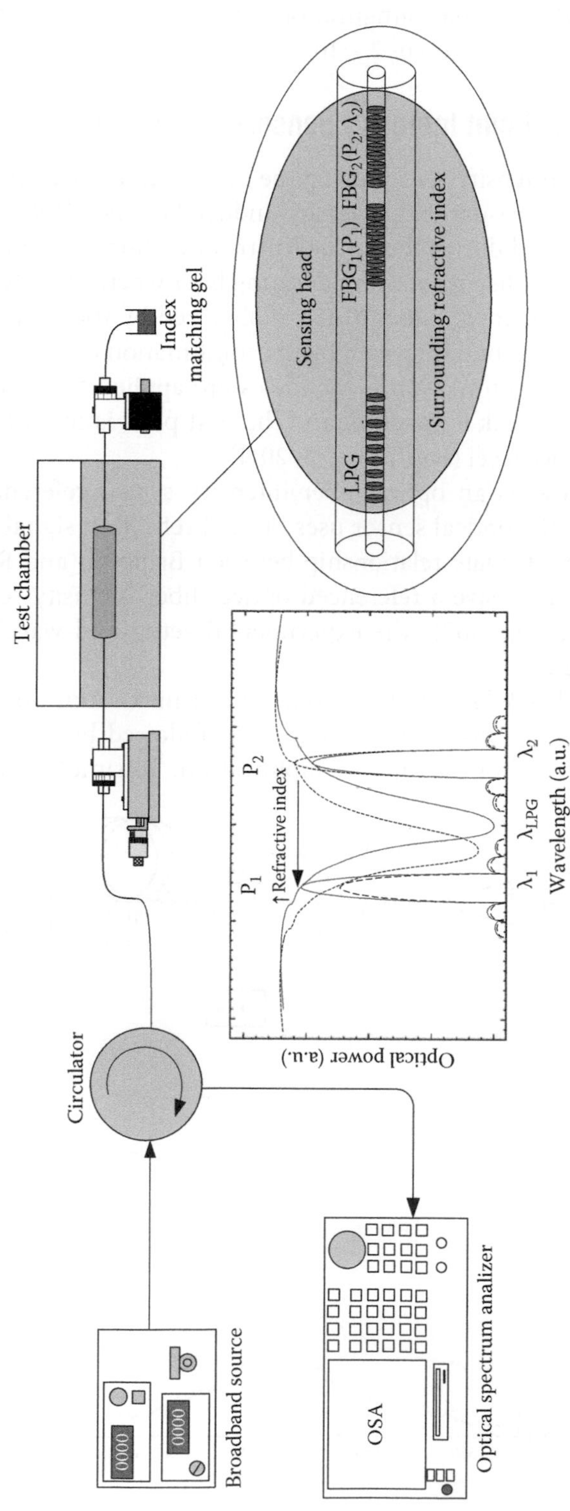

FIGURE 13.21 Experimental setup, showing the sensing head and the principle of sensor operation.

shift, and thus to the external refractive index, but is independent of any other optical power fluctuations. The sensing configuration exhibited a linear response with sensitivity of 4%/10^{-3} RIU and resolution of 2×10^{-5}.

13.5.2 Fiber-Optic Point Intensity Sensors Using Nonlinear Optical Effects

A Raman laser point intensity sensor using the multipath interference produced inside a ring cavity is able to measure the power loss induced by the displacement of a moving taper. The laser is created due to the virtual distributed mirror formed by the Rayleigh scattering produced in a dispersion compensating fiber when pumped by a Raman laser. Two laser peaks were formed: one of them is obtained by the Raman gain (1555 nm) inside the ring and the second is created by the combination of the Raman gain and the Rayleigh scattering (1565 nm). When the losses are applied to the taper via displacement, the second laser peak is reduced, and the first peak is maintained constant and can be used as reference level (Baptista et al. 2011).

Another work presents an optical fiber intensity sensor referenced by stimulated Brillouin scattering. The optical sensor uses Fresnel reflection signal at the sensor fiber end and employs an adequate relationship between Brillouin and Rayleigh scattering and Fresnel reflection to have a referenced optical fiber intensity sensor addressed in reflection (Baptista et al. 2007). The experimental setup and wavelength spectra are shown in Figure 13.22.

A distributed feedback laser (DFB) diode with a maximum power of 50 mW and central wavelength λ_c = 1554.15 nm is used. It is followed by an erbium-doped fiber amplifier (EDFA) of 1 W of maximum amplification. To simulate optical fluctuations

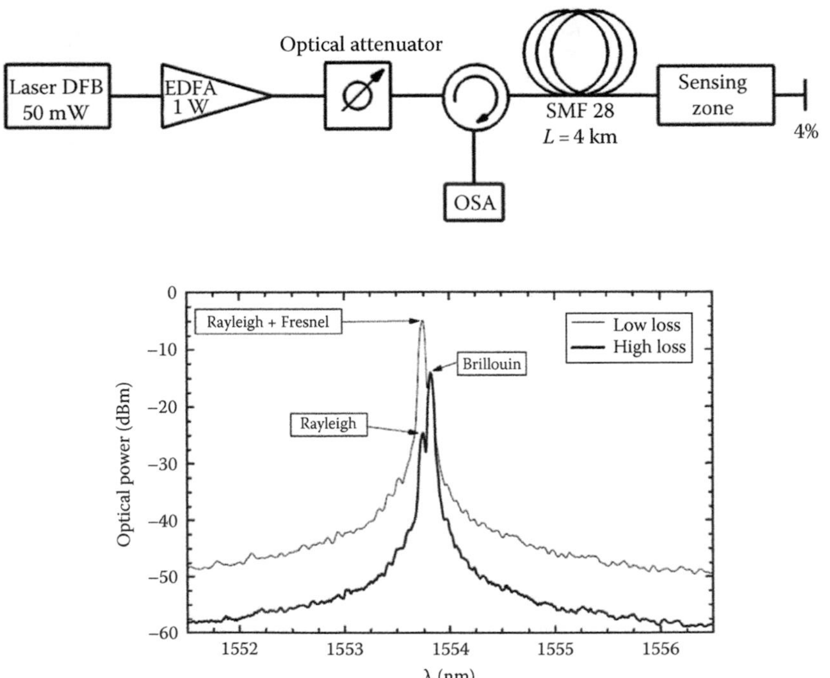

FIGURE 13.22 Experimental setup and wavelength spectra.

of the optical source, an optical attenuator is inserted between the EDFA and the optical circulator. A standard single-mode fiber roll, SMF 28, with length $L = 4$ km is used. The far end of the fiber is cleaved inducing the corresponding Fresnel reflection of ≈4%. The Brillouin and the Rayleigh scattering and Fresnel reflection signals are collected by an optical spectrum analyzer through the optical circulator. The sensing zone is implemented by an optical attenuator, and it is placed at the end of length L, just before the cleaved fiber end, in order not to affect the generation of the stimulated Brillouin scattering. This attenuator emulates the intensity sensing head. As it can be seen, the referencing is possible using the Brillouin peak.

13.5.3 Fiber-Optic Point Intensity Sensors Operated in Conjunction with Electrical or Virtual Techniques

Self-referencing fiber-optic intensity sensors based on RF electro-optical finite impulse response configuration have been reported in different configurations, as is the case of the one proposed by Abad et al. (2002). In a later development (Montalvo et al. 2009), the long fiber delay coil in each measuring point is replaced by a compact and reconfigurable electronic delay in the processing unit, as shown in Figure 13.23. This approach provides enhanced flexibility, compact design, and single-point reconfiguration of

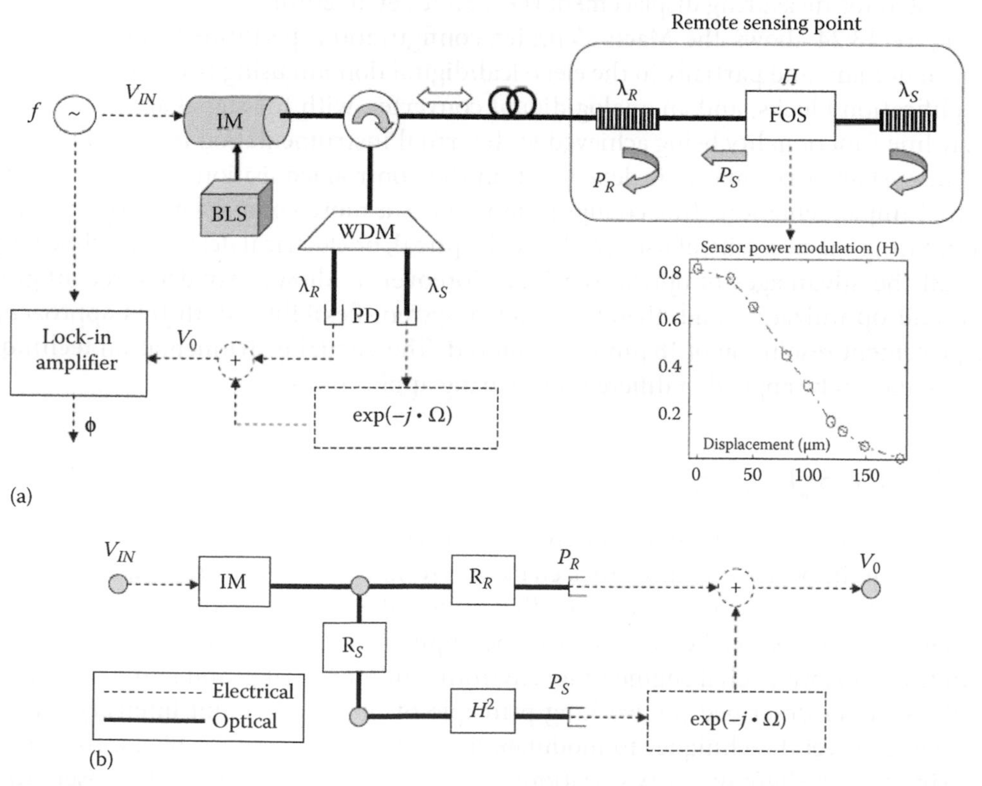

FIGURE 13.23 (a) Electro-optical configuration and (b) filter model for the proposed self-referencing technique without delay fiber coil. AOM, acousto-optic modulator; FOS, fiber-optic sensor; PD, photodetector.

Chapter 13

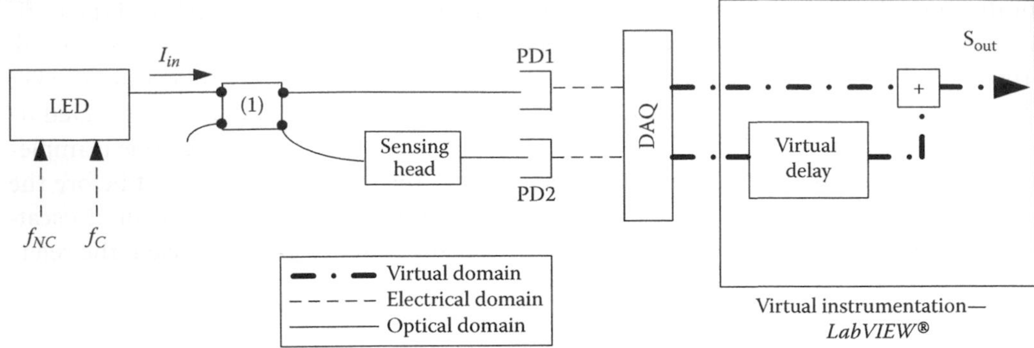

FIGURE 13.24 Implementation of a fiber-optic point intensity sensor referenced in frequency, based on a Mach–Zehnder configuration using virtual instrumentation.

multiple sensors. Long fiber delay coils of hundreds of meters are avoided, and the operating point of the topology can be adjusted just by fixing the appropriated electrical phase-shift of an RF signal at 2 kHz in a simple and flexible way. More recently, this concept has been extended to the virtual instrumentation domain in an implementation of a fiber-optic point intensity sensor referenced in frequency. In this case, the sensor was tested for measuring displacement (Fernandes et al. 2010).

Figure 13.24 shows the Mach–Zehnder configuration, performed partially in the optical domain and partially in the electrical/digital domain using two photodetection/amplification blocks, and an analog/digital converter, with the signal aggregation and delay line functionality being achieved with virtual instrumentation techniques.

The system performance is dependent on the control/acquisition parameters of the virtual implementation. This configuration allows a more compact fiber-optic sensor, eliminating the necessity of using a physical, optical, or electrical delay line, while keeping all the advantages of optical sensing. Moreover, it allows easy delay reconfiguration and optimization and, therefore, higher system flexibility. With this approach, a displacement resolution of 18 μm was achieved. The virtual instrumentation technique described can be applied in different measuring applications.

13.6 Conclusion

Fiber-optic point intensity sensors were the first type of fiber-optic sensors to be studied. Beyond the important features that characterize fiber-optic sensors in general, fiber-optic point intensity sensors are normally associated with simple concepts and low cost implementation. Nevertheless, they can be applied in very different areas associated with transportation, civil engineering, environment monitoring, and many others.

The chapter described the working principle of fiber-optic point intensity sensors, presenting several techniques to modulate light intensity in optical fibers. In order to discriminate the light intensity variations only due to the parameter under observation, the most common referencing methods were also addressed. Finally, the chapter gives an overview of the current state of the art in fiber-optic point intensity sensors, illustrating the most recent techniques that can be applied to this type of sensors.

The continuous technological development in the optics communications arena is a reality due to many reasons, certainly one of them being the fierce competition between telecom operators worldwide. This competition is expected to continue to drive the decline in prices of optical components and systems bringing, therefore, very positive consequences to the optical fiber sensor systems. Moreover, the growing deployment of optical fiber networks, such as to the households via the fiber-to-the-home technology, may allow new interesting applications to the fiber-optic sensor technology. Thus, a promising future for the sensing technology supported by the optical fiber platform and, within this domain, to the ones based on intensity modulation by the reasons outlined in this chapter is foreseen.

References

Abad S., M. López-Amo, F. M. Araújo et al. 2002. Fiber Bragg grating-based self-referencing technique for wavelength multiplexed intensity sensors. *Optics Letters* 27(4): 222–224.

Adamovsky G. 1987. Referencing in fiber optic sensing systems. *Proceedings of SPIE* 787: 17–23.

Adamovsky G. and N. Piltch. 1986. Fiber-optic thermometer using temperature dependent absorption, broadband detection, and time domain referencing. *Applied Optics* 25(23): 4439–4443.

Arregui F. J., C. Fernandez-Valdivieso, I. Ilundain et al. 2000. A Ph sensor made using cellulosic coating on a biconically tapered singlemode optical fiber. *Proceedings of the 14th International Conference on Optical Fibre Sensors*, Venice, Italy, pp. 464–467.

Baptista J. M., C. Correia, M. B. Marques et al. 2011. A Raman laser intensity sensor induced by cooperative Rayleigh scattering in a ring configuration. *Laser Physics* 21(5): 928–930.

Baptista J. M., J. M. Marques, O. Frazão et al. 2007. Stimulated Brillouin scattering as the referencing mechanism of an optical fibre intensity sensor. *Optics Communications* 271: 224–227.

Baptista J. M., J. L. Santos, and A. S. Lage. 2000. Mach-Zehnder and Michelson topologies for self-referencing fiber optic sensors. *Optical Engineering* 39(6): 1636–1644.

Baptista J. M., S. F. S. Santos, G. Rego et al. 2006. Micro-displacement or bending measurement using a long-period fibre grating in a self-referenced fibre optic intensity sensor. *Optics Communications* 260(1): 8–11.

Bariárin C., I. R. Matías, F. J. Arregui et al. 1998. Experimental results towards development of humidity sensors by using a hygroscopic material on biconically tapered optical fibre. *Proceedings of SPIE* 355: 95–105.

Bergougnoux L., J. Misguich-Ripaultand, and J. Firpo. 1998. Characterization of an optical fiber bundle sensor. *Review of Scientific Instruments* 69(5): 1985–1990.

Berthold J. W. 1988. Overview of fiber-optic intensity sensors for industry. *Proceedings of SPIE* 838: 2–8.

Berthold J. W. 1995. Historical review of microbend fiber-optic sensors. *Journal of Lightwave Technology* 13(7): 1193–1999.

Brenci M., G. Conforti, A. Mencaglia et al. 1988. Fibre-optic position sensor array. *International Journal of Optoelectronics* 3(6): 473–480.

Extance P., G. D. Pitt, B. J. Scott et al. 1983. Intelligent turbidity monitoring using fiber optics. *Proceedings of the 1st International Conference on Optical Fibre Sensors*, London, U.K., pp. 109–113.

Fernandes A. J. G., C. Jesus, P. A. S. Jorge et al. 2011. Fiber optic intensity sensor referenced with a virtual delay line. *Optics Communications* 284(24): 5665–5668.

Fernandez-Vallejo M., R. A. Pérez-Herrera, C. Elosúa et al. 2010. Stable multiwavelength fiber laser for referencing intensity sensor networks using multiple amplified ring resonators. *Proceedings of SPIE* 7653: 76533V-1–76533V-4.

Frazão O., L. A. Ferreira, F. M. Araújo et al. 2006. Intensity-referenced sensing system refractive index measurement. *Proceedings of the 18th International Conference on Optical Fiber Sensors*, Cancun, MX, pp. TuE50-1–TuE50-4.

Hwuang D., L. V. Nguyen. D. S. Moon et al. 2009. Intensity-based optical fiber strain sensor using long period fiber gratings and core mode blocker. *Measurement and Science Technology* 20: 034020.

Ilev I. K. and R. W. Waynant. 1999. All-fiber-optic sensor for liquid level measurement. *Review of Scientific Instruments* 70(5): 2551–2554.

Jesus C., P. Caldas, O. Frazão et al. 2009. Simultaneous measurement of refractive index and temperature using a hybrid fiber Bragg grating/long-period fiber grating configuration. *Fiber and Integrated Optics* 28: 440–449.

Jones B. E., R. S. Medlock, and R. C. Spooncer. 1989. Intensity and wavelength-based sensors and optical actuators. In *Optical Fibre Sensors: Systems and Application*, Vol. II, B. Culshaw and J. Dakin (eds.), pp. 431–474. Norwood, MA: Artech House.

Kharaz A., B. Jones, K. Hale et al. 2000. Optical fibre relative humidity sensor using a spectrally absorptive material. *Proceedings of the 14th International Conference on Optical Fibre Sensors*, Venice, Italy, pp. 370–373.

Koyama Y., M. Nishiyama, and K. Watanabe. 2009. Multipoint real-time displacement sensing method employing multi-wavelength by intensity-based hetero-core fiber optics sensors. *Proceedings of SPIE* 7503: 75031X-1–75031X-4.

Krohn D.A. 1986. Intensity modulation fiber optic sensors: Overview. *Proceedings of SPIE* 718: 2–11.

Lagakos N., J. H. Cole, and J. A. Bucaro. 1987. Microbend fiber-optic sensor. *Applied Optics* 26(11): 2171–2180.

Lagakos N., T. Litovitz, P. Macedo et al. 1981. Multimode optical fiber displacement sensor. *Applied Optics* 20(2): 167–168.

Leng J., A. K. Asundi, S. Du et al. 1999. Vibration measurement of smart composite structures using a new intensity-modulated optical fiber sensor. *Proceedings of SPIE* 3541: 110–115.

Medlock R. S. 1987. Fibre optic intensity modulated sensors. *NATO Advanced Study Institute* 466: 131–134.

Miao Y., Y. Liu, B. Liu et al. 2011. Intensity-modulated temperature sensor based on the photonic crystal fibers filled with magnetic fluid. *Proceedings of SPIE* 7753: 775347-1–775347-4.

Montalvo J., F. M. Araújo, L. A. Ferreira, et al. 2008. Electrical FIR filter with optical coefficients for self-referencing WDM intensity sensors. *IEEE Photonics Technology Letters* 20(1): 45–47.

Murtaza G. and J. M. Senior. 1994a. Wavelength selection strategies to enhance referencing in LED based optical sensors. *Optics Communications* 112(3–4): 201–213.

Murtaza G. and J. M. Senior. 1994b. Referenced intensity-based optical fibre sensors. *International Journal of Optoelectronics* 9(4): 339–348.

Murtaza G. and J. M. Senior. 1995. Dual wavelength referencing of optical fibre sensors. *Optics Communications* 120(5–6): 348–357.

Nakamura K. 2000. A two-dimensional optical fiber sensor array with matrix-style data readout. *Proceedings of the 14th International Conference on Optical Fibre Sensors*, Venice, Italy, pp. 268–273.

Passaro D., M. Fernández, R. A. Pérez-Herrera et al. 2009. Intensity sensors multiplexing using a multiwavelength ring fiber laser with hybrid serial-tree configuration. *Proceedings of SPIE* 7503: 75031V-1–75031V-4.

Pérez-Herrera R. A., O. Frazão, J. L. Santos et al. 2009. Frequency modulated continuous wave system for optical fiber intensity sensors with optical amplification. *IEEE Sensors Journal* 9(12): 1647–1653.

Pham V. H., H. Bui, C. D. Hoang et al. 1998. Novel fiber-optic sensor equipment for directly measuring the sugar contents inside sugar canes. *Proceedings of SPIE* 355: 121–126.

Rahnavardy K., V. Arya, A. Wang et al. 1997. Investigation and application of the frustrated-total-internal-reflection phenomenon in optical fibers. *Applied Optics* 36(10): 2183–2187.

Ramos R. T. and E. J. Fordham. 1999. Oblique-tip fiber-optic sensors for multiphase fluid discrimination. *Journal of Lightwave Technology* 17(8) 1392–1400.

Shlyagin M. G., R. M. Manuel, and O. Esteban. 2009. Wide range index sensor using a twin-grating interferometer for intensity reference. *Proceedings of SPIE* 7503: 75031J-1–75031J-4.

Spenner K., H. Schulte, H. J. Boehnel et al. 1983. Experimental investigations on fiber optic liquid level sensors and refractometers. *Proceedings of the 1st International Conference on Optical Fibre Sensors*, London, U.K., pp. 96–99.

Spillman, W. B. and D. H. McMahon. 1983. Multimode fiber optic sensors. *Proceedings of the First International Conference on Optical Fibre Sensors*, London, U.K., pp. 160–163.

Vergara M. C., I. Khanina, G. W. Baxter et al. 2000. Demonstration of fluorescence-intensity-ratio based optical-fibre temperature sensing of window glass during fires. *Proceedings of the 14th International Conference on Optical Fibre Sensors*, Venice, Italy, pp. 214–217.

Wade S. A., J. C. Muscat, S. F. Collins et al. 1999. Nd^{3+}-doped optical fiber temperature sensor using the fluorescence intensity ratio technique. *Review of Scientific Instruments* 70(11): 4279–4282.

Wang A., M. S. Miller, A. J. Plante et al. 1996. Split spectrum intensity-based optical fiber sensors for measurement of micro-displacement, strain, and pressure. *Applied Optics* 35(15): 2595–2601.

Wang H. 1996a. Effects of fibre geometry on the modulation function of a reflective intensity modulation sensor. *Journal of Modern Optics* 43(11): 2355–2366.
Wang H. 1996b. Reflective fibre optical displacement sensors for the inspection of titled objects. *Optical and Quantum Electronics* 28(11): 1655–1668.
Xiao-Jun F., W. Anbo, H. Shi et al. 1992. A compensation technique for amplitude optical fibre sensors. *International Journal of Optoelectronics* 7(4): 547–552.
Zhang Z. Y., K. T. V. Grattan, A. W. Palmer et al. 1998. Characterization of erbium-doped intrinsic optical fiber sensor probes at high temperatures. *Review of Scientific Instruments* 69(8): 2924–2929.
Zhao Z., W. S. Lau, A. C. K. Choi et al. 1995. Modulation function of a reflective fiber sensor with random fiber arrangement based on a pair model. *Optical Engineering* 34(10): 3055–3061.

14. Point Sensors

Interferometric Sensors

William N. MacPherson
Heriot-Watt University

Handbook of Optical Sensors. Edited by José Luís Santos and Faramarz Farahi © 2015 CRC Press/Taylor & Francis Group, LLC. ISBN: 9781439866856.

14.1 Introduction

Optical interference is readily observed in the natural world; the colorful patterns of oil films on water or swirling patterns seen in the delicate walls of soap bubbles vividly demonstrate the effect of the interference of light. In such examples, the color is not a result of material absorption or pigmentation, but rather, it is due to the interference of light as it interacts with a thin optical film, in these cases, the oil or soapy water, respectively. The colors observed depend upon the *optical thickness* of the film and are evident even for physical film thicknesses that are less than the wavelength of the illuminating light. This ability to gather information on subwavelength dimensions demonstrates the measurement resolution achievable with interferometry. This subject has seen extensive application both in the well-controlled laboratory environment and in many real-world applications, particularly in situations where more conventional electrical sensors are unsuitable. When applied to scientific and engineering measurement problems, it has proven to be a powerful tool for precision metrology.

Interest in interferometry as a measurement tool arises predominantly from the inherent measurement resolution afforded by interference of light: visible and near infrared light, with wavelengths in the region of 0.4 μm to few μm, makes an excellent *measuring stick* with which we can measure some characteristic length, or change in length, of the optical paths in the interferometer. Appropriate analysis of the interference yields subwavelength resolution and accuracy, with nanometer scale measurements commonplace. Optical techniques have a number of often cited advantages that can be particularly relevant for challenging measurement environments. In addition to measurement resolution, these include inherent safety, good noise immunity, and ease of sensor multiplexing. Optical measurements can also offer an advantage through non-contact measurements. This is exploited in cases where it is important to eliminate mass loading of the structure under observation, or for use in harsh environments in which a contact sensor lifetime would be limited.

Despite the often cited advantages of optical interferometry as a measurement tool, it is only with the development of optical fibers that it has become practical to apply this technique to many real-world engineering applications. Deploying traditional free-space optical techniques into harsh engineering environments poses many challenges such as maintaining optical component cleanliness, optical access, alignment, and stability. These issues can be addressed, or at least partly addressed, by exploiting optical fibers to simplify the implementation of a sensing scheme. However, it should be noted that fiber systems do not always solve all the engineering difficulties, and indeed can bring some of their own such as difficulty in producing robust packaging, fiber strain-temperature cross-sensitivity, and bend induced effects (Culshaw 2004). Therefore, the practical implementation of any interferometer depends critically upon the intended application and the desired parameter to be measured (commonly, referred to as the measurand).

In this chapter, we restrict interest to interferometer configurations suited to measurements at a *single point* in space. Here we define a single point as a region in which one measurement gives a good representation of the measurand of interest at that location. We consider the merits of different interrogation techniques and different interferometer configurations with examples from recent work in this field, but we should bear in mind that advances in materials, fabrication techniques, and optical components mean that this sensing approach continues to see new advances.

14.2 Optical Interferometers: Point or Distributed Sensors?

The classic textbook interferometer is often illustrated as a single beam of light that is amplitude split into two parts, commonly called the two *arms* of the interferometer. These are then recombined, at which point the effect of interference is observed. If we examine a typical Michelson interferometer, as shown in Figure 14.1, then the optical path length may be modulated by changes along *any part* of the interferometer arms: for example, due to the change in refractive index in a gas cell within one arm of the interferometer. Alternatively, the optical path can be varied by physical displacement of one of the mirrors along the axis of the beam. These examples illustrate two different ways of modulating the optical path: a change of the refractive index over an extended region, or a change of the path length. The former is an integrated effect in which the interferometer observes a change due to the summation of all effects along the affected optical paths. The latter could be considered as a point measurement, in-so-much that the interferometer is monitoring the displacement of a single point, the mirror. Strictly speaking, this description of a *point* measurement relies upon an integrated effect, since the interferometer must always integrate all changes along the optical path. However, for practical purposes, the distinction that we are able to make is that the mirror displacement

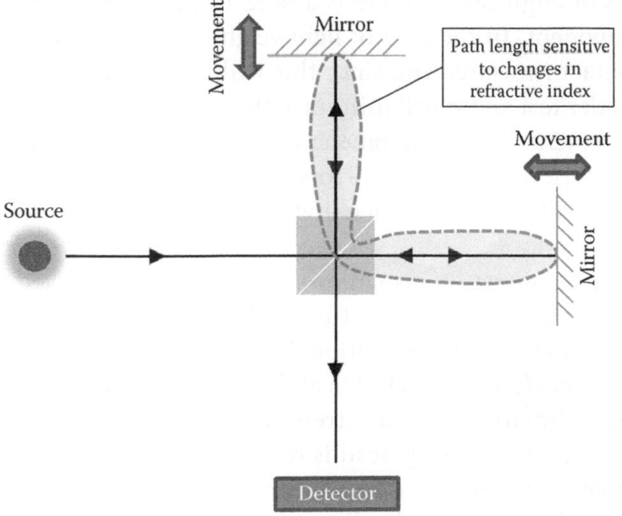

FIGURE 14.1 Modulating the optical path in a Michelson interferometer—either by moving the interferometer mirrors or by varying the index along the arms of the interferometer.

has a single discrete value, as opposed to an integral measurement arising from summation of effects over an extended path.

As a result, the interferometer can be used as a point sensor or as a sensor that integrates the effect of interest over an extended region depending upon the configuration. Often *point measurement* is taken to mean a small sensing element in which the small sensor dimension achievable by exploiting optical fibers is an advantage. However, in applications where submillimeter spatial resolution is not required, sensor configurations may be physically larger in size. For example, a mandrel wound hydrophone incorporates a length of fiber wound round a suitable cylinder which is used to observe the acoustic signal, that is, it is actually an integrating type of sensor element, but in comparison with the measurand field, it can be considered as a *point* measurement. In both cases, the criteria are the same: that the sensor must be small compared with any spatial distribution in the parameter of interest. Therefore, a *point* sensor could be defined as one in which, by any means, a measurement is made that is representative of a single location in the measurand field, that is, a single *point* measurement. However, since one great benefit of optical fibers is their compact size, examples in this chapter concentrate upon physically small sensors on the scale of the fiber itself rather than inherently larger *integrating* configurations as commercially used in the optical fiber hydrophone or gyroscope.

14.3 Advantages of Optical Interferometry

The prospect of high measurement resolution is the obvious benefit afforded by using interferometry. However, for many applications, there are further benefits that strengthen the case for deploying optical fiber point sensors in preference to competing technologies.

14.3.1 Noncontact Operation

In many branches of engineering, there is a need to monitor the deformation or vibration of a test component. In the case of lightweight or fragile objects, there is clearly a benefit for noncontact measurement, since this will not adversely perturb the motion of the component under test and it will minimize the possibility of damage to the test piece during the measurement. An interferometer can be configured such that the test surface forms one of the interferometer arm mirrors; therefore, it is possible to monitor the test surface position without physically contacting the surface as long as suitable optical access is available.

Noncontact remote operation also allows sensors to be deployed for measurements on high temperature structures, or structures in hazardous environments both of which pose challenges for deploying conventional measurement techniques. For example, sensing at high temperatures with electrical devices is difficult due to reduced component lifetime, or inability to operate at extreme elevated temperatures. By using noncontact optical techniques, the sensing head is remote from the measurement point and it is therefore possible to protect the optical components from extremes of temperature, thereby permitting vibrations and displacements to be monitored safely and reliably in such harsh environments (Carolan et al. 1997).

14.3.2 All-Optical Sensing for Enhanced Safety

Assuming that active components such as light sources, modulators, and detectors are suitably located away from the measurement region, then the all-optical nature of interferometry offers benefits in terms of enhanced safety for sensor deployment in hazardous environments. Flammable and explosive atmospheres require great care when deploying and using electrical sensors where the outcome of an unexpected electrical discharge could prove disastrous. The all-optical nature of interferometry minimizes the risk of the sensor causing an incident by eliminating possible electrical discharge sources. Typical sensing interferometers use low optical powers that will not act as heat or ignition sources. Therefore, it is possible to conceive of intrinsically safe instruments for use in such environments.

Similarly, in regions of high electric fields, the remote sensing ability of interferometry mean that sensors can be deployed while maintaining safe electrical isolation between the senor and the readout unit.

14.3.3 Absolute Measurement Capability

In addition to the high measurand resolution achievable using interferometry, it is possible to realize high *absolute accuracy* by referencing against a traceable *standard*. The length of an optical cavity, or imbalance in an interferometer, is related to the free spectral range (FSR) of the cavity according to Equation 14.1:

$$\text{FSR} \approx \frac{\lambda_0^2}{n\Delta L} \tag{14.1}$$

where n is the average refractive index over the path imbalance ΔL. For example, in the case of the Fabry–Perot cavity, $\Delta L = 2 \times$ physical cavity length. Therefore, a traceable measure of the FSR will give a measure of the cavity length. By the appropriate use of optical standards when determining the FSR, it will be the accuracy of these standards that defines the overall measurement accuracy.

In optics, we are fortunate that good standards are readily available by exploiting atomic transitions—either via a stabilized laser source, for example, the stabilized HeNe laser, or by reference to gas absorption spectra such as the iodine cell or hydrogen cyanide gas cell for use in the visible or near infrared regions, respectively. Typical frequency stability is better than 1 in 10^7 over long timescales (several hours) for commercial stabilized HeNe lasers and 1 in 10^{11} for iodine cell stabilized lasers over many hours.

Such stable references are exploited in many traditional optical spectrum analyzers for wavelength calibrating and also in Fourier transform spectroscopy to calibrate the scan mechanism. Using such instruments, the interferometer spectrum is obtained under illumination by a broadband source and from this spectrum, the FSR can be calculated.

Optical cavity length measurement with resolution better than 1 part in 10^5 has been reported using a technique based upon Fourier transform spectrometry (Jiang 2008). Fourier transform techniques have also been used for analysis of fiber Bragg grating (FBG) spectra (Davis and Kersey 1995) where Bragg wavelength resolution on the scale of picometers is demonstrated.

14.4 Drawbacks of Optical Interferometry

Despite the many benefits of using interferometry, there are also drawbacks associated with employing optical interferometers as sensors. These tend to be related to the inherent sensitivity of the whole interferometer to the surrounding environment, cross-sensitivity to parameters other than that of interest, and what appear to be more complicated interrogation system requirements when compared to other optical and nonoptical sensors.

14.4.1 Interferometer Sensitivity and Crosstalk

The excellent measurand resolution also alludes to one potential difficulty that may be encountered when using interferometry for real-world applications. In addition to the measurand, the interferometer is sensitive to any perturbation of *any part* of the interferometer. Vibration, movement, and temperature changes are all capable of inducing changes in the interferometer phase, and this is indistinguishable from phase changes due to the desired measurand. For a system that is designed to operate in the *real world* away from the controlled laboratory environment, these effects must be accounted for, or eliminated altogether. Appropriate choice of interferometer architecture, covered in Section 14.6, can go some way to addressing this, while careful choice of the sensor design and materials can act to minimize crosstalk, as discussed in relation to some practical sensors in Section 14.7.

14.4.2 Phase Ambiguity

Interferometry in its simplest form suffers an inherent drawback due to the periodic relationship between the transmitted (or reflected) intensity and the optical cavity phase when illuminated by monochromatic light. This results in a phase ambiguity that limits the unambiguous range of operation: for example, a Michelson interferometer operating at 633 nm, which is used to monitor vibration would be limited to an unambiguous vibration range of only ±79 nm centered around one of the quadrature points, as shown in Figure 14.2.

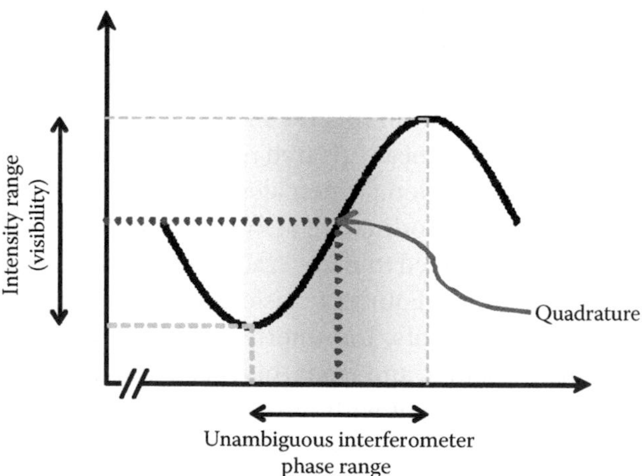

FIGURE 14.2 Interferometer operating at quadrature is limited in phase range before reaching a point (the turning points) at which the phase is not single-valued for a given measured intensity. This severely limits the use of simple interferometers for many industrial applications.

This is of limited practical use in real engineering applications where much larger vibration amplitudes are common. Techniques to address this include multiple wavelength illumination (Creath 1988, Ezbiri and Tatam 1996, 1997), extending to broadband illumination and low coherence interferometry (Li et al. 1995) in order to extend the unambiguous range. An alternative approach requires active phase control to deliberately modulate the cavity phase in a known manner (Sasaki and Takahashi 1988), or alternatively to lock the phase at a given value from which the *error* signal required to lock the phase represents the measurand (Jackson et al. 1980).

The final choice of interrogation technique is further complicated by the temporal requirement of the measurement system. A high temporal bandwidth generally precludes the use of a large number of wavelengths due to a combination of hardware limitations, such as the scan rate of a spectrometer or tunable wavelength source and the challenges associated with communicating, storing, and processing the associated quantities of data. As a result, high bandwidth interrogation systems often use a small number of discrete illumination wavelengths. However, it should be noted that spectrometer and swept laser technology is improving rapidly and systems operating in the 10^3–10^4 Hz regime are becoming commercially available, and in due course, this could offer the benefits of broadband interrogation for high bandwidth measurement systems.

Using a restricted number of interrogation wavelengths has the disadvantage of a potential loss of absolute phase measurement ability. In the extreme case of single wavelength interrogation, there are issues with *unknown* optical losses between the interrogator and the sensor, which may vary as a function of time. This sensitivity to system losses often precludes the use of a single wavelength approach in many practical implementations, since this approach cannot discriminate between intensity change due to interferometer phase change and that due to varying system losses. As a consequence of this, it is common to use multiple wavelength interrogation schemes, a selection of which is described in Section 14.6.5.

14.4.3 Optical Fibers: An Enabling Technology

Interferometry offers great potential in terms of measurement accuracy, resolution, and ability to realize noncontact operation. Optical fibers offer a practical means to take these advanced optical techniques out of the well-controlled laboratory environment and apply them to real-world applications.

14.5 Optical Fiber Interferometers

Optical fibers offer a means to control the route of the light, thereby relaxing the physical *beam path* restrictions imposed by conventional free-space optics. To a first approximation, the fiber acts as a low loss guide that allows the optical path to be routed as desired without concern about geometrical alignment (for small bend radii, the bend-loss becomes an issue; however, in most cases, it is possible to avoid this problem with appropriate cable routing). Development of optical fiber components for modern telecommunication networks, in which component lifetime of 25+ years is expected, has provided optical components with a demonstrated ability for reliable long-term usage at competitive prices. Even in relatively harsh environments, fibers may be deployed without overdue concern by exploiting the range of cable protection that has been developed for electrical power and communication systems, although the often cited benefit of small fiber dimensions is somewhat lost if this becomes necessary. Therefore, optical fibers offer a practical route to exploit interferometry outside the well-controlled laboratory environment.

Optical fibers can provide a well-defined optical path for the interferometer arms and, as discussed earlier, the entire path is sensitive to external perturbation, thereby offering potential for a spatially integrating sensor. It should be noted that this is somewhat different to a distributed sensor scheme such as those based upon optical time domain reflectometry in which the measurand is determined as a function of position. By exploiting the potential of optical fibers to *miniaturize* the interferometer, we can produce devices that begin to deliver the benefit of point sensors; for example, a short submillimeter fiber Fabry–Pérot cavity attached to a *downlead fiber* offers a point sensor capable of remote interrogation. In this case, the connecting fiber, or downlead, acts only to deliver the light to and from the sensing interferometer and does not form part of the sensing interferometer itself. Here, any downlead perturbation should not result in a change in the interferometer phase. In such configurations, we note the advantage of using optical fibers to produce miniaturized sensors easily capable of submillimeter dimensions. There are many applications in which sensor miniaturization is vital, for example, in vivo measurements in which the measurement volume is clearly limited, or aerodynamic measurements in which the sensor probe must be miniaturized to reduce disturbance to the air-flow of interest.

As we can see, the use of optical fibers in fabricating and connecting to sensing interferometers offers great benefits in terms of routing flexibility and sensor miniaturization, and allows a range of interferometer configurations to be considered for point sensing.

14.6 Interrogation Schemes

There are a wide range of interferometer configurations, although not all are obviously suited for use as point sensors. The most common are the Michelson, Mach–Zehnder, Sagnac, and Fabry–Pérot interferometers. In the case of low reflectivity cavity mirrors forming a Fabry–Pérot cavity, it is more correctly described as a Fizeau interferometer, which can be considered as a 2-beam implementation of the Fabry–Pérot. However, in many publications, this detail is often ignored and it is common to see the 2-beam case referred to as a Fabry–Pérot cavity. These configurations are relatively easily implemented using single mode optical fibers and fiber components using, for example, bidirectional couplers to replace traditional *cube beamsplitters* as shown in Figure 14.3.

14.6.1 Michelson Interferometer

The Michelson interferometer, shown in Figure 14.4, has two discrete arms, both are sensitive to environmental perturbation and the optical phase integrates the effect of the overall interaction. If a point measurement is required, for example, monitoring the position of the *signal mirror* at the end of one of the interferometer arms, then great

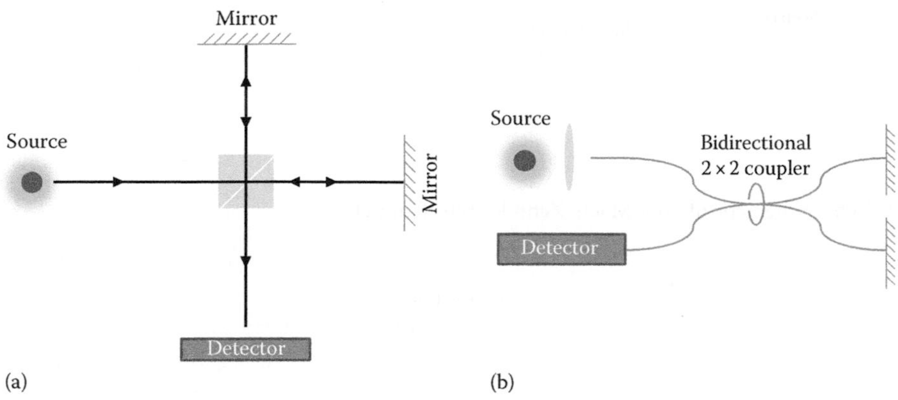

FIGURE 14.3 Schematic of bulk optic Michelson interferometer (a) and fiber optic Michelson interferometer (b) using a fused taper 2 × 2 coupler. The reflectors at the end of the fibers could be due to Fresnel reflectivity, coatings, or external mirrors.

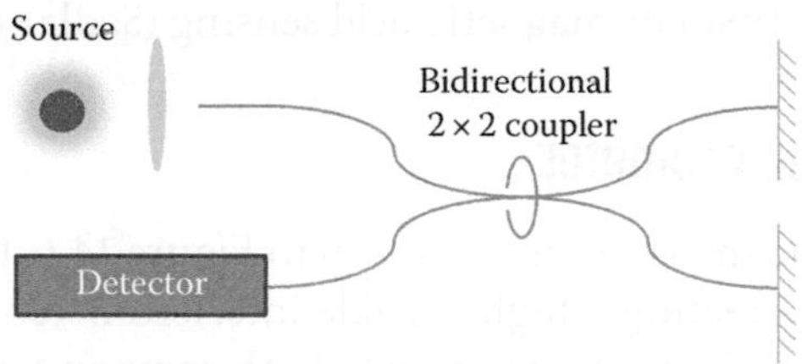

FIGURE 14.4 Schematic of fiber Michelson interferometer.

care must be taken to ensure there is no phase change over the remainder of the signal and reference arms. This can be challenging for optical fiber systems as it will require stability of the fiber route as well as careful control of the temperature of these fibers. These restrictions can be limiting for engineering applications of the fiber Michelson interferometer, unless the whole arm becomes the *point sensor*—for example, by wrapping a long length of fiber onto a sensing element. This has been successfully demonstrated for acoustic pressure wave detection in hydrophone arrays where a length of fiber wound onto a sensing mandrel can be used to detect acoustic signals in the surrounding media (De Freitas 2011). If the whole arm is not the sensing element, then the difficulty associated with stabilizing the interferometer often prevents use of this configuration.

14.6.2 Mach–Zehnder Interferometer

The Mach–Zehnder interferometer, shown in Figure 14.5, has similar issues as that of the Michelson interferometer in that both arms integrate the effect of any perturbations, and as such suffers the same challenge in practical implementation. The Mach–Zehnder has been successfully demonstrated for acoustic pressure wave detection in hydrophone arrays by exposing a fiber coil to the acoustic signal (Bucaro et al. 1977) or by wrapping one arm onto a sensing mandrel (De Freitas 2011). The integrating effect has also been

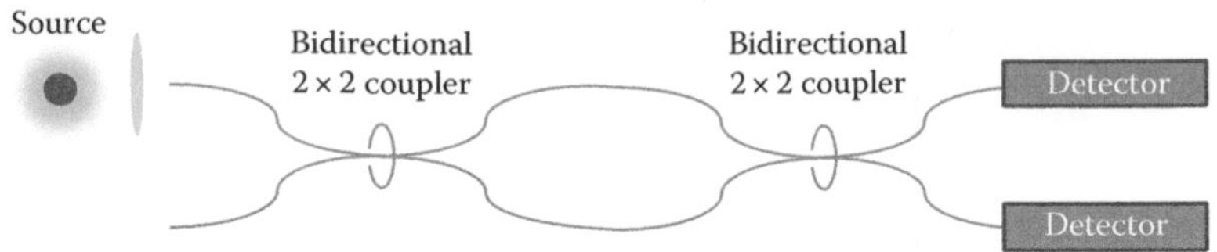

FIGURE 14.5 Schematic of fiber Mach–Zehnder interferometer.

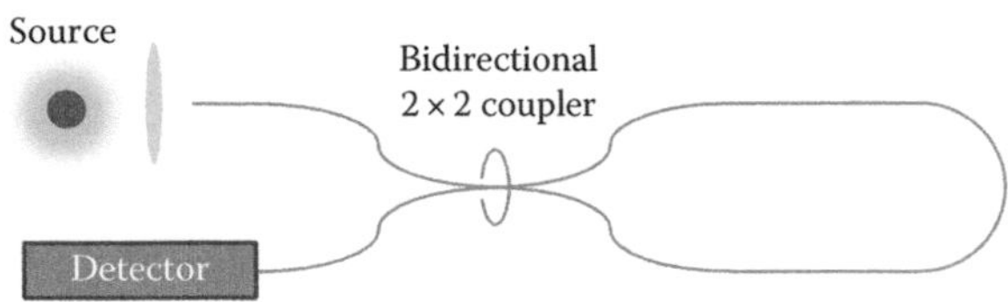

FIGURE 14.6 Schematic of fiber Sagnac interferometer.

used for sensors based upon sensitized fiber jacket materials, for example, for pressure (Hughes and Jarzynski 1980) or magnetic field sensing (Sedlar et al. 2000).

14.6.3 Sagnac Interferometer

In the case of the Sagnac interferometer, shown in Figure 14.6, the two arms are formed along a common path, creating a highly stable interferometer—for example, temperature gradients around the loop are common to both arms and, therefore, do not modify the interferometer phase. This configuration has been widely exploited in the optical fiber gyroscope where the rotation of the entire loop results in an observed phase change. These have been commercially successful, and bearing drifts better than 0.1°h^{-1} are reported (Culshaw 2006, Ferreira et al. 2007).

14.6.4 Fabry–Pérot Interferometer

The Fabry–Pérot interferometer, shown in Figure 14.7, holds promise for point sensors due to the common path taken by signal and reference beam right up to the location of a partially transmitting mirror. This partially reflecting mirror and a secondary mirror define the sensing cavity. Outside this cavity, all effects are common mode, and therefore, the fiber connecting the sensor to the interrogation system, commonly referred to as the *downlead* or *addressing fiber*, is insensitive to external perturbations, leaving

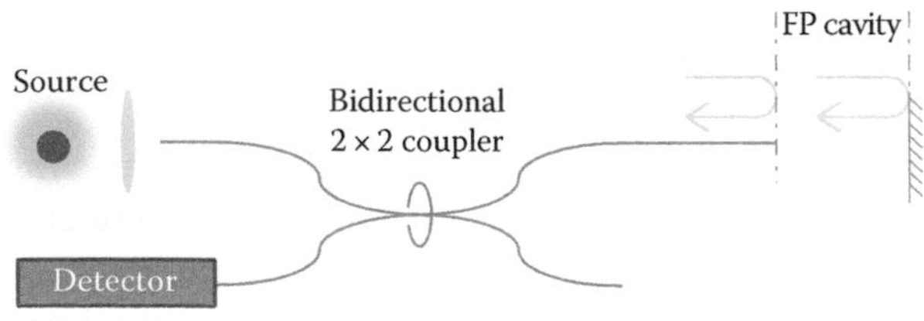

FIGURE 14.7 Schematic of fiber Fabry–Pérot interferometer.

only the Fabry–Pérot cavity sensitive to perturbation. This scheme lends itself to both measurements under reflection and transmission conditions, although in many cases, it is convenient to consider the cavity under reflection as this offers a minimally invasive sensor at the end of a single downlead fiber.

14.6.5 Phase Recovery

Consider a simple 2-beam interferometer. The optical path difference between the two arms of the interferometer is given by $L_{OPD} = nl$ where n is the mean refractive index over the path difference l. This is more commonly expressed in terms of optical phase, ϕ, given by

$$\phi(\lambda) = 2\pi \frac{nl}{\lambda} \tag{14.2}$$

From this equation, we obtain a relationship for the measured intensity

$$I(\lambda) = k\left[1 - V\cos(\phi)\right] \tag{14.3}$$

where

- V is the visibility of the interferometer
- k is a constant that includes detail of the optical source intensity, coherence, and system losses

In general, both V and k may be wavelength dependent and may be time varying; however, they are commonly assumed to be constant for purposes of data analysis.

The inspection of Equation 14.3 exemplifies the phase ambiguity issue discussed in Section 14.4.2. Since the phase, ϕ, varies as a result of the applied measurand, monotonic changes in the measurand will result in the observed intensity cycling between the maximum and minimum values. This leaves uncertainty in the absolute interferometer phase and, hence, uncertainty in the absolute value of the measurement. If we follow this argument further, then, for an instrument that is not in continuous operation, it is not possible to state an absolute value of the measurand. This is because there may be changes of the phase (i.e., the measurand) exceeding 2π that will be unaccounted for during the powered-off period.

This issue of phase ambiguity has received a great deal of attention, and some approaches to address this are discussed.

14.6.5.1 Single Wavelength Interrogation: Quadrature Detection

If we again consider Equation 14.3, and assume that k and V are constant, then we note that the greatest sensitivity of I on ϕ is observed when $\phi = \pm\pi/2$. These are commonly referred to as the quadrature positions. If we construct our interferometer so that we can actively control the phase in one arm of the interferometer using an error signal generated from the deviation of I from the quadrature value, then we can *lock* the interferometer at the quadrature operating point. For example, in the case of a free-space

Michelson interferometer, we might wish to monitor the displacement of one of the mirrors and achieve this quadrature lock by mounting the second mirror onto a piezo-electric translation stage and arranging the feedback such that the piezo-controlled mirror tracks the displacement of the test mirror to keep the interferometer in quadrature. In this case, the error signal is proportional to the measurand.

It is noteworthy that as long as the movement is slow enough that the feedback system (detector and associated electronics) and mirror translation mechanism is fast enough, this system can track mirror displacements that exceed 2π phase changes, that is, over multiple fringes, that would otherwise be problematic for a single wavelength system due to phase ambiguity. Therefore, this approach avoids the issue of phase ambiguity as long *as the rate of change can be followed and that the system is in continuous operation*. Unfortunately when the system is powered off, interferometer phase changes are unknown, and therefore, the absolute phase value cannot be determined without appropriate calibration during start-up.

Phase modulation can be realized in a number of different ways. Movement of a mirror, as suggested earlier, is a simple and effective method of inducing a phase change. Commercial piezo-electric actuators are capable of subnanometer resolution and have typical resonant frequencies in the range of tens to hundreds of kilohertz, allowing submillisecond response times. The disadvantage of this approach is the requirement for a free space optical beam. Free space beams introduce the risk of external vibration sensitivity, optical loss, and unwanted reflections (which in turn cause phase noise). Unless great care is taken when implementing this approach, these concerns may preclude the use of extrinsic phase modulation.

Piezo-electric actuators can also be used to stretch an optical fiber to form an in-fiber phase modulator (Martini 1987). This has the advantage of keeping the light inside the waveguide, does not suffer from misalignment errors, and minimizes unwanted reflections that can be associated with free space optics. In practice, care must be taken to ensure the fiber is under tension *at all times* and to prevent slippage or slack fiber distorting the relationship between the applied voltage and the induced phase change. Although this approach still suffers from the bandwidth limit associated with mechanical movement, operation up to tens of kHz has been reported (Jackson et al. 1980). A potential issue of this approach, especially when long lengths of fiber are wrapped around a piezo-electric cylinder, arises due to bend induced birefringence. Techniques to address this have been reported (Luke et al. 1995). Alternatively, linear stretching of a length of fiber avoids this issue but with the penalty of a less convenient geometry.

An alternative to mechanical stretching relies upon changing the refractive index of a waveguide element, via the electro-optic effect. Waveguide phase modulators, such as lithium niobate phase modulators, offer a route to very high bandwidth phase modulation (GHz). However, they tend to be limited in the magnitude of the phase change possible, with only a few radians common.

14.6.5.2 Passive Quadrature Interrogation

If we consider again an interferometer, with single wavelength illumination chosen such that it is operating at the quadrature point, assuming that system losses remain constant and that source intensity fluctuations are negligible, then the reflected intensity will be proportional to the interferometer phase change, assuming that the phase change is small. In cases where these criteria are met, this offers a simple, but effective way to measure interferometer phase with sub-mrad phase resolution.

14.6.5.3 Two Wavelength Interrogation

Single wavelength interrogation is susceptible to error due to uncompensated power fluctuations. While it is relatively straightforward to compensate for source intensity variation by making a reference intensity measurement, it is less obvious to compensate for connector loss and fiber bendloss as these are likely to be unknown, and potentially time varying.

Furthermore, the passive single wavelength interrogation approach restricts the range of phase that can be measured to a maximum of π radians, that is, to avoid ambiguity the phase must not pass through a turning point of the transfer function. By observing more than one point on the transfer function, it is possible to address the difficulties encountered with the passive single wavelength approach. Multiple points can be observed by deliberate modulation of the interferometer phase by a known amount. In general, it is possible to achieve the desired phase shift by actively modulating one of the arms of the interferometer. However, this is impractical in the case of a short Fabry–Pérot interferometer, since the active modulation element would have to be included within the sensing cavity itself. An alternate approach uses multiple wavelength illumination to access two different points on the transfer function, with the phase difference between these points set by the chosen wavelengths. It is common that the wavelength separation is chosen such that the two interrogation wavelengths are in quadrature in order to simplify the phase recovery procedure, that is, the phase difference between the two wavelengths is $\pi/2$. This concept is illustrated in Figure 14.8.

Two wavelength interrogation can be realized by using two discrete optical sources (with spectral separation or temporal modulation to allow the two intensities to be measured independently), or by suitable modulation of a single optical source to switch between two slightly different wavelengths. Alternatively, a broadband source and wavelength filters can be used. This approach has been demonstrated using interference filters (Furstenau and Schmidt 1998) and FBGs as filters (Lo and Sirkis 1997). Using these techniques, it is possible to measure the reflected signal from two points

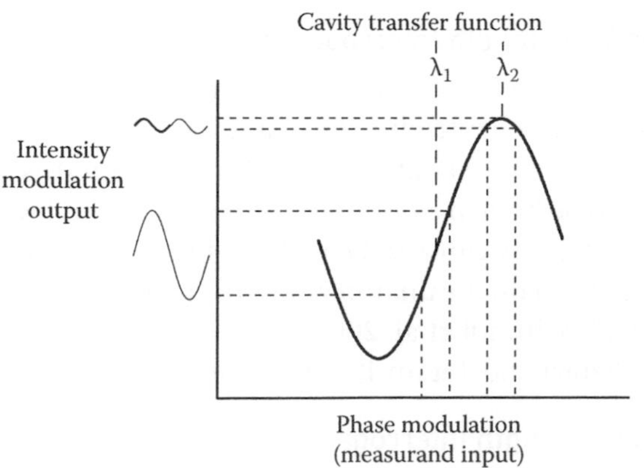

FIGURE 14.8 Principle of two wavelength quadrature interrogation.

Chapter 14

on the transfer function, and assuming that these are chosen to be in quadrature, then the normalized intensities S_1 and S_2 are given by

$$S_1 = \frac{I_1}{I_0} = k\left(1 + V\sin\phi\right)$$
$$S_2 = \frac{I_2}{I_0} = k\left(1 + V\cos\phi\right) \tag{14.4}$$

where
I_0 is the incident intensity
I_1 and I_2 are the interferometer intensities for λ_1 and λ_2, respectively
k includes a component due to bend loss, and is assumed to be common mode for both interrogation wavelengths

Solving for ϕ requires knowledge of either k or V. While k is unknown, it is possible to measure V experimentally, assuming that it can also be considered a constant. In this case, the cavity phase is recovered:

$$\phi = \cos^{-1}\left(\frac{S_2 - S_1}{\pm V\sqrt{S_1^2 + S_2^2}}\right) + \tan^{-1}\left(-\frac{S_2}{S_1}\right) \tag{14.5}$$

The benefit of this approach is that we have effectively eliminated the unknown loss contained in k, assuming that bendloss is *wavelength invariant*. While this assumption is not strictly true, it has been shown experimentally that for bend radii greater than ~20 mm, it is reasonable to assume that bend loss is common-mode for wavelengths separated by up to 40 nm (Faustini and Martini 1997).

This solution has an important feature: the argument of the arc cosine is multivalued due to the presence of the root. This results in multiple solutions for the recovered phase value. This can be addressed by inspecting the absolute values of S_1 and S_2, although this is susceptible to error particularly when close to the transition point between these two solutions.

14.6.5.4 Multiple Wavelength Interrogation

While the two wavelength approach eliminates some of the issues with monitoring the phase of the interferometer, the introduction of additional wavelengths further improves the robustness of the phase recovery algorithm. More wavelengths can be used to remove the requirement for V to be known as long as it is the same for all wavelengths. Reduction of the mathematical ambiguity can also be realized by using multiple wavelengths; however, there remains a 2π limit on the unambiguous measurement range. Three-wavelength (MacPherson et al. 1997, Schmidt et al. 2001), four-wavelength (Ezbiri and Tatam 1996), and five-wavelength (Ezbiri and Tatam 1997) algorithms have been reported.

14.6.5.5 Broad Wavelength Interrogation

Further extending the multiple wavelength approach ultimately leads to broadband interrogation. Here the interferometer is illuminated with a very broad spectrum of light, typically tens of nanometers wide, with the interferometer output recorded as a function of wavelength. A typical configuration might be to use a superluminescent LED as an optical source and an optical spectrum analyzer as the *detector.* Alternatively, a single detector and swept wavelength source such as tunable laser, or broadband source and tunable filter can be used.

If we again consider Equations 14.2 and 14.3, then combining these yields

$$I(\lambda) = k\left(1 - V\cos\left(\frac{2\pi nl}{\lambda}\right)\right) \tag{14.6}$$

A plot of $I(\lambda)$ against $1/\lambda$ results in a cosinusoidal function where the frequency is related to the interferometer path length, nl. This frequency can be obtained by Fourier analysis of $I(\lambda)$ as a function of $1/\lambda$ and provides an *absolute measure* of the interferometer phase. The resolution of this is related to the length of data used in the Fourier analysis, which, in turn, is related to the source spectral width and the spectral resolution with which the interferogram is recorded.

This approach has the clear benefit of being able to report the interferometer phase without requiring continuous operation; however, for a scanning spectrometer (or scanning tunable source), this approach can be significantly limited in terms of temporal resolution. Typical optical spectrum analyzers have scan rates of the order of a few Hz, although higher bandwidth systems have been developed (e.g., high speed tunable laser sources for FBG interrogation), although these higher bandwidth systems often achieve this at the expense of wavelength range.

An alternative to using scanning systems is to exploit interrogation systems based upon spectrometers incorporating detector arrays. Here the dispersed spectrum is incident upon an array of photodetectors, and now it is the photodetector array that defines the temporal and the wavelength resolution. This is an area of technology that has seen significant development in recent years, and although commercial systems still cannot compete with scanning optical spectrum analyzers in terms of combined range and resolution, this may not always be the case. Unfortunately, operation at communications wavelengths remains costly with this type of spectrometer due to the high cost associated with large InGaAs detector arrays; however, lower cost systems are commercially available for operation below ~1100 nm.

14.6.6 Sensor Multiplexing

For many applications, a compelling advantage of optical sensor technologies is the ease of sensor multiplexing. The FBG sensor is an excellent example where many sensors can be addressed by single optical fiber. In comparison with electrical temperature and strain sensors, which typically require 2–4 wires per sensor, the FBG sensor has significant cost and weight savings where many sensors are required. This multiplexing capability is realized by selecting different grating wavelengths for each sensor, with sufficient separation to prevent grating spectra overlapping as the measurands vary. In this way, each sensor occupies its own region in the wavelength domain, known as wavelength division multiplexing. Unfortunately, interferometers have a very broad spectral fingerprint and as a result, wavelength division multiplexing does not offer an easy route to sensor multiplexing. Despite this, interferometer multiplexing has been realized by using wavelength filters to allow some level of spectral encoding (Rao 2006). Other approaches rely upon using arrays of fibers, by using coherence, and most commonly by using time division multiplexing (Cranch et al. 2003). The issue of sensor multiplexing is considered in more detail in Chapter 15.

14.7 Encoding the Measurand

For any fiber sensor, we have two requirements of the *encoding* mechanism: (1) we desire high sensitivity to the measurand, and (2) we desire minimal sensitivity to other effects. For example, a temperature sensor might exploit both the thermal expansion and the thermo-optic effect to change the optical path length in one of the interferometer arms, but might be constructed so that it is isolated from any potential strain effects that could be applied to it by its surroundings. This highlights the importance of considering optical effects such as thermo-optic or strain-optic effects as well as mechanical effects such as thermal expansion or mechanical strain when designing a sensor element and its housing.

While we may choose to maximize the sensitivity of a sensor by appropriate sensor design and material choice, issues with cross-sensitivity often remain. Temperature cross-sensitivity poses a significant problem since a temperature change can affect both material optical properties as well as physical dimensions via thermal expansion. Therefore, there is often a need to eliminate or compensate for any thermal effects. This is commonly done by appropriate sensor design to minimize thermal sensitivity. If this cannot be realized, then an independent temperature measurement can be made to compensate for thermal drift during the data processing stage.

14.7.1 Fabry–Pérot Strain Sensing

Strain measurement is typically required with microstrain resolution to be useful for monitoring the effect of loading on engineering and civil structures. Typical electrical strain gauges are based upon a thin metal film resistive element patterned onto a compliant backing such that the electrical resistance varies as a function of the applied strain. These are *point* sensors with spatial resolution on the scale of several millimeters. The main disadvantage of electrical strain gauges lies in their deployment with each normally requiring four wires to connect to a standard signal conditioning unit. This quickly becomes cumbersome for large sensor arrays. The four wire approach is used to help mitigate against noise and compensate for the resistance of the connecting wires. Despite this, it is still challenging to measure the small change in resistance for microstrain resolutions, typically <0.25 mΩ per microstrain for a 120 Ω strain gauge. In the case of long connecting cables, great care must taken when shielding the cables to prevent the electrical noise dominating the signal. By comparison, optical fiber sensors do not suffer from the same level of electromagnetic pickup and only require a single connecting cable. Depending upon the intended application, low bandwidth measurements may be sufficient, and in this case, FBG-based strain gauge is often the chosen technology. FBGs have seen significant application in this area, exploiting their ability to multiplex many FBG sensors on a single fiber; however, there have also been studies on interferometer-based strain sensing. High bandwidth and high temperature measurements of strain are less straightforward with FBGs, and interferometers retain some benefit in such cases.

Strain sensing requires the measure of a change in length, ΔL, of a test section length L. One approach to measure this length change is to monitor displacement of one part of the structure with respect to another. This could be done using a Michelson interferometer; however, care must be taken to ensure that a well-defined region of the interferometer observes the elongation due to strain and that all other parts of the interferometer remain unaffected by any uncompensated strain or temperature. Practical implementation of this approach has been demonstrated by monitoring displacement of one part of the test structure with respect to a reference point on the structure (Harrison et al. 2005, Chen et al. 2010).

A more convenient approach is to use a Fabry–Pérot configuration as a sensing element, as shown in Figure 14.9, where the change on optical cavity length can be used to infer the strain, effectively forming the fiber optic equivalent of the electrical foil strain gauge.

This approach, where a short sensor length is bonded to the test structure, has been realized by gluing two fibers into an alignment capillary (Murphy et al. 1992), splicing a hollow fiber section between two single-mode fibers (Sirkis et al. 1993), hydrofluoric acid (HF) etching and splicing to form an in-fiber cavity (Cibula and Donlagic 2007), and femtosecond laser machining to directly fabricate an in-fiber cavity (Rao et al. 2007). Two examples of this approach are shown in Figure 14.10.

In the case where the cavity is air-filled, the thermal cross-sensitivity of the cavity media can be considered negligible. However, there remains a residual intrinsic effect due to the thermal expansion of the materials involved in the sensor body. This is not always a problem: in cases where the sensor is securely bonded to a large test structure, it will be the thermal expansion coefficient of the structure that dominates. For example,

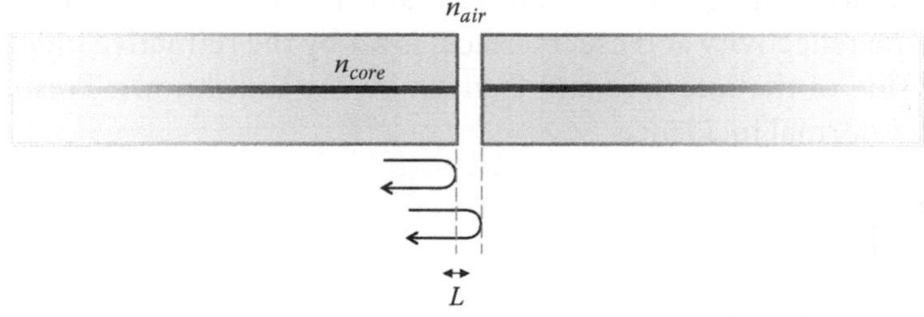

FIGURE 14.9 Interferometer formed by the gap between two fibers. The interferometer phase is a function of the gap length.

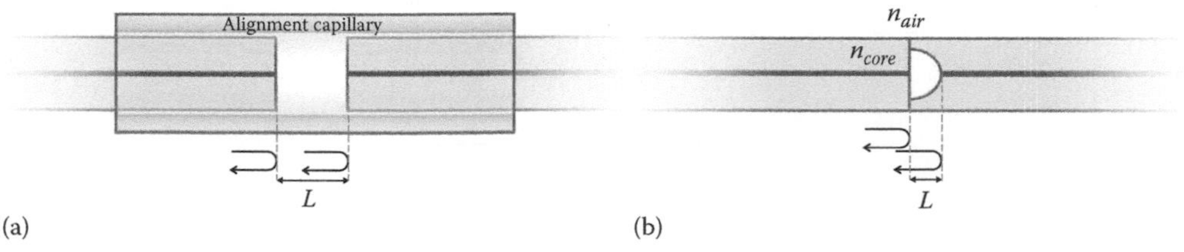

FIGURE 14.10 (a) Fibers held in alignment capillary; (b) cavity formed by splicing etched fiber onto cleaved fiber.

consider an all-silica sensor bonded to a large metal test object; the thermal expansion coefficient of fused silica is ~5.5×10^{-7} compared to >1×10^{-5} for typical metals. Although the thermal expansion of the structure dominates, an appropriate compensation must still be applied to account for any apparent strain arising due to thermal expansion of the structure. Unfortunately, if the structure is too small, then there is concern that the presence of the sensor acts to modify the overall structure properties, essentially the fiber acts as a reinforcing element, and in such cases, the strain measurement itself may be compromised. In extreme examples, the attached *strain gauge* sensor may not be appropriate, and noncontact displacement monitoring may become essential.

Displacement monitoring can be exploited to measure a wide range of parameters by appropriate choice of the transduction mechanism. Pressure sensing, discussed in subsequent sections, is an example of this where the deflection of a pressure sensitive diaphragm is measured interferometrically. Hydrogen sensing has been demonstrated by monitoring the deflection of a palladium foil as it *expands* during hydrogen absorption (Maier et al. 2007), and magnetic field sensing has been reported using a cavity structure fabricated from magnetostrictive amorphous metallic wire (Oh et al. 2004) such that the cavity length becomes a function of the applied magnetic field.

A similar configuration can also be used to determine refractive index. By considering a fixed cavity length and changing the media inside the cavity, this approach has demonstrated refractive index measurement with a resolution better than 10^{-6} predicted (Ran et al. 2009).

14.7.2 Thin Optical Film Sensors

Miniaturized optical cavities can be fabricated by deposition of thin optical films onto the cleaved end of an optical fiber. In this case, the optical cavity arises due to reflections associated with the glass–film interface and the film–air interface as shown in Figure 14.11. The reflectivity at these interfaces is set by the refractive index of the materials on each side of the interface and is given by the well-known Fresnel reflectivity equation under normal incidence

$$R = \left(\frac{n_1 - n_2}{n_1 + n_2}\right)^2 \tag{14.7}$$

where n_1 and n_2 are the refractive indices either side of the interface. Clearly, it is important that there is sufficient index difference between the fiber and the film to form an optical cavity.

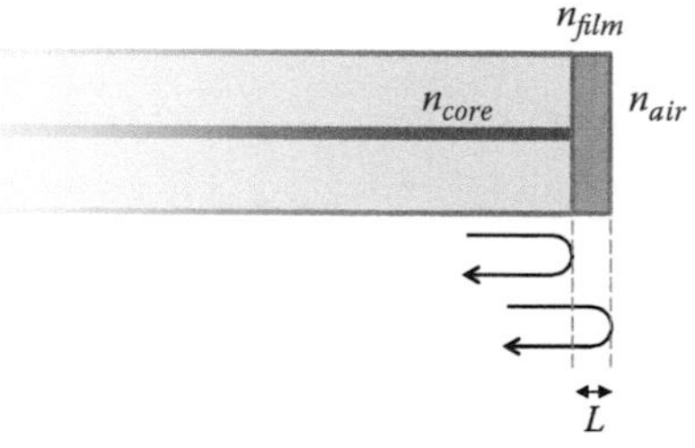

FIGURE 14.11 Optical thin film cavity on the end of cleaved fiber.

For many optical materials, the reflectivity at each interface is sufficiently low that the cavity operates in the low reflectivity *2-beam approximation* regime, that is, a Fizeau cavity. While this affords relatively straightforward data processing, issues can arise if any contamination is deposited onto the end of the sensor. Contamination, such as a thin film of liquid, will affect the cavity visibility by modifying the end face reflectivity and may itself form an additional optical cavity which will significantly affect the recovered phase. To address this, it is often necessary to optically isolate the cavity from the environment by application of suitable outer coatings such as an opaque end coating, for example, a metallic film. The resulting increased reflectivity will change the visibility and finesse. In some cases, it is necessary to balance this increase with an increased reflectivity at the fiber–cavity interface, perhaps by additional of another metal layer. With careful selection, the sensitivity ($dI/d\phi$) can be improved, at least over a limited range of ϕ, albeit at the expense of a less straightforward relationship between measured intensities and optical phase over the complete 2π range (Santos et al. 1992, Kilpatrick et al. 2000).

By appropriate selection of the cavity material, the sensor can be chosen to be sensitive to one of many measurands; for example, a polymer coating may be pressure sensitive via the strain optic effect and physical changes in length (Beard et al. 1998) and temperature sensitivity can be selected by choosing materials with strong thermo-optic and/or thermal expansion effects (Kidd et al. 1995). As always, it is important to minimize measurand cross-sensitivity and therefore, appropriate material selection is critical to a practical sensor design if active compensation is to be avoided.

As an example of a temperature sensor based upon this technique, consider a 1.5 μm thick zinc selenide layer with mirror coatings of ~10 nm nickel on the fiber–film interface and 100 nm aluminum on the film end face. In this case, the optical phase change, $\Delta\phi$, associated with a change, ΔT, of the mean cavity temperature is given by

$$\Delta\phi = \frac{4\pi(n\alpha + \beta)l}{\lambda}\Delta T \tag{14.8}$$

where

- λ is the illumination wavelength
- n is the mean cavity refractive index
- l is the cavity length
- α and β are the thermal expansion coefficient and thermo-optic coefficient of the film, respectively

Calibration and published values of n, α, and β yield a sensitivity of 2.2 mrad K^{-1}. In a practical sensor, 0.1 K temperature resolution at 10 kHz measurement bandwidth has been demonstrated (Kilpatrick et al. 2002). Here, the purpose of the 100 nm aluminum layer is primarily to optically isolate the cavity for use in an environment where oil deposits are expected on the end of the sensor. The increased reflectivity due to the aluminum film is *balanced* at the fiber–ZnSe interface by the presence of the nickel film. This results in a higher finesse cavity which is exploited here for simple measurement of sub-mrad phase by biasing a single wavelength interrogation part way up the steepest part of the optical transfer function.

Sensors based upon optical thin films are physically small and have the potential to be mass-produced, but their response to various measurands is limited by the practical choice of materials.

14.7.3 Miniature Pressure Sensors

In principle, a pressure sensor could be fabricated by using a pressure sensitive compliant thin film coating and operated in a similar manner to the thin film temperature sensor described previously. However, limited pressure sensitivity and thermal cross-sensitivity remain a concern with this approach and it is more common to measure pressure by detecting the deflection of a diaphragm as shown in Figure 14.12.

Miniature pressure sensors are of interest for a wide range of applications from niche research measurements to widespread biomedical use. Many fields of engineering research have the desire to be able to instrument scale models to determine fluid flow without the instrumentation affecting overall flow conditions; hence, small sensor size is important. Biomedical applications have the obvious requirement for miniaturized sensors especially for in vivo deployment. In such cases, the miniaturization offered by optical fibers is important and this has seen significant interest both academically and, more recently, commercially.

If we consider a circular diaphragm clamped around its circumference, then the deflection, y_{center}, at the center of the diaphragm is given by Di Giovanni (1992)

$$y_{center} = \frac{3(1-\mu)Pa^4}{16Eh^3} \tag{14.9}$$

where

a is the diaphragm radius

h is the thickness

E and μ are the Young's modulus and Poisson's ratio of the diaphragm material, respectively

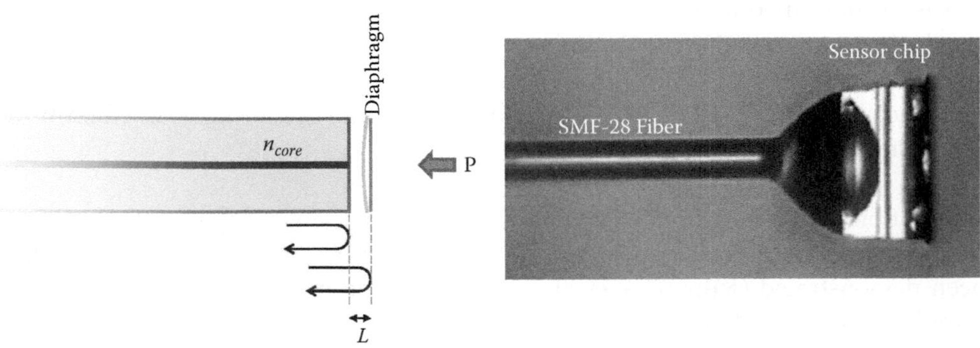

FIGURE 14.12 Schematic of Fabry–Pérot pressure sensor. A pressure sensitive diaphragm deflects due to external pressure change, and this deflection is observed as a change in cavity length.

A measure of the deflection of the center of the diaphragm is, for small deflections, proportional to the pressure difference across the diaphragm; hence, a measure of y_{center} can be used to determine the pressure acting across the diaphragm.

The fabrication of these devices can be challenging due to the dimensions involved; inspection of Equation 14.9 reveals that, for diaphragm with dimensions of the order of the diameter of an optical fiber, the diaphragm thickness is only a few micrometers for a pressure range of a few hundred kPa. This diaphragm must be attached in a manner that it remains flat, since any buckling will distort the pressure–phase relationship and may introduce unwanted hysteresis. The attachment must also be leak-free to ensure repeatable and reliable long-term operation as a pressure sensor. To realize these requirements, a variety of fabrication techniques have been exploited, including silicon micromachining technology (Watson et al. 2006b), laser machining (Watson et al. 2006a), and microassembly (Totsu et al. 2005, Wang et al. 2010). High temperature operation is possible if adhesives are eliminated from the design, and components with low thermal expansion coefficients are used; for example, an all-silica pressure sensor (Xu et al. 2005) has been demonstrated for use up to 700°C. At this temperature, many electrical pressure transducers, particularly those based upon semiconductor technology, would fail to operate reliably.

Silicon micromachining offers a route to producing miniature sensor *chips* that can be fixed onto the end of the optical fiber, an example is shown in Figure 14.12b. Conventional integrated circuit lithography processes are well able to realize the resolution and tolerance required for fabrication of precision components suitable for attachment to optical fibers; however, they suffer from a limited processing depth. With care, deep reactive ion etching can realize the high aspect ratios desired for a sensor chip incorporating a hole to locate the fiber as well as diaphragm-supporting structures. Using this approach, sensor chips incorporating a cavity *end stop* to help define the cavity length have been developed. These have been demonstrated to be capable of ~100 Pa resolution over a range of 0–1 MPa, with a high bandwidth operation extending into the hundreds of kilohertz regime (Watson et al. 2006b).

Further miniaturization is offered by attaching the diaphragm directly to the fiber itself. Fiber end face laser machining (Ran et al. 2009, 2011) and differential etching using HF acid (Donlagic and Cibula 2005, Zhu and Wang 2005) have both been demonstrated for producing a cavity recess in the end of the fiber, onto which a diaphragm can be attached.

Other novel pressure sensor designs have been reported, including the use of glass spheres (Dakin et al. 2009) as sensor elements where the interferometer is formed by the wall of the sphere as illustrated in Figure 14.13. The microsphere is attached using UV curing adhesive to allow the reflected interferogram to be optimized during the

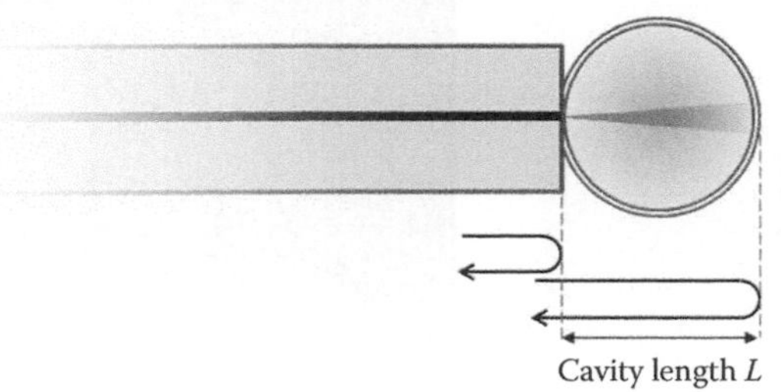

FIGURE 14.13 Microsphere pressure sensor concept.

fabrication process before curing the adhesive and fixing the sphere in place. In this case, the external pressure compresses the microsphere, resulting in a reduced cavity length and a sensitivity of 1000 Pa is reported using a broadband interrogation technique.

14.7.4 Fiber Top Cantilevers

With recent developments in areas of material processing and micromachining, it is becoming possible to fabricate elaborate structures directly onto the end of the fiber with dimensions on the micron scale. Focused ion beam milling has been used to fabricate miniature cantilevers onto the end of the fiber itself (Melngailis 1987). In such examples (Figure 14.14), the cantilever end has a pointed *tip* for use as fiber-end atomic force microscope, which has then been used to measure the Casimir force (Zuurbier et al. 2011). Different approaches to monitoring the cantilever deflection have been reported including measurement of the length of the optical cavity formed between the end of the fiber and the cantilever itself.

In the experiments of Zuurbier et al., a single wavelength interrogation was chosen such that it was near the quadrature point when the cantilever was in its neutral position. Therefore, small cantilever deflections will result in a linear change in measured intensity, with the exact proportionality constant determined experimentally. In this case, it is assumed that the source and downlead losses are stable over the duration of the calibration and measurement. Operating in this manner, a minimum detectable cantilever tip deflection of ~80 pm has been demonstrated (Iannuzzi et al. 2007).

Microcantilevers are increasingly being investigated for environmental sensing (Bashir et al. 2002) and biosensing applications (Pei et al. 2004, Wenfeng and Lee 2009), and the prospect of producing cantilevers on the end of fibers for use in miniaturized sensing probes is attractive.

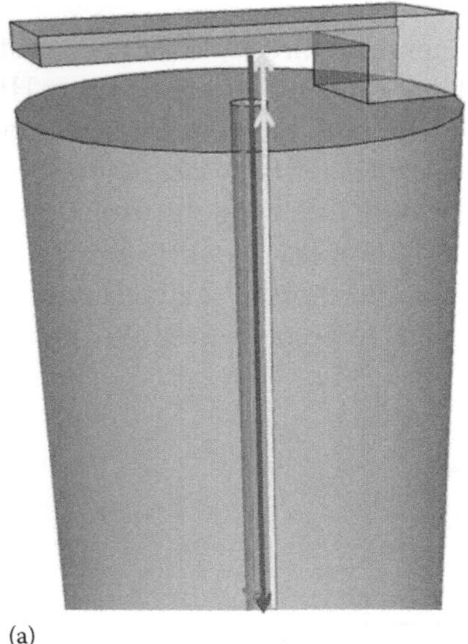

(a)

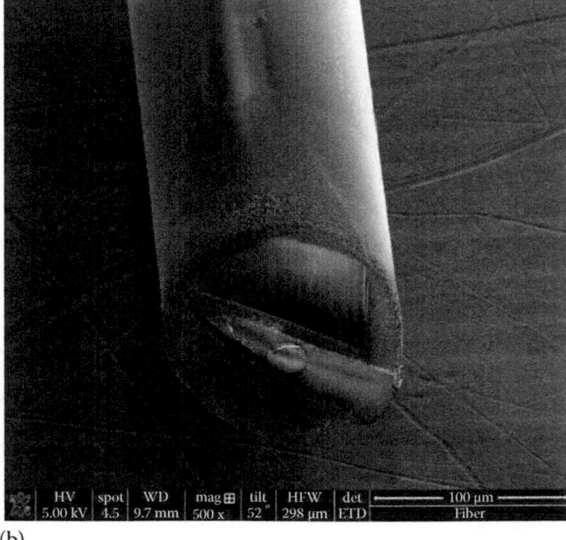

(b)

FIGURE 14.14 (a) Concept microcantilever on the end of a 125 μm diameter fiber. (b) Example cantilever sensor fabricated on SMF-28 fiber.

Unfortunately, fabrication using focused ion beam machining is a slow and expensive approach, with a typical sensor taking several hours to manufacture. Published work (Deladi et al. 2006, Said et al. 2008) suggests that a removal rate of 5000–4000 μm³/h in fused silica is possible with an appropriate assist gas. Although this approach is acceptable for low-volume high-value sensor manufacture, for research purposes, for example, there is little scope of this becoming an affordable sensor fabrication process for more widespread use. Researchers have investigated alternatives such as fabrication using laser machining (Said et al. 2008) and self-aligned UV photolithography (Petrušis et al. 2009). Both of these have potential to dramatically increase the production rate, but with the possible loss of the high resolution and fine optical finish achievable using focused ion beam machining techniques.

14.7.5 Modal Interferometers

The majority of fiber interferometer sensors exploit single-mode fiber to provide a convenient means of defining optical paths in a well-known and predictable manner. However, by allowing more than one mode to propagate over a given length of fiber, it is possible to construct an interferometer where the two arms of the interferometer are formed by light propagating in two different modes. This has been demonstrated using microstructured fibers (Aref et al. 2009), few moded fiber (Kumar et al. 2001), and by using fiber tapers to couple light between core and cladding modes (Amaral et al. 2011). The latter approach is shown in Figure 14.15, where the interferometer is illuminated with a two wavelength source formed by combining a broadband source with FBG filters. Here wavelength division multiplexing is used to measure the two wavelengths separately.

14.8 Concluding Remarks

Despite advances in other fiber sensor technologies, point measurement using fiber interferometers remains of interest for niche applications where it may be inappropriate to use more convenient techniques such as FBGs or distributed measurements. This includes high temperature applications, at which conventional FBGs may become unreliable, high spatial resolution measurements beyond that achievable using conventional FBGs and distributed measurements, and high temporal bandwidth measurements.

In many cases, it is clear that the potential advantage of sensor miniaturization is one of the driving forces behind fiber interferometers. The realization of

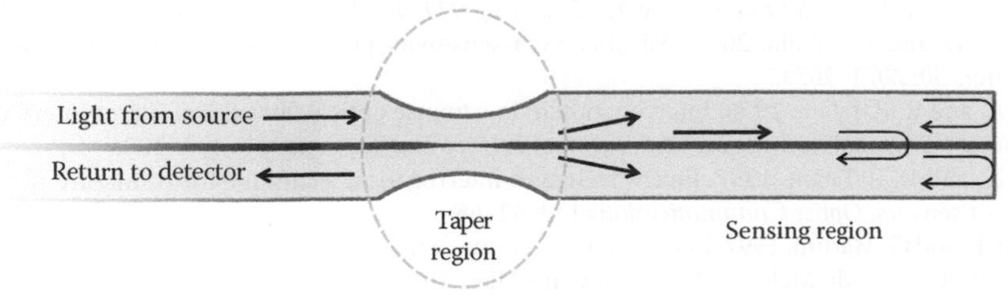

FIGURE 14.15 Fiber taper sensor design to couple light from core into cladding modes.

such miniaturized sensors has required exploitation of advanced manufacturing processes—laser material processing, semiconductor processing, lithography, and ion-beam milling. In some cases, even these approaches are unsuitable either because of technical limitations, or cost implications, and new fabrication routes are being explored that may prove to be suitable for precision manufacturing on a micron scale, but with mass manufacture potential.

References

Amaral, L. M. N., O. Frazão, J. L. Santos et al. 2011. Fiber-optic inclinometer based on taper Michelson interferometer. *IEEE Sensors Journal* 11: 1811–1814.

Aref, S. H., O. Frazão, P. Caldas et al. 2009. Modal interferometer based on ARROW fiber for strain and temperature measurement. *IEEE Photonics Technology Letters* 21: 1636–1638.

Bashir, R., J. Z. Hilt, O. Elibol et al. 2002. Micromechanical cantilever as an ultrasensitive pH microsensor. *Applied Physics Letters* 81: 3091–3093.

Beard, P. C., A. Hurrell, E. Van Den Elzen et al. 1998. Comparison of a miniature ultrasonic optical fibre hydrophone with PVDF hydrophone technology. *Proceedings IEEE Ultrasonics Symposium*, Sendai, Japan, pp. 1881–1884.

Bucaro, J. A., H. D. Dardy, and E. F. Carome. 1977. Fiber-optic hydrophone. *Journal of the Acoustic Society of America* 62: 1302–1304.

Carolan, T. A., R. L. Reuben, J. S. Barton et al. 1997. Fiber-optic Sagnac interferometer for noncontact structural monitoring in power plant applications. *Applied Optics* 36: 380–385.

Chen, J.-H., X.-G. Huang, J.-R. Zhaon et al. 2010. Fabry–Perot interference-based fiber-optic sensor for small displacement measurement. *Optics Communications* 283: 3315–3319.

Cibula, E. and D. Donlagic. 2007. In-line short cavity Fabry–Perot strain sensor for quasi distributed measurement utilizing standard OTDR. *Optics Express* 15: 8719–8730.

Cranch, G. A., P. J. Nash, and C. K. Kirkendall. 2003. Large-scale remotely interrogated arrays of fiber-optic interferometric sensors for underwater acoustic applications. *IEEE Sensors Journal* 3: 19–30.

Creath, K. 1988. Phase-measurement interferometry techniques. In *Progress in Optics XXVI*, E. Wolf (ed.). Amsterdam, the Netherlands: Elsevier, pp. 349–393.

Culshaw, B. 2004. Optical fiber sensor technologies: Opportunities and-perhaps-pitfalls. *Journal of Lightwave Technology* 22: 39–50.

Culshaw, B. 2006. The optical fibre Sagnac interferometer: An overview of its principles and applications. *Measurement Science and Technology* 17: R1–R16.

Dakin, J. P., W. Ecke, K. Schroeder et al. 2009. Optical fiber sensors using hollow glass spheres and CCD spectrometer interrogator. *Optics and Lasers in Engineering* 47: 1034–1038.

Davis, M. A. and A. D. Kersey. 1995. Application of a fiber Fourier transform spectrometer to the detection of wavelength-encoded signals from Bragg grating sensors. *Journal of Lightwave Technology* 13: 1289–1295.

De Freitas, J. M. 2011. Recent developments in seismic seabed oil reservoir monitoring applications using fibre-optic sensing networks. *Measurement Science and Technology* 22: 052001.

Deladi, S., D. Iannuzzi, V. J. Gadgil et al. 2006. Carving fiber-top optomechanical transducers from an optical fiber. *Journal of Micromechanics and Microengineering* 16: 886–889.

Di Giovanni, M. 1992. *Flat and Corrugated Diaphragm Design Handbook*. New York: Marcel Dekker.

Donlagic, D. and E. Cibula. 2005. All-fiber high-sensitivity pressure sensor with SiO_2 diaphragm. *Optics Letters* 30: 2071–2073.

Ezbiri, A. and R. P. Tatam. 1996. Interrogation of low finesse optical fibre Fabry–Pérot interferometers using a four wavelength technique. *Measurement Science and Technology* 7: 117–120.

Ezbiri, A. and R. P. Tatam. 1997. Five wavelength interrogation technique for miniature fibre optic Fabry–Pérot sensors. *Optics Communications* 133: 62–66.

Faustini, L. and G. Martini. 1997. Bendloss in single-mode fibers. *Journal of Lightwave Technology* 15: 671–679.

Ferreira, E. C., F. F. de Melo, and J. A. Siqueira Dias. 2007. Precision analog demodulation technique for open-loop Sagnac fiber optic gyroscopes. *Review of Scientific Instruments* 78: 024704.

Furstenau, N. and M. Schmidt. 1998a. Fiber-optic extrinsic Fabry–Perot interferometer vibration sensor with two-wavelength passive quadrature readout. *IEEE Transactions on Instrumentation and Measurement* 47: 143–147.

Furstenau, N. and M. Schmidt. 1998b. Interferometer vibration sensor with two-wavelength passive quadrature readout. *IEEE Transactions on Instrumentation and Measurement* 47: 143–147.

Harrison, P. B., R. R. J. Maier, J. S. Barton et al. 2005. Component position measurement through polymer material by broadband absolute distance interferometry. *Measurement Science and Technology* 16: 2066–2071.

Hughes, R. and J. Jarzynski. 1980. Static pressure sensitivity amplification in interferometric fiber-optic hydrophones. *Applied Optics* 19: 98–107.

Iannuzzi, D., K. Heeck, M. Slaman et al. 2007. Fibre-top cantilevers: Design, fabrication and applications. *Measurement Science and Technology* 18: 3247–3252.

Jackson, D. A., R. Priest, A. Dandridge et al. 1980. Elimination of drift in a single-mode optical fiber interferometer using a piezoelectrically stretched coiled fiber. *Applied Optics* 19: 2926–2929.

Jiang, Y. 2008. Fourier transform white-light interferometry for the measurement of fiber-optic extrinsic Fabry–Perot interferometric sensors. *IEEE Photonics Technology Letters* 20: 75–77.

Kidd, S. R., J. S. Barton, and J. D. C. Jones. 1995. Demonstration of optical fiber probes for high bandwidth thermal measurements in turbomachinery. *Journal of Lightwave Technology* 13: 1335–1339.

Kilpatrick, J. M., W. N. MacPherson, J. S. Barton et al. 2000. Phase-demodulation error of a fiber-optic Fabry–Perot sensor with complex reflection coefficients. *Applied Optics* 39: 1382–1388.

Kilpatrick, J. M., W. N. MacPherson, J. S. Barton et al. 2002. Measurement of unsteady gas temperature with optical fibre Fabry–Perot microsensors. *Measurement Science and Technology* 13: 706–712.

Kumar, A., N. K. Goel, and R. K. Varshney. 2001. Studies on a few-mode fiber-optic strain sensor based on LP01-LP02 mode interference. *Journal of Lightwave Technology* 19: 351–358.

Li, T., A. Wang, K. Murphy et al. 1995. White-light scanning fiber Michelson interferometer for absolute position-distance measurement. *Optics Letters* 20: 785–787.

Lo, Y.-L. and J. S. Sirkis. 1997. Fabry–Perot sensors for dynamic studies using spectrally based passive quadrature signal processing. *Experimental Mechanics* 37: 119–125.

Luke, D. G., R. McBride, J. G. Burnett, A. H. Greenaway, and J. D. C. Jones. 1995. Polarization maintaining single-mode fibre piezo-electric phase modulators. *Optics Communications* 121(4–6): 115–120. doi: 10.1016/0030-4018(95)00569-4.

MacPherson, W. N., S. R. Kidd, J. S. Barton et al. 1997. Phase demodulation in optical fibre Fabry–Perot sensors with inexact phase steps. *IEE Proceedings on Optoelectronics* 144: 130–133.

Maier, R. R. J., B. J. S. Jones, J. S. Barton et al. 2007. Fibre optics in palladium-based hydrogen sensing. *Journal of Optics A: Pure and Applied Optics* 9: S45–S59.

Martini, G. 1987. Analysis of a single-mode optical fibre piezoceramic phase modulator. *Optical and Quantum Electronics* 19: 179–190.

Melngailis, J. 1987. Focused ion beam technology and applications. *Journal of Vacuum Science and Technology B* 5: 469.

Murphy, K. A., M. F. Gunther, A. M. Vengsarkar et al. 1992. Fabry–Perot fiber-optic sensors in full-scale fatigue testing on an F-15 aircraft. *Applied Optics* 31: 431–433.

Oh, K. D., A. Wang, and R. O. Claus. 2004. Fiber-optic extrinsic Fabry–Perot dc magnetic field sensor. *Optics Letters* 29: 2115–2117.

Pei, J., F. Tian, and T. Thundat. 2004. Glucose biosensor based on the microcantilever. *Analytical Chemistry* 76: 292–297.

Petrušis, A., J. H. Rector, K. Smith et al. 2009. The align-and-shine technique for series production of photolithography patterns on optical fibres. *Journal of Micromechanics and Microengineering* 19(4): 047001.

Ran, Z., Z. Liu, Y. Rao et al. 2011. Miniature fiber-optic tip high pressure sensors micromachined by 157 nm laser. *IEEE Sensors Journal* 11: 1103–1106.

Ran, Z., Y. Rao, J. Zhang, Z. Liu, and B. Xu. 2009. A miniature fiber-optic refractive-index sensor based on laser-machined Fabry–Perot interferometer tip. *Journal of Lightwave Technology* 27(23): 5426–5429. doi: 10.1109/JLT.2009.2031656.

Rao, Y. J. 2006. Recent progress in fiber-optic extrinsic Fabry–Perot interferometric sensors. *Optical Fiber Technology* 12: 227–237.

Rao, Y.-J, M. Deng, D.-W. Duan et al. 2007. Micro Fabry–Perot interferometers in silica fibers machined by femtosecond laser. *Optics Express* 15: 14123–14128.

Said, A. A., M. Dugan, S. de Man et al. 2008. Carving fiber-top cantilevers with femtosecond laser micromachining. *Journal of Micromechanics and Microengineering* 18: 035005.

Santos, J. L., A. P. Leite, and D.A. Jackson. 1992. Optical fiber sensing with a low-finesse Fabry–Perot cavity. *Applied Optics* 31: 7361–7366.

Sasaki, O. and K. Takahashi. 1988. Sinusoidal phase modulating interferometer using optical fibers for displacement measurement. *Applied Optics* 27: 4139–4142.

Schmidt, M., B. Werther, N. Fuerstenau et al. 2001. Fiber-optic extrinsic Fabry–Perot interferometer strain sensor with <50 pm displacement resolution using three-wavelength digital phase demodulation. *Optics Express* 8: 475–480.

Sedlar, M., V. Matejec, and I. Paulicka. 2000. Optical fibre magnetic field sensors using ceramic magnetostrictive jackets. *Sensors and Actuators A: Physical* 84: 297–302.

Sirkis, J. S., D. D. Brennan, M. A. Putman et al. 1993. In-line fiber étalon for strain measurement. *Optics Letters* 18: 1973–1975.

Totsu, K., Y. Haga, and M. Esashi. 2005. Ultra-miniature fiber-optic pressure sensor using white light interferometry. *Journal of Micromechanics and Microengineering* 15: 71–75.

Wang, W., N. Wu, Y. Tian et al. 2010. Miniature all-silica optical fiber pressure sensor with an ultrathin uniform diaphragm. *Optics Express* 18: 9006–9014.

Watson, S., M. J. Gander, W. N. MacPherson et al. 2006a. Laser-machined fibers as Fabry–Perot pressure sensors. *Applied Optics* 45: 5590–5596.

Watson, S., W. N. MacPherson, J. S. Barton et al. 2006b. Investigation of shock waves in explosive blasts using fibre optic pressure sensors. *Measurement Science and Technology* 17: 1337–1342.

Wenfeng, X. and C. Lee. 2009. Nanophotonics sensor based on microcantilever for chemical analysis. *IEEE Journal of Selected Topics in Quantum Electronics* 15: 1323–1326.

Xu, J., G. Pickrell, X. Wang et al. 2005. A novel temperature-insensitive optical fiber pressure sensor for harsh environments. *IEEE Photonics Technology Letters* 17: 870–872.

Zhu, Y. and A. Wang. 2005. Miniature fiber-optic pressure Sensor. *IEEE Photonics Technology Letters* 17: 447–449.

Zuurbier, P., S. de Man, G. Gruca et al. 2011. Measurement of the Casimir force with a ferrule-top sensor. *New Journal of Physics* 13: 023027.

15. Fiber-Optic Sensor Multiplexing Principles

Geoffrey A. Cranch
US Naval Research Laboratories

Chapter 15

Handbook of Optical Sensors. Edited by José Luís Santos and Faramarz Farahi © 2015 CRC Press/Taylor & Francis Group, LLC. ISBN: 9781439866856.

15.1 Introduction

The development of fiber-optic sensors began in the late 1970s and progressed in parallel to the development of fiber-optic-based telecommunication systems. This exciting new technology promised many benefits over existing sensor technology such as the immunity of optical fiber to electromagnetic interference (EMI) and the potential to apply a vast range of optical effects to the detection of various phenomena. Although these techniques were well understood, many of them (such as linear/nonlinear scattering, electro-optic, and magneto-optic effects) were impractical when applied in bulk silica glass due to the short interaction lengths. Research efforts increased considerably through the 1980s and 1990s as practical light sources and fiber-based components (such as connectors, couplers, multiplexers, modulators, switches, and detectors) became available.

Many fiber-optic sensors developed in the early 1980s used multimode fiber, which was also the standard for telecommunication systems at the time. Multimode optical fiber supports many transverse modes due to the core diameter being more than 30–50 times larger than the wavelength, which also makes them easy to splice and launch light into. Ray tracing can effectively represent the propagation, where total internal reflection results in guidance along the fiber for rays above the critical angle. Each ray reflects at the core–cladding interface at a characteristic angle. As a result, propagation in multimode fiber is highly dispersive due to the large variation in transit time of each ray through the fiber (known as modal dispersion). The polarization of each reflected ray also depends on the plane of incidence, such that although each mode exhibits its own well-defined polarization properties, the polarization behavior of multiple modes is very complex. For these reasons, multimode fibers are only suitable for implementing incoherent sensing techniques, such as intensity sensors. These limitations forced the community to adopt the use of single mode fiber some time before its adoption in telecommunication system research in order to exploit other sensing modalities (Hecht 1999). In single-mode fiber, the core diameter is typically up to 10 times the wavelength, such that guidance of only one mode is supported. These fibers can transmit radiation at the lowest loss region of silica optical fiber (less than 0.2 dB/km at 1550 nm) and by increasing the difference in refractive index between the core and cladding are extremely resistant to bending (due to the high mode confinement) (Harris and Castle 1986). These fibers can be used to implement coherent optical sensors using interferometry as well as polarimetric sensors. In the last decade, a new class of optical fiber known as microstructured or photonic crystal fiber (Russell 2006) has attracted considerable interest. The inclusion of air holes within the cladding or core makes possible many new sensing concepts, particularly for gas and chemical sensing (Monro et al. 2001).

Although fiber-optic sensors can be found in many highly specialized sensing applications (e.g., biomedical, food quality control), individual sensors are less commonly found in widespread use. Ever since the first demonstrations in the late 1970s and up until the present day, fiber-optic sensors (and probably most optical sensors) often struggle to compete commercially due to the relatively high cost of components, when other more conventional (and usually lower cost) sensor technologies are available. Two notable exceptions, which are probably the most commercially successful individual fiber-optic sensor, are the fiber-optic gyroscope and the fiber-optic current sensor. These owe their success to being able to outperform other sensing technologies in a reliable, compact sensing unit. In the case of the fiber-optic current sensor, its ability to perform reliably in very high electromagnetic field environments is a significant advantage. It was, however, recognized early on that one of the inherent benefits of fiber-optic sensors over other technologies is the considerable potential for multiplexing. The high cost of individual components can be circumvented with the efficiency by which, in some cases, hundreds of sensors can be multiplexed onto a single optical fiber. Consequently, the majority of other successful fiber-optic sensors are highly multiplexed systems for military surveillance, petrochemical subsea monitoring, and large-scale monitoring systems for civil infrastructure and perimeter security.

Multiplexing methods were investigated very soon after the acceptance of fiber-optic sensors as a viable sensing technology in the early 1980s. New techniques are still being developed today illustrating the huge scope of methods available. To a newcomer in the field, the sheer number of reported techniques most likely appears somewhat daunting, especially when one desires simply to identify a suitable technique for a given application. However, as this chapter attempts to demonstrate, many of these techniques have strong underling similarities and are not unique to fiber-optic sensors. Their origins can be traced to fields as diverse as radar, telecommunications, laser ranging, and transmission line analysis.

This chapter will discuss the basic concepts for multiplexing fiber-optic sensors. Its focus will be on multiplexing discrete sensors, although some of the concepts are directly applicable to distributed fiber-optic sensors described in the next chapter. The objective is to provide a basic understanding of a broad range of multiplexing methods and emphasize their similarities. In Section 15.2, the basic multiplexing architecture is introduced and relevant terminology used to characterize multiplexed systems is defined. Section 15.3 describes components found in such systems. Section 15.4 reviews sensing techniques and is intended to serve as a brief summary, as many of these techniques have been described in detail in previous chapters. These will enable us to show how these techniques can be incorporated into multiplexed systems. Section 15.5 describes and compares the basic multiplexing topologies upon which the majority of multiplexed sensor systems are based. Section 15.6 covers specific multiplexing techniques, dividing them up into spatial-, time-, frequency-, wavelength-, and coherence-based systems. Finally, by way of illustrating in more detail the operating principle of a prototype system, Section 15.7 describes an acoustic surveillance array and shows how its performance can be analyzed. Data taken during a field trial is presented to further illustrate the performance of such a system. Other useful references to book chapters on the subject can be found in the Further Reading section at the end.

15.2 Basic Multiplexing Concepts and Definitions

15.2.1 Basic Concepts

Figure 15.1 illustrates two types of multiplexed sensor arrays: (a) discretely multiplexed sensors and (b) a fully distributed sensor array. In the former case, the sensor discretely samples the local measurand field, X_n, at a fixed spacing. For fields with a rapid spatial

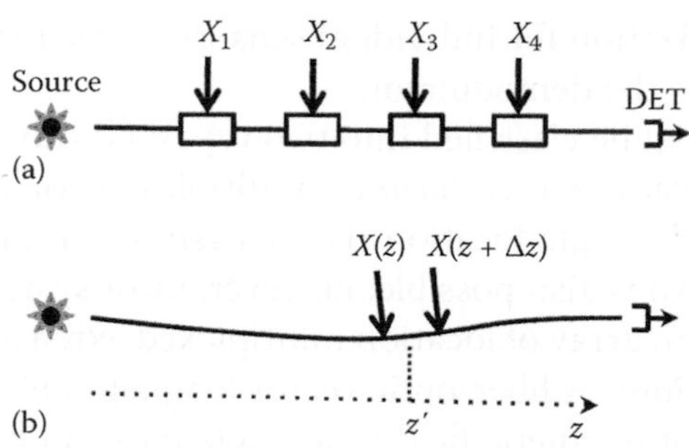

FIGURE 15.1 (a) Discrete sensor multiplexing and (b) fully distributed sensor multiplexing.

variation, the sensor will integrate the measurand over its active surface area. In a fully distributed system, the sensor is continuous along the length of the fiber. The sensor responds to the measurand field, $X(z)$, integrated over a distance given by the spatial resolution, Δz, such that

$$\overline{X}(z') = \int_{z'-\Delta z/2}^{z'+\Delta z/2} X(z)dz \tag{15.1}$$

In practice, the spatial resolution is determined by the interrogation method and scattering mechanism utilized (e.g., Rayleigh, Raman, Brillouin). Although fully distributed sensor systems offer considerable flexibility and high spatial resolution, discretely multiplexed sensors generally offer higher sensitivity and dynamic range.

A system can be considered multiplexed if a single source or detector is shared by two or more sensors. A generalized multiplexed configuration, shown in Figure 15.2, illustrates the basic components found in a typical system. Here a laser source is injected into a modulator (MOD) that applies an encoding to the optical signal, enabling addressing of individual sensors. For example, an intensity modulator may produce optical pulses for implementation of time division multiplexing (TDM). The output of the modulator is distributed to the sensor array(s) through a multiplexer (MUX), such as a directional coupler (i.e., fiber equivalent of the beam splitter) or an optical switch. The return signals from the array are decoded (DECODER) to convert the optical property into an intensity modulation and detected (DET). Finally, the demodulator (DEMOD) removes the modulation imposed by the modulator and

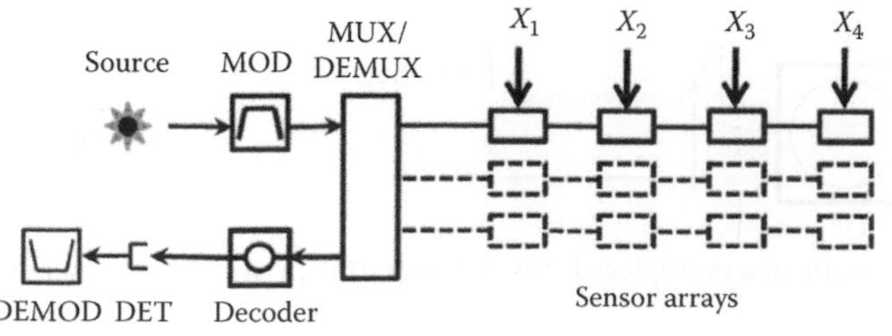

FIGURE 15.2 Generalized configuration of a multiplexed sensor system.

enables response linearization for individual sensors. Some form of signal analysis or recording device follows the demodulator.

In general, systems can be classified into two types: (a) those that measure a single quantity at multiple locations (i.e., *location* multiplexed) or (b) those that measure multiple parameters at a single location (i.e., *measurand* multiplexed). Of course, a combination of these two is also possible; however, most systems fall into the former category. For example, an array of location multiplexed extensometers measure strain at multiple spatial locations. A fiber-optic vector magnetometer is capable of measuring three components of magnetic field in a single location by frequency multiplexing their signal (Bucholtz et al. 1995), and a fiber-optic laser Doppler velocimeter is capable of measuring three components of fluid velocity in a single location (James et al. 1996).

An appropriate starting point is to define the terminology that will be used in the following discussions. A fiber-optic sensor operates by converting the *measurand* (i.e., the quantity to be measured) into a change in a measurable property of the light (e.g., intensity, phase, or polarization). For example, a hydrostatic pressure or inertial force may induce an axial strain in the optical fiber, which modulates the phase of the propagating light. The mechanism implemented to induce this change is known as the transduction mechanism. The *transducer* refers to the sensor head that converts the measurand into a change in an optical property of the light. The *responsivity* (defined formally in the following text) of the transducer quantifies the effectiveness of the transduction mechanism. Sensors may be either *intrinsic* or *extrinsic* in design. In the former case, light does not exit the fiber and thus forms an intrinsic part of the transduction mechanism. This is a highly desirable configuration, since it avoids excess losses and intensity fluctuations associated with coupling to/from the fiber. In the latter case, light may exit the fiber and be either coupled into a second fiber or reflected back into the original fiber after passing through the sensor. Some examples are extrinsic Fabry–Pérot sensors (Claus et al. 1992) and intensity sensors based on a reflecting membrane (Bucaro et al. 2005). In some cases, the fiber simply acts to deliver the light to the sensing mechanism and does not necessarily form part of the transduction mechanism.*

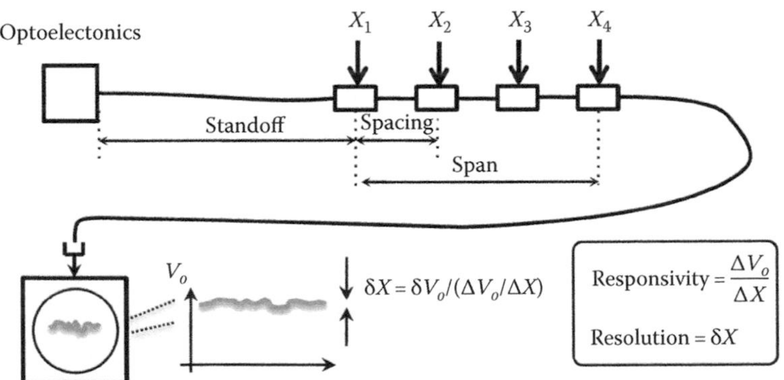

FIGURE 15.3 Schematic of a multiplexed system illustrating performance parameters.

* This distinction can become ambiguous in some cases such as evanescent field sensors where light does not fully exit the fiber but does interact with an external optical material.

15.2.2 System Characteristics and Specifications

The performance of a sensor system can be defined by a number of parameters. These can be separated into parameters describing the multiplexed system and those of the individual sensors. A selection of these parameters is indicated schematically in Figure 15.3.

15.2.2.1 System Characteristics

1. The *multiplexing gain* is defined as the number of sensors that can be multiplexed onto a single or pair of optical fibers. In many situations, it is of interest to know how many sensors can be accommodated by a given number of fibers. Although it is possible to multiplex hundreds of sensors onto a single optical fiber, in some cases, this may not always be the most effective approach. Most fiber-optic cables carry several fibers, and for reasons of redundancy, it may be preferable to multiplex fewer sensors onto multiple fibers.
2. The *sensor system span* typically refers to the total extent of the sensing array. In many practical situations, a sensor array will cover a large physical area. For example, a condition monitoring system may be required to measure several parameters at different points along a length of pipeline or an intruder detection system may cover a large perimeter or fence.
3. The *standoff distance* (sometimes called download length) typically refers to the distance between the interrogation system and the first sensor in the array. An inherent benefit of fiber-optic systems is their ability to transmit sensor information over very large distances, due to the very low transmission loss in optical fiber (<0.2 dB/km at 1550 nm in single mode fiber or about 3 dB/km at 1 μm in multimode fiber). Consequently, the interrogation system can be located far from the sensor array, which in some cases can be up to 100 km.
4. It is often desirable to know the maximum and minimum permissible *sensor spacing*. Typically, it is required to measure either a single parameter or multiple parameters at various spatial locations. In some cases, such as condition or structural monitoring, this spacing may not be even. In others such as security systems or acoustic sensing arrays, this spacing will be constant. The minimum sensor spacing also defines the spatial resolution.

15.2.2.2 Sensor Characteristics

1. The *transducer responsivity* is the change in the optical property, carrying the sensor information, produced per unit change in the parameter to be measured. The term *sensitivity* is sometimes used with the same meaning; however, this can lead to confusion due to its association with the resolution.
2. The *resolution* defines the minimum resolvable change in the measurand.
3. The *accuracy* is the difference in the measurand as measured by the sensor and the true value of the measurand.
4. The *measurement bandwidth* refers to the maximum electronic bandwidth over which a measurement can be made. It also indicates the maximum rate of change of the measurand that can be resolved by the sensor.

5. The *linearity* is a measure of the deviation of the sensor response from an exactly linear response. It is generally desirable to design a sensor with an inherently linear response (this is important for ac sensors). However, a nonlinear (but monotonic) response can be linearized with suitable calibration for static measurements.
6. The *dynamic range* is defined as the ratio of the maximum value to the minimum value of the measurand that can be measured within the specified accuracy or linearity. The minimum is usually determined by the system noise. High levels of background noise can further degrade the dynamic range. The maximum level may be defined by a gain compression value or predefined deviation from linearity.
7. *Crosstalk* in a sensor array refers to the appearance of signals injected onto one sensor on other sensors. Crosstalk may arise from internal optical effects or external mechanical effects in smaller, closely spaced sensors arrays. It is often defined in terms of the ratio of a narrowband signal injected on one sensor to that appearing on another sensor.

15.2.2.3 Other Useful Characteristics

1. The *power budget* is not considered a performance metric; however, it is often used to determine many of the other array properties such as multiplexing gain, sensor system span, standoff distance as well as the resolution and dynamic range. It quantifies the change in optical power at consecutive locations in the sensor system. Usually written in tabular form, the power budget quantifies the optical losses and gain in the system, allowing the minimum power level incident on the detector to be determined.
2. The *duty cycle* applies specifically to time division multiplexed systems. Here, a pulse of width, τ, interrogates an individual sensor at a rate $1/\tau_1$ s^{-1}. The duty cycle is given by $\tau/(\tau_1-\tau)$. The interrogation rate $1/\tau_1$ sets an upper limit for the sensor bandwidth.
3. The *array insertion loss* quantifies the power loss due to splitting of the light between sensors in an array and is used to calculate the power budget. A common design trade-off involves minimizing the insertion loss without increasing complexity or compromising performance of the sensor system. The sensitivity is ultimately limited by the received power level.
4. The *extinction ratio* is used to quantify the ability of an optical switch to extinguish the transmitted light. It can be defined as the ratio of transmitted power in the *on* state to that in the *off* state.

These parameters broadly define the performance of most systems. Caution should be applied when comparing performance specifications of systems from different commercial suppliers as a lack of standard terminology results in variation in the definition of some parameters.

15.3 Components for Multiplexing

Implementing optical sensing techniques requires generating, manipulating, and modulating the light, which requires fiber-coupled optical components. A brief description of commonly used components is now given. More detailed descriptions of the physics of these devices can be found in popular books (e.g., Rogers, *Principles of Optoelectronics*, and Davis, *Lasers and Electro-Optics*).

15.3.1 Sources of Optical Radiation

The following sections describe optical sources commonly used in fiber-optic sensing systems. Selected properties of these sources are later tabulated in Table 15.1.

15.3.1.1 *p–n* Junction Devices

A simple luminescent light source is the *surface-emitting light-emitting diode* (SELED) consisting of a *p–n* junction in a direct-bandgap semiconductor such as gallium arsenide (GaAs) or gallium phosphide (GaP). Injection of current under forward bias into the diode will cause emission of photons with energy equal to the bandgap. Variation in the bandgap energy leads to emission over wavelength ranges of ~10–30 nm. These are low cost, rugged, and reliable devices capable of being easily coupled to multimode fiber. Recent interest in light-emitting diodes (LEDs) for general lighting applications has driven the development of devices covering the visible wavelength range through the infrared. Another type of LED is the *edge-emitting LED* (ELED), which emits radiation with a smaller beam divergence, enabling higher coupling efficiencies into single-mode optical fiber and over a smaller wavelength range. LEDs produce output powers of a few milliwatts coupled into a multimode optical fiber and are available over a very broad wavelength range (GaAs or AlGaAs for 850 nm, InGaAsP, and InP for 1310 and 1550 nm emission). Another related device is the *superluminescent diode* (SLED), which operates with the same principle as the LED but in a superluminescent mode. Applying a forward bias to a highly doped *p–n* junction generates amplified spontaneous emission (ASE), which by design of the junction covers a broad range of wavelengths. These devices emit much higher powers than LEDs and ELEDs but are quite sensitive to feedback, and their spectra also depend strongly on the junction temperature. Emission bandwidths can range from 5 to 100 nm centered from 400 to 1700 nm with output power of several milliwatts for injection current of several hundred mA.

A *semiconductor laser diode* (SLD) can be formed by adding reflective facets to a SLED, to obtain a population inversion and hence lasing. Single-mode laser diodes can be made from distributed feedback structures and can emit around 50 mW with several hundred mA injection current. External cavity laser diodes provide a linewidth narrowed emission for coherent applications as well as tunability.

All of these devices can be current modulated, yielding both frequency and amplitude modulation. Frequency modulation is most efficient at low frequencies due to a thermal tuning mechanism. Amplitude modulation at frequencies exceeding a GHz is possible.

Table 15.1 Characteristics of Selected Optical Sources

Type	Spectral Bandwidth	Fiber-Coupled Power (mW)	Cost
SELED	10–30 nm	Few	<$10
ELED	10–30 nm	Few	<$'00
SLED	≫20 nm	Up to 10	$'00s
SLD	<1 MHz	Up to 50	~$1k
Fiber Laser	<1 kHz	Up to 100	$10–20k
EDFA	~35 nm	≫10	Few $k up
Nonplanar Ring Laser	<1 kHz	>100	>$20k

15.3.1.2 Fiber-Based Sources

Fiber sources of radiation are made by integrating a gain medium within the core of the optical fiber. This can be achieved by adding a suitable dopant to the preform prior to drawing the fiber. For example, optical radiation around 1.0–1.1 μm is produced with ytterbium, 1.25–1.6 μm with erbium (Er), or erbium and ytterbium codoping (Er:Yb), 1.3 μm with praseodymium, 1.32–1.35 μm with neodymium, and 1.7–2 μm with thulium doping. For optical fiber sensors, erbium fiber sources are most common due to their wide availability and relatively low cost. Such devices can be used as either sources of radiation or amplifiers, known as erbium doped fiber amplifiers (EDFA). Erbium-doped fiber fluoresces over a 35 nm bandwidth centered on ~1542 nm with output powers greatly exceeding 10 mW and can be conveniently pumped with laserdiodes at 980 or 1480 nm. Incorporating feedback elements enables lasing action at either single or multiple frequencies. Single frequency fiber lasers (similar to the fiber laser sensors described in Section 15.6.4.3) offer extremely narrow linewidth (~1 kHz) and hence very long coherence lengths, making them particularly well suited for implementing coherent sensors based on interferometry. Erbium fiber sources cannot be directly modulated due to the long lifetime of the excited state (~12 ms); thus, external modulation is generally required. However, it is possible to achieve low frequency (~kHz) modulation of the emission frequency of fiber lasers by incorporating a piezoelectric element into the cavity.

Another class of fiber laser that finds utility in sensor multiplexing is mode-locked lasers (MLLs). In this laser, the many longitudinal modes are forced to be phase synchronous (many mechanisms exist to achieve this, descriptions of which can be found in texts on laser physics). In this regime, the laser emission may be transform limited where the temporal emission is the Fourier transform of the emission spectrum and in general comprises a periodic pulse train. MLLs with long cavities and broad emission bandwidth are well suited to interrogating fiber-optic sensors by time and wavelength division multiplexing (WDM). One particular benefit of using MLLs to generate pulses of radiation is their high efficiency, as all the stored energy within the resonant cavity is emitted within a single pulse. One such MLL is capable of generating pulse powers of several Watts over an optical bandwidth of 65 nm, for pulse durations of 10 ns and repetition rates of 300 kHz, powered from a single laserdiode drawing ~0.4 W of electrical power (Matsas et al. 1992, Putnam et al. 1998).

15.3.1.3 Solid-State Lasers

Another class of high-performance laser is the Nd:YAG nonplanar ring laser (NPRO) operating at 1319 nm. This laser source exhibits a very narrow linewidth (<1 kHz) with powers greater than 100 mW and is used only for very high-performance applications due to its relatively high cost. It is generally considered a benchmark for high-performance lasers. This exceptional performance has resulted in its use in gravitational wave sensing (Heurs et al. 2004).

15.3.2 Optical Modulators

15.3.2.1 Amplitude (or Envelope) Modulation

The p–n-junction-based optical sources can be amplitude modulated by direct current modulation. External modulation can be achieved with an integrated optic modulator made from an electro-optic crystal such as lithium niobate. Forming a Mach–Zehnder interferometer (MZI) from this material and applying an external electric field across one arm causes a change in the refractive index of the arm through the *Pockels* effect (a linear electro-optic effect). Intensity modulation can thus be achieved by switching the phase in one arm by π rad, causing the output intensity to flip from maximum to minimum when the bias phase is 0 modulo 2π. Switching voltages, V_π around a few volts are possible and switching times can exceed GHz. Extinction ratios (i.e., ratio of on to off) are typically around 30 dB and can be increased with multistage devices. Amplitude modulation can also be achieved with *mechanical switches* based on microelectromechanical system (MEMS) devices. These are capable of high extinction ratios; however, their switching time is limited to greater than ~1 ms, restricting their use to low speed sensor systems.

Acousto-optic modulators (AOMs) can be used for both frequency modulation and switching. In this device, an acoustic wave is excited in an optical medium, forming a volume diffraction grating or hologram. An incoming optical signal is diffracted by the grating at an angle related to the laser wavelength, acoustic wavelength, and angle of incidence. The hologram thus acts as a beam deflector and yields very high extinction ratio switching (due to the light being physically shifted to a different position). The switched light also experiences a Doppler shift equal to the acoustic frequency, due to its interaction with a traveling acoustic wave. Fiber-coupled devices are commercially available with insertion losses less than 5 dB and drive frequencies up to 100 MHz. Switching times are on the order of 50 ns limited by the bulk acoustic velocity of the medium. High electrical drive powers (several Watts) and availability of low phase noise radio frequency (RF) sources make these devices more challenging to operate.

An elegant device for high speed, high extinction ratio switching over a broad optical bandwidth is the *semiconductor optical amplifier (SOA)*. This device is broadly similar to the *p–n* junction laser diode; however, fiber coupling at each end of the junction allows an optical signal to be coupled in and out of the device with negligible feedback. With a forward biased junction, an optical signal will experience gain due to stimulated emission. In the absence of an electrical bias, the optical signal will be strongly attenuated due to absorption. Thus, driving the junction with a switched injection current will yield amplitude modulation with a high extinction (>50 dB for peak injection currents of a few 100 mA). The broad emission bandwidth of the SOA also permits broadband operation without compromising the extinction ratio. Switching times are much less than 1 ns.

15.3.2.2 Frequency/Phase Modulation

Frequency is the rate of change of the phase; thus, frequency and phase modulation are closely related. A phase modulation given by $\Delta\phi(t) = \phi_0\sin(\omega_m t)$ yields a frequency modulation, $\Delta\nu(t) = \phi_0\omega_m\cos(\omega_m t)$, where the frequency modulation depth $\nu_0 = \phi_0\omega_m$. Thus, an appreciable frequency modulation depth at optical frequencies (~10^{14} Hz) implemented through phase modulation is achieved either with a very large phase modulation or a high modulation frequency. Consequently, frequency modulation is more efficiently achieved through direct modulation of the laser cavity. However when this is not possible, one method to achieve narrowband modulation is by applying a large sinusoidal strain to a length of optical fiber. This can be achieved by wrapping the fiber around a large cylindrical piezoelectric tube and applying a sinusoidal voltage at a frequency equal to the mechanical resonance of the first radial mode. The resulting phase modulation can be large enough to achieve appreciable frequency modulation. Placing a Faraday rotation mirror (FRM)

Table 15.2 Characteristics of Selected Modulators

Type	Effect	Responsivity	Bandwidth	Insertion Loss	Cost
Phase (Small wrapped PZT)	Phase	Few rad/V	~20 kHz	Very low	<$100
Phase (Large wrapped PZT)	Phase	~50 rad/V (×100 at f_{res})	f_{res} ~ 18 kHZ	Very low	$'00s
Phase ($LiNbO_3$)	Phase	~1 rad/V	GHz	~5 dB	Few $k
Intensity ($LiNbO_3$)	Intensity	—	GHz	~5 dB	Few $k
Acousto-optic	Frequency, intensity	—	MHz	~5 dB	Few $k
Semiconductor optical amplifier	Intensity	—	GHz	Gain	Few $k

on one fiber end and operating the device in reflection reduces the effect of birefringence modulation that accompanies the phase modulation (Bush et al. 1996). For higher frequency phase modulation, an integrated optic phase modulator similar to that described earlier but comprising a single lithium niobate waveguide can be used.

For low frequency phase modulation, fiber-wrapped piezoelectric cylinders are highly effective with modulation depths of a few rad/V achievable from a meter of wrapped fiber on a 25 mm diameter cylinder (Jackson et al. 1980) over a bandwidth of a few kHz. Characteristics of the modulators discussed earlier are summarized in Table 15.2.

15.3.3 Other Components

Other components commonly found in fiber-optic sensor systems perform basic functions such as splitting and combining light signals to/from several fibers, connecting two fibers, and separating spectral regions in an optical signal (i.e., filtering). The directional coupler is the fiber-optic equivalent of the beam splitter and can be fabricated by fusing two optical fibers together at high temperature. For single mode fiber, the length of the fused region determines the split ratio and the wavelength range over which the coupler can operate. These devices can be fabricated with split ratios ranging from 50% to a few percent, with polarization dependent loss less than 0.1 dB and excess loss less than 0.05 dB. Single and multimode fibers can be fusion spliced or connected using a variety of polished connector types designed for multiple use (i.e., ST or FC/PC) or low back-reflection (i.e., angle polished FC/APC).

Wavelength selective coupling can also be achieved with directional couplers; however, more precise filtering can be achieved with fiber-coupled thin film interference filters, fiber Bragg grating–based (FBG) devices (see Section 15.4.4.1), or arrayed waveguide gratings (Smit and Dam 1996).

15.4 Sensing Techniques

A linearly polarized electromagnetic wave is represented by, $E = E_0\exp(i\omega t)$, which at optical frequencies oscillates at a frequency greater than 10^{14} Hz. An electronic detector is not capable of responding directly to an oscillation rapidly and thus responds to the time-averaged intensity defined by, $I = \langle E \cdot E^* \rangle$, where "*" represents the complex conjugate.

Thus, to measure a property of the optical signal such as frequency, phase, or polarization, these must first be converted to changes in optical intensity, which *can* be readily measured with a photodiode. It has already been shown, in previous chapters, how sensors can be implemented utilizing these optical properties, so a brief summary is now given.

15.4.1 Intensity-Modulated Sensors

As detailed in Chapters 4 and 13, probably, the simplest optical sensor is based on direct intensity modulation, illustrated in Figure 15.4a. Here, the transmitted power of an optical signal is modulated by the measurand, which is measured with a photodetector. Intensity sensors have many benefits:

- They generally are relatively low cost due to their simple implementation utilizing multimode fiber, LED light sources, and direct detection methods.
- Are generally immune to polarization-related effects and effects of source coherence.
- Can be sensitive, although not as sensitive as interferometric methods.
- However, they require a referencing technique to remove effects of source power fluctuations, which increases complexity.

A method to induce propagation loss in an optical signal is illustrated in Figure 15.4b. Here, two fibers are separated by a small distance. One is fixed and the other is allowed to move in response to the measurand, changing the coupled light intensity into the fixed fiber. One implementation of this concept consists of a fiber bundle with a single launch fiber surrounded by multiple return fibers. These capture the light reflected from a deflecting membrane coupled to the measurand (Bucaro et al. 2005). Another implementation utilizes microbending losses, illustrated in Figure 15.4c, where the fiber is periodically perturbed with a serrated surface (Berthold 1995). Light can be made to couple between modes in a multimode fiber when the propagation constants of the modes satisfy the phase matching condition,

$$\beta - \beta_n = \frac{\pm 2\pi}{\Lambda} \tag{15.2}$$

where

β and β_n are the propagation constants of the initial mode and coupled mode and $\beta = 2\pi n_e/\lambda$, where n_e is the effective index of the respective mode

Λ is the period of the perturbation

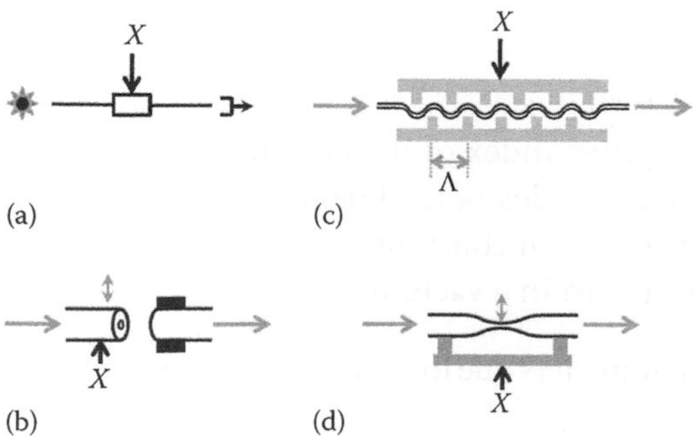

FIGURE 15.4 Intensity-modulated sensor implementations: (a) basic concept, (b) two fiber separated by a gap, (c) microbend sensor, and (d) tapered fiber.

Incorporating a short section of multimode fiber between two single mode fibers causes a change in transmission loss through the device. For the case of a periodically perturbed single mode fiber, microbending losses cause coupling from the fundamental LP_{01} mode to one of the discrete cladding modes, which can be optimized by choice of the period of the microbend (Arya et al. 1995). More recently, an intensity sensor based on changes in transmission of a tapered fiber sensor has been demonstrated, as illustrated in Figure 15.4d (Matías et al. 2003).

Although a simple and potentially low cost concept, intensity-based sensors have some limitations. They are sensitive to changes in the source power and transmission loss in the connecting fiber and therefore require referencing methods to enable accurate and unambiguous measurement. Also, their effect on transmitted power also restricts available multiplexing topologies. Despite these limitations, their low cost combined with relatively high sensitivity and dynamic range makes them attractive for many applications.

15.4.2 Phase-Modulated Sensors

A very sensitive detection mechanism is through modulation of the phase of the optical signal by applying an axial strain to an optical fiber or by applying a transverse stress through the stress-optic effect. As detailed in Chapters 5 and 14, phase modulation–based sensors have several benefits:

- Phase can be measured independent of optical power.
- They achieve very high sensitivity and dynamic range.
- They provide great flexibility for implementing sensors (in terms of gage length and a large choice of interferometric implementations) as well as multiplexing.
- However, this comes with many added complexities in terms of advanced components, complex demodulation techniques and requires sophisticated implementations for dc measurements. Furthermore, interferometric sensors must generally be implemented in single-mode fiber.

Despite these added complexities, interferometric sensors are probably the most developed; thus, their operation will be discussed in more detail.

15.4.2.1 Optical Phase in Fibers

Considering a length of fiber, l, the total phase of the optical signal propagating through the fiber, ϕ, is given by,

$$\phi = n_e k l \tag{15.3}$$

where

n_e is the effective refractive index of the fiber mode (which can be roughly approximated to the refractive index of the fiber core)

k is the free space propagation constant equal to $2\pi/\lambda_0$

λ_0 is the optical wavelength in a vacuum

A change in these three parameters due to the measurand, X, yields a phase shift, $\Delta\phi$, given by

$$\frac{\Delta\phi}{\phi} \simeq \left(\frac{1}{k}\frac{\partial k}{\partial X} + \frac{1}{n_e}\frac{\partial n_e}{\partial X} + \frac{1}{l}\frac{\partial l}{\partial X} \right)\Delta X \tag{15.4}$$

The terms on the right hand side of Equation 15.4 represent a change in phase due to a change in the laser wavelength, change in refractive index, and change in length, respectively. For the case of an axial strain, $\Delta\varepsilon = \Delta l/l$, applied to the fiber, the phase change simplifies to

$$\frac{\Delta\phi_\varepsilon}{\phi} = (0.78)\Delta\varepsilon \tag{15.5}$$

where the factor (0.78) is the stress-optic coefficient for single-mode silica fiber and accounts for a change in the refractive index associated with an applied strain (Butter and Hocker 1978) (the laser wavelength is unaffected by the applied strain). A change in temperature will also cause a phase change due to the thermo-optic effect. The thermally induced phase shift due to a temperature change, ΔT, is given by (Hocker 1979)

$$\frac{\Delta\phi_T}{\phi} = \left(\frac{1}{n_e} \frac{\partial n_e}{\partial T}\bigg|_\rho + \alpha_e \right) \Delta T \tag{15.6}$$

where the term in brackets represents contributions from the temperature-induced refractive index change at constant density and the thermal expansion coefficient, α_e. These are collectively known as the thermo-optic coefficient, which has a value of ~5×10^{-5} K^{-1} for silica optical fiber. The sensitivity of the optical fiber length to temperature makes dc or very low frequency measurements of strain difficult. Such a sensor requires either an independent measure of temperature or carefully matched and co-located isothermal optical paths.

15.4.2.2 Interferometric Methods

An optical phase change can be measured with interferometry. Probably, the simplest configuration is the *MZI*, illustrated in Figure 15.5a. Light from the source is split into the two arms of the interferometer using a fiber directional coupler and recombined using a second coupler. Assuming the coupler power splitting ratio, $\kappa = 50\%$, ignoring losses and taking the launched power as $2P$ then the photocurrent is

$$i_{ph} = r_d P\left[1 \pm V\cos(\phi_s - \phi_r)\right] \tag{15.7}$$

where

- r_d is the photodiode responsivity in AW^{-1} and $r_d = e\eta_d/h\nu$, where e is the electron charge, η_d is the photodiode quantum efficiency (ratio of the number of electron–hole pairs generated for one photon absorption), h is the Planck constant, and ν is the optical frequency
- ϕ_r and ϕ_s represent the phase of the signals from the reference and signal arms, respectively, such that $\phi_s - \phi_r = |\Delta\phi_s| = 2\pi n_e \Delta L/\lambda$, where ΔL is the length difference between the two arms

The second term in Equation 15.7 is the interferometric term, and its amplitude depends on the fringe visibility, V. For a coherent light source, this is determined

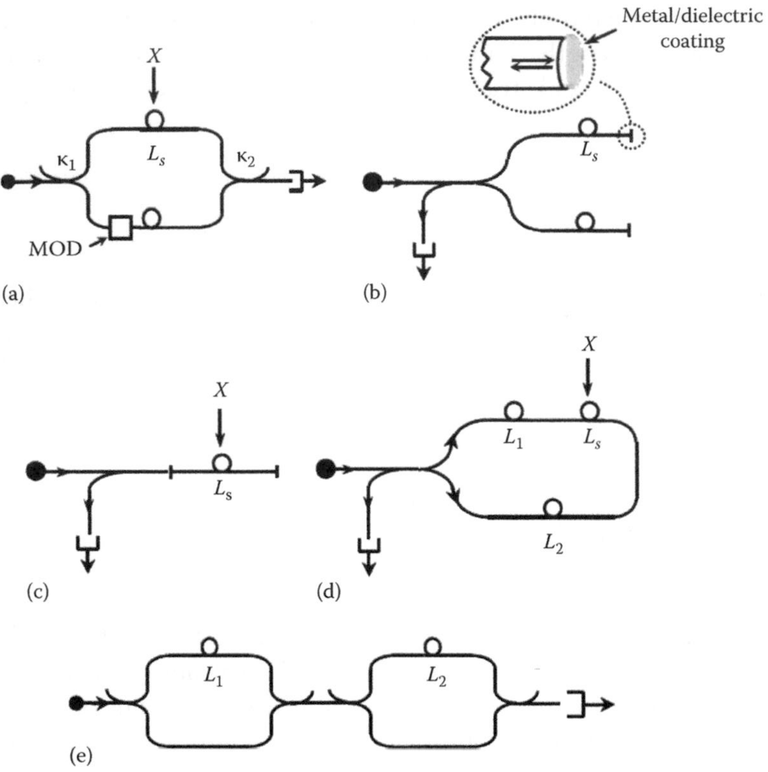

FIGURE 15.5 Interferometric sensors: (a) Mach–Zehnder; (b) Michelson; (c) Fabry–Pérot; (d) Sagnac; and (e) tandem interferometry.

by the relative orientation of the polarization states of the recombined optical signals. For a low coherence light source, the fringe visibility is also related to the path-imbalance in the interferometer and the coherence properties of the source. The sign of the interferometric signal depends on which output of the directional coupler is used.

A similar interferometer configuration known as the *Michelson* interferometer is shown in Figure 15.5b. Here the light is reflected at each fiber end and recombined through the input coupler. The reflectors can be formed by cleaved ends (which yields a ~4% reflection) or with mirrored coatings applied to the fiber ends. Alternatively, mirrors can be formed with fiber loop reflectors (Mortimore 1988), which have the advantage of achieving close to 100% reflectance, FBGs, or FRMs. This latter device acts as an ortho-conjugate mirror (Martinelli 1994), removing the effect of birefringence on the polarization state of the beam propagating in the interferometer arm and therefore maximizing the fringe visibility. This interferometer has the same transfer function (TF) as the MZI given by Equation 15.7. However, the phase in each arm is now doubled due to the forward and return transit through each arm. In this case, $\phi_s - \phi_r = |\Delta\phi_s| = 4\pi n_e \Delta L/\lambda$.

The *fiber Fabry–Pérot* (FFP) interferometer, shown in Figure 15.5c, is a multiple beam interferometer. An interference signal is generated by the reflection from two series reflectors. If we assume that the interferometer is illuminated by a monochromatic source and there is no birefringence or attenuation in the sensor fiber, then in reflection, the FFP produces Fizeau fringes and the power response is given by (Vaughan 1989)

$$P(\phi) = P_{inc}\left[1 - \frac{(1-R_1)(1-R_2)}{1+R_1R_2 - 2\sqrt{R_1R_2}\cos(\phi_s)}\right] \tag{15.8}$$

where
R_1, R_2, are the mirror reflectivities
P_{inc}, is the incident power
$\phi_s = 4\pi n_e L_s/\lambda$

For the case of R_1 and R_2 close to unity, this function exhibits narrow resonant peaks, as illustrated in Figure 15.6a, enabling resonant locking techniques to be employed to track the phase of the cavity as described in Section 15.4.2.4. The parameters $\Delta\nu_c$ and $\Delta\nu_{fsr}$ define the linewidth and free-spectral range, respectively, of the cavity. An interesting situation arises as $R \approx R_1 \approx R_2$ become small, where Equation 15.8 reduces to

$$P \approx 2RP_{inc}(1-\cos(\phi_s)) \tag{15.9}$$

and resembles the response of the two beam interferometer (described by Equation 15.7). This type of interferometer can be implemented with weak FBG reflectors.

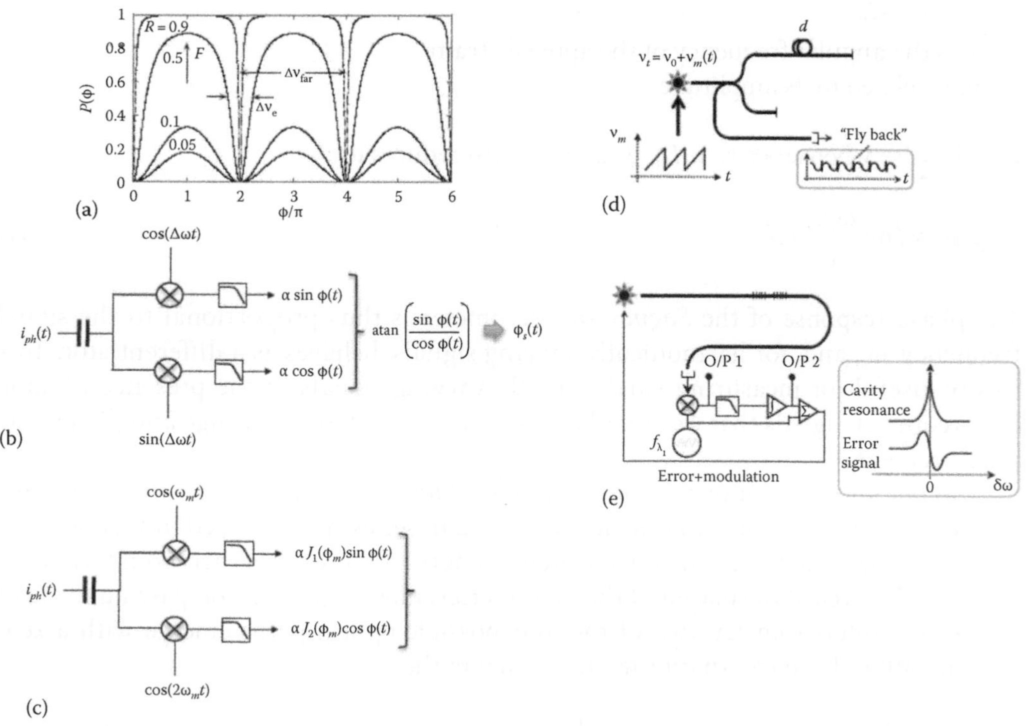

FIGURE 15.6 (a) Response in reflection of the Fabry–Pérot interferometer; (b) signal processing steps for heterodyne interrogation; (c) signal processing steps for phase-generated-carrier interrogation; (d) implementation of pseudo-heterodyne interrogation; (e) and the PDH interrogation technique (inset) PDH error signal for low modulation frequency.

Figure 15.5d illustrates another two beam interferometer known as the *Sagnac* interferometer. Here the light is split by a coupler and travels in opposite directions around the same closed loop to be recombined at the same coupler. If the loop is subjected to a rotation, the relative phase of the two beams in a loop of this type will experience a shift due to the Sagnac effect. It is this mechanism that forms the basis of the fiber-optic gyroscope. However, this interferometer can also be used to measure time-varying strain applied to the loop, providing the strain is applied asymmetrically (i.e., not at the center of the loop). Considering the configuration, if the loop contains two delay coils of length L_1 and L_2 and a sensor coil of length L_s, then a static perturbation acting on the complete loop will have no net effect on the phase delay between the two beams, because both beams will be affected equally. However, a dynamic signal acting on the sensor, but not on the delay coils, will introduce a phase difference between the two counter-propagating beams due to the fact that one beam is in a different part of the loop when the other beam is passing through the sensor. Analysis of the Sagnac interferometer requires separate integration of each beam around the loop. It can be shown that the difference in phase between the two beams is given by

$$|\Delta\phi_s| = 2L_s k\delta n \left| \operatorname{sinc}\left(\frac{\omega_s n_e}{2c} L_s\right) \sin\left(\frac{\omega_s n_e}{2c} L_d\right) \right| \tag{15.10}$$

where

$L_d = L_2 - L_1$

ω_s is the angular frequency of the applied strain

δn is related to its amplitude

For the case when $\omega_s \ll \pi c/n_e L_d$, Equation 15.10 reduces to

$$|\Delta\phi_s| \approx k\delta n \frac{\omega_s n_e}{c} L_s L_d \tag{15.11}$$

The phase response of the *Sagnac* interferometer is thus proportional to the signal frequency, ω_s, and, for harmonically varying signals, behaves as a differentiator. This can be useful for measuring small, rapidly varying signals in the presence of large low frequency signals that would otherwise exceed the dynamic range and distort the measurement.

Another implementation is known as *tandem interferometry*. Consider the case of two MZIs with equal optical path imbalances placed in series, as illustrated in Figure 15.5e. If the path imbalance exceeds the coherence length of the source, then no interference signal is observed at the output of the first interferometer. However, on passing through the second interferometer, two of the four possible optical paths emerge with a zero optical path imbalance. An interference signal of the form

$$i_{ph} = \frac{r_d P}{2}\left[1 \pm \frac{V}{2}\cos\left(\phi_{MZI1} + \phi_{MZI2}\right)\right] \tag{15.12}$$

is observed at the output. As only two paths produce an interference signal, the fringe visibility is reduced by a factor of 2; the other paths exhibit an effective path imbalance that exceeds the coherence length of the optical source. The total phase difference accumulated through both interferometers is represented by $\phi_{MZI1,2}$. The visibility is given by $V = \exp(-\tau^2/(4\tau_c^2))$, where $\tau = n_e L_{21}/c$ is the differential time delay in the interferometer pair ($L_{21} = L_2 - L_1$ is the differential fiber path imbalance) and τ_c is the coherence time of the source. In practice, the source coherence must be kept sufficiently low to avoid residual coherence of the imbalanced optical paths, generating high levels of frequency-induced phase noise.

15.4.2.3 Measurement of Interferometric Phase in Fiber Interferometers

The measurement of the intensity at the output of the interferometer provides a measure of the interferometric phase, which is equal to the phase difference between the recombined light beams. However, as well as the actual phase value it is usually desired to determine the order of the fringe and the phase change direction, preferably independent of the received power and fringe visibility. This can be achieved by determining quadrature values of the phase (i.e., two identical signals that are phase shifted by $\pi/2$ and are expressed in terms of $\sin(\phi)$ and $\cos(\phi)$) from which the phase can be obtained from the arctangent of their ratio. Furthermore, by recording quadrant changes of the phase using these quadrature components, the fringe order can be tracked indefinitely.

One method to accomplish this is to shift the frequency of the light in one arm relative to the other, such that a *heterodyne* beat frequency is generated on the detector, yielding a photodiode current of the form

$$i_{ph} = r_d P\left[1 \pm V\cos\left(\Delta\omega \cdot t + \phi_s - \phi_r\right)\right] \tag{15.13}$$

where $\Delta\omega = \omega_s - \omega_r$ is the *heterodyne* frequency difference. In this situation, the *heterodyne* frequency is phase modulated by the interferometric phase. High pass filtering to remove the dc term in Equation 15.13 and mixing with quadrature components of the unmodulated *heterodyne* frequency (i.e., $\sin(\Delta\omega t)$ and $\cos(\Delta\omega t)$, which can be generated electronically) followed by low pass filtering to remove the harmonic component of $\Delta\omega$ yields the required quadrature phase components. This procedure is illustrated in Figure 15.6b.

Another method involves sinusoidal modulation of the phase in one arm of the interferometer, which yields a photocurrent of the form

$$i_{ph} = r_d P\left[1 \pm V\cos\left(\phi_m \sin\left(\omega_m t\right) + \phi_s - \phi_r\right)\right] \tag{15.14}$$

where

ϕ_m is the modulation depth

ω_m is the angular modulation frequency

The necessary processing of this signal is less obvious in this case. However, if we note the Fourier series

$$\cos\left(\phi_s \sin\left(\omega_m t\right)+\phi_s-\phi_r\right)=\cos\left(\phi_s-\phi_r\right)\left\{J_0\left(\phi_m\right)+2\sum_{k=1}^{\infty}J_{2k}\left(\phi_m\right)\cos\left(2k\omega_m t\right)\right\}$$

$$-\sin\left(\phi_s-\phi_r\right)\left\{2\sum_{k=0}^{\infty}J_{2k+1}\left(\phi_m\right)\sin((2k+1)\omega_m t)\right\} \qquad (15.15)$$

then Equation 15.14 comprises a harmonic series of tones, beginning at the modulation frequency and weighted by their respective Bessel functions. It can be seen that consecutive harmonics are proportional to the quadrature components of the phase. So if Equation 15.14 is high pass filtered to remove the dc term and the result separately mixed with unmodulated reference frequencies at ω_m and $2\omega_m$ (again these are generated electronically), the resulting signals, after low pass filtering to remove ω_m and its harmonics, yield the quadrature components of the phase. These steps are illustrated in Figure 15.6c. This is a similar procedure to processing the *heterodyne* signal and is known as *phase-generated carrier* (Cole et al. 1982, Dandridge et al. 1982). It should be noted that to balance the required harmonic components in Equation 15.14, the modulation depth ϕ_m has to be correctly set (e.g., if using ω_m and $2\omega_m$ components, then set ϕ_m = 2.6 rad where $J_1(\phi_m) \approx J_2(\phi_m)$). The required phase modulation can be generated using an interferometer with a path imbalance, d, and a frequency-modulated laser. The required frequency modulation depth is given by $\Delta\nu_m = \phi_m c/(2\pi n_e d)$. This implementation is known as *frequency-modulated phase-generated carrier* (FM-PGC). Alternatively, the required phase modulation can be directly applied to one of the arms of the interferometer using a phase modulator.

A third method involves applying a linear ramp to the laser frequency with an excursion rate of Ψ Hz/s, such that the photodiode current becomes

$$i_{ph}=rP\left[1\pm V\cos\left(\omega' t+\phi_s-\phi_r\right)\right] \qquad (15.16)$$

where $\omega' = (2\pi n_e d/c)\Psi$. A carrier frequency is only generated with a nonzero path-imbalance, d. The photocurrent resembles a heterodyne signal leading to its designation as a pseudo-heterodyne method (Jackson et al. 1982), and the processing to extract the interferometric phase follows as described earlier (see Figure 15.6d). In practice, it is not possible to generate an indefinite linear increase in the laser frequency, and a practical alternative is to modulate with a periodic ramp (i.e., serrodyne signal), as illustrated in Figure 15.6d. The photodiode current deviates from the true heterodyne signal during the *fly back*, which can introduce measurement error in certain circumstances (Jorge et al. 2000). Also, linearity of the laser frequency sweep can affect the spectral purity of the carrier. The phase ramp can also be applied through a phase modulator placed in one of the interferometer arms.

Other methods of phase recovery are also possible such as phase-stepping, where a series of samples are taken over an interference fringe from which the phase can be extracted (Hariharan 1987), and passive methods (i.e., modulation free) using the 3 × 3 directional coupler (Todd et al. 2002).

The phase recovery techniques described earlier are capable of achieving phase resolutions approaching 1 μrad/Hz$^{1/2}$ although in many situations, other noise sources such as laser frequency noise degrade this resolution somewhat.

15.4.2.4 Measurement of Displacements in Passive Optical Cavities: The Pound–Drever–Hall Method

Another way of measuring length changes in optical fiber involves incorporating the fiber into a high-finesse cavity of the type shown in Figure 15.5c and locking a single-frequency laser to one of the cavity resonances. This technique is a form of *frequency modulation spectroscopy.* Changes in cavity length will be observed as changes in the laser frequency. One implementation known as the Pound–Drever–Hall (PDH) method (Black 2001) is illustrated in Figure 15.6e. Here, the output of a frequency-modulated laser is injected into a FFP cavity. The modulated signal consists of a carrier and modulation sidebands, which experience different interactions with the narrow cavity resonance. A change in the cavity resonance (due to a change in cavity length) is then observed in the transmitted light. Synchronous detection of the transmitted signal with the modulation frequency generates an error signal. The general behavior of the interaction also depends on the relative magnitudes of the modulation frequency, ω_m, and the cavity resonance linewidth, $\Delta\nu_c$. For the case when $(\omega_m/2\pi) \ll \Delta\nu_c$ the error signal as a function of the laser offset frequency from the cavity resonance, $\delta\omega$, takes the form illustrated in the inset of Figure 15.6e. Close to the cavity resonance, the error signal is given by (Black 2001)

$$\varepsilon \approx -\frac{4}{\pi}\sqrt{P_c P_s}\,\frac{\delta\omega}{\Delta\nu_c} \tag{15.17}$$

where P_c, P_s are the power in the carrier and sideband, respectively. A linear relationship exists between ε and $\delta\omega$, and the error signal can be used as a control signal to lock the laser frequency to the cavity resonance. This is implemented by adding the integrated error signal to the modulation signal and applying this to the laser frequency modulation input. The same can be achieved when $(\omega_m/2\pi) > \Delta\nu_c$, although a slightly different error signal is observed. The feedback loop can be set up in two configurations. The bandwidth of the feedback loop can be made low such that the laser tracks slow variations in the cavity length. The signal is then observed at the output of the photodiode (marked O/P 1 in Figure 15.6e). If the loop bandwidth is made large (i.e., covering the signal bandwidth), then the signal is read from the output of the integrator (marked O/P 2 in Figure 15.6e).

When operating in this linear region, the relationship between the error signal and laser frequency fluctuation, δf, or cavity length fluctuation, δL, is given by (Black 2001)

$$\varepsilon \simeq -8\sqrt{P_c P_s}\,\frac{2LF}{\lambda}\left(\frac{\delta f}{f}+\frac{\delta L}{L}\right) \tag{15.18}$$

where the *finesse* of the cavity, F is defined as $F \equiv \Delta\nu_c/\Delta\nu_{fsr}$. Thus, the slope of the error signal increases with increasing cavity *finesse.* Both fractional changes in laser frequency fluctuations and cavity length fluctuations (i.e., the bracketed terms) contribute in equal proportion to the error signal. This is identical behavior to that observed in conventional two-beam interferometers. The error signal response is however enhanced by the cavity finesse, which is useful for overcoming other noise sources that are not related to the cavity *finesse* (such as thermal noise from the amplifier following the detector, laser intensity noise, and multipath interference). Shot noise limited performance is also more readily achievable using this technique.

This technique is capable of resolving very small changes in the cavity length. Short cavity (~3 cm) strain sensors achieving strain resolutions less than 1 pm/Hz$^{1/2}$ at frequencies below 1 kHz have been demonstrated although coherent lasers such as fiber lasers or stabilized semiconductor lasers are required (Chow et al. 2005). This performance can only be matched by conventional interferometry using much longer fiber lengths.

15.4.3 Polarization-Modulated Sensors

Polarimetric sensing techniques rely on the interference between two polarization resolved signals generated by the same light source. They have several benefits:

- They can be used to implement single fiber interferometers.
- They can be made to exhibit low sensitivity to temperature and hence measure quasi-dc parameters.
- Simple demodulation is achieved using a polarizer to convert polarization rotation to intensity modulation.
- Near balancing of optical path lengths provides immunity from laser frequency noise.

When an optical signal propagates in a conventional single mode optical fiber, its state of polarization (SOP) will evolve on transmission through the fiber, due to low levels of intrinsic birefringence. Birefringence causes light of different polarization states to propagate at different group velocities. Linear birefringence can be represented by two orthogonal eigenaxes with refractive indices denoted by n_x and n_y. The group velocity is given by $v_g = c/n$; thus, the axis with the lower index is known as the *fast* axis with the other denoted the *slow* axis. The birefringence can be perturbed by applying transverse stress or by twisting the fiber, which will change the polarization state of the propagating signal. Transverse stress induces linear birefringence, whereas twist induces circular birefringence. A linear polarizer placed at the output of the fiber will convert the polarization change of the light exiting the fiber into changes in transmitted intensity. With conventional fiber, this effect provides a very limited degree of control due to the intrinsic birefringence and its environmentally induced variability, causing the output SOP to change unpredictably over time. However, if the optical fiber has a large amount of linear birefringence, then light in one polarization axis cannot couple to the other eigenaxis and is guided in the same polarization state.* If a linearly polarized light source is injected with its polarization azimuth at 45° to the eigenaxes of the fiber such that

* These fibers are known as polarization maintaining (PM) fibers and can be made by placing two parallel stress rods centered around the core, which induces a high degree of transverse stress and hence linear birefringence, or by making the core elliptical.

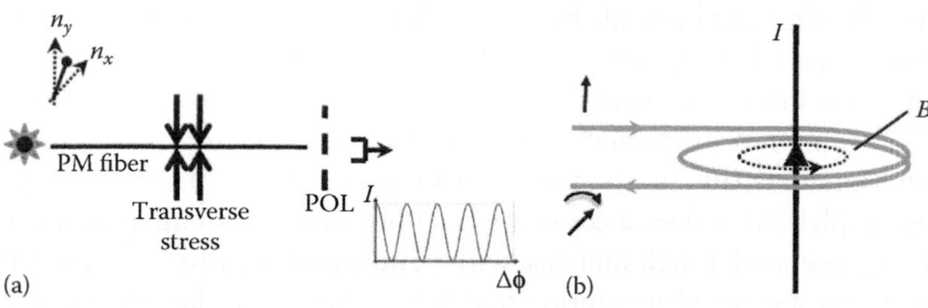

FIGURE 15.7 Polarization-modulated sensor: (a) basic configuration and (b) current sensor principle.

each axis is equally exited, then each polarization eigenstate will accumulate a phase determined by the refractive index of the eigenaxis, as illustrated in Figure 15.7a, generating an elliptical state at the output. This is passed through a linear polarizer (POL) also with its azimuth at 45° to the eigenaxes. Each eigenstate will be equally resolved along the polarizer axis and interfere on a detector generating a signal closely resembling that of a two-beam interferometer

$$i_{ph} = r_d P\left[1 \pm V\cos\left(\phi_x - \phi_y\right)\right] \tag{15.19}$$

where ϕ_x and ϕ_y are the phases of each eigenstate. Following the same analysis used in Section 15.4.2.1, a change in differential phase can be expressed as

$$\Delta\left(\phi_x - \phi_y\right) = k\left(\left(n_x - n_y\right)\frac{\partial L}{\partial X} + L\frac{\partial\left(n_x - n_y\right)}{\partial X}\right)\Delta X \tag{15.20}$$

where the first term on the right hand side is the effect on the fiber length due to the measurand and the second term is the effect of the measurand on the index difference. Thus, a two beam interferometer can be implemented in a single PM optical fiber.

Another polarimetric sensor involves the interaction of the light with an external magnetic field through the *Faraday effect.* If a magnetic field is applied to a medium in a direction parallel to the optical path, the light will experience a polarization rotation. The magnetic field induces an effective circular birefringence in the medium. If we consider the idealized case of an optical fiber wound around a current carrying conductor with no intrinsic birefringence, as illustrated in Figure 15.7b, the total rotation, ρ, will be given by the line integral of the field around the loop

$$\begin{aligned}\rho &= NV_m \oint \underline{H} \cdot d\underline{l} \\ &= NV_m I\end{aligned} \tag{15.21}$$

where

N is the number of loops
$\underline{H}$ is the magnetic field strength
V_m is the Verdet constant (~0.5 rad/(T · m) at 1550 nm (Cruz et al. 1996))

Further simplification can be made by noting that the line integral of the magnetic field around a loop is equal to the current in the conductor, *I*. A linearly polarized beam propagating in the fiber will experience a rotation of its polarization azimuth given by Equation 15.21. Thus, measurement of the polarization rotation yields a measurement of the current and it is this mechanism that forms the basis of the fiber-optic current sensor. As we explained earlier, a conventional fiber exhibits small amounts of intrinsic linear birefringence, which will increase with bending of the fiber and contribute to the total polarization rotation. One solution to this is a fiber with a large amount of intrinsic circular birefringence,* which dominates the total birefringence (i.e., linear plus circular). In the presence of a magnetic field, the total birefringence then becomes the sum of the intrinsic circular birefringence and the magnetically induced circular birefringence. One can then use the nonreciprocity of the magnetically induced circular birefringence to remove the effect of the (reciprocal) intrinsic circular birefringence by reflecting the light back through the fiber (i.e., the magnetically induced rotation does not depend on the propagation direction). This essentially untwists the effect of the intrinsic circular birefringence. The output SOP is then dependent only on the magnetically induced polarization rotation (Bohnert et al. 2002).

15.4.4 Wavelength-/Frequency-Modulated Sensors

Another sensing method is through modulation of the wavelength or frequency of the optical radiation. Wavelength-modulated sensors have several benefits:

- They are robust to power fluctuations and fiber birefringence.
- Wavelength encoding provides an absolute measurement reference.
- They are inherently suited to WDM.
- Wavelength-modulated fiber laser sensors achieve the highest possible sensitivity limited by thermodynamic noise.

We now consider specific examples of wavelength or frequency encoded sensors in single mode fiber.

15.4.4.1 Fiber Bragg Grating Sensors

As described in detail in Chapter 17, the FBG consists of a periodic modulation of the refractive index in the core of optical fiber, illustrated in Figure 15.8a. FBGs can be inscribed in silica optical fiber, where the core is doped with germanium, by exposure to a periodic UV radiation pattern generated by a phase mask. This creates a permanent change in the refractive index of the core. Each period of the modulation forms an index plane perpendicular to the axis of the optical fiber. Efficient diffraction occurs when light scatters from each index plane, and the wavelength of the incident radiation and the period of the grating meet the same phase matching condition as for the microbending sensor given by Equation 15.2. In this situation, back-reflected light waves from

* Circular birefringence can be induced by spinning an asymmetric perform during the drawing process of the fiber or by twisting the fiber.

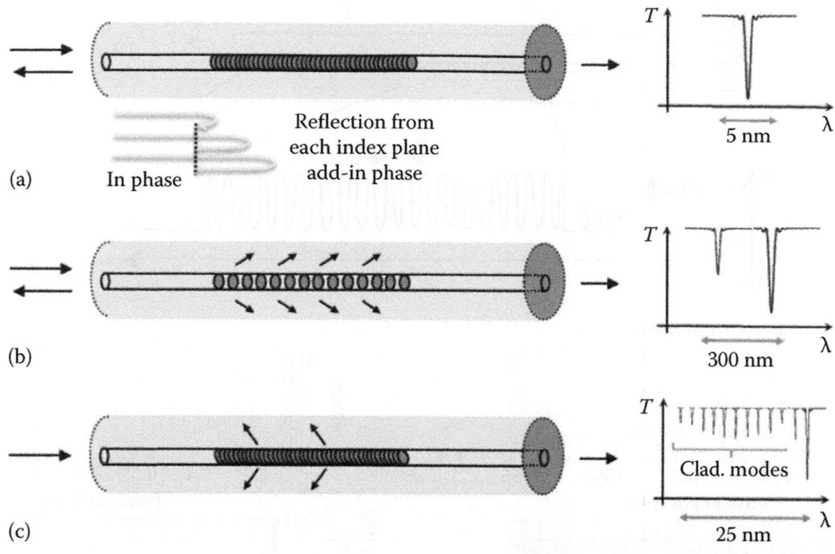

FIGURE 15.8 Wavelength-/frequency-modulated sensor: (a) FBG sensor; (b) LPG sensor; and (c) tilted FBG sensor.

each plane combine in phase. For coupling to the reverse propagating mode, $\beta = -\beta_n$, the condition reduces to

$$\lambda_B = 2n_e\Lambda \tag{15.22}$$

where

- λ_B is the wavelength at which coupling occurs, known as the Bragg wavelength
- Λ is the pitch of the grating

This criterion is only exact for an infinitely long, ideal FBG. For finite length, FBGs, λ_B, is not uniquely defined and coupling occurs over a range of wavelengths (although one usually defines a wavelength of peak coupling). Applying a strain to the FBG will change the pitch, Λ, causing a change in λ_B. Many techniques are available for modeling the interaction of optical radiation with an arbitrary grating structure such as coupled mode theory. However, for our purposes here, we shall consider the simplifying case of weak gratings, where multiple reflections within the grating structure can be ignored.

Figure 15.9a illustrates the spatial properties of an arbitrary FBG (a uniform FBG is illustrated for clarity). The properties of the grating, Δn_{ac}, Δn_{dc}, and Λ may all be functions of z. In coupled mode theory, a quantity known as the complex coupling coefficient, q, represents the spatial variation in strength of the FBG and is slowly varying compared with the FBG period. It is defined as

$$|q(z)| = \frac{\eta_m \pi}{\lambda_B} \Delta n_{ac}(z) \tag{15.23}$$

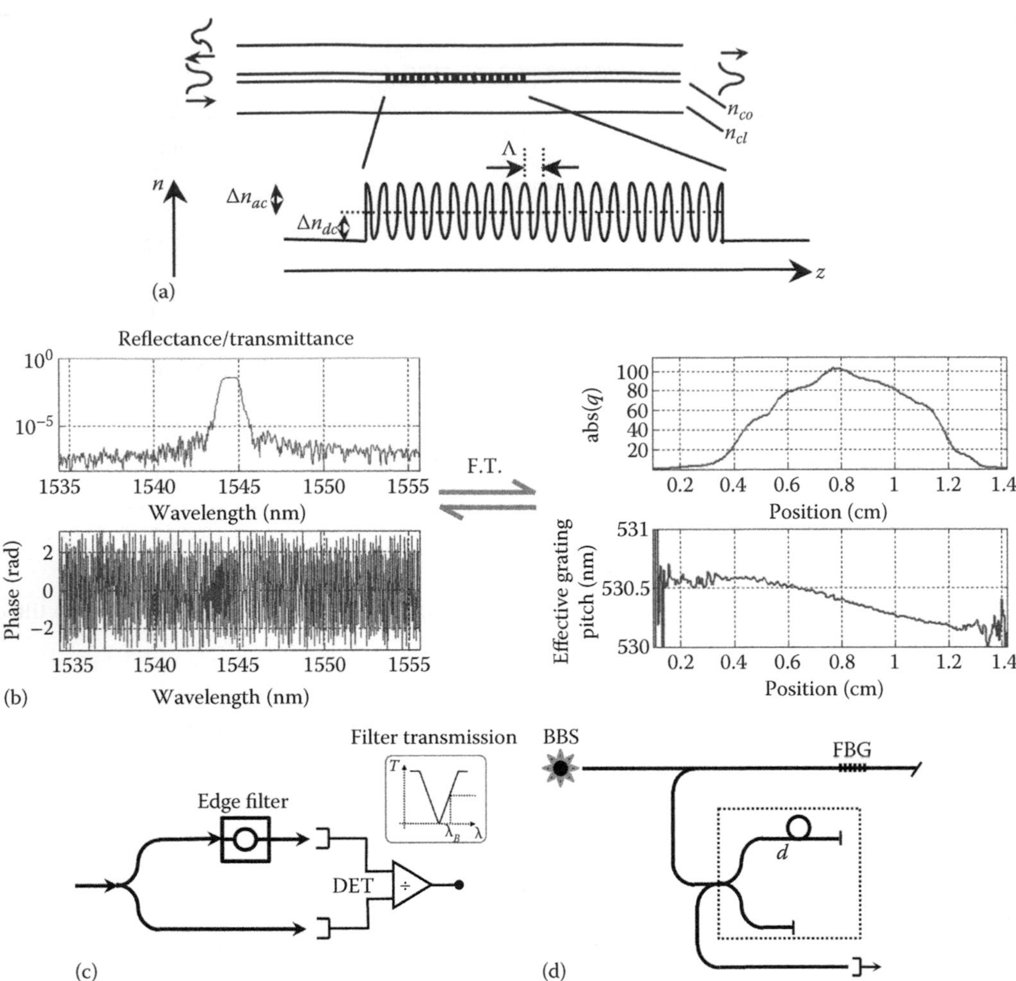

FIGURE 15.9 (a) Physical parameters of the FBG; (b) example of complex reflectivity and complex coupling coefficient (amplitude and phase); (c) edge filter with ratiometric output-based demodulator (c); and (d) interferometric decoding.

$$\arg(q(z)) = \theta'(z) - 2\eta_m k \int_0^z \Delta n_{dc}(z')dz' + \frac{\pi}{2} \tag{15.24}$$

where

η_m is a mode overlap factor

$\theta'(z)$ represents chirp in the grating

The amplitude of the coupling coefficient is proportional to the amplitude of the rapidly varying ac index modulation. The phase of the coupling coefficient represents the excess optical phase accumulated on propagation through the grating. It incorporates variation in the slowly varying dc index and any variation in the spatial frequency of the index modulation (i.e., chirp). It can be shown that for the case of a weak grating, the complex reflectivity of the FBG and the coupling coefficient form a Fourier transform pair such that (Skaar 2000)

$$r(\delta) = -\frac{1}{2}\int_{-\infty}^{\infty} q^*\left(\frac{z}{2}\right)\exp\left(i\delta z\right)dz \tag{15.25}$$

or

$$r(\delta) \overset{\text{FT}}{\longleftrightarrow} -\frac{1}{2}q^*\left(\frac{z}{2}\right) \tag{15.26}$$

where we have introduced the normalized frequency parameter, $\delta = 2\pi n_e((1/\lambda)-(1/\lambda_B))$, known as the detuning. The derivation of Equation 15.25 follows from the weakly interacting approximation of the coupled mode equations and is rather complex; however, its implications are very significant. If we assume the complex reflectivity can be measured in some way, then the coupling coefficient can be deduced from its Fourier transform. According to Equation 15.24, the phase of the coupling coefficient contains information on the local pitch of the grating; hence, it can reveal perturbations of the grating pitch (i.e., local strain applied to the grating or spatially varying temperature). The effective grating pitch is related to the phase of the coupling coefficient by

$$\Lambda_{eff}(z) = \Lambda\left(1+\frac{\Lambda}{2\pi}\frac{d\arg(q(z))}{dz}\right)^{-1} \tag{15.27}$$

The spatial variation in $\Lambda_{eff}(z)$ thus provides information on strain variation along the fiber. We will show in Section 15.6.2.6 how the complex reflectivity can be measured with optical frequency-domain reflectometry (OFDR). The reflection spectra and complex coupling coefficient of a 1 cm long, chirped Bragg grating are shown in Figure 15.9b. The spatial variation in the effective grating pitch is clearly visible in the lower right hand plot.

The wavelength shift due to an axial strain or temperature change can be derived in the same way as the dependence of optical phase on these parameters (recall from Section 15.4.2.1) and for Ge doped silica fiber is given by

$$\frac{\Delta\lambda_\varepsilon}{\Delta\varepsilon} \simeq 1.2 \text{ pm}/\mu\varepsilon \tag{15.28}$$

$$\frac{\Delta\lambda_T}{\Delta T} \simeq 10 \text{ pm/K} \tag{15.29}$$

Thus, a 10 $\mu\varepsilon$ signal produces a similar wavelength shift to a 1 K temperature change, indicating that a high degree of cross-sensitivity to temperature exists in these devices.

Many methods have been investigated to overcome this cross-sensitivity to provide discrimination between temperature and strain (Farahi et al. 1990).

The potential of the FBG as a discrete strain sensor, with performance comparable to the piezoresistive strain gage, has driven considerable efforts to develop methods to decode their wavelength shifts. A simple method, illustrated in Figure 15.9c, uses an edge filter. Here the transmitted intensity through the filter is proportional to the wavelength. A parallel reference path provides invariance to signal power fluctuations. For very high resolution measurement of wavelength shifts, interferometric decoding can be used, as illustrated in Figure 15.9d. If the FBG is illuminated with a broad band source (BBS) and the reflected signal from the BBS is injected into an imbalanced interferometer, then according to Equation 15.4, a change in the FBG wavelength will generate a phase shift. The phase shift at the interferometer output is given by

$$\Delta\phi = \frac{2\pi n_e d}{\lambda^2}\Delta\lambda = \frac{2\pi n_e d}{c}\Delta\nu \tag{15.30}$$

where d is the path imbalance in the interferometer. The imbalance must be smaller than the coherence length of the light returned from the grating. The resolution of most wavelength detection schemes is on the order of 1–10 με for quasi-dc signals (<10 Hz). Interferometry is more suited to measurement of ac strains, unless a temperature reference is incorporated (Kersey et al. 1993a), and is capable of strain resolutions approaching 1 nε/Hz$^{1/2}$ (Kersey et al. 1992). Other methods for decoding wavelength shifts suitable for multiplexed sensors will be described in Section 15.6.4.1.

The FBG can also be used as a wavelength selective mirror for forming fiber-based Michelson and Fabry–Pérot interferometers, as well as for implementing multiplexed interferometric sensor arrays.

15.4.4.2 Long-Period Fiber Grating Sensors

A long-period grating (LPG) consists of a periodic perturbation in the optical fiber that causes coupling of a forward propagating core mode to a forward propagating mode that propagates in the cladding, as illustrated in Figure 15.8b. The perturbation period ranges from tens of micrometers to several hundreds of micrometers (compared with ~500 nm in FBGs at 1550 nm). Coupling occurs when a core mode becomes phase matched to a cladding mode and satisfies the condition given by Equation 15.2. This can be expressed as

$$\lambda = \left(n_e(\lambda) - n_{cl}^i(\lambda)\right)\Lambda \tag{15.31}$$

where

Λ is the pitch of the LPG
$n_e(\lambda)$ is the effective index of the core mode
$n_{cl}^i(\lambda)$ is the effective index of the ith cladding mode

Efficient coupling is only observed when good spatial overlap exists between core and cladding modes. This is only found to be the case for core and circularly symmetric cladding modes of odd order. Since the phase matching condition is met over a much broader range of wavelengths compared with typical FBGs, the spectral width of LPG resonances is much broader than those observed in FBG sensors. Also, the coupling interaction in LPGs is only observable in transmission. These differences clearly restrict the interrogation and multiplexing methods applicable to LPG sensors. However, despite these limitations, LPGs exhibit many useful properties.

The response of the LPG to a measurand depends on the period of the LPG, the order of the resonance being observed, and the chemical composition of the fiber. Therefore, LPGs can be designed to respond with a positive sensitivity, negative sensitivity, or no sensitivity. The response of the LPG to strain is (Bhatia 1999, James and Tatam 2003)

$$\frac{d\lambda}{d\varepsilon} = \frac{d\lambda}{d(\delta n_e)}\left(\frac{dn_e}{d\varepsilon} - \frac{dn_{cl}}{d\varepsilon}\right) + \Lambda\frac{d\lambda}{d\Lambda} \tag{15.32}$$

and to temperature is

$$\frac{d\lambda}{dT} = \frac{d\lambda}{d(\delta n_e)}\left(\frac{dn_e}{dT} - \frac{dn_{cl}}{dT}\right) + \Lambda\frac{d\lambda}{d\Lambda}\frac{1}{L}\frac{dL}{dT} \tag{15.33}$$

where $\delta n_e = n_e - n_{cl}$. In both cases, by tailoring the response of the core and cladding indices to the measurand, they can be made to counteract or reinforce the dispersion term, $d\lambda/d\Lambda$, enabling control of the LPG response. The cladding index also depends on the refractive index of the material surrounding the fiber and therefore responds to changes in external refractive index which can be a useful sensing mechanism (Patrick et al. 1998).

LPGs can be made by exposing the fiber to UV radiation in the same way as FBGs; however, they can also be inscribed by exposure to a CO_2 laser or an electric arc, which causes a thermally induced change in the refractive index. This produces a structure with much higher thermal resistance and thus can be used at high temperatures.

15.4.4.3 Tilted Gratings

Tilted FBGs are a general form of the FBG described earlier, where the index planes are tilted or blazed relative to the fiber axis. The tilt of the index planes enhances the coupling from the forward propagating core mode to the reverse propagating cladding mode and reduces the coupling to the reverse propagating core mode. These devices are interrogated in transmission and exhibit the core mode resonance with the cladding mode resonances (typically greater than 50) on the short wavelength side, as illustrated in Figure 15.8c. The cladding mode resonances exhibit a width similar to the main Bragg resonance with a strength related to the tilt angle. The cladding mode is guided by the outer cladding surface and thus exhibits an effective index dependent on the outer material. The Bragg wavelength for the tilted grating is given by Equation 15.22; however, the grating pitch must be replaced with $\Lambda = \Lambda_g\cos(\theta)$, where Λ_g is the actual pitch of the grating and θ is the tilt angle. The cladding mode resonance obeys the phase matching condition given by Equation 15.2, which can be expressed as

$$\lambda_{cl}^{i} = \left(n_{e}^{i} + n_{cl}^{i}\right)\Lambda \tag{15.34}$$

where

- λ_{cl}^{i} is the wavelength of the ith cladding mode
- n_{e}^{i} is the effective index of the core mode at λ_{cl}^{i}
- n_{cl}^{i} is the effective index of the ith cladding mode

Such devices can be used as refractometers with reduced sensitivity to temperature (Chan et al. 2007) or compact fiber-optic spectrometers (Wielandy and Dunn 2004). An excellent review is given by Albert et al. (2013).

15.4.5 Noise Sources

The performance of a fiber-optic sensor system can be determined by considering the various noise sources present in the system and enables the signal resolution and dynamic range to be calculated. A brief review of noise sources is now given (this section serves as adjunct material and can be omitted on first reading).

15.4.5.1 Shot Noise and Amplifier Noise

Quantum theory represents light as a flow of discrete wave packets called photons. At a detector, they arrive at discrete, random time intervals, giving rise to a random variation in the photocurrent. This variation is observed as a current noise known as *shot noise* and for an incident power of P is given in A^2/Hz by

$$\overline{i_{sh}^{2}} = \frac{2e^{2}\eta_{d}P\Delta f}{h\nu} \tag{15.35}$$

where Δf is the electrical measurement bandwidth and the other parameters were defined in Section 15.4.2.2. After generation of the photocurrent, a transimpedance amplifier is often used to convert the current into a voltage. This amplifier contains a feedback resistor, which generates *Boltzmann noise* expressed as an equivalent noise current referred to the input of the amplifier

$$\overline{i_{th}^{2}} = \frac{4k_{B}T_{e}\Delta f}{R_{L}} \tag{15.36}$$

where

- k_B is Boltzmann constant
- $T_e = T + T_a$
- T is the ambient temperature
- T_a is the effective noise temperature of the amplifier (Yariv 1991)
- R_L is the photodiode load resistance

In the simplest case, the optical power can be increased such that $\overline{i_{sh}^2} \gg \overline{i_{th}^2}$ yielding shot noise limited detection.

15.4.5.2 Source Intensity and Frequency Noise

Power fluctuations of the source can cause measurement error in intensity sensors. However, in other systems, source *intensity noise* generates a noise current at the detector limiting the SNR. This noise current is given by

$$\overline{i_{RIN}^2} = (r_d P)^2 \cdot RIN(f) \tag{15.37}$$

where $RIN(f)$ is the relative intensity noise spectrum of the source.

As indicated before, in interferometric systems, *laser frequency noise* generates phase noise in imbalanced interferometers. The physical mechanisms that cause frequency noise (or line broadening) vary for different lasers. Some lasers such as single frequency fiber lasers and the NPRO laser described in Section 15.3.1.3 exhibit very low frequency noise (<10 Hz/Hz$^{-1/2}$ at 1 kHz) and are thus well suited to interrogating interferometric sensors. The phase noise due to input frequency fluctuations can be calculated using Equation 15.30.

15.4.5.3 Noise from Optical Amplification

In most cases, signal amplification is not possible without addition of some noise. In the case of amplification in an inverted optically active medium, amplification is accompanied by the addition of noise due to ASE. This generates photons with wavelengths that cover the gain spectrum of the medium and for the case of the erbium-doped fiber amplifier (EDFA) corresponds to 1525–1560 nm. When detected, these spontaneously generated photons beat both with themselves and with the signal to generate four new noise currents. These are:

1. Beating between the signal and spontaneous emission, $\overline{i_{s-sp}^2}$ (*s–sp*)
2. Beating between the spontaneous emission and itself, $\overline{i_{sp-sp}^2}$ (*sp–sp*)
3. Beating between the spontaneous emission and the shot noise, $\overline{i_{sp-sh}^2}$ (*sp–sh*)
4. Beating between the signal and the shot noise, $\overline{i_{s-sh}^2}$ (*s–sh*)

The noise currents for each noise source are given by (Olsson 1989)

$$\overline{i_{s-sp}^2} = 4r_d^2 G P_{in} P_{ASE} \Delta f \tag{15.38}$$

$$\overline{i_{sp-sp}^2} = 4r_d^2 P_{ASE}^2 \Delta\nu_{opt} \Delta f \tag{15.39}$$

$$\overline{i_{sp-sh}^2} = 4r_d e P_{ASE} \Delta\nu_{opt} \Delta f \tag{15.40}$$

$$\overline{i_{s-sh}^2} = 2r_d eGP_{in}\Delta f \tag{15.41}$$

where

r_d is the detector responsivity

G is the small signal amplifier gain

$\Delta\nu_{opt}$ is the optical bandwidth

the ASE power in one polarization mode is given by $P_{ASE} = n_{sp}(G-1)h\nu$. Here, n_{sp} is the inversion parameter*

In most instances, both $\overline{i_{s-sh}^2}$ and $\overline{i_{sp-sh}^2}$ can be neglected. Furthermore, $\overline{i_{s-sp}^2}$ can be made to dominate $\overline{i_{sp-sp}^2}$ by making $\Delta\nu_{opt}$ sufficiently small.

15.4.5.4 Other Complex Noise Sources

Other more complex noise sources also exist in multiplexed sensor systems. For example, long fiber length reflective systems can give rise to significant levels of Rayleigh scattering. This generates an additional photocurrent and can be very significant in highly coherent systems. In time division multiplexed systems (see Section 15.6.2.2), poor extinction ratio switching gives rise to leakage light, which can generate excess noise and increase crosstalk. In high vibration environments, perturbations applied to the connecting leads can cause undesirable modulation of the polarization, generating excess phase noise in interferometric sensor arrays. Sideband noise on RF oscillators, used to drive AOMs, can also contribute to phase noise. Oscillators with phase noise less than −100 dB/Hz over the measurement bandwidth are generally required.

15.4.5.5 Calculating the Sensor Resolution

The sensor resolution is calculated by determining the signal-to-noise ratio (SNR). For simple direct detection systems, where the detected power is proportional to the measurand, the SNR is given by

$$SNR = \frac{(r_d P)^2}{\overline{i_n^2}} \tag{15.42}$$

The minimum detectable change in optical power is determined by assuming a $SNR = 1$. The minimum detectable signal is then obtained from the transducer responsivity. For systems employing a carrier frequency, where the measurand modulates the carrier power (i.e., AM),† it is more appropriate to calculate the carrier-to-noise power ratio (CNR). In this case, the noise contributions must be calculated at frequencies about the carrier frequency. Finally, for systems where the signal modulates the phase or frequency of the carrier (i.e., PM, FM), the relationship between the carrier power and sideband power must be determined. We consider here the case of phase-modulated heterodyne, where the photocurrent is given by

$$i_{ph} \propto \left[1 + V\cos(\Delta\omega t + \phi_s(t))\right] \tag{15.43}$$

* Note that $\overline{i_{s-sp}^2}$ is calculated using the ASE power in the same polarization mode as the signal.

† An example may be an intensity sensor with an amplitude-modulated source. The signal can be extracted with synchronous detection using a lock-in amplifier.

For a harmonic signal with a peak modulation depth of ϕ_o where $\phi_s(t) = \phi_o \sin(\omega_s t)$ and dropping the dc term yields,

$$\begin{aligned} i_{ph\_ac} &\propto V\cos\left(\Delta\omega t + \phi_o \sin\left(\omega_s t\right)\right) \\ &\propto V\left\{\cos\left(\Delta\omega t\right)\cos\left(\phi_o \sin\left(\omega_s t\right)\right)\cdots \right. \\ &\qquad \left. -\sin\left(\Delta\omega t\right)\sin\left(\phi_o \sin\left(\omega_s t\right)\right)\right\} \end{aligned} \tag{15.44}$$

Here we have also expanded the cosine term. Finally, for the case of small modulation depth ($\phi_o \ll 1$) Equation 15.44 becomes

$$i_{ph\_ac} \propto V\left\{\cos\left(\Delta\omega t\right) + \frac{\phi_o}{2}\left[\cos\left(\left(\Delta\omega + \omega_s\right)t\right) - \cos\left(\left(\Delta\omega - \omega_s\right)t\right)\right]. \right. \tag{15.45}$$

The signal exists as sidebands at $\Delta\omega \pm \omega_s$ about the carrier at $\Delta\omega$. It follows that the carrier to sideband ratio (CSBR) is given by

$$\begin{aligned} CSBR &= \left(\frac{2}{\phi_0}\right)^2 \\ &= \left(\frac{\sqrt{2}}{\delta\phi}\right)^2, \end{aligned} \tag{15.46}$$

where $\delta\phi$ is the root-mean-square (RMS) modulation depth. We will show in Section 15.7.2 how this expression can be related to the CNR in terms of the optical power and the various noise contributions described earlier enabling the sensor resolution in a multiplexed system to be calculated.

15.5 Multiplexing Topologies

A convenient method of classifying multiplexing architectures is by their topology. Sensors can be multiplexed in a serial, parallel, or ladder topology. Multiplexing sensors requires a method by which the signals from different sensors can be distinguished. For example, if sensors are located at difference positions, then their signals could be distinguished by the difference in time taken for the probing signal to travel to and return from the sensor; this is known as *TDM* and can be implemented using direct time gating methods or with frequency-domain methods where the Fourier relationship between the frequency and time domain is utilized. If sensors are designed to be wavelength selective, then their signals can be identified by their respective wavelength and can be *wavelength division multiplexed* (WDM). For the case of interferometric sensors, it has been shown that the phase of the interferometer is dependent on the laser frequency when a path imbalance is present in the interferometer. Thus, using the technique described in Section 15.4.2.3, sweeping the laser frequency generates a phase carrier with a frequency dependent on the path imbalance. Thus, sensors with different imbalances can be identified by their respective carrier frequencies. This is known as *frequency division multiplexing* (FDM) and is distinguished from WDM by the frequencies involved, which tend to be less than 1 MHz (compared with frequencies much greater than 1 GHz for WDM). Finally, another multiplexing method utilizes the coherence properties of the source and is only applicable to coherent interferometric systems. Individual sensors can be addressed by ensuring that only their signal is coherent at the detector. This is known as *coherence division multiplexing* (CDM).

We shall return to these basic multiplexing techniques in the next section, where we examine specific implementations in more detail. First, the generalized topologies of multiplexed systems are considered.

15.5.1 Parallel Topologies

Parallel topologies, as the name implies, involves addressing the sensors in parallel, as illustrated in Figure 15.10a through e. In the configuration shown in (a), the output of the source is coupled into a parallel array of sensors through a multiplexer, which directs the signal to each sensor. The reflected signal from each sensor is then routed to the detector. The multiplexer may be an optical switch that sequentially switches from one sensor to the next. Alternatively, it may be a wavelength multiplexer that routes a specific wavelength band to each sensor and then recombines the return signals onto a detector. Some form of signal decoding is required in this case to separate the wavelengths at the detector.

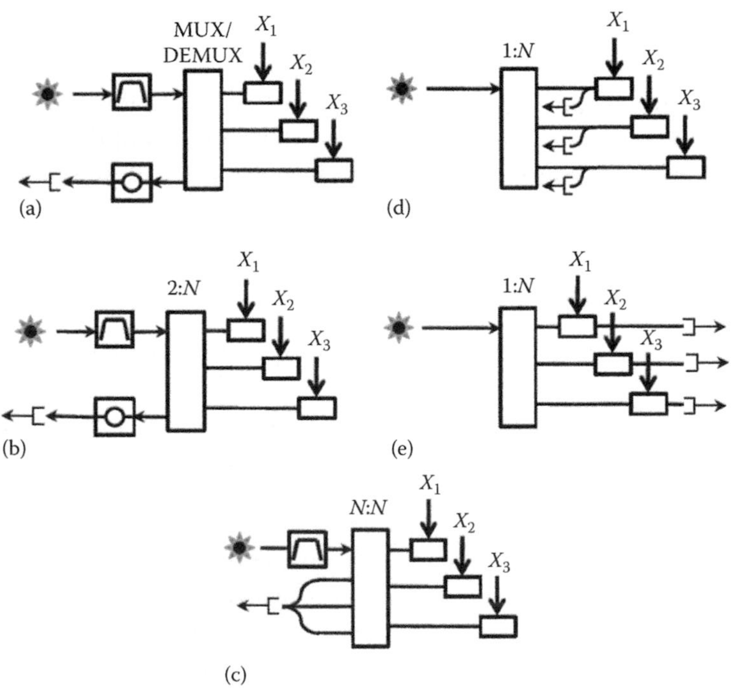

FIGURE 15.10 (a–e) Parallel topologies.

The configuration shown in (b) is similar to configuration (a); however, the multiplexer is replaced with a 2:*N* splitter. The splitter divides the input light between *N* sensors, and the reflected light from each sensor is combined onto a single fiber and routed to a detector. Some form of modulation and decoding is required to distinguish between sensor signals. For example, the sensors could be designed to generate a specific carrier frequency at the detector (i.e., FDM), or they could be designed to respond to a specific wavelength (this would require a broad band optical source). One disadvantage of this configuration is the high splitting loss. For the case of a *N*:*N* coupler where two input ports are utilized, then in the absence of other excess losses, the received power per sensor is given by

$$P_r = \frac{1}{N^2} P_l \tag{15.47}$$

where P_l and P_r are the launched and received power, respectively. An improvement in power budget is obtained by combining the input fibers of the splitter onto a single detector as illustrated in configuration (c). Here, the received power is given by

$$P_r = \frac{(N-1)}{N^2} P_l \tag{15.48}$$

which for a large *N* is an improvement by a factor of *N* compared with configuration (b). Alternatively, the sensor signals can be individually measured in reflection (configuration (d)) or transmission (configuration (e)). In transmission, the received power is

$$P_r = \frac{1}{N} P_l \tag{15.49}$$

In reflection, the received power is reduced by a further factor of ¼ due to the additional 50:50 coupler.

15.5.2 Ladder Topologies

Parallel topologies provide considerable flexibility in sensor location, as each sensor has its own dedicated fiber. However, for applications that require sensors to be linearly or closely spaced, a more efficient multiplexing configuration is the ladder topology. Figure 15.11a through c illustrates three such topologies. Here a telemetry fiber distributes the light sequentially to sensors by coupling off a small fraction. The output of each sensor is then recombined onto a second telemetry fiber which routes the return signals to a detector. Configurations (a) and (c) illustrate reverse coupled ladders and configuration (b) illustrates a forward coupled configuration. In order to ensure that equal power is received from each sensor, the power split ratio of each coupler must be tailored depending on its position in the array. Assuming no excess losses, then equal power is received from each sensor for a coupler split ratio given by

$$\kappa_j = \frac{1}{N-j+1} \tag{15.50}$$

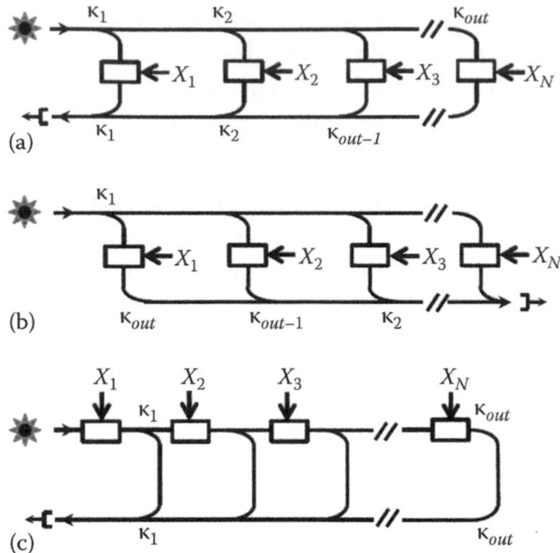

FIGURE 15.11 (a–c) Ladder topologies.

where j indicates the position of the coupler. These multiplexing configurations are well suited to TDM, where a pulse of light is injected into the array. Delay coils must be inserted between each rung of the ladder for configurations (a) and (b) such that pulses returned from each sensor do not overlap. The power returned from an individual sensor is given by

$$P_r = \frac{1}{N^2} P_I \tag{15.51}$$

which is identical to the parallel configuration shown in Figure 15.10b. In practice, fabrication tolerances in coupler split ratios limit the number of sensors that can be multiplexed in these parallel topologies to between 8 and 10.

15.5.3 Serial Topologies

A further simplification in the multiplexing telemetry can be achieved with transmissive or reflective serial topologies, illustrated in Figure 15.12a and b, respectively. In both topologies, sensors are multiplexed such that the input signal passes through each sensor and the sensor output is detected either in transmission or reflection. An evident benefit of such a configuration is that only a single fiber is required to multiplex sensors. Serial topologies also generally require the lowest number of components (such as directional couplers) and can take advantage of wavelength selective components and broadband optical sources to provide considerable flexibility in multiplexing a variety of sensors. The transmissive topology has the added advantage of being immune to Rayleigh backscatter that can become significant with long fiber lengths and coherent sources.

The reflective topology is often used to multiplex interferometric, FBG and intensity-based sensors. The transmissive topologies can be used to multiplex LPG and high finesse Fabry–Pérot sensors.

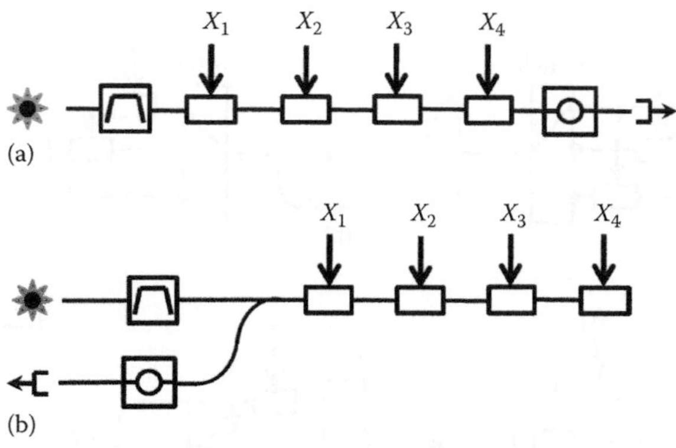

FIGURE 15.12 Serial topologies: (a) transmissive and (b) reflective.

15.6 Multiplexing Methods

The implementation of a multiplexing method can be further classified by the technique employed to address individual sensors. These are closely related to the sensing techniques described in Section 15.4: spatial, time, frequency, wavelength, coherence, and combinations of these methods. We shall explain the topologies used for each implementation and describe selected demonstrations.

15.6.1 Spatial Multiplexing

Spatial multiplexing is probably the simplest multiplexing method. Its implementation conforms to a parallel topology where light from a single source is delivered to each sensor through a dedicated fiber. Sensors are placed at different locations, where each one can be described by a geometric coordinate (Paton 1990). The source, fibers, and detectors can all be located at different coordinates. Figure 15.13 illustrates how a combination of a splitter and switch can be used to separate signals from individual fibers. Such a configuration is well suited to multiplexing intensity sensors. The benefits of this technique are the small number of components required for implementation and that only simple modulation is required for separating sensor signals (even further simplification is possible if an individual detector is used for each sensor output). The gate is necessary only if the switch is configured to switch rapidly from one sensor to the next. The system can be further simplified with a reflective configuration, illustrated in Figure 15.13b. Here the sensors are operated in reflection and the switch directs the input and return signal to each sensor consecutively.

Another configuration is shown in Figure 15.13c, where a splitter is used to direct light to different sensors. Sensors can be operated in either transmission or reflection, depending on the placement of the detectors. Such configurations are well suited to multiplexing Mach–Zehnder interferometric sensors or Michelson interferometers operating in reflection.

In applications where a large number of optical fibers are available, for example, when a multifiber cable is to be used or when an existing fiber network is in place, spatial multiplexing has many favorable attributes. Since each fiber carries information for only a single sensor, crosstalk is likely to be low (arising only in the optical switch or electronic

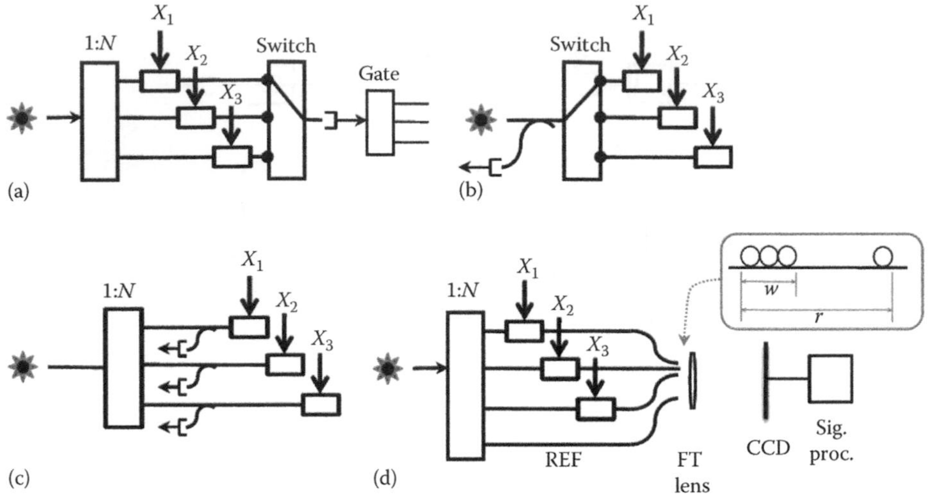

FIGURE 15.13 Basic implementations of spatial multiplexing: (a) using a splitter and switch in transmission; (b) a single switch in reflection; (c) transmission and reflection configurations with multiple detectors; and (d) spatial multiplexing of interferometric sensors with a frequency-domain read-out.

gate if used). It is also unlikely that the multiplexing implementation will have any detrimental effect on the sensor performance, as each sensor is operated almost independently. Spatially multiplexed systems also distribute optical power very efficiently to each sensor. Referring to Figure 15.13, configuration (a) exhibits a splitting loss of $1/N$ (assuming the switch is lossless). Configuration (b) exhibits a splitting loss of 1/4 due to the directional coupler, whereas configuration (c) exhibits a splitting loss of $1/(4N)$. A multiplexing configuration of the type illustrated in Figure 15.13c imposes no additional restriction on the measurement bandwidth. The sensors are interrogated continuously, and the measurement bandwidth is ultimately determined by the response time of either the transducer or, in the case of interferometric sensors, the frequency response of the interferometer. For configurations (a) and (b), the measurement bandwidth is limited by the switching speed.

An elegant implementation of spatially multiplexed interferometric sensors is illustrated in Figure 15.13d. Here the output of a laser is injected into a series of fibers through a 1:N splitter. The fibers are brought together in a bundle and imaged onto a CCD array. One of the fibers is separated from the others and acts as a reference. The position of this reference fiber in the bundle is also located at a distance from the others, whose end faces are positioned on a line parallel to a linear CCD array (inset of Figure 15.13d). If a Fourier transforming (FT) lens is placed between the fiber and the CCD, then the image formed on the CCD array will be the Fourier transform of the fields on the fiber end faces (this situation also occurs if the distance between the fibers and CCD is made large, i.e., the far-field). The interference between the light from a sensor fiber and the reference will produce a unique spatial frequency on the CCD array related to their physical separation, which can be extracted by taking the fast Fourier transform (FFT) of the spatial interference pattern.* The amplitude and phase for each spatial

* The two single-mode fibers act as point emitters of spherical waves. Recalling from elementary optics, the spatial interference pattern formed by two such emitters will exhibit fringes with a pitch related to their separation.

frequency in the FFT then reveals the amplitude and phase of the light from each sensor fiber (Hu and Chen 1995). The technique requires very few optical components and has no moving parts. It also does not require additional modulation techniques to recover the interferometric phase. Sensor information is carried on a unique spatial frequency; thus, crosstalk is expected to be low as each spatial frequency can be uniquely resolved. However, measurement bandwidth is likely to be limited by the processing time of the FFT and the readout speed of the CCD array. This concept of assigning a specific frequency (either spatial or temporal) to a sensor forms the underlying basis of FDM and will be discussed in more detail in Section 15.6.3.

15.6.2 Time Division Multiplexing

In TDM, sensors are individually addressed by the difference in arrival times of their respective signals at a detector. This is implemented by essentially performing a measurement of the *impulse response* (IR) of the sensor network (i.e., the response of the network to a unit impulse). The various techniques differ by the amount of information in the input pulse and whether the measurement is performed in the time or frequency domain. Each sensor is allocated a unique time slot during which its respective information is transmitted to the detector. Provided that information from different sensors does not overlap within the same time slot, a sensor's information can be uniquely decoded, free from crosstalk. TDM can be implemented either directly in the time domain, by interrogating a sensor with a short optical pulse, or in the frequency domain, where a frequency sweep is used to identify transit time differences between sensors by observing the phase shift of the reflected light at each frequency relative to zero delay reference. In the latter case, a Fourier transform is applied to the received signal to transform back to the time domain. Due to the various methods available to implement TDM, it is a highly flexible multiplexing technique and as a result is also highly developed. The direct implementation of TDM using optical pulses is essentially the same as laser range finding that simply measures time of flight of optical pulses reflected from targets. Optical time-domain reflectometry (OTDR), used as an optical fiber diagnostic tool to measure variations in Rayleigh backscatter along a fiber and hence identify regions of high loss, also employs the same principle as TDM. Furthermore, the frequency-domain implementation of TDM can be traced to early radar systems that measured a beat frequency generated by a swept RF signal mixed with a delayed version of itself (i.e., the received *echo*) (Hymans and Lait 1960).

TDM can be implemented in parallel, series, and ladder topologies. It is capable of multiplexing large numbers of sensors (hundreds) over long distances (tens of kilometers) with low crosstalk (<−40 dB). Furthermore, frequency-domain implementations can overcome spatial resolution limitations experienced with conventional TDM. For example, conventional TDM cannot resolve sensors separated by fiber lengths less than a few meters (without the use of ultra-short pulses and very high speed detectors). However, frequency-domain implementations are capable of spatial resolutions less than 100 μm with only low frequency detection electronics.

The ease with which TDM can be combined with WDM enables a further dramatic increase in number of sensors addressable without compromising performance.

15.6.2.1 Time-Domain Multiplexing of Intensity Sensors

An implementation of a TDM ladder topology of intensity sensors is illustrated in Figure 15.14. Here the light from a pulsed source is launched into a ladder array of recirculating optical loop sensors. Each sensor consists of two directional couplers spliced together such that an injected pulse recirculates in a fiber loop. On each transit of the loop, light is coupled out of the second coupler and returned to a detector. The return signal from each sensor consists of a series of decaying pulses corresponding to multiple transits of the input pulse in a single loop. Decaying pulse trains are returned from each sensor at different times, enabling temporal separation of signals from different sensors. It was shown that the power ratio between consecutive pulses in a sensor pulse train is independent of losses in the connecting fibers but dependent on losses within the loop (Spillman and Lord 1987). An intensity sensor of the type described in Section 15.4.1 can be inserted into each fiber loop. To reduce the fiber lengths required in each sensor, the input pulse width must be short ($\ll$1 ns). The loop length must meet the condition, $L_l > c\tau/n_e$, where c is the vacuum light velocity and τ is the pulse width, to ensure separation of recirculated pulses. An array of three sensors was demonstrated using multimode fiber and employing a pulsed laser diode producing pulses with τ = 135 ps. The loop length must therefore be greater than ~2.7 cm with fiber delay lengths between sensors an order of magnitude greater to minimize overlap of pulse trains from consecutive sensors. Some residual crosstalk will always be present as each sensor generates an infinite pulse train of rapidly diminishing pulse power. The measurement bandwidth of a sensor multiplexed in this configuration is limited by the pulse repetition rate of the laser. If we assume that 10 sensors are multiplexed in a ladder and the temporal separation between return signals from each sensor is 10τ, then the maximum interrogation rate is ~50 MHz, which greatly exceeds the response time of any mechanical sensor.

15.6.2.2 Time Division Multiplexing of Interferometric Sensors

TDM of interferometric sensors has received considerable attention, due to their application in acoustic sensing arrays for military surveillance (Davis et al. 1997, Cranch et al. 2004) and oil reservoir monitoring systems (www.tgs.com, Nakstad and Kringlebotn 2008). Implementations can take the form of both serial and ladder topologies operated in reflection or transmission. Parallel topologies using mechanical optical switches do not enable sufficiently rapid switching between sensors to achieve typically desired sensor bandwidths ($\gg$1 kHz) due to the slow switching times (~1–10 ms).

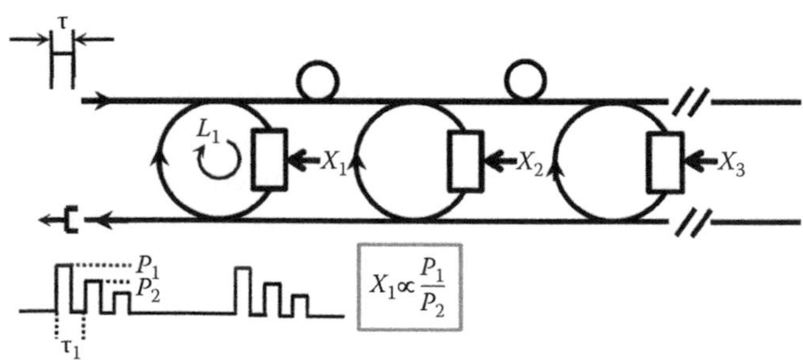

FIGURE 15.14 TDM of self-referenced intensity sensors.

Figure 15.15 shows four multiplexing implementations for interferometric sensors: (a) forward coupled MZIs, (b) forward coupled Michelson, (c) reflectometric/in-line Michelson (using directional couplers or FBG reflectors), and (d) in-line Michelson incorporating WDM. Implementations (a) and (b) are ladder topologies and (c) and (d) are serial topologies. These implementations all operate on the same principle that a single optical pulse injected into the array will generate returned pulses from each sensor in a unique time slot. However, their differences result in slight variations in performance, ease of fabrication, and possibly construction cost. One can find various other implementations of these in the literature (Brooks et al. 1987, Kersey and Dandridge 1989,

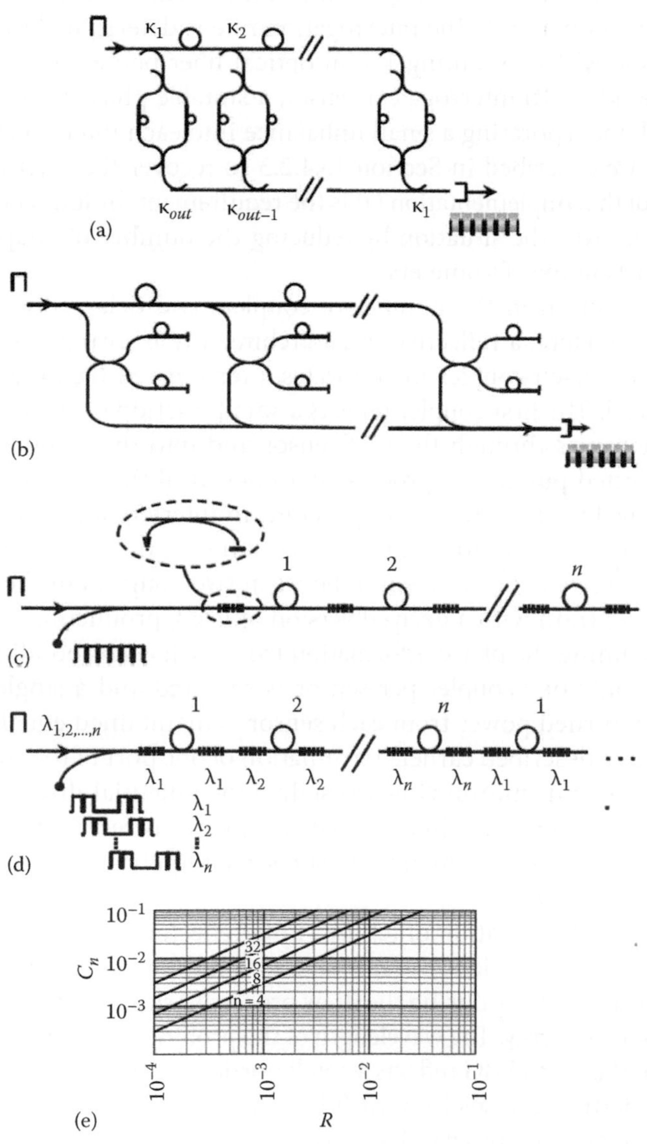

FIGURE 15.15 TDM implementations of interferometric sensors: (a) forward coupled Mach–Zehnder; (b) forward coupled Michelson; (c) in-line Michelson; (d) WDM/TDM inline Michelson; (e) and crosstalk in in-line Michelson interferometers.

Kersey et al. 1989a) as well as hybrid architectures incorporating interferometric and intensity sensors (Lobo Ribeiro et al. 1995).

Referring to Figure 15.15a, if an optical pulse of width τ is launched into the input fiber, a small fraction is coupled into the first *rung*. This passes through a MZI producing two overlapping pulses at the output coupler, which are recombined through a second coupler onto a return fiber. The remainder of the first pulse passes through a delay coil to the second *rung* and the process repeats for all *rungs*. The received signal is a series of overlapping pulses from each Mach–Zehnder in the ladder. The power split ratios of the couplers forming the Mach–Zehnder are 50:50; however, to ensure equal power returned from each rung, the power split ratio of the telemetry coupler must be tailored as described in Section 15.5.2. The interrogation rate is determined by the transit time to the final sensor, which assuming 100 m optical fiber per sensor with 10 sensors is found to be ~200 kHz. To interrogate a sensor, a suitable phase measurement method must be selected. Incorporating a small imbalance into each interferometer enables the use of FM-PGC we described in Section 15.4.2.3 to recover the interferometric phase. A disadvantage of this implementation (a) is the requirement for four couplers per sensor. Figure 15.15b improves the situation by reducing the number of couplers to three per rung using Michelson interferometers.

A dramatic reduction in the number of couplers and delay coils is achieved with implementation (c). Here, a reflective serial architecture is formed by directional couplers (shown inset). Each coupler incorporates a terminated (i.e., index matched) end and a mirrored end. The first coupler reflects a small fraction of the incident pulse. The remaining pulse passes through the first sensor and into the second coupler producing a second reflected pulse. This process continues until the final sensor, producing a pulse train returned to the detector. To generate an interference signal, the pulse train is passed through a compensating interferometer (not shown). This has a path length imbalance equal to twice the path length between two couplers in the array and interferes with the pulse train with a delayed version of itself, producing a series of interference signals containing the phase information from each corresponding sensor. In this implementation, only one coupler per sensor is required and a single fiber addresses all sensors. The returned power from each sensor is maintained equal by tailoring the coupler split ratio as described earlier. Termination of one port of the coupler suppresses multiple reflections, thus minimizing crosstalk. One potential disadvantage is that all fiber between couplers forms the interferometer, which may restrict the maximum separation of sensors. However, for many acoustic sensor applications, this architecture is well suited (Lin et al. 2000).

An alternative implementation of (c) uses FBG reflectors with identical Bragg wavelengths in place of the directional couplers. This is advantageous since it is possible to inscribe FBGs in optical fiber during the draw process (Askins et al. 1994), thus reducing the assembly cost of an array. The wavelength of the probing laser is matched to the Bragg wavelength, such that each FBG reflects a small portion of the incident pulse. Reflections are generated in both directions from each FBG resulting in crosstalk. In such an array, the time-averaged crosstalk observed in sensor n is given by (Kersey et al. 1989b)

$$C_n = \frac{(n-1)R}{(1-R)} \tag{15.52}$$

where R is the power reflectivity of the FBG.* Equation 15.52 is plotted in Figure 15.15e for increasing array size, n. Crosstalk can thus be minimized by reducing R; however, this also comes at the expense of an increased insertion loss of the array.

One solution to reducing crosstalk in an FBG-based array, without increasing the array insertion loss, is shown in Figure 15.15d. Here, each sensor is formed by a matched pair of FBGs. The wavelength of each pair of FBGs is incremented and each wavelength group is repeated along a single fiber. Pulses at each wavelength are launched into the array, and each corresponding FBG pair produces two reflected pulses. The large temporal separation of pulses from consecutive FBG pairs at the same wavelength allows sufficient decay of multiple reflections, resulting in low crosstalk. Thus, the reflectivities of the FBGs do not have to be very low; however, to prevent rapid attenuation of the incident pulse, they should be around a few percent. Of course this improvement in performance comes at the cost of increased complexity of multiplexed laser sources and wavelength demultiplexers.

The insertion loss of these multiplexing topologies can be calculated from the ratio of the peak power returned from a single sensor to the input pulse power. For the case of Figure 15.15c with directional couplers, the insertion loss is

$$IL = \frac{P_o}{P_i} = \frac{0.25}{(N+1)^2} \tag{15.53}$$

The factor of 0.25 accounts for the power loss of the directional coupler separating the forward and return signals, and the factor in the denominator corresponds to two passes through the first coupler. Typical insertion losses are around −27 dB for an eight sensor array including excess losses.

Suitable methods for generating the necessary optical pulses, which are of duration 10–500 ns, are dual-stage Mach–Zehnder modulators or SOA switches. High extinction ratio switching is also necessary, particularly when highly coherent laser sources are used. Low levels of leakage light emitted when the modulator is in the *off* state give rise to excess noise (due to multipath interference) and crosstalk.

15.6.2.3 Combined Time and Wavelength Multiplexing of Fiber Fabry–Pérot Sensors

TDM can be applied to multiplexing FBG and short cavity Fabry–Pérot strain sensors. One such implementation is illustrated in Figure 15.16. FBGs are multiplexed in series along an optical fiber with incrementing wavelength from λ_1 to λ_n. This group of n is then repeated m times to form a serial array with $m \times n$ sensors in total. If a pulse is launched into this serial array that has a broad optical spectrum covering the wavelength range of the FBGs contained within a group (see inset of Figure 15.16), then each FBG will reflect a small portion of the incident pulse. Consecutive FBGs at the same wavelength will each reflect a small portion of the input pulse, which are separated in time at the detector, providing the transit time between the two FBGs exceeds the input pulse width (i.e., $\tau < n_e l_1/c$, where l_1 is the separation between FBGs at the same wavelength or group

* This can be characterized in practice by applying identical signals to all sensors except for sensor n and observing the resulting crosstalk term on sensor n. To express this quantity in dB, one would take $20\log_{10}(C_n)$.

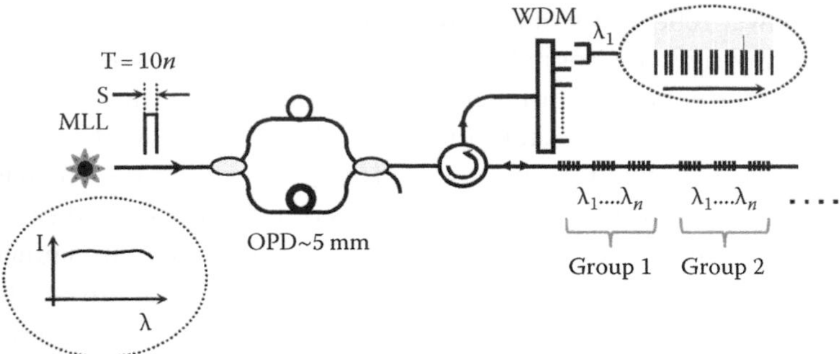

FIGURE 15.16 TDM/WDM of FBGs.

separation). For a 10 ns input pulse, this corresponds to a physical separation between each group of 2 m. The physical separation between FBGs in a group is then determined by their wavelength separation. For example, if the optical bandwidth of the input pulse is 60 nm and the FBGs in a group are spaced by 2 nm, then around 30 sensors can be multiplexed in a group, corresponding to a physical separation of ~7 cm.

This technique is capable of multiplexing large numbers of sensors onto a single fiber (Dennis et al. 1997). However, some careful design constraints must be observed. The FBGs must have a low reflectivity to minimize the amplitudes of multiple reflections between FBGs with the same wavelength. These multiple reflections give rise to cross-talk as pulses generated by secondary reflections arrive at the detector in the same time slot as the primary reflection from later FBGs. Generating a pulse with the desired characteristics requires an advanced optical source. One such source capable of generating these pulses is a mode-locked erbium-doped ring fiber laser (Putnam et al. 1998). This laser generates multiple optical frequencies covering the wavelength range of the gain medium, which is approximately 35 nm for the case of erbium-doped fiber. The spacing of the frequencies is ~100 kHz (due to the very long cavity length) and as such, the emission can be considered continuous. A passive mode-locking method, based on nonlinear polarization switching, forces the laser frequencies to be phase locked, resulting in the output consisting of a train of square pulses. The high intensity within the laser cavity generates Raman scattering, which broadens the spectral width to around 60 nm. This type of laser source is a highly efficient means of generating broadband laser pulses as all the energy stored within the laser cavity is emitted within a pulse. Peak pulse powers exceeding several Watts are possible with modest pump powers of 50 mW at 980 nm at a repetition rate of ~300 kHz.

A demonstration of this multiplexing architecture used interferometric interrogation to convert wavelength shifts of the FBGs into a phase shift (Cranch et al. 2005), as described in Section 15.4.2.1. The interferometer is placed at the input to the FBG array, and the reflected signal is detected and processed in the same way as signals from TDM-based interferometric sensors described in Section 15.6.2.2. The imbalance in the interferometer must be less than the coherence length of the light reflected by an FBG (typically a few mm). Each reflected pulse then contains an interference fringe, with a phase determined by the wavelength of the FBG. Such a system demonstrated strain resolutions on the order of 20–60 $n\varepsilon/Hz^{1/2}$. An alternative configuration replaces the FBG sensors with matched pairs of weak FBGs, each forming a Fabry–Pérot cavity.

A matched interferometer is placed at the input, with a path imbalance equal to the Fabry–Pérot cavity length, typically a few centimeters. Such a configuration is an implementation of the tandem interferometer, described in Section 15.4.2.2, and achieved a strain resolution of 2 nε/Hz$^{1/2}$ from 1 Hz to 1 kHz.

15.6.2.4 Code Division Multiplexing of Interferometric Sensors and FBGs

A special implementation of TDM is known as code division multiplexing (CDMA) and is a spread spectrum technique. These techniques have been used extensively in covert telecommunications systems in multiple-domains including optical, RF, and acoustic. Applying a spread spectrum technique to a signal dramatically broadens its bandwidth, making it indistinguishable from ambient noise and hence ideal for secure communications. One such multiplexed sensor implementation is illustrated in Figure 15.17a. Here, the output of a coherent laser is intensity modulated with a pseudo-random bit sequence (PRBS), which is injected into a ladder array of interferometers. Each interferometer returns a copy of the modulation envelope at a particular time delay, τ_n, determined by the delay coil length between each rung. This relative delay allows addressing of individual sensors. A sensor signal can be recovered by mixing the received signal and the modulating envelop with an appropriate time delay applied (which must be an integer number of bit lengths). The electronic power spectrum of the signal at the detector contains information from all sensors, which is spread over a very broad bandwidth due to the modulation by the PRBS. If the source power is P_i, then the output of the MOD is $P_i \cdot s(t)$, where $s(t)$ represents the PRBS, and the output of the detector can be expressed as

$$i_{ph}(t) \propto \left[P_i \cdot s(t-\tau_1)\right]X_1(t) + \left[P_i \cdot s(t-\tau_2)\right]X_2(t) + \left[P_i \cdot s(t-\tau_3)\right]X_3(t) \tag{15.54}$$

where $X_n(t)$ is the optical power modulating function of each rung and incorporates the sensor information and coupler losses. The time delay for each sensor can be extracted from the cross-correlation, $R(\tau)$, between $i_{ph}(t)$ with $s(t)$ given by

$$R(\tau) = \int_{-\infty}^{\infty} i_{ph}(t) \cdot s(t-\tau)d\tau \tag{15.55}$$

This function, illustrated in the inset of Figure 15.17a, contains a series of correlation peaks corresponding to each sensor's time delay, τ_n. One method to recover a sensor signal is to multiply $i_{ph}(t)$ by $s(t-\tau_n)$ and low-pass filter above the measurement bandwidth, which will recover $P_i X_n(t)$, as illustrated in the inset of Figure 15.17a.

Benefits of this technique are as follows: (1) improved power budget and sensor bandwidth compared to TDM (due to the sensor being interrogated continuously) and (2) improved spatial resolution as the sensor spacing is now limited by the bit length (which may be <100 ps). Early demonstrations yielded encouraging results, but demonstrated high phase noise (~1 mrad/Hz$^{1/2}$) associated with residual coherence from other optical paths in the system and high crosstalk (Al-Raweshidy 1994). Later demonstrations improved on both these aspects achieving a phase resolution of 100 μrad/Hz$^{1/2}$ and crosstalk less than −60 dB from an array of eight sensors (Kersey et al. 1992). This technique

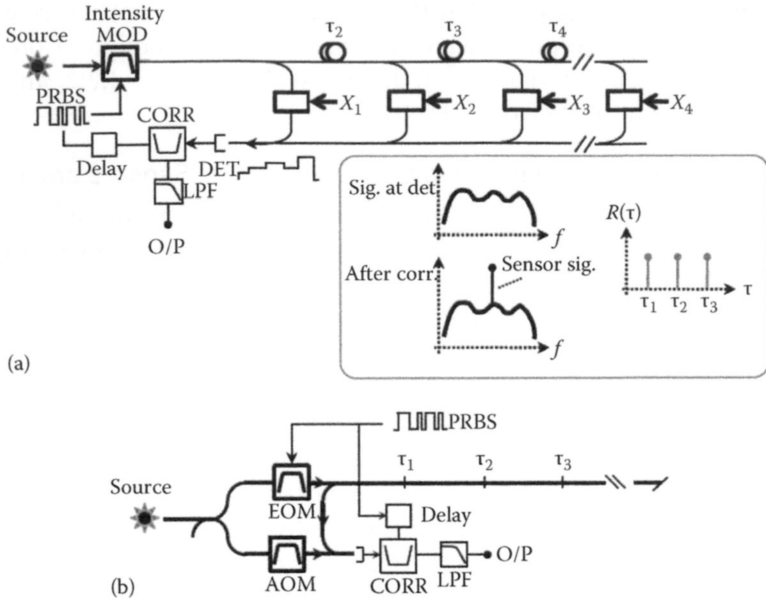

FIGURE 15.17 Code division multiplexing implementation: (a) through source intensity modulation and (b) through interferometric phase modulation.

has also been applied to interrogation of FBG strain sensors, enabling close wavelength spacing without compromising dynamic range (Koo et al. 1999). In conventionally multiplexed FBG sensors using WDM, the wavelengths of each FBG cannot overlap. Using CDMA, the wavelengths from each FBG can now overlap each other, as the signal from each wavelength can still be uniquely reconstructed due to the time delay between FBG sensor signals.

Another implementation of this concept has been demonstrated more recently (Shaddock 2007, Wuchenich et al. 2011) and is illustrated in Figure 15.17b. Here, the interferometric phase (rather than the intensity) is modulated with the PRBS. An electro-optic phase modulator (EOM) is incorporated into the sensing arm of a Michelson interferometer. Reflections are generated at different points along the sensing fiber, which are combined with light from the reference arm through a second coupler. An acoustic-optic modulator (AOM) is placed in the reference arm to generate a carrier for demodulation purposes. The optical field from the sensor arm is given by

$$E_s = E_1 \cdot e^{-i\phi_1} \cdot s(t-2\tau_1) + E_2 \cdot e^{-i\phi_2} \cdot s(t-2\tau_2) + E_3 \cdot e^{-i\phi_3} \cdot s(t-2\tau_3) + \cdots \quad (15.56)$$

where

E_n is the optical field reflected from each feature
ϕ_n and τ_n are the corresponding phase and time delay, respectively

If E_r is the optical field from the reference arm, then the photocurrent is given by

$$i_{ph} \propto \left\langle (E_s + E_r) \cdot (E_s + E_r)^* \right\rangle \quad (15.57)$$

The detected signal is processed by performing a correlation with the PRBS for differential time-delays corresponding to the reflecting features. This involves mixing the detected signal with a suitably delayed version of the PRBS and applying a low pass filter, cut-off at the desired measurement bandwidth. This technique thus permits the amplitude and phase of the optical signal from the reflecting region to be isolated from the signals originating from other features in the sensing fiber. Crosstalk was measured to be −52 dB, and displacement resolutions on the order of 10^{-10} m/Hz$^{1/2}$ at 1 Hz were demonstrated, which corresponds to a phase resolution of ~1 mrad/Hz$^{1/2}$. The spatial resolution of such a technique is limited by the bit length of the PRBS, and the range is limited by the total length of the PRBS. A code frequency of 10 GHz would yield a spatial resolution of ~2 cm.

15.6.2.5 Amplified (Active) Arrays

Optical amplification became a practical technique for fiber-optic systems after the invention of the EDFA (Desurvire et al. 1987, Mears et al. 1987). Amplification of a weak optical signal around 1550 nm by greater than 30 dB is now easily achievable, requiring a pump power of ~100 mW. Furthermore, developments in high power fiber lasers have enabled Raman amplification to become a practical technique, being particularly well suited to low noise distributed amplification. EDFAs have been incorporated into TDM arrays to overcome high telemetry losses. They can also boost the launch power and improve the receiver sensitivity (Lin et al. 2000, Cranch et al. 2003a). Careful design of the system prevents these benefits from being negated by the addition of excess noise from ASE. Figure 15.18a illustrates such a configuration with a remotely pumped power-EDFA and pre-EDFA. The pump for the remote pre-EDFA can be delivered through the

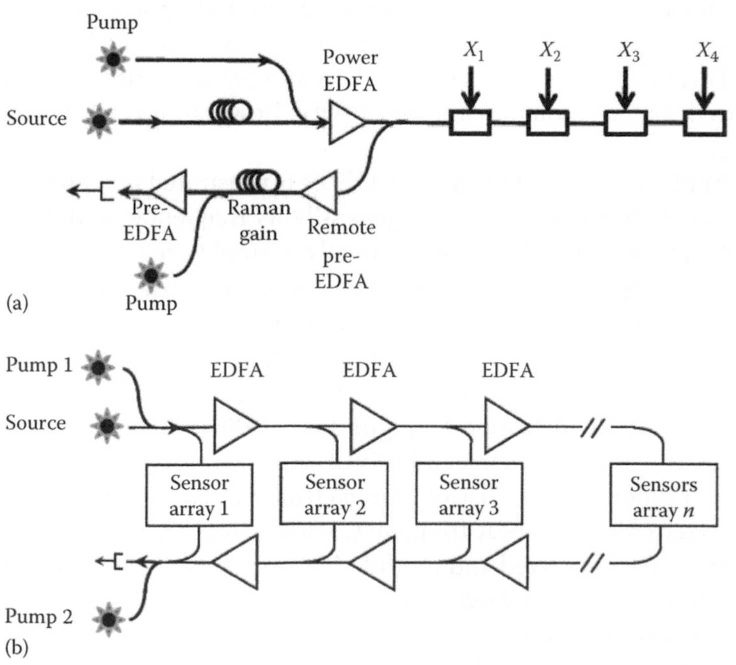

FIGURE 15.18 (a) EDFA/Raman amplified interferometric and FBG arrays and (b) ladder array with amplified telemetry.

return fiber, providing additional Raman gain to the return signals. Standoff distances much greater than 50 km have been demonstrated (Huh et al. 2011).

Figure 15.18b illustrates a ladder topology incorporating EDFAs between the ladder rungs to overcome splitting losses. In such a system, a significant source of noise will be ASE generated by the amplifiers. An appropriate means of characterizing the addition of ASE to the signal is by the noise figure defined as

$$F = \frac{OSNR_{in}}{OSNR_{out}} \tag{15.58}$$

where *OSNR* is the optical SNR. Analyses have shown that despite the introduction of additional noise due to ASE, an improvement in electrical SNR can be obtained with this type of architecture (Sæther and Bløtekjær 1997, Hodgson et al. 1998a,b). A 12 dB improvement in noise figure with a 10-element ladder topology incorporating 20 Er:Yb-doped fiber amplifiers has been reported (Hodgson et al. 1997, Vakoc et al. 1998) compared to an array without optical amplification.

15.6.2.6 Time-Domain Multiplexing with Incoherent Optical Frequency-Domain Multiplexing

TDM can also be implemented in the frequency domain using techniques such as OFDR. These methods are generalized implementations of the frequency-domain continuous wave (FMCW) technique, which will be described in Section 15.6.3.1. However, their inclusion within the TDM section is due to their ability to measure the complex IR of a sensor array. There are two implementations of OFDR: based on modulation of the optical intensity (incoherent OFDR) or sweeping the optical frequency (coherent-OFDR). In incoherent OFDR, the information is encoded on an RF carrier, whereas in coherent OFDR, the information is encoded on an optical carrier. The latter technique offers improved spatial resolution and sensitivity; however, it is slightly more complex to implement.

The basic implementation of incoherent OFDR is illustrated in Figure 15.19. This particular implementation is known more generally as frequency-modulated continuous wave (FMCW) incoherent OFDR, since it involves amplitude modulating the envelope (or intensity) of the laser emission with a periodic waveform (MacDonald 1981, Venkatest and Dolfi 1990). In this figure, a laser diode is amplitude modulated with a swept frequency from a few kHz to a few GHz and the output is launched into a fiber. Replicas of the modulated signal are backscattered due to Rayleigh scattering and reflections from discontinuities. A simply detection configuration is illustrated in Figure 15.19a where the backscattered signal is detected and mixed with the (undelayed) modulation signal driving the laser diode. For the case of a single backscattering feature, the signal generated by the mixer contains a beat frequency, with a frequency dependent on the time delay between the backscattered and modulation signals and whose amplitude depends on the strength of the scattered light and accumulated loss in the sensor network. If a single discrete discontinuity is located at a position, L_d, then the returned signal will be a replica of the modulation signal, delayed by approximately, $\tau_d = 2L_d n_e/c$. The instantaneous frequency of the modulation and backscattered signal is shown in the inset of Figure 15.19. Multiplying this with the modulation signal and low pass filtering will

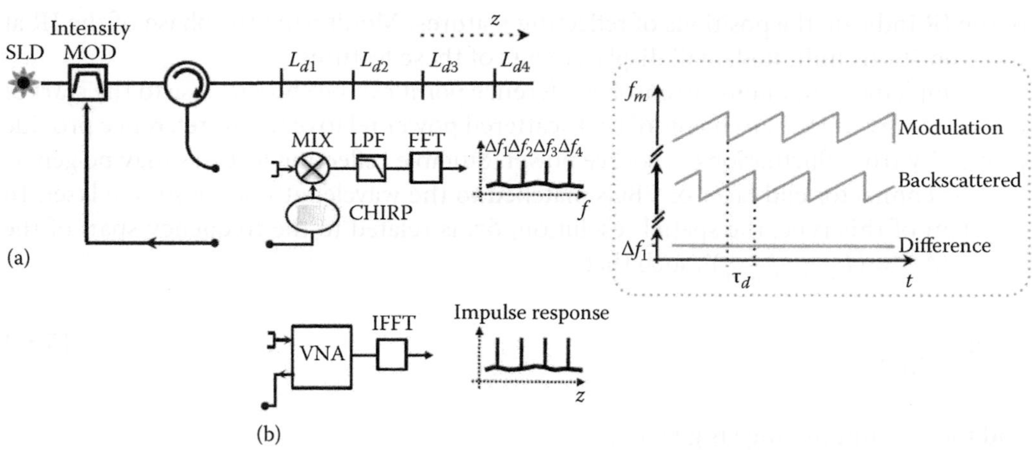

FIGURE 15.19 TDM of intensity sensors using incoherent OFDR implemented (a) with simply synchronous detection and (b) by measuring the TF of the sensor array.

produce a difference frequency, Δf_1, which is related to the time delay between the two signals, also shown in the inset of Figure 15.19 (lower trace), and hence the position of L_d.* Introducing other discontinuities will generate more reflected signals, each producing its own prominent unique beat frequency. It is useful to note the similarity between the received signal with this technique and that obtained with the CDMA method described in Section 15.6.2.4. The received signal is of the form given by Equation 15.54, with $s(t)$ now representing the chirp modulation. In the CDMA method, the sensor information is spread over a very broad bandwidth, due to its modulation by the PRBS. In the case of incoherent OFDR, the sensor information is also spread over a broad bandwidth. However, mixing the received signal with the modulating signal generates a unique carrier frequency determined by the time delay. Each sensor signal is therefore assigned to a specific frequency by the modulating signal. Taking the Fourier transform of this received time-domain signal produces a frequency *map* of discontinuities within the fiber, where the amplitude of each frequency component is related to the power reflected from the discontinuity and the frequency is proportional to the time delay. This signal is closely related to the IR of the sensor array. The implementation described in Venkatest and Dolfi (1990) uses an optical modulator to perform the mixing between the received signal and modulation in the optical domain, removing the need for high speed detection electronics.

An alternative method of processing the received signal can yield far more detailed information by retaining both the amplitude and phase information from the received signal. If the complex frequency response or TF is measured between the received signal from the sensor array and the modulation signal, then the inverse Fourier transform of the TF yields the complex IR of the sensor array.† In Liehr and Krebber (2010), the TF was measured with a vector network analyzer and the inverse fast Fourier transform (IFFT) is taken to obtain the IR, as illustrated in Figure 15.19b. The peaks in the amplitude

* One can also think of this as an RF interferometer with the delay implemented in the optical domain and a swept frequency input.

† This relationship forms the basis of linear systems theory.

of the IR indicate the positions of reflecting features. Monitoring the phase of the IR at these positions indicates small displacements of these features.

To implement an intensity sensor, a reference point can be chosen close to the start of the sensor array. Measurements of backscattered power relative to this reference provide immunity from fluctuations in source power. Suitable reflecting features may be generated by connector end faces or FBGs matched to the wavelength of the source laser. In a system of this type, the spatial resolution, δz, is related to the frequency span of the sweep ($\Delta f_m = (f_{m,\max} - f_{m,\min})$), such that

$$\delta z = \frac{c}{2n_e \Delta f_m} \tag{15.59}$$

and the maximum range is given by

$$z_{\max} = \frac{c}{2n_e \delta f_m} \tag{15.60}$$

where δf_m is the frequency step interval. Assuming a continuous frequency sweep, the frequency step interval is limited by the sample rate of the received signal. Thus, for $\Delta f_m = 2$ GHz and assuming 10,000 points are sampled within a sweep to yield $\delta f_m = 200$ kHz, then we find $\delta z \sim 5$ cm and $z_{\max} \sim 500$ m. If one sweep occurs per second, the sample rate is 10 kHz. One can see from Equations 15.59 and 15.60 that there is a direct trade-off between spatial resolution and range if the measurement time is to be kept constant. This technique has been implemented to measure the local increases in backscatter in polymer optical fiber due to applied strain (Liehr et al. 2010). Also, by measuring differences in frequencies between two discrete reflecting points, the change in length can be measured. Length changes less than 1 μm at measurement frequencies up to 2 kHz have been demonstrated (Liehr et al. 2010).

15.6.2.7 Time Division Multiplexing with Coherent Optical Frequency-Domain Reflectometry

Coherent OFDR is implemented by sweeping the optical frequency of the laser instead of modulating its envelope or intensity. A simplified implementation is shown in Figure 15.20a.

Here, the output of a frequency swept external cavity laser diode is launched into a Michelson interferometer. The light is split into two arms; one arm is terminated with a mirror at a position, L_{ref}, acting as the reference and the other arm is the sensor with a series of reflecting elements, described by an amplitude reflection coefficient, r_n. For the case of a single discrete element at a distance, L_d, the effective path imbalance in the interferometer is, thus, $L_d - L_r$, yielding a photocurrent (similar to Equation 15.16) given by

$$i_{ph}(t) \propto \left(A + B|r_1|\cos(\omega' t)\right) \tag{15.61}$$

where

A and B are constants

ω' is proportional to the imbalance and the frequency sweep rate

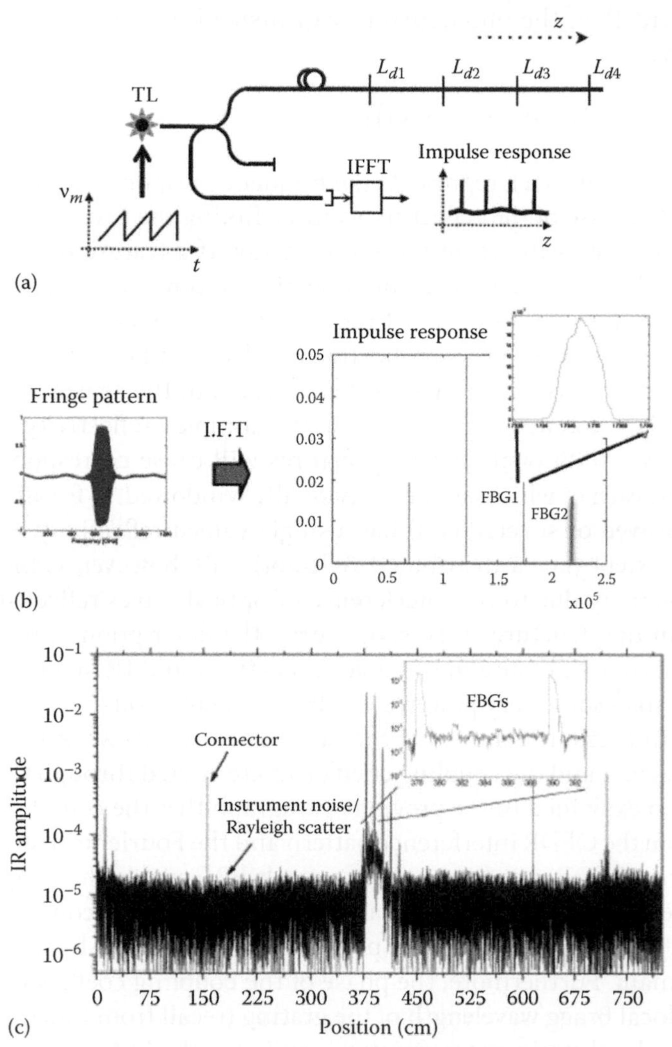

FIGURE 15.20 (a) TDM using coherent OFDR; (b) signal processing steps to extract the IR of an FBG sensor; and (c) example IR of an 8 m optical fiber illustrating signals from connectors and FBGs.

The interferometric portion of the photocurrent comprises a time varying term whose frequency is proportional to the position of the reflecting point and amplitude is proportional to the strength of the reflector. Measurement of ω' yields L_d–L_r, since the sweep rate is known. Generalizing the representation of the spectrum of the reflecting feature to exhibit a complex reflectivity as a function of optical frequency, ν, given by, $\tilde{r}(\nu) = |r(\nu)|\exp(i\varphi(\nu))$, the photocurrent is now given by

$$i_{ph}(t) \propto \left(A(\nu) + B|r(\nu)|\cos\left(\omega' t + \varphi(\nu)\right)\right) \tag{15.62}$$

Thus, the interferometric (second) term contains all the information on the complex reflectivity of the reflecting point. Assuming the frequency sweep of the laser is exactly linear,* then the photocurrent can instead be expressed as a function of laser frequency:

$$i_{ph}(\nu) \propto \left(A(\nu) + B|r(\nu)|\cos(\omega' t + \varphi(\nu)) \right) \tag{15.63}$$

The complex reflectivity, $r(\nu)$, represents the frequency response discussed in the previous section, in this case of the reflecting feature. Taking the inverse Fourier transform of Equation 15.63 yields the *IR* of the sensor array. If a feature of interest exists at a spatial position, L_d, then applying a time-domain window, centered on the time delay corresponding to $(L_d - L_r)n_e/c$, to the IR removes the components associated with $A(\nu)$ (this term also incorporates terms proportional to $|r(\nu)|$ and is thus no longer constant) as well as components associated with other features in the sensor array. The Fourier transform of this windowed function yields the complex reflectivity of the reflecting feature of interest. Additional reflecting features will cause corresponding features to appear in the IR, each of which can be individually windowed. A feature with no spatial extent (i.e., a cleaved or silvered end) has a single valued reflectivity. A structure with a finite spatial extent (i.e., a distributed reflector) will, however, exhibit a frequency-dependent reflectivity due to the interference of optical waves reflected from different positions within the structure. This is, of course, the description of the FBG. Thus, we have described how to measure the complex reflectivity and IR of a FBG. We shall now see how this enables some very powerful multiplexing methods.

We showed in Section 15.4.4.1 that for the special case of weak FBGs, the complex reflectivity spectrum and the coupling coefficient are related through the Fourier transform. It was also explained in the previous paragraph that the complex reflectivity can be obtained from the OFDR interference pattern and the Fourier transform of this yields the IR. Thus, the OFDR measurement of a weak FBG yields the coupling coefficient of the FBG. We have therefore devised a way of measuring the coupling coefficients of FBGs inscribed in the sensor fiber arm (provided they are weak), by separating their IRs in the time domain. Furthermore, the phase of the coupling coefficient contains information on the local Bragg wavelength of the grating (recall from Equation 15.27), which can be related to local strain or temperature applied to the FBG.

In summary, a single swept frequency measurement yields the IR of the sensor arm. If the sensor arm contains weak FBGs, then the respective regions of the IR are directly proportional to the coupling coefficients of the FBGs. The processing steps are illustrated in Figure 15.20b. Here, the fringe pattern measured as a function laser frequency is shown. The Fourier transform of this pattern yields the IR showing four features. Two of the features are FBGs, with the inset showing an expanded view of the IR for FBG1. This is the same FBG shown in the right side plot in Figure 15.9b.

It is now possible to compare the performance of this technique with the incoherent OFDR. Firstly, it is evident from Equation 15.63 that the interferometric part of the photocurrent is proportional to the amplitude of the reflection coefficient (and hence the amplitude of the reflected electric field). In incoherent OFDR, the photocurrent is proportional to the backscattered power; thus, the dynamic range of coherent OFDR

* In practice, the frequency sweep of the laser is not linear. An unbalanced reference interferometer is used to either measure the instantaneous frequency of the tunable laser or act as a trigger for digital sampling of the photocurrent $i_{ph}(t)$, yielding an evenly sampled interferogram with optical frequency.

is expected to be much larger than incoherent OFDR by up to a power of two. This also results in coherent OFDR being more sensitive than incoherent OFDR. Coherent OFDR is subject to the same fundamental limitations of spatial resolution and range as incoherent OFDR; however, the sweep range in coherent-OFDR is much larger. If the laser frequency sweep is 20 nm (Δf_m = 2.5 THz), then using Equation 15.59 yields a spatial resolution of 40 μm. If the laser sweeps at a rate of 20 nm/s and the detector signal is sampled at 10 kHz, then the wavelength step size is 2 pm ($\delta\nu$ ~ 250 MHz). Using Equation 15.60 yields a range of only 40 cm; however, in practice, the sample rates can be very high (≫1 MHz), dramatically increasing the range. Therefore, coherent OFDR yields a dramatic improvement in spatial resolution and higher sensitivity, but with a more limited range. Multiplexing hundreds of FBGs has been demonstrated with this technique (Childers et al. 2001).

Recent developments in rapidly tunable, coherent lasers, operating around 1550 nm, have enabled the full potential of coherent OFDR to be realized. Its primary advantages over conventional OTDR are spatial resolution (tens of micrometers compared with several meters) and its ability to enable distributed strain and temperature measurement. An example IR from an 8 m optical fiber measured with coherent OFDR is shown in Figure 15.20c. Here, the reflecting features corresponding to a connector and two FBGs (observed as two prominent peaks at 375 cm) are clearly visible. The measurement noise is due to a combination of instrument noise and Rayleigh scatter from the optical fiber. The three features observed following the two FBGs are due to multiple reflections between the FBGs. The inset shows a magnified version of the FBGs IRs.

One fascinating extension of this technique is its application to measuring distributed strain by the effect it has on the local spectra of Rayleigh backscatter in single-mode silica fiber (Froggatt and Moore 1998, Lunatechnologies 2013). Measurement of local strain and temperature has been demonstrated by measuring changes in the spectral distribution of the Rayleigh scattering with spatial resolutions on the order of centimeters and strain and temperature resolutions of 1 με and 0.1 K. Such a technique has been applied to structural monitoring in nuclear power reactors (Sang et al. 2008). One limitation of these swept frequency methods is their relatively low speed. Measurement time is limited by the sweep speed of the tunable laser, which is typically on the order of 1 s, although this can be increased by reducing the sweep range. Consequently, these measurement techniques are usually limited to static and quasi-dc sensing; however, development of rapidly tunable laser diodes has increased this measurement rate somewhat.

15.6.3 Frequency Division Multiplexing

The OFDR techniques described earlier could be classified as FDM methods as each sensor is distinguished by a specific spatial frequency. However, we also showed how the received signal can be processed to retrieve the IR of the sensor array, hence its inclusion under TDM. This illustrates that clear distinction between frequency-domain TDM methods and other common FDM methods is difficult. Other derivatives of these more general OFDR are also possible. In fact, FDM was one of the first multiplexing techniques developed for fiber-optic sensors due to the availability of necessary modulators for its implementation. FDM systems operate on the principle that a sensor can be identified by a unique carrier frequency and can be implemented in serial, parallel, and ladder topologies. As well as being easy to implement with phase modulators or by direct laser diode modulation, FDM techniques involve continuous interrogation of the sensors, potentially yielding higher SNR. However, since sensors are each assigned a region of electronic bandwidth, restrictions on total available electronic bandwidth result in lower multiplexing gains than that achievable with TDM.

15.6.3.1 FMCW and FM-PGC Applied to Interferometric Sensors and FBGs

FDM can be applied to interferometric sensors using pseudo-heterodyne or its derivative FM-PGC techniques as explained in Section 15.4.2.3. Figure 15.21a illustrates an implementation of FMCW. Here a laser diode is frequency modulated with a serrodyne ramp, the output of which is launched into an array of three MZIs in a ladder topology. Each MZI has a different path-imbalance, such that at the detector, a carrier frequency dependent on the imbalance and ramp rate is generated in accordance with Equation 15.16. The phase information for each interferometer appears as modulation sidebands on each carrier frequency. The bandwidth of a sensor is thus determined by the spacing of the carrier frequencies.

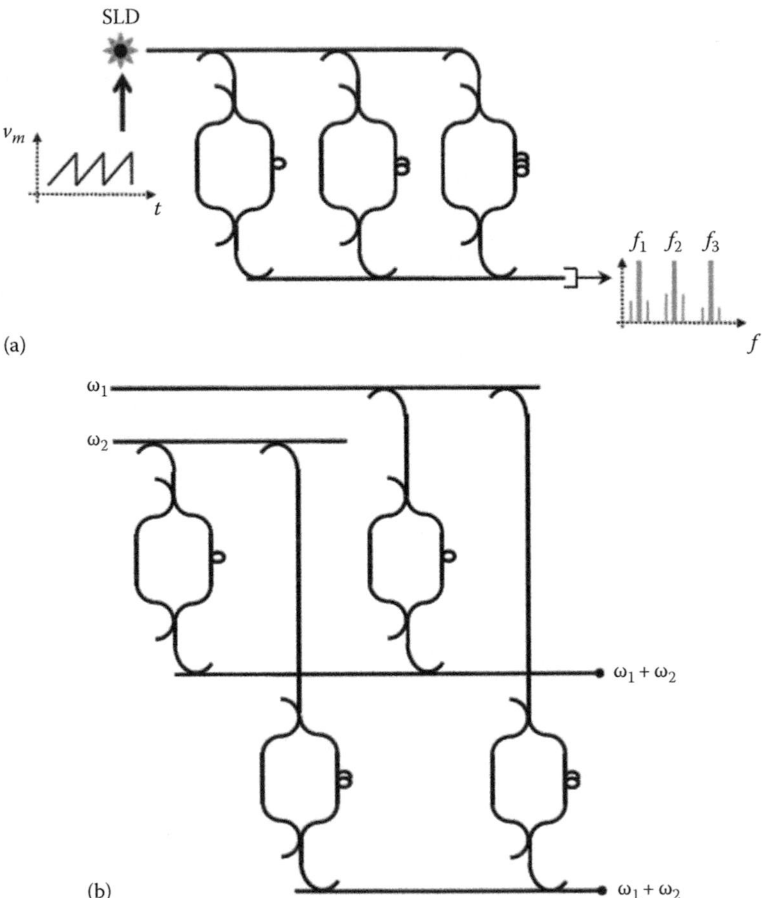

FIGURE 15.21 Interferometric sensor multiplexing with: (a) pseudo-heterodyne interrogation and (b) FM-PGC.

The performance of systems based on pseudo-heterodyne has never matched other techniques such as phase-generated carrier and heterodyne. This is partly due to the difficulties associated with generating perfectly linear frequency ramps and conversion of laser frequency noise to phase noise. These can be avoided using sinusoidal modulation of the laser frequency (i.e., FM-PGC) (Dandridge et al. 1987), a technique described in Section 15.4.2.3. Here, a frequency modulation is applied to the source laser (say by current modulating a laser diode) and interrogates an imbalanced interferometer with a modulation depth set to ensure balancing of the amplitudes of the detected carrier harmonics corresponding to the sine and cosine. If additional laser diodes are each frequency modulated at a different frequency, then FDM can be employed to combine signals from multiple interferometers, each with a different carrier frequency, onto a single fiber. Figure 15.21b illustrates this concept for multiplexing four interferometers. The number of sensors that can be interrogated scales as the square of the number of source lasers. Modulation frequencies and receiver bandwidths must be carefully chosen to allow separation of the carrier spectral components to enable PGC demodulation with adequate dynamic range. Reported demonstrations multiplexed around eight sensors per fiber with a sensor bandwidth around 10 kHz (Dandridge et al. 1994). This technique has also been successfully implemented in hydrophone arrays (Dandridge et al. 1994). The high bandwidth achievable with this technique compared with coherent-OFDR arises from the possibility to sinusoidally modulate, either by direct current injection of an SLD or using an external modulator, at frequencies exceeding 1 MHz.

Other variations of this basic method have also been reported for interrogating a serial array of MZIs with different path imbalances (Sakal et al. 1987) as well as extrinsic displacements sensors (Zheng 2005, 2007). Frequency multiplexing techniques have been applied to FBG sensors using the incoherent OFDR technique illustrated in Figure 15.19a. FBGs can be multiplexed both in series and parallel, separated with by a suitable delay. The processed signal will comprise a unique carrier frequency for each FBG, enabling individual sensor addressing (Chan et al. 1999, 2000).

15.6.4 Wavelength Division Multiplexing

WDM (WDM) of fiber-optic sensors owes much of its success to its rapid development and subsequent adoption in optical fiber telecommunication systems. This has driven the development of low-cost, high-performance components necessary for implementing the technique. Devices based on thin film filter and FBGs are available for multiplexing and demultiplexing wavelengths with spacing less than 0.4 nm (50 GHz at 1550 nm). Wavelength multiplexing is generally not capable of matching the multiplexing gain achieved with TDM; however, in combination, they can achieve very large sensor counts exceeding many other methods (with the possible exception of coherent OFDR, although one can argue that this is a form of time and wavelength multiplexing). WDM utilizes the very large bandwidth available in optical fiber (tens of terahertz) to encode sensor signals. Due to the large optical frequency separation between sensor signals, demultiplexing is carried out in the optical domain, in contrast to FDM where demultiplexing is usually carried out in the RF domain. Analysis of wavelength multiplexed systems therefore requires use of an optical spectrum analyzer as well as an electronic spectrum analyzer. A benefit of WDM systems is that crosstalk is typically very low due to the ease with which wavelengths can be efficiently separated in the optical domain. It was shown in Section 15.6.2.3 how FBG sensors can be efficiently multiplexed with time and WDM using a MLL that generated a short pulse with a broad optical bandwidth. Interferometry was used to decode the wavelength shift of the FBG. There are many other techniques available for multiplexing and decoding FBG signals. These will now be discussed.

15.6.4.1 Wavelength Multiplexing of FBG Sensors

FBG sensors are inherently well suited to WDM. However, since the FBGs' wavelength both uniquely identifies the sensor and encodes the sensor information, decoding methods have to be carefully designed. A trade-off exists between the dynamic range of the sensor and the minimum permissible wavelength spacing between adjacent FBGs. Figure 15.22 illustrates a serial array of FBG sensors interrogated by a broadband source. Such a source may be an ELED, SLED, or EDFA (described in Section 15.3.1).

The output of a broadband source is injected into the serial array of FBG sensors. Each FBG reflects a unique wavelength that is returned to a detector through a circulator. In its simplest form, an optical spectrum analyzer can be used to monitor the positions of the reflected peaks, illustrated in Figure 15.22a. These typically achieve a spectral resolution ~10 pm. However, dedicated spectrometers have been developed for improved performance. These operate by reflecting the received light from a dispersive grating and imaging this onto a linear CCD array, illustrated in Figure 15.22b. This provides the ability for high speed measurement (~kHz) with high spectral resolution (Askins et al. 1995, Chen et al. 1998). The best combinations of speed and resolution are currently achieved with CCD arrays operating in the 850 nm region;

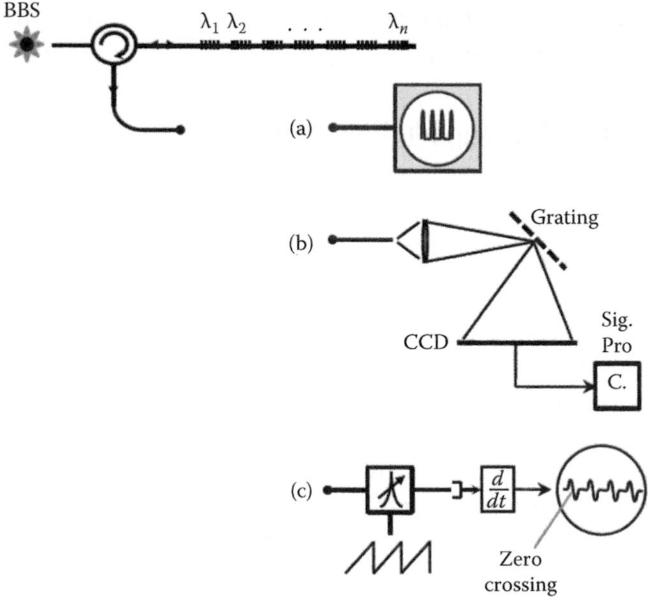

FIGURE 15.22 Multiplexing of FBG sensors using: (a) optical spectrum analyzer; (b) CCD spectrometer; and (c) scanning Fabry–Pérot.

however, these require more specialized single-mode components operating in this wavelength region. Full spectral information is available, with this method allowing sophisticated peak tracking algorithms to be applied to monitor shifts in the FBG center wavelength (Lucas et al. 2011). Techniques such as these can achieve quasi-static strain resolutions less than 1 $\mu\varepsilon/Hz^{1/2}$ corresponding to a spectral resolution of 1.5 $pm/Hz^{1/2}$ at 1550 nm. This method can also be used to interrogate multiplexed LPG sensors (Balzhiev et al. 2011).

Another method utilizes a scanning Fabry–Pérot filter that demultiplexes the FBG signals. An example is illustrated in Figure 15.22c. Here a tunable, high-finesse Fabry–Pérot filter is driven by a voltage ramp, causing its transmission resonance to scan across a wavelength range covering the FBG sensor signals. The output of the filter consists of the optical spectrum from the sensor array convolved with the transmission spectrum of the filter. A series of peaks are observed corresponding to each FBG. This signal is repeated at a rate given by the repetition rate of the applied voltage ramp. The wavelength separation of the sensor signals has been translated into a temporal separation by the scanning filter. Small changes in the FBG wavelength will be observed as changes in the positions of the peaks measured at the output of the tunable filter (TF). These can be accurately tracked by differentiating this signal and monitoring the points of zero-crossing (Kersey et al. 1993). This method has also been combined with interferometric detection to achieve high strain resolution (few $n\varepsilon/Hz^{1/2}$) at frequencies above a few Hertz over a bandwidth of several kHz (Johnson et al. 2000). The maximum dynamic range for the case of 35 FBG sensors spaced by 1 nm over the C-band (extending from 1530–1565 nm), will be ~60 dB, assuming a strain resolution of 1 $\mu\varepsilon/Hz^{1/2}$.

15.6.4.2 Wavelength Multiplexing of Intensity Sensors

Multiplexing of intensity sensors generally requires incorporating an intensity referencing method to provide immunity from variations in the source power and losses in the connecting lead. Figure 15.23 illustrates a method of multiplexing intensity sensors with WDM and a frequency modulation–based self-referencing technique (Abad et al. 2002).

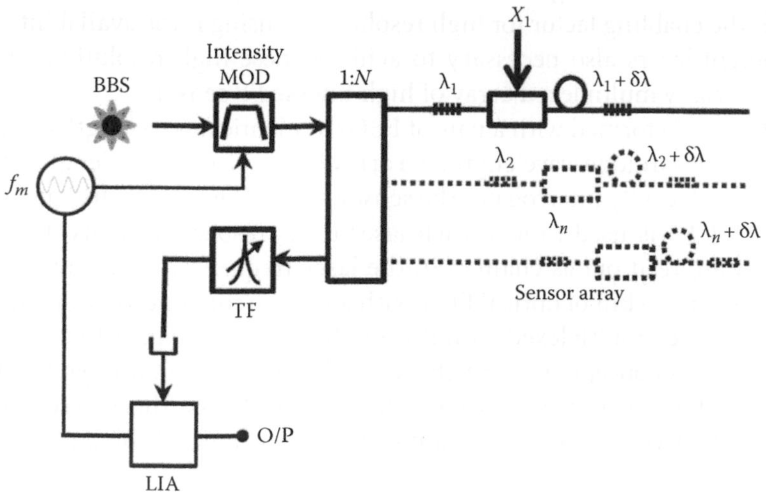

FIGURE 15.23 Self-referenced multiplexing of intensity sensors.

The output of a broad-band source is intensity modulated at a frequency f_m, which is launched into a parallel array of intensity sensors though a 1:N splitter. An intensity sensor is placed between two FBGs with slightly different Bragg wavelengths to prevent coherent interference of the reflected signals. The reflections from each FBG pair produce a signal on the detector, which is the sum of two amplitude-modulated signals. The amplitude of the signal reflected from the first FBG is affected only by losses preceding the sensor. The amplitude of the signal from the second FBG depends on these losses plus the losses from the intensity sensor. Crucially, a phase difference also exists between these two signals due to the optical path delay between the two FBGs. Variations in the sensor loss will change the relative power between these two signals. This will be observed as a change in phase of the signal at the detector, which can be measured with a lock-in amplifier.* A TF can be used to separate signals from different intensity sensors; however, later demonstrations have used a wavelength division demultiplexer with multiple detectors (Montalvo et al. 2009). Encoding the sensor signals as a phase shift significantly reduces the sensitivity to fluctuations in the source power or losses in the connecting leads. Appropriate choice of the modulation frequency (~100 kHz) and delay length between the FBGs yields a near-linear dependence of the measured phase shift on the sensor loss; however, it is necessary for the delay length to be relatively large (~300 m). This technique has also been adapted to multiplex FBG and intensity sensors in series (Abad et al. 2003).

15.6.4.3 Multiplexing High-Finesse Fabry–Pérot Interferometers with Laser Frequency Locking

Interferometric sensors can achieve very high strain resolution (~10^{-14} ε) and high bandwidths (up to MHz). However, many of the configurations discussed in Section 15.4.2.2 require either large numbers of directional couplers (at least 1 per sensor) and/or long fiber lengths (≫10 m per sensor) to achieve a high resolution. A technique capable of achieving high strain resolution with short cavity (few cm) Fabry–Pérot sensors has been recently demonstrated based on frequency locking a laser to the optical cavity (Chow et al. 2005). Although the basic concepts of laser frequency locking are well established, the enabling factor for high resolution sensing is the availability of low cost, highly coherent lasers also necessary to achieve these high resolutions. Figure 15.24 illustrates a serially multiplexed array of high finesse FP sensors.

Each FP cavity is formed with a pair of FBGs at identical wavelengths, such that each cavity operates at a unique wavelength. An array of lasers with wavelengths matched to a resonance of each cavity interrogates the sensor array. The PDH technique we described in Section 15.4.2.3 is used to lock each laser to its respective cavity. Changes in the cavity length are read out as changes in the laser frequency. An array of four sensors interrogated over 100 km of optical fiber with a measurement resolution on the order of 1 pε/$Hz^{1/2}$ have been multiplexed with this method (Littler et al. 2010).

This sensor array concept can be further extended by incorporating gain sections within each cavity, yielding an array of wavelength multiplexed lasers. The gain section comprises a short length (few centimeters) of erbium-doped fiber in which Bragg grating reflectors

* Recall that adding two sine waves of the same frequency, but with a phase difference, will cause the phase of the resulting signal to change if the relative amplitudes of the two initial sine waves are varied.

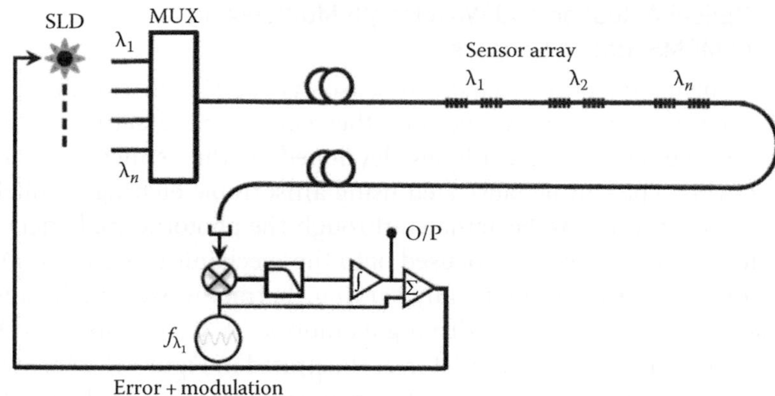

FIGURE 15.24 Wavelength multiplexed high-finesse Fabry–Pérot sensors using the PDH laser locking method.

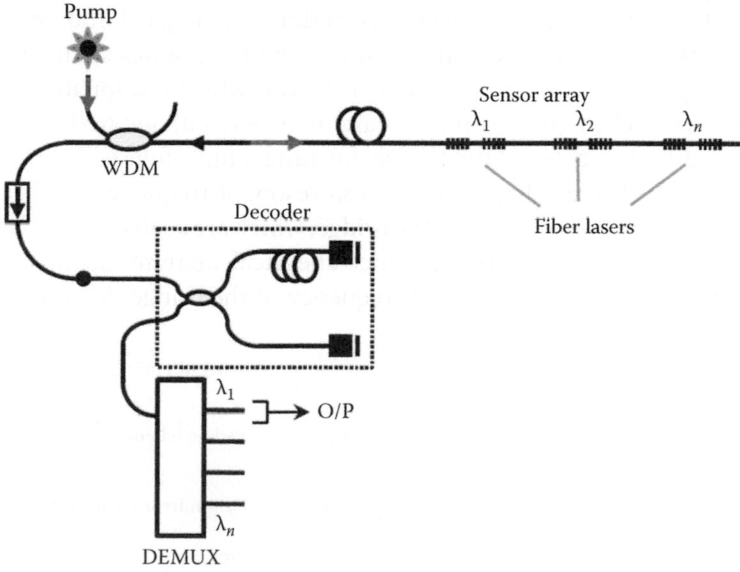

FIGURE 15.25 Fiber laser strain sensors with interferometric decoding.

are written. A SLD is used to optically pump the sensor array, as illustrated in Figure 15.25. Changes in the cavity length of each laser are now encoded on the respective laser frequency, which can be decoded using the interferometric techniques described in Section 15.4.2.3 (Cranch et al. 2008). One benefit of this technique is that only a single, low cost laserdiode (known as a pump diode) is required to pump the fiber lasers. The strain resolution of such a laser is found to be limited by thermodynamic noise within the laser cavity, yielding fundamentally limited strain resolutions less than 10^{-13} $\varepsilon/\text{Hz}^{1/2}$. These sensors have been utilized in several high-performance applications including magnetometers (Cranch et al. 2011) and hydrophone arrays (Goodman et al. 2009). Longer cavity (~1 m) fiber lasers have demonstrated strain resolutions approaching 10^{-14} $\varepsilon/\text{Hz}^{1/2}$ (Cranch and Miller 2011).

15.6.4.4 Optical Actuation and Wavelength Multiplexing of MEMS-Based Sensors

Another implementation of a mechanical sensor is the silicon cantilever or bridge. Silicon-based sensors are attractive, because they can be manufactured to a high precision in large volume using techniques developed in the semiconductor industry. A structure of this type can be fabricated using anisotropic etching (Wolfelschneider et al. 1987). The cantilever can be actuated through the photothermal effect, where an intense modulated optical beam is focused onto the mechanical oscillator. Absorption of the radiation generates a harmonically varying thermal wave, which causes a periodic expansion of the silicon wafer, driving its motion. A second fiber can be used to read out the displacements of the oscillator, as illustrated in Figure 15.26a. Sweeping the frequency of the optical excitation allows the frequency response of the oscillator to be characterized. A measurand can be configured to change the resonance frequency of the mechanical oscillator by changing its effective stiffness or mass. Pressure sensors based on this concept have been demonstrated (Lammerink and Gerritsen 1987).

This sensing concept has evolved dramatically since these early demonstrations into the field of MEMS and applied to the detection of biological and chemical species (Lavrik et al. 2004). A recently demonstrated device incorporates a silicon microbridge with a rib-waveguide formed perpendicular to the MEMS resonator, illustrated in Figure 15.26b. Two etched air/silicon FBGs are incorporated, one within the waveguide and one attached to the center of the bridge, forming a high-finesse Fabry–Pérot cavity. Displacements of the bridge change the optical resonant frequency of the Fabry–Pérot cavity. Four paddles are also attached to the bridge, which are coated with chemo-selective polymers. These respond only to a distinct chemical analyte, absorption of which changes its mass and hence the resonant frequency of the bridge. Multiplexing of these

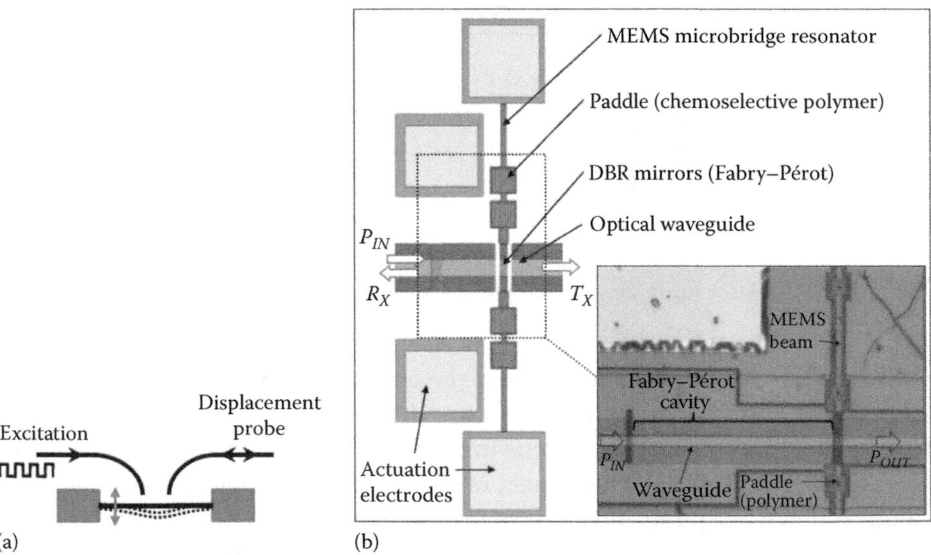

FIGURE 15.26 (a) Optical excitation of a MEMS resonator and (b) MEMS resonator incorporated within a Fabry–Pérot microcavity. (Reprinted with permission from Pruessner, M.W., Stievater, T.H., Rabinovich, W.S. et al., Chip-scale wavelength-division multiplexed integrated sensor arrays, *IEEE Sensor Conference*, Waikoloa, HI, Copyright 2010 IEEE.)

devices can be achieved by ensuring that the optical resonances of each Fabry–Pérot cavity occur at different regions of the optical spectrum. An individual sensor can be read out by wavelength tuning the interrogating laser. Two sensors multiplexed onto a single chip were demonstrated using this technique (Pruessner et al. 2010). Such a technique is capable of incorporating multiple sensors configured to measure different chemical species, enabling broad functionality on a single chip.

15.6.5 Coherence Division Multiplexing

Coherence multiplexing, along with FDM, was one of the earliest methods investigated but, unlike FDM, was not adapted from other engineering fields. The concept of signal coherence has more impact in optical systems than other electronic systems operating in lower regions of the electromagnetic spectrum. Early laser diode sources exhibited coherence lengths typically less than 1 m, so adapting this property to enable sensor multiplexing was a natural progression.

Although coherence multiplexing cannot generally match the multiplexing gain of TDM and WDM, it can be implemented without specialized components (i.e., amplitude modulators, wavelength mux/demux). As we shall show, the use of coherence properties of the laser source to selectively probe multiplexed sensors provides some powerful capabilities particularly in terms of spatial resolution.

15.6.5.1 Optical Low Coherence Interferometry

One of the earliest implementations of coherence multiplexing, known as *path-matched differential interferometry* (PMDI), is shown in Figure 15.27. Here the output of a laser diode is injected into a serial array of MZIs, each with a different path-imbalance that is longer than the coherence length of the laser source. The output of this array is launched into a parallel array of MZIs. The optical path difference (OPD) in each parallel MZI is matched to one of the serial MZIs, such that a differential optical path through the matched interferometers exists with a (near) zero path-imbalance.

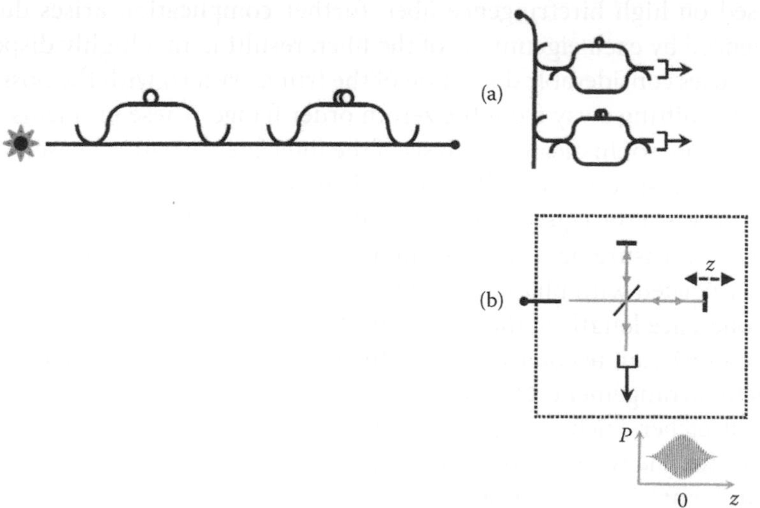

FIGURE 15.27 Multiplexing with PMDI.

Each sensor in the array operates as a tandem interferometer of the type described in Section 15.4.2.2. The sensor output takes the same form as that given by Equation 15.12 but incorporating much higher splitting losses associated with the additional optical paths and directional couplers.

Early demonstrations comprising two sensors demonstrated phase resolutions around 1 mrad/Hz$^{1/2}$ and crosstalk on the order of −40 dB (Brooks et al. 1985). The phase resolution is limited by residual coherence of the unbalanced optical paths, generating high levels of frequency-induced phase noise, and can be increased by accurate path-balancing and reducing the coherence length of the laser source (Wentworth 1989, Pendock and Sampson 1997). Alternatively, noise reduction by two orders of magnitude is possible by applying a large frequency modulation to the laser source, which selectively upconverts this excess phase noise outside the signal bandwidth (Kersey and Dandridge 1986) (interference signals corresponding to a near-zero imbalance are unaffected while those with a large imbalance are upconverted onto sideband of the modulation frequency). Application of PMDI systems to static strain sensors also requires careful design to avoid anomalous sensitivity of the connecting leads due to the combined effects of partially polarized light sources and birefringence within the interferometers (Cranch et al. 2005). More recent implementations using path-scanned receiving interferometers have demonstrated multiplexing six sensors (Yang et al. 2007) as well as a three-sensor system for humidity sensing (McMurtry et al. 2000). One benefit of using a path-scanned interferometer instead of a fixed path interferometer is the ability to precisely identify the zero OPD point, providing the source exhibits a well-behaved (e.g., Gaussian) lineshape. Figure 15.27b illustrates such a scanning system with an example fringe pattern showing the zero OPD point at the peak of the fringes, where the fringe visibility and hence the SNR is maximized. If the source coherence length is sufficiently short, the rapid decrease in visibility with OPD also enables the fringe order to be determined, hence increasing the unambiguous measurement range.

In practice, the source spectrum is not ideal, resulting in asymmetry of the fringe pattern. Furthermore, when this technique is applied to interrogation of polarimetric sensors based on high birefringence fiber, further complication arises due to dispersion experienced by each eigenmode of the fiber, resulting in a highly dispersive group delay. This causes considerable distortion of the fringe pattern with the possibility of the peak visibility shifting away from the zeroth order fringe. These problems can be overcome using a more sophisticated analysis of the interference pattern based on dispersive Fourier transform spectroscopy (Flavin et al. 1998), which measures the difference in group delay between the eigenmodes and is unaffected by the group delay dispersion or source spectra. Measurement of differential group delay also avoids the measurement ambiguity associated with phase tracking.

As the coherence length of the input light to the scanning interferometer increases, the required OPD scan becomes too large to be practically implemented in a free-space interferometer arrangement. One solution is to apply large strain to a long fiber length with a fiber stretcher. Such a device was implemented as a Fourier transform spectrometer for decoding the wavelengths of a serial array of FBGs (Davis and Kersey 1995), as illustrated in Figure 15.28. Here, a near-balanced Michelson interferometer has ~200 m of fiber from one arm wrapped around a lead zirconate titanate (PZT) stretcher. FRMs placed at the end of each arm prevent polarization fading, and a reference wavelength,

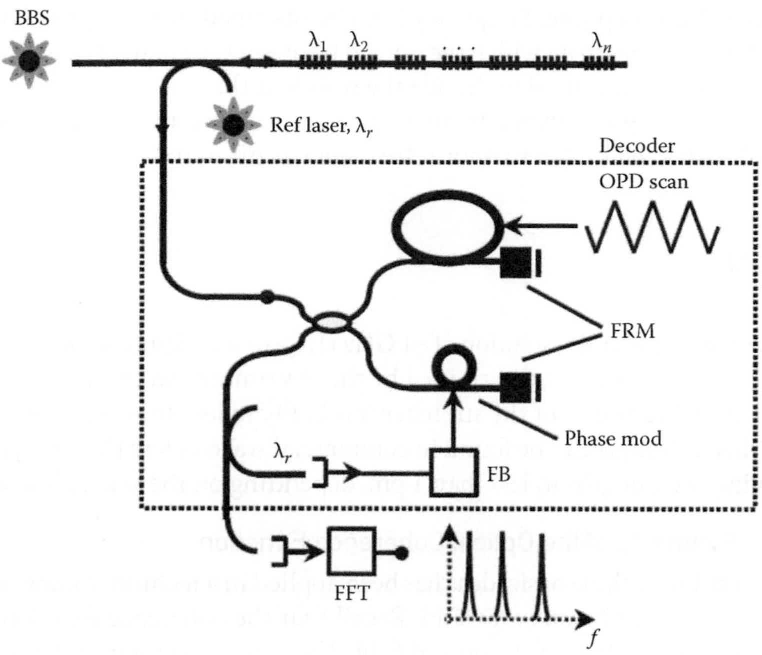

FIGURE 15.28 Wavelength decoding of FBG sensors with a fiber-optic Fourier transform spectrometer. (Reprinted with permission from Davis, M.A. and Kersey, A.D., Application of a fiber Fourier Transform spectrometer to the detection of wavelength encoded signals from fiber Bragg grating sensors, *J. Lightwave Technol.*, 13, 1289–1295, Copyright 1995 IEEE.)

λ_r, is injected into the same interferometer to enable stabilization of the optical paths that will otherwise drift due to environmental effects. A high voltage signal applied to the PZT was capable of inducing an OPD of ~30 cm in one arm of the interferometer.

We know from Equation 15.4 that the interferometric phase is dependent on the optical path imbalance and the laser frequency such that

$$\Delta\phi = \frac{2\pi(2n_e\Delta L)\nu}{c} \tag{15.64}$$

Therefore, if a fixed optical frequency, ν, is injected into the interferometer, then a linear sweep of the path imbalance will produce sinusoidal fringes (according to Equation 15.7) with a frequency given by

$$f = \frac{2V_m\nu}{c} \tag{15.65}$$

where V_m is the effective velocity of one of the mirrors and for the fiber-optic interferometer is

$$V_m = n_e\frac{\Delta l}{\Delta t} \tag{15.66}$$

where $\Delta l/\Delta t$ is the sweep rate. Frequency f can be obtained by taking the Fourier transform of the fringe signal from which the optical frequency, ν is calculated using Equation 15.65. This technique was used to decode the wavelengths of a serial array of three FBG sensors, by measuring the fringe frequency corresponding to each FBG. The spectral resolution of such a technique in optical frequency is given by*

$$\delta\nu = \frac{c}{2n_e \Delta L} \tag{15.67}$$

For a 10 cm scan, a spectral resolution of ~1 GHz (15 pm) was demonstrated. The measurement time of such a system is determined by the maximum sweep rate, which is limited by the mechanical response of the stretcher (probably to less than a few Hz). Free-space versions of this technique can be found in commercial wavemeters that are typically capable of resolving wavelengths to less than 1 pm, depending on the source linewidth.

15.6.5.2 Synthesis of the Optical Coherence Function

A further extension of these basic ideas has been applied in a technique known as *synthesis of the optical coherence function* (SOCF). Recall that the coherence function, $\gamma(\tau)$, is the normalized autocorrelation of the optical field. The autocorrelation, $c(\tau)$, is related to the power spectral density of the optical field, $S_\nu(\omega)$, through the Fourier transform according to the Wiener–Khinchin theorem. Thus, $\gamma(\tau)$ can be expressed as (Rogers 1997)

$$\gamma(\tau) = \frac{c(\tau)}{c(0)} = \frac{\int_{-\infty}^{\infty} S_\nu(\omega)\exp(i\omega\tau)\,d\omega}{\int_{-\infty}^{\infty} S_\nu(\omega)\,d\omega} \tag{15.68}$$

For the case of a monochromatic source, $\gamma(\tau)$ is unity (i.e., independent of time delay). Therefore, two waves from this source will interfere with maximum efficiency, regardless of their relative time delay. Conversely, for an infinitely broad source, $\gamma(\tau)$ is a Dirac delta function resulting in no interference for any nonzero differential time delay. One can therefore imagine intermediate situations where the coherence function falls off gradually or is periodic. The former situation occurs naturally (due to Gaussian linewidth sources), and the latter situation can be contrived or *synthesized* by appropriate frequency modulation of the source.

Figure 15.29 illustrates a simplified method of implementing this technique for reflectometry. Here, a laser diode is current modulated with a sinusoid, generating a sinusoidal frequency modulation (Hotate et al. 2004). This is injected into an optical fiber containing reflecting features. The reflected signals are separated through a directional coupler and mixed with the direct output of the laser diode. An AOM is also incorporated to generate an RF carrier. An interference signal is only generated when the differential time delay between the reflected and reference signals corresponds to a peak in the coherence function. Therefore, this technique enables selective probing

* Note the similarity with Equation 15.60 for the spatial resolution of OFDR.

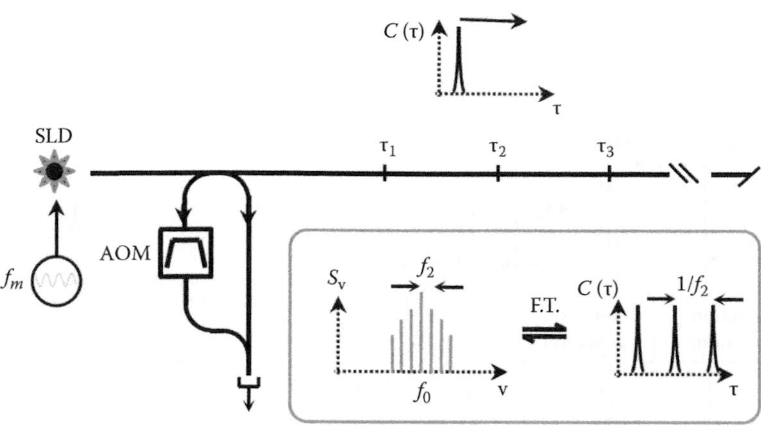

FIGURE 15.29 SOCF concept. (Reprinted with permission from Hotate, K. and Arai, H., Enlargement of measurement range of simplified BOCDA distributed strain sensing system using a temporal gating scheme, *17th International Conference on Optical Fiber Sensors*, Bruges, Belgium, SPIE Vol. 5855, pp. 184–187, Copyright 2006 IEEE.)

of a limited spatial region within the test fiber. If one considers the case of a sinusoidal frequency modulation of the form

$$\nu = f_0 + f_1 \sin(2\pi f_2 t) \tag{15.69}$$

where

f_0 is the mean laser frequency

f_1 is the frequency modulation depth

f_2 is the modulation frequency, then the coherence function will be the Fourier transform of this optical field, which consists of a series of peaks separated in time by $1/f_2$

Therefore, the spatial region being probed within the test fiber can be changed by adjusting the modulation frequency, f_2, and can be swept along the test fiber. For unambiguous measurement, only a single coherence peak should exist within the test fiber, which determines the choice of f_2. Using this implementation, the spatial resolution is found to be $\Delta z = 4.9/f_1$ (Hotate et al. 2004).

Other methods have been implemented to synthesize different coherence functions such as those based on linearly sweeping the laser frequency in a stepwise fashion, which yields a cyclical pattern of peaks in the coherence function (Hotate and He 2006). By synchronously modulating the phase of the receiving interferometer, the positions of the coherence peaks can be changed or swept. Any coherence function can be synthesized by simultaneously modulating both the frequency and intensity. The periodicity of the correlation peaks restricts the measurement range in some circumstances; however, techniques such as time gating have been developed to alleviate this restriction and have been applied to Brillouin-based distributed sensing systems (Hotate and Arai 2005).

An implementation of SOCF to interrogate a serial array of FBGs is shown in Figure 15.30. Here, a laserdiode is sinusoidally modulated, generating a periodic coherence function. An FBG is interrogated by overlapping a single coherence peak with the FBG. Applying a linear ramp to the mean frequency enables the FBG

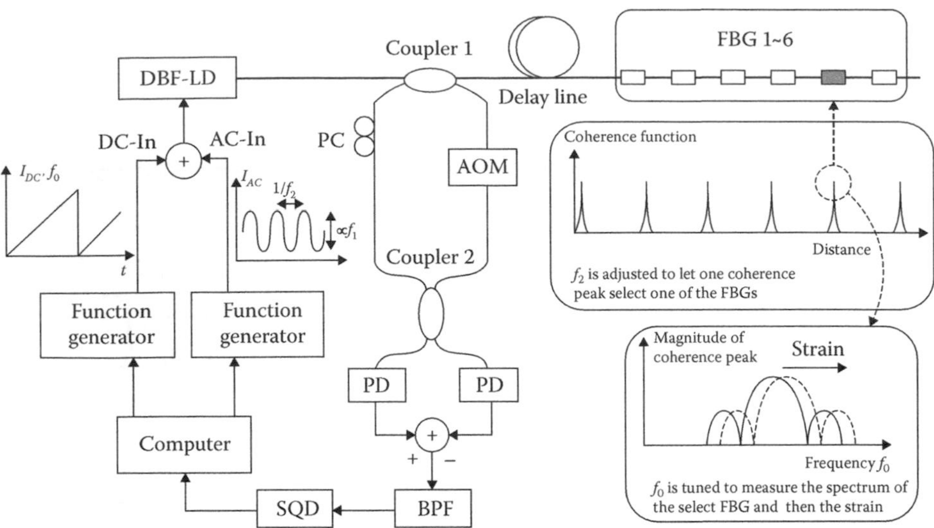

FIGURE 15.30 Interrogation of FBG array using SOCF. (Reprinted with permission from Hotate, K. and Arai, H., Enlargement of measurement range of simplified BOCDA distributed strain sensing system using a temporal gating scheme, *17th International Conference on Optical Fiber Sensors*, Bruges, Belgium, SPIE Vol. 5855, pp. 184–187, Copyright 2006 IEEE.)

reflectivity to be mapped out. Performing this measurement on each FBG in the array enables the FBG wavelengths to be tracked (Hotate and He 2006).

A benefit of the SOCF technique is that no mechanical scanning is required, making it robust and potentially high speed in certain implementations. Furthermore, no significant signal processing is required to decode the output signals.

It is interesting to compare the SOCF technique with the coherent CDMA method described in Section 15.6.2.4. Both techniques operate with a continuous wave probing laser and utilize a form of correlation to selectively probe a limited spatial region of a sensor array. In SOCF, the coherence properties of the probing laser are manipulated such that only spatial regions with the appropriate differential delay yield a signal. A sensor signal is extracted only when high correlation exists between the reflected field and the reference field, in this case, in the optical domain. In coherent CDMA, although the signal probing the sensor array is rapidly modulated, the reflected signal from the sensor array remains coherent with the reference signal to produce an interference signal carrying the sensor information (providing the sensor array is shorter than the coherence length of the source). However, an individual sensor signal is only extracted by coherently mixing the interference signal with the appropriately delayed modulation signal in the electronic domain. Thus, in this case, the correlation is performed in the RF domain. One can conclude that these two techniques employ the same basic principle of operation. That is, they use correlation properties of a modulated carrier to selectively probe a region of optical fiber. The difference lies in which domain this correlation is performed in.

15.6.6 Comparison of Techniques

This section has described specific methods used to multiplex fiber-optic sensors. These were classified as spatial, time, frequency, wavelength, and coherence multiplexing.

It was shown that time- and frequency-based multiplexing methods are closely related. Both methods utilize the optical transit time as the key distinguishing factor between different sensors. Some multiplexing methods such as swept frequency interferometry can fall under either category. Wavelength and frequency multiplexing are based on similar concepts (i.e., whereby a unique frequency is assigned to a sensor) but with differing implementations. Wavelength multiplexing is implemented by tailoring the response of the sensor in the optical domain, such that it responds only to a narrow portion of the optical spectrum. Frequency multiplexing is implemented at radio frequencies in the electronic domain. Coherence multiplexing utilizes the coherence properties of the light returned from a sensor and can be clearly distinguished from the other methods. However, the coherence of the sensor signal is measured relative to a reference signal and therefore also depends on the transit time of the optical signal. It was discussed earlier how the implementations of SOCF and CDMA, described in Section 15.6.2.4, are based on the same underlying concept. This leaves spatial multiplexing as being a unique multiplexing method, as it is implemented by interrogating sensors through physically separated paths. This also makes it the simplest to implement, but lacking in the flexibility and efficiency of the other techniques. Table 15.3 summarizes key properties of these multiplexing techniques. Here, low measurement speed corresponds to less than a few Hz, medium up to a few kHz, high up to 100 kHz, and very high greater than 100 kHz. It is interesting to note that the differences in performance of the various systems do not necessarily arise from fundamental differences in operating principle, rather by limitations imposed by components used in a specific implementation.

Table 15.3 Summary of Key Properties of Multiplexing Techniques

Technique	Measurement Speed	Multiplexing Gain	Spatial Resolution	Span	Sensor Type
Spatial	Very high	Low	—	km	1–5
TDM pulsed	High	High	10^0 m	km	1, 2, 4
TDM/WDM	High	Very high	10^{-2} m	km	1, 2, 4
Swept frequency (coherent OFDR)	Low	Very high	10^{-5} m	<km	2, 4
Swept frequency (incoherent OFDR)	Med	Med	10^{-3} m	km	1
CMDA	Very high	Med	10^{-2} m	km	1, 2, 4
FMCW (serrodyne)	Med	Med	—	—	1, 2, 4
FM-PGC	High	Med	—	—	2, 5
WDM only	Very high	Med	—	Km	1, 2, 4, 5
PMDI (path scanned)	Low	Low	—	—	1, 3
SOCF	Med	Med	10^{-2} m	m–km	1, 2, 4

1, Intensity; 2, interferometric; 3, polarimetric; 4, FBG; 5, LPG.

15.6.7 General Comments on System Performance

We complete this section with some general comments on obtaining optimum performance in multiplexed systems. Self-noise in fiber-optic sensors is usually generated by the laser source or receiving electronics, rather than in the transducer itself. A few exceptions are transducers with very long lengths of fiber generating thermal noise or very high responsivity accelerometers where thermomechanical noise in the mass-spring transducer mechanism can be significant (Thomas and Garrett 1988). In order to achieve low noise performance in a fiber-optic sensor system, the same principle must be applied to the system design as for any other multielement system. That is, every component must be viewed as a potential noise source and therefore should be characterized individually. A single poorly performing component can severely degrade the entire system performance, affecting both the sensor resolution and dynamic range. For example, a poorly chosen oscillator used for the AOM drive signal in an interferometric sensor may significantly reduce the phase resolution of every sensor. The reduction in phase resolution is also accompanied with an equal reduction in dynamic range. In fiber-optic hydrophone arrays that operate in relatively benign environments (such as sea-bed arrays), laboratory-demonstrated performance can generally be achieved in field-test systems. However, the performance of systems operating in more extreme environments is more difficult to characterize without rigorous field-testing. High vibration environments may induce both excess frequency noise in the laser and excess phase noise in the launch and receive optics. Coherent systems with long stand-off fiber links may also be susceptible to polarization-induced phase noise caused by birefringence modulation in the link fiber being converted to phase-shift in the interferometer (Kersey et al. 1988). For systems of these types, careful attention to aspects such as sensor and fiber packaging becomes essential in obtaining acceptable performance. Active control of device performance when exposed to extreme conditions, using feedback techniques, may also be required to ensure the required performance is maintained.

Another important design principle is to reduce stray photons. In systems with optical amplifiers and long fiber links, the use of bandpass optical filters and isolators to reduce ASE power residing away from the signal wavelength and backscattered power due to Rayleigh scattering safeguards against degrading amplifier performance and removes unwanted noise sources. Unwanted reflections from poor splices and connectors can also generate parasitic optical cavities. Uncontrolled interferometers of this type can be a major source of noise, particularly when high coherence lasers are used, since on detection, the noise fields add coherently (Dagenais et al. 1993).

15.7 Multiplexed Fiber-Optic Sensors in Action: Fiber-Optic Hydrophone Array

In this section, we bring together many of the concepts introduced in preceding sections by examining a prototype multiplexed fiber-optic sensor system. This system represents a highly developed multiplexed fiber-optic sensor and comprises 96 hydrophones combined using TDM and WDM. This system was field tested during a sea trial in 2002 and used to track synthetic acoustic targets. Results from this trial are described.

15.7.1 System Operating Principle

A fiber-optic hydrophone operates by converting the acoustically induced strain in an optical fiber placed within the acoustic field into a phase-shift in the light propagating in the fiber. This phase-shift is converted into intensity modulation by incorporating the fiber into one arm of a fiber-optic interferometer. To enhance the strain induced in the fiber, an amplification mechanism is incorporated. The technique used in this system is based on a fiber-wrapped air-backed mandrel design (Nash and Keen 1990). The physical change in the diameter of a plastic mandrel, around which the fiber is wrapped, under the influence of a time varying pressure field induces a strain in the fiber. The air backing increases the compliance of the structure and therefore increases the induced strain in the fiber. An omni-directional hydrophone response is obtained when the acoustic wavelength is much greater than the maximum hydrophone dimension. Multiplexing is achieved by serial concatenation of sensors and using the time of flight of injected optical pulses to sequentially address each sensor. WDM is incorporated with TDM to permit signals from several time division multiplexed sensor arrays to be combined onto a single optical fiber. In the prototype system, 16 sensors are multiplexed with time (expandable to 64) and 6 wavelengths are used to allow 96 hydrophones to be interrogated through 2 optical fibers. To achieve high phase resolution from the interferometric sensor, a high coherence laser is required that emits a stable single optical frequency. Six erbium-doped distributed feedback fiber lasers are used in this system with emission wavelengths ranging from 1541.35 to 1549.32 nm spaced by 1.6 nm. The TF of the interferometer is cosinusoidal, and therefore, an interrogation method is required to linearize the response of the sensor. A heterodyne-based method of the type described in Section 15.4.2.3 is employed, whereby the frequency of the light from each arm of the interferometer is shifted. When the two beams interfere on the detector, a beat frequency equal to the difference in frequency between the two beams is generated. A strain imposed on the fiber will modulate the phase of the light, which will appear as phase modulation sidebands around the beat frequency. The beat frequency is then mixed with a phase locked local oscillator, and the phase information of interest is retrieved using a trigonometric method. A schematic of the system layout is shown in Figure 15.31. It can be separated into three subsystems: the launch optics, hydrophone array, and the receive optics and electronics. The launch and receive optics are usually co-located. The launch optics generates and amplifies two pulses at each wavelength. The maximum launch power at each wavelength is limited by stimulated Brillouin scattering, which occurs at about 10 mW of average power in conventional single-mode optical fiber. The hydrophone array contains the sensors and wavelength multiplexers. Each wavelength is coupled off sequentially using optical add-drop multiplexers and injected into a reflective in-line Michelson array, of the type illustrated in Figure 15.15c using directional couplers. The return pulses are combined onto a return fiber. To allow the stand-off distance to be increased, a remotely pumped EDFA is incorporated into the array, providing around 20 dB of gain to the return signals. The 1480 nm pump, λ_p, for this EDFA is provided through a separate fiber. The receive electronics contains an EDFA to amplify the return signals before separating them through a wavelength demultiplexer followed by detection. The detector consists of a polarization diversity

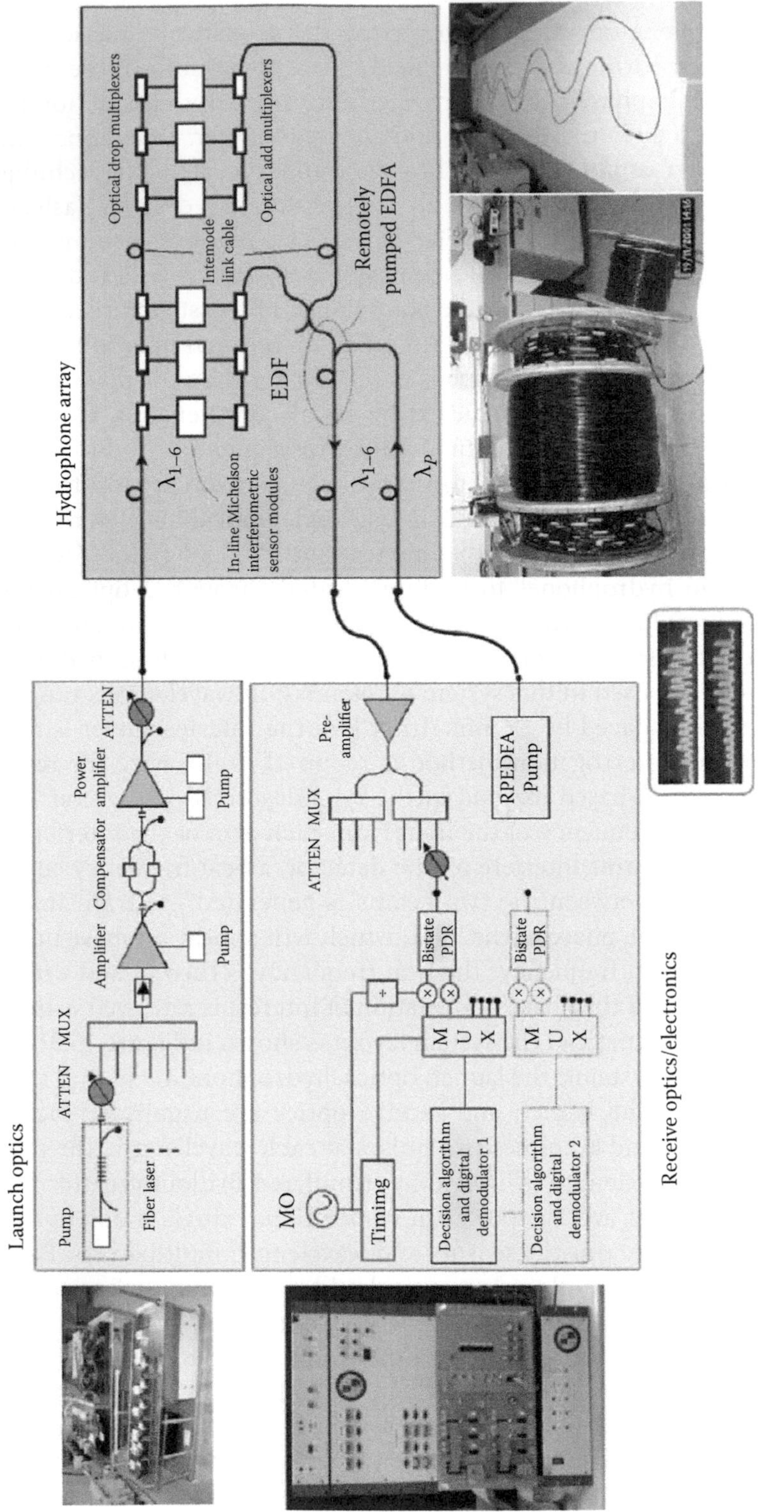

FIGURE 15.31 System arrangement of multiplexed array. (Adapted with permission from Cranch, G.A., Kirkendall, C.K., Daley, K. et al. Large-scale remotely pumped and interrogated fiber-optic interferometric sensor array, *Photon. Technol. Lett.*, 15, 1579–1581, Copyright 2003 IEEE.)

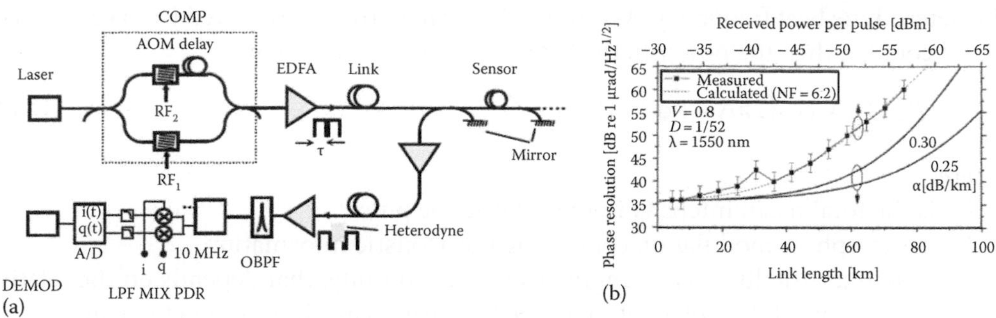

FIGURE 15.32 (a) Configuration of a single interferometric hydrophone in a multiplexed array and (b) dependence of phase resolution on received optical power. (Reproduced from Cranch, G.A. et al., *J. Acoust. Soc. Am.*, 115, 2848. Copyright 2004, Acoustical Society of America. With permission.)

receiver (PDR), which overcomes the problem of polarization-induced signal fading (Frigo et al. 1984) and the remaining electronics performs the trigonometric calculations to recover the phase from each sensor.

Despite the apparent complexity of this system, the arrangement for a single sensor can be deduced for analysis purposes and is shown in Figure 15.32a.

The optical emission from the laser is injected into a path-imbalanced interferometer or compensator (COMP). AOMs in each arm frequency shift and amplitude modulate the light when driven by a pulsed RF source. The output from one AOM is delayed with a fiber delay line before being combined with the other arm. The compensator thus generates two delayed pulses with a frequency difference, $\Delta f = RF_2 - RF_1$, which are repeated at a frequency, f_{rep}. The pulses are amplified with an EDFA and launched into the sensor array. The sensing region of the fiber is defined by splicing two reflective directional couplers at each end. The fiber length in-between the coupler forms one arm of a Michelson interferometer, and is set to give an optical path length equal to half the optical path length of the delay line in the compensator. Therefore, the reflection of the first pulse from the mirror after the sensor arrives back at the detector at the same time as the reflection of the second pulse from the mirror before the sensor. The total phase of this interferometer configuration is the difference in phase between the compensator and the Michelson interferometer sensor

$$\phi_{tot} = \left(\frac{2\pi n_e}{\lambda}\right) \cdot \left(L_{COMP} - 2L_{MI}\right) \tag{15.70}$$

where L_{COMP} and L_{MI} are the fiber path imbalances in the compensator and Michelson interferometer, respectively. In the ideal case $(L_{COMP} - 2L_{MI}) = 0$ and $\phi_{tot} = 0$; however, imperfect matching of optical fiber paths and thermally induced fluctuations in the fiber lengths results in a small effective imbalance and hence a nonzero total phase. At the detector, the two pulses reflected from each directional coupler overlap and generate the heterodyne beat frequency. Assuming that equal power is reflected from each directional coupler, then the power in the heterodyne pulse is given by

$$P = P_{inc}\left(1 + V\cos\left(2\pi\Delta ft + \phi_s(t)\right)\right) \tag{15.71}$$

where

P_{inc} is the total mean intensity in the heterodyne pulse
$\phi_s(t)$ is the phase modulation containing the acoustic information
V is a fringe visibility term, ranging from zero to unity, that depends on the relative orientations of the polarization of the light from each interferometer arm

Removing the dc component from Equation 15.71 and mixing with in-phase (i) and quadrature (q) components of a stable local oscillator yields

$$i(t) = A\cos\left(\phi_s(t)\right) \tag{15.72}$$

$$q(t) = A\sin\left(\phi_s(t)\right) \tag{15.73}$$

where A is proportional to the fringe visibility, photodiode responsivity, received power, and gain in the mixing process. The signal phase of interest is then obtained by taking the arctangent of the ratio of Equations 15.72 and 15.73. Prior to detection, the pulse train is amplified by a second EDFA. An optical bandpass filter (OBPF) is placed at the output to remove ASE, generated by the EDFAs, that lies outside the optical bandwidth of the signal interrogating the sensor. The effect of polarization-induced signal fading is alleviated with a polarization diversity receiver (PDR). The two outputs of the PDR are each mixed with the local oscillator, low-pass filtered (LPF) and digitally sampled (A/D). The phase measurement and tracking is then performed digitally by the demodulator (DEMOD). Phase locking the local oscillator to the oscillators driving the AOMs allows the phase of the interferometer to be tracked. Although the phase will drift due to thermally induced fluctuations in the fiber lengths and laser wavelength, these effects are usually very slow, and in most cases, very low frequency environmental pressure fluctuations ($\ll$1 Hz) can be resolved.

15.7.2 Noise Sources

The phase resolution of the sensor can be determined from the CNR. The carrier in this case is the heterodyne beat frequency, whose power is given by the interferometric term in Equation 15.13. The CNR is given by

$$CNR = \frac{\left(r_d V P_{inc}/\sqrt{2}\right)^2}{\overline{i_n^2}} \tag{15.74}$$

where $\overline{i_n^2}$ is the noise current and may comprise contributions from all the noise sources described in Section 15.4.5. Assuming the various noise sources are statistically independent, then the total noise is the rms sum of the individual noise sources

$$\overline{i_n^2} = \overline{i_{sh}^2} + \overline{i_{th}^2} + \overline{i_{RIN}^2} + \overline{i_{s-sp}^2} + \overline{i_{sp-sp}^2} + \overline{i_{sp-sh}^2} + \overline{i_{s-sh}^2} \tag{15.75}$$

The equivalent phase noise is calculated by equating the CSBR given by Equation 15.46, which we derived for phase-modulated signals, with Equation 15.74. The dependence of the phase resolution as a function of input power to the EDFA proceeding the detectors is shown in Figure 15.32b. At high optical powers, the phase resolution is weakly dependent on optical power. The limiting noise in this region is a combination of laser frequency noise and possibly multipath noise. As the power decreases (i.e., as the stand-off distance or link length increases), the phase resolution decreases more rapidly. In this region, the limiting noise is due to beating between the spontaneous emission and itself (i.e., $\overline{i_{sp-sp}^2}$), according to Equation 4.38. Using Equations 15.39 and 15.74, the phase noise at low received optical power is given by (Cranch et al. 2003)

$$\delta\phi_{sp-sp} = \frac{2Fh\nu}{VP_{inc}}\sqrt{\frac{\Delta\nu_{opt}\Delta f}{D}} \tag{15.76}$$

where

P_{inc} is the total power in the heterodyne pulse at the input to the EDFA preceding the detector

F is its noise figure

The duty cycle, D, takes into account the effect of aliasing spontaneous emission noise, which occupies a bandwidth much larger than the sampling rate. Also shown in Figure 15.32b is the dependence of the phase resolution on the standoff distance for fiber attenuations of 0.3 and 0.25 dB/km. The performance of these systems can be accurately modeled using these simple noise analyses, which are described in more detail in the literature (Cranch et al. 2003).

15.7.3 Acoustic Performance

The 96 hydrophones are separated into 2 arrays of 48 hydrophones. These were deployed a few kilometers from the coast of a major military and commercial shipping port. A 3 km fiber-optic cable joined each array and a 5 km cable connected array 1 to the shore station. For the majority of the tests, an extra 35 km of optical fiber wound onto spools was added to the input and output fibers, to increase the effective array standoff to 40 km. Array 1 was deployed at a depth of 57 m and array 2 at a depth of 73 m. The ambient acoustic induced phase noise was found to be approximately 20 dB higher than the mean sensor self noise at ~488 Hz for array 1 and up to 10 dB higher than the sensor self-noise for array 2. The temporal outputs of hydrophones from array 1 without high pass filtering are shown in Figure 15.33a. The vertical striation represents the slow pressure variations that propagate along the hydrophone array due to surface wave motion. This plot illustrates the capability of fiber-optic hydrophones to resolve very low frequency variation in ambient pressure.

The characterization of the beam forming performance of the array was achieved with a towed acoustic source. The source transmitted a number of discrete frequencies, and since the angular resolution of an array increases with frequency, tracking is most accurately achieved using the highest of the discrete frequencies, which is 445 Hz.

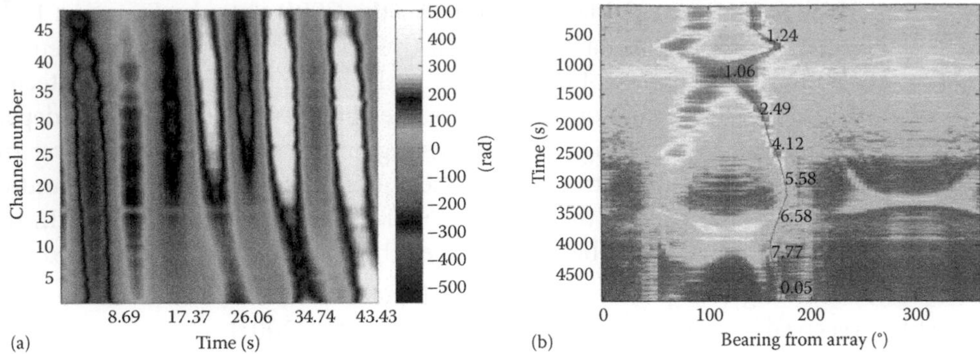

FIGURE 15.33 (a) Temporal output from 48 sensor array and (b) bearing time plot from a 48 element array. (Reproduced from Cranch, G.A. et al., *J. Acoust. Soc. Am.*, 115, 2848. Copyright 2004, Acoustical Society of America. With permission.)

To show the capability of the system to track the towed source, only the section of the beam pattern at 445 Hz is used, and this is combined with the same section from subsequent beam patterns. This enables an image to be created that shows how the bearing and strength of the signal at 445 Hz varies as the towed source moves. Such an image, shown in Figure 15.33b, has been created from 4900 s of data from array 1.

At times up to 2500 s, the strongest signal detected is from the towed source; however, as it moves further away, its signal strength decreases and the signal from another nearby vessel becomes higher.

There are normally two maxima on the trace due to the left/right ambiguity. When the towed source moves near the end fire position (which happens around 1200 s), the angular resolution of the beam pattern becomes worse and the two traces merge. The right hand trace is narrower and more intense, which shows that this represents the true bearing. The dotted white line on this figure shows the actual bearing of the towed source calculated from the log of its GPS positions. The numbers beside this line show the range in kilometers of the towed source from the array at that time. There is clearly very good agreement between the actual and measured bearing out to a range of around 5.5 km (3000 s) after which time the signal from another vessel temporarily masks the signal from the towed source. After 4200 s, when the other vessel has moved away, there is still a faint trace that follows the dotted line, which shows that the towed source is being detected to a range of ~9 km (Cranch et al. 2004).

15.8 Concluding Remarks

This chapter has illustrated the vast potential of fiber-optic sensors to impact a broad range of fields in science, engineering, and metrology. However, their uptake has occurred primarily in niche areas where limited alternative sensor options exist. Perhaps a reason for this is the conservative nature of many existing industries. The introduction of fiber-optics into fields unfamiliar with this technology brings with it many unknowns in terms of installation, handling, and maintenance. Engineering fiber-optic sensors into existing systems can present major challenges that require dedicated and highly specialized engineering practices to be developed, implemented, and tested as well as additional associated cost and commitment in order to ensure successful introduction of the technology. Furthermore, sensor technology based on glass fiber can raise unfounded skepticism. For example, a common misconception of optical fiber is its fragility compared to copper wire. However, silica fiber exhibits a higher tensile strength than copper wire and deforms elastically up to several percent. In contrast, copper wire is ductile and when subject to high transient pressure or strain does not return to its pristine condition. Although optical fiber can be made resistant to bending, conventional optical fiber (i.e., SMF-28) will exhibit high attenuation for bend diameters less than a few centimeters and all optical fiber must generally be wound around uniform structures, to avoid microbending losses. Failure to adopt these basic handling practices can result in catastrophic failure of a system.

Many companies that specialize in fiber-optic sensors have addressed these issues by focusing heavily on easing installation burden through engineered sensor packages to simplify installation as well as low maintenance and user-friendly interrogation units. The development of standards for fiber-optic sensors has also provided a consistent foundation on which to design and market sensor systems.

At the time of writing, multiplexed fiber-optic hydrophone arrays have been adopted by the US Navy (Dandridge et al. 1994) and in the petrochemical industry for well-monitoring applications (Nakstad and Kringlebotn 2008). FBG-based sensors have been developed for down-well monitoring systems as well as for infrastructure monitoring. As the technology continues to demonstrate reliability, one can expect significant further uptake in new applications areas.

References

Abad, S., F. M. Araújo, L. A. Ferreira et al. 2003. Transparent network for hybrid multiplexing of fiber Bragg gratings and intensity-modulated fiber-optic sensors. *Applied Optics* 42: 5040–5045.

Abad, S., M. López-Amo, F. M. Araújo et al. 2002. Fiber Bragg grating-based self-referencing technique for wavelength-multiplexed intensity sensors. *Optics Letters* 27: 222–224.

Albert, J, L. Y. Shao, and C. Caucheteur. 2013. Tilted fiber Bragg grating sensors. *Laser & Photonics Reviews* 7: 83–108.

Al-Raweshidy, H. S. 1994. Hybrid CDMA/WDM multiplexing strategy for an interferometric optical fibre sensor network. *Proceedings of the 10th International Conference on Optical Fiber*, Glasgow, U.K.

Arya, V., K. A. Murphy, A. Wang et al. 1995. Microbend losses in singlemode optical fibers: Theoretical and experimental investigation. *Journal of Lightwave Technology* 13: 1998–2002.

Askins, C. G., M. A. Putnam, and E. J. Friebele. 1995. Instrumentation for interrogating many-element fiber Bragg grating arrays. *Smart Structures and Materials* (*SPIE* 2444): 257–266.

Askins, C. G., M. A. Putnam, G. M. Williams et al. 1994. Stepped-wavelength optical-fiber Bragg grating arrays fabricated in-line on a draw tower. *Optics Letters* 19: 147–149.

Balzhiev, P., W. J. Bock, T. A. Eftimov et al. 2011. A spectrally and spatially multiplexed LPG sensor system using an INGAAs CCD linear array. *Proceedings of the 21st International Conference on Optical Fiber Sensors*, Ottawa, Ontario, Canada, SPIE Vol. 7753, paper 9L.

Berthold III, J. W. 1995. Historical review of microbend fiber-optic sensors. *Journal of Lightwave Technology* 13: 1193–1199.

Bhatia, V. 1999. Applications of long-period gratings to single and multi-parameter sensing. *Optics Express* 4: 457–466.

Black, E. D. 2001. An introduction to Pound–Drever–Hall laser frequency stabilization. *American Journal of Physics* 69: 79–87.

Bohnert, K., P. Gabus, J. Nehring et al. 2002. Temperature and vibration insensitive fiber-optic current sensor. *Journal of Lightwave Technology* 20: 267–276.

Brooks, J. L., B. Moslehi, B. Y. Kim et al. 1987. Time-domain addressing of remote fiber-optic interferometric sensor arrays. *Journal of Lightwave Technology* LT-5: 1014–1023.

Brooks, J. L., R. H. Wentworth, R. C. Youngquist et al. 1985. Coherence multiplexing of fiber-optic interferometric sensors. *Journal of Lightwave Technology* LT-3: 1062–1072.

Bucaro, J. A., N. Lagakos, B. H. Houston et al. 2005. Miniature, high performance, low-cost fiber optic microphone. *Journal of the Acoustical Society of America* 118: 1406–1413.

Bucholtz, F., C. A. Villarruel, A. R. Davis et al. 1995. Multichannel fiber-optical magnetometer system for undersea measurements. *Journal of Lightwave Technology* 13: 1385–1395.

Bush, J., C. A. Davis, F. McNair et al. 2013. Low cost fiber optic interferometric sensor. *Proc, SPIE 2872*, Second Pacific Northwest Fiber Optic Sensor Workshop, 48, Troutdale, OR, (August 5, 1996); doi:10.1117/12.245584.

Butter, C. D. and G. B. Hocker. 1978. Fiber optics strain gauge. *Applied Optics* 17: 2867–2869.

Chan, C. F., C. Chen, A. Jafari et al. 2007. Optical fiber refractometer using narrowband cladding-mode resonance shifts. *Applied Optics* 46: 1142–1149.

Chan, P. K. C., W. Jin, and M. S. Demokan. 2000. FMCW multiplexing of fiber Bragg grating sensors. *IEEE Journal on Selected Topics in Quantum Electronics* 6: 756–763.

Chan, P. K. C., W. Jin, J. M. Gong et al. 1999. Multiplexing of fiber Bragg grating sensors using an FMCW technique. *Photonics Technology Letters* 11: 1470–1472.

Chen, S., Y. Hu, L. Zhang et al. 1998. Multiplexing of large-scale FBG arrays using a two-dimensional spectrometer. *Conference on Sensory Phenomena and Measurement Instrumentation for Smart Structures and Materials*, San Diego, CA, SPIE Vol. 3330, pp. 245–252.

Childers, B. A., M. E. Froggatt, S. G. Allison et al. 2001. Use of 3000 Bragg grating strain sensors distributed on four eight-meter optical fibers during static load tests of a composite structure. *Proceedings of SPIE* 4332: 133–142.

Chow, J. H., D. E. McClelland, M. B. Gray et al. 2005. Demonstration of a passive subpicostrain fiber strain sensor. *Optics Letters* 30: 1923–1925.

Claus, R. O., M. F. Gunther, A. Wang et al. 1992. Extrinsic Fabry–Perot sensor for strain and crack opening displacement measurements from −200 to 900 degrees C. *Modeling and Simulation in Materials Science and Engineering* 1: 237–242.

Cole, J. H., B. A. Danver, and J. A. Bucaro. 1982. Synthetic-heterodyne interferometric demodulation. *Journal of Quantum Electronics* QE-18: 694–697.

Cranch, G. A., R. Crickmore, C. K. Kirkendall et al. 2004. Acoustic performance of a large-aperture, seabed, fiber-optic hydrophone array. *Journal of the Acoustical Society of America* 115: 2848–2858.

Cranch, G. A., G. M. H. Flockhart, and C. K. Kirkendall. 2005. Efficient fiber Bragg grating and fiber Fabry–Pérot sensor multiplexing scheme using a broadband pulsed mode-locked laser. *Journal of Lightwave Technology* 23: 3798–3807.

Cranch, G. A., G. M. H. Flockhart, and C. K. Kirkendall. 2008. Distributed feedback fiber laser strain sensors. *IEEE Sensors Journal* 8: 1161–1172.

Cranch, G. A., C. K. Kirkendall, K. Daley et al. 2003a. Large-scale remotely pumped and interrogated fiber-optic interferometric sensor array. *Photonics Technology Letters* 15: 1579–1581.

Cranch, G. A. and G. A. Miller. 2011. Fundamental frequency noise properties of extended cavity erbium fiber lasers. *Optics Letters* 36: 906–909.

Cranch, G. A., G. A. Miller, and C. K. Kirkendall. 2011. Fiber laser sensors: Enabling the next generation of miniaturized, wideband marine sensors. *SPIE Defense, Security, and Sensing*, Orlando, FL, SPIE Vol. 8028, paper 17.

Cranch, G. A., P. J. Nash, and C. K. Kirkendall. 2003b. Large-scale remotely interrogated arrays of fiber-optic interferometric sensors for underwater acoustic applications. *IEEE Sensors Journal* 3: 19–30.

Cruz, J. L., M. V. Andres, and M. A. Hernandez. 1996. Faraday effect in standard optical fibers: Dispersion of the effective Verdet constant. *Applied Optics* 35: 922–927.

Dagenais, D. M., K. P. Koo, and F. Bucholtz. 1993. Effects of parasitic Fabry–Perot cavities in fiber-optic interferometric sensors. *Optics Letters* 18: 388–390.

Dandridge, A., A. B. Tveten, and T. G. Giallorenzi. 1982. Homodyne demodulation scheme for fiber optic sensors using phase generated carrier. *Journal of Quantum Electronics* QE-18: 1647–1653.

Dandridge, A., A. B. Tveten, A. D. Kersey et al. 1987. Multiplexing of interferometric sensors using phase carrier techniques. *Journal of Lightwave Technology* LT-5: 947–952.

Dandridge, A., A. B. Tveten, and C. K. Kirkendall. 2004. Development of the fiber optic wide aperture array: From initial development to production. NRL Review, Optical Sciences Division, Naval Research Laboratory Washington, DC (available at www.nrl.navy.mil).

Davis, A. R., C. K. Kirkendall, A. Dandridge et al. 1997. 64-channel all-optical deployable acoustic array. *OSA Technical Digest Series* 16: 616–621.

Davis, M. A. and A. D. Kersey. 1995. Application of a fiber Fourier Transform spectrometer to the detection of wavelength encoded signals from fiber Bragg grating sensors. *Journal of Lightwave Technology* 13: 1289–1295.

Dennis, M. L., M. A. Putnam, J. U. Kang et al. 1997. Grating sensor array demodulation by use of a passively mode-locked fiber laser. *Optics Letters* 22: 1362–1364.

Desurvire, E., J. R. Simpson, and P. C. Becker. 1987. High-gain erbium-doped traveling-wave fiber amplifier. *Optics Letters* 12: 888–890.

Farahi, F., D. J. Webb, J. D. C. Jones et al. 1990. Simultaneous measurement of temperature and strain: Cross-sensitivity considerations. *Journal of Lightwave Technology* 8: 138–142.

Flavin, D. A., R. McBride, and J. D. C. Jones. 1998. Demodulation of polarimetric optical fibre sensors by dispersive Fourier transform spectroscopy. *Optics Communications* 156: 367–373.

Frigo, N. J., A. Dandridge, and A. B. Tveten. 1984. Technique for elimination of polarisation fading in fibre interferometers. *Electronics Letters* 20: 319–320.

Froggatt, M. and J. Moore. 1998. High-spatial-resolution distributed strain measurement in optical fiber with Rayleigh scatter. *Applied Optics* 37: 1735–1740.

Goodman, S., S. Foster, J. Van Velzen, and H. Mendis. 2009. Field demonstration of a DFB fibre laser hydrophone seabed array in Jervis Bay, Australia. *Proceedings of SPIE* 7503: 75034L.

Hariharan, P. 1987. Digital phase stepping interferometry: Effects of multiply reflected beams. *Applied Optics* 26: 1987.

Harris, A. J. and P. F. Castle. 1986. Bend loss measurements on high numerical aperture single-mode fibers as a function of wavelength and bend radius. *Journal of Lightwave and Technology* LT-4: 34–40.

Hecht, J. 1999. *City of Light: The Story of Fiber Optics*. Oxford University Press, New York.

Heurs, M., V. M. Quetschke, B. Willke et al. 2004. Simultaneously suppressing frequency and intensity noise in a Nd:YAG nonplanar ring oscillator by means of the current-lock technique. *Optics Letters* 28: 2148–2150.

Hocker, G. B. 1979. Fiber-optic sensing of pressure and temperature. *Applied Optics* 18: 1445–1448.

Hodgson, C. W., M. J. F. Digonnet, and H. J. Shaw. 1997. Large-scale interferometric fiber sensor arrays with multiple optical amplifiers. *Optics Letters* 22: 1651–1653.

Hodgson, C. W., J. L. Wagener, M. J. F. Digonnet et al. 1998a. Optimization of large-scale fiber sensor arrays incorporating multiple optical amplifiers—Part I: Signal-to-noise ratio. *Journal of Lightwave Technology* 16: 218–223.

Hodgson, C. W., J. L. Wagener, M. J. F. Digonnet et al. 1998b. Optimization of large-scale fiber sensor arrays incorporating multiple optical amplifiers—Part II: Pump power. *Journal of Lightwave Technology* 16: 224–231.

Hotate, K. and H. Arai. 2005. Enlargement of measurement range of simplified BOCDA distributed strain sensing system using a temporal gating scheme. *17th International Conference on Optical Fiber Sensors*, Bruges, Belgium, SPIE Vol. 5855, pp. 184–187.

Hotate, K., M. Enyama, S. Yamashita et al. 2004. A multiplexing technique for fibre Bragg grating sensors with the same reflection wavelength by the synthesis of optical coherence function. *Measurement Science and Technology* 15: 148–153.

Hotate, K. and Z. He. 2006. Synthesis of optical-coherence function and its applications in distributed and multiplexed optical sensing. *Journal of Lightwave Technology* 24: 2541–2557.

Hu, Y. and S. Chen. 1995. Electronic scanning spatial multiplexing technique for interferometric optical fiber sensor arrays. *IEEE Photonics Technology Letters* 7: 673–675.

Huh, J. H., Y. M. Chang, and J. H. Lee. 2011. Performance comparison of Raman/EDFA hybrid amplification-based long distance fiber Bragg grating sensor system. *Applied Optics* 51: 348–355.

Hymans, A. J. and J. Lait. 1960. Analysis of a frequency modulated continuous-wave ranging system. *Proceedings of IEE* 107B: 365–372.

Jackson, D. A., A. D. Kersey, M. Corke et al. 1982. Pseudoheterodyne detection scheme for optical interferometers. *Electronics Letters* 18: 1081–1083.

Jackson, D. A., R. Priest, A. Dandridge et al. 1980. Elimination of drift in a single-mode optical fiber interferometer using a piezoelectrically stretched coiled fiber. *Applied Optics* 19: 2926–2929.

James, S. W., R. A. Lockey, D. A. Egan et al. 1996. 3D fiber optic laser Doppler velocimeter. *SPIE* 2839: 323–326.

James, S. W. and R. P. Tatam. 2003. Optical fibre long-period grating sensors: Characteristics and applications. *Measurement Science and Technology* 14: R49–R61.

Johnson, G. A., M. D. Todd, B. L. Althouse et al. 2000. Fiber Bragg grating interrogation and multiplexing with a 3 × 3 coupler and a scanning filter. *Journal of Lightwave Technology* 18: 1101–1105.

Jorge, P. A. S., L. A. Ferreira, and J. L. Santos. 2000. Analysis of the flyback effects on the serrodyne interferometric demodulation of fiber optic Bragg grating sensors. *Optical Engineering* 39: 1399–1404.

Kersey, A. D., T. A. Berkoff, and W. W. Morey. 1992. High-resolution fibre-grating based strain sensor with interferometric wavelength-shift detection. *Electronics Letters* 28: 236–238.

Kersey, A. D., T. A. Berkoff, and W. W. Morey. 1993a. Fiber-optic Bragg grating strain sensor with drift-compensated high-resolution interferometric wavelength-shift detection. *Optics Letters* 18: 72–74.

Kersey, A. D., T. A. Berkoff, and W. W. Morey. 1993b. Multiplexed fiber Bragg grating strain-sensor system with a fiber Fabry–Perot wavelength filter. *Optics Letters* 18: 1370–1372.

Kersey, A. D. and A. Dandridge. 1986. Phase-noise reduction in coherence-multiplexed interferometric fibre sensors. *Electronics Letters* 22: 616–618.

Kersey, A. D. and A. Dandridge.1989. Multiplexed Mach-Zehnder ladder array with ten sensor elements. *Electronics Letters* 25: 1298–1299.

Kersey, A. D., A. Dandridge, and K. L. Dorsey. 1989a. Transmissive serial interferometric fiber sensor array. *Journal of Lightwave Technology* 7: 846–854.

Kersey A. D., K. L. Dorsey, and A. Dandridge. 1989b. Cross talk in a fiber-optic Fabry–Perot sensor array with ring reflectors. *Optics Letters* 14: 93–95.

Kersey, A. D., M. J. Marrone, and A. Dandridge. 1988. Observation of input-polarization induced phase noise in interferometric fiber-optic sensors. *Optics Letters* 13: 847–849.

Koo, K. P., A. B. Tveten, and S. T. Vohra. 1999. Dense wavelength division multiplexing of fibre Bragg grating sensors using CDMA. *Electronics Letters* 35: 165–167.

Lammerink, T. S. J. and S. J. Gerritsen. 1987. Fiber-optic sensors based on resonating mechanical structures. *Fiber Optic Sensors II*, The Hague, the Netherlands, SPIE Vol. 798.

Lavrik, N. V., M. J. Sepaniak, and P. G. Datskos. 2004. Cantilever transducers as a platform for chemical and biological sensors. *Review of Scientific Instruments* 75: 2229–2253.

Liehr, S. and K. Krebber. 2010. A novel quasi-distributed fibre optic displacement sensor for dynamic measurement. *Measurement Science and Technology* 21: 075205.

Liehr, S., N. Nöther, and K. Krebber. 2010. Incoherent optical frequency domain reflectometry and distributed strain detection in polymer optical fibers. *Measurement Science and Technology* 21: 017001.

Lin, W., S. Huang, J. Tsay et al. 2000. System design and optimization of optically amplified WDM-TDM hybrid polarization-insensitive fiber-optic Michelson interferometric sensor. *Journal of Lightwave Technology* 18: 348–359.

Littler, I. C. M., M. B. Gray, T. T. Lam et al. 2010. Optical-fiber accelerometer array: Nano-g infrasonic operation in a passive 100 km loop. *IEEE Sensors Journal* 10: 1117–1124.

Lobo Ribeiro, A. B., R. F. Caleya, and J. L. Santos. 1995. Progressive ladder network topology combining interferometric and intensity fiber-optic-based sensors. *Applied Optics* 34: 6481–6488.

Lucas, H. N., J. K. Hypolito, and A. S. Paternoa. 2011. Benchmark for standard and computationally intelligent peak detection algorithms for fiber Bragg grating sensors. *Proceedings of the 21st International Conference on Optical Fiber Sensors*, Ottawa, Ontario, Canada, SPIE Vol. 7753, paper 77537F.

Luna Technologies. 2013. Optical backscatter reflectometer (OBR). www.lunatechnologies.com.

MacDonald, R. I. 1981. Frequency domain optical reflectometer. *Applied Optics* 20: 1840–1844.

Martinelli, M. 1994. Time-reversal of the polarization state in optical fiber circuits. *Proceedings of the Optical Fiber Sensor Conference 10*, SPIE Vol. 2360, Glasgow, UK, pp. 312–318.

Matías, I. R., C. Fernández-Valdivielso, F. J. Arregui et al. 2003. Transmitted optical power through a tapered single-mode fiber under dynamic bending effects. *Fiber and Integrated Optics* 22: 173–187.

Matsas, V. J., T. P. Newson, and M. N. Zervas. 1992. Self-starting passively mode-locked fibre ring laser exploiting nonlinear polarisation switching. *Optics Communications* 92: 61–66.

McMurtry, S., J. D. Wright, and D. A. Jackson. 2000. Multiplexed low coherence interferometric system for humidity sensing. *Sensors and Actuators B: Chemical* 67: 52–56.

Mears, R. J., L. Reekie, I. M. Jauncie et al. 1987. Low-noise erbium-doped fibre amplifier operating at 1.54 μm. *Electronics Letters* 23: 1026–1027.

Monro, T. M., W. Belardi, K. Furusawa et al. 2001. Sensing with microstructured optical fibres. *Measurement Science and Technology* 12: 854–858.

Montalvo, J., O. Frazão, J. L. Santos et al. 2009. Radio-frequency self-referencing technique with enhanced sensitivity for coarse WDM fiber optic intensity sensors. *Journal of Lightwave Technology* 27: 475–482.

Mortimore, D. A. 1988. Fiber loop reflectors. *Journal of Lightwave Technology* 6: 1217–1224.

Nakstad, H. and J. T. Kringlebotn. 2008. *Nature Photonics, Technology Focus*. dio:10.1038/nphoton.2018.18, March.

Nash, P. J. and J. Keen. 1990. Design and construction of practical optical fibre hydrophones. *Proceedings of the Institute of Acoustics* 12: 201–212.

Olsson, N. A. 1989. Lightwave systems with optical amplifiers. *Journal of Lightwave Technology* 7: 1071–1082.

Paton, B. E. 1990. Low cost fibre optic sensing systems using spatial division multiplexing. *Proceedings of SPIE—The International Society for Optical Engineering*, San Diego, CA, doi:10.117/12.51093, p. 446.

Patrick, H. J., A. D. Kersey, and F. Bucholtz. 1998. Analysis of the response of long period fiber gratings to external index of refraction. *Journal of Lightwave Technology* 16: 1606–1612.

Pendock, G. J. and D. D. Sampson. 1997. Noise in coherence-multiplexed optical fiber systems. *Applied Optics* 36: 9536–9540.

Pruessner, M. W., T. H. Stievater, W. S. Rabinovich et al. 2010. Chip-scale wavelength-division multiplexed integrated sensor arrays. *IEEE Sensor Conference*, Waikoloa, HI.

Putnam, M. A., M. L. Dennis, I. N. Duling III et al. 1998. Broadband square-pulse operation of a passively mode-locked fiber laser for fiber Bragg grating interrogation. *Optics Letters* 23: 138–140.

Rogers, A. 1997. *Essentials of Optoelectronics*. Chapman & Hall, London, U.K., Section 5.2.

Russell, P. S. J. 2006. Photonic-crystal fibers. *Journal of Lightwave Technology* 24: 4729–4749.

Sakai, I., R. C. Yongquist, and G. Parry. 1987. Multiplexing of optical fiber sensors using a frequency-modulated source and gated output. *Journal of Lightwave Technology* LT-5: 932–940.

Sang, A. K., M. E. Froggatt, D. K. Gifford et al. 2008. One centimeter spatial resolution temperature measurements in a nuclear reactor using Rayleigh scatter in optical fiber. *IEEE Sensors Journal* 8: 1375–1380.

Sæther, J. and K. Bløtekjær. 1997. Optical amplifiers in time domain multiplexed sensor systems. *Proceedings of the 12th International Conference on Optical Fiber Sensors*, Williamsburg, VA.

Shaddock, D. A. 2007. Digitally enhanced heterodyne interferometry. *Optics Letters* 32: 3355–3357.

Skaar, J. 2000. Synthesis and characterization of fiber Bragg gratings. PhD thesis, The Norwegian University of Science and Technology, Trondheim, Norway.

Smit, M. K. and C. V. Dam. 1996. Phasar based WDM devices: Principles, design and applications. *Journal of Selected Topics in Quantum Electronics* 2: 236–250.

Spillman Jr., W. B. and J. R. Lord. 1987. Self-referencing multiplexing technique for fiber-optic intensity sensors. *Journal of Lightwave Technology* LT-5: 865–869.

Stingraygeo.com. www.stingraygeo.com. http://www.tgs.com/products-and-services/reservior/stingray-systems/

Thomas, J. H. and S. L. Garrett. 1988. Thermal noise in a fiber optic sensor. *Journal of the Acoustical Society of America* 84: 471–475.

Todd, M. D., M. Seaver, and F. Bucholtz. 2002. Improved, operationally-passive interferometric demodulation method using 3 × 3 coupler. *Electronics Letters* 38: 784–786.

Todd, M. D., M. Seaver, and F. Bucholtz. 2003. Erratum: Improved, operationally-passive interferometric demodulation method using 3 × 3 coupler. *Electronics Letters* 39: 1873.

Vakoc, B. J., C. W. Hodgson, M. J. F. Digonnet et al. 1998. Phase-sensitivity measurement of a 10-sensor array with erbium-doped fiber amplifier telemetry. *Optics Letters* 23: 1313–1315.

Vaughan, J. M. 1989. *The Fabry–Perot Interferometer: History, Practice and Applications*. IOP Publishing, Bristol, U.K., Chapter 3.

Venkatest, S. and D. W. Dolfi. 1990. Incoherent frequency modulated cw optical reflectometry with centimeter resolution. *Applied Optics* 29: 1323–1326.

Wentworth, R. H. 1989. Theoretical noise performance of coherence-multiplexed interferometric sensors. *Journal of Lightwave Technology* 7: 941–956.

Wielandy, S. and S. C. Dunn. 2004. Tilted superstructure fiber grating used as a Fourier-transform spectrometer. *Optics Letters* 29: 1614–1616.

Wolfelschneider, H., R. Kist, G. Knoll et al. 1987. Optically excited and interrogated micromechanical silicon cantilever structure. *Fiber Optic Sensors II*, The Hague, the Netherlands, SPIE 798.

Wuchenich, D. M. R., T. T. Lam, J. H. Chow et al. 2011. Laser frequency noise immunity in multiplexed displacement sensing. *Optics Letters* 36: 672–674.

Yang, J., L. Yuan, and W. Jin. 2007. Improving the reliability of multiplexed fiber optic low-coherence interferometric sensors by use of novel twin-loop network topologies. *Review of Scientific Instruments* 78: 055106.

Yariv, A. 1991. *Optical Electronics*, 4th edn. Saunders College Publishing, Section 11.7.

Zheng, J. 2005. Multiplexed reflectometric fiber optic frequency-modulated continuous-wave interferometric displacement sensor. *Optical Engineering* 44: 084401.

Zheng, J. 2007. Triple-sensor multiplexed frequency-modulated continuous-wave interferometric fiber-optic displacement sensor. *Applied Optics* 46: 2189–2196.

Further Reading

Abad, S., M. López-Amo, and I. Matías. 2002. Active fiber optic sensor networks, in *Handbook of Optical Fibre Sensing Technology*, J.M. López-Higuera, ed. John Wiley & Sons Ltd.

Dandridge, A. and C. Kirkendall. 2002. Passive fiber optic sensor networks, in *Handbook of Optical Fibre Sensing Technology*, J.M. López-Higuera, ed. John Wiley & Sons Ltd.

Jones, J. D. C. and R. McBride. 1998. Multiplexing optical fiber sensors, in *Optical Fiber Sensor Technology: Devices and Technology*, K.T.V. Grattan and B.T. Meggitt, eds., Vol. 2. Chapman & Hall.

Kersey, A. 1991. Distributed and multiplexed fiber optic sensors, in *Fiber Optic Sensors: An Introduction for Engineers and Scientists*, E. Udd, ed. Wiley Interscience.

Kist, R. 1989. Point sensor multiplexing principles, in *Optical Fiber Sensors: Systems and Applications*, B. Culshaw and J. Dakin, eds., Vol. 2. Artech House.

16. Distributed Fiber-Optic Sensors Based on Light Scattering in Optical Fibers

Xiaoyi Bao
University of Ottawa

Wenhai Li
University of Ottawa

Liang Chen
University of Ottawa

Handbook of Optical Sensors. Edited by José Luís Santos and Faramarz Farahi © 2015 CRC Press/Taylor & Francis Group, LLC. ISBN: 9781439866856.

16.1 Introduction

Distributed fiber optic sensor technology is one of the most promising candidates among the numerous sensor technologies that are adopted for structural health monitoring (SHM) (Measures 2001). This is due to its inherent fiber-optic properties such as light in weight, small in size, noncorrosive, and the immunity to electromagnetic (EM) interference (Kersey 1996). Distributed fiber-optic sensors can be used to measure the physical property changes at any position along a single optical fiber, with a gauge length of 2 m over hundreds of kilometers sensing length (Bao and Chen 2011a). The state-of-the-art technology for distributed strain measurement is 1–2 cm with sensing length of a few kilometers using the Brillouin scattering mechanism (Foaleng et al. 2010, Dong et al. 2012). In applications where monitoring is required at a large number of points on a large structure, distributed fiber-optic sensors provide particularly attractive solutions as cost- and space-effective tool.

The basic principle of distributed fiber-optic sensors is optical time-domain reflectometry (OTDR), which was first introduced to monitor fiber attenuation (Barnoski and Jensen 1976). In OTDR technique, the system is probed by a narrow pulse using the optical radar concept, Rayleigh backscatter from the fiber material is detected, and the changes can be seen due to different properties of fiber such as discontinuities in the fiber or coupler, and the location information is acquired by mapping position through time of flight of a pulse of light traveling to and from the sensing location. For this reason, the spatial resolution is determined by the duration of the optical pulse. The spatial resolution can be improved as the pulses are shortened at the expense of the broadened bandwidth and increased noise level and consequently in the reduction of dynamical range (a well-known trade-off between dynamical range and spatial resolution). As a result, the resolution of measured parameters, such as temperature, strain, vibration, and acoustic wave, will be compromised. OTDR-based sensors usually use direct detection to measure local intensity changes.

Another approach is the optical frequency-domain reflectometry (OFDR) with the frequency scanning, in which the probe signal is a continuous frequency-modulated optical wave instead of a pulsed signal as in OTDR (Kingsley and Davies 1985). In contrast to OTDR, the OFDR systems have relatively higher-power-per-frequency components; more dynamic range is obtained. The spatial resolution is determined by the tuning range of the laser, which can be a few millimeters for sensing of temperature or strain; the sensing length is limited by the coherent length of the laser to about 100 m, typically. The demodulation scheme of OFDR is interferometric, so it measures relative phase changes associated with temperature or strain variations.

Based on OTDR technique, a number of different distributed sensors have been explored based on Rayleigh, Raman, and Brillouin scatterings. OFDR systems require a coherent source and they are mainly deployed with Rayleigh and Brillouin mechanisms due to the small spectrum linewidth, while Raman scattering has bandwidths of THz, which makes it difficult to recover the location information and the change of the measured parameters, as both of them are in the same frequency range and it is hard to differentiate them. This chapter reviews the various types of distributed sensors that have been developed based on Rayleigh, Brillouin, and Raman scatterings, their working principles, as well as some applications. It is arranged as follows: the introduction to distributed sensing as in Section 16.1; the theory and working principle of spontaneous Rayleigh, Brillouin, and Raman scatterings and their mechanisms for measuring strain and temperature as in Section 16.2; the applications of the Stokes and anti-Stokes ratio in Raman scattering for distributed temperature sensing as in Section 16.3; the Rayleigh scattering–based OTDR and OFDR and system performance as in Sections 16.4 and 16.5; and the Brillouin scattering–based distributed sensors as in Section 16.6, which have been intensely studied for the last 20 years, resulting into the most successful commercially available distributed measurement systems, with important applications in several fields, particularly in the monitoring of large civil engineering structures.

Detailing more Section 16.6, it starts with the history of the Brillouin optical time-domain reflectometry (BOTDR) and Brillouin optical time-domain analysis (BOTDA) developments. Then the following topics will be addressed: theory of the combined Brillouin gain and loss to form parametric gain to make distributed sensors, the phase-matching conditions for the gain-and-loss process and their different Brillouin frequencies due to the chromatic dispersion (CD) and polarization mode dispersion (PMD), and the potential applications to measure CD and PMD using the off-resonance Brillouin spectrum. The recent developments of the Brillouin grating, differential Brillouin gain, and Brillouin *echo* have been summarized in this section as well, which briefly described the principle of Brillouin scattering–based distributed sensing in frequency domain. Section 16.7 gives the discussion on the limitations for sensing length and spatial and temperature or strain resolution, and it provides a performance chart of different sensing systems based on different mechanisms. In Section 16.8, we discuss the challenges of distributed sensors in applications involving structural monitoring. Finally, Section 16.9 summarizes and concludes this chapter.

16.2 Spontaneous Scattering in Optical Fibers

When a light wave propagates in a medium, it interacts with the constituent atoms and molecules, and if its wavelength is far from a medium resonance, the electric field induces a time-dependent polarization dipole. The induced dipole generates a secondary EM wave, and this is so-called light scattering. Because the distances between scattering centers (particles) are smaller than the wavelength of light in optical fibers, the secondary light waves are coherent. Hence, the resulting intensity is the addition of the scattered fields.

When the medium is perfectly homogeneous, the phase relationship of the emitted waves only allows the forward scattered beam. The medium can then be considered as continuous and the scattered light is coherent. The optical fiber is an inhomogeneous medium; scattering arises from microscopic or macroscopic variations in density, composition, or structure of a material through which light is passing. The random ordering of the molecules and the presence of dopants cause localized variations in density

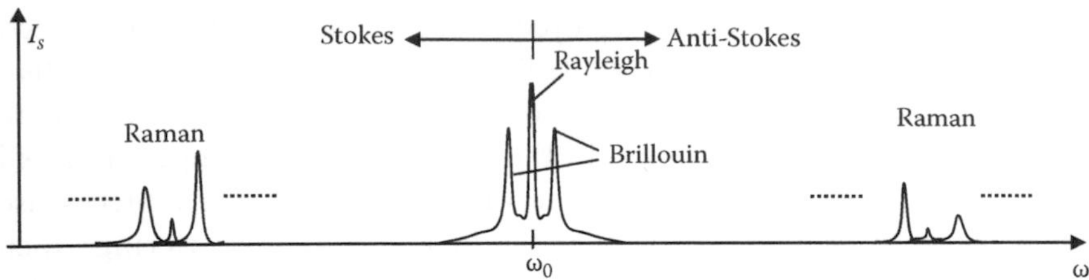

FIGURE 16.1 Typical spontaneous scattering spectrum from a solid-state matter.

(and therefore refractive index). These give rise to Rayleigh scattering that causes attenuation of the forward propagating signal (and creation of a backward propagating wave) that is proportional to $1/\lambda^4$. Rayleigh scattering is a linear scattering process in that the scattered power is simply proportional to the incident power. Also, no energy is transferred to the glass in Rayleigh scattering; therefore, there is no change in frequency of the scattered light comparing with the wavelength of the incident light, so-called elastic scattering. It is attributed to nonpropagating density fluctuations (Boyd 2008). In Figure 16.1, the two lines appearing on both sides of the Rayleigh peak are the Brillouin lines. They are contributed by the scattering of sound waves moving in opposite directions. The left peak with a downshifted frequency is called the Stokes peak, while the right one with an upshifted frequency is called the anti-Stokes line. Raman lines are contributed by the interaction of the light wave with molecular vibrations in the medium. Both the Brillouin and Raman scatterings are *inelastic* scatterings because they are associated with some frequency shifts. The last mechanism that can be observed is the Rayleigh wing scattering attributed to fluctuations in the orientation of anisotropic molecules. Raman spectra usually contain many sharp bands where separations between bands, corresponding to the electronic vibrations, and each bandwidth result from molecular rotation or reorientation excitations.

As long as the input light is scattered without strongly altering the property of the medium, we will say that the scattering is spontaneous. When the light intensity increases to a level such that the optical property of the medium is modified and then the scattered light also changes its property, the regime becomes stimulated. In other words, the evolution from spontaneous to stimulated scattering corresponds to a transition of the medium behavior from linear to nonlinear regime.

16.2.1 Impact of Incident Light on Medium and Scattered Light

In a homogeneous and isotropic dielectric medium such as glass, the response of the medium to an electric field ($\boldsymbol{E}$) is described by the polarization vector ($\boldsymbol{P}$) defined as

$$\boldsymbol{P} = \varepsilon_0 \chi \boldsymbol{E} \tag{16.1}$$

where

ε_0 is the vacuum dielectric constant
χ is the medium susceptibility

We have seen that scattering is caused by fluctuations $\Delta\varepsilon$ of the medium dielectric constant ε. These fluctuations can then induce a small polarization $\boldsymbol{P}_S$ such that the displacement vector $\boldsymbol{D}_S$ of the scattered field can be expressed as a linear combination of scattered ($\boldsymbol{E}_S E_S$) and input fields ($\boldsymbol{E}_p$) (Jackson 1999):

$$\boldsymbol{D}_S = \varepsilon_0 \varepsilon \boldsymbol{E}_S + \Delta \boldsymbol{E}_p = \varepsilon_0 \varepsilon \boldsymbol{E}_S + \boldsymbol{P}_S \tag{16.2}$$

$\Delta\varepsilon$ is a tensor even for an isotropic medium, which can be separated into three components: a scalar scattering, a symmetric scattering, and an antisymmetric scattering. The spatial (x, y, z) and temporal (t) dependence of the scattered wave can be described by a perturbed wave equation:

$$\nabla^2 \boldsymbol{E}_S - \left(\frac{n}{c_0}\right)^2 \frac{\partial^2 \boldsymbol{E}_S}{\partial t^2} = -\mu_0 \frac{\partial^2 \boldsymbol{P}_S}{\partial t^2} \tag{16.3}$$

where

μ_0 is the magnetic permeability of vacuum
c_0 is the velocity of light in vacuum
n is the refractive index of the medium

Fluctuations of pressure, temperature, entropy, and density are at the origin of scalar scattering, that is, Brillouin and Rayleigh scatterings. Symmetric scattering gives to Rayleigh and Brillouin scatterings, antisymmetric scattering is associated with the reorientation (Rayleigh wing scattering), and optical polarizability change of molecules gives Raman scattering in the presence of an electric field. For the case of the Rayleigh scattering, $\Delta\epsilon$ is only spatially varying.

16.2.2 Rayleigh Scattering

Rayleigh scattering is an elastic scattering where the scattered light frequency is the same as that of the incident light. On the microscopic level, the molecules making up any ordinary matters are immersed in a violent internal EM environment, in spite of the macroscopic charge neutrality as is true for most of the macroscopic materials. Those violent EM environments are constantly causing the molecule to readjust its electron clouds. By its own changing of the electron cloud configuration, this molecule is contributing to the changing environment for other neighboring molecules in a perpetual cycle. Therefore, on a relatively small spatial scale (order of tens of molecular sizes), one would observe fluctuations in terms of local charge density or local temperature or even strain values. Without incident light, such kind of short-range fluctuations would not produce measurable macroscopic effect at the far distance, as they are mutually incoherent and thus cancelled out. In this case, the macroscopic EM fields inside any material are zero. However, with external light incident on a material, this EM field $\boldsymbol{E}$ would reorient the originally incoherent random fluctuating molecular clouds to have a tendency to respond collectively the same way on a small spatial scale covering a small fraction of the wavelength of the EM field. Such collective tendency to respond to an

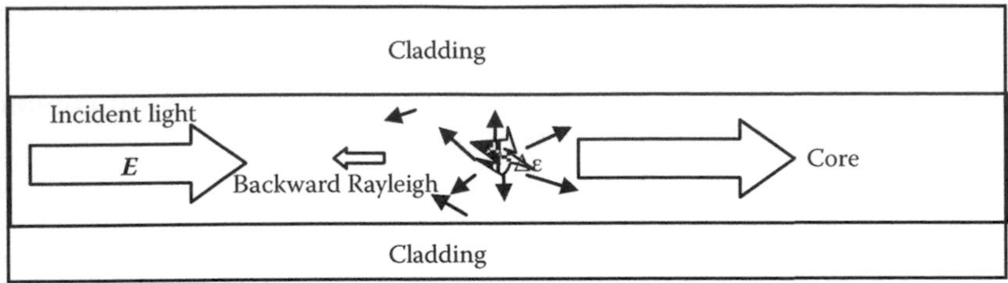

FIGURE 16.2 Schematic diagram for the spontaneous Rayleigh scattering process.

EM field would result in macroscopic polarization that is proportional to the external electric field $\boldsymbol{E}$, $\boldsymbol{P} = \epsilon_0 \chi \boldsymbol{E}$. The parameter χ is a material status–dependent quantity characterizing the collective response; the value χ possesses a randomly fluctuating portion associated with $\Delta\varepsilon(t,z)$. This fluctuating dielectric parameter $\Delta\varepsilon$ gives a fluctuating polarization-induced light emission in all directions as cartooned in Figure 16.2. Some of the scattered Rayleigh lights are recaptured by the waveguide, and they were sent in the backward direction. This backward propagating Rayleigh scattered light has a time delay that can be used to refer the spatial location of the scattering event provided that it does not suffer further scatterings (i.e., multiple scattering events), since multiple scattered light would lose its spatial location information. The Rayleigh scattering is generally weak; this makes single scattering assumption easily satisfied in most of the experimental applications. Hence, OTDR trace can be used to locate the fiber components in a network.

Assuming simple linearity of the polarization for a nonmagnetic material like the fiber, we write the following electric displacement field vector:

$$\boldsymbol{D} = \epsilon_0 (1 + \chi) \boldsymbol{E} = (\varepsilon + \Delta\epsilon) \boldsymbol{E} \tag{16.4}$$

Here, ε is a constant, while $\Delta\epsilon$ describes the locally fluctuating physical mechanism, that is, spontaneous scattering. For the case of the Rayleigh scattering, one assumes that $\Delta\epsilon$ varies spatially. Using Maxwell equation, we can get the following electric field $\boldsymbol{E}$:

$$\mu_0 \varepsilon \frac{\partial^2 \boldsymbol{E}}{\partial t^2} - \nabla^2 \boldsymbol{E} - \nabla \left[\boldsymbol{E} \cdot \nabla \ln (\varepsilon + \Delta\epsilon) \right] + \mu_0 \frac{\partial^2 (\Delta\epsilon \boldsymbol{E})}{\partial t^2} = 0 \tag{16.5}$$

The first two terms in Equation 16.5 describe the ordinary coherent propagation process, while the third and the fourth terms describe the random spontaneous scattering terms caused by the fluctuation $\Delta\epsilon$ that is both time and spatially dependent. To further simplify the physics, we could assume $\Delta\epsilon$ to be time independent (i.e., consider the Rayleigh scattering only), and then one can replace the partial time derivative using the transformation $\partial/\partial t \to -i\omega$ (assuming a time variation of the complex form $e^{-i\omega t}$ for the incident $\boldsymbol{E}$ field). Hence, the Maxwell equation can be further simplified:

$$\nabla^2 \boldsymbol{E} + \nabla \left[\boldsymbol{E} \cdot \nabla \ln (\varepsilon + \Delta\epsilon) \right] + \mu_0 \varepsilon \omega^2 \left(1 + \frac{\Delta\epsilon}{\varepsilon} \right) \boldsymbol{E} = 0 \tag{16.6}$$

For the case of optical fiber, we could neglect the lateral dependence and consider only the longitudinal dependence on z. Furthermore, by making the transverse wave approximation (i.e., neglecting the $\boldsymbol{E}$ field projection in the direction of propagation), we get a simpler looking differential equation (i.e., 1D plane wave approximation):

$$\frac{\partial^2 \boldsymbol{E}}{\partial z^2} + \mu_0 \varepsilon \omega^2 \left(1 + \frac{\Delta\epsilon(z)}{\varepsilon}\right) \boldsymbol{E} = 0 \tag{16.7}$$

Equation 16.7 can be viewed as a scalar differential equation of the following type:

$$\frac{\partial^2 E}{\partial z^2} + \beta^2 \left(1 + \frac{\Delta\epsilon(z)}{\varepsilon}\right) E = 0 \tag{16.8}$$

Here, $\beta = \omega\sqrt{\mu_0 \varepsilon}$ is the propagation constant. Thus, one could propose a solution composed of forward and backward traveling waves (Froggatt and Moore 1998):

$$E = E_0 e^{i\beta z} + \Psi(z,\beta) e^{-i\beta z} \tag{16.9}$$

The differential equation for backward scattered wave is then

$$\frac{\partial^2 \Psi}{\partial z^2} - 2i\beta \frac{\partial \Psi}{\partial z} + \beta^2 \frac{\Delta\epsilon(z)}{\varepsilon} E_0 e^{2i\beta z} + \beta^2 \frac{\Delta\epsilon(z)}{\varepsilon} \Psi = 0 \tag{16.10}$$

Considering the case of a weak Rayleigh scattering by neglecting the second-order derivative and the last term (i.e., $|\Psi| \ll E_0$), we can find an approximate solution for the backward scattered Rayleigh light due to the random spatial variation of the permittivity:

$$\Psi(z,\beta) - \Psi(0,\beta) \approx \frac{\beta E_0}{2i} \int_0^z \frac{\Delta\epsilon(\zeta)}{\varepsilon} e^{2i\beta\zeta} d\zeta \tag{16.11}$$

For most of the time, the detected signal is $\Psi(0,\beta)$, which is seen to be related to the end face reflection amplitude, $\Psi(z = L,\beta)$. Hence, the Rayleigh backscattered signal is a type of Fourier transform of the random permittivity fluctuation. Fiber attenuation α can be easily incorporated by the replacement of $\beta \rightarrow \beta + i\alpha$. The applications of the Rayleigh scattering in the fiber sensor are relatively wide; it can be used to sense local temperature or strain. It can also be used to sense acoustic vibrations.

Although we emphasized that Rayleigh scattering is a type of scattering without changing of frequency, that is, elastic scattering, there is other scattering arising from larger scattering centers such as dust particles that gives no frequency change as well. Such processes are called the Mie scattering, in which their scattering strength is controlled by the size of the scattering particles and its refractive index relative to the scattering medium. Although the Mie scattering can be used to detect dust particle size and it is widely used in biomedical sensing, it cannot be used to sense temperature like Rayleigh scattering can in the fiber.

16.2.3 Spontaneous Brillouin Scattering in a Single-Mode Optical Fiber

For a perfectly symmetric waveguide, the fiber supports two modes, but they are degenerate. The radial intensity follows essentially a Gaussian distribution characterized by the spot size r_a defined as the $1/e$ intensity width. The electric field propagating in the fiber is then considered as a plane wave with a Gaussian radial distribution.

The Brillouin scattering represents collective acoustic oscillations of the glass. From the microscopic point of view, the intermolecular interaction in glass makes it favorable for molecules to stay at some stable distance away from each other. There is an energy penalty when the intermolecular distance is either farther apart or closer than some given stable distances. This microscopic existence of some balanced intermolecular distances would set a new collective motion. Imagine if a neighboring molecule was getting closer than the stable separation, then it will be pushed away toward the stable separation distance; however, when it reaches the stable separation, it will not stop; rather, it will overshoot passing the stable separation distance. Once it is farther away, it will experience an attraction to pull it back toward the stable separation distance; however, it will again overshoot when it returns. Such a repeating cycle forms a collective motion called acoustic phonons. To describe this process, we need to use macroscopic parameters like the density (ρ), entropy (s), and pressure (P) of the matter as well as the temperature of the matter. Recall these parameters are all macroscopic thermodynamic quantities that can be directly related to the macroscopic Maxwell equations. It makes perfect physical sense to expect that material polarizability to be proportional to material density if one traces the steps of macroscopic Maxwell equation. Furthermore, as the local density ρ is changed, one can also expect local pressure changes as well. As we are interested in Δε variations induced by thermodynamic quantities, we first consider ρ and T as independent thermodynamic variables and write the dielectric constant as (Boyd 2008)

$$\Delta\varepsilon = \left(\frac{\partial\varepsilon}{\partial\rho}\right)_T \Delta\rho + \left(\frac{\partial\varepsilon}{\partial T}\right)_\rho \Delta T \tag{16.12}$$

According to Barnoski and Jensen (1976), the second term can be neglected with error of 2% because density fluctuations affect the dielectric constant significantly more than temperature fluctuations. Also,

$$\Delta\rho = \left(\frac{\partial\rho}{\partial p}\right)_S \Delta p + \left(\frac{\partial\rho}{\partial s}\right)_p \Delta s \tag{16.13}$$

The first term corresponds to adiabatic density fluctuations, which are pressure waves or acoustic waves, that is, Brillouin scattering. The second term is entropy or temperature fluctuations, that is, Rayleigh scattering. The dielectric constant fluctuation density can be expressed as

$$\Delta\varepsilon = \left(\frac{\partial\varepsilon}{\partial\rho}\right)_T \left(\frac{\partial\rho}{\partial p}\right)_S \Delta p = \frac{\gamma_e}{\rho_0}\left(\frac{\partial\rho}{\partial p}\right)_S \Delta\tilde{p} \tag{16.14}$$

where the electrostriction constant, γ_e, is defined as $\gamma_e = \rho_0(\partial\varepsilon/\partial\rho)_T$. Here, ρ is the average density of the fiber material. The acoustic wave is captured in the following wave equation describing the pressure wave with the local pressure variation parameter, $\Delta\tilde{p}$:

$$\frac{\partial^2\Delta\tilde{p}}{\partial t^2} - \Gamma'\nabla^2\frac{\partial\Delta\tilde{p}}{\partial t} - V_a^2\nabla^2\Delta\tilde{p} = 0 \tag{16.15}$$

Γ' is a damping parameter related to the local viscosity of the material, while V_a is the sound velocity. From this equation, one can get the solution for Brillouin frequency of Stokes and anti-Stokes wave solution as the following (Bao and Chen 2011):

$$\Omega_B \cong \frac{2n(\omega)\omega}{(c/V_a) \mp n_g(\omega)} \approx 2\pi\frac{2n(\omega)V_a}{\lambda}\left(1 \pm n_g(\omega)\frac{V_a}{c}\right) \tag{16.16}$$

Here, $n_g(\omega) = d[n(\omega)\omega]/d\omega$ is the group refractive index, wherein the upper sign is for anti-Stokes and lower sign is for the Stokes side resonance, respectively. For a given light, its corresponding Stokes and anti-Stokes Brillouin resonance frequency difference is

$$\delta\Omega_B = \Omega_B^{AS} - \Omega_B^S \cong 2\pi\frac{2n(\omega)V_a}{\lambda}\delta n_g(\omega)\frac{V_a}{c} \tag{16.17}$$

Due to the geometry of the fiber, both cladding and core acoustic modes propagate. If the optical wave and the acoustic modes do not significantly overlap, as the case of step-index fibers, the scattering due to variations of the density along the optical fiber is weak, and it is position dependent due to the density and/or shape nonuniformity. Such a process is related to the well-known effect of PMD (Gordon and Kogelnik 2000). Such variations are induced from manufacturing process or environmental perturbations. This means even in a single-mode fiber (SMF), two orthogonally polarized modes are nondegenerate, that is, we should take separate effective refractive indexes n_x and n_y and $n_x \neq n_y$ due to the variation in fiber core shape and anisotropic stress along the fiber length (these indexes also change with the position). Hence, Equation 16.17 should be generalized to represent two axes for the effective modal index for both Stokes and anti-Stokes frequencies:

$$\Omega_{S,x,y}(z) = 2\pi\frac{2n_{x,y}(z,\omega_S)V_a}{\lambda}\left[1 - n_{gx,gy}(z,\omega_S)\frac{V_a}{c}\right] \tag{16.18a}$$

$$\Omega_{AS,x,y}(z) = 2\pi \frac{2n_{x,y}(z,\omega_{AS})V_a}{\lambda}\left[1 + n_{gx,gy}(z,\omega_{AS})\frac{V_a}{c}\right] \tag{16.18b}$$

Because of the mode coupling in SMF, the modal Brillouin frequency varies and it introduces the uncertainty between each measurement, as the input state of polarization (SOP) of the pump wave changes. Therefore, the local SOP changes, and even the Stokes and anti-Stokes frequencies can be different due to their different modal index at two different frequencies. This is in addition to the wavelength-induced difference.

16.2.4 Combined Brillouin Gain and Loss via Stimulated Brillouin Scattering

The spontaneous Brillouin scattering can be generated from the thermodynamic noises. So long as a Stokes or anti-Stokes light wave is generated, they interact with the external light wave to enhance this process via an effect called electrostriction.

A molecule with a permanent dipole moment $\boldsymbol{p}$ in the external electric field $\boldsymbol{E}$ has the potential energy $U = -\boldsymbol{p} \cdot \boldsymbol{E}$. This molecule would then experience a force that is proportional to the external field gradient:

$$\boldsymbol{F} = -\nabla U = \nabla(\boldsymbol{p} \cdot \boldsymbol{E}) \tag{16.19}$$

As a result, the molecules will reorganize themselves according to the external field both spatially and temporally. This effect is called electrostriction. The dielectric constant ε of a material depends on the mass density ϱ of the material. Following the treatment of Boyd (2008), the increase of the field energy density that is associated with the change of dielectric constant is attributed by the work done by local strictive pressure p_{st} (per unit volume) to the material. Considering the high optical frequency of the electric field, average values are taken as follows:

$$\Delta U = \frac{1}{2}\varepsilon_0 \langle E^2 \rangle \Delta\epsilon = \frac{1}{2}\varepsilon_0 \langle E^2 \rangle \left(\frac{\partial \epsilon}{\partial \varrho}\right)\Delta\varrho = \Delta w = p_{st}\frac{\Delta V}{V} = -p_{st}\frac{\Delta\varrho}{\varrho} \tag{16.20}$$

Therefore, we could obtain an expression for external force via the mechanical strictive pressure:

$$\mathbf{f} = \nabla p_{st}, \quad p_{st} = -\frac{1}{2}\varepsilon_0 \gamma_e \langle \boldsymbol{E} \cdot \boldsymbol{E} \rangle \tag{16.21}$$

The nonlinear polarization contribution is induced by the electrostriction effect associated with the mass density change $\boldsymbol{P}_{NL}$:

$$\boldsymbol{P}_{NL} = \varepsilon_0 \Delta\epsilon \boldsymbol{E} = \varepsilon_0 \varrho_0 \left(\frac{\Delta\epsilon}{\Delta\varrho}\right)\frac{\Delta\varrho}{\varrho_0}\boldsymbol{E} = \frac{\varepsilon_0 \gamma_e}{\varrho_0}\Delta\varrho \boldsymbol{E} \tag{16.22}$$

The equation governing the density change takes the same form as the pressure wave Equation 16.14 of the spontaneous Brillouin scattering except the external field term (Boyd 2008):

$$\frac{\partial^2(\Delta\varrho)}{\partial t^2} - \Gamma'\nabla^2\frac{\partial(\Delta\varrho)}{\partial t} - V_a^2\nabla^2(\Delta\varrho) = \nabla\cdot\mathbf{f} \tag{16.23}$$

The maximum SBS interaction is achieved when polarizations of the electric fields are matched; thus, we will assume that this is satisfied (although in reality, it is hard to achieve total polarization alignment, especially in SMFs due to PMD as discussed earlier). Hence, instead of vector equation for the electric fields, we would treat them as scalar quantities. Second, for the combined Brillouin gain-and-loss-based distributed sensors, one can input both Stokes and anti-Stokes waves simultaneously on an incident laser light in the fiber (Li et al. 2010). In this case, a given incident light is simultaneously involved in the gain (with an anti-Stokes light) and loss (with a Stokes light) process via two acoustic waves that move in opposite directions in the fiber.

Figure 16.3 shows the schematic diagram for the simultaneous gain-and-loss SBS configuration.

The 1D *stimulated Brillouin scattering* (SBS) modeling equations in the fiber are given as follows (Boyd 2008):

$$\frac{\partial^2(\Delta\varrho)}{\partial t^2} - \Gamma'\frac{\partial^3(\Delta\varrho)}{\partial z^2\partial t} - v^2\frac{\partial^2(\Delta\varrho)}{\partial z^2} = \frac{\partial^2\left[-(1/2)\varepsilon_0\gamma_e E^2(z,t)\right]}{\partial z^2} \tag{16.24}$$

$$\frac{\partial^2 E}{\partial z^2} - \mu_0\varepsilon_0\epsilon\frac{\partial^2 E}{\partial t^2} = \mu_0\frac{\varepsilon_0\gamma_e}{\varrho_0}\frac{\partial^2[\Delta\varrho E]}{\partial t^2} \tag{16.25}$$

In the following, we write down the scalar electric fields for this process:

$$E(z,t) = \underbrace{\left[A_{AS}(z,t)e^{i(k_{AS}z-\omega_{AS}t)} + c.c.\right]}_{\text{Anti-Stokes wave}} + \underbrace{\left[A_S(z,t)e^{i(k_S z-\omega_S t)} + c.c.\right]}_{\text{Stokes wave}} + \underbrace{\left[A(z,t)e^{i(-kz-\omega t)} + c.c.\right]}_{\text{Simultaneous gain-loss wave}} \tag{16.26}$$

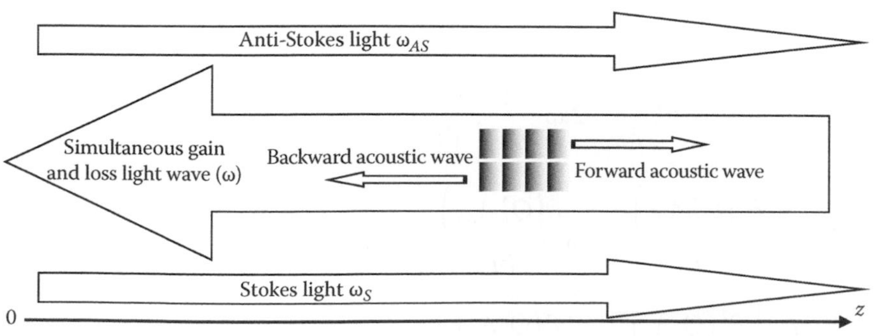

FIGURE 16.3 Schematic diagram for simultaneous gain-and-loss SBS configuration in fiber.

Both the anti-Stokes and Stokes waves are propagating in the positive z-direction, and the simultaneous gain-and-loss light is propagating in the negative z-direction. The dispersion and energy conversation relationships are satisfied for the three light waves:

$$k_{AS}^2 = \left(\frac{n}{c}\omega_{AS}\right)^2, \quad k_S^2 = \left(\frac{n}{c}\omega_S\right)^2, \quad k^2 = \left(\frac{n}{c}\omega\right)^2 \tag{16.27}$$

$$\omega_{AS} - \omega = \Omega_f, \qquad \omega - \omega_S = \Omega_b \tag{16.28}$$

The forward and backward moving acoustic waves are related to density variation:

$$\Delta\varrho(z,t) = \underbrace{\left[Q_f(z,t)e^{i(q_f z - \Omega_f t)} + c.c.\right]}_{\text{Forward acoustic wave}} + \underbrace{\left[Q_b(z,t)e^{i(-q_b z - \Omega_b t)} + c.c.\right]}_{\text{Backward acoustic wave}} \tag{16.29}$$

In the simplified SBS modeling process, we only equate the same time-varying terms, thus capturing the essential physics. For the terms having a time variation $e^{-i\omega_{AS}t}$, we get

$$\begin{aligned}
&\frac{\partial^2 A_{AS}}{\partial z^2} - \frac{1}{(c/n)^2}\frac{\partial^2 A_{AS}}{\partial t^2} + 2ik_{AS}\frac{\partial A_{AS}}{\partial z} + 2i\omega_{AS}\frac{1}{(c/n)^2}\frac{\partial A_{AS}}{\partial t} = \\
&= \frac{\gamma_e}{c^2\varrho_0}\left[\frac{\partial^2(Q_f A)}{\partial t^2} - 2i\omega_{AS}\frac{\partial(Q_f A)}{\partial t} - \omega_{AS}^2 Q_f A\right]e^{-i(k_{AS}+k-q_f)z}
\end{aligned} \tag{16.30}$$

The coefficient of the terms having a time variation $e^{-i\omega_S t}$ is similarly found:

$$\begin{aligned}
&\frac{\partial^2 A_S}{\partial z^2} - \frac{1}{(c/n)^2}\frac{\partial^2 A_S}{\partial t^2} + 2ik_S\frac{\partial A_S}{\partial z} + 2i\omega_S\frac{1}{(c/n)^2}\frac{\partial A_S}{\partial t} = \\
&= \frac{\gamma_e}{c^2\varrho_0}\left[\frac{\partial^2\left(Q_b^* A\right)}{\partial t^2} - 2i\omega_S\frac{\partial\left(Q_b^* A\right)}{\partial t} - \omega_S^2 Q_b^* A\right]e^{-i(k_S+k-q_b)z}
\end{aligned} \tag{16.31}$$

The coefficient of the terms with time variation $e^{-i\omega t}$, which represents combined gain-and-loss terms, is

$$\begin{aligned}
&\frac{\partial^2 A}{\partial z^2} - \frac{1}{(c/n)^2}\frac{\partial^2 A}{\partial t^2} - 2ik\frac{\partial A}{\partial z} + 2i\omega\frac{1}{(c/n)^2}\frac{\partial A}{\partial t} = \\
&= \frac{\gamma_e}{c^2\varrho_0}\left[\frac{\partial^2\left(Q_f^* A_{AS}\right)}{\partial t^2} - 2i\omega\frac{\partial\left(Q_f^* A_{AS}\right)}{\partial t} - \omega^2 Q_f^* A_{AS}\right]e^{i(k_{AS}+k-q_f)z} \\
&+ \frac{\gamma_e}{c^2\varrho_0}\left[\frac{\partial^2(Q_b A_S)}{\partial t^2} - 2i\omega\frac{\partial(Q_b A_S)}{\partial t} - \omega^2 Q_b A_S\right]e^{i(k_S+k-q_b)z}
\end{aligned} \tag{16.32}$$

The coefficients of a time variation $e^{-i\Omega_f t}$ give the following equation:

$$\left(V_a^2 q_f^2 - \Omega_f^2 - i\Omega_f \Gamma' q_f^2\right) Q_f + \left(\Gamma' q_f^2 - 2i\Omega_f\right)\frac{\partial Q_f}{\partial t} - 2\left(\Omega_f \Gamma' q_f + iV_a^2 q_f\right)\frac{\partial Q_f}{\partial z} + \frac{\partial^2 Q_f}{\partial t^2}$$
$$+\left(i\Omega_f \Gamma' - V_a^2\right) - 2i\Gamma' q_f \frac{\partial^2 Q_f}{\partial z \partial t} - \Gamma' \frac{\partial^3 Q_f}{\partial z^2 \partial t} \frac{\partial^2 Q_f}{\partial z^2}$$
$$= \varepsilon_0 \gamma_e \left[\left(k + k_{AS}\right)^2 A_{AS} A^* - 2i\left(k + k_{AS}\right)\frac{\partial\left(A_{AS} A^*\right)}{\partial z} - \frac{\partial^2\left(A_{AS} A^*\right)}{\partial z^2}\right] e^{i\left(k_{AS} + k - q_f\right)z} \tag{16.33}$$

Lastly, for the coefficients of a time variation of the form $e^{-i\Omega_b t}$, we obtain

$$\left(V_a^2 q_b^2 - \Omega_b^2 - i\Omega_b \Gamma' q_b^2\right) Q_b + \left(\Gamma' q_b^2 - 2i\Omega_b\right)\frac{\partial Q_b}{\partial t} + 2\left(\Omega_b \Gamma' q_b + iV_a^2 q_b\right)\frac{\partial Q_b}{\partial z} + \frac{\partial^2 Q_b}{\partial t^2}$$
$$+\left(i\Omega_b \Gamma' - V_a^2\right)\frac{\partial^2 Q_b}{\partial z^2} + 2i\Gamma' q_b \frac{\partial^2 Q_b}{\partial z \partial t} - \Gamma' \frac{\partial^3 Q_b}{\partial z^2 \partial t}$$
$$= \varepsilon_0 \gamma_e \left[\left(k + k_S\right)^2 A_S^* A + 2i\left(k + k_S\right)\frac{\partial\left(A_S^* A\right)}{\partial z} - \frac{\partial^2\left(A_S^* A\right)}{\partial z^2}\right] e^{-i\left(k_S + k - q_b\right)z} \tag{16.34}$$

Considering that the nonlinear Brillouin scattering effect–induced index change is orders of magnitude smaller than the ordinary linear interactions in the refractive index, it is adequate to apply the slowly varying amplitude approximation. Therefore, any changes caused by the nonlinear interaction will be treated as a small perturbation on the amplitudes of the ordinary linear propagating waves. Hence, we can remove all the terms with more than second-order differentiation with respect to space z or time t, so we get the following approximate equation for the anti-Stokes wave:

$$2ik_{AS}\frac{\partial A_{AS}}{\partial z} + 2i\omega_{AS}\frac{1}{(c/n)^2}\frac{\partial A_{AS}}{\partial t} = \frac{\gamma_e}{c^2 \varrho_0}\left[-2i\omega_{AS}\frac{\partial\left(Q_f A\right)}{\partial t} - \omega_{AS}^2 Q_f A\right] e^{-i\left(k_{AS} + k - q_f\right)z} \tag{16.35}$$

The Stokes wave equation is approximated accordingly:

$$2ik_S\frac{\partial A_S}{\partial z} + 2i\omega_S\frac{1}{(c/n)^2}\frac{\partial A_S}{\partial t} = \frac{\gamma_e}{c^2 \varrho_0}\left[-2i\omega_S\frac{\partial\left(Q_b^* A\right)}{\partial t} - \omega_S^2 Q_b^* A\right] e^{-i\left(k_S + k - q_b\right)z} \tag{16.36}$$

The simultaneous gain-and-loss wave equation becomes

$$-2ik\frac{\partial A}{\partial z}+2i\omega\frac{1}{(c/n)^2}\frac{\partial A}{\partial t}=\frac{\gamma_e}{c^2\varrho_0}\left[-2i\omega\frac{\partial\left(Q_f^{*}A_{AS}\right)}{\partial t}-\omega^2 Q_f^{*}A_{AS}\right]e^{i\left(k_{AS}+k-q_f\right)z}$$

$$+\frac{\gamma_e}{c^2\varrho_0}\left[-2i\omega\frac{\partial\left(Q_b A_S\right)}{\partial t}-\omega^2 Q_b A_S\right]e^{i\left(k_S+k-q_b\right)z} \tag{16.37}$$

For the forward propagating acoustic wave equation, we get

$$\left(V_a^2 q_f^2-\Omega_f^2-i\Omega_f\Gamma' q_f^2\right)Q_f+\left(\Gamma' q_f^2-2i\Omega_f\right)\frac{\partial Q_f}{\partial t}-2\left(\Omega_f\Gamma' q_f+iV_a^2 q_f\right)\frac{\partial Q_f}{\partial z}$$

$$=\varepsilon_0\gamma_e\left[\left(k+k_{AS}\right)^2 A_{AS}A^{*}-2i\left(k+k_{AS}\right)\frac{\partial\left(A_{AS}A^{*}\right)}{\partial z}\right]e^{i\left(k_{AS}+k-q_f\right)z} \tag{16.38}$$

For the backward propagating acoustic wave, we obtain the following approximation:

$$\left(V_a^2 q_b^2-\Omega_b^2-i\Omega_b\Gamma' q_b^2\right)Q_b+\left(\Gamma' q_b^2-2i\Omega_b\right)\frac{\partial Q_b}{\partial t}+2\left(\Omega_b\Gamma' q_b+iV_a^2 q_b\right)\frac{\partial Q_b}{\partial z}$$

$$=\varepsilon_0\gamma_e\left[\left(k+k_S\right)^2 A_S^{*}A+2i\left(k+k_S\right)\frac{\partial\left(A_S^{*}A\right)}{\partial z}\right]e^{-i\left(k_S+k-q_b\right)z} \tag{16.39}$$

The right-hand side terms in the preceding equations involve first-order spatial or time derivatives, and they are much smaller than the other terms in the bracket. Therefore, we neglect those terms and consider the attenuation of the optical fibers:

$$\frac{\partial A_{AS}}{\partial z}+\frac{n}{c}\frac{\partial A_{AS}}{\partial t}=i\frac{\omega_{AS}\gamma_e}{2nc\varrho_0}Q_f A e^{-i\left(k_{AS}+k-q_f\right)z}-\frac{1}{2}\alpha A_{AS} \tag{16.40}$$

$$\frac{\partial A_S}{\partial z}+\frac{n}{c}\frac{\partial A_S}{\partial t}=i\frac{\omega_S\gamma_e}{2nc\varrho_0}Q_b^{*}Ae^{-i\left(k_S+k-q_b\right)z}-\frac{1}{2}\alpha A_S \tag{16.41}$$

$$-\frac{\partial A}{\partial z}+\frac{n}{c}\frac{\partial A}{\partial t}=i\frac{\omega\gamma_e}{2nc\varrho_0}Q_f^{*}A_{AS}e^{i\left(k_{AS}+k-q_f\right)z}+i\frac{\omega\gamma_e}{2nc\varrho_0}Q_b A_S e^{i\left(k_S+k-q_b\right)z}-\frac{1}{2}\alpha A \tag{16.42}$$

$$\left[V_a^2 q_f^2 - (\omega_{AS} - \omega)^2 - i(\omega_{AS} - \omega)\Gamma' q_f^2\right] Q_f + \left[\Gamma' q_f^2 - 2i(\omega_{AS} - \omega)\right]\frac{\partial Q_f}{\partial t}$$

$$-2\left[(\omega_{AS} - \omega)\Gamma' q_f + iV_a^2 q_f\right]\frac{\partial Q_f}{\partial z} = \varepsilon_0 \gamma_e (k + k_{AS})^2 A_{AS} A^* e^{i(k_{AS} + k - q_f)z} \tag{16.43}$$

$$\left[V_a^2 q_b^2 - (\omega - \omega_S)^2 - i(\omega - \omega_S)\Gamma' q_b^2\right] Q_b + \left[\Gamma' q_b^2 - 2i(\omega - \omega_S)\right]\frac{\partial Q_b}{\partial t} +$$

$$-2\left[(\omega - \omega_S)\Gamma' q_b + iV_a^2 q_b\right]\frac{\partial Q_b}{\partial z} = \varepsilon_0 \gamma_e (k + k_S)^2 A_S^* A e^{-i(k_S + k - q_b)z} \tag{16.44}$$

Special cases can be considered for either the anti-Stokes wave to be zero and the forward acoustic wave $Q_f = 0$ or the Stokes wave to be zero and the backward acoustic wave $Q_b = 0$. Then the five aforementioned equations become the ordinary three coupled wave equations as described in Boyd (2008), except one further assumption is needed: sound speed is much smaller than that of light. Hence, the spatial derivative terms $\partial Q_f/\partial z$ and $\partial Q_b/\partial z$ are neglected. Under the condition of simultaneous gain-and-loss process (Li et al. 2010), the depletion of the Stokes and anti-Stokes wave is much smaller than that of the Brillouin gain or Brillouin loss process (Minardo et al. 2009), that is, the $\partial Q_f/\partial z$ and $\partial Q_b/\partial z$ should be kept.

The choice of acoustic moment of q_f and q_b is constrained by the SBS process, so we introduce two momentum mismatch variables, first, for the *gain process* (i.e., the anti-Stokes light, forward moving acoustic wave, and the gain light):

$$\Delta q_G = q_f - k_{AS} - k \tag{16.45}$$

Then, we can rewrite the forward acoustic wave momentum:

$$q_f = k_{AS} + k + \Delta q_G = \frac{1}{c}\left[n(\omega_{AS})\omega_{AS} + n(\omega)\omega\right] + \Delta q_G \tag{16.46}$$

Note the frequency dependence of refractive index n takes into account of the CD and modal index of the fiber waveguide properties, which reflects the fast and slow axes of the optical fiber and PMD (Gordon and Kogelnik 2000). We are considering only one of the principal axes in developing the earlier model since polarization matching has been assumed at the start.

The momentum mismatch for the *loss process* (i.e., the loss light, back moving acoustic wave, and the Stokes light) takes

$$\Delta q_L = q_b - k_S - k \tag{16.47}$$

Then, we can rewrite the backward acoustic wave momentum:

$$q_b = k + k_S + \Delta q_L = \frac{1}{c}\left[n(\omega)\omega + n(\omega_S)\omega_S\right] + \Delta q_L \tag{16.48}$$

Apparently, the momentum conservation varies with fiber position, as $n(z,\omega)$ is a function of position (due to PMD) and frequency (due to the chromatic dispersion). There are two gratings moving at the same speed but in opposite directions, and they are created by the Brillouin gain and Brillouin loss processes. The middle frequency wave will experience simultaneous gain (from the anti-Stokes light) and loss (from the Stokes light) via two oppositely moving acoustic waves excited by the two coupled SBS processes. The coupling of two SBS processes is controlled by ω_{AS},ω, and ω_S and Δq_G and Δq_L. The maximum coupling is determined by optimizing the acoustic waves, which is $\Delta q_G = 0$ for the forward moving acoustic wave and $\Delta q_L = 0$ for the backward moving acoustic wave. Using those conditions, we can define a Brillouin frequency for the gain side that varies with the modal index (PMD) and fiber position:

$$\Omega_B^G(z) = \omega_{AS} - \omega = \frac{V_a}{c}\left[n\left(\omega_{AS}, z\right)\omega_{AS} + n(\omega, z)\omega\right] \tag{16.49}$$

Similarly, we can define a Brillouin frequency for the loss side:

$$\Omega_B^L(z) = \omega - \omega_S = \frac{V_a}{c}\left[n(\omega, z)\omega + n\left(\omega_S, z\right)\omega_S\right] \tag{16.50}$$

The Brillouin frequency for the gain side Ω_B^G is not the same as the Brillouin frequency for the loss side Ω_B^L, mainly due to the CD; furthermore, fiber birefringence would of course make the measured Brillouin frequency having statistical feature, rather than being governed only by the signal-to-noise ratio (SNR) of the sensor system (Bao and Chen 2011). At positions of equal SOPs for all three optical waves, the depletion of central wave (simultaneous gain and loss) is zero, and then $\Omega_B^L = \Omega_B^G$; this means the gain-and-loss spectrum is symmetric. Such a situation can be found in polarization-maintaining fiber (PMF) (Li et al. 2010). However, in SMF, due to the local SOP variation, the power of the carrier frequencies varies due to the phase mismatching attributed by the SOP change with the position. This concept can be used to measure the beat length of SMF by measuring the Ω_B^G and Ω_B^L simultaneously. It is clear why the Brillouin frequency is sensitive to the local temperature and strain as it is a function of local acoustic speed V_a and of the refractive modal indexes of the two light waves that excite the acoustic wave. Both acoustic speed and the refractive indexes are functions of temperature and strain.

In SBS-based distributed sensor, the input SOP of the probe and pump waves is critically related to the local Brillouin gain, as SOP of the fiber varies in position due to the PMD. The PMD represents two parts: (1) local SOP changes and (2) differential group delay (DGD) (Gordon and Kogelnik 2000), although the introduction of polarization scrambler could reduce the Brillouin gain fluctuation from SOP contribution of PMD. However, it does not remove the fluctuation in the Brillouin frequency measurement, as the Brillouin frequency as stated in the spontaneous Brillouin scattering is associated with effective refractive index, that is, phase velocity, rather than SOP. In addition, the polarization scrambler itself also has PMD, which reduces the minimum measureable stress or temperature resolution.

16.2.5 Spontaneous Raman Scattering

When light is scattered from an atom or a molecule, most photons are elastically scattered (Rayleigh scattering), such that the scattered photons have the same energy (frequency) and wavelength as the incident photons. However, a small fraction of the scattered light (approximately 1 in 10 million photons) is scattered by an excitation, with the scattered photons having a frequency different from the incident photons. The interaction of light with matter in a linear regime allows the absorption and emission of a photon precisely matching the difference in energy levels of the interacting electron or electrons.

The light and molecular interaction process can be described by the behavior of the induced electric dipole moment $\boldsymbol{P}$ and the dynamic electric field $\boldsymbol{E}$ that produces it. A monochromatic plane wave $\boldsymbol{E}$ is incident on the molecule, and the total induced molecular polarization field can be written as the sum of a series of time-dependent electric dipole moment vectors $\boldsymbol{P}^{(1)},\boldsymbol{P}^{(2)},\boldsymbol{P}^{(3)}\ldots$ as follows:

$$\boldsymbol{P} = \boldsymbol{P}^{(1)} + \boldsymbol{P}^{(2)} + \boldsymbol{P}^{(3)} + \cdots \tag{16.51}$$

We have $\boldsymbol{P}^{(1)} \gg \boldsymbol{P}^{(2)} \gg \boldsymbol{P}^{(3)}\cdots$, and this is a very fast-converging series. Their relationships with incident $\boldsymbol{E}$ field have the following forms:

$$\begin{aligned} \boldsymbol{P}^{(1)} &= \epsilon_0 \chi^{(1)} \cdot \boldsymbol{E} \\ \boldsymbol{P}^{(2)} &= \epsilon_0 \chi^{(2)} : \boldsymbol{EE} \\ \boldsymbol{P}^{(3)} &= \epsilon_0 \chi^{(3)} \vdots \boldsymbol{EEE} \end{aligned} \tag{16.52}$$

where

$\chi^{(1)}$ is the polarizability tensor (second rank)
$\chi^{(2)}$ is the first hyperpolarizability tensor (third rank)
$\chi^{(3)}$ is the second hyperpolarizability tensor (fourth rank)

These tensors have time-dependent nature according to the molecule's characteristic oscillation mode frequency ω_M, where $\chi^{(1)}(t)$ can be written as (with $\chi^{(1)}_{Ray}$ is the elastic Rayleigh scattering polarizability and $\chi^{(1)}_M$ is the polarizability of mode ω_M)

$$\chi^{(1)}(t) = \chi^{(1)}_{Ray} + \sum_M \chi^{(1)}_M \cos\omega_M t \tag{16.53}$$

Assume the incident $\boldsymbol{E}$ field has angular frequency ω_i, that is,

$$\boldsymbol{E}(t) = \boldsymbol{E}_0 \cos\omega_i t \tag{16.54}$$

We then have the first-order electric dipole moment vector:

$$\boldsymbol{P}^{(1)} = \epsilon_0 \chi^{(1)}_{Ray} \cdot \boldsymbol{E}_0 \cos\omega_i t + \frac{1}{2}\sum_M \chi^{(1)}_M \cdot \boldsymbol{E}_0 \left[\cos(\omega_i + \omega_M)t + \cos(\omega_i - \omega_M)t\right] \tag{16.55}$$

The time dependence of the first-order dipole moment has three angular frequencies: ω_i and $\omega_i \mp \omega_M$. The unchanging frequency ω_i corresponds to the Rayleigh scattering line, the $\omega_i - \omega_M$ component corresponds to the Stokes line, and the $\omega_i + \omega_M$ component corresponds to the anti-Stokes line. The same procedure can be repeated for the hyperpolarizability tensor $\boldsymbol{P}^{(2)}$ for the new frequency components of 0, $2\omega_i$, and $2\omega_i \mp \omega_M$. The angular frequency $2\omega_i$ is called the hyper-Rayleigh scattering line (or second-harmonic generation), the angular frequency $2\omega_i - \omega_M$ is called the hyper-Stokes line, and the angular frequency $2\omega_i + \omega_M$ is called the hyper-anti-Stokes line. The remaining *zero* frequency component is in fact corresponding to a constant polarization that is associated with a constant internal electric displacement vector. This phenomenon is called optical rectification. If one does the analysis for the second hyperpolarizability tensor $\boldsymbol{P}^{(3)}$, then one would see the third-harmonic generation $3\omega_i$ (also called second hyper-Rayleigh line), the second hyper-Stokes line $3\omega_i - \omega_M$, as well as the second hyper-anti-Stokes line $3\omega_i + \omega_M$. In summary, one expects to observe the Stokes and anti-Stokes lines equally spaced on the side of the Rayleigh line on an angular frequency axis corresponding to different orders. These spacings are unique for gaseous molecules. On the other hand, these spacing features are influenced by the environment if the molecules find themselves in a liquid- and especially in solid-state matter, which forms the foundation for Raman scattering–based sensors, particularly temperature and pressure. One of the wide applications is the distributed temperature sensing based on the spontaneous Raman scattering. As for the pressure sensing, it is mostly in the field of high-pressure physics.

In quantum mechanics (QM) domain, a harmonic oscillator oscillates at an angular frequency ω_M with quantized energy levels:

$$E_n = \left(n + \frac{1}{2}\right)\hbar\omega_M, \quad n = 0, 1, 2, \ldots \tag{16.56}$$

According to the statistical mechanics, if such a quantum oscillator is in contact with a thermal reservoir of temperature T, then this oscillator has probability P_n being in the energy level E_n, given by

$$P_n = \frac{\exp\left(-\dfrac{\left(n + \frac{1}{2}\right)\hbar\omega_M}{k_B T}\right)}{\sum_{n'=0}^{\infty} \exp\left(-\dfrac{\left(n' + \frac{1}{2}\right)\hbar\omega_M}{k_B T}\right)} \tag{16.57}$$

where

k_B is the Boltzmann constant

$\hbar = h/2\pi$, with h the Planck constant

Furthermore, according to the QM, the dipole transition strength from energy level $E_n \to E_{n+1}$ is found proportional to quantum number n: $|\,\boldsymbol{p}\,|_{n,n+1} \propto \sqrt{n+1}$. Now, we can evaluate the Stokes line strength from an ensemble of N identical oscillators connected to a thermal bath at temperature T (Long 2002):

$$N\sum_{n=0}^{\infty}\left(\sqrt{n+1}\right)^2 P_n = \frac{N}{1-\exp\left(-\hbar\omega_M/k_BT\right)} \tag{16.58}$$

Conversely, the dipole transition strength from energy level $E_{n+1} \to E_n$ is found in the following proportion $|\,\boldsymbol{p}\,|_{n+1,n} \propto \sqrt{n}$. For the anti-Stokes line strength of an ensemble of N identical oscillators,

$$N\sum_{n=0}^{\infty}\left(\sqrt{n}\right)^2 P_n = \frac{N}{\exp\left(\hbar\omega_M/k_BT\right)-1} \tag{16.59}$$

We could write down the strength of the Raman Stokes line λ_S from an ensemble of identical QM oscillators at ω_M that is dominated by the induced electric dipole radiation:

$$I_S = I_0\left(\frac{\ell}{\lambda_S}\right)^4 \frac{1}{1-\exp\left(-\hbar\omega_M/k_BT\right)} \tag{16.60}$$

where

ℓ is the length scale

I_0 the intensity scale proportional to the incident light strength

At the same time, the associated anti-Stokes Raman line λ_{AS} has the strength:

$$I_{AS} = I_0\left(\frac{\ell}{\lambda_{AS}}\right)^4 \frac{1}{\exp\left(\hbar\omega_M/k_BT\right)-1} \tag{16.61}$$

We thus can derive the well-known expression that served as the basis for the distributed spontaneous Raman temperature sensor:

$$\frac{I_{AS}}{I_S} = \left(\frac{\lambda_S}{\lambda_{AS}}\right)^4 \exp\left(-\frac{\hbar\omega_M}{k_BT}\right) \tag{16.62}$$

The preceding formula is derived with the assumption that each molecule is independent in the system and their mutual interactions are only represented by a statistical temperature parameter T. The first distributed Raman scattering is based on earlier relation (Dakin et al. 1985a,b). In this work, 1 km sensing length with 3 m spatial resolution and 10°C temperate resolution was demonstrated. Now the commercial system

(York Sensors DTS80) is able to measure the temperature at accuracy of ±1°C with a spatial resolution of 1 m over the length of 10 km in times of minutes (John and Dakin 1995). It uses Q-switched fiber laser source to create strong pulses with grade index multimode fiber.

16.3 Application of Stokes and Anti-Stokes Ratio–Based Raman Temperature Sensing

In the early demonstration of Stokes and anti-Stokes ratio to measure the temperature with Raman scattering, the difference of the fiber attenuation at the Stokes line λ_S and anti-Stokes line λ_{AS} was not counted. In reality, because of the large wavelength difference of Stokes and anti-Stokes lines, it can be 200 nm or larger at 1550 nm depending on the type of the fiber; typically, the SMF has attenuation of 0.2 dB/km at 1550 nm and 0.4 dB/km at 1310 nm. When we convert it to the loss, it means the loss of 5%/km and 0.95%/10 m. The temperature difference of the Stokes and anti-Stokes ratio, which is 0.8%/°C at room temperature in SMF28, is comparable to the fiber loss difference at two wavelengths. Hence, a few methods have been proposed to automatically correct this error.

In the spontaneous Raman scattering distributed fiber temperature sensor, one necessarily uses a high-intensity optical pulse to induce the reflected Stokes and anti-Stokes signal to be detected. The reflected Stokes and anti-Stokes light passes the same fiber length with different attenuations; therefore, if one were simply using their reflected ratio to decode the temperature, an error will be introduced.

In Figure 16.4, we illustrate a dual-end Raman scattering method to automatically correct the error caused by the attenuation dependence on the wavelength.

Let us define a function $f(\lambda,z)$ that describes the intensity attenuation of a light at wavelength λ when it propagates a distance z inside the fiber. Hence,

$$f(\lambda,z)=\exp\left\{-\int_0^z \alpha(\lambda,\xi)\,d\xi\right\} \tag{16.63}$$

where $\alpha(\lambda,\xi)$ is the local attenuation function of wavelength λ at the fiber location ξ. It is assumed that the incident laser wavelength is λ, its Raman Stokes scattering is at the wavelength λ_S, and anti-Stokes scattering is at the wavelength λ_{AS} at the same location z and that their reflected signals are being detected at the origin $z = 0$. Suppose the Raman scattering vibration angular frequency is ω_M, then the detected Raman anti-Stokes to Stokes line intensity ratio for location z is expressed by

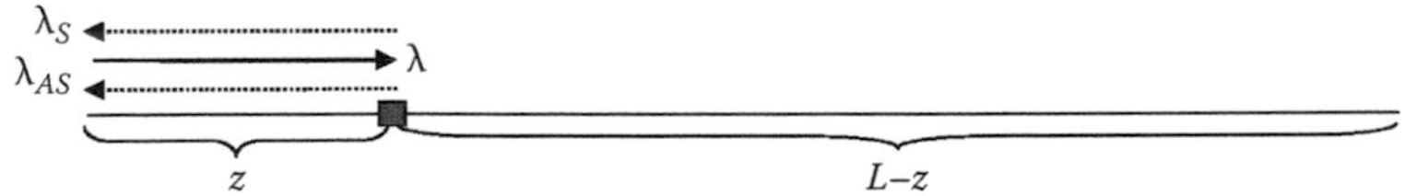

FIGURE 16.4 Schematic of the dual-end Raman scattering method.

$$R(z,T)=\frac{I_{AS}(z)_{\text{detected reflection}}}{I_S(z)_{\text{detected reflection}}}=\frac{f(\lambda,z)f(\lambda_{AS},z)}{f(\lambda,z)f(\lambda_S,z)}\left(\frac{\lambda_S}{\lambda_{AS}}\right)^4\exp\left(-\frac{\hbar\omega_M}{k_BT(z)}\right) \tag{16.64}$$

Obviously, this ratio cannot be directly used to find the temperature $T(z)$ if the ratio $f(\lambda_{AS},z)/f(\lambda_S,z)$ is unknown. Note the same factor $f(\lambda,z)$ appearing both in the numerator and denominator emphasizes the fact that the exciting laser field also experiences attenuation as it goes to the location z to induce the Raman Stokes and anti-Stokes scattering signal. In the dual-end method (Fernandez et al. 2005), it is proposed that one measures the reflected Raman signals using the same laser parameters at the other end $z = L$ and evaluates the similar detected reflection ratio as follows:

$$\tilde{R}(z,T)=\frac{I_{AS}(L-z)_{\text{detected reflection}}}{I_S(L-z)_{\text{detected reflection}}}=\frac{f(\lambda,L-z)f(\lambda_{AS},L-z)}{f(\lambda,L-z)f(\lambda_S,L-z)}\left(\frac{\lambda_S}{\lambda_{AS}}\right)^4\exp\left(-\frac{\hbar\omega_M}{k_BT(z)}\right) \tag{16.65}$$

The idea of the dual-end Raman scattering method is to find a geometric mean between the two ends at the same location z:

$$R^{DE}(z,T)=\sqrt{R(z,T)\tilde{R}(z,T)}=\left(\frac{\lambda_S}{\lambda_{AS}}\right)^4\exp\left(-\frac{\hbar\omega_M}{k_BT(z)}\right)\sqrt{\frac{f(\lambda_{AS},z)f(\lambda_{AS},L-z)}{f(\lambda_S,z)f(\lambda_S,L-z)}} \tag{16.66}$$

Note that the value under the square root is now almost a constant since it describes the ratio of the entire fiber, that is,

$$\frac{f(\lambda_{AS},z)f(\lambda_{AS},L-z)}{f(\lambda_S,z)f(\lambda_S,L-z)}\equiv\frac{f(\lambda_{AS},L)}{f(\lambda_S,L)} \tag{16.67}$$

In this way, one can add a reference ratio for temperature ϑ at any location ξ inside the fiber:

$$R^{DE}(\xi,\vartheta)=\left(\frac{\lambda_S}{\lambda_{AS}}\right)^4\exp\left(-\frac{\hbar\omega_M}{k_B\vartheta}\right)\sqrt{\frac{f(\lambda_{AS},\xi)f(\lambda_{AS},L-\xi)}{f(\lambda_S,\xi)f(\lambda_S,L-\xi)}}\equiv R^{DE}(\vartheta) \tag{16.68}$$

We could then solve the temperature at z as follows:

$$T(z)=\left\{\frac{1}{\vartheta}-\frac{k_B}{\hbar\omega_M}\ln\frac{R^{DE}(z,T)}{R^{DE}(\vartheta)}\right\}^{-1} \tag{16.69}$$

This dual-end method essentially eliminates the error aforementioned due to the variance of attenuation between the Stokes and anti-Stokes lines. Here, we would like to mention as well a subtle dependence of the ratio $f(\lambda_{AS},L)/f(\lambda_S,L)$ for the entire fiber length. If this global ratio is a sensitive function of the overall fiber status, then the dual-end method

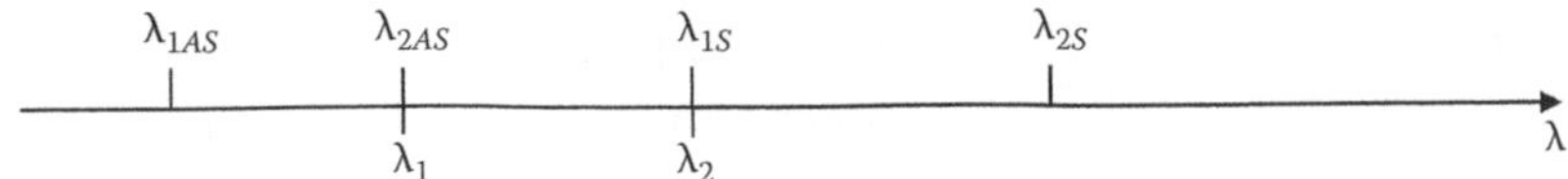

FIGURE 16.5 The schematic relationships between wavelengths in the double light source method.

discussed earlier would not be as accurate since one needs to use a reference temperature ϑ. Fortunately, for most of the temperature measurement, this global ratio is almost identical. The disadvantage of the dual-end method is to have access to both fiber ends at the same time; this would essentially reduce the usable fiber sensing length to half.

Another method is the double light source method (Suh and Lee 2008). Figure 16.5 shows the relationships between the two light sources: the first light source at the wavelength λ_1 is the primary light source; it generates a Raman Stokes line at λ_{1S} and an anti-Stokes line at λ_{1AS} by a common molecular vibration mode ω_M. The secondary light source has the wavelength λ_2 that matches the Stokes line of the primary light source, that is, $\lambda_2 = \lambda_{1S}$. As such, the anti-Stokes line λ_{2AS} of the secondary light source via the same molecular vibration mode ω_M becomes the same as the primary light source, that is, $\lambda_{2AS} = \lambda_1$.

Now the input intensity of the primary light source at the fiber input is I_{01}, and that of the secondary light source is I_{02}. In this arrangement, we will only need to measure the Stokes line intensity of the primary light source at location z as the following:

$$I_{1S}(z) = I_{01} f(\lambda_1, z) f(\lambda_{1S}, z) \left(\frac{\ell}{\lambda_{1S}}\right)^4 \frac{1}{1 - \exp(-\hbar\omega_M / k_B T(z))} \tag{16.70}$$

The reflected anti-Stokes intensity by the secondary light source can be written as

$$I_{2AS}(z) = I_{02} f(\lambda_2, z) f(\lambda_{2AS}, z) \left(\frac{\ell}{\lambda_{2AS}}\right)^4 \frac{1}{\exp(\hbar\omega_M / k_B T(z)) - 1} \tag{16.71}$$

Taking a ratio of the secondary anti-Stokes line with that of the primary Stokes line at ω_M,

$$R^{DL}(z) = \frac{I_{2AS}(z)}{I_{1S}(z)} = \frac{I_{02}}{I_{01}} \frac{f(\lambda_2, z) f(\lambda_{2AS}, z)}{f(\lambda_1, z) f(\lambda_{1S}, z)} \left(\frac{\lambda_{1S}}{\lambda_{2AS}}\right)^4 \exp\left(-\frac{\hbar\omega_M}{k_B T(z)}\right) \tag{16.72}$$

Because $\lambda_1 = \lambda_{2AS}$ and $\lambda_2 = \lambda_{1S}$, then we have the cancellation related to the attenuations:

$$R^{DL}(z) = \frac{I_{2AS}(z)}{I_{1S}(z)} = \frac{I_{02}}{I_{01}} \left(\frac{\lambda_2}{\lambda_1}\right)^4 \exp\left(-\frac{\hbar\omega_M}{k_B T(z)}\right) \tag{16.73}$$

For a calibration of known temperature ϑ with the same double light source operation, we have a reference ratio:

$$R^{DL}(\vartheta) = \frac{I_{2AS}}{I_{1S}} = \frac{I_{02}}{I_{01}} \left(\frac{\lambda_2}{\lambda_1}\right)^4 \exp\left(-\frac{\hbar\omega_M}{k_B \vartheta}\right) \tag{16.74}$$

Under such a calibration, we have the following temperature $T(z)$:

$$T(z)=\left\{\frac{1}{\vartheta}-\frac{k_B}{\hbar\omega_M}\ln\frac{R^{DL}(z)}{R^{DL}(\vartheta)}\right\}^{-1} \tag{16.75}$$

Comparing with the dual-end method, this double light source method is in principle more robust as it achieves the exact cancellation for the attenuation factors with two specific wavelengths. However, it requires the wavelength stability of the two light sources; hence, it adds extra cost for practical applications. Anyway, it does offer higher precision for better temperature resolution, and therefore a compromise is asked for between performance and cost.

Both dual-end and double light source methods need to measure both Stokes and anti-Stokes lines. Therefore, an alternative solution is to use a single light source to measure either Stokes or anti-Stokes lines instead of two lines (Hwang et al. 2010).

One drawback of Raman-based systems is the low Raman scattering coefficient, which results in Raman scattered powers some 30 dB less than Rayleigh scattering and 10 dB less than Brillouin scattering. For this reason, high-powered lasers and long acquisition times are needed. Even so a temperature accuracy of 1°C with 1 m resolution on a 10 km sensing length is possible (Bao and Chen 2011). The best spatial resolution is achieved with the photon-counting technique to give a few centimeters (Feced et al. 1997). Currently available commercial systems provide performances of 5 m for distances up to 30 km. Measurement times are of the order of tens of seconds (Rogers 1999).

The application of the distributed temperature sensors includes the power supply industry, in which fiber sensors are inserted into high-power transformers to detect hot spots and determine their temperature. As the highest temperature determines the lifetime of the insulation materials, so does the equipment itself. The advantage of continuous monitoring of hot spots can prevent the failure of the transformers. They can also be used in thermal power stations to monitor high-pressure steam pipes for leaks (Hartog 1996), as well as in pipeline temperature monitoring to search for leaks in fluids, because cooling will be generated from expansion of gas leaving the pipes, and for the monitoring of heating materials to reduce viscosity. However, excess heating can degrade the materials that are being transported and increase the costs of heating.

Storage vessels in industry require leak monitoring; this is particularly true for liquefied natural gas, and hence an optical fiber can be used to monitor the temperature change. More applications can be found in process industry to monitor the long-term thermal curing, in drying processes, and in fire alarms in tunnels and buildings.

16.4 Rayleigh Scattering–Based OTDR

Rayleigh OTDR was developed for fault detection in telecommunication cables, and later it has been used for a number of applications. These all involve enhancing the effects of the measurand on the loss of the fiber, therefore allowing that measurand to be profiled along the fiber length. One source of loss is microbending loss and a distributed sensor system to measure lateral pressure on a fiber has been developed (Oscroft 1997) that places a fiber inside a spiral sheath that induces microbending in the fiber when a lateral force is applied. The original idea was developed for mechanical sensing based on microbending-induced loss from small and sharp bend on fiber (Marcuse 1976), which increases local attenuation using SMF. The extra loss is detected through standard OTDR techniques.

A Rayleigh scattering–based temperature sensor is possible, at least in principle, through the dependence of the Rayleigh scattering intensity with temperature. For normal glass fibers, the dependence is too weak to result in an effective sensor; however, a successful approach is to use liquid-core fibers to measure temperature with an accuracy of ±1°C with a spatial resolution of a few meters (Hartog 1983). The practical application of the sensor is limited by the need for special liquid-core fibers.

The use of the temperature variation of the attenuation coefficient of doped glass fibers has also been studied. These sensors rely on the absorption bands of the dopants that shifted wavelength with temperature. By monitoring the loss near the edge of an absorption band, changes with temperature can be seen as changes in attenuation. Results obtained with Nd-doped fiber showed 2°C accuracy and 15 m resolution over a 200 m sensing length (Farries et al. 1986). Using Ho-doped fiber, these figures were improved to 1°C and 3.5 m (Farries et al. 1987).

Another factor that can change the fiber loss is the polarization, as the backscattering Rayleigh scattering includes local SOP evolution. Therefore, if any disturbance occurs in one location, the SOP will be modulated. Based on this approach, polarization OTDR was demonstrated (Rogers 1981), and it monitors the spatial distribution of the fiber's polarization properties that can be modulated by pressure, strain, temperature, electric field, and magnetic field. The negative part of the sensitivity to so many parameters is that it cannot distinguish them, unless such a sensor is used for a dynamic measurement, in which measurement is taken within 1 s or less as a distributed vibration sensor (Zhang and Bao 2008). In this case, 2 km sensing length has been demonstrated with 10 m spatial resolution and 5 kHz vibration frequency.

Another direction that has been pursued actively with polarization dependent OTDR (POTDR) is to measure the PMD, which is a major limiting factor for high-speed fiber communication systems due to the PMD-induced pulse broadening, and it is very difficult to compensate as PMD changes with time (Cameron et al. 1998). Normally, high PMD occurs in one or two sections of the fiber due to either the fiber manufacturing process or local environmental changes, such as high-temperature gradient, strong wind, or sun radiation–induced temperature gradients on the cable fiber (Waddy et al. 2001). It is important to identify and locate the high-PMD section through spatial distribution of the fiber's polarization properties. Using POTDR, such a task can be realized (Corsi et al. 1998, Shatalin and Rogers 2006).

Conventional OTDR technique, including polarization OTDR, uses broadband frequency laser of ~0.1 nm bandwidth. Recently, phase OTDR sensor was proposed (Juarez and Taylor 2005). It used KHz linewidth laser as a source and a large pulse, which is equivalent to the spatial resolution of a few hundred meters, to measure the intruder over 12 km with direct detection. Because of coherent Rayleigh scattering, it allows accurate location of the intruder event, unlike polarization OTDR, which acts as an alarm system for locating starting point without ending point due to the continuous SOP change in optical fiber from disturbance point.

By introducing the coherent detection to the phase OTDR, the SNR has been improved significantly, which allows the spatial resolution improvement to 10 m with frequency detection up from Hz level to kHz over the sensing length of kilometers (Lu et al. 2010), thanks to the signal processing scheme of moving average for SMF. By changing to PMF configuration, the spatial resolution has been improved to 1 m, and the frequency range has been increased to over 2 kHz. Furthermore, the disturbance point can be as far as 18 cm from vibration point with metal plate as transducer to cover the area (Qin et al. 2011).

16.5 Rayleigh Scattering–Based Optical Frequency-Domain Reflectometry

The drive for higher performance in distributed fiber sensor systems has pushed the interest in coherent detection. An important goal of distributed optical fiber system-based OFDR is to have high spatial resolution (not limited by the pulse length or digitizer speed as in OTDR technique (Eickhoff and Ulric 1981)), combined with high strain and temperature resolution.

OFDR was originally proposed to measure the fiber loss. In the original setup, a highly monochromatic light frequency ω was tuned by $\Omega = (z/V_g)(d\omega/dt)$; here, $V_g = d\beta/d\omega$ is the group velocity of the fiber, and β is the propagation constant, which depends on the refractive index profile of the fiber, and z is the distance from light source to the fiber point under consideration. Interference fringes are generated as the laser frequency is linearly tuned. Those fringes are detected and related to the optical power, to the phase response of the system, and to the Rayleigh backscattering signal from the fiber under test. This interference signal can be measured via a Fourier transform of the Rayleigh scattering to the space domain. The coherent detection was introduced by mixing a coherent reference wave with a Rayleigh scattering light at a delayed time of $2z/V_g$ in a detector. Hence, a map of the Rayleigh scattering as a function of fiber location on the fiber can be constructed. Because the refractive index can be varied by temperature, pressure, and strain, this means OFDR can be used for measurement of these parameters (Froggatt and Moore 1998, Froggatt et al. 2004, Soller et al. 2005). The setup of a typical OFDR system is shown in Figure 16.6.

As the Rayleigh scattering in optical fibers is caused by fluctuations of the dielectric constant, it is associated with the thermodynamic change of the density and temperature, as shown in Section 16.2.1. When polarized light is launched into the fiber, the SOP of the Rayleigh scattering changes continuously along the fiber (Zhang and Bao 2008). The first-order scatter is guided in SMF that can maintain certain SOP; however, the density

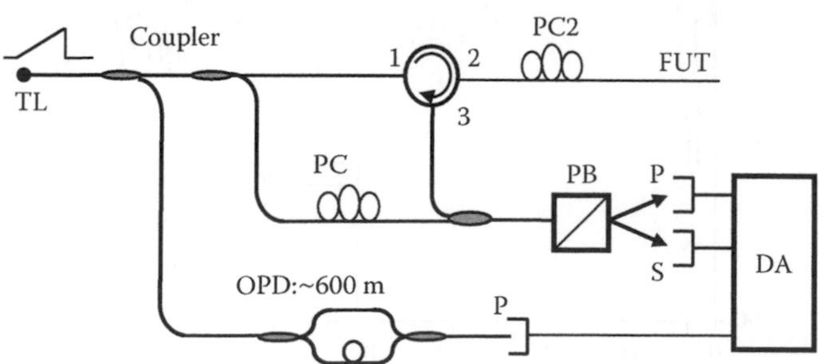

FIGURE 16.6 Setup of an OFDR system (TL, tunable laser; PC, polarization controller; PB, polarizing beam splitter; P and S, detectors; DA, processing; OPD, optical path difference).

fluctuations in each location result in different group velocities via the refractive index changes, which leads to a relative phase shift in different locations between the local fast and slow axes of SMF in detected Rayleigh scattering. Although the Rayleigh scattering has weak dependence on the SOP as the maximum of the polarized OTDR keeps constant, this allows the coherent polarization diversity (PD) detection for continuous wave (CW) Rayleigh signal at sensitivity of 100 dB to remove SOP dependence. This dynamical range does not depend on the spatial resolution, which is in the range of millimeters without the need of expensive tens of GHz/s sampling rate card and detectors, over a few hundreds of meters of fiber. A typical OFDR trace is shown in Figure 16.7 with p and s components and their vector sum over a length of 150 m SMF-28 fiber.

A distributed sensor, for example, temperature or strain, is realized with the use of OFDR method via two steps: (1) the Rayleigh scatter profile of the fiber under test at an ambient temperature in loose condition is measured as reference signal; (2) under strain or a temperature perturbation, Rayleigh scattering is measured again. Then the two profiles are separated to many segments with length Δs for performing Fourier transforms to frequency domain. A cross correlation is performed for each segment to determine the spectral shift between the reference and perturbed profiles. The spectral difference between the shifted peak and the unshifted peak is directly proportional to the temperature or strain changes in each segment.

The nonlinearity of the tunable laser is corrected by an unbalanced auxiliary Mach–Zehnder interferometer (AMZI), which is used as a trigger signal to sample the Rayleigh scattering signal. The maximum measurement length, L_{max}, is approximately determined by the Nyquist sampling criterion:

$$L_{max} = \frac{c\tau_g}{4n} \tag{16.76}$$

where τ_g is the differential delay in the AMZI, used as indicated to calibrate the tunable laser-induced nonlinear wavelength sweep. The factor of four in the denominator

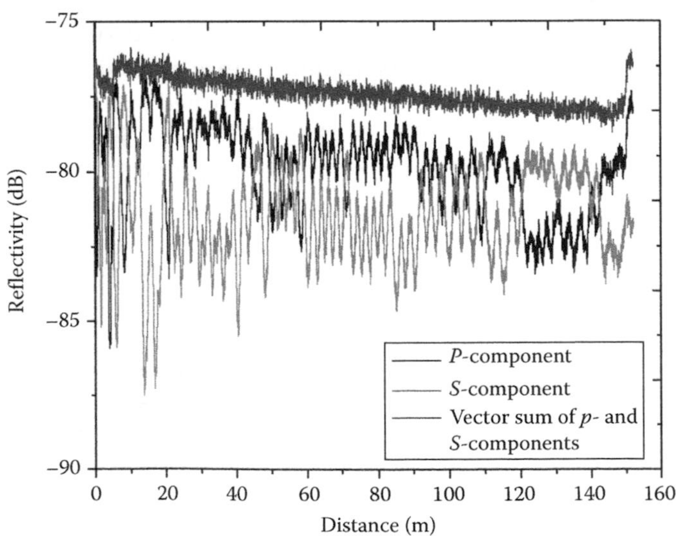

FIGURE 16.7 Typical reflectivity display of an OFDR system.

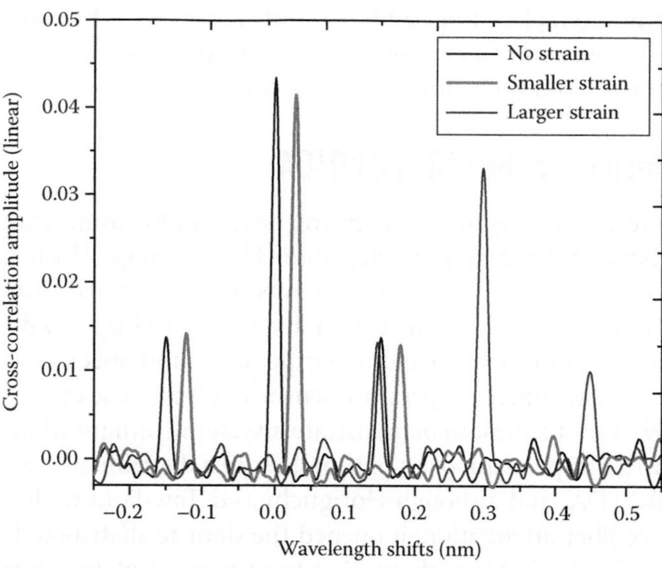

FIGURE 16.8 Typical reflectivity display of an OFDR system. The shifts of cross-correlation peaks indicate stresses at particular locations of the fiber under test.

is due to the sampling theorem and the double-pass nature of the measurement interferometer. The coherence length of the light source also limits the maximum measurement range. Using phase-reconstruction schemes, the maximum length can be improved to 1 km.

The recovered temperature and strain information through *auto- and cross correlation* are shown in Figure 16.8 for PMF, in which two axes have different temperature and strain coefficients. This means using PMF, one can get simultaneous temperature and strain sensing, so long as the position dependence birefringence change on temperature and strain can be neglected and calibration is crucial for different kinds of PMFs.

The spatial resolution of the measurement, Δz, is directly related to the resolution in the time domain and is determined by the optical frequency sweep range ΔF. It is given by

$$\Delta z = \frac{c}{2n_g \Delta F} \tag{16.77}$$

This represents the minimum separation necessary to resolve two reflective events. This resolution can be affected by environmental noise, insufficient linear laser tuning, and excess of unbalanced dispersion in the measurement arm. It has been demonstrated that a spatial resolution of 1–2 cm with strain and temperature errors of 12.5 ppm and 0.5°C can be achieved, respectively.

16.6 Brillouin Scattering–Based Distributed Sensors

While there are a variety of distributed fiber-optic sensors, research on distributed strain sensing has focused almost exclusively on Brillouin scattering–based sensors. For over two decades, distributed optical fiber sensors based on Brillouin scattering have gained much interest for their potential capabilities of monitoring temperature and strain in large infrastructures to replace thousands of point sensors. This kind of sensors can find applications in civil structures, environmental monitoring, aerospace industry, power generator monitoring, and geotechnical engineering.

16.6.1 Development of BOTDR and BOTDA

The first work leading directly to strain sensing based on Brillouin scattering was done in 1989 by the Nippon Telegraph & Telephone (NTT) group (Horiguchi et al. 1989). They found the Brillouin shift of optical fiber was linearly related to applied strain. At about the same time, researchers at the University of Kent in England demonstrated that Brillouin shift was also linearly related to temperature and speculated that this could form the basis of a distributed temperature sensor (Culverhouse et al. 1989). Horiguchi and Tateda (1989) were the first to demonstrate a system configuration capable to measure the Brillouin gain spectrum of an optical fiber in a distributed manner. They called this technique BOTDA, and although Horiguchi and Tateda initially used their setup simply to measure fiber attenuation, it opened the door to distributed strain and temperature sensing. The BOTDA configuration used two counterpropagating lasers and took advantage of the Brillouin amplification to provide a very strong signal.

In 1990, Kurashima et al. (1990a) performed their own investigation of the temperature dependence of the Brillouin shift of optical fiber and later in the same year demonstrated distributed temperature measurement using their BOTDA setup (Kurashima et al. 1990b). They achieved a 3°C temperature accuracy with a spatial resolution of 100 m over a sensing length of 1.2 km. Later, a new configuration of BOTDR was proposed (Kurashima et al. 1991). This system had the advantage of requiring access to only one end of the sensing fiber. Performance was also slightly improved over the earlier BOTDA configuration, with the total sensing length being increased to 11 km with similar spatial resolution and accuracy. One year later, a further improved version was demonstrated with a similar performance but requiring only one laser (Shimizu et al. 1993).

With the optimization of the pump depletion and probe wave saturation for more uniform Brillouin gain, a significant improvement in sensing length and spatial resolution has been reported in 1993 by Bao et al. (1993b), enabling an accuracy of 1°C for temperature measurements with a spatial resolution of 10 m and a total sensing length of 22 km.

This performance was improved upon quickly by the introduction of the Brillouin loss BOTDA configuration, which also attained a temperature measurement accuracy of 1°C with a spatial resolution of 5 m and a total length of 32 km (Bao et al. 1993a). Eventually, the total sensing length of this configuration was pushed to over 50 km (Bao et al. 1995).

In addition to working on extremely long-distance temperature sensing, distributed strain measurements were made by the Kent group. A strain measurement accuracy of 20 μm per meter fiber was demonstrated by Bao et al. (1995) with a spatial resolution of 5 m and a total sensing length of 22 km using the same Brillouin loss setup. This group was also the first to report the simultaneous measurement of both strain and temperature using a single fiber (Bao et al. 1994). This was accomplished by isolating half of the length of the fiber from strain so that it was only sensitive to temperature, while the remaining half was sensitive to both parameters. By laying the two sections of fiber in parallel, it was possible to separate the effects of both parameters by comparing the Brillouin shift occurring in each section of fiber. Accuracies of 20 μm per meter fiber and 2°C were obtained, with a spatial resolution of 5 m over a total fiber length of 22 km.

Unlike the single-ended BOTDR system from NTT, Swiss Federal Institute of Metrology introduced a novel single laser system that made use of Brillouin amplification in much the same way as a BOTDA configuration. This system has the performance of 45 m spatial resolution per 1.4 km sensing length (Niklès et al. 1996). It was eventually developed to the point that slightly better than 1 m spatial resolution was attained in 1997 (Fellay et al. 1997), though the NTT researchers reached that milestone several years earlier (Horiguchi et al. 1995). Both groups reported that they felt this was the limit of spatial resolution for Brillouin scattering–based sensors due to the observed broadening of the spectral linewidth.

Around 1996/1997, several research groups started investigating the possibility of developing a method of making simultaneous distributed temperature and strain measurements using a single fiber rather than the two parallel fiber sections. Wait and Newson (1996) used the Landau–Placzek ratio to measure temperature in a distributed fashion that could theoretically be combined with data from a regular Brillouin scattering–based distributed sensor to determine both strain and temperature. While their initial results were not practically useful, having achieved only a 10°C accuracy with a spatial resolution of 600 m, further development has improved their spatial resolution to 10 m, which could be used for some applications (Lees et al. 1998). This technique has been refined with Raman fiber amplifiers and coherent detection, which allows 150 km sensing length with temperature resolution of 5.2°C and 50 m spatial resolution (Alahbabi et al. 2005).

In 1997, Parker et al. (1997) also investigated simultaneous temperature and strain measurements by determining the temperature dependence of Brillouin gain. They demonstrated simultaneous measurement using a single fiber, achieving 100 με and 4°C accuracy with a spatial resolution of 40 m and a total sensing length of 1.2 km. The measurement time of their system was on the order of 1 h, making it very impractical, but it did demonstrate that simultaneous measurement of temperature and strain was possible by observing only the Brillouin spectrum of a fiber. In 1999, Smith et al. (1999) demonstrated a simultaneous measurement system using polarization-maintaining sensing fiber that achieved significantly better performance. A strain resolution of 128 μm/m was obtained, with a temperature resolution of 3.9°C at a spatial resolution of 3.5 m, an order of magnitude better than that reported by Parker et al. (1997).

The first demonstration of a Brillouin scattering sensor system with spatial resolution substantially better than 1 m was demonstrated in 1998 (DeMerchant et al. 1998). Spatial resolution of 50 cm was demonstrated for strain measurements that were the first to report on realistic structural testing rather than simply measuring stretched sections of fiber. This was further improved to 10 cm spatial resolution (Brown et al. 1999).

The best performance of a differential Brillouin gain signal system has been a 2 km sensing length with 2 cm spatial resolution, for a Brillouin frequency shift resolution of 2 MHz that is equivalent to 2°C temperature resolution (Bao and Chen 2011). Figure 16.9a shows the measured stress section of 1 cm length using 8/8.2 ns pulse pair

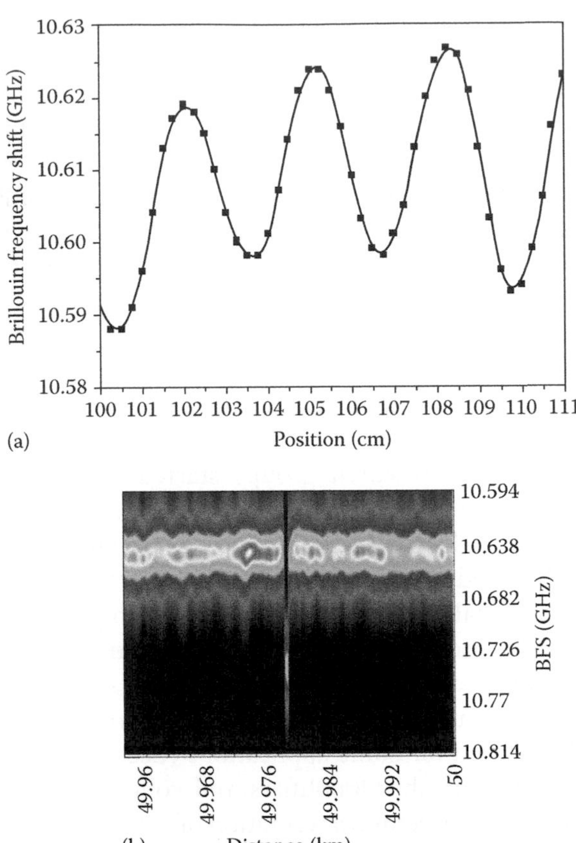

FIGURE 16.9 (a) 1.5 cm stress section is measured with 8/8.2 ns pulse pair in a BOTDA; (b) 50 cm stressed section is measured with RZ-coded pulse of 60 ns/55 ns pulse pair. (Reprinted with kind permission from Springer Science + Business Media: *Sensors Journal of Photonic Sensors*, Recent progress in optical fiber sensors based on Brillouin scattering at University of Ottawa, 1(2), 2011b, 102, Bao, X. and Chen, L. Copyright 2011 Springer-Verlag.)

in a BOTDA with 40 GHz/s (2.5 mm/point) sampling rate digitizer. This kind of performance is achieved with a differential pulse-width pair Brillouin optical time-domain analysis (DPP-BOTDA) for high-spatial-resolution sensing, still using the prepumping idea to measure the differential Brillouin gain signal instead of the Brillouin gain itself (Li et al. 2008). This technique uses two different pulses of nearly identical duration, having high extinction ratios of 40–50 dB to remove the pre-DC effect. Two separate measurements are implemented with respect to the individual pulses, and the differential Brillouin gain signal is then obtained by subtraction of the two Brillouin signals.

For long sensing lengths of 50 km, it achieved 50 cm spatial resolution and 0.7 MHz equivalent Brillouin frequency shift. For temperature, it means 0.7°C and for strain it means 15 μm per meter fiber (Hao et al. 2010). Figure 16.9b shows 50 cm stressed fiber section at the end of 50 km.

Such a performance level is achieved with large effective area fiber (LEAF) fiber, which is a dispersion-shifted fiber with three Brillouin peaks so that it allows higher pulsed power to be used. To avoid the gain saturation, return-to-zero (RZ)-coded pulses have been used for relative lower power and to reduce the nonlinear effect (SBS)–induced bit pattern dependence at the end of long fiber length. In addition, the CW beam must be as low as possible to avoid large depletion-induced Brillouin spectrum distortion. As a trade-off of the low Brillouin interaction, lower gain is expected, so does the lower SNR. To overcome this problem, DPP-BOTDA technique is introduced.

The longest sensing length of BODA in the report is 150 km with the spatial resolution of 2 m and Brillouin frequency accuracy of 1.5 MHz at 1.5°C/30 με. It was achieved with three different fiber sections of different Brillouin frequencies at normal CD for the purposes of (1) avoiding intensive Brillouin interaction over the entire sensing length, which will induce the Brillouin spectrum distortion, that is, depletion effect, and (2) avoiding the onset of modulation instability (MI) (Tai et al. 1986), which is a process that the amplitude and phase modulation of the wave grow as a result of interplay between nonlinearity and anomalous dispersions. In the frequency domain, MI leads to the generation of sidebands symmetrically placed about the pump frequency; hence, the energy in the high-power pump or probe waves will be transferred to those sidebands instead of contributing to the Brillouin gain or loss process. The sensing fiber includes two identical spans of fibers being amplified by erbium doped fiber amplifiers (EDFAs) between them.

16.6.2 Frequency-Domain Distributed Brillouin Sensor

Unlike the time-domain BOTDA or BOTDR, a Brillouin optical frequency-domain analysis (BOFDA) was developed in 1997 (Garus et al. 1997), which measures the complex baseband transfer function given by the ratio of the Fourier transforms of the pump and Stokes intensities at the end of the fiber (nearest to the Stokes laser). From this, the inverse Fourier transform may be taken to give the temporal pulse response that can be converted to a spatial response by using the relationship $t = 2nz/c$. The intensities of the two beams are detected and fed into a network analyzer that determines the baseband transfer function. This in turn is fed into a digital signal processor that takes the inverse fast Fourier transform (IFFT), giving the pulse response of the fiber at the given laser beat frequency. Initial results performed using this system showed an accuracy of 1.5°C and 40 μm/m fiber with 1.4 m spatial resolution on an 11 km long fiber.

An application of the BOFDA sensor that has been reported is its use for beat length measurements by locating the areas of maximum and minimum Brillouin loss. This was done using the BOFDA method, and beat lengths in the range of 50–60 m were observed on a 10 km length of fiber (Golgolla and Krebber 2000).

More recently, another technique Brillouin optical correlation-domain analysis (BOCDA) was proposed and demonstrated (Hotate and Hasegawa 2000). The spatial resolution of the BOCDA system is determined by the modulation parameters (amplitude and frequency) of a light source, rather than by the decay time of an acoustic wave. In this approach, two CW light waves with a Brillouin frequency difference are identically frequency modulated. SBS occurs at the correlation peak position, where the two light waves are highly correlated. The correlation peak width determines the spatial resolution. The signal processing of correlation between different positions provides a sharp peak for the matched Brillouin frequency, and the spatial resolution of BOCDA can be as high as 1 cm for the short sensing length of tens of meters (Hotate and Tanaka 2002). The most recent progress has improved this sensor system to 1 km length of 7 cm spatial resolution (Hotate 2011).

16.6.3 Differential and Parametric Brillouin Gain

As explained in Section 16.6.1, the acoustic wave has a lifetime of ~10 ns, which means the Brillouin gain in BOTDA or the Stokes signal in BOTDR will be very weak if the optical pulse is less than 10 ns, which is equivalent to 1 m spatial resolution. In order to achieve submeter spatial resolution sensors, various complex techniques have to be used in order to enhance nonlinear Brillouin interaction portion, so-called preexcitation (Afshar et al. 2003) by a small portion of DC leakage of the pulse. The prepumping principle have been further explored by different techniques: (1) exploiting Brillouin echoes (Thévenaz and Mafang 2008), (2) BOCDA system as explained in Section 16.6.2, as well as (3) optical coded BOTDA with DPP schemes with RZ coding of submeter (50 cm) spatial resolution over 50 km sensing length (Hao et al. 2010) without amplifier and for direct BOTDA of RZ-coded pulse with preamplifier at detection end for the spatial resolution of 3 m over 120 km sensing length (Soto et al. 2011).

The idea of π-phase-shifted pulse added to non-phase-shifted pulse is similar as that of a bright pulse being added to a dark base (Wang et al. 2008) and dark pulse (Brown et al. 2005); they are based on intensity modulation rather than being added in phase modulation format. The dark pulse used the bright pulse as prepumping and dark pulse to give the spatial resolution. In the π-phase-shifted pulse pair, two pulses share the same pulse width except that the last portion of the second pulse is phase inverted (π phase shift). As a result, the spatial resolution is limited by the fall time of the pump modulation and the phenomenon of secondary *echo* signals, as was previously proposed in the magnetic resonance imaging (MRI).

On the other hand, the DPP simply uses two pulses with different pulse widths. It involves subtraction of two Brillouin signals originating from two light pulses with different pulse widths; the subtraction process degrades the SNR, especially at the end of long sensing ranges. Hence, pulse coding techniques could be efficiently used, providing both a sensing range and resolution enhancement factor linked to the coding gain.

One of the drawbacks with the two previous time-domain techniques is the gain saturation of the large pulse in both *echo* and DPP-BOTDA techniques. Hence, the depletion could become a problem for long sensing length. To overcome the depletion-induced Brillouin spectrum distortion, coherent interaction of the Brillouin gain and loss is applied (Minardo et al. 2009, Li et al. 2010). The working principle of this parametric Brillouin gain process is described in Section 16.2.4. Because of the introduction of Stokes and anti-Stokes wave, as well as carrier wave, there are two acoustic waves being created, which means two phase-matching conditions are required for two counterpropagating acoustic waves. Such a process is different from pure Brillouin gain or Brillouin loss process that is governed by one acoustic wave, in which phase-matching condition can be met easily. However, the interaction of the combined gain-and-loss process is further complicated due to two different SOPs, that is, the PMD, of the gain-and-loss process unless the PMF is used (Li et al. 2010). In the experiment, both the Stokes and anti-Stokes pulsed signals are used; the center of the carrier wave is a CW signal. The two identical sidebands

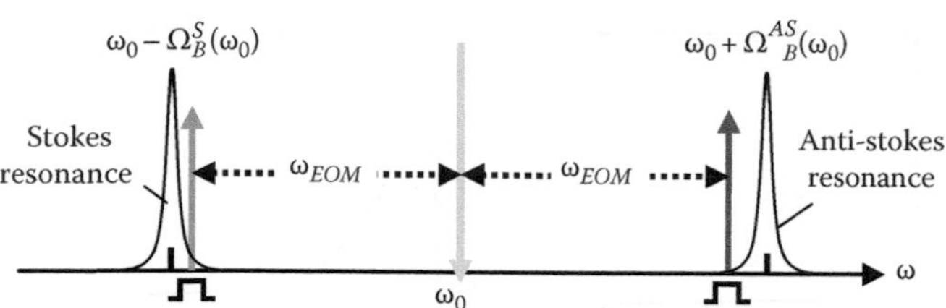

FIGURE 16.10 Schematic relationships in the experimental light frequencies.

are created by electro-optic modulator (EOM). Because of the unequal Brillouin frequency due to their different effective index at different frequencies (i.e., CD), both of the Brillouin gain and Brillouin loss are not in-resonance simultaneously as stated in Section 16.2.4 and illustrated in Figure 16.10.

The monitored CW beam experiences simultaneous Brillouin gain (energy is transferred from the sideband at $\omega_0 + \omega_{EOM}$ via an acoustic grating moving in the direction of the pulsed beam due to the SBS) and loss (energy is given to the sideband at $\omega_0 - \omega_{EOM}$ via another acoustic grating moving in the direction of the CW beam due to the SBS). Therefore, as one scans the ω_{EOM} (sidebands of the CW beam), the CW spectrum varies from a loss-dominated behavior (when $\omega_{EOM} = \Omega_B^S(\omega_0)$) to a gain-dominated behavior (when $\omega_{EOM} = \Omega_B^{AS}(\omega_0)$). Somewhere in the middle, we expect to observe an exact balance between the gain and the loss. In fact, because of the small difference between the Brillouin frequency of the anti-Stokes domain and that of the Stokes domain, we expect to directly observe a type of *off-resonance* behavior of the conventional Brillouin resonance spectrum as one scans the ω_{EOM}.

Figure 16.11a is the simulation of the off-resonance Brillouin spectrum for the usual case $\Omega_B^{AS}(\omega_0) > \Omega_B^S(\omega_0)$ based on equations in Section 16.2.4; more specifically we took $\Omega_B^{AS}(\omega_0) = 12{,}801$ MHz and $\Omega_B^S(\omega_0) = 12{,}799$ MHz. Figure 16.11b is the simulation result for the anomalous case of $\Omega_B^{AS}(\omega_0) < \Omega_B^S(\omega_0)$; more specifically we took $\Omega_B^{AS}(\omega_0) = 12{,}799$ MHz and $\Omega_B^S(\omega_0) = 12{,}801$ MHz. It is seen that the quadrature points (maximum slope points) have opposite signs and they represent a gain-or-loss-dominated process controlled by the phase of the two coupled acoustic waves.

It is noted that the aforementioned simulation results works for the ideal situation where polarizations of the Stokes and anti-Stokes waves are perfectly aligned with the CW that is being monitored, which requires PMF (and yet without CD). In a real fiber system, there is another unavoidable complication, namely, the PMD or polarization effect. Under the influences of PMD, the perfect polarization alignments between the three light waves can never be achieved; therefore, the monitored CW signal shows even more rich features. The experimental condition is reported in Li et al. (2010). This can be shown by the experiments that are conducted with Stokes and anti-Stokes pulse width of 25 ns, peak pulsed power of 30 mW, and CW power of ~1 mW. Figure 16.12 shows the balance of two phase-matching conditions in PMF. Because of the fixed SOP for the Brillouin gain/loss processes, the quadrature point, which is maximum slope of the off-resonance Brillouin spectrum, maintains the same sign, although the slope changes with the position, while in SMF, this slope changes sign along the fibers as seen in Figure 16.12b. The amplitude of Figure 16.12a for the Brillouin gain process (when the

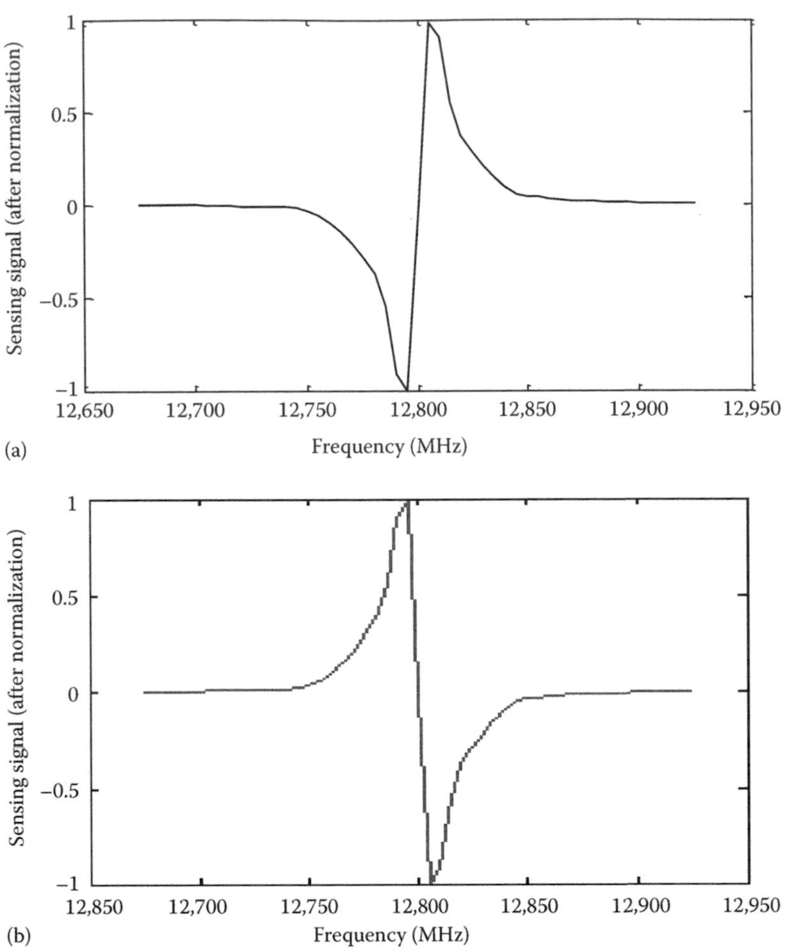

FIGURE 16.11 Simulated off-resonance Brillouin spectrum for (a) $\Omega_B^{AS}(\omega_0) > \Omega_B^S(\omega_0)$ and (b) $\Omega_B^{AS}(\omega_0) < \Omega_B^S(\omega_0)$.

anti-Stokes wave is absent) is about 75 times larger than that of the off-resonance spectra, which represents parametric gain of the coupling of the two acoustic waves, created by three waves in Figure 16.10. The amplitude of the off-resonance spectra varies with position, and the slope of the quadrature point maintains the same orientation, as SOP of three waves is kept the same.

For the LEAF fiber, with certain SOPs, that is, the principal SOPs, we can find similar off-resonance spectra as that of PMF, as shown in Figure 16.13. The time trace of LEAF while sweeping the Stokes and anti-Stokes frequency is seen in Figure 16.13a through c, for three different input SOPs. For (a) and (b), we have the gain-or-loss-dominated condition remained in the 90 m LEAF without changing the slopes at quadrature point. Hence, the phase-matching condition does not change in position for both gain and loss cases. The period of the off-resonance spectra amplitude change is 5 m, which should represents the beat length, as it is related to the cycle of the maximum Brillouin gain changes to minimum value along the fiber length due to the SOP change of the

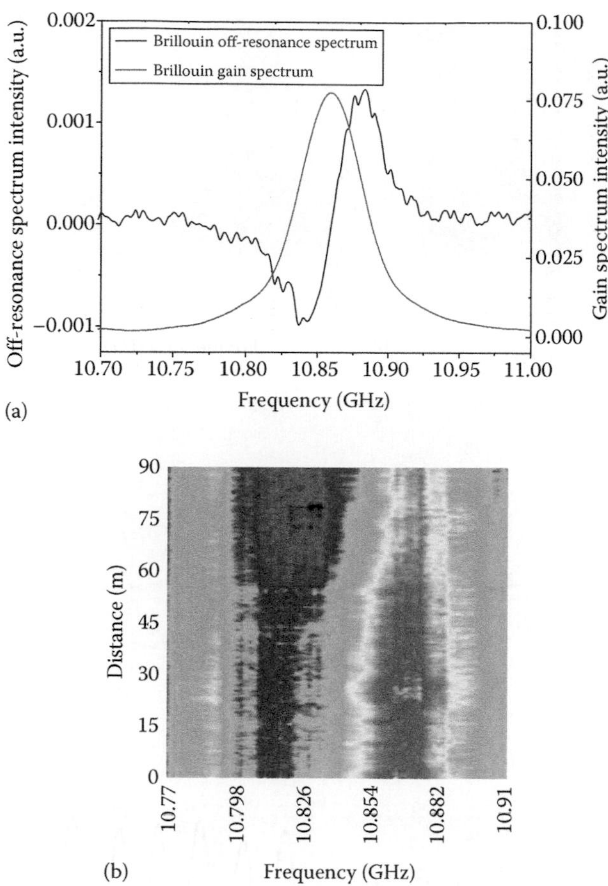

FIGURE 16.12 (a) The off-resonance Brillouin and Brillouin gain spectrum in PMF; (b) the off-resonance Brillouin spectrum versus position in PMF.

Stokes or anti-Stokes wave. Such a cycle is the same for the Brillouin gain or loss process, as illustrated in Figure 16.13a. For the normal input SOP (Figure 16.13c), the intensity of off-resonance spectrum reduced to 10% comparing with launching to SOP condition, and the slopes change the signs along the positions. Furthermore, the period varies in time domain.

Figure 16.14 shows the off-resonance of SMF 28 at two different locations and their variation along 425 m fiber. For SMF 28, we can't find a SOP to give a fixed period as that shown in LEAF fiber; in addition, the slope changes signs with position. This means the beat length in SMF changes at different locations due to PMD (LEAF is dispersion-shifted fiber and it has minimum CD at 1550 nm). From Figure 16.14b, the gain-and-loss-dominated condition alternates in the fiber. Hence, the phase-matching condition changes in position, so does the slope of the off-resonance Brillouin spectrum around peak frequency of the Brillouin gain. The relative amplitude of the Brillouin gain and off-resonance spectrum has a ratio of 4–10 depending on input SOPs. The two spectra at off-resonance condition represent two different locations. This difference is much smaller than that of the PMF. Because the Brillouin gain in SMF is much smaller than that in PMF due to the mismatched phase condition in the fiber, that is, the SOP of the

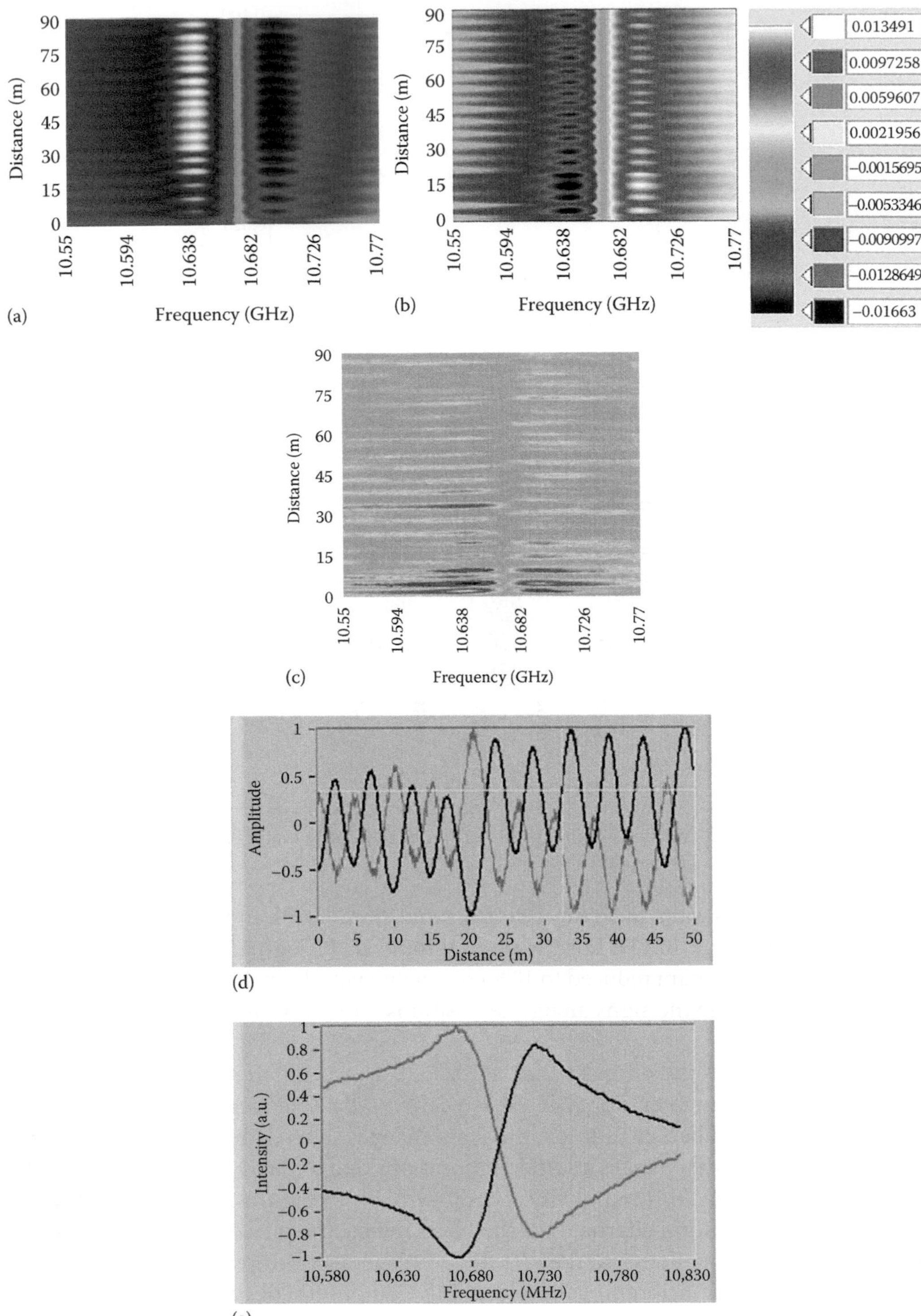

FIGURE 16.13 The off-resonance Brillouin and Brillouin gain spectrum in 90 m LEAF at three different input SOPs (a–c) at different driver frequencies of EOM for SOPs of (a) and (b). Time trace for the first 50 m (d). The spectra of (a) and (b) are shown in (e).

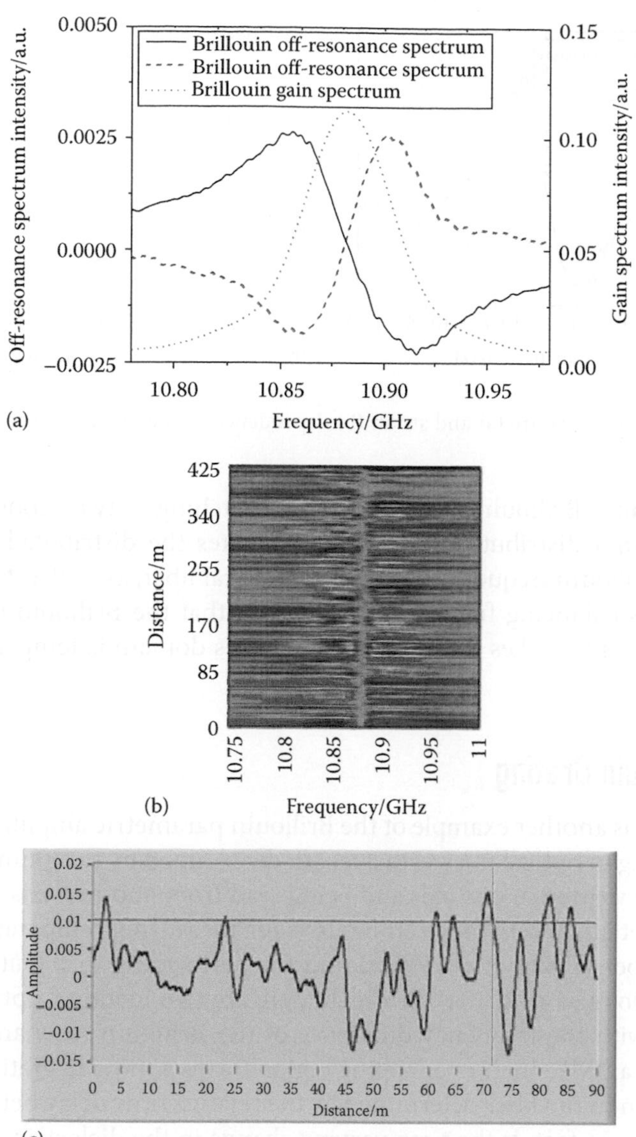

FIGURE 16.14 (a) The off-resonance Brillouin and Brillouin gain spectrum in SMF28, (b) the off-resonance Brillouin spectrum over 425 m fiber, and (c) time trace of first 90 m fiber.

pump and that of the probe are different in every position due to PMD as shown in Figure 16.14; hence, the Brillouin gain is reduced significantly.

PMF is a high-birefringence fiber, and birefringence changes with temperature and strain, so does the maximum slope points, as illustrated in Figure 16.15.

The amplitude of the off-resonance signal varies with position as illustrated in Figures 16.12 through 16.14. There is no peak in those spectra. As a result, the detuned Brillouin gain-and-loss process via carrier and its two sidebands have promoted two similar strength acoustic gratings. Such a process has been verified by a new five coupled wave equations, which account for the coupling of two acoustic gratings and their phase-matching conditions.

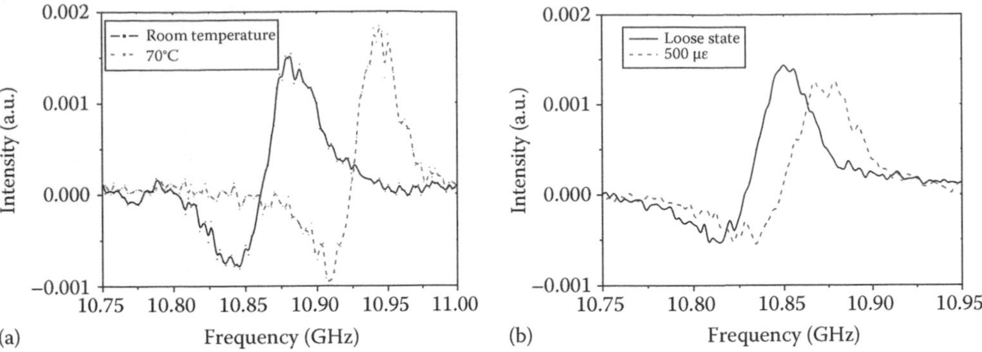

FIGURE 16.15 The temperature (a) and strain (b) dependence of the off-resonance Brillouin spectrum in PMF.

The off-resonance Brillouin spectrum (phase matching of two acoustic gratings) has been measured in a distributed way, which indicates the distributed refractive index change within Brillouin frequency range in the optical fiber, as well as the time response around gain–loss balancing frequency. It is shown that the Brillouin frequency difference between the anti-Stokes domain and the Stokes domain is temperature and strain dependent.

16.6.4 Brillouin Grating

Brillouin grating is another example of the Brillouin parametric amplification by writing a Brillouin grating in optical fiber with a length determined by two pump pulses in PMF, as grating can be written in one axis and being read from another axis via birefringence coupling (Song et al. 2008). They can be used for measuring temperature via birefringence in PCF fiber (Dong et al. 2009a) and for performing distributed birefringence measurement (Dong et al. 2010a), in which there are two identical optical pulses counterpropagating with the frequency difference of the Brillouin resonance in one of the principal axes of a PMF. Their overlapping region forms a moving grating, which can be set at any position in the fiber determined by the relative time delay between two pulses. This moving *FBG*, in fact, is the time-varying change in the dielectric constant induced by beating two pump beams through the electrostriction effect, namely, SBS. The size of the grating is determined by the pulse length. The lifetime of this Brillouin grating is limited by the phonon lifetime (~10 ns). The grating spectrum is a convolution of the pulse and Brillouin gain spectra. Such a Brillouin grating can be probed with a third pulse launched on the orthogonal axis; the optimum frequency of the third pulse is related to the local birefringence between two axes of PMF. This means the grating feature is affected by the temperature and strain, reflected in the Brillouin frequency and the local birefringence from two axes of PMF. This local transient Brillouin grating (TBG) (Zou et al. 2009, Dong et al. 2010b) can be used for the purpose of sensing. When we change the relative delay between two pump pulses, distributed sensing can be realized for the entire fiber. Because the large frequency shift associated with birefringence change is on the order of 40–50 GHz, the measurement accuracy of temperature and strain can be much higher than direct Brillouin gain spectrum measurement. It has been reported that

0.08°C and 3 με can be achieved with a continuous wave (without location information) (Zou et al. 2009) and with a pulse of 2 ns, equivalent to 20 cm of spatial resolution and temperature and strain accuracy of 0.4°C and 9 με, respectively (Dong et al. 2010b). For BFS measurement, the DPP-BOTDA is used to realize a high spatial resolution. A high spatial resolution of 20 cm is achieved by using a pulse-width difference of 2 ns, and a narrowband Brillouin gain spectrum is obtained from a pulse pair of 30/28 ns. For birefringence-induced frequency shift (BireFS) measurement, two short pump pulses (2 ns) are used to generate a local Brillouin grating and thus get a high spatial resolution of 20 cm. The temperature and strain range can be up to 700°C and 14 mε due to the short pulse (2 ns)–based TBG.

16.7 Limitations for Sensing Length, Spatial Resolution, and Temperature/Strain Resolution

Generally speaking, scattering-based OTDR sensor is capable for long sensing length. Among Rayleigh, Raman, and Brillouin scattering–based sensors, Brillouin scattering–based sensors have shown the best performance in terms of sensing length, over 150 km, and high spatial resolution of a few centimeters. The limitations of sensing length are the fiber losses over the long sensing length, which can be compensated by either Raman scattering or EDFAs.

For a high-performance distributed sensor, it requires (1) long sensing length and (2) high spatial resolution. On the detection end, it means larger dynamic range, broadband, and minimum detection level. Those cannot be achieved at the same time, as the detection noise is proportional to the bandwidth. For instance, if we need 1 cm spatial resolution, the detection bandwidth shall be 10 GHz, the same requirement for the digitizer. With broadband detection, one can choose either AC- or DC-coupled detector. At present, a high precision 2 m spatial resolution (equivalent to Brillouin frequency shift of 1–2 MHz) over hundred km (<200 km) for temperature or strain measurement has been demonstrated. On the other hand, for the BOTDR or Raman OTDR, because of the weak spontaneous scattering, although 150–200 km sensing length is possible, the spatial resolution has been compromised significantly to a value that is in the order of 50–100 m, an important limitation unless the applications are tolerable to such a degraded performance level (Bao and Chen 2011).

The advantage of DC detection is to preserve the relative frequency response for the full frequency range, and it provides a baseline for the power conversion for absolute measurement. While AC detection removes the low-frequency components, which allows only the measurement of the relative changes, it is a good choice for small signal measurement, as it removes the large DC-based noises. With a long sensing length, this can be a problem, as the low-frequency filter tends to alter constant strain or temperature sections in long sensing length (Dong et al. 2009b). The relaxation time of the capacitors (τ_{relax}) in AC-coupled detector would change the measured relaxation time, as the duration time of the long strain or temperature could be comparable or longer than τ_{relax}, which is equivalent to an AC filter.

For the short fiber length (~100 m), a broadband AC-coupled detector is good to achieve the highest spatial resolution, the probe pulse power should be a few mW, and the

Table 16.1 Main Characteristics of Optical Fiber–Based Distributed Sensing Techniques

	Distributed Sensing Technique				
Characteristics	BOTDPP–BOTDA	Brillouin Grating	BOTDR	OFDR (Rayleigh)	OTDR (Rayleigh)
Spatial resolution	2 cm (2 km) 2 m (150 km)	1–2 cm	~1 m	~1 mm	~1 m
Sensing range	150–200 km	200 m	Kilometers	~50 m	1–2 km
Measurement time	Minutes	Minutes	Minutes	Seconds	<1 ms
Temperature and strain	Yes	Yes	Yes	Yes	No
Temperature accuracy	1°C	0.3°C	2°C–3°C	0.2°C	No
Strain accuracy	20 με	5 με	60 με	2 με	No
Dynamic measurement	Yes	No	No	No	Yes
Calibration	One time	One time	One time	Before every measurement	No

pump power should be in the same range to achieve the highest contrast. For a long sensing length, a broadband DC-coupled low-noise detector is required for the high spatial and strain resolution, and the probe pulsed power should be much larger than the pump power, which should be as small as possible to have minimum depletion and to achieve the highest modulation depth (contrast) due to the pump and probe wave interaction.

For high spatial and temperature/strain resolution over long sensing length (>30 km), both of DPP-BOTDA and parametric Brillouin gain-based BOTDA are relatively simpler and cheaper (Dong et al. 2009b) than the Brillouin grating technique, which requires three lasers to be locked together, including one tunable laser, while for BOTDA technique, it can be realized with two lasers or one laser plus EOM driven by a tunable and stable microwave generator.

With super–high spatial resolution (<1 mm), the frequency-domain distributed sensor is much more cost-effective than OTDR, as the broadband electronics and digitizer requirements have made OTDR-based system very expensive, especially Brillouin grating-type sensor, that requires three lasers to be located together on the top of the PMF as sensing fiber, which makes system complicated and expensive. The advantage of the Brillouin scattering–based distributed sensor versus the Rayleigh scattering–based sensor systems is ground on the fact it measures the Brillouin frequency change, which is an absolute change, rather than the intensity or phase change, which is a relative change from one reference time. Furthermore, such a reference must be determined in every test or measurement.

Table 16.1 gives the state-of-the-art main characteristics of the several optical fiber–based distributed sensing techniques.

16.8 Challenges in Applications of Distributed Sensors

For a large-scale structure, the number of point sensors needed to generate complete strain information can grow rapidly. Distributed sensors offer an advantage over point sensors for global strain measurements. The thousands of sensing points that the distributed sensor provides enable mapping of strain distributions in two or even three dimensions. Thus, real measurements can be used to reveal the global behavior of a structure rather than extrapolation from a few point measurements. Such a process is SHM, which has been used to identify early signs of potential problems in civil structures, to prevent disasters, and to conduct needed repairs at the appropriate time, avoiding unnecessary costs and reducing economic burden. Thus, it is important to have accurate and real-time monitoring on the safety assessment of civil structures, such as bridges, dams, and pipelines. Currently, such evaluations are carried out by engineers trained in visual inspection, which sometimes can be inaccurate due to differences in their personal experience with safety condition assessment. To increase the inspection efficiency and accuracy, fiber-optic sensors are one of the most promising candidates, due to their features of durability, stability, small size, and insensitivity to external EM perturbations, which makes them ideal for the long-term health assessment of civil structures. Optical fibers can cover the large areas of civil structures, enabling safely access and monitoring the status of these structures. Hence, distributed fiber sensor systems will play important roles in the monitoring and diagnostics of critical structures, as they have the advantage of a long sensing range and the capability of providing strain or temperature at every spatial resolution along the entire sensing fiber, imbedded in or attached to the structures, using the fiber itself as the sensing medium. The first temperature-compensated structural test with BOTDA was demonstrated in 2001 on a steel beam (Bao et al. 2001), and other field tests with BOTDA have been reviewed in Bao and Chen (2008). The industrial applications of BOTDR have been reviewed in Ohno et al. (2001), in which the results of three field tests are reported, namely, as a damage-detection system for America's Cup yachts, as an optical fiber sensor for detecting changes in river levees, and as a strain-sensing optical fiber embedded in concrete structures.

One of the challenges in field applications of the distributed sensors is the interface problem, that is, ensuring an optimized interaction between the fibers' system and the measurand field, which is uniform and constant both in space and in time. The demanding measurand, in this respect, is temperature and strain; since all materials come naturally to thermal equilibrium with their varying environment, the separation of temperature and strain in field is difficult, although some of the work has been demonstrated with LEAF fiber of different peaks. However, the difference of their strain and temperature coefficients is small, which means that from the recovered strain and temperature accuracy, it is difficult to meet the engineering monitoring requirement.

For field demonstrations, the choice of the glue for embedding the fiber is critical and depends on the material parameters of the structures. The choice of glue affects the strain transferring matrix. Normally harder glue applied in thinner layers is good for static strain measurement. For dynamic measurement, relatively soft epoxy glue is preferred as it is not easy to break fibers with repeated bending and stretching, especially for the large extension measurement of a steel pipe. However, for concrete and plastic pipes, due to small thermal extension, hardened glue gives better strain transferring ability.

A strain calibration pretest is needed to verify if the interface between structures and fibers works well, which means the strain can be transferred linearly. The signal processing approach is critical to monitor these changes in various structures subjected to heavy loads or environmental conditions. The measured strain distribution of both the concrete columns and pipes provided detailed information on the structure's health at the local and global level, associated with deformations, cracks, and buckling. Such critical information is crucial to field engineers for making decisions pertaining to repairs and maintenance in order to protect public safety.

16.9 Summary and Conclusion

Distributed fiber sensors are one of the most exciting developments in the field of optical fiber sensors in recent years. As they offer a number of distinguishing advantages, great advancements have been achieved in recent years in terms of long sensing length (150 km), high spatial resolution (mm to 1 cm), and high strain and temperature resolution (less than 1 με and less than 0.1°C temperature resolution). This has proven them to be one of the most promising candidates for fiber-optic smart structures.

This chapter presents a comprehensive and systematic overview of the history of distributed sensor technology regarding many aspects including sensing principles, properties, and their performance, system limitations, and applications. It is anticipated that many more distributed sensor systems will be commercialized and widely applied in practice in the near future due to the maturity of many new technologies and the availability of cost-effective instrumentation techniques.

Acknowledgment

The authors acknowledge the contributions of many of their graduate students, postdoctoral fellows, and research associates described in part in this review and the supports of the National Science and Engineering Research Council of Canada via Discovery Grants, Equipment Grants and Strategic Grants, and Intelligent Sensing for Innovative Structures (ISIS), Canada, as well as Canada Research Chair Program. Special thanks goes to Hao Liang for the measurement of the off-resonance of the Brillouin scattering in various optical fibers.

References

Afshar, S., G. Ferrier, X. Bao et al. 2003. The impact of finite extinction ratio of EOM on the performance of the pump-probe Brillouin sensor system. *Optics Letters* 28: 1418–1420.

Alahbabi, M. N., Y. T. Cho, and T. P. Newson. 2005. 150-km-range distributed temperature sensor based on coherent detection of spontaneous Brillouin backscatter and in-line Raman amplification. *Journal of Optical Society of America B* 22: 1321–1324.

Bao, X. and L. Chen. 2008. Development of the distributed Brillouin sensors for health monitoring of civil structures. *Optical Waveguide Sensing and Imaging*. NATO Science for Peace and Security Series B: Physics and Biophysics, Bock, W.J., Gannot, I., and Tanev, S. (Eds.), pp. 101–125. Springer, New York.

Bao, X. and L. Chen. 2011a. Recent progress in Brillouin scattering based fiber sensors. *Sensors Review* 11: 4152–4187.

Bao, X. and L. Chen. 2011b. Recent progress in optical fiber sensors based on Brillouin scattering at University of Ottawa. *Sensors Journal of Photonic Sensors* 1(2): 102–117.

Bao, X., M. DeMerchant, A. Brown et al. 2001. Tensile and compressive strain measurement in the lab and field with the distributed Brillouin scattering sensor. *Journal of Lightwave Technology* 19: 1698–1704.

Bao, X., J. Dhliwayo, N. Heron et al. 1995a. Experimental and theoretical studies on a distributed temperature sensor based on Brillouin scattering. *Journal of Lightwave Technology* 13: 1340–1348.

Bao, X., D. J. Webb, and D. A. Jackson. 1993a. 32 km distributed temperature sensor based on Brillouin loss in an optical fiber. *Optics Letters* 18: 1561–1563.

Bao, X., D. J. Webb, and D. A. Jackson. 1993b. 22-km distributed temperature sensor using Brillouin gain in an optical fiber. *Optics Letters* 18: 552–554.

Bao, X., D. J. Webb, and D. A. Jackson. 1994. Combined distributed temperature and strain sensor based on Brillouin loss in an optical fiber. *Optics Letters*19: 141–143.

Bao, X., D. J. Webb, and D. A. Jackson. 1995b. 22 km distributed strain sensor using Brillouin loss in an optical fiber. *Optics Communications* 104: 298–302.

Barnoski, M. K. and S. M. Jensen. 1976. Fiber waveguides: A novel technique for investigating attenuation characteristics. *Applied Optics* 15: 2112–2115.

Boyd, R. W. 2008. *Nonlinear Optics*, 3rd edn. Academic Press, San Diego, CA.

Brown, A., M. DeMerchant, X. Bao et al. 1999. Spatial resolution enhancement of a Brillouin scattering based distributed sensor. *Journal of Lightwave Technology* 17: 1179–1183.

Brown, A. W., B. G. Colpitts, and K. Brown. 2005. Distributed sensor based on dark-pulse Brillouin scattering. *Photonics Technology Letters* 17: 1501–1503.

Cameron, J., L. Chen, X. Bao et al. 1998. Time evolution of polarization mode dispersion in optical fibers. *Photonics Technology Letters* 10: 1265–1267.

Corsi, F., A. Galtarossa, and L. Palmieri. 1998. Polarization mode dispersion characterization of single-mode optical fiber using a backscattering technique. *Journal of Lightwave Technology* 16: 1832–1843.

Culverhouse, D., F. Farahi, C. Pannell et al. 1989. Potential of stimulated Brillouin scattering as sensing mechanism for distributed temperature sensors. *Electronics Letters* 25: 913–915.

Dakin, J. P., D. J. Pratt, G. W. Bibby et al. 1985a. Distributed optical fibre Raman temperature sensor using a semiconductor light source and detector. *Electronics Letters* 21: 569–570.

Dakin, J. P., D. J. Pratt, J. N. Ross et al. 1985b. Distributed anti-Stokes Raman thermometry. *Proceedings Conference on Optical-Fibre Sensors 3*, San Diego, CA.

DeMerchant, M. D., A. W. Brown, X. Bao et al. 1998. Automated system for distributed sensing. *Proceedings of SPIE* 3330: 315–322.

Dong, Y., X. Bao, and L. Chen. 2009a. Distributed temperature sensing based on birefringence effect on transient Brillouin grating in a polarization-maintaining photonic crystal fiber. *Optics Letters* 34: 2590–2592.

Dong, Y., X. Bao, and W. Li. 2009b. Using differential Brillouin gain to improve the temperature accuracy and spatial resolution in a long-distance distributed fiber sensor. *Applied Optics* 48: 4297–4301.

Dong, Y., L. Chen, and X. Bao. 2010a. Truly distributed birefringence measurement of polarization-maintaining fibers based on transient Brillouin grating. *Optics Letters* 35: 193–195.

Dong, Y., L. Chen, and X. Bao. 2010b. High-spatial-resolution simultaneous strain and temperature sensor using Brillouin scattering and birefringence in a polarization-maintaining fibre. *Photonics Technology Letters* 22: 1364–1366.

Dong, Y., H. Zhang, L. Chen, and X. Bao. 2012. A 2-cm-spatial-resolution and 2-km-range Brillouin optical fiber sensor using a transient differential pulse pair. *Applied Optics* 51(9): 1229–1235.

Eickhoff, W. and R. Ulric. 1981. Optical frequency domain reflectometry in single-mode fiber. *Applied Physics Letters* 39: 693–695.

Farries, M. C., M. E. Fermann, R. I. Laming et al. 1986. Distributed temperature sensor using Nd^{3+} doped fiber. *Electronics Letters* 22: 418–419.

Farries, M. C., M. E. Fermann, S. B. Poole et al. 1987. Distributed temperature sensor using holmium 3^+ doped fiber. *Proceedings of OFC'87*, Reno, NV, 170pp.

Feced, R., M. Farhadiroushan, V. A. Handerek et al. 1997. Advances in high resolution distributed temperature sensing using the time-correlated single photon counting technique. *Proceedings of the IEE Opto-Electronics* 144: 183–188.

Fellay, A., L. Thevanez, M. Facchini et al. 1997. Distributed sensing using SBS: Towards ultimate resolution. *OSA Technical Digest Series* 16: 324–327.

Fernandez, A. F., P. Rodeghiero, B. Berghmans et al. 2005. Radiation-tolerant Raman distributed temperature monitoring system for large nuclear Infrastructures. *IEEE Transactions on Nuclear Science* 52: 2689–2694.

Foaleng, S. M., M. Tur, J. Beugnot, and L. Thévenaz. 2010. High spatial and spectral resolution long-range sensing using Brillouin echoes. *Journal of Lightwave Technology* 28: 2993–3003.

Froggatt, M. and J. Moore. 1998. High-spatial-resolution distributed strain measurement in optical fiber with Rayleigh scatter. *Applied Optics* 37: 1735–1740.

Froggatt, M., B. Soller, D. Gifford et al. 2004. Correlation and keying of Rayleigh scatter for loss and temperature sensing in parallel optical networks. *OFC Technical Digest*, Paper PDP17.

Garus, D., T. Golgolla, K. Krebber et al. 1997. Brillouin optical frequency-domain analysis for distributed temperature and strain measurements. *Journal of Lightwave Technology* 15: 654–662.

Golgolla, T. and K. Krebber. 2000. Distributed beat length measurements in single-mode optical fibers using stimulated Brillouin-scattering and frequency-domain analysis. *Journal of Lightwave Technology* 18: 320–328.

Gordon, J. P. and H. Kogelnik. 2000. PMD fundamentals: Polarization mode dispersion in optical fibers. *Proceedings of the National Academy of Sciences of the United States* (*PNAS*) 97(9): 4541–4550.

Hao, L., W. Li, N. Linze et al. 2010. High resolution DPP-BOTDA over 50 km fiber using return to zero coded pulses. *Optics Letters* 35: 1503–1505.

Hartog, A. H. 1983. A distributed temperature sensor based on a liquid-core optical fibre. *Journal of Lightwave Technology* 1: 498–509.

Hartog, A. H. 1996. Distributed fiber-optic sensors. In: *Optical Fiber Sensor Technology*, K. T. V. Grattan and B. T. Meggitt (eds.), Chapter 11. Chapman & Hall, London, U.K.

Horiguchi, T., T. Kurashima, and M. Tateda. 1989. Tensile strain dependence of Brillouin frequency shift in silica optical fibers. *Photonics Technology Letters* 1: 107–108.

Horiguchi, T., K. Shimizu, T. Kurashima et al. 1995. Development of a distributed sensing technique using Brillouin scattering. *Journal of Lightwave Technology* 13: 1296–1302.

Horiguchi, T. and M. Tateda. 1989. Optical-fiber-attenuation investigation using stimulated Brillouin scattering between a pulse and a continuous wave. *Optics Letters* 14: 408–410.

Hotate, K. 2011. Brillouin scattering accompanied by acoustic grating in an optical fiber and applications in fiber distributed sensing. *Proceedings of SPIE* 7753: 7–10.

Hotate, K. and T. Hasegawa. 2000. Measurement of Brillouin gain spectrum distribution along an optical fiber with a high spatial resolution using a correlation-based technique—Proposal, experiment and simulation. *IEICE Transactions Electronics* E83-C: 405–411.

Hotate, K. and M. Tanaka. 2002. Distributed fiber Brillouin strain sensing with 1 cm spatial resolution by correlation-based continuous-wave technique. *Photonics Technology Letters* 14: 179–181.

Hwang, D., D. Yoon, I. Kwon et al. 2010. Novel auto-correction method in a fiber-optic distributed-temperature sensor using reflected anti-Stokes Raman scattering. *Optics Express* 18: 9747–9754.

Jackson, J. D. 1999. *Classical Electrodynamics*, 3rd edn. John Wiley & Sons, Inc., New York.

John, P. and J. P. Dakin. 1995. Distributed optical fiber sensor. In: *Fiber Optic Smart Structures*, E. Udd (ed.), Chapter 14. Wiley Series in Pure and Applied Optics. John Wiley & Sons, New York.

Juarez, J. C. and H. F. Taylor. 2005. Polarization discrimination in a phase-sensitive optical time-domain reflectometer intrusion-sensor system. *Optics Letters* 24: 3284–3286.

Kersey, A. D. 1996. A review of recent developments in fiber optic sensor technology. *Optical Fiber Technology* 2: 291–317.

Kingsley, S. A. and D. E. N. Davies. 1985. OFDR diagnostics for fibre and integrated-optic systems. *Electronics Letters* 21: 434–435.

Kurashima, T., T. Horiguchi, H. Izumita et al. 1991. Brillouin OTDR. *Proceedings of the International Quantum Electronics Conferences*, Vienna, Austria, pp. 42–44.

Kurashima, T., T. Horiguchi, and M. Tateda. 1990a. Distributed-temperature sensing using stimulated Brillouin scattering in optical silica fibers. *Optics Letters* 15: 1038–1040.

Kurashima, T., T. Horiguchi, and M. Tateda. 1990b. Thermal effects on the Brillouin frequency shift in jacketed optical silica fibers. *Applied Optics* 29: 2219–2222.

Lees, G. P., P. C. Wait, M. J. Cole et al. 1998. Advances in optical fiber distributed temperature sensing using the Landau–Placzek ratio. *Photonics Technology Letters* 10: 126–128.

Li, W., X. Bao, Y. Li et al. 2008. Differential pulse-width pair BOTDA for high spatial resolution sensing. *Optics Express* 16: 21616–21625.

Li, Y., L. Chen, Y. Dong et al. 2010. A novel distributed Brillouin sensor based on optical differential parametric amplification. *Journal of Lightwave Technology* 28: 2621–2626.

Long, D. A. 2002. *The Raman Effect*. John Wiley & Sons Ltd., Chichester, U.K.

Lu, Y., T. Zhu, L. Chen et al. 2010. Distributed vibration sensor based on coherent detection of phase-OTDR. *Journal of Lightwave Technology* 28: 3243–3249.

Marcuse, E. 1976. Microbending losses of single-mode, step index and multimode parabolic-index fibers. *Journal Bell Systems Technology* 55: 937–955.

Measures, R. 2001. *Structural Monitoring with Fiber Optic Technology*. Academic Press, San Diego, CA.

Minardo, A., R. Bernini, and L. Zeni. 2009. A simple technique for reducing pump depletion in long-range distributed Brillouin fiber sensors. *IEEE Sensors Journal* 9: 633–634.

Niklès, M., L. Thévenaz, and P. A. Robert. 1996. Simple distributed fiber sensor based on Brillouin gain spectrum analysis. *Optics Letters* 21: 758–760.

Ohno, H., H. Naruse, M. Kihara et al. 2001. Industrial applications of the BOTDR optical fiber strain sensor. *Optical Fiber Technology* 7: 45–64.

Oscroft, G. 1997. Intrinsic fiber optic sensors. *Journal of Optical Sensors* 2: 269–282.

Parker, R., M. Farhadiroushan, V. A. Handerek et al. 1997. A fully distributed simultaneous strain and temperature sensor using spontaneous Brillouin backscatter. *Photonics Technology Letters* 9: 979–981.

Qin, Z., T. Zhu, and X. Bao. 2011. High frequency response distributed vibration sensor based on all polarization-maintaining configurations of phase-OTDR. *Photonics Technology Letters* 23: 1091–1093.

Rogers, A. 1999. Distributed optical fiber sensing. *Measurement Science and Technology* 10: R75–R99.

Rogers, A. J. 1981. Polarization-optical time domain reflectometry: A technique for the measurement of field distributions. *Applied Optics* 20: 1060–1074.

Shatalin, S. V. and A. J. Rogers. 2006. Location of high PMD sections of installed system fiber. *Journal of Lightwave Technology* 24: 3875–3881.

Shimizu, K., T. Horiguchi, Y. Koyamada et al. 1993. Coherent self-heterodyne detection of spontaneously Brillouin-scattered light waves in a single-mode fiber. *Optics Letters* 18: 185–187.

Smith, J., M. DeMerchant, A. Brown et al. 1999. Simultaneous distributed strain and temperature measurement. *Applied Optics* 38: 5372–5377.

Soller, B. J., D. K. Gifford, M. S. Wolfe et al. 2005. High resolution optical frequency domain reflectometry for characterization of components and assemblies. *Optics Express* 13: 666–674.

Song, K. Y., W. Zou, Z. He et al. 2008. All-optical dynamic grating generation based on Brillouin scattering in polarization-maintaining fiber. *Optics Letters* 33: 926–928.

Soto, M. A., G. Bolognini, and Di F. Pasquale. 2011. Long-range simplex-coded BOTDA sensor over 120 km distance employing optical pre-amplification. *Optics Letters* 36: 232–234.

Suh, K. and C. Lee. 2008. Auto-correction method for differential attenuation in a fiber-optic distributed-temperature sensor. *Optics Letters* 16: 1845–1847.

Tai, K., A. Hasegawa, and A. Tomita. 1986. Observation of modulational instability in optical fibers. *Physics Review Letters* 56: 135–138.

Thévenaz, L. and S. F. Mafang. 2008. Distributed fiber sensing using Brillouin echoes. *Proceedings of SPIE* 7004: 1–4.

Waddy, D. S., P. Lu, L. Chen et al. 2001. Fast state of polarization changes in aerial fiber under different climatic conditions. *Photonics Technology Letters* 13: 1035–1037.

Wait, P. C. and T. P. Newson. 1996. Landau Placzek ratio applied to distributed fibre sensing. *Optics Communications* 122: 141–146.

Wang, F., X. Bao, L. Chen et al. 2008. Using pulse with dark base to achieve high spatial and frequency resolution for the distributed Brillouin sensor. *Optics Letters* 33: 2707–2709.

Zhang, Z. and X. Bao. 2008. Distributed optical fiber vibration sensor based on spectrum analysis of polarization-OTDR system. *Optics Express* 16: 10240–10247.

Zou, W., Z. He, and K. Hotate. 2009. Complete discrimination of strain and temperature using Brillouin frequency shift and birefringence in a polarization-maintaining fiber. *Optics Express* 17: 1248–1255.

17. Fiber Bragg Grating Sensors

David Webb
Aston University

Handbook of Optical Sensors. Edited by José Luís Santos and Faramarz Farahi © 2015 CRC Press/Taylor & Francis Group, LLC. ISBN: 9781439866856.

17.1 Introduction

An optical fiber Bragg grating (FBG) usually takes the form of a spatially periodic modulation of the refractive index along the core of a section of optical fiber. As we shall see, such a structure has the property of reflecting light of essentially one wavelength determined by the period of the modulation and the mean index of the fiber. Periods of around half a micron reflect light in the important 1550 nm spectral window of silica optical fiber. Such devices have found numerous applications in optical communications systems, for example, as add-drop multiplexers, dispersion compensators, or gain equalizers (Giles 1997). However, the focus of this chapter is on sensing applications, where the key point is that the reflected wavelength—the Bragg wavelength—is found to depend on the temperature of the fiber or any strain to which the fiber is subjected. Thus, by illuminating the fiber with light of a broadband spectrum and measuring the wavelength reflected by the grating, it is possible to infer what is happening to the fiber in the region of the grating.

17.2 Historical Development

FBGs were discovered in the late 1970s (Hill et al. 1978, Kawasaki et al. 1978) but remained a curiosity for about 10 years until a means was discovered of arranging for the reflected wavelength to lie in a convenient spectral region (Meltz et al. 1989). Since the first demonstration of FBG sensor elements at the start of the 1990s, the field has grown so that since 2005, publications on FBG sensors have amounted to about a third of all the fiber sensor papers, as shown in Figure 17.1. As well as being a fertile topic for research, FBG sensors have also become a commercial success story, partly perhaps as a result of photonics start-up companies caught out by the bursting of the telecoms bubble at the turn of the millennium, switching their technology focus from optical communications to sensing.

There is already a significant amount of review literature on the subject. There are three textbooks specifically dedicated to FBGs (Kashyap 1999, Othonos and Kalli 1999, Webb and Kalli 2011) and other books that contain significant chapters on FBG sensors (Ansari 1998, López-Higuera 2002). There are review papers purely dedicated to general sensing applications of FBG (Kersey et al. 1997, Rao 1997), applications in medicine (Mishra et al. 2011), structural health monitoring (Majumder et al. 2008, Todd et al. 2007), and microstructured FBGs (Cusano et al. 2009) as well as others

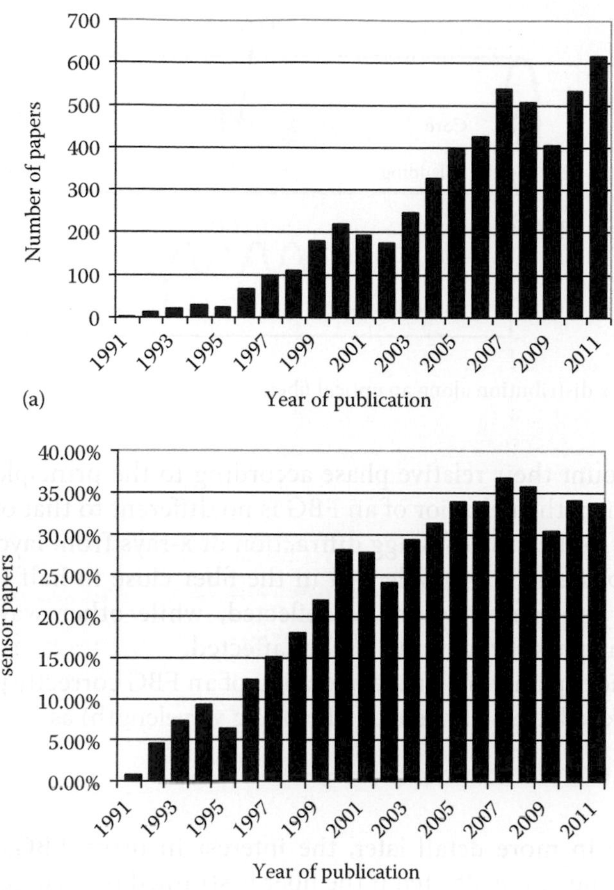

FIGURE 17.1 Bibliographic information relating to FBG sensors. (a) Number of papers each year, retrieved from a search of Web of Science with search term *Bragg grating (sensor OR sensing) (fiber OR fiber)*. (b) Percentage of FBG sensor papers of all optical fiber sensor papers retrieved with search term *optical (sensor OR sensing) (fiber OR fiber)*.

of more general nature (Bennion et al. 1996, Hill et al. 1993, Hill and Meltz 1997, Othonos 1997, Russell et al. 1993, Vasil'ev et al. 2005).

17.3 Optical Properties

17.3.1 Uniform Index Modulation

In its simplest incarnation, an FBG takes the form of a sinusoidal modulation of the refractive index along the core of a section of optical fiber, see Figure 17.2. Intuitively, the behavior of such a grating can be understood by considering it to consist of a series of weakly reflecting planes situated at the positions of the peaks in the refractive index. Each plane reflects a small amount of the incident light, and the resultant back-reflected signal is obtained by summing the contributions from all the planes,

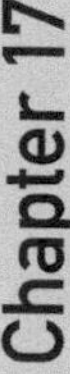

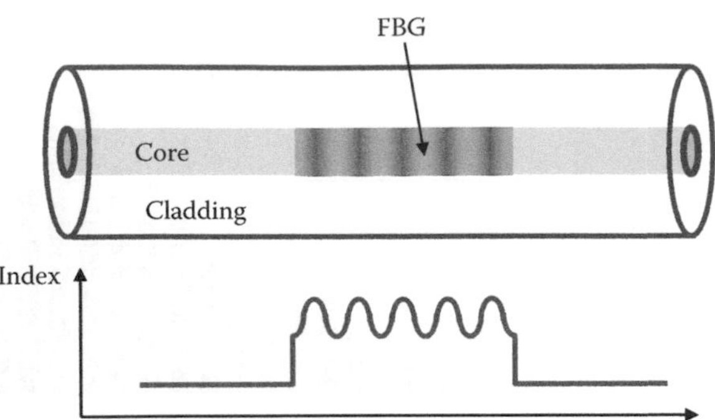

FIGURE 17.2 Index distribution along an optical fiber.

taking into account their relative phase according to the principle of superposition. Conceptually then, the behavior of an FBG is no different to that of a multilayer mirror and similar to the case of Bragg diffraction of x-rays from layers of atoms. Back-reflected light possessing a wavelength in the fiber close to half the grating period, $\Lambda/2$, interferes constructively and is reflected, while other wavelengths interfere destructively and pass through the FBG unaffected.

This simplistic explanation for the operation of an FBG correctly provides an expression for the peak reflected wavelength (the Bragg wavelength) as

$$\lambda_B = 2n\Lambda. \tag{17.1}$$

As we shall see in more detail later, the interest in using FBGs as sensors arises, because both n and Λ are affected if the fiber is strained or its temperature changed,* and hence, these parameters can be monitored using a measurement of the precise wavelength reflected by the grating, but a more rigorous analysis is needed to extract further information. A common approach is to use coupled wave theory to obtain an analytic solution (Kogelnik 1976). Here we shall simply quote the major results. The reflectivity of a grating of length l with an index modulation of constant amplitude is given by

$$R = \frac{\kappa^2 \sinh^2 \psi l}{\Delta k^2 \sinh^2 \psi l + \psi^2 \cosh^2 \psi l}, \tag{17.2}$$

where $\Delta k = k - \pi/\lambda$ is called the detuning wave vector and provides a measure of how much the Bragg condition is violated, $\psi = (\kappa^2 - \Delta\kappa^2)^{1/2}$, with κ being the coupling constant given by

$$\kappa = \frac{\pi \Delta n \eta}{\lambda}, \tag{17.3}$$

* For completeness, it should be stated that the refractive index, and hence the Bragg wavelength, is also intrinsically sensitive to the magnetic field along the fibre axis, due to the Faraday effect, though realistic fields only give a small wavelength shift.

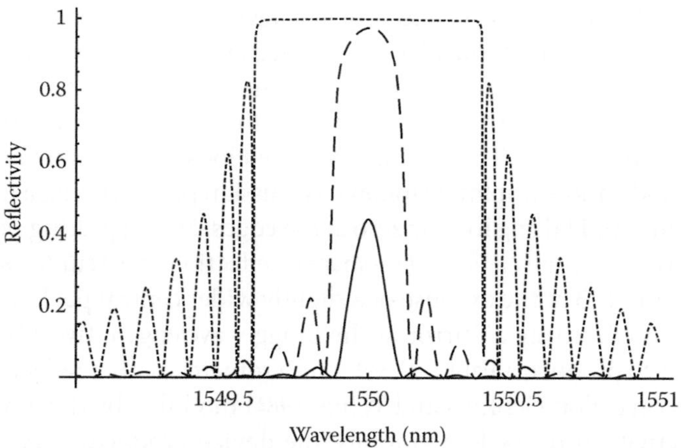

FIGURE 17.3 Theoretical reflectivity of 1 cm gratings with a Bragg wavelength of 1550 nm and with coupling constant $\kappa = 80$ (solid), 250 (dashed) and 1000 (dotted).

where Δn is the amplitude of the index modulation. The quantity η is a measure of the grating efficiency and represents how much of the guided power is actually in the core overlapping the grating. In terms of the fiber parameters, η is given by Russell et al. (1993)

$$\eta \approx 1 - \left(\frac{\lambda_0}{2\pi a}\right)^2 \frac{1}{\left(n_1^2 - n_2^2\right)}, \tag{17.4}$$

where

λ_0 is the free space wavelength
a the fiber core radius
n_1 the core index
n_2 the cladding index

If we turn our attention to Equation 17.2, if the incident light is Bragg matched so that $\Delta k = 0$, the reflectivity simplifies to

$$R = \tanh^2(\kappa \mathrm{l}), \tag{17.5}$$

so the reflectivity increases asymptotically toward unity as either the length or the strength (Δn) of the grating is increased. It turns out that strong gratings may have a high reflectivity over a wide range of wavelengths, as shown in Figure 17.3. For sensing applications, weaker gratings are usually used (reflectivity 95% or less) to avoid the broadening of the reflection peak displayed by stronger gratings, which could reduce the accuracy of the Bragg wavelength determination.

17.3.2 More Complex Grating Structures

Most sensing applications of Bragg gratings use devices with properties quite close to the ideal model just described; however, there are more complex structures that can bring benefits to some applications. Analysis of the behavior of these structures is often not possible analytically, in which case the coupled wave equations may be solved numerically or the transfer matrix method (Skaar and Waagaard 2003) may be used, for example.

The simplest modification to the uniform grating described earlier is to apodize the index profile. Apodization most generally refers to the control of the modulation amplitude along the grating length. The term is often used loosely to mean Gaussian apodization, where the modulation amplitude follows a Gaussian profile, being a maximum in the center of the grating and falling to near zero at the edges. Such a grating possesses (at least when the reflectivity is small) a Gaussian-shaped reflection spectrum, avoiding the sidebands seen in Figure 17.3 at the expense of a slightly wider central peak (Malo et al. 1995).

Since an FBG is effectively a mirror at the Bragg wavelength, two FBGs can be combined in one fiber to produce a Fabry–Pérot interferometer, with a reflection profile given by that of a classical etalon (Jenkins and White 1981) modified by the grating reflectivity. With high reflectivity and low loss gratings, the device produces very narrow spectral peaks enabling the length of the cavity to be monitored with high precision, at the expense of a more complicated signal processing scheme than would be the case with a single grating. In contrast to a simple FBG sensor where the sensitivity to temperature or strain does not depend on the grating length, the sensitivity of an interferometer is proportional to the cavity length. The very highest resolution optical fiber sensors have been produced by combining an interferometric cavity with amplifying fiber to produce a narrow linewidth fiber laser; wavelength shifts of just a few Hz can be monitored (Koo and Kersey 1995)—corresponding to a resolution in wavelength measurement of better than 1 part in 10^{13}!

A variation on the Fabry–Pérot cavity is the π-shifted grating, which can be considered as a Fabry–Pérot cavity where the grating separation has been reduced to just a quarter of a wavelength (Canning and Sceats 1994). Such a structure has a single, narrow peak in the center of the grating's transmission band. While these devices are not usually employed directly as sensors, they can be used to create a single longitudinal mode fiber laser, simplifying the demodulation of the fiber laser sensors (Asseh et al. 1995). These gratings may be fabricated in several ways, such as by using a blank UV beam to raise the index in the central region of a uniform FBG (Asseh et al. 1995) or by superimposing two gratings with slightly different periods to record part of the resulting beat pattern; the latter structure is known as a Moiré grating (Reid et al. 1990).

A chirped grating is one in which the Bragg wavelength varies with position along the grating. This can be achieved by arranging for either the period of the grating or its mean index to vary with position. This can result in a broad reflection profile covering many nanometers. Chirped grating usually find application in communications systems as broadband mirrors or for dispersion compensation (Oullette 1991); however, they do have some specialized sensing applications, permitting high spatial resolution recovery of information from along the sensor length (LeBlanc et al. 1996).

17.3.3 Long-Period Gratings

The final grating structure to be discussed is rather different to those that have gone before, as it is the only one that cannot be readily operated in reflection. As its name suggests, a long-period grating (LPG) has a period much greater than is typical for a Bragg grating, which is around half a micron when the Bragg wavelength is in the 1550 nm spectral region. With typical periods of a few hundred microns, an LPG acts to couple light from the core mode to forward traveling cladding modes. To understand how this occurs, consider that the Bragg condition for diffraction may be written in the form

$$\mathbf{k}_2 = \mathbf{k}_1 + \mathbf{K}; \quad \mathbf{k}_1 = \frac{2\pi n_{\text{core}}}{\lambda}, \quad \mathbf{k}_2 = \frac{2\pi n_{\text{cladding}}}{\lambda}, \tag{17.6}$$

where
- $\mathbf{k}_1$ is the propagation constant of the incident light in the fiber core
- $\mathbf{k}_2$ is that of the diffracted light
- $\mathbf{K}$ is the grating vector equal to $2\pi/\Lambda$
- λ is the vacuum wavelength
- n_{core} is the effective core index
- n_{cladding} is the effective index of the relevant cladding mode

In the case of the standard FBG, $\mathbf{k}_1$ and $\mathbf{k}_2$ are equal but of opposite sign so that $\mathbf{K} = 2\mathbf{k}_1$, and hence, Λ is small. For the LPG, $\mathbf{k}_1$ and $\mathbf{k}_2$ are in the same direction and not too dissimilar; so $\mathbf{K}$ is small and Λ is therefore large. Note that the resonance condition (Equation 17.6) is very sensitive to changes in the difference between the core and cladding indices.

Because the coupling in LPGs is to forward travelling cladding modes that suffer high attenuation, the devices cannot be operated in reflection. Instead, the shift in the notch in the transmission profile in response to the measurand is monitored. A given LPG may couple light into a range of cladding modes, depending on the wavelength. Thus, the transmission spectrum consists of a set of attenuation bands, as shown in Figure 17.4.

LPG sensors represent a field in their own right (James and Tatam 2003), and hence, we will not dwell on them in this chapter. We will just briefly survey how their behavior differs from that of FBG sensors. First, as already noted, they can only be operated in transmission. Second, the multiple spectral resonances displayed by a typical LPG and

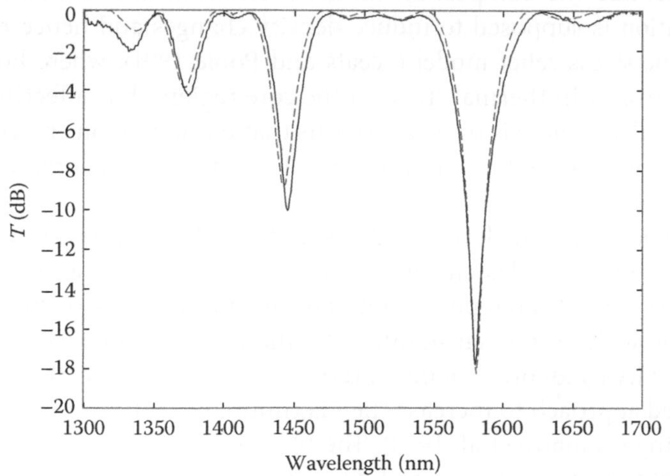

FIGURE 17.4 Typical long-period Bragg grating transmission spectrum. Solid curve—experimental, dashed curve theoretical. (Courtesy of L. Zhang and I. Bennion, Photonics Research Group, Aston University, U.K.)

the larger resonance bandwidths tend to limit the amount of wavelength multiplexing that can be practically employed. Third, in addition to strain and temperature sensitivity, their resonant wavelengths are also affected by bending and the refractive index surrounding the cladding. Fourth, whereas the sensitivity of FBGs to strain and temperature varies very little between one silica fiber and another, because the sensitivity of the LPG depends on the difference between the core and cladding indices, the sensitivities vary significantly, both from fiber to fiber and even between different cladding mode resonances of the same fiber. Importantly, the sensitivity of the LPG to the various parameters (strain, temperature, index, and bending) has a degree of independence, so that it is possible, for example, to find one fiber with reasonable strain sensitivity that is almost insensitive to temperature (Dobb et al. 2004). With an appropriate choice of fiber, it then becomes possible in principle to maximize the sensitivity to a desired parameter while minimizing the cross-sensitivity to the other parameters.

17.4 Photosensitivity

The photosensitivity that permits the inscription of FBGs is a complex phenomenon. More accurately, it is a complex range of phenomena as there are various mechanisms that give rise to a refractive index change, depending on the dopants used in the core of the fiber and the wavelength, intensity, and pulse duration of the light source used for inscription. We start by looking at sensitivity to relatively low intensities* of continuous wave UV light. It is thought that in the case of standard germania-doped SMF-28 optical fiber, the photosensitivity arises due to oxygen-deficiency centers in the glass (Skuja 1998), which leads to absorption bands centered around 244 and 330 nm. An incident photon exciting such a center bleaches out the absorption, which may in turn affect the refractive index in the NIR through a variety of mechanisms. One invokes the Kramers–Kronig relationship, relating the absorption change in the UV to a change in index in the near infrared (Hand and Russel 1990). Other models that are still under investigation include the compaction model (Bernardin and Lawandy 1990), where the UV irradiation is supposed to induce density changes and hence refractive index changes, and the stress-relief model (Sceats and Poole 1991), where bonds broken by the UV reduce frozen-in thermal stress in the core region, thus affecting the index via the stress-optic effect. There is some consensus that more than one mechanism is likely to be involved, with their relative importance depending on the precise fiber type and illumination conditions.

With the germania-doping levels used in standard telecommunications fiber (around 3 mol%), peak refractive index changes of about 3×10^{-5} are obtained; however, index changes up to 10^{-3} have been reported with heavily doped fiber (10–20 mol%) (Xie et al. 1993). Codoping with boron can significantly increase the index change and improve the photosensitivity by an order of magnitude (Williams et al. 1993).

An often used approach to increase the maximum refractive index change is that of hydrogen loading (Lemaire et al. 1993). The fiber is placed in high-pressure hydrogen (typically 150 atmospheres) for several days at room temperature (or slightly above). The fiber then typically exhibits a maximum index modulation up to two orders of

* By low intensity we are referring here to situations where we can ignore multiphoton interactions.

magnitude greater than for untreated fiber, with modulations over 0.01 being achievable. The presence of hydrogen within the fiber increases the refractive index and therefore the Bragg wavelength compared to an untreated fiber with the same grating spacing. Following inscription, there is a gradual diffusion of unreacted hydrogen out of the fiber, which leads to a slowly decreasing Bragg wavelength (Malo et al. 1994), though this drift problem can be avoided by thermally annealing the fiber before use (Lemaire 1995).

Archambault et al. discovered a new form of recording process that occurs when short pulses are used above a certain threshold energy density (Archambault et al. 1993). In their experiment, increasing the pulse energy from 20 to 40 mJ resulted in an increase in the index modulation of three orders of magnitude to 0.006. This process is associated with physical damage on one side of the core-cladding boundary. Such gratings are characterized by high reflectivity (approaching 100%), good thermal stability (over 800°C), and wide reflecting bandwidth (~several nm). In transmission, they are observed to couple wavelengths shorter than the Bragg wavelength into the cladding. As a result of them being written using short laser pulses, these were the first type of gratings capable of being written as the fiber is being formed on the pulling tower (Dong et al. 1993), thus opening the way to cost-effective mass production of gratings (Lindner et al. 2011).

These damage gratings were originally labeled as type II gratings to distinguish them from the original (Type I) variety. Since that time, several other inscription mechanisms have been developed and a complex terminology of grating classification arose. An admirable attempt has been recently made by Canning to bring some order to this field by renaming the categories of gratings based on what we now know about their physical origin (Canning et al. 2009). Here we just mention two inscription methods that are currently the subjects of much research.

The first is the regenerated grating, classified by Canning as type R. These are gratings that appear when type I gratings are being annealed out at high temperature. They tend to be quite weak and exhibit a significant red shift of a few nanometers on return to room temperature. These gratings are of considerable practical interest as they can survive repeated cycling to high temperatures (>1000°C), permitting sensors to be developed for harsh environments (Zhang and Kahrizi 2007).

The second inscription method is the fabrication of gratings using very high intensity, short laser pulses, typically around 100 fs in length. Multiphoton absorption processes allow the inscription of gratings in materials that do not exhibit conventional photosensitivity and furthermore permit the recording of Bragg gratings one plane at a time, offering precise control over grating structure (Martinez et al. 2004).

17.5 Inscription

Standard FBGs require the production of a roughly half-micron period index modulation along the fiber core. The first, and possibly the most flexible, approach to do this was the holographic technique shown in Figure 17.5, where the precise period is controlled by the angular separation of the two interfering beams. The technique has a number of disadvantages though; the main one is the difficulty in maintaining the optical paths of the two UV laser beams, which are derived from a single laser using a beam splitter, constant to within a small fraction of a wavelength. Any variation greater than this will tend to move

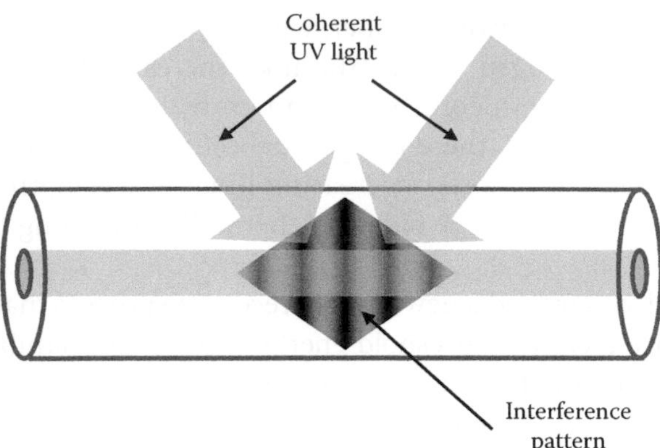

FIGURE 17.5 Holographic approach to grating inscription, with beams incident from the side of the fiber.

the interference pattern significantly along the fiber axis *washing out* the grating, reducing the resulting index modulation, and consequently reducing the reflectivity. In practice, this means that a high-quality optical table is required for the recording apparatus to prevent movements of the mirrors and beam splitter used to generate the two beams, and care must be taken to prevent air currents in the free space paths. The demands placed by this approach on the UV source can also be significant as a reasonable temporal coherence is required to avoid having to match the two optical paths precisely. Spatial coherence needs to be taken into consideration but is not a particular requirement, so long as the beams in the two arms of the interferometer are made to suffer the same number of reflections, so that similar parts of the two beams overlap in the fiber core.

More stable designs exist that use a wavefront splitting interferometer; a recording geometry using a Lloyds mirror wavefront splitting interferometer is illustrated in Figure 17.6 (Limberger et al. 1993). In this approach, interference occurs between directly transmitted light and part of the beam that is deflected from the mirror. Despite the advantage of mechanical stability, this sort of system is rarely used for several reasons: first, the amount of wavelength tuning is limited by the system geometry and

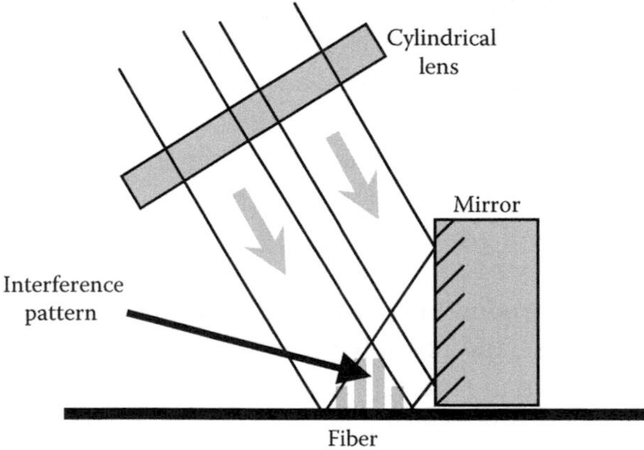

FIGURE 17.6 Lloyds mirror approach to grating fabrication. The cylindrical lens focuses the light onto the core of the fiber to maximize the intensity and hence minimize the recording time.

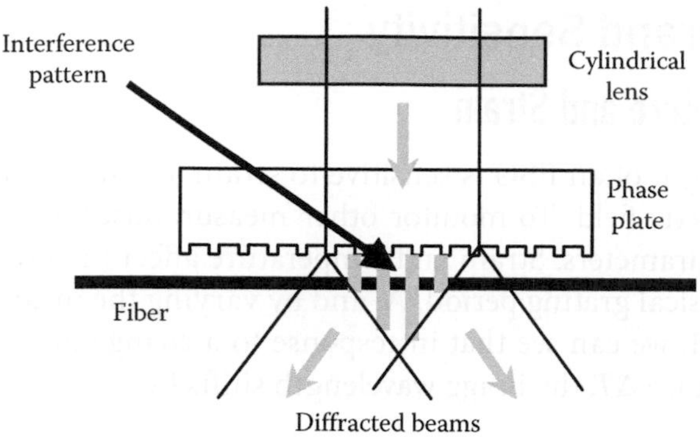

FIGURE 17.7 FBG recording using a phase mask.

second, because different parts of the beam are required to interfere, the laser beam must possess high spatial coherence.

By far, the most common approach for recording gratings is the use of a phase mask (Hill et al. 1993), illustrated in Figure 17.7. The phase mask is a diffracting surface relief structure, usually formed on a high-quality silica optical flat: essentially, it is a transmission diffraction grating. It is placed close to, or in contact with, the fiber and interference occurs between usually the +1 and −1 diffracted order to produce the required intensity pattern. Operation is optimized by selecting a surface profile that in the case shown simultaneously suppresses the zero order transmitted beam (to a few percent) and maximizes the power in the first order beams (often more than 35% each). Because the mask is in contact with the fiber, stability requirements are quite easily met. Furthermore, the close proximity of the core to the mask means that the spatial and temporal coherence of the laser beam does not even have to be particularly good.

Phase masks are usually produced using one of two main techniques: Holographic recording can be used to produce a grating with very uniform period over several centimeters; complex grating structures are not easily produced though. Alternatively, the grating can be written using electron beam lithography (Anderson et al. 1993), allowing arbitrary profiles to be written but usually with nonuniformities in the period of the grating. The major disadvantage of the phase mask is the difficulty of adjusting significantly the Bragg wavelength, although adjustment over a few nanometers is possible by either straining the fiber prior to recording (Campbell and Kashyap 1991), tilting the fiber so that it makes an angle to the surface of the phase mask (Othonos and Lee 1995), or by using a magnifying lens in front of the mask (Prohaska et al. 1993).

A rather different method of grating inscription is the point-by-point approach. As the name suggests, this involves directing a tightly focused beam of UV light onto the core for a brief period, stepping the fiber along by the grating period, and then repeating the process many times. This process has been used to record gratings for use in the 1550 nm band (Malo et al. 1993) using an excimer laser by making use of the second- and third-order reflections from gratings with spacings two and three times that satisfy the Bragg condition at 1550 nm. The point-by-point approach has more recently been used with some success for grating inscription with short pulse (fs) laser systems, as described in the previous section. With these sources, first order gratings may be recorded.

17.6 Measurand Sensitivity

17.6.1 Temperature and Strain

The Bragg wavelength of an FBG is sensitive to strain and temperature and to a small degree axial magnetic field. To monitor other measurands, they must be transduced into one of these parameters. Strain and temperature affect the Bragg wavelength both by altering the physical grating period, Λ, and by varying the mean refractive index, n. From Equation 17.1, we can see that in response to a change in axial strain $\Delta\varepsilon_z$ and a change in temperature ΔT, the Bragg wavelength shifts by

$$\Delta\lambda = 2\left(\Lambda\frac{\partial n}{\partial\varepsilon_z} + n\frac{\partial\Lambda}{\partial\varepsilon_z}\right)\Delta\varepsilon_z + 2\left(\Lambda\frac{\partial n}{\partial T} + n\frac{\partial\Lambda}{\partial T}\right)\Delta T$$

$$= \lambda\left(\frac{1}{n}\frac{\partial n}{\partial\varepsilon_z} + 1\right)\Delta\varepsilon_z + \lambda(\varsigma + \alpha)\Delta T, \tag{17.7}$$

where

$$\alpha = \frac{1}{\Lambda}\frac{\partial\Lambda}{\partial T} \tag{17.8}$$

is the thermal expansion coefficient of the fiber and

$$\varsigma = \frac{1}{n}\frac{\partial n}{\partial T} \tag{17.9}$$

is the thermo-optic coefficient. In practice, the temperature sensitivity of a silica fiber FBG is mainly a result of the variation in the refractive index, rather than the change in physical length, the former being more than 10 times the latter.

The variation of refractive index with axial strain may be shown to be (Butter and Hocker 1978):

$$\frac{1}{n}\frac{\partial n}{\partial\varepsilon_z} = -\frac{n^2}{2}\left(p_{12} - \nu(p_{11} + p_{12})\right), \tag{17.10}$$

where

ν is Poisson's ratio

p_{11} and p_{12} are components of the strain-optic tensor

This equation reflects the fact that not only does the axial strain modify the refractive index of the guided mode directly, but it also causes a radial strain (through Poisson's ratio), which in turn also affects the refractive index due to the tensorial nature of the strain-optic effect.

Table 17.1 Typical Sensitivities of Silica FBGs at Various Wavelengths

Grating Wavelength (nm)	Strain Coefficient (pm/με)	Temperature Coefficient (pm/°C)	References
789	0.60	6.6	Anderson et al. (1993)
1300	1.0	10	Rao (1998)
1561	1.20	11.9	Brady et al. (1997)

The right hand side of Equation 17.10 turns out to be negative, meaning that tension in the fiber leads to a reduction in the refractive index and hence the Bragg wavelength. However, this is more than compensated for by the physical elongation of the grating, so tensile strain leads to a net increase in Bragg wavelength. Table 17.1 lists the strain and temperature sensitivities of typical Bragg gratings operating in the three main telecommunication windows.

For many applications involving slowly varying strain or temperature, the typical resolutions required are in the region of 1 με or 0.1°C. In either case, this implies a measurement of the Bragg wavelength with a precision of the order of 1 pm, or roughly 1 part in 10^6. This is quite a demanding task and a great deal of effort has been devoted to the design of suitable interrogation schemes, which are described in Section 17.7.

17.6.2 Pressure

When FBGs are deployed as strain sensors, often, the fiber is fixed to the device being monitored in some way at two points either side of the grating itself, which leads to an imposed axial strain on the fiber, the radial strain then being determined by Poisson's ratio. FBGs have also been used as pressure sensors, which leads to an isotropic strain in the fiber causing a refractive index variation given by

$$\frac{1}{n}\frac{\partial n}{\partial P} = -\frac{n^2}{2E}(1-2\nu)(p_{11}+2p_{12}), \tag{17.11}$$

where E is the Young's modulus of the glass. Although this expression only strictly applies to static pressure, because of the small size of the FBG sensor, the equation is a good approximation for most acoustic frequencies of interest. The exception is MHz ultrasound that is relevant to medical applications (Fisher et al. 1998) where the wavelength may become comparable to the grating length, limiting omnidirectionality, or even to the fiber diameter at which point an anisotropic strain is set up (DePaula et al. 1982).

Unfortunately, the pressure sensitivity is not large, for example, around -3.6×10^{-9} nm/Pa in the 1550 nm window (Hill and Cranch 1999). For acoustic sensing, a common target specification is that of deep sea state zero (DSS0), which corresponds roughly to the acoustic noise level in the quieter parts of the oceans. DSS0 is defined by

$$\text{DSS0} = 90 - 16.65\log_{10}(\text{frequency})\ \text{dB relative to } 1\ \mu\text{Pa}/\sqrt{\text{Hz}}. \tag{17.12}$$

To detect pressures as low as this, a wavelength measurement resolution of 3.6×10^{-13} nm/$\sqrt{\text{Hz}}$ is required or around 2 parts in 10^{16}. To put this further in perspective, the equivalent strain resolution would be about 1 fε. We have already seen that many requirements for quasi-static monitoring demand a resolution of around 1 pm, which is readily achievable. For signal frequencies above around 100 Hz, much better resolutions are available, due to the much reduced influence of 1/*f* noise, with many systems operating in the 1–10 nε/√Hz region; detecting DSS0 requires an improvement on this of about 7 orders of magnitude. While this level of performance does not appear to be achievable using standard gratings, it can be approached using a fiber laser where the cavity is defined using Bragg gratings (Ball and Glenn 1992).

To achieve the highest pressure sensitivities, it is possible to use mechanical amplification by using a compliant coating surrounding the fiber. The correct choice of the Young's modulus and Poisson's ratio of the material* results in a significant increase in the Bragg wavelength shift: an enhancement of 25 dB has been reported with a 5 mm diameter coating of polyurethane (APT-Flex F17) (Hill et al. 1999). Housing the grating in a glass bubble has also been shown to increase the pressure sensitivity (Xu et al. 1996).

17.6.3 Magnetic Field

Although not usually sensitive enough for practical use, for completeness, here we discuss the FBGs sensitivity to magnetic fields. The transduction mechanism here is the Faraday effect (Wilson and Hawkes 1983), where the axial component of the magnetic field induces circular birefringence within the fiber. As a result of this, the Bragg wavelengths for left and right circularly polarized light change in the presence of a field, and hence, by monitoring the difference, the field strength may be deduced. The magnitude of the induced birefringence is proportional to the axial field strength, the interaction length, and the Verdet constant of the material. Kersey and Marrone have demonstrated with standard silica fiber a resolution of 4 gauss in a 0.15 Hz bandwidth using a high-sensitivity interferometric interrogation scheme (Kersey and Marrone 1994).

17.6.4 Cross-Sensitivity

While it may be the case that for temperature sensing applications, it is often possible to arrange for an FBG to be isolated from strain, the reverse is usually not possible and the cross-sensitivity of FBG strain sensors to temperature is a long-standing issue. The literature contains many ingenious solutions to this problem, though all unfortunately with flaws, which means there is no ideal solution. The general approach is to have two gratings experience the same temperature and strain environment but to somehow arrange for them to have different temperature and strain sensitivity coefficients.

In response to changes in temperature and strain of ΔT and $\Delta\varepsilon$, respectively, the wavelength shifts experienced by the two gratings can be written as

$$\frac{\Delta\lambda_1 = \Theta_1 \Delta T + S_1 \Delta\varepsilon}{\Delta\lambda_2 = \Theta_2 \Delta T + S_2 \Delta\varepsilon}, \tag{17.13}$$

* For hydrostatic boundary conditions, both Young's modulus and Poisson's ratio should be as small as possible.

where Θ and S are the temperature and strain coefficients for gratings 1 and 2. We therefore have two simultaneous equations in two unknowns, which are easily solved so long as the determinant

$$\begin{vmatrix} \Theta_1 & S_1 \\ \Theta_2 & S_2 \end{vmatrix} \tag{17.14}$$

is nonzero. The first demonstration of this technique made use of two FBGs at 850 and 1300 nm recorded in the same region of fiber (Xu et al. 1994). A later approach involved the use of a single grating interrogated at the Bragg wavelength (1561 nm) and at its first submultiple (789 nm*) (Brady et al. 1997). An alternative that avoids the need to work in widely different wavelength regions where the fiber may not always be single mode involves recording two gratings on either side of a splice between fibers of different diameters (James et al. 1996), or using two different types of FBG (Simpson et al. 2004).

The effectiveness of these different approaches to temperature/strain discrimination depends on how well conditioned is the system of equations (17.13). This concept can be understood graphically by viewing equations (17.13) as defining two straight lines in strain/temperature space (all other parameters taken as constant). The actual strain and temperature is determined by the intersection of these two lines. The system of equations is said to be ill-conditioned if the two lines have similar gradients and subtend a small angle; in this case, small inaccuracies in the determination of the strain and temperature coefficients or the wavelength shifts can produce a large error in the strain or temperature.

Unfortunately, none of the schemes just described results in particularly well-conditioned equations. Furthermore, they are all demanding in terms of either the sophistication of the signal processing or the assembly of the system. The pragmatic solution to temperature–strain discrimination when space is not at a premium is often simply to collocate two gratings, one of which experiences temperature and strain, the other being isolated from the effects of strain and yet still subject to temperature changes (Haran et al. 1998).

The development of photonic crystal (or microstructured) fiber has opened up new possibilities to deal with temperature cross-sensitivity problems. The holes in the fiber may be filled with a material whose properties counteract those of the glass, for example, a polymer, resulting in a much-reduced temperature sensitivity (Sorensen et al. 2006). Alternatively, if it is acceptable to use LPGs, then it has been shown that LPGs fabricated in photonic crystal fiber have a very small temperature sensitivity (Humbert et al. 2003).

17.7 Interrogation Systems

As discussed earlier, the need to resolve changes of temperature or strain of 0.1 K or 1 με requires the measurement of the Bragg wavelength to better than 1 part in a million. The desire to do this in a field-portable instrument has led to the development of a large number of signal processing schemes, each with its own advantages and disadvantages. Here, we describe the main classes of grating interrogation schemes.

* Note that due to dispersion, the two wavelengths do not differ by a factor of exactly 2.

17.7.1 Filters

17.7.1.1 Fabry–Pérot Filter

The Fabry–Pérot interferometer (Hecht and Zajac 1974) offers the highest optical resolution of all the interferometers and can be implemented in a variety of formats. The scanning Fabry–Pérot filter (SFPF) used both in optical fiber communications and sensors consists of two closely spaced highly reflective flat mirrors, one of which can be moved along the optical axis using a piezoelectric actuator. In a typical SFPF, light is guided to the input mirror by a monomode fiber and collimated with a miniature lens before entering the SFPF, the output light from the SFPF is then launched into another monomode fiber. The diameter of the collimated beam entering the fiber will typically be less than a millimeter, and so the flatness over the beam diameter and the consequent finesse and resolution can be very high. In most sensing and communications applications, the *free spectral range* (FSR) is chosen to be approximately the same as the bandwidth of the illuminating source, typically of the order of 50 nm, in order to maximize the resolution, while preventing measurement errors caused by the presence of two transmission peaks in the source bandwidth. SFPFs with low loss coatings have been reported with resolutions of ~0.01 nm. The simplest configuration for using an SFPI to interrogate an FBG is shown in Figure 17.8.

Light from the broadband source would typically have a bandwidth of the order of 50 nm. After reflection by the grating, the light returned through the coupler to the SFPI will have a bandwidth determined by the grating, typically around 0.2 nm. The SFPI is then used to monitor the Bragg wavelength in one of two ways. Either the SFPI transmission peak is repeatedly scanned over the FSR of the interferometer, in which case a signal is detected by the photodetector each time the scanned wavelength matches that of the grating, or the SFPI pass band is locked to the grating utilizing the dither signal shown in Figure 17.8, leading to submicrostrain resolution (Kersey et al. 1993).

Note that the behavior of the system in Figure 17.8 in the spectral domain can be understood as a series of multiplications: the resulting spectrum is obtained by multiplying the source spectrum by the reflectivity profile of the FBG followed by the transmission spectrum of the SFPI. Because the order of terms in a multiplication of this kind does not affect the answer, we can deduce that the SFPI can be placed before the FBG if desired (e.g., it could be placed between the source and the coupler) without affecting the outcome. As we shall see later, alternative arrangement opens up possibilities for sensor multiplexing.

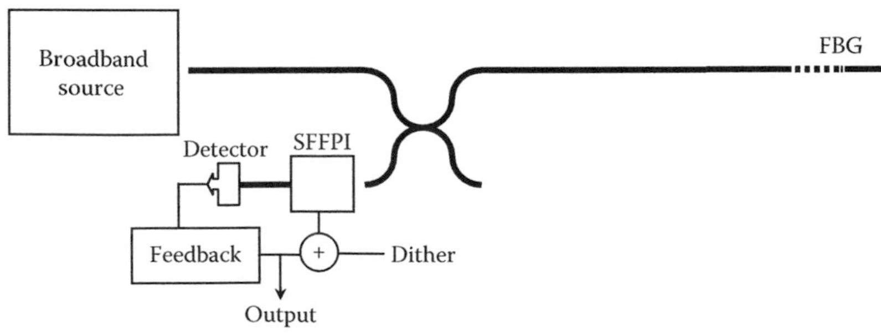

FIGURE 17.8 The use of an SFPI to interrogate a single FBG sensor.

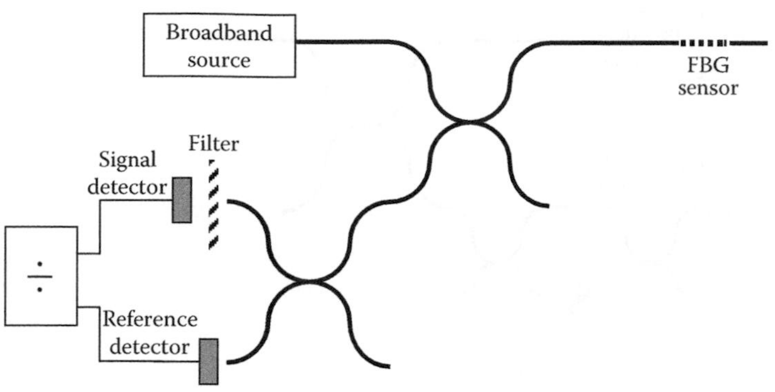

FIGURE 17.9 Edge filter interrogation with intensity referencing.

17.7.1.2 Edge Filter

The idea here is to convert a shift in grating wavelength into a change in intensity by passing the light reflected back by the grating through a filter where ideally, the transmission is a linear function of wavelength (Melle et al. 1992). Typically, this is achieved by using the edge of a filter stop band—hence the name edge filter demodulation. One difficulty with this approach is that the output is sensitive to changes in intensity caused by source power fluctuations and varying losses in the system, and not just the grating wavelength; this is effectively abandoning one of the key features of FBG sensors: that they are not normally affected by intensity changes. This problem can however be overcome by referencing the filtered signal against the power received before the filter. An example is shown in Figure 17.9, where a second coupler is used to sample the light returned by the grating before it reaches the filter. The ratio of the two detector signals gives a normalized signal that depends only on the instantaneous wavelength of the light reflected from the FBG.

Using the edge filter approach, James et al. were able to monitor strain transients in a gun barrel with sample times of less than 100 μs (James et al. 1999). A variation on this approach removes the need for the discrete filter by using the wavelength response of the coupler itself. Davis and Kersey (1994) reported a system in which a broadband superfluorescent fiber source was used to illuminate an FBG and a coupler with a monotonic variation between the coupling ratio in the range 1500–1570 nm; the system provided a strain resolution of about ±5 με.

17.7.2 Interferometric Wavelength Shift Detection

An unbalanced two-beam interferometer, such as a Michelson or Mach–Zehnder design, can be used to provide high-resolution wavelength recovery. An example arrangement for this is shown in Figure 17.10.

The FBG reflects back a slice of the source spectrum to the interferometer. We then consider two scenarios. First, let us assume the FBG bandwidth is narrow, in that the coherence length of the light reflected from the grating is larger than the imbalance in the interferometer. Interference therefore occurs between the light in the two arms of the interferometer, and the emerging power in one of the output fibers is a sinusoidal

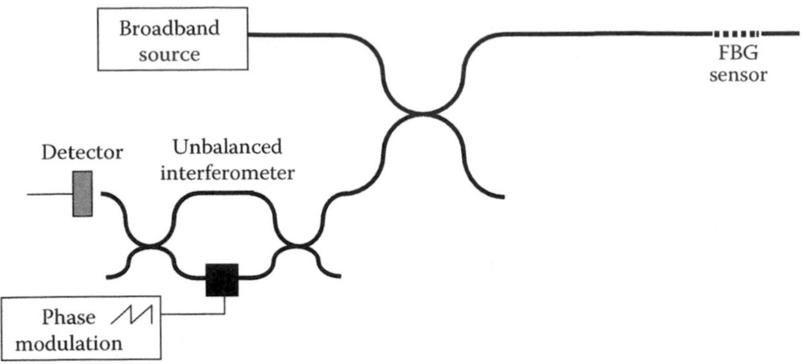

FIGURE 17.10 Interferometric wavelength shift detection.

function of the phase of the interferometer, which in turn depends on the Bragg wavelength of the grating, λ_B.

The relationship between the interferometer phase, $\Delta\varphi$, and the optical path difference in the interferometer, nL, where n is the refractive index of the fiber, is given by

$$\varphi = \frac{2\pi nL}{\lambda_B}. \tag{17.15}$$

In response to a shift in the Bragg wavelength of $\Delta\lambda_B$, the magnitude of the change in phase is obtained by differentiating Equation 17.15 to give

$$\Delta\varphi = \frac{2\pi nL\Delta\lambda_B}{\lambda_B^2}. \tag{17.16}$$

It is readily seen that the sensitivity to very small wavelength changes can be improved by increasing L. An upper limit on this is provided by the coherence length of the light reflected by the Bragg grating, determined by the grating's bandwidth.

The interferometric phase is not directly measured; instead, we observe the power resulting from the two-beam interference:

$$I = I_0(1 + V\left(\cos\left(\varphi_0 + \Delta\varphi\right)\right) = I_0(1 + V\left(\cos\left(\varphi_0 + k\Delta\lambda_B\right)\right), \tag{17.17}$$

where

- I_0 is a constant related to source power
- V is the fringe visibility
- $k = 2\pi nL/\lambda_B^2$
- φ_0 is an arbitrary phase

To recover $\Delta\lambda_B$ from a measurement of I when the wavelength shift is small, it is possible to adjust the interferometer to a quadrature point ($\varphi_0 = \pi/2 + n\pi$, where n is an integer), in which case ($\cos(\varphi_0 + k\Delta\lambda_B) \approx k\Delta\lambda_B$) and the output power is proportional to $\Delta\lambda_B$.

In addition to requiring $k\Delta\lambda_B \ll 1$, this approach makes wavelength demultiplexing multiple sensors difficult. A common approach is to arrange to linearly scan the interferometer phase using a phase modulator or acousto-optic modulator (Webb et al. 1997). If the phase is scanned over 2π in time T, the power output from the interferometer takes the form

$$I = I_0(1+V(\cos(\omega t + k\Delta\lambda_B))), \quad (17.18)$$

where $\omega = 2\pi f = 2\pi/T$. After photodetection, we have an electric carrier at frequency f that is phase modulated by the measurand-induced wavelength shift. It is then possible to recover the measurand using conventional phase demodulation techniques, such as through the use of a phase-locked loop. When detecting periodic signals with characteristic frequencies above 100 Hz, resolutions tend to be in the order of 1 nε/√Hz (Kersey et al. 1992). Quasi-static signals are more problematic as it is difficult to prevent slow drifts within the interferometer, which are indistinguishable from changes in the Bragg wavelength. It is however possible to use a stabilized reference grating to compensate for the drift (Kersey et al. 1993).

17.7.3 Spectrometers

17.7.3.1 Wavelength Domain Spectrometry

In the laboratory, the measurement of the Bragg wavelength is often done using an optical spectrum analyzer (OSA). These are very expensive, bulky, and usually somewhat fragile items of test equipment, ill suited to measurements in the field. Rather more rugged and cheaper miniature spectrometers are available from a number of suppliers, and these have been successfully utilized for grating interrogation (Askins et al. 1995). These spectrometers do not have such high resolution as an OSA; however, it has been shown that by curve fitting or the use of a centroid algorithm, resolutions around 100th of the pixel-to-pixel wavelength shift are achievable (Ezbiri et al. 1998). Disadvantages of this approach are the expense of CCD arrays operating at wavelengths above 1 μm and the relatively low data rate normally achievable. This latter disadvantage is starting to disappear; Ibsen Photonics currently produce a model capable of operation up to 17 kHz (Ibsen 2012).

17.7.3.2 Fourier Transform Spectroscopy

Fourier transform spectroscopy (FTS) is a complementary approach to wavelength domain spectroscopy. Typically, FTS is based on a Michelson two-beam interferometer. Light reflected by the sensing grating passes through the interferometer and onto a detector. The path difference in the interferometer is scanned and this results in a periodic variation in received intensity—a fringe pattern—with the period determined by the mean grating wavelength. Multiple gratings at different wavelengths can be interrogated, in which case a complex fringe pattern is obtained, the Fourier transform of which reveals the location of each grating in wavelength space. FTS has a signal-to-noise ratio advantage over a scanned wavelength domain spectrometer since all wavelengths are detected simultaneously and hence no light is *thrown away* (Fellgate 1951).

Initially based on a scanned bulk optic Michelson, all-fiber versions have been demonstrated where the path difference is scanned by stretching the fiber using a PZT

(Davis and Kersey 1995, O'Mahoney et al. 2009). The resolution measured by analyzing the light back-reflected from an FBG illuminated by a superfluorescent fiber source was 0.07 cm^{-1} equivalent to a wavelength shift of 16 pm, which corresponds to a strain resolution of about 12 με. More recently, a grating interrogation system based on an FTS without any scanning element has been realized. This makes use of a bulk optic Michelson interferometer where one of the mirrors is tilted so that the optical path difference varies across a linear CCD array, on which the fringes are observed. A spectral resolution of 0.025 nm was obtained for a spatially scanned optical path delay of only 200 μm (Murphy et al. 2001).

17.7.3.3 Acousto-Optic Tunable Filter

The acousto-optic tunable filter (AOTF) utilizes a MHz frequency traveling acoustic wave to couple light between two polarization states in a crystal. Only light close to a well-defined wavelength determined by the precise acoustic frequency is efficiently coupled and allowed to leave the modulator. The device thus acts as a tunable filter and may be used in a similar fashion to the SFPI in the system shown in Figure 17.8, when a resolution better than 1 με was obtained (Geiger et al. 1995). The AOTF is a relatively fast device and wavelength changes can be made in microseconds enabling high-bandwidth measurands to be followed. Excitation at multiple frequencies allows several gratings to be interrogated simultaneously (Norman and Webb 2007).

17.7.4 Tunable Optical Source

All the approaches described so far have used a broadband light source, coupled with some kind of spectral filter to measure the Bragg wavelength. It is also possible to use a tunable optical source, either to sweep through the wavelength space of the system or to lock directly to the Bragg wavelength (Lissak et al. 1998), where a strain resolution of 45 pε/√Hz at 3 kHz was obtained.

17.7.5 FBG Lasers

If a pair of gratings (or a single π-shifted grating) are recorded in an amplifying fiber, they can act as cavity mirrors, allowing a fiber laser to be created. Such lasers can possess very narrow linewidths of 10 kHz or less (Sejka et al. 1995), which is more than a factor of 10^{10} less than the optical frequency, allowing the optical frequency to be monitored with great precision. This is usually carried out using interferometric wavelength shift detection (see Section 17.7.2) (Koo and Kersey 1995). Remember that the sensitivity of this approach is proportional to the OPD in the interferometer and this is limited by the coherence length of the light. When the effective light source is a standard Bragg grating with a bandwidth of around 0.2 nm, the coherence length would only be about a centimeter. Because we now have such a narrow linewidth source, OPDs of hundreds of meters become feasible. Indeed a demodulation system with an OPD of 100 m was shown to provide the sensor with a strain resolution of just $5 \times 10^{-14}/\sqrt{Hz}$, achieved by monitoring wavelength shifts of 8×10^{-11} nm/√Hz (Koo and Kersey 1995).

17.8 Demultiplexing

In order to simplify the descriptions of some of the interrogation techniques described in the previous section, we have mostly concentrated on systems with just one sensor. However, the ability to incorporate many sensors in one system is a key feature of the technology and helps to reduce the per-sensor cost of a complete system. There are three generic approaches.

17.8.1 Wavelength Division Multiplexing

Most of the interrogation techniques that incorporate a broadband source can be used to interrogate multiple sensors of different wavelengths on a single fiber. For example, the scanning Fabry–Pérot filter or spectrometer-based approaches cope very well with multiple sensors. A key parameter with such systems is the number of sensors that can be interrogated and the important point here is that it is essential to ensure that the wavelengths of any two sensors never overlap, which would render it impossible to determine which one is which. An upper limit on the number of sensors is obtained by dividing the source bandwidth by the working range of each sensor. For example, if we take 40 nm as a typical bandwidth of a broadband source in the 1550 nm region and want our sensors to each be capable of measuring ±2000 με, corresponding to a roughly 4 nm wavelength range, then we can cope with a maximum of 10 sensors in the system. In practice, it may be prudent to leave a slightly larger gap between sensor wavelengths, reducing the number slightly.

17.8.2 Spatial Division Multiplexing

The previous section has shown that wavelength division multiplexing alone cannot generally permit more than around a dozen sensors to be used. When there is sufficient optical power, the number can be extended through the use of spatial division multiplexing. This approach utilizes the fact that with the techniques that use some kind of tunable filter, such as the SFPI interrogation approach shown in Figure 17.8, the filter can be placed between the source and the grating, rather than between the grating and the detector. By doing this, it is possible to split the power coming from the filter, directing it into a number of fibers each containing a string of WDM gratings. Each string comprises gratings of different wavelengths, but each string contains gratings with nominally the same wavelengths as any other string. What stops the gratings being confused is that light from each string is directed back to a different photodetector. Figure 17.11 shows one example where 64 sensors were addressed providing 32 temperature-compensated measurement points (Henderson et al. 1999). The number of strings that can be addressed in this way depends on the source power and desired signal-to-noise ratio but can result in an order of magnitude more sensors than from a simple WDM system.

In situations where the number of sensors is limited by the available source power, instead of using a coupler tree to distribute power among the strings of gratings, it is possible to use optical switches to distribute the power. In one example, a 1×4 and a 1×2 switch were used to distribute power to 60 FBGs, each providing a resolution of 1 με (Davis et al. 1996). This approach is generally only suitable for applications that

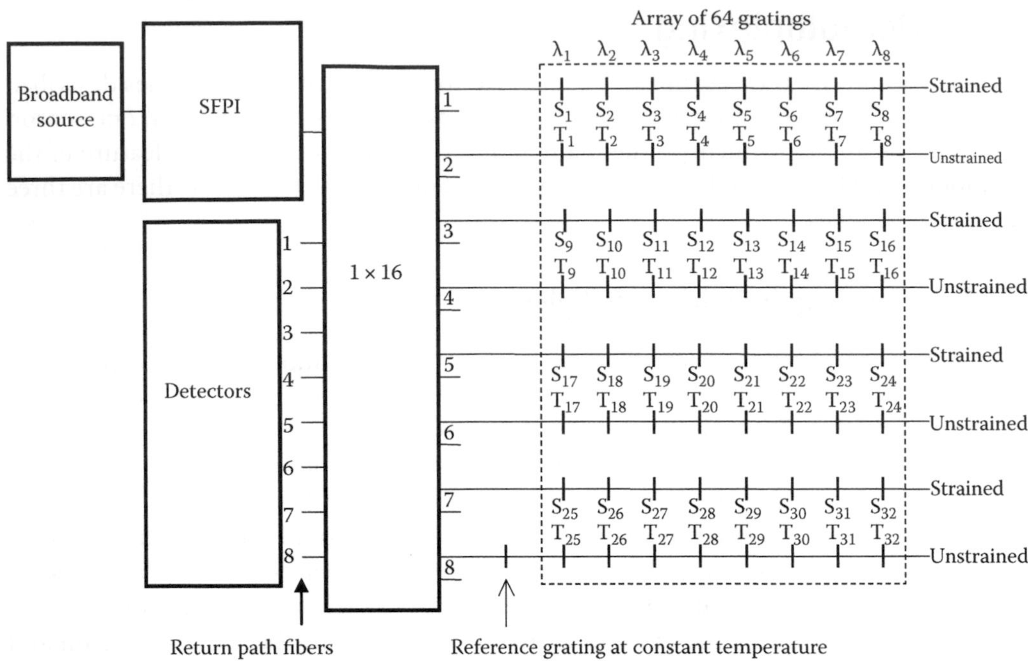

FIGURE 17.11 Combination of WDM and SDM addressing 64 sensors with an additional reference grating.

require a very slow switching speed; otherwise, operation over weeks and months can exceed the lifetime number of switching operations guaranteed for the switch.

17.8.3 Time Division Multiplexing

When the optical source emits short pulses, it is possible to discriminate between similar FBGs based on the difference in arrival time at the detector of pulses of light reflected from each grating (Berkoff et al. 1995). This approach allows sets of FBGs with similar wavelengths to be distributed along a single optical fiber, though in this case, low-reflectivity gratings are required to prevent nearer gratings from *stealing* all the light from those of a similar wavelength that are further away. Although such systems are highly efficient in that only a single sensing fiber is required, this brings with it the disadvantage that a break in the fiber close to the interrogation unit could mean the loss of all sensors in the system. In contrast, with the SDM approach described earlier, the breakage of one sensing fiber can at most result in the failure of one sensor string.

An interesting variant on the TDM approach illustrated in Figure 17.12 was commercialized by Indigo Photonics, and is currently used by Moog (Lloyd et al. 2004). All the low-reflectivity (~4%) sensing gratings have nominally the same wavelength. The system operates by essentially mode locking a laser cavity formed by a chosen sensor grating, the semiconductor optical amplifier, and a broadband reflecting grating, with the light being directed to a spectrometer for readout (the commercialized version uses a bespoke miniature spectrometer). Different sensing gratings are selected by adjusting the frequency at which the current to the semiconductor optical amplifier (SOA) is

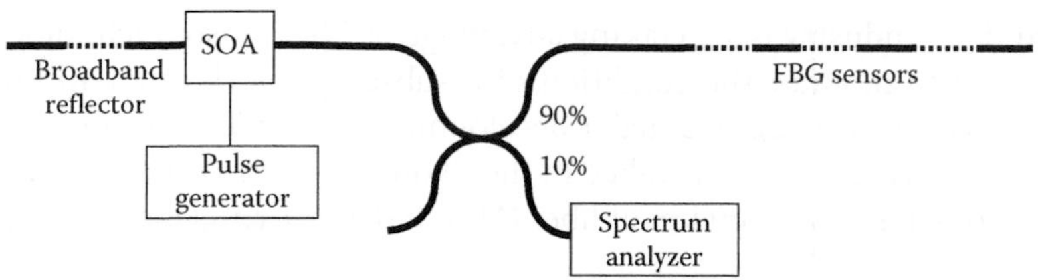

FIGURE 17.12 Indigo Photonics interrogation system based on mode locking a laser cavity using a SOA.

pulsed to provide optical amplification. Grating separation needs to be at least around a meter to avoid cross-talk. Tens of sensors can be operated over several kilometers with a resolution of around a microstrain.

17.8.4 Combined Systems and Other Demultiplexing Techniques

Where the largest possible number of sensors must be addressed, the different multiplexing approaches just described can be combined; for example, spatial, wavelength, and time division multiplexing have all been employed in a single system (Rao et al. 1996).

17.9 Applications

FBG sensors have some unique features and applications that take advantage of these tend to be the ones with most chance of success. The key properties of FBG sensors are

- Low fiber loss
- Multiplexing capability
- Small size
- Immunity to electromagnetic interference

We will look at each of these properties in turn and see how they are utilized in sensing systems.

17.9.1 Low Fiber Loss and Multiplexing Capability

We combine these two advantages as they often feature together when FBGs are used to monitor large engineering structures using tens or even hundreds of sensors (Majumder et al. 2008). The low fiber loss permits sensors to be located many kilometers from the interrogation system, without the need for intermediate amplification, such as would be required with conventional electrical strain gauges. This feature has been exploited for a number of civil and marine infrastructure projects. The world's largest combined road and rail suspension bridge—the 1377 m long Tsing Ma bridge in Hong Kong—has been instrumented with 40 FBG sensors, installed on the hanger cables, rocker bearings, and truss girders (Chan et al. 2006). Also in Hong Kong, FBG strain sensors are in use, monitoring the passage of trains on the 36 km Kowloon-Canton Railway Corporation East Rail Link (East Rail Link 2012). FBG strain sensors have also been used to monitor dams (Bronnimann et al. 1998).

The oil and gas industry is also taking advantage of FBG sensor technology. Gratings have been used to monitor the condition of a subsea pump located 10 km from the platform (SmartFibers 2012). In a major development, Optoplan has installed a seismic sensing array covering 60 km^2 of seabed in the Ekofisk oil field comprising 24,000 FBG-based interferometers and 3500 km of fiber (Nakstad and Kringlebotn 2008).

17.9.2 Small Size

The small size of the optical fiber—typically 125 μm in diameter, though smaller fibers are available—allows the sensors to be embedded in other materials with little effect on the material's properties. This permits the creation of smart structures whose behavior can be easily monitored without the risk of damage that might occur with surface-mounted sensors. This feature is considered particularly important when applied to modern composite materials (Measures 1992). Sensors have been used for assessing composite cure (O'Dwyer et al. 1998) and monitoring strain in yacht masts (Read et al. 1999), ship hulls (Johnson et al. 2000), and concrete strengthening wraps (Seim et al. 1999). Although slow to be taken up because of the very conservative nature of the industry, FBG sensors seem likely to have a big future in monitoring aircraft structures, which are increasingly made of composites. Here the sensors bring the additional advantages of low weight and immunity to electromagnetic interference, particularly from lightning strikes. In addition to their use with modern structural materials, FBG sensors have also been embedded in concrete by attachment to the rebars (Maaskant et al. 1997).

The small size of the optical fiber suggests that FBG sensors should be suitable for in vivo sensing, for example, during minimally invasive surgery. The first demonstration of in vivo sensing involved monitoring the heat inside an organ during thermal therapy (Webb et al. 2000), and FBG sensors have also been proposed as candidates for thermodilution measurements of blood flow (Rao et al. 1997).

A final advantage of the small fiber size is that the sensors can be very inconspicuous; this property has been taken advantage of in a project to sense the movement of historic buildings, where it is important that the sensors do not distract from the visual aesthetics of the structure (Lima et al. 2008).

17.9.3 Immunity to Electromagnetic Interference

Optical fiber is composed of a dielectric material, and it is therefore nonconductive and essentially immune to electromagnetic interference. This last point deserves a little qualification, as already noted, light traveling in a fiber is subject to the Faraday effect in the presence of an axial magnetic field (Wilson and Hawkes 1983); however, this small sensitivity never seems to pose any problem in practice. FBG sensors have been used to detect hot spots in the stator of an electricity generator (Theune et al. 2000). In addition to providing immunity from the intense electromagnetic fields present, the dielectric fiber prevents the possibility of an electrical connection from the control room to the high voltages present in the generator. FBG sensors have also been used in a magnetic resonance imaging environment (Park et al. 2010) and for sensing strain on a MAGLEV train (Kang and Chung 2009). The absence of any electrical current in an FBG sensor means that such devices are well suited to use in potentially explosive environments such as fuel tanks or for monitoring hydrogen (Caucheteur et al. 2008).

17.10 Commercial Landscape

FBG sensors were first commercialized around 1995 by 3 M and Photonetics. Since that time, the commercial activity in this field has grown significantly in three broad areas. First, there are the sensors themselves, available either as simple fiber devices or else packaged for ease of use or to render them sensitive to parameters other than strain or temperature. For example, SmartFibres Ltd produces a range of strain, temperature, and pressure probes for various applications (SmartFibres 2012). Different versions of the strain sensors may be glued or welded to the underlying structure.

The second area of commercial activity relates to the readout units required to interrogate the sensors. As described earlier, a variety of ways of measuring the Bragg wavelength of one or more gratings have been developed and many of these have been commercialized. At one end of the scale are sophisticated—and consequently expensive—units, often based on tunable filters or sources capable of addressing many sensors. An example here is the Micron Optics SM125 unit, which can address 16 channels each with many sensors (Micron Optics 2012). At the other end and with a price more than an order of magnitude cheaper are devices capable of interrogating just one or two gratings, such as the FCG transceiver from Redondo Optics (Redondo Optics 2012).

The third area of commercial activity relates to companies selling complete sensing solutions for particular applications. An example here is Moog's wind turbine rotor monitoring system (Mooginfo 2012). Such complete system installations can appear attractive to clients who simply want to obtain the key information needed to run their facility optimally, but have no interest in what technology is used to generate that information and indeed may well have no interest in developing technical capability with that technology.

It is difficult to find reliable data quantifying the size of the FBG sensor market. A review of activity back in 2007 estimated the market as being around 100 million USD and growing at 20% (Mendez 2007), so it is probably a safe bet to conclude that in 2012, the market should be over 200 million USD.

17.11 Conclusions and Current Research Directions

In this chapter, we have surveyed a technology that has grown over the last 20 years to produce a significant level of commercial activity that has been increasing at a rate that has significantly outstripped general economic growth. Despite this increasing commercial maturity, as was described in the introduction, FBG sensors also continue to grow as a research field.

While there are now many publications describing industrial applications and field trials of sensing systems, there continues to be a sizable number of papers dealing with more fundamental aspects. There are some very important sensing requirements, which currently cannot adequately be met due to the harsh nature of the environment, particularly concerning high temperature operation, for example, in aircraft engines or process control applications. It has recently been discovered that when *standard* type I gratings are annealed at high temperature, a new grating arises resulting from structural transformations to the glass caused by the inscription of the original grating (Canning et al. 2009). These regenerated, or type R, gratings can be used at temperatures in excess of 1000°C and are currently a *hot topic* in Bragg grating research; they can even be made on the draw tower as the fiber is produced (Lindner et al. 2011).

Another area enjoying a high level of research activity concerns the use of ultra short (fs) laser pulses for grating inscription. The high intensities typically provided by these pulses facilitate multiphoton interactions, allowing modification of any material, not just those displaying conventional photosensitivity. FBGs have been written in pure silica photonic crystal fiber (Mihailov et al. 2004), which offers another approach to creating sensors useable up to around 1000°C. For even higher temperature applications, well beyond the softening point of silica, crystalline *fibers* may be employed. Gratings fabricated by an fs laser in thin sapphire rods were usable up to 1500°C (Grobnic et al. 2004).

There are further advantages to be gained by recording gratings in materials other than silica. As an example, gratings can now be written in both step-index (Xiong et al. 1999) and microstructured (Dobb et al. 2005) polymer optical fiber (Webb and Kalli 2010). Such fiber can easily survive very high strains and is much more compliant than silica, therefore perturbing any flexible structure to which it is affixed much less. Some polymers have an affinity for water, which has been shown to open up the possibility of the direct detection of water in fuel using an FBG sensor (Zhang et al. 2009).

References

Anderson D. Z., V. Mizrahi, T. Erdogan et al. 1993. Production of in-fibre gratings using a diffractive optical element. *Electronics Letters* 29: 566–568.

Ansari F. (ed.). 1998. *Fiber Optic Sensors for Construction Materials and Bridges*. Technomic Publishing Co. Inc., Lancaster, PA.

Archambault J.-L., L. Reekie, and P. St. J. Russell. 1993. 100% reflectivity Bragg reflectors produced in optical fibres by single excimer laser pulses. *Electronics Letters* 29: 453–454.

Askins C. G., M. A. Putna, and E. J. Friebele. 1995. Instrumentation for interrogating many-element fiber Bragg grating arrays embedded in fiber/resin composites. *Proceedings SPIE* 2444: 257–266.

Asseh A., H. Storoy, J. T. Kringlebotn et al. 1995. 10 cm Yn^{3+} DFB fiber laser with permanent phase-shifted grating. *Electronics Letters* 31: 969–970.

Ball G. A. and W. H. Glenn. 1992. Design of a single-mode linear-cavity erbium fiber laser utilizing Bragg reflectors. *IEEE Journal of Lightwave Technology* 10: 1338–1343.

Bennion I., J. A. R. Williams, L. Zhang et al. 1996. UV-written in-fibre Bragg gratings. *Optical and Quantum Electronics* 28: 93–135.

Berkoff A., M. A. Davis, D. G. Bellemore et al. 1995. Hybrid time and wavelength division multiplexed fiber grating array. *Proceedings of Smart Structures and Materials, SPIE* 2444: 288–295.

Bernardin J. P. and N. M. Lawandy. 1990. Dynamics of the formation of Bragg gratings in germanosilicate optical fibres. *Optics Communications* 79: 194–199.

Brady G. P., K. Kalli, D. J. Webb et al. 1997. Simultaneous measurement of strain and temperature using the first- and second-order diffraction wavelengths of Bragg gratings. *IEE Proceedings-Optoelectronics* 144: 156–161.

Bronnimann R., P. M. Nellen, P. Anderegg et al. 1998. Application of optical fiber sensors on the power dam of Luzzone. *International Conference on Applied Optical Metrology*, Balatonfured, Hungary.

Butter C. D. and G. B. Hocker. 1978. Fiber optics strain gauge. *Applied Optics* 17: 2867–2869.

Campbell R. J. and R. Kashyap. 1991. Spectral profile and multiplexing of Bragg gratings in photosensitive fiber. *Optics Letters* 16: 898–900.

Canning J., S. Bandyopadhyay, M. Stevenson et al. 2009. Regenerated gratings. *Journal of the European Optical Society-Rapid Publications* 4: 09052.

Canning J. and M. G. Sceats. 1994. Pi-phase shifted periodic distributed structures in optical fibers by UV post-processing. *Electronics Letters* 30: 1344–1345.

Caucheteur C., M. Debliquy, D. Lahem et al. 2008. Catalytic fiber Bragg grating sensor for hydrogen leak detection in air. *Photonics Technology Letters* 20: 96–98.

Chan T. H. T., L. Yu, T. Ling et al. 2006. Fiber Bragg grating sensors for structural health monitoring of Tsing Ma bridge: Background and experimental observation. *Engineering Structures* 28: 648–659.

Cusano A., D. Paladino, and A. Iadicicco. 2009. Microstructured fiber Bragg gratings. *Journal of Lightwave Technology* 27: 1663–1697.

Davis M. A., D. G. Bellemore, M. A. Putman et al. 1996. A 60 element fibre Bragg grating sensor system. *11th International Conference on Optical Fiber Sensors*, Sapporo, Japan, pp. 100–103.

Davis M. A. and A. D. Kersey. 1994. All fiber Bragg grating strain sensor demodulation technique using a wavelength division coupler. *Electronics Letters* 30: 75–76.

Davis M. A. and A. D. Kersey. 1995. Application of a fibre Fourier transform spectrometer to the detection of wavelength encoded signals from Bragg grating sensors. *Journal of Lightwave Technology* 13: 1289–1295.

DePaula R. P., L. Flax, J. H. Cole et al. 1982. Single-mode fiber ultrasonic sensor. *Journal of Quantum Electroncics QE* 18: 680–683.

Dobb H., K. Kalli, and D. J. Webb. 2004. Temperature-insensitive long period grating sensors in photonic crystal fibre. *Electronics Letters* 40: 657–658.

Dobb H., D. J. Webb, K. Kalli et al. 2005. Continuous wave ultraviolet light-induced fiber Bragg gratings in few- and single-mode microstructured polymer optical fibers. *Optics Letters* 30: 3296–3298.

Dong L., J.-L. Archambault, L. Reekie et al. 1993. Single pulse Bragg gratings written during fibre drawing. *Electronics Letters* 29: 1577–1578.

East Rail Link. http://www.micronoptics.com/uploads/library/documents/FBGs_Rail_Monitoring.pdf. Accessed April 10, 2012.

Ezbiri A., S. E. Kanellopoulos, and V. A. Handerek. 1998. High resolution instrumentation system for fibre-Bragg grating aerospace sensors. *Optics Communications* 150: 43–48.

Fellgate P. 1951. On the theory of infra-red sensitivities and its application to the investigation of stella radiation in the near infra-red. PhD thesis, Cambridge University, Cambridge, U.K.; also see 1958, *Journal of Physics and Radiation* 19: 187.

Fisher N. E., D. J. Webb, C. N. Pannell et al. 1998. Ultrasonic hydrophone based on short in-fiber Bragg gratings. *Applied Optics* 37: 8120–8128.

Geiger H., M. G. Xu, N. C. Eaton et al. 1995. Electronic tracking system for multiplexed fibre grating sensors. *Electronics Letters* 31: 1006–1007.

Giles C. R. 1997. Lightwave applications of fiber Bragg gratings. *Journal of Lightwave Technology* 15: 1391–1404.

Grobnic D., S. J. Mihailov, C. W. Smelser et al. 2004. Sapphire fiber Bragg grating sensor made using femtosecond laser radiation for ultrahigh temperature applications. *Photonics Technology Letters* 16: 2505–2507.

Hand D. P. and P. St. J. Russel. 1990. Photoinduced refractive index changes in germanosilicate fibres. *Optics Letters* 15: 144–146.

Haran F. M., J. K. Rew, and P. D. Foote. 1998. A strain-isolated fibre Bragg grating sensor for temperature compensation of fibre Bragg grating strain sensors. *Measurement Science and Technology* 9: 1163–1166.

Hecht E. and A. Zajac. 1974. *Optics.* Addison-Wesley, Reading, MA, Chapter 9.

Henderson P. J., D. A. Jackson, L. Zhang et al. 1999. Highly-multiplexed grating-sensors for temperature-referenced quasi-static measurement of strain in concrete bridges. *13th International Conference on Optical Fiber Sensors*, Kyongju, Korea, Vol. 3746, pp. 320–323.

Hill D. J. and G. A. Cranch. 1999. Gain in hydrostatic pressure sensitivity of coated fibre Bragg grating. *Electronics Letters* 35: 1268–1269.

Hill D. J., P. J. Nash, D. A. Jackson et al. 1999. A fiber laser hydrophone array. *Proceedings of SPIE* 3860: 55–66.

Hill K. O., Y. Fujii, D. C. Johnson et al. 1978. Photosensitivity on optical fiber waveguides: Application to reflection filters fabrication. *Applied Physics Letters* 32: 647–649.

Hill K. O., B. Malo, F. Bilodeau et al. 1993a. Bragg gratings fabricated in monomode photosensitive optical fiber by UV exposure through a phase mask. *Applied Physics Letters* 62: 1035–1037.

Hill K. O., B. Malo, F. Bilodeau et al. 1993b. Photosensitivity in optical fibers. *Annual Review Material Science* 23: 125–157.

Hill K. O. and G. Meltz. 1997. Fiber Bragg grating technology fundamentals and overview. *Journal Lightwave Technology* 15: 1263–1276.

Humbert G., A. Malki, S. Fevrier et al. 2003. Electric arc-induced long-period gratings in Ge-free air-silica microstructure fibres. *Electronics Letters* 39: 349–350.

Ibsen. http://www.ibsen.dk/im/I-MON%20256%20High%20Speed. Accessed April 9, 2012.

James S. and R. Tatam. 2003. Optical fibre long-period grating sensors: Characteristics and application. *Measurement Science and Technology* 14: R49–R61.

James S. W., M. L. Dockney, and R. P. Tatam. 1996. Simultaneous independent temperature and strain-measurement using in-fiber Bragg grating sensors. *Electronics Letters* 32: 1133–1134.

James S. W., R. P. Tatam, S. R. Fuller et al. 1999. Monitoring transient strains on a gun barrel using fibre Bragg-grating sensors. *Measurement Science and Technology* 10: 63–67.

Jenkins F. A. and H. E. White. 1981. *Fundamentals of Optics*. International Edition, McGraw-Hill, New York, pp. 301–302.

Johnson G. A., K. Pran, G. Sagvolden et al. 2000. Surface effect ship vibro-impact monitoring with distributed arrays of fiber Bragg gratings. *Proceedings IMAC-XVIII: A Conference on Structural Dynamics*, San Antonio, TX, pp. 1406–1411.

Kang D. and W. Chung. 2009. Integrated monitoring scheme for a maglev guideway using multiplexed FBG sensor arrays. *NDT and E International* 42: 260–266.

Kashyap R. 1999. *Fiber Bragg Gratings*. Academic Press, San Diego, CA.

Kawasaki B. S., K. O. Hill, D. C. Johnson et al. 1978. Narrow-band Bragg reflectors in optical fibers. *Optics Letters* 3: 66–68.

Kersey A. D., T. A. Berkoff, and W. W. Morey. 1992. High resolution fiber Bragg grating based strain sensor with interferometric wavelength shift detection. *Electronics Letters* 28: 236–237.

Kersey A. D., T. A. Berkoff, and W. W. Morey. 1993a. Fiberoptic Bragg grating strain sensor with drift-compensated high resolution interferometric wavelength-shift detection. *Optics Letters* 18: 72–74.

Kersey A. D., T. A. Berkoff, and W. W. Morey. 1993b. Multiplexed fiber Bragg grating strain-sensor system with a fiber Fabry-Peror wavelength filter. *Optics Letters* 18: 1370–1372.

Kersey A. D., M. A. Davis, H. J. Patrick et al. 1997. Fiber grating sensors. *IEEE Journal Lightwave Technology* 15: 1442–1462.

Kersey A. D. and M. J. Marrone. 1994. Fiber Bragg grating high-magnetic-field probe. *Tenth International Conference on Optical Fibre Sensors*, Glasgow, U.K. *Proceedings of SPIE*, Vol. 2360, pp. 53–56.

Kogelnik H. 1976. Filter response of nonuniform almost-periodic structures. *Journal Bell Systems Technology* 55: 109–126.

Koo K. P. and A. D. Kersey. 1995a. Bragg grating based laser sensors systems with interferometric interrogation and wavelength division multiplexing. *Journal of Lightwave Technology* 13: 1243–1249.

Koo K. P. and A. D. Kersey. 1995b. Fiber laser sensor with ultrahigh strain resolution using interferometric interrogation. *Electronic Letters* 31: 1180–1182.

LeBlanc M., S. Huang, M. Ohn et al. 1996. Distributed strain measurement based on a fiber Bragg grating and its reflection spectrum analysis. *Optics Letters* 21: 1405–1407.

Lemaire P. J. 1995. Enhanced UV photosensitivity in fibres and waveguides by high-pressure hydrogen loading. *Conference on Optical Fibre Communication*, San Diego, CA. *Technical Digest* WN5, pp. 162–163.

Lemaire P. J., R. M. Atkins, V. Mizrahi et al. 1993. High pressure H_2 loading as a technique for achieving ultrahigh UV photosensitivity and thermal sensitivity in GeO_2 doped optical fibres. *Electronic Letters* 29: 1191–1193.

Lima H. F., R. S. Vicente, R. N. Nogueira et al. 2008. Structural health monitoring of the church of Santa Casa da Misericordia of Aveiro using FBG sensors. *Sensors Journal* 8: 1236–1242.

Limberger H. G., P. Y. Fonjallaz, P. Lambelet et al. 1993. Photosensitivity and self-organization in optical fibers and waveguides. *Proceedings of SPIE* 2044: 272–283.

Lindner E., J. Canning, C. Chojetzki et al. 2011. Regenerated draw tower grating (DTG) temperature sensors. *21th International Conference on Optical Fiber Sensors*, Ottawa, Ontario, Canada, Vol. 7753-37.

Lindner E., J. Moerbitz, C. Chojetzki et al. 2011. Draw tower fiber Bragg gratings and their use in sensing technology. *Conference on the Fiber Optic Sensors and Applications VIII*, Orlando, FL.

Lissak B., A. Arie, and M. Tur. 1998. Highly sensitive dynamic strain measurements by locking lasers to fibre Bragg gratings. *Optics Letters* 23: 1930–1932.

Lloyd G. D., L. A. Everall, K. Sugden et al. 2004. Resonant cavity time-division-multiplexed fiber Bragg grating sensor interrogator. *Photonics Technology Letters* 16: 2323–2325.

López-Higuera J. M. (ed.). 2002. *Handbook of Optical Fiber Sensing Technology*. John Wiley & Sons, New York.

Maaskant R., T. Alavie, R. M. Measures et al. 1997. Fiber-optic Bragg grating sensors for bridge monitoring. *Cement & Concrete Composites* 19: 21–33.

Majumder M., T. K. Gangopadhyay, A. K. Chakraborty et al. 2008. Fibre Bragg gratings in structural health monitoring—Present status and applications. *Sensors and Actuators A-Physical* 147: 150–164.

Malo B., J. Albert, K. O. Hill et al. 1994. Effective index drift from molecular hydrogen diffusion in hydrogen-loaded optical fibres and its effect on Bragg grating fabrication. *Electronic Letters* 30: 442–444.

Malo B., K. O. Hill, F. Bilodeau et al. 1993. Point by point fabrication of micro-Bragg gratings in photosensitive fiber using single excimer pulse refractive index modification techniques. *Electronics Letters* 29: 1668–1669.

Malo B., S. Theriault, D. C. Johnson et al. 1995. Apodised in-fiber Bragg grating reflectors photoimprinted using a phase mask. *Electronics Letters* 31: 223–225.

Martinez A., M. Dubov, I. Khrushchev et al. 2004. Direct writing of fibre Bragg gratings by femtosecond laser. *Electronics Letters* 40: 1170–1172.

Measures R. M. 1992. Smart composite structures with embedded sensors. *Composites Engineering* 2: 597–618.

Melle S. M., K. Li, and R. M. Measures. 1992. A passive wavelength demodulation system for guided-wave Bragg grating sensors. *Photonics Technology Letters* 4: 516–518.

Meltz G., W. W. Morey, and W. H. Glenn. 1989. Formation of gratings in optical fiber by a transverse holographic method. *Optics Letters* 14: 823–825.

Mendez A. 2007. Fiber Bragg grating sensors: A market overview. *Third European Workshop on Optical Fibre Sensors*, Naples, Italy, Vol. 661905.

Micronoptics. http://micronoptics.com/sensing_instruments.php. Accessed April 10, 2012.

Mihailov S. J., C. W. Smelser, D. Grobnic et al. 2004. Bragg gratings written in all-SiO_2 and Ge-doped core fibers with 800-nm femtosecond radiation and a phase mask. *Journal of Lightwave Technology* 22: 94–100.

Mishra V., N. Singh, U. Tiwari et al. 2011. Fiber grating sensors in medicine: Current and emerging applications. *Sensors and Actuators A-Physical* 167: 279–290.

Mooginfo. http://www.mooginfo.com/RMS/case-study--2-304RR-9259Z.html Accessed April 10, 2012.

Murphy D. F., D. A. Flavin, R. McBride et al. 2001. Interferometric interrogation of in-fiber Bragg grating sensors without mechanical path length scanning. *Journal of Lightwave Technology* 19: 1004–1009.

Nakstad H. and J. T. Kringlebotn. 2008. Realisation of a full-scale fibre optic ocean bottom seismic system. *19th International Conference on Optical Fiber Sensors*, Perth, Australia, Vol. 7004, p. 36.

Norman D. C. C. and D. J. Webb. 2007. Fibre Bragg grating sensor interrogation using an acousto-optic tunable filter and low-coherence interferometry. *Measurement Science and Technology* 18: 2967–2971.

O'Dwyer M. J., G. M. Maistros, S. W. James et al. 1998. Relating the state of cure to the real-time internal strain development in a curing composite using in-fibre Bragg gratings and dielectric sensors. *Measurement Science and Technology* 9: 1153–1158.

O'Mahoney K. T., R. P. O'Byrne, S. V. Sergeyev et al. 2009. Short-scan fiber interferometer for high-resolution Bragg grating array interrogation. *Sensors Journal* 9: 1277–1281.

Othonos A. 1997. Fiber Bragg gratings. *Review Scientific Instruments* 68: 4309–4341.

Othonos A. and K. Kalli. 1999. *Fiber Bragg Gratings: Fundamentals and Applications in Telecommunications and Sensing*. Artech House, Boston, MA.

Othonos A. and X. Lee. 1995. Novel and improved methods of writing Bragg gratings with phase masks. *IEEE Photonics Technology Letters* 7: 1183–1185.

Oullette F. 1991. All-fiber filter for efficient dispersion compensation. *Optics Letters* 16: 303–305.

Park Y.-L., S. Elayaperumal, B. Daniel et al. 2010. Real-time estimation of 3-D needle shape and deflection for MRI-guided interventions. *Transactions on Mechatronics* 15: 906–915.

Prohaska J. D., E. Snitzer, S. Rishton et al. 1993. Magnification of mask fabricated fibre Bragg gratings. *Electronics Letters* 29:1614–1615.

Rao Y. J. 1997. In-fibre Bragg grating sensors. *Measurement Science and Technology* 8: 355–375.

Rao Y. J. 1998. Fiber Bragg grating sensors: Principles and applications. In *Optical Fiber Sensor Technology*, Grattan, K. T. V. and Meggitt, B. T. (eds.). Chapman and Hall, London, U.K.

Rao Y. J., A. B. L. Ribeiro, D. A. Jackson et al. 1996. Simultaneous spatial, time and wavelength division multiplexed in-fibre grating sensing network. *Optics Communications* 125: 53–58.

Rao Y. J., D. J. Webb, D. A. Jackson et al. 1997. In-fiber Bragg-Grating temperature sensor system for medical applications. *Journal of Lightwave Technology* 15: 779–785.

Read I. J., P. D. Foote, D. Roberts et al. 1999. Smart carbon-fibre mast for a super-yacht. *Structural Health Montoring* 2000: 276–286.

Redondo Optics. http://www.redondooptics.com/FBGT_060209.pdf. Accessed April 10, 2012.

Reid D. C. J., C. M. Ragdale, I. Bennion et al. 1990. Phase-shifted Moire grating fiber resonators. *Electronics Letters* 26: 10–11.

Russell P. St. J., J.-L. Archambault, and L. Reekie. 1993. Fibre gratings. *Physics World*, pp. 41–46.

Sceats M. G. and S. B. Poole. 1991. Stress-birefringence reduction in elliptical-core fibers under ultraviolet irradiation. *Australian Conference Optical Fiber Technology*, Adelaide, Australia, p. 302.

Seim J., E. Udd, W. Schulz et al. 1999. Composite strengthening and instrumentation of the Horsetail Falls Bridge with long gauge length fiber Bragg grating strain sensors. *13th International Conference on Optical Fiber Sensors & Workshop on Device and System Technology Toward Future Optical Fiber Communication and Sensing*, Kyongju, Korea, pp. 196–199.

Sejka M., P. Varming, J. Hübner et al. 1995. Distributed feedback Er^{3+} fibre laser. *Electronics Letters* 31: 1445–1446.

Simpson G., K. Kalli, K. M. Zhou et al. 2004. Blank beam fabrication of regenerated type IA gratings. *Measurement Science & Technology* 15: 1665–1669.

Skaar J. and O. H. Waagaard. 2003. Design and characterization of finite-length fiber gratings. *IEEE Journal of Quantum Electronics* 39: 1238–1245.

Skuja L. 1998. Optically active oxygen-deficiency-related centers in amorphous silicon dioxide. *Journal Non-Crystalline Solids* 239: 16–48.

Smartfibres. http://www.smartfibres.com/Attachments/SFref671.pdf. Accessed April 10, 2012.

Smartfibres. http://www.smartfibres.com/FBG-sensors. Accessed April 10, 2012.

Sorensen H. R., J. Canning, J. Laegsgaard et al. 2006. Control of the wavelength dependent thermo-optic coefficients in structured fibres. *Optics Express* 14: 6428–6433.

Theune N. M., M. Kaufmann, J. Kaiser et al. 2000. Fiber Bragg gratings for the measurement of direct copper temperature of stator coil and bushing inside large electrical generators. *14th International Conference on Optical Fiber Sensors*, Venice, Italy, *Proceedings of SPIE*, Vol. 4185.

Todd M. D., J. M. Nichols, S. T. Trickey et al. 2007. Bragg grating-based fibre optic sensors in structural health monitoring. *Philosophical Transactions of the Royal Society A-Mathematical Physical and Engineering Sciences* 365: 317–343.

Vasil'ev S. A., O. I. Medvedkov, I. G. Korolev et al. 2005. Fibre gratings and their applications. *Quantum Electronics* 35: 1085–1103.

Webb D. J., M. W. Hathaway, D. A. Jackson et al. 2000. First in-vivo trials of a fiber Bragg grating based temperature profiling system. *Journal of Biomedical Optics* 5: 45–50.

Webb D. J. and K. Kalli. 2010. Polymer fibre Bragg gratings. In *Fiber Bragg Grating Sensors: Thirty Years from Research to Market*, A. Cusano (ed.). Bentham eBooks.

Webb D. J. and K. Kalli. 2011. Polymer fiber Bragg gratings. In *Fiber Bragg Gratings Sensors: Thirty Years from Research to Market*. Cusano, A., A. Cutolo, and J. Albert (eds.). Bentham eBooks.

Webb D. J., S. F. G. O'Neill, N. Fisher et al. 1997. Signal recovery technique for in-fibre Bragg grating and interferometric sensors using a Mach-Zehnder interferometer incorporating a fibre-pigtailed acousto-optic modulator. *OSA Technical Digest* 16: 508–511.

Williams D. L., B. J. Ainslie, J. R. Armitage et al. 1993. Enhanced UV photosensitivity in boron codoped germanosilicate fibres. *Electronics Letters* 29: 457–458.

Wilson J. and J. F. B. Hawkes. 1983. *Optoelectronics: An Introduction*. Prentice-Hall International, London, U.K., pp. 109–111.

Xie W. X., P. Niay, P. Bernage et al. 1993. Experimental evidence of two types of photorefractive effects occurring during photoinscriptions of Bragg gratings within germanosilicate fibres. *Optics Communications* 104: 185–195.

Xiong Z., G. D. Peng, B. Wu et al. 1999. Highly tunable Bragg gratings in single-mode polymer optical fibers. *IEEE Photonics Technology Letters* 11: 352–354.

Xu M. G., J. L. Archambault, L. Reekie et al. 1994. Discrimination between strain and temperature effects using dual-wavelength fibre grating sensors. *Electronics Letters* 30: 1085–1087.

Xu M. G., H. Geiger, and J. P. Dakin. 1996. Fibre grating pressure sensor with enhanced sensitivity using a glass-bubble housing. *Electronics Letters* 32: 128–129.

Zhang B. and M. Kahrizi. 2007. High-temperature resistance fiber Bragg grating temperature sensor fabrication. *IEEE Sensors Journal* 7: 586–591.

Zhang C., X. Chen, D. J. Webb et al. 2009. Water detection in jet fuel using a polymer optical fibre Bragg grating. *20th International Conference on Optical Fiber Sensors*, Edinburgh, Scotland, *SPIE* 7503, p. 750380.

18. Optical Fiber Chemical Sensor Principles and Applications

Tong Sun
City University

T.H. Nguyen
City University

Kenneth T.V. Grattan
City University

Handbook of Optical Sensors. Edited by José Luís Santos and Faramarz Farahi © 2015 CRC Press/Taylor & Francis Group, LLC. ISBN: 9781439866856.

18.1 Introduction

Chemical sensing deals with the task to measure the concentration or activity of a chemical species in a sample of interest. This measurement involves a functionalized sensing head, a communication channel, and a powering/processing unit that energizes the system and displays the information about the target chemical parameter using the most convenient format. Ideally, the measurement shall be performed in a continuous and reversible way and in remote conditions if necessary, looking to the ideal chemical sensing system with its ability to provide in real time the spatial and temporal distributions of a particular molecular or ionic species.

Fiber-optic chemical sensing is a subclass of optical chemical sensing in which the radiation guided by the optical fiber is modulated in one or more of its characteristics by the measurand, with the modulated light carried out to the detection and processing unit by the same or a second fiber. Therefore, this sensing approach eliminates the need of dedicated power and telemetry channels. Typically, two sensing approaches are considered. In one of them, a spectroscopically physical property of the analyte can be measured directly through its interaction with the fiber-guided optical field, not requiring a specific chemical recognition phase. In another approach, a chemical recognition phase is used to generate an analyte-dependent spectroscopically detectable signal within the sensing region of the fiber. The chemical changes induced by the interactions of the analyte with the immobilized reagents are measured spectroscopically by analyzing the radiation that returns from the sensing head. A third approach to chemical sensing, based on the physical changes of the sensor materials rather than chemical reactions, is feasible and indeed widely explored in recent years.

This chapter addresses fiber-optic-based chemical sensing, summarizing in its first part the basic principles associated with spectroscopic sensing and underpinning chemistry, while in the second part, a detailed description of specific applications will highlight the critical issues associated with this sensing technology.

18.2 Spectroscopic Sensing

In this section, the basic principles of spectroscopic sensing will be outlined. First, absorption-based spectroscopy is summarily described, followed by the description of the essence of luminescence-based spectroscopy. Finally, the topic of underpinning chemistry for spectroscopic sensors is introduced.

18.2.1 Absorption-Based Spectroscopy

Absorption-based optical sensors can be *colorimetric* or *spectroscopic* in nature. Colorimetric sensors are based upon detection of an analyte-induced color change in the sensor material, while spectroscopic absorption-based sensors rely on detection of the analyte by probing its intrinsic molecular absorption (McDonagh et al. 2008, Wolfbeis 2008).

Absorption of optical energy arises from transitions in the electronic, vibrational, and/or rotational energy states of the atoms and molecules and occurs only if the difference in the energy states involved matches exactly the same amount of energy of the exciting photons. Visible and ultraviolet (UV) radiations induce electronic excitation, infrared (IR) radiation promotes vibrational excitation, and microwave radiation gives rise to rotational transitions.

Thus, absorption leads to a loss of the power of the radiation as it passes through the target sample. Therefore, after encountering a number of absorbing species, a light beam of initial intensity, I_0, will be transmitted by the sample with a reduced intensity I. It should be noted that only those frequencies that are absorbed will be attenuated, and all other frequencies will pass through with no power loss. The decrease in the light intensity is determined by the number of absorbing species in the light path and is related to the concentration, c, of the absorbing species through the Beer–Lambert equation (Narayanaswamy 1993):

$$A_b = \log\left(\frac{I_0}{I}\right) = \varepsilon l c \qquad (18.1)$$

where

A_b is the absorbance

l is the length of the light path

ε is the molar absorptivity, which is characteristic of the analyte substance at a given wavelength

Optical fiber chemical sensors can be designed to enable the guided light in the fiber to interact with the target analyte, therefore inducing either the color change on the sensor material (e.g., indicator) or the change of intensity of the light as a result of the spectroscopic absorption expressed in Equation 18.1.

18.2.2 Luminescence-Based Spectroscopy

Luminescence represents one of the energy release formats when atoms or molecules are excited to a higher energy state after the absorption of energy from exciting photons. The deexcitation process, when the excited atoms or molecules return to lower energy states, involves the energy release via several pathways. One of the relaxation modes is luminescence, by which radiation at a lower frequency is emitted.

Deactivation through luminescence can occur either from a singlet state, in which case the emission is called *fluorescence*, or from a triplet state, in which case the emission is called *phosphorescence*. Fluorescence is extremely rapid and occurs within 1–100 ns after excitation. Phosphorescence, however, has a longer decay period (10^{-3}–10s) and persists after the removal of the excitation source. In both cases, the emitted radiation is of a different frequency to that of the exciting radiation, and its intensity, I_L, is dependent on the intensity of the exciting radiation, I_0, and the concentration, c, of the luminescent species.

For weakly absorbing species, that is, $A_b < 0.05$, the intensity of luminescence can be expressed by the following equation:

$$I_L = k'I_0\theta\varepsilon lc \tag{18.2}$$

where

l is the length of the light path in the sample
ε is the molar absorptivity
θ is the quantum efficiency of the luminescence
k' is the fraction of the emission that can be measured

When the excitation radiation I_0 is a constant, Equation 18.2 can be simplified to be

$$I_L = k_L c \tag{18.3}$$

where $k_L = k'I_0\theta\varepsilon l$.

In the presence of some species (e.g., oxygen), the luminescence decay of an activated species could compete with a collisional quenching decay mode. The mean lifetime of the activated species is decreased and the luminescence intensity is reduced. In this case, the luminescence intensity, I_L, is related to the concentration of the quenching species, c_q, by the Stern–Volmer equation:

$$\frac{I_0}{I_L} = 1 + K_{SV}c_q \tag{18.4}$$

where

I_0 is the luminescence intensity in the absence of the quencher
K_{SV} is the Stern–Volmer constant

In addition to the luminescence quenching mechanism discussed earlier, excited species can also be generated through a chemical reaction. The measured light that is emitted as the excited species returns to the ground state is then known as chemiluminescence, which can be quantitatively related to the concentration of the analyte species.

In light of the above, a fluorescence-based optical fiber sensor can be realized through the immobilization of analyte-sensitive fluorescence dye onto the fiber platform. One of the possible sensor designs is to replace a portion of a fiber cladding with a solid matrix that contains the analyte-sensitive fluorescent compound. When the evanescent wave (*leaked* from the section where the fiber cladding is removed) interacts with the target analyte, the fluorescence signal will change accordingly.

Alternatively, the analyte-sensitive fluorescent compound can be immobilized directly onto the fiber end surface to form a fiber sensor tip. When the sensor probe interacts with the target analyte, the fluorescence signal generated can be directly collected by the fiber, which also sends the *excitation* light. Both the pH and cocaine sensors discussed later in the chapter are fluorescence-based sensors. A detailed description of the principles and techniques of fluorescent measurement can be found in Chapter 6.

In addition to the earlier absorption and luminescence-based spectroscopic techniques, it is worth mentioning that there are some other chemical sensors that are based on the physical changes of the sensor materials rather than chemical reactions. For example, the humidity sensor described in one of the following sections is based on the swelling effect of a polymer sensor material, which induces the strain effect on the fiber Bragg grating (FBG), thus causing the wavelength shift.

18.2.3 Underpinning Chemistry for Spectroscopic Sensors

The general chemistry governing the spectroscopic sensing mechanisms is briefed here (Narayanaswamy 1993). When a sensing reagent (R) reacts with an analyte species (A), a product (AR) is formed, thus the process can be represented as

$$A + R \leftrightarrow AR \tag{18.5}$$

where R or AR is usually absorbing or luminescent to enable the target analyte A to be measured in an optical form. The reagent (sensor material) employed is required to be both analyte selective and analyte sensitive so that it can produce a distinctive optical signal change for the given analyte species.

The chemical transduction is normally based on the equilibrium established during the chemical reaction between A and R, and this equilibrium can be described as

$$K = \frac{[AR]}{[A][R]} \tag{18.6}$$

where

K is the equilibrium constant

the square brackets indicate the equilibrium concentration of the species involved

The change in absorption or luminescence signal can be caused either by the decrease of the reagent R consumed in the chemical reaction or by the formation of product AR, which increases its absorption or fluorescence. Under both circumstances, the change in optical property of R or AR can be related to their concentration and, in turn, related to the concentration of the analyte A causing changes in the measured optical property.

Assuming the reagent is consumed during the reaction, its total initial concentration (c_R) at any time will be given by

$$c_R = [AR] + [R] \tag{18.7}$$

Chapter 18

Thus, from Equations 18.6 and 18.7, the analyte A concentration can be expressed as

$$[A] = \frac{1}{K} \times \left(\frac{c_R}{[R]} - 1 \right) \tag{18.8}$$

If the optical property of AR is being measured, the analyte concentration can thus be related to the concentration of AR as follows:

$$\frac{1}{[A]} = K \times \left(\frac{c_R}{[AR]} - 1 \right) \tag{18.9}$$

The concentrations of R in Equation 18.8 and of AR in Equation 18.9 are related to the measured optical property.

Based on Equations 18.5, 18.8, and 18.9, the underpinning sensing mechanism is based on the use of a direct indicator (reagent R), which requires an appropriate equilibrium constant for the measurement of the desired analyte concentration range. The response of the sensor, however, is also dependent on the total amount of the indicator used. As a result, any uncontrolled variable that affects the equilibrium constant will be a potential source of error. For example, for any reaction involving ions, variations in ionic strength (IS) will affect the K values.

Indicators showing reversible reactions with analytes are generally preferred for use in optical sensors because they can provide continuous and unperturbed measurements. The equilibrium response of indicators does not depend on mass transfer; however, the response time (i.e., the time required to reach equilibrium) is dependent on mass transfer. Therefore, the sensors that do not involve chemical reactions to enable the reversible interactions of analytes with indicators usually provide much shorter response time.

If the measured optical parameter is dependent on the ratio of the concentrations of the two forms of the indicator, that is, $[AR]/[R]$, the response no longer depends on the total amount of the indicator, although the dependence on the equilibrium constant remains. As a result, measurements of the ratio of optical intensities at two wavelengths can be used to determine the analyte concentration. Such ratio measurements are inherently more stable and thus more favored for long-term measurements as it is able to minimize the drift caused by the light source fluctuation or the environmental influence. This type of ratio measurements can be described as follows by rearranging Equation 18.6:

$$[A] = \frac{1}{K} \times \frac{[AR]}{[R]} \tag{18.10}$$

A range of *precalibrated* sensors based on ratiometric intensity measurements have been widely reported for various applications.

18.3 Sensor Design and Applications

In this section, illustrative examples of new chemical sensor designs and their respective underpinning sensing technologies projected to fulfill specific application requirements are presented. It includes fiber-optic sensing systems designed specifically for the measurement of humidity, pH, and cocaine.

18.3.1 Optical Fiber Humidity Sensor

Masonry is widely recognized as an adaptable and sustainable construction material, with a low carbon signature, and is a key element of the world's tangible cultural heritage. Weathering of stone appears to be one of the major reasons for the damage of stone masonry structures, through a complex and interlinked chemical, physical, and biological process. The moisture content or *humidity* plays an important role in this process as water is an effective transport medium of all these reactive elements; therefore, monitoring effectively the moisture ingress would help for a better understanding of the deterioration processes of building stone.

The most commonly used techniques for measuring moisture in building materials are gravimetric, electrical and mechanical, and chilled mirror hygrometric methods (Lee and Lee 2005). Some of these are labor intensive, suitable only for laboratory measurements, and often constrained by the cost rather than by the performance. The electrical technique, based upon monitoring the electrical resistance (ER) change as a function of the moisture variation, is widely used for monitoring the presence of moisture in building stone, due to the advantages such as ease of deployment, real-time monitoring, and direct measurement. This technique, however, is influenced by the presence of dissolved salts and other chemical changes in the porous structure of building stone; thus, the sensor signal change is not uniquely correlated to the level of moisture being present.

The fiber-optic sensing system described here is aimed to address the earlier and other limitations of current techniques by providing novel solutions to achieve a *direct* measurement of the moisture content in stonework through the development and evaluation of a minimally invasive optical fiber humidity sensor. The approach taken has been designed to allow comparison with the performance of existing sensing technologies and indeed to better their performance, when applied to moisture ingress/egress monitoring.

18.3.1.1 Sensing Principle

An optical FBG is used in this work as the basis of a moisture ingress monitoring system. The Bragg wavelength (λ_B) that satisfies the Bragg condition is thus given by Equation 18.11:

$$\lambda_B = 2n_{eff}\Lambda \tag{18.11}$$

where

n_{eff} is the effective refractive index of the fiber core

Λ is the grating period, where *both* are affected by strain and temperature variations, a feature that is reflected in the sensor design

The shift in Bragg wavelength due to the change in strain or thermal effect is thus given by

$$\frac{\Delta\lambda_B}{\lambda_B} = (1 - P_e)\varepsilon + \left[(1 - P_e)\alpha + \zeta\right]\Delta T \tag{18.12}$$

where

P_e is the photoelastic constant of the fiber
ε is the strain induced on the fiber
α is the fiber thermal expansion coefficient
ζ is the fiber thermal-optic coefficient

The first term of Equation 18.12 represents the longitudinal strain effect on the FBG, and the second term represents the thermal effect, which comprises a convolution of thermal expansion of the material and the thermal-optic effect.

The humidity sensing concept used in this sensor exploits the strain effect induced in an FBG through the swelling of a thin layer of applied polymer coating. The swelling of the polymer coating, arising from the absorption of moisture, changes the Bragg wavelength of the FBG, where this can be calibrated to give a direct indication of the humidity level. Thus, the shift in the Bragg wavelength in Equation 18.12 for the polymer-coated FBG can be modified as follows:

$$\frac{\Delta\lambda B}{\lambda B} = (1 - P_e)\alpha_{RH} \cdot \Delta RH + \left[(1 - P_e)\alpha_T + \zeta\right]\Delta T \tag{18.13}$$

where α_{RH} and α_T are the moisture expansion coefficient and the thermal expansion coefficient of the coated FBG.

The detailed discussions of the fabrication of the FBGs used, the coating material chosen, the coating thickness, and the resulting humidity sensor response time have been reported by some of the authors elsewhere (Yeo et al. 2005, 2006).

Due to the sensitivity of the FBG that forms the key sensor element to both strain *and* temperature, it is important to compensate the temperature effect when the coated FBG is used for humidity measurement as this depends only on the strain induced in the coating.

18.3.1.2 Sensor Design, Fabrication, and Packaging

The FBGs used in this work were inscribed in boron–germanium (B/Ge) codoped photosensitive optical fibers using a phase mask fabrication technique. In order to produce a stable sensor that could be used over a wide temperature range, the grating was first annealed for more than 7 h at 200°C prior to the polymer coating being applied. Subsequently, a thin layer of moisture-sensitive polyimide (PI) with a coating thickness of 24 μm was coated onto the FBG using an automated dip coating machine. Achieving temperature compensation is important as such grating-based devices are temperature sensitive, and thus a second grating element is used to create the complete sensor system. To do so, a bare FBG is also included in the sensor design. Figure 18.1 shows, respectively, the schematic diagram and a picture of the humidity sensor probe design, in which both grating elements can be seen—a bare FBG without coating is used for temperature measurement and for temperature compensation of the coated humidity sensor.

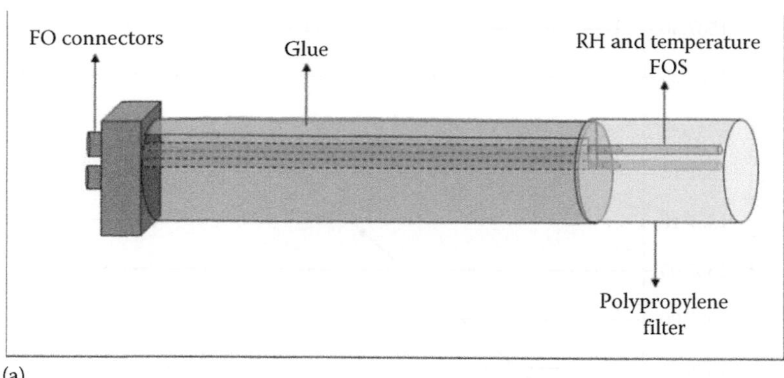

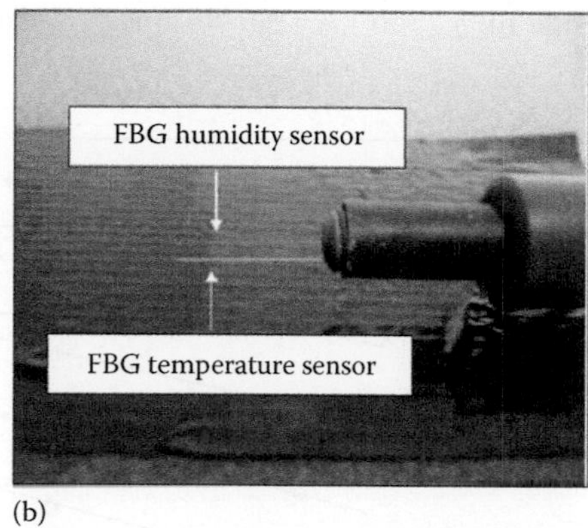

FIGURE 18.1 (a) Schematic diagram of the sensor design, showing a coated grating as an RH sensor and a bare grating as a temperature sensor; (b) picture of the packaged sensor probe.

A critical aspect of the work in this research has been to tailor the design of the sensor system to the requirements of the measurement *in the field* and for the stonework itself: this involving developing a specific design of its protection and packaging for use in the harsh environments experienced. As a result, the dual-parameter sensor elements have been protected in this special design created to allow monitoring of the key parameters involved in the deterioration of stone masonry structures, as prescribed by the users of the devices.

18.3.1.3 Sensing System and Its Calibration

After the construction of the dual-grating sensor probe illustrated earlier, an effective humidity sensing system can thus be created, requiring the wavelength-encoded output from the sensor probe to be determined, using a commercial FBG interrogation system. Such an interrogation device can either include a broadband light source, a F-P filter and a photodetector to capture the Bragg wavelength, or include a swept laser source, and a photodetector to capture the peak wavelength.

In order to show that the sensor was ready for use with exterior masonry, the sensor was first tested and calibrated in the laboratory. This involved determining the Bragg wavelength change, separately with both temperature and humidity changes. Figure 18.2 shows the calibration data obtained from the specially designed and packaged humidity

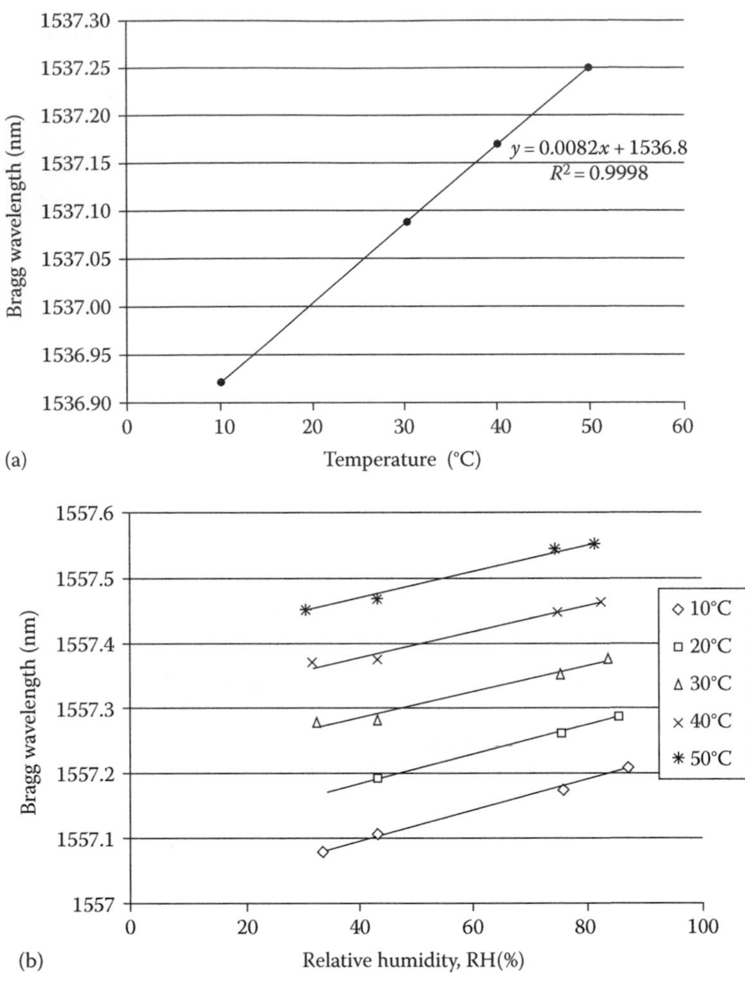

FIGURE 18.2 (a) Temperature calibration curve of the optical fiber RH sensor at a constant RH level; (b) the RH calibration curves of the RH humidity sensor at a series of known temperatures.

sensor probe as a function of both (a) temperature and (b) relative humidity (RH), while the other parameter is kept constant.

Such an accurate calibration is essential and, for the humidity calibration, this was undertaken using a saturated salt solution technique (which can provide a known and constant RH value), with the probe being placed in an enclosed container with sequentially a series of different saturated salt solutions, which then provide a series of standard RH levels—all this was done at a series of known temperature (Greenspan 1977) representative of the range to be used in actual monitoring work.

To achieve a well-referenced calibration, the humidity inside the container was also monitored by using a reference hygrometer, to allow comparative measurements with a commercial probe to be made. The temperature inside the container was varied from 10°C to 50°C at each specific humidity level so that the influence of temperature on the humidity measurement could be seen from the sensor output. Thus, from the humidity calibration graph, the temperature coefficient may be calculated as this is required to provide a correction to the humidity data in situations where the temperature is not constant, thus allowing a correction factor to the apparent RH measurements to be applied. For convenience of operation, the temperature and RH coefficients obtained from the calibration curves were entered into the LabVIEW-based software, allowing the user to have a direct readout of the temperature-corrected RH value from the output of the instrument.

18.3.1.4 Evaluation of Sensor Performance

18.3.1.4.1 Stone Sample Preparation To facilitate the evaluation of the sensor system for in situ testing, a limestone block of the type used in building conservation of dimensions 150 mm × 150 mm × 80 mm had been specifically fabricated for the planned laboratory tests. Before placing the sensors in the carefully drilled holes in the sample, the block was dried at 50°C—this being determined when the weight of the sample reached a constant value. The specially designed optical fiber temperature/humidity probe was placed at a depth of 30 mm below the surface, as shown in Figure 18.3. In order to validate and provide a cross comparison of the measurements from the fiber-optic sensor (FOS)-RH probe, both commercially available capacitance-based RH probe and ER-based moisture sensors are placed together in the same stone block, and all their outputs were closely monitored during the tests carried out.

As shown in Figure 18.3, the capacitance-based RH sensor was placed in a position parallel to the fiber-optic RH probe and with the active element at the same depth (30 mm below the surface). The ER sensors were placed in drilled holes at four different depths of 0.5, 1, 2, and 5 cm, respectively, from the exposed surface.

18.3.1.4.2 Simulated Capillary Tests Prior to the simulated capillary test in the laboratory, the limestone sample was placed on a tray supported with two stainless steel rods where the tray was filled with water maintained at 20°C and to a level such that the base

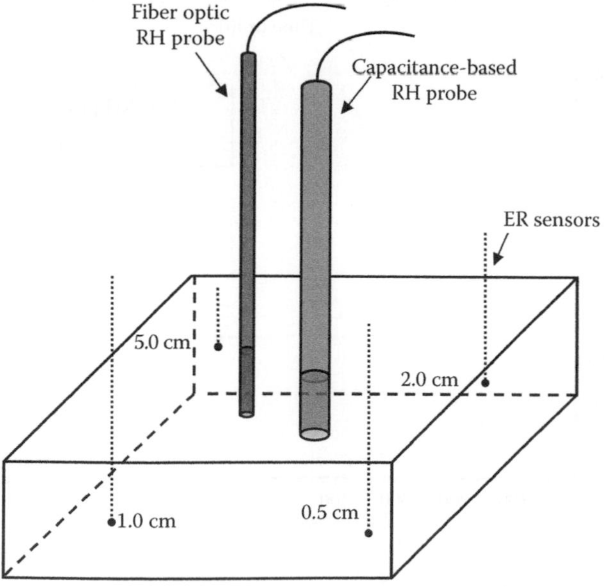

FIGURE 18.3 The layout of the sensors in a limestone block.

of the block is immersed in the water to a depth of 3 mm. The whole setup was positioned in a climatic cabinet maintained at 20°C and 65% RH. The water level in the tray was monitored carefully and kept constant by adding any necessary additional water throughout the test. The moisture and temperature changes inside the block were also monitored continuously at an interval of 1 min, and the data obtained were collected and recorded using a PC.

The changes in RH due to capillary rise of water, as measured by the FOS and capacitance-based sensor, are shown in Figure 18.4a. As the waterfront rises to the sensors located at 30 mm from the surface, the FOS-RH sensor responded more rapidly and gave values of RH, which were seen to stabilize as would be expected, after 400 min. The measurements made by capacitance sensor, however, showed a gradual change in RH and stabilization was reached after 600 min. The delay demonstrated by the capacitance sensor is likely due to the larger volume of the sensor probe and the location of the opening allowing moisture into the sensing part of the probe, being placed on the side rather than at the tip

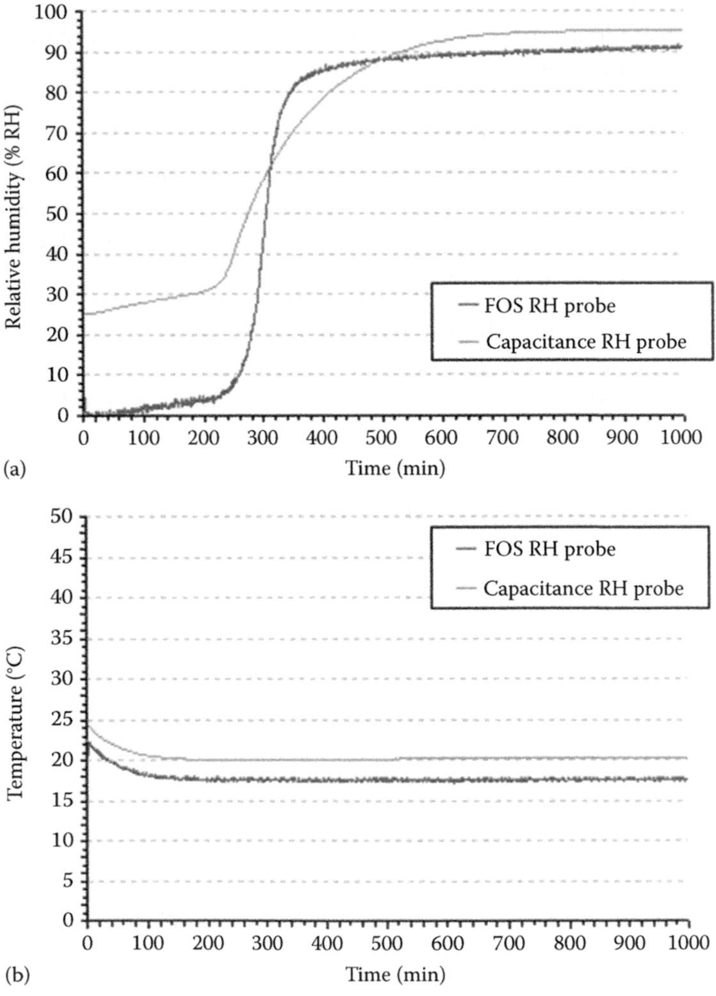

FIGURE 18.4 (a) Changes in RH at 30 mm depth of stone with capillary rise of water; (b) changes in temperature at 30 mm depth of stone with capillary rise of water.

of the probe. The temperature measurements made by both RH probes are closely related (agreeing within experimental error) and follow the same trend, as shown in Figure 18.4b.

After the limestone block was fully saturated with moisture as a result of the capillary rise test, the block was subsequently subjected to a slow heating process, allowing the drying of the block and with that the change of the moisture content to be monitored. The heating of the sample was performed in an oven maintained at 40°C. The changes in RH and temperature were measured using both the FOS-RH sensor and the capacitance-based sensor when the limestone block was drying. The results obtained from the sensors are shown in Figure 18.5. In this experiment, the RH and temperature changes inside the stone were monitored continuously for 3 days. It can be observed from Figure 18.5a that the RH values, as measured by FOS probe, decreased gradually with time, showing a clear correlation between the response of this humidity sensor and the limestone block drying process with the expected decrease in RH due to the expulsion of moisture from the block.

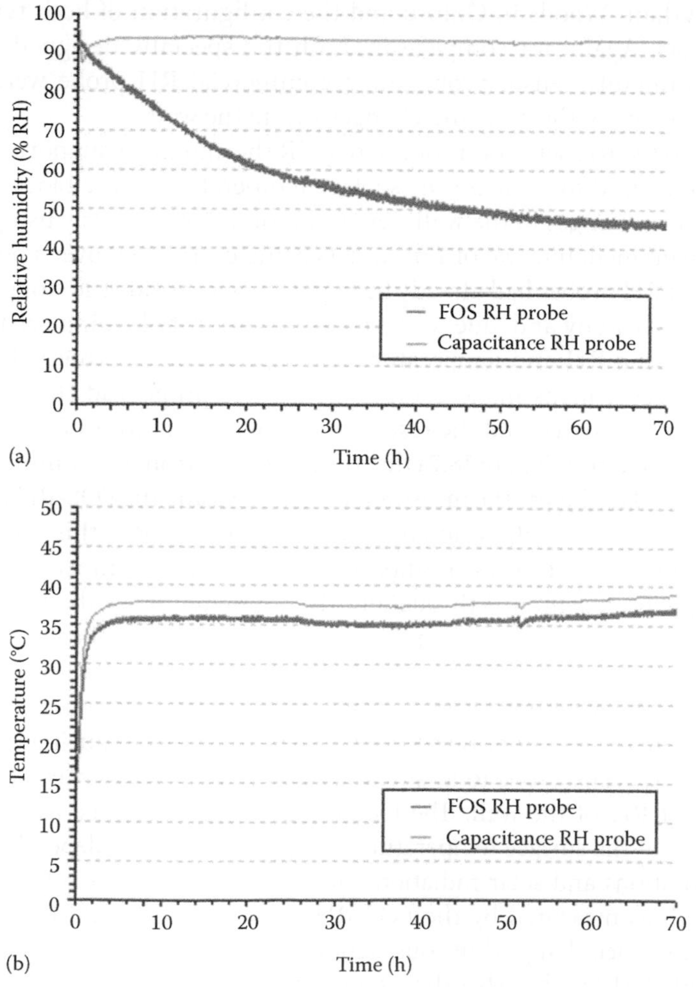

FIGURE 18.5 (a) Changes in RH at 30 mm depth of stone with drying of the stone block; (b) changes in temperature at 30 mm depth of stone with drying of the stone block.

However, the contrast was seen in the RH measurement made by the capacitance-based sensor, which shows only the ongoing saturation of the sensor during the experiment—the reading obtained indicated a minimal decrease with the initial few hours of drying, but the saturated level is maintained throughout the drying period. This effect was likely due to the condensation of moisture on the sensing part of capacitance-based RH probe and reflects that this sensor is not actually measuring the condition of the stonework, as required. Figure 18.5b confirms again the good agreement seen in the temperature measurements made with the optical probe.

18.3.1.5 Field Tests of Humidity Sensors

Further to the aforementioned simulated tests in the laboratory, an extensive series of field tests was planned and subsequently carried out on a test wall in Oxford, United Kingdom. To allow for controlled measurements, the wall itself was built specifically using methods similar to those now employed in conservation and which were similar to those used over many centuries. Figure 18.6 shows the pictures of the test wall constructed at Wytham Woods in Oxford and the configuration of both types of sensors used in the tests. Again, for comparison, both the specially designed, temperature-compensated humidity sensor probe and a commercial RH probe were used in this study for monitoring of the moisture changes inside the wall.

In addition to the RH sensors, a commercial IR thermometer supported using a truss arrangement was used to measure the surface temperature of the block. The temperature and moisture changes in the wall were monitored continuously using the FOS-RH probe at a measurement interval of 1 min, tests being carried out over a period of 2 days. The commercial RH probes had already been put in place inside the stone blocks several months before this study and, due to the more limited data logging capacity, had been used for monitoring RH and temperature continuously at 1 h intervals. The RH and temperature measurements made using the FOS-RH probe and the commercial RH probe were cross-compared and the results are shown in Figure 18.7.

It can be observed from Figure 18.7a that the indication from the commercial RH probe was always 100% RH, whereas the measurements made using the FOS-RH probe showed a variation in RH of less than 100% and between 90% and 93% RH, this varying with time. As with the laboratory tests, these results also show that the commercial RH sensor element had simply been saturated throughout the tests, and hence the RH measurements were continuously reading an apparent RH of 100%—indicating the failure of the sensor for monitoring accurately the true conditions of the wall itself. This indicates a clear drawback in this conventional monitoring approach, and indeed many commercial RH sensors that have a high mass fail to dry out properly when wet initially during use. By contrast, the optical fiber RH sensor, due to its small size and low mass, has been able to follow the actual change of RH of the wall. The temperature measurements, as shown in Figure 18.7b, indicate that the surface temperature changes rapidly, this depending on the environmental conditions and solar radiation. On the other hand, the temperature changes at 50 mm depth as measured by the FOS-RH probe are, as would be expected, rather slower than the surface changes but consistent with the surface temperature measurement undertaken with the IR probe. Also, the response of the temperature measurements made by the FOS-RH probe was more rapid, in comparison to the temperature determined using the instrumentation within the commercial RH probe. It can be observed from Figure 18.7b

(a)

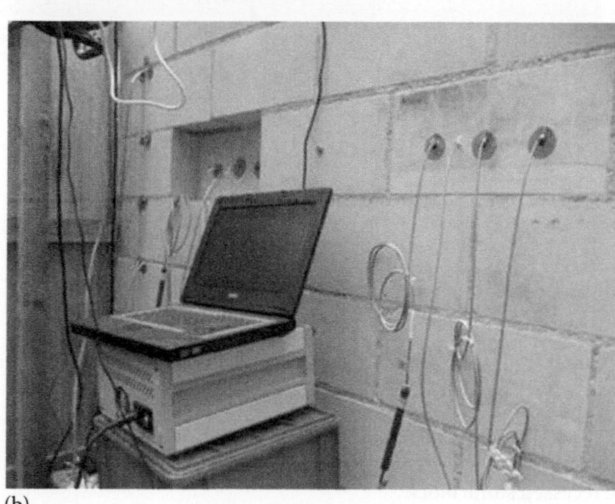
(b)

FIGURE 18.6 (a) Front face of test wall showing the monitoring surface of the nonrecessed stone block using IR thermometer; (b) rear side face of the test wall showing the fiber-optic monitoring setup.

that the surface temperature of the stone block changes rapidly, recording a more substantial temperature difference (as would be expected due to the mass of the material) than was observed at a depth of 50 mm from the surface.

18.3.1.6 Comments

The results presented indicate that with careful design of the sensor packaging to meet the application requirements, the optical RH sensors have demonstrated their robustness and been shown to be more accurate, with additionally a faster response, when their outputs are compared to those from commercial capacitance sensors, the more so when they are subjected to cycles of wet/dry conditions over several days. This is due to the fact that, unlike most conventional RH sensors, the optical fiber RH probes discussed here are not influenced by condensation of moisture within or on the sensor itself.

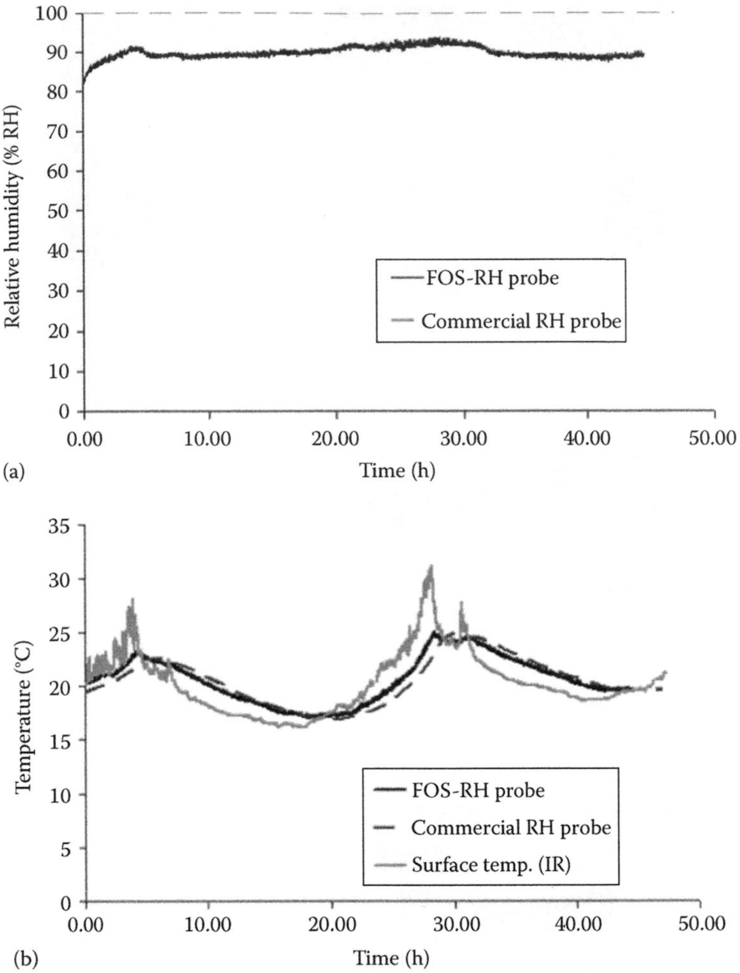

FIGURE 18.7 (a) Changes in RH at 50 mm depth of the recessed stone block; (b) changes in temperature at 50 mm depth of the recessed stone block.

Therefore, the optical fiber temperature/humidity sensors can ideally complement the currently available temperature and RH sensors for enhanced long-term monitoring of moisture changes in building stone and indeed show superior performance when issues such as rapid changes in RH and temperature are concerned.

18.3.2 Optical Fiber pH Sensor

Optical fiber pH sensors have been actively investigated because of their importance for both in situ and in vivo pH measurements in various aspects of scientific research and in a range of practical applications, in particular those where available conventional glass electrodes are not suitable (Grant et al. 2001, Wolfbeis 2002, Vasylevska et al. 2007). As indicated before, optical fiber–based sensors have shown many advantageous characteristics, such as small size, immunity to electromagnetic (EM) interference, remote sensing capability, resistance to chemicals, and biocompatibility (Lee et al. 2001). Most reported pH optrodes (the optical analog of electrode) function through monitoring the changes in the absorbance or fluorescence properties of certain pH-sensitive indicators, which are immobilized on/in proton-permeable solid substrates (Liu et al. 2005). A majority of the reported optical fiber pH sensors were constructed to operate in the physiological or near neutral pH region, and very limited research has been undertaken to explore the pH measurements at both extremes, that is, either at the low or high pH regions where the pH response of most glass pH electrodes is imperfect (Carey et al. 1989, Safavi and Bagheri 2003).

This limitation may lie in the selection of pH indicators and their subsequent immobilization techniques used, as they usually govern the sensor lifetime and stability. Poor immobilization tends to result in dye leaching and consequently a drifting of the calibration of the probe, which leads to the gradual breakdown of its useful sensing ability (Uttamlal et al. 2002). Among several widely used immobilization methods are included absorption or entrapment (Fujii et al. 1993, Arregui et al. 2002), layer-by-layer (LbL) electrostatic self-assembly (Egawa et al. 2006, Goicoechea et al. 2008), and covalent bonding (Ensafi and Kazemzadeh 1999, Baldini et al. 2007). The covalent bonding method can produce more reliable and durable sensors, as the indicators are virtually bonded to the substrate; therefore, they are unlikely to leach out under normal conditions, although the fabrication process is relatively complicated and time consuming (Lin 2000).

When sensors are considered to be used under extreme conditions, that is, very low pH or very high pH conditions, the sensor reliability and durability are determined not just by the immobilization method used but also by other factors, such as the stability of the pH indicators themselves and the stability of the substrates and the linking bonds between the fiber substrate and the sensor material. For example, the commonly used ester linkage and acid amide linkage are not very stable in acidic or alkaline aqueous conditions (Peterson et al. 1980).

Here is a typical example through a novel intrinsic pH sensor design, created specifically for low pH measurements, which is aimed to overcome the limitations of the existing sensor technologies highlighted earlier, thus creating more stable and therefore more useful devices.

18.3.2.1 Operational Principle of Fluorescence-Based pH Measurements

The development of the present pH optrode is basically based on the fluorometric determination of pH. It makes use of a fluorescent dye as an indicator, *HA*, to induce pH-sensitive changes in the measured fluorescence intensity. In aqueous solution, the following equilibrium can be reached:

$$HA \leftrightarrow H^{+} + A^{-}$$

The relationship between the protonation state of the indicator and the pH is governed by the Henderson–Hasselbalch equation:

$$\text{pH} = \text{p}K_{\text{a}} + \log\frac{[A^{-}]}{[HA]} \qquad (18.14)$$

where

$[A^-]$ and $[HA]$ are the concentrations of the dissociated and undissociated forms of the indicator

$\text{p}K_{\text{a}}$ is the acid–base constant

$[A^-]$ and $[HA]$ are related to fluorescence intensities by $[A^-] = F - F_{max}$ and $[HA] = F_{min} - F$ where F is a measured fluorescence intensity of the system, F_{max} is the fluorescence intensity of the fully protonated system, and F_{min} is the fluorescence intensity of the deprotonated system. The expressions are then substituted into Equation 18.14 to provide Equation 18.15:

$$\mathrm{pH} = \mathrm{p}K_a + \log\frac{F - F_{max}}{F_{min} - F} \tag{18.15}$$

Thus, Equation 18.15 can be rewritten in terms of F to give

$$F = \frac{F_{max} + F_{min} \times 10^{(\mathrm{pH}-\mathrm{p}K_a)}}{10^{(\mathrm{pH}-\mathrm{p}K_a)} + 1} \tag{18.16}$$

This results in an *S-shaped* relation of the fluorescence intensity versus pH graph, centered on the $\mathrm{p}K_a$ value. Equation 18.16 is used as a model for a nonlinear fitting method to calculate the $\mathrm{p}K_a$ value, which is the pH where 50% of the dye population in solution is protonated.

18.3.2.2 Intrinsic pH Sensor Probe Design

As discussed earlier, the sensor reliability and durability are determined by many factors, which include the immobilization method used, the stability of the pH indicators themselves, and the stability of the substrates and the linking bonds between the fiber substrate and the sensor material. Thus, in the following new pH sensor design, all these aspects have been taken into account.

18.3.2.2.1 Selection of Fluorescent Dyes A novel polymerizable coumarin dye, bearing a carboxylic acid group, was designed and synthesized for low pH measurement. Coumarins have been chosen for this application as they are widely used as laser dyes for single-molecule fluorescence, and so they are *tried and tested* in terms of the key property of being photostable (Drexhage 1976, Eggeling et al. 1998). The dissociation of the carboxylic acid group allows for the determination of pH in the acidic region of the pH scale, which makes it suitable for gastric measurements (Wiczling et al. 2005) and acidic soil measurements (Simek et al. 2002) as well as the measurement of pH in certain chemical reactors. The dye was covalently bonded to the fiber surface by polymerization, in an approach similar to the method reported (Sloan and Uttamlal 2001), but in this work, allyltriethoxysilane (ATES) was used to functionalize the fiber surface with polymerizable groups rather than 3-(trimethoxysilyl) propyl methacrylate to avoid the unstable ester linkage. The fluorescence detection method was employed rather than the simpler commonly used method based on colorimetric measurements as fluorescent sensors are usually more precise and offer higher sensitivity than their colorimetric counterparts (Staneva and Betcheva 2007).

18.3.2.2.2 Synthesis of a Fluorescent Dye All chemicals used in the sensor fabrication process were of analytical grade, purchased from Sigma-Aldrich, and were used without further purification. All solvents used were of high-performance liquid chromatography (HPLC) grade from Fisher Scientific. All aqueous solutions were prepared using distilled water. ^1H and ^{13}C NMR spectra were recorded on a Bruker Avance 500 spectrometer. Mass spectra were run by negative ion electrospray (ES) mode on a Waters LCT Premier XE mass spectrometer. IR spectra were recorded using a 8700 Shimadzu Fourier transform infrared spectrophotometer. Melting points were recorded on a Stuart SMP30 melting point apparatus and were uncorrected. Elemental analyses were carried out at the Microanalytical Laboratory, Department of Chemistry, University College London, United Kingdom. Absorption and fluorescence measurements of aqueous solutions containing fluorophores were carried out on a PerkinElmer Lambda 35 spectrophotometer and a Horiba Jobin Yvon FluoroMax-4 spectrofluorometer system with FluorEssence™ as driving software, respectively. Refractive indices were measured on an Abbe refractometer. Quantum yields of fluorescence were determined using quinine sulfate as the standard ($\Phi = 0.55$) (Eaton 1988).

The detailed procedure for the synthesis of the fluorescent dye is as follows. First of all, 7-vinylbenzylaminocoumarin-4-carboxylic acid (7-VBACC) was prepared by hydrolyzing methyl 7-vinylbenzylaminocoumarin-4-carboxylate, which was synthesized from methyl 7-aminocoumarin-4-carboxylate (Besson et al. 1991) and 4-vinyl benzyl chloride as shown in Figure 18.8.

18.3.2.2.3 pH Sensor Probe Design and Fabrication Building on the successful synthesis of the fluorescent dye shown in Figure 18.8, the next step in the development of the sensor was the creation of an appropriate pH sensing probe incorporating the dye developed. This requires a multistep process and the fabrication of the pH sensing probe used in the work is shown schematically in Figure 18.9.

The distal end of a 1000 μm diameter UV multimode fiber was polished in succession with 5, 3, and 1 μm polishing pads and washed with acetone to create a clean, polished surface. The distal end was then immersed in 10% KOH in isopropanol for 30 min with subsequent rinsing in copious amounts of distilled water and dried with compressed nitrogen. After that, it was treated in a 30:70 (v/v) mixture of H_2O_2 (30%) and

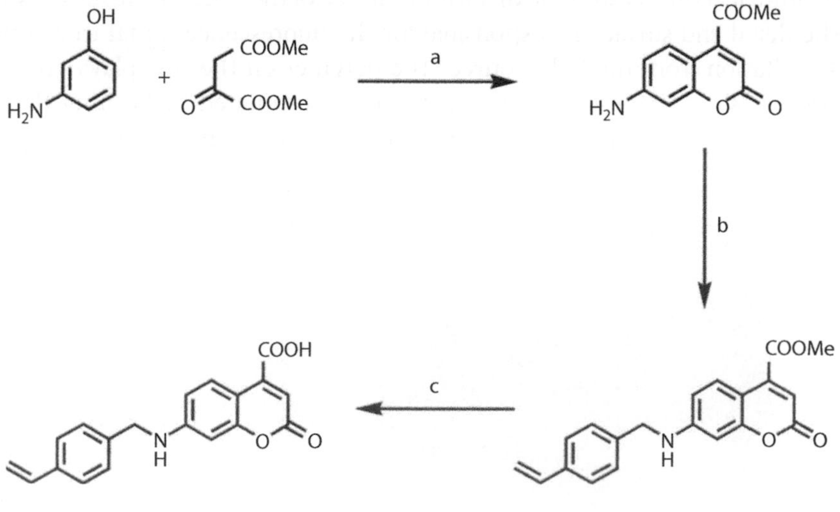

FIGURE 18.8 Preparation of fluorescent monomer 7-VBACC: (a) 110°C, 2 h, 42%; (b) $CH_2{=}CHC_6H_4CH_2Cl$, K_2CO_3, KI, MeCN, 85°C, 50 h, 44%; (c) 1 M NaOH, EtOH/THF (2:1), 50°C, 12 h, 94%.

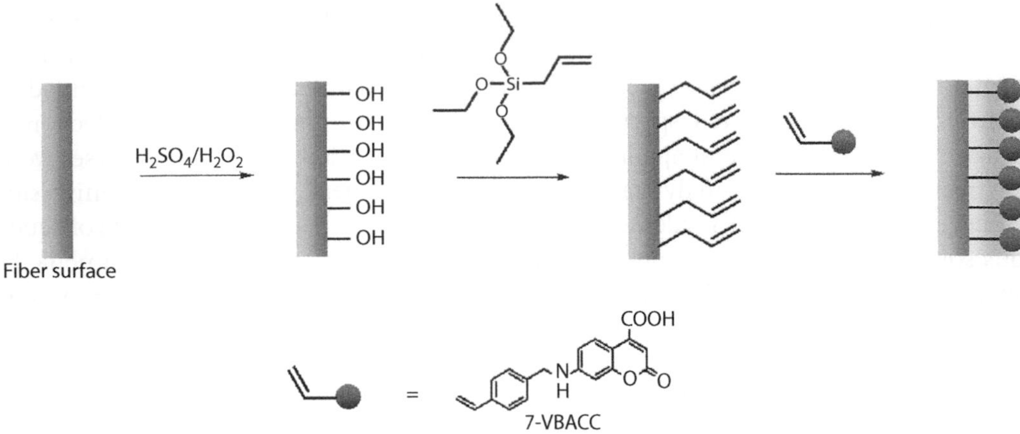

FIGURE 18.9 Preparation of a pH sensor probe: schematic of the processes involved.

H_2SO_4 (conc.) (piranha solution) for 60 min, rinsed in distilled water for 15 min, and dried in an oven at 100°C for 30 min. This procedure leaves the surface with exposed hydroxyl groups, which facilitate bonding of ATES.

The fiber surface was then modified by silanizing for 2 h in a 10% solution of ATES in ethanol. The fiber was washed with methanol and distilled water, respectively, in an ultrasonic bath. Subsequently, it was dried in an oven at 60°C for 2 h. This procedure functionalizes the fiber surface with polymerizable allyl groups.

The monomer stock solution was prepared by dissolving 7-VBACC (6.4 mg, 0.02 mmol), 1,4-bis(acryloyl)piperazine cross-linker (19.4 mg, 0.1 mmol), acrylamide comonomer (2.8 mg, 0.04 mmol) and AIBN initiator (4 mg) in 200 μL dimethylformamide (DMF). The stock solutions were purged thoroughly with argon for 10 min. A small volume of the solution was placed into a capillary tube via syringe and the distal end of the fiber was inserted. They were sealed quickly with PTFE tape and polymerized in an oven at 80°C for 18 h. This procedure forms a polymer layer of the dye, which is covalently bound to both the cylindrical surface and the distal end surface of the fiber. However, only the polymer on the distal end surface is responsible for the fluorescence signal that is produced by direct excitation from the light source. The polymer on the side plays no role in the sensing process since evanescent wave excitation is eliminated by keeping the cladding of the fiber intact. A typical pH probe prepared by this procedure is shown in Figure 18.10

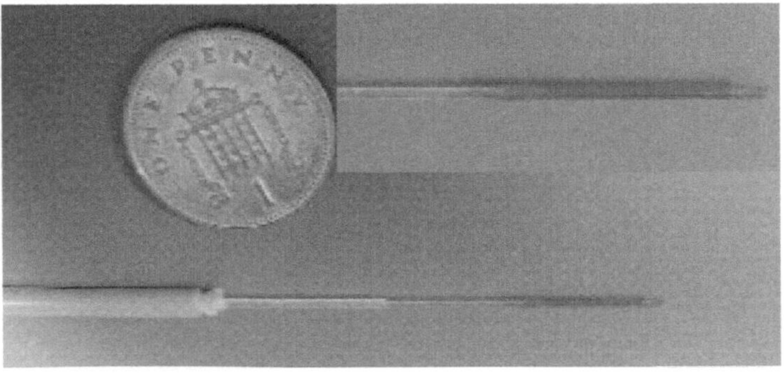

FIGURE 18.10 Typical pH sensor tip prepared in this work showing the active distal end of the sensor.

where the distal end of the probe shows a distinctive coloration due to the presence of the dye. The sensor tip was placed in pH 7 buffer for 24 h to remove all unreacted materials and the excess amount of polymer formed that was not directly bound to the fiber. The probe was then stored in a cool and dark place until use.

18.3.2.3 Setup of a pH Sensor System

Further to the completion of the sensor fabrication, Figure 18.11 shows the setup of a low-pH sensor system. As shown in the figure, light from a LED, emitting at a center wavelength of 400 nm, is coupled through a multimode UV/visible fiber (with hard polymer cladding, 1000 μm silica core, and numerical aperture *NA* of 0.37), using collimation and focusing lenses, into one branch of a 2 × 1 multimode fiber coupler to illuminate the active sensing region being located at the other end of the fiber coupler. Following pH interaction with the active region, a portion of the total light emitted from the sensing layer is collected and guided through the other branch of the fiber coupler to an Ocean Optics USB2000 spectrometer, with the output being displayed on a computer screen.

18.3.2.4 Evaluation of the Intrinsic pH Sensor System

18.3.2.4.1 Properties of the Fluorescent Dye in Solution The absorption spectrum of 7-VBACC shows only one main absorption band in the UV region (Figure 18.12). The dye exhibits a large Stokes shift (the difference in wavelength between the absorption and the fluorescence spectral peaks) of 150 nm, which is very important for the sensor system design to minimize the interference of the excitation light with the fluorescence emission.

The quantum yield of 7-VBACC is reasonably good in ethanol and much higher in H_2O (details are shown in Table 18.1).

In aqueous solution, there is equilibrium between the protonated and deprotonated forms of the dye as shown in Figure 18.13. The deprotonated form is fluorescent and the protonated form is much less so. Therefore, the fluorescence intensity of the dye is higher at higher pH values. To determine the pK_a value for the free dye, a series of pH titration experiments was carried out using 50 mM citrate buffer solutions with different pH values.

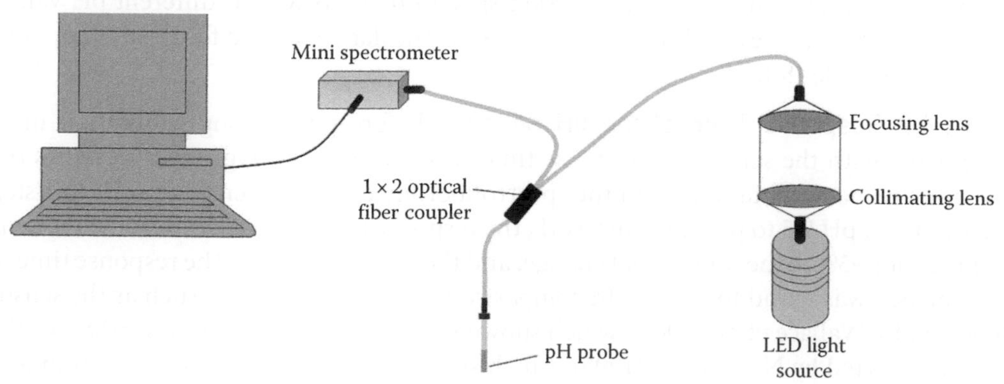

FIGURE 18.11 Low-pH sensor system setup.

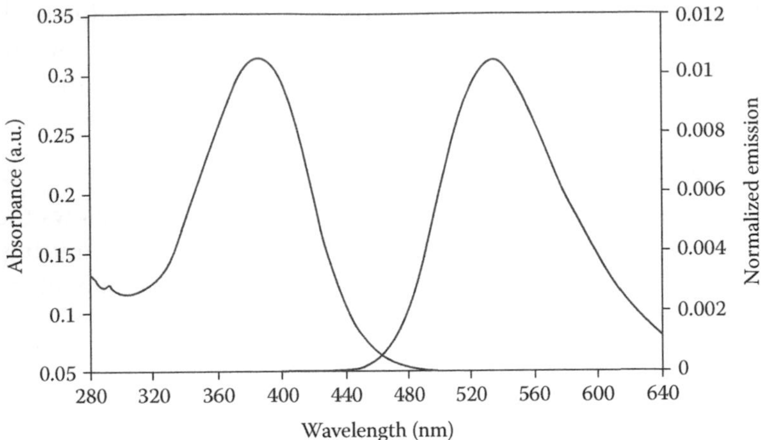

FIGURE 18.12 Absorption (a) and emission (b) spectra of 7-VBACC (14 μM) in H_2O; emission spectra recorded with $\lambda_{ex} = 350$ nm.

Table 18.1 Special Data and pK_a Values of the Fluorescent Dye

Solvent	UV Max (nm)	Emission Max (nm)	Quantum Yield	pK_a
EtOH	384	496	0.565	—
H_2O	387	534	0.146	2.40

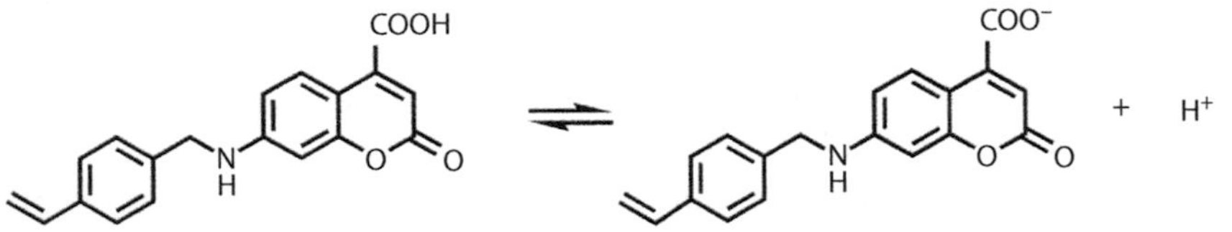

FIGURE 18.13 Equilibrium between the protonated and deprotonated forms of the dye in solution.

In the titration, 50 μL of 0.8 mM dye stock solution was added to 3 mL of buffer in a cuvette, followed by measurement of emission spectra.

The calculation of the pK_a value was performed using a nonlinear fitting method according to Equation 18.14. The emission spectra of 7-VBACC at different pH values and the titration curve are shown in Figure 18.14. The data obtained for the dye are summarized in Table 18.1.

18.3.2.4.2 Response Time of the pH Sensor Before performing calibration measurements with the sensor, its response time was investigated. Figure 18.15 shows the dynamic response obtained from the spectrofluorometer of the sensing probe to a step change from pH 0.5 to pH 6. In this work, the response time is considered to be the time required for 95% of the total signal change, and the measurement of the response time of the optrode was found to be 25 s. In comparison to other pH sensors such as the sensor reported by Wallace et al. (2001), which showed a response time of around 500 s, or the device reported by Netto et al. (1995), which showed a response time of a few minutes, this pH sensor responds much more rapidly. This is likely due to its key design features: both the relatively low thickness of the polymer film and its hydrophilicity. All this has

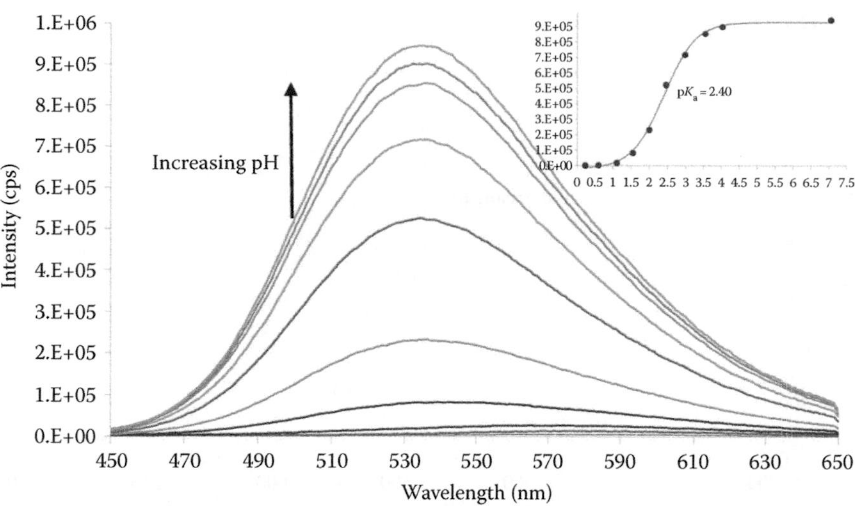

FIGURE 18.14 Emission spectra of 7-VBACC at pH from 0.2 to 7.0. The inset shows the titration plots at 535 nm (λ_{ex} = 380 nm).

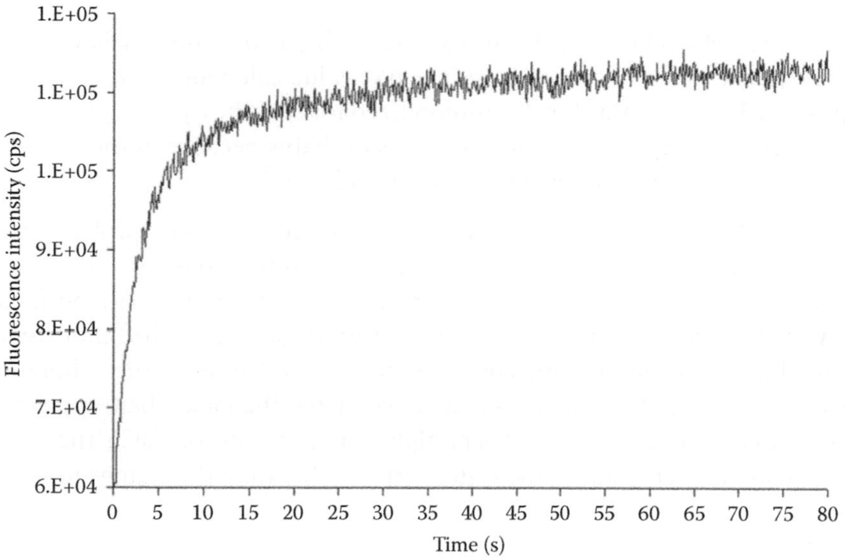

FIGURE 18.15 Dynamic response of the VBACC pH sensor showing the 25 s response time (to 95%).

obvious advantages where a rapid change of pH is to be monitored and a real-time measurement to be achieved.

18.3.2.4.3 Response of the Sensor to Different pH Values The calibration measurements of the sensor characteristics were performed in 50 mM citrate buffer solutions at different pH values (note: citrate does not act as a good buffer at pH higher than 6 and lower than 3; however, citrate was used for all pH values to avoid any differences in fluorescence caused by the difference in buffer composition). The titration curve is shown in Figure 18.16. The sensor probe exhibited an increase in fluorescence intensity

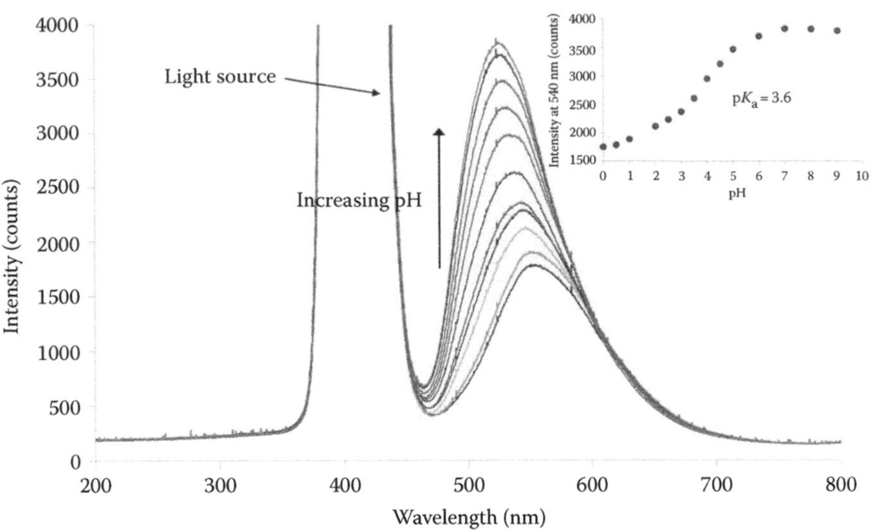

FIGURE 18.16 The evolution of fluorescence spectra of the sensor probe with pH in the range from 0.5 to 7. Inset shows the titration plots at 540 nm.

with increasing pH in the range from 0.5 to 6.0, which conveniently is wider than the dynamic response range of the free dye. The pK_a value calculated using Equation 18.14 for the probe is 3.6. This value for the immobilized form of the dye is slightly higher than that for its free form in solution, and this arises probably because of the decrease in the polarity of the microenvironment (Vasylevska et al. 2007).

18.3.2.4.4 Effect of Ionic Strength Sensitivity to IS can be a serious problem in the cases of optical fiber sensors as it affects pK_a values, thus resulting in errors in pH determination. The effect of IS was investigated with the prepared pH 4 and pH 7 buffer solutions adjusted with NaCl to different ISs ranging from 10 to 1000 mM. The fluorescence intensity obtained for each solution was converted to a pH value using the calibration curve and this is presented in Table 18.2. As can be seen from the table, there appears to be no sensitivity to IS for the sensor, even at very high concentrations of NaCl. The insignificant errors caused are probably due to the system error rather than the change in IS.

Table 18.2 pH Measurements of pH 4 and pH 7 Buffer Solutions with Different IS

IS (mM)	pH 4.00 Buffer Solution	Difference	pH 7.00 Buffer Solution	Difference
0	4.05	0.05	7.02	0.02
10	4.06	0.06	7.11	0.11
50	4.01	0.01	7.09	0.09
100	4.11	0.11	7.04	0.04
200	4.02	0.02	7.04	0.04
500	4.05	0.05	7.07	0.07
1000	4.07	0.07	7.10	0.10
2000	4.10	0.10	7.07	0.07

18.3.2.4.5 Reproducibility and Photostability of the pH Sensor The stability of the probe in terms of storage, its susceptibility to error due to intense irradiation of the sample, and its reproducibility in use are all very critical to the successful application of the system. An evaluation of these parameters was made in order to understand better the performance of the sensor and establish its suitability for industrial applications. The stability of the sensor was tested by calibrating it with buffer solutions at different pH values ranging from 0.5 to 7.0 and recalibrating it after 24 h and then after 5 months. After each calibration, the probe was washed thoroughly with a pH 7.0 buffer, followed by the same procedure with distilled water and then it was stored in the dark until next use. No significant difference was observed between the measurements and the pK_a values calculated: it is very pleasing to note that these are essentially the same even after several months, illustrating the high stability of the sensor scheme produced (Figure 18.17).

Photostability is one of the critical properties of fluorescent indicators and thus of the dye used in this sensor application. In order to test the photostability of the dye, the probe was coupled into the fluorimeter through a dichroic mirror using a fiber bundle. The excitation light (at a wavelength of 400 nm) was launched to the distal end of the probe illuminating the sensing material with light from the intense, high-power Xe lamp of the fluorimeter continuously for 1 h. The fluorescence intensity data from the probe were collected over that period and displayed. As can be seen from Figure 18.18, no photobleaching was observed over the time investigated and with the high flux of photons onto the probe. When compared to the results of other materials, this offers excellent performance: the decrease observed in the fluorescence intensity was 65% for carboxyfluorescein and 10%–13% for iminocoumarin derivatives, again after 60 min of continuous illumination using a mercury lamp. Thus, an important conclusion is that the material prepared using the coumarin fluorophore and synthesized specifically for this application in this work possesses superior photostability, a feature that is critically important with excitation of sensor probes by high-intensity solid-state sources.

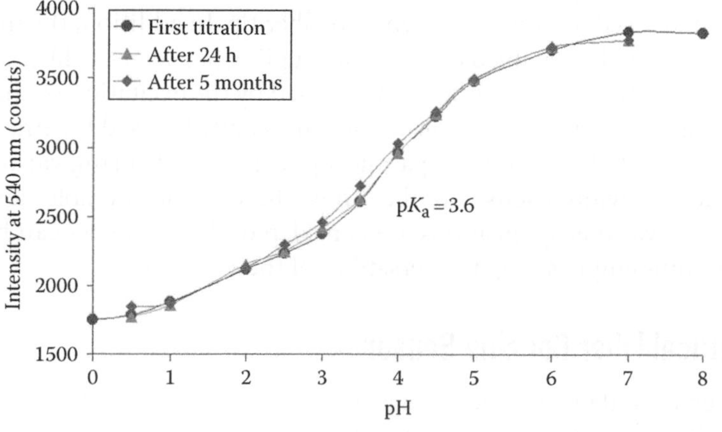

FIGURE 18.17 Titration curves for the sensor probe obtained between 24 h interval and after 5 months.

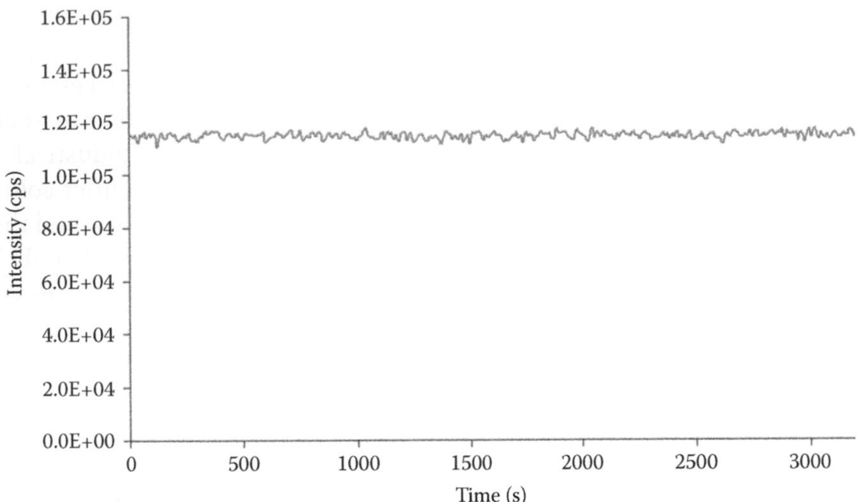

FIGURE 18.18 Fluorescence intensity of the probe at the emission wavelength as function of time during 60 min of continuous illumination by light from a high-power Xe lamp.

18.3.2.5 Comments

The example presented earlier of a fiber-optic intrinsic chemical sensor based on a novel and highly effective approach used for the measurement of low pH values indicates the potential of this technology, in the case expressed in a superior performance compared with conventional solutions (fast response and increase in fluorescence intensity in response to pH in the range from 0.5 to 6.0 with very good stability over a period of several months, in addition to the characteristics of robustness, compactness, and portability).

A further important feature of this type of sensors is that it is potentially inexpensive to produce in quantity and the large Stokes shift shown allows for more accurate measurements due to the minimum level of interference between light source and fluorescence signals generated. For industrial applications, it is necessary to develop a suitable *packaging* to withstand use by inexperienced operators. However, prior work with optical fiber RH sensors, as discussed in Section 18.3.1, has shown an effective design that could be employed to protect the sensitive fiber tip. In addition, the intrinsic sensor design discussed here has enabled direct light coupling between the fiber and the sensor material; therefore, there is a minimum loss caused by the excitation or the fluorescence signal collection. Also, sensors of this design can readily be used together, multiplexed along a single optical fiber or along a parallel optical network, using various techniques to identify each individual sensor probe. Thus, there is considerable flexibility in the approach, and as various applications are considered, these sensors can be tailored for different uses, thus emphasizing the versatility of their design.

18.3.3 Optical Fiber Cocaine Sensor

The total value of trade from illicit drugs is at an estimated \$321 bn per year, which is higher than the cumulative GDP of 88 countries across the world (UN report 2005). Illicit drug use in the United Kingdom remains very high and costs run to £15.4 billion each year (Murray and Tinsley 2006), in areas ranging across policing, detection, related crime, medical care, and family support. Cocaine is one of the most commonly abused drugs and this has led to extensive investigative research efforts for its detection, due to the adverse health effects and related dangers associated with its illicit use (Gilloteaux and Ekwedike 2010).

There are several major analytical methods available for the analysis of cocaine and its metabolites including gas chromatography mass spectrometry (GC/MS) (Popa et al. 2009), HPLC (Trachta et al. 2004), thin-layer chromatography (Antonilli et al. 2001), voltammetry (Oiye et al. 2009), radioimmunoassay (Baumgartner et al. 1982), and enzyme-linked immunosorbent assay (ELISA) (Verebey and Depace 1989). These traditional methods, despite having achieved very good results, are generally expensive, time consuming, and cumbersome for real-time measurements outside the laboratory; some of which also require sample cleanup and derivatization of cocaine prior to analysis.

Biosensors, which rely on the specificities of the binding sites of receptors, enzymes, antibodies, or DNA as biological sensing elements, have been considered as alternative analytical devices due to their specificity, portability, speed, and low cost (Wang 2006). Biosensors for cocaine based on monoclonal antibodies (Meijler et al. 2005) and especially aptamers (Li et al. 2007, Zhang and Johnson 2009) have been developed in recent years. However, these sensors suffer from certain limitations in light of their potential practical applications in the field due to the fragile and unstable nature of the biological recognition elements. Therefore, the development of stable, compact, and portable chemical sensing systems, which are capable of real-time detection of the target drug, remains a compelling goal. The system described in this section illustrates how the fiber-optic sensing technology can contribute to this objective.

18.3.3.1 Sensing Methodology

Molecular imprinting has been extensively demonstrated over the last three decades as a versatile technique for the preparation of synthetic molecular receptors capable of the selective recognition of given target molecules. The approach is based on the self-assembly of a template molecule with polymerizable monomers possessing functional group(s) interacting with the template (Haupt and Mosbach 2000). After polymerization, the template is removed, leaving vacant recognition sites that are complementary in shape and functional groups to the original template. Molecularly imprinted polymers (MIPs) provide an exciting alternative to biological receptors as recognition elements in chemical sensors (Alexander et al. 2006).

Here, an example of a robust fiber-optic chemical sensor for cocaine detection based on the combination of molecular imprinting (as a method for generating chemically selective binding sites) and fluorescence modulation (as a means of signaling the presence and concentration of the analyte) is presented. The attraction of this approach lies in the advantages offered both by the optical fiber in terms of small size, immunity to EM interference, remote sensing capability, resistance to chemicals, and biocompatibility (Lee et al. 2001) and by the synthetic polymer receptor in terms of robustness, thermal and chemical stabilities, low cost, and long shelf life (Haupt and Mosbach 2000).

The MIP receptor that is selective for cocaine was covalently bonded to the distal end of the optical fiber, which facilitated rapid and highly sensitive detection. Acrylamidofluorescein (AAF) was used as fluorescent functional monomer interacting

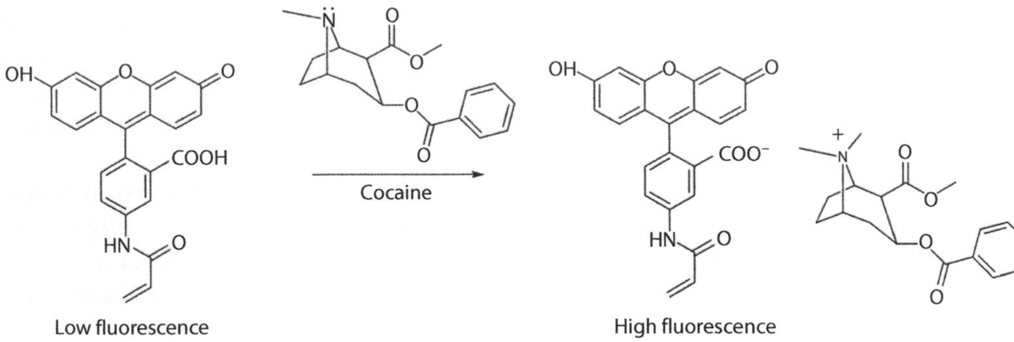

FIGURE 18.19 Interaction between AAF and cocaine.

with the template cocaine. The sensing mechanism depends on changes in the frontier orbitals of fluorescein, which occur when it is deprotonated by a base. The deprotonated form is fluorescent and the protonated form is much less so. In the presence of cocaine, the carboxylate group of AAF is deprotonated. Cocaine acts as a base in the ion pair complex, accepting a proton from AAF and leading to an increase in the observed fluorescence intensity (Figure 18.19).

The imprinting and sensing strategy is illustrated in Figure 18.20. A complex is formed between the functional group –COOH on the fluorophore and the amine group on the template/analyte. The complex is copolymerized with a cross-linking monomer and comonomer on the surface of the fiber, which has been functionalized with polymerizable groups. Then, the template/analyte is extracted from the polymer. The resulting MIP formed on the fiber contains recognition sites incorporating the fluorophore and exhibits an increase in fluorescence intensity selectively in the presence of the template/analyte. As a result, the selectivity of the sensor has been designed to arise from the functional group of the fluorophore and from the shape of the cavity.

18.3.3.2 Design of a Cocaine Sensor System

18.3.3.2.1 Sensor Probe Fabrication All chemicals were of analytical grade, purchased from Sigma-Aldrich, and were used without further purification, except for ethylene glycol dimethacrylate (EDMA), which was distilled under reduced pressure prior to use. AAF was prepared from fluoresceinamine according to the literature procedure (Munkholm et al. 1990) as shown in Figure 18.21. All solvents used were of HPLC grade from Fisher Scientific. Dry ethanol and dry acetonitrile for probe fabrication were taken from sealed bottles under argon. All aqueous solutions were prepared using distilled water. Absorption and fluorescence measurements of aqueous solutions containing fluorophore were carried out on a PerkinElmer Lambda 35 spectrophotometer and a Horiba Jobin Yvon FluoroMax-4 spectrofluorometer system with FluorEssence as driving software, respectively.

The fabrication of the cocaine sensing probe requires a multi-step process, which is described below. The distal end of a 1000 μm diameter UV multimode fiber was polished in succession with 5, 3, and 1 μm polishing pads and washed with acetone. The distal end was then immersed in 10% KOH in isopropanol for 30 min with subsequent rinsing in copious amounts of distilled water and dried with compressed nitrogen. After that, it

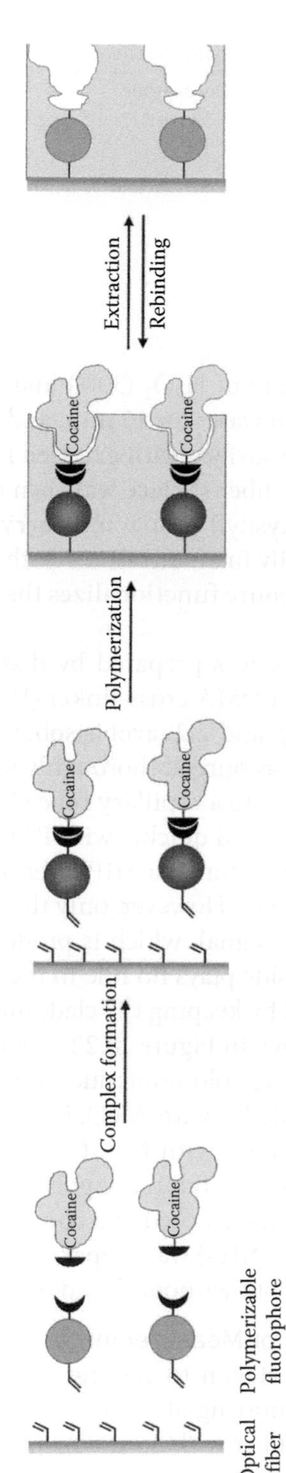

FIGURE 18.20 The preparation of a cocaine sensing MIP on the surface of the optical fiber, which exhibits fluorescence changes upon template binding.

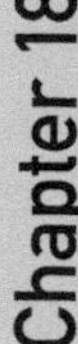

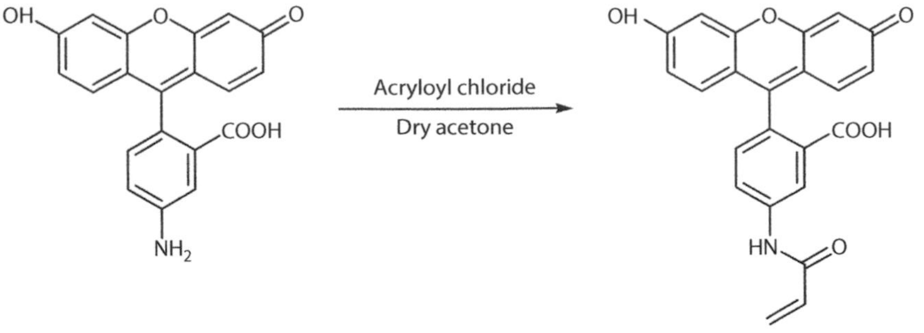

OH, O, O, COOH, NH2 — Acryloyl chloride / Dry acetone → OH, O, O, COOH, HN, O

FIGURE 18.21 Preparation of AAF.

was treated in a 30:70 (v/v) mixture of H_2O_2 (30%) and H_2SO_4 (conc.) (piranha solution) for 30 min, rinsed in distilled water for 15 min, and dried in an oven at 100°C for 30 min. This procedure leaves the surface with exposed hydroxyl groups, which facilitate bonding of a silane agent. The fiber surface was then modified by silanizing for 2 h in a 10% solution of 3-(trimethoxysilyl) propyl methacrylate in dry ethanol. The fiber was washed with ethanol repeatedly in an ultrasonic bath. Subsequently, it was dried in an oven at 70°C for 2 h. This procedure functionalizes the fiber surface with polymerizable acrylate groups.

The prepolymerization mixture was prepared by dissolving cocaine (6.1 mg, 0.02 mmol), AAF (4.0 mg, 0.01 mmol), EDMA cross-linker (150.9 μL, 0.8 mmol), acrylamide comonomer (10.0 mg, 0.14 mmol), and 2,2′-azobisisobutyronitrile initiator (1.1 mg) in 222 μL dry MeCN. The solution was purged thoroughly with argon for 10 min. A small volume of the solution was placed into a capillary tube via syringe and the distal end of the fiber was inserted. They were sealed quickly with PTFE tape and polymerized in an oven at 70°C for 16 h. This procedure forms a MIP layer on both the cylindrical surface and the distal end surface of the fiber. However, only the MIP on the distal end surface is responsible for the fluorescence signal, which is produced by direct excitation from the light source. The MIP on the side plays no role in the sensing process since evanescent wave excitation is eliminated by keeping the cladding of the fiber intact. The probe prepared by this procedure is shown in Figure 18.22, where it can be seen that the distal end of the probe shows a distinctive coloration due to the presence of the fluorophore. The sensor tip was washed repeatedly with MeOH–AcOH (8:2, v/v) in an ultrasonic bath, followed by the same procedure with MeOH alone to remove the template and all unreacted materials and the excess amount of polymer formed that was not directly bound to the fiber. The probe was then stored in a cool and dark place until use. A control probe (nonimprinted polymer [NIP]) was prepared at the same time under identical conditions, using the same recipe but without the addition of the template cocaine.

18.3.3.2.2 Sensor System Setup for Measurements The cocaine sensor system setup used for the measurements undertaken to calibrate the probe is presented in Figure 18.23, where light from a LED, emitting at a center wavelength of 375 nm, is coupled through a multimode UV/visible fiber with hard polymer cladding, 1000 μm silica core, and numerical aperture (*NA*) of 0.37, using collimation and focusing lenses, into a 2 × 1 Y fiber coupler, made using two multimode UV/visible fibers with hard polymer cladding, 600 μm silica core, and 0.37 *NA*, which is connected to the sensor probe with

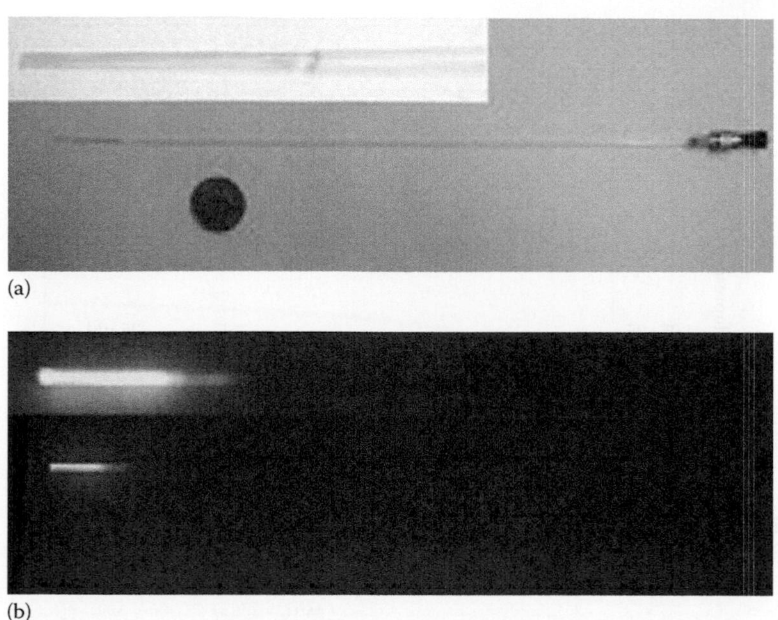

FIGURE 18.22 Cocaine probe prepared in this work showing the active distal end of the sensor (a) under normal conditions and (b) when 375 nm UV light was launched to the end of the fiber.

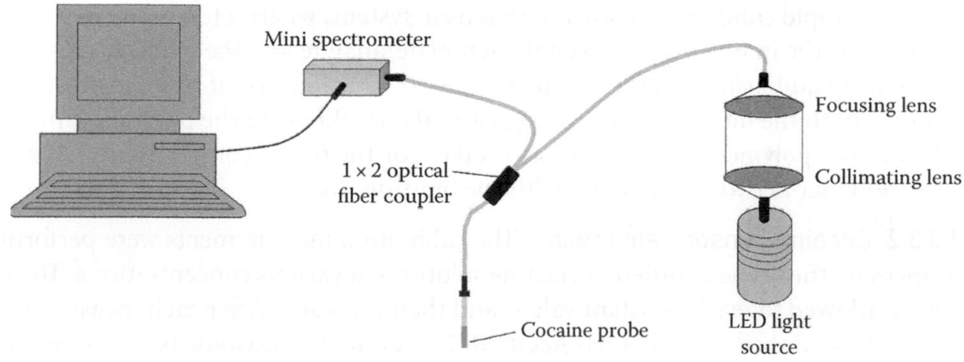

FIGURE 18.23 Experimental setup used in the evaluation of the performance of the probe designed.

the active sensing region being located at the distal end of the fiber. Following interaction of cocaine with the active region, a portion of the total light emitted from the sensing layer is collected and guided through the other end of the fiber coupler to an Ocean Optics USB2000 spectrometer, the output from which is then displayed on a computer screen.

18.3.3.3 Implementation and Evaluation of the Cocaine Sensor System

18.3.3.3.1 Response Time of the Cocaine Sensor Before performing measurements to calibrate the sensor, its response time was investigated. Figure 18.24 shows the dynamic response of the sensor obtained from the spectrofluorometer to a step change from no cocaine present (0 μM) to 25 μM and to 250 μM cocaine in H_2O/MeCN 9:1.

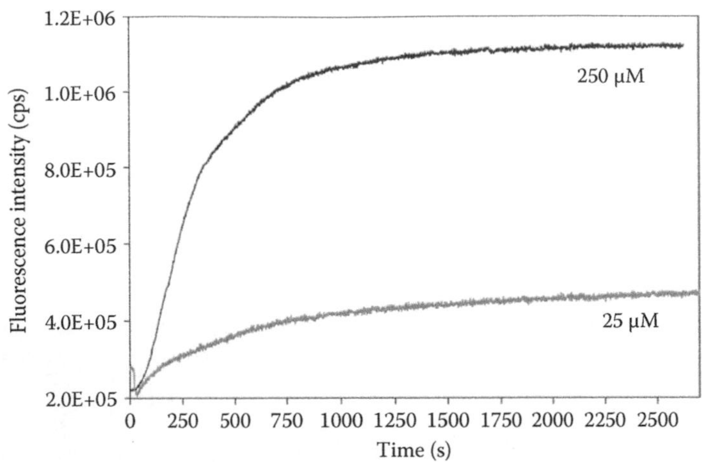

FIGURE 18.24 Dynamic response of the sensor probe at 515 nm (excitation at 375 nm) showing the 15 min response time (to 95%).

Although around 70% of the total signal change occurred within 5 min, it took around 15 min for the sensor to attain equilibrium (to 95%) in 250 μM cocaine and 20 min in 25 μM cocaine. The higher concentration of cocaine appeared to give a slightly quicker response time. However, the difference was not significant. This response time is considered to be rapid compared to other MIP sensor systems where a few hours incubation is required for the interaction between the template/analyte and the binding sites in the MIP to reach equilibrium (Turkewitsch et al. 1998). This important result is most probably due to both the intrinsic sensor design and the thickness of the polymer film since the thicker the polymer layer, the longer it takes for the target compound to penetrate into the polymer network to interact with the binding sites.

18.3.3.3.2 Cocaine Sensor Calibration The calibration measurements were performed by immersing the probe in different cocaine solutions at various concentrations. The signals were allowed to reach constant values and then recorded. After each measurement, the probe was washed with MeOH–AcOH (8:2, v/v) in an ultrasonic bath, followed by the same procedure with MeOH alone to remove bound cocaine. Initially, experiments were carried out in MeCN/H_2O 9:1. MeCN was used because the MIP was prepared in MeCN, so its recognition properties would be expected to be best in MeCN (since this should result in no loss of selectivity due to MIP swelling) (Kempe and Mosbach 1991). H_2O was added at 10% (v/v) in order to reduce nonspecific binding. The sensor exhibited an increase in fluorescence intensity with increasing cocaine concentration in the range from 0 to 250 μM (Figure 18.25a). At higher concentrations of cocaine, no further change of intensity was observed due to the saturation of all available binding sites. It was also interesting to see if the sensor could work in aqueous media where biological recognition mainly occurs. Measurements were carried out in a manner similar to those of Figure 18.7a, but the solvent system was replaced by H_2O/MeCN 9:1 (MeCN was added to solubilize the analyte). The sensor showed a greater increase in fluorescence in the aqueous than in the organic solution (Figure 18.7b), which is attributed to the difference between the photophysical properties of the fluorophore in aqueous and in organic

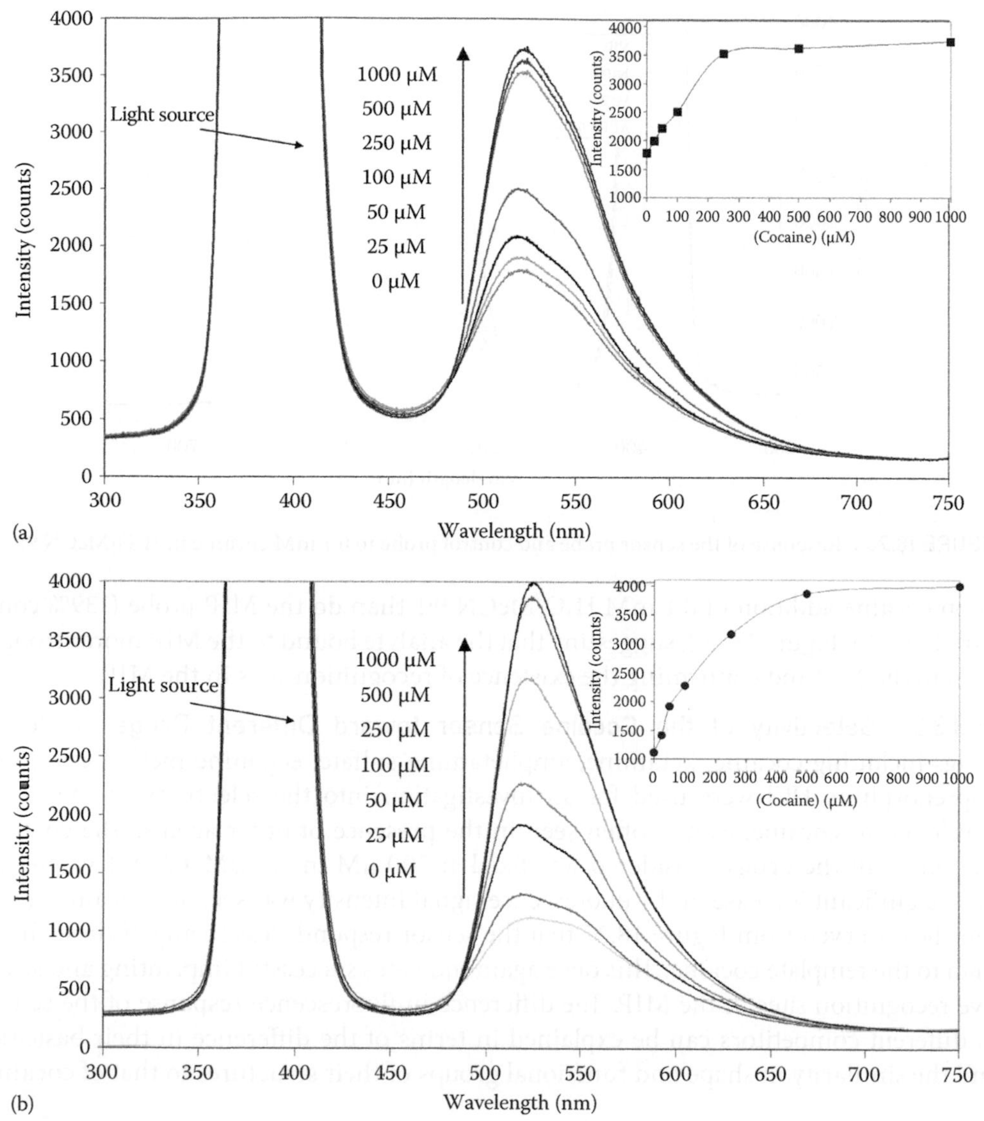

FIGURE 18.25 Response of the sensor to cocaine in the concentration range from 0 to 1000 μM in (a) $MeCN/H_2O$ 9:1 and (b) in $H_2O/MeCN$ 9:1. Insets show the dependence of emission maximum on cocaine concentration.

media. The dynamic response range of the sensor in aqueous solution is also wider, from 0 up to 500 μM. This arises because noncovalent interactions between cocaine and the functional groups in the MIP were weaker in H_2O, and thus the available binding sites were not fully occupied until higher concentrations of cocaine were used.

The lower limit of detection of the system may vary since it depends on the type and sensitivity of detector used. With the Ocean Optics minispectrometer used in this work, the lowest concentration of cocaine that can cause a distinguishable change in fluorescence intensity is 2 μM. The response of the control probe (NIP) to cocaine was also studied, and it was observed that the NIP probe showed a lesser increase in fluorescence

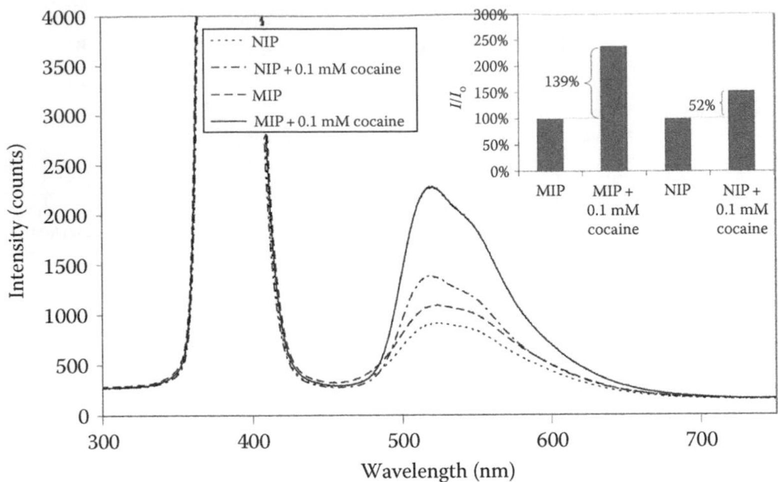

FIGURE 18.26 Response of the sensor probe and control probe to 0.1 mM cocaine in H_2O/MeCN 9:1.

upon cocaine addition of 0.1 mM H_2O/MeCN 9:1 than do the MIP probe (139% compared to 52%, Figure 18.26), suggesting that the analyte bound to the MIP more strongly than to the NIP and confirming the existence of recognition sites in the MIP.

18.3.3.3.3 Selectivity of the Cocaine Sensor toward Different Drugs Different drugs including cocaine, ketamine, amphetamine sulfate, ecgonine methyl ester, and buprenorphine HCl were used for an investigation into the selectivity of the probe developed to cocaine, as it is often seen in the presence of other agents. The concentration of all the drugs considered was fixed at 500 μM in H_2O/MeCN 9:1 where the most significant increase in the fluorescence signal intensity was seen for cocaine. It can thus be observed from Figure 18.27 that the sensor responds less to any of these drugs than to the template cocaine. This once again indicates successful imprinting and selective recognition sites in the MIP. The difference in fluorescence response of the sensor to different competitors can be explained in terms of the difference in their basicities and the similarity in shape and functional groups of their structures to that of cocaine.

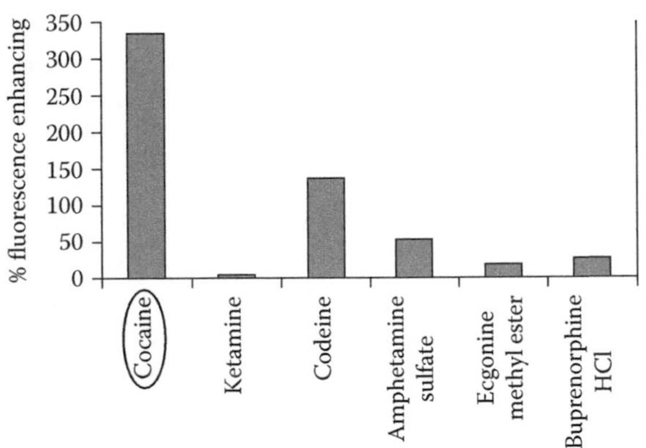

FIGURE 18.27 Response of the sensor probe to different drugs with concentrations of 500 μM in H_2O/MeCN 9:1.

Significantly higher reactivity of the sensor for codeine compared to that for other competitors may also be due to the availability of more functional groups on the codeine molecule, which are able to interact noncovalently with the binding sites in the MIP. It should also be noted that some of the drugs tested were in the salt forms, not free bases, and the presence of acids might affect the test results obtained.

18.3.3.3.4 Reproducibility and Photostability The stability of the probe in terms of storage, its susceptibility to error due to intense irradiation of the sample, and its reproducibility in use is very critical to the successful application of the system. A preliminary evaluation of these parameters was made in order to understand better the performance of the sensor. The stability of the sensor was tested by calibrating it with different cocaine concentrations ranging from 0 to 500 μM and recalibrating it after 24 h and then 1 month. After each calibration, the probe was washed thoroughly with MeOH–AcOH (8:2, v/v) in an ultrasonic bath, followed by the same procedure with MeOH alone to remove bound cocaine and then it was stored in the dark until next use. No significant difference was observed between the measurements, and the results obtained were found to be fairly reproducible even after 1 month. In order to test the photostability of the sensor, it was coupled into the fluorimeter through a dichroic mirror using a fiber bundle. The excitation light at 375 nm was launched to the distal end of the probe consisting of the sensing material by the high-power Xe lamp of the fluorimeter continuously for 1 h. The fluorescence intensity of the probe was dynamically collected. As can be seen from Figure 18.28, very little photobleaching (less than 1%) was observed over the time investigated. Compared to the decrease in fluorescence intensity by 65% observed for carboxyfluorescein or by 10%–13% observed for iminocoumarin derivatives on their free forms in solution after 60 min of continuous illumination using a mercury lamp, the MIP prepared in this work possesses superior photostability, a feature that is critically important with excitation by high-intensity solid-state sources.

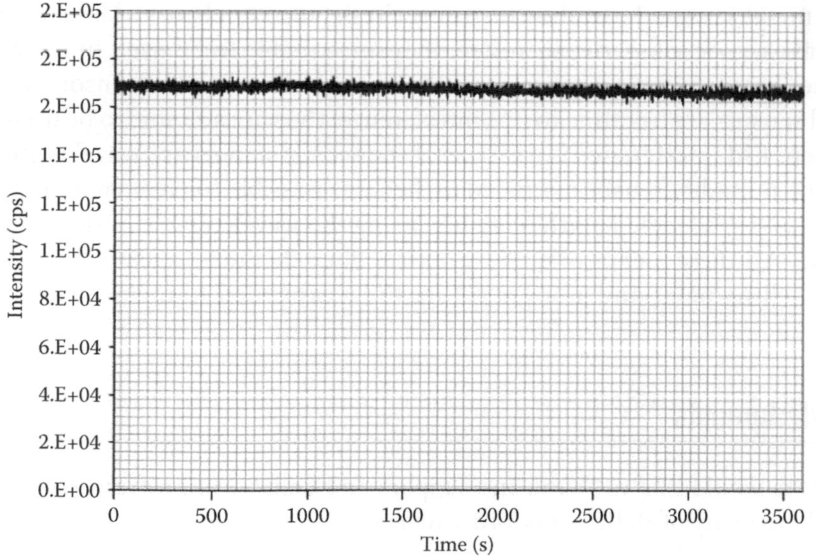

FIGURE 18.28 Fluorescence intensity of the probe at the emission wavelength as function of time during 60 min of continuous illumination by a high-power Xe lamp.

18.3.3.4 Comments

The fiber-optic sensing structure described earlier is an example of an effective approach using MIP technology for the development of a specific drug sensor for cocaine detection, which shows both superior performance and fast response. Indeed, the cocaine sensor developed has demonstrated an increase in fluorescence intensity in response to cocaine in the concentration range of 0–500 μM in aqueous acetonitrile mixtures with good selectivity and reproducibility over 1 month. Once its performance is further refined, this type of sensors will potentially make a significant impact on the homeland security enhancement as it can provide technical evidence on the spot with minimum invasion. Based on the earlier successful evaluation of this novel, robust, compact, and portable cocaine sensor, the MIP technology has shown promise both for the expansion of drug detection and for next-stage commercial exploitation.

18.4 Conclusions and Future Outlook

This chapter has addressed generic principles of fiber-optic chemical sensing, as well as some chemical sensor core designs using an optical fiber as a generic sensor substrate (based upon a generic spectroscopic sensor design protocol), illustrated with examples derived from City University London R&D in this field. The underpinning sensing mechanism has varied from the induced physical change of an FBG to the chemical interaction between the target molecules and the sensor materials, thus inducing the fluorescence signal change. The corresponding spectral signal change, as a result, has been demonstrated through either the spectral shift or the fluorescence intensity variation.

These optical fiber chemical sensors developed have shown a huge potential to address various industrial challenges. Compared to electrochemical sensors, optical fiber chemical sensors have shown stronger resistance to chemical attacks due to the use of glass sensor substrate; therefore, they are suitable to be used in harsh environment where contamination may pose a threat to electrochemical sensors. In addition, chemical sensors are immune to EM interference, thus are suitable to be used in conditions where EM interference is an issue. If the sensor signals are wavelength encoded, such as the humidity sensor discussed earlier, it is possible to create a multiplexed sensor array within a single length of a fiber, using the wavelength-division-multiplexing technique. All these merits offered by optical fiber chemical sensors have given them a unique selling point and a bright future to be expanded to meet various industrial needs.

Acknowledgments

The authors would like to thank the support and collaboration from colleagues from Queen's University of Belfast, Oxford University, and Home Office Centre for Applied Science and Technology.

The authors would also like to thank the U.K. Engineering and Physical Sciences Research Council for the funding support via a number of research projects.

References

Alexander, C., H. S. Andersson, L. I. Andersson et al. 2006. Molecular imprinting science and technology: A survey of the literature for the years up to and including 2003. *Journal of Molecular Recognition* 19: 106–180.

Antonilli, L., C. Suriano, M. C. Grassi et al. 2001. Analysis of cocaethylene, benzoylecgonine and cocaine in human urine by high-performance thin-layer chromatography with ultraviolet detection: A comparison with high-performance liquid chromatography. *Journal of Chromatography B* 751: 19–27.

Arregui, F. J., M. Otano, C. Fernandez-Valdivielso et al. 2002. An experimental study about the utilization of Liquicoat solutions for the fabrication of pH optical fiber sensors. *Sensors and Actuators B: Chemical* 87: 289–295.

Baldini, F., A. Giannetti, and A. A. Mencaglia. 2007. Optical sensor for interstitial pH measurements. *Journal of Biomedical Optics* 12: 024024.

Baumgartner, W. A., C. T. Black, P. F. Jones et al. 1982. Radioimmunoassay of cocaine in hair—Concise communication. *Journal of Nuclear Medicine* 23: 790–792.

Besson, T., G. Coudert, and G. Guillaumet. 1991. Synthesis and fluorescent properties of some heterobifunctional and rigidized 7-aminocoumarins. *Journal of Heterocyclic Chemistry* 28: 1517–1523.

Carey, W. P., M. D. Degrandpre, and B. S. Jorgensen. 1989. Polymer-coated cylindrical wave-guide absorption sensor for high acidities. *Analytical Chemistry* 61: 1674–1678.

Drexhage, K. H. 1976. Fluorescence efficiency of laser-dyes. *Journal of Research of the National Bureau of Standards Section A—Physics and Chemistry* 80: 421–428.

Eaton, D. F. 1988. Reference materials for fluorescence measurement. *Pure and Applied Chemistry* 60: 1107–1114.

Egawa, Y., R. Hayashida, and J. I. Anzai. 2006. Multilayered assemblies composed of brilliant yellow and poly(allylamine) for an optical pH sensor. *Analytical Sciences* 22: 1117–1119.

Eggeling, C., J. Widengren, R. Rigler et al. 1998. Photobleaching of fluorescent dyes under conditions used for single-molecule detection: Evidence of two-step photolysis. *Analytical Chemistry* 70: 2651–2659.

Ensafi, A. A. and A. Kazemzadeh. 1999. Optical pH sensor based on chemical modification of polymer film. *Microchemical Journal* 63: 381–388.

Fujii, T., A. Ishii, Y. Kurihara et al. 1993. Multiple fluorescence-spectra of fluorescein molecules encapsulated in the silica xerogel prepared by the sol-gel reaction. *Research on Chemical Intermediates* 19: 333–342.

Gilloteaux, J. and N. N. Ekwedike. 2010. Cocaine causes atrial Purkinje fiber damage. *Ultrastructural Pathology* 34: 90–98.

Goicoechea, J., C. R. Zamarreno, I. R. Matias et al. 2008. Optical fiber pH sensors based on layer-by-layer electrostatic self-assembled Neutral Red. *Sensors and Actuators B: Chemical* 132: 305–311.

Grant, S. A., K. Bettencourt, P. Krulevitchm et al. 2001. In vitro and in vivo measurements of fiber optic and electrochemical sensors to monitor brain tissue pH. *Sensors and Actuators B: Chemical* 72: 174–179.

Greenspan, L. 1977. Humidity fixed points of binary saturated aqueous solutions. *Journal of Research of the National Bureau of Standards Section A—Physics and Chemistry* 81A: 89–96.

Haupt, K. and K. Mosbach. 2000. Molecularly imprinted polymers and their use in biomimetic sensors. *Chemical Reviews* 100: 2495–2504.

Kempe, M. and K. Mosbach. 1991. Binding-studies on substrate-and enantio-selective molecularly imprinted polymers. *Analytical Letters* 24: 1137–1145.

Lee, C. Y. and G. B. Lee. 2005. Humidity sensors: A review. *Sensor Letters* 3: 1–14.

Lee, S. T., J. Gin, V. P. N. Nampoori et al. 2001. A sensitive fiber optic pH sensor using multiple sol-gel coatings. *Journal of Optics A—Pure and Applied Optics* 3: 355–359.

Li, Y., H. L. Qi, Y. Peng et al. 2007. Electrogenerated chemiluminescence aptamer-based biosensor for the determination of cocaine. *Electrochemistry Communications* 9: 2571–2575.

Lin, J. 2000. Recent development and applications of optical and fiber-optic pH sensors. *Trends in Analytical Chemistry* 19: 541–552.

Liu, Z. H., J. F. Liu, and T. L. Chen. 2005. Phenol red immobilized PVA membrane for an optical pH sensor with two determination ranges and long-term stability. *Sensors and Actuators B: Chemical* 107: 311–316.

McDonagh, C., C. S. Burke, and B. D. MacCraith. 2008. Optical chemical sensors. *Chemical Review* 108: 400–422.

Meijler, M. M., G. F. Kaufmann, L. W. Qi et al. 2005. Fluorescent cocaine probes: A tool for the selection and engineering of therapeutic antibodies. *Journal of the American Chemical Society* 127: 2477–2484.

Munkholm, C., D. R. Parkinson D. R. Walt et al. 1990. Intramolecular fluorescence self-quenching of fluoresceinamine. *Journal of the American Chemical Society* 112: 2608–2612.

Murray, R. and L. Tinsley (eds). 2006. Measuring different aspects of problem drug use: Methodological developments. Home Office Online Report 16/06, 2006. London, U.K.: Home Office.

Narayanaswamy, R. 1993. Optical chemical sensors: Transduction and signal processing. *Analyst* 118: 317–322.

Netto, E. J., J. I. Peterson, M. McShane et al. 1995. A fiber-optic broad-range pH sensor system for gastric measurements. *Sensors and Actuators B: Chemical* 29: 157–163.

Oiye, E. N., N. B. de Figueiredo, J. F. de Andrade et al. 2009. Voltammetric determination of cocaine in confiscated samples using a cobalt hexacyanoferrate film-modified electrode. *Forensic Science International* 192: 94–97.

Peterson, J. I., S. R. Goldstein, R. V. Fitzgerald et al. 1980. Fiber optic pH probe for physiological use. *Analytical Chemistry* 52: 864–869.

Popa, D. S., L. Vlase, S. E. Leucuta et al. 2009. Determination of cocaine and benzoylecgonine in human plasma by LC-MS/MS. *Farmacia* 57: 301–308.

Safavi, A. and M. Bagheri. 2003. Novel optical pH sensor for high and low pH values. *Sensors and Actuators B: Chemical* 90: 143–150.

Simek, M., L. Jisova, and D. W. Hopkins. 2002. What is the so-called optimum pH for denitrification in soil? *Soil Biology & Biochemistry* 34: 1227–1234.

Sloan, W. D. and M. Uttamlal. 2001. A fiber-optic calcium ion sensor using a calcein derivative. *Luminescence* 16: 179–186.

Staneva, D. and R. Betcheva. 2007. Synthesis and functional properties of new optical pH sensor based on benzo[de]anthracen-7-one immobilized on the viscose. *Dyes and Pigments* 74: 148–153.

Trachta, G., B. Schwarze, B. Sagmuller et al. 2004. Combination of high-performance liquid chromatography and SERS detection applied to the analysis of drugs in human blood and urine. *Journal of Molecular Structure* 693: 175–185.

Turkewitsch, P., B. Wandelt, G. D. Darling et al. 1998. Fluorescent functional recognition sites through molecular imprinting. A polymer-based fluorescent chemosensor for aqueous cAMP. *Analytical Chemistry* 70: 2025–2030.

UN report puts world's illicit drug trade at estimated $321b. Boston.com. June 30, 2005, cited from www.wikipedia.com.

Uttamlal, M., W. D. Sloan, and D. Millar. 2002. Covalent immobilization of fluorescent indicators in photo- and electropolymers for the preparation of fiber-optic chemical sensors. *Polymer International* 51: 1198–1206.

Vasylevska, A. S., A. A. Karasyov, S. M. Borisov et al. 2007. Novel coumarin-based fluorescent pH indicators, probes and membranes covering a broad pH range. *Analytical and Bioanalytical Chemistry* 387: 2131–2141.

Verebey, K. and A. Depace. 1989. Rapid confirmation of enzyme multiplied immunoassay technique (emit) cocaine positive urine samples by capillary gas-liquid-chromatography nitrogen phosphorus detection (Glc Npd). *Journal of Forensic Sciences* 34: 46–52.

Wallace, P. A., N. Elliott, M. Uttamlal et al. 2001. Development of a quasi-distributed optical fiber pH sensor using a covalently bound indicator. *Measurement Science and Technology* 12: 882–886.

Wang, J. 2006. Electrochemical biosensors: Towards point-of-care cancer diagnostics. *Journal of Biosensors & Bioelectronics* 21: 1887–1892.

Wiczling, P., M. J. Markuszewski, M. Kaliszan et al. 2005. Combined pH/organic solvent gradient HPLC in analysis of forensic material. *Journal of Pharmaceutical and Biomedical Analysis* 37: 871–875.

Wolfbeis, O. S. 2002. Fiber-optic chemical sensors and biosensors. *Analytical Chemistry* 74: 2663–2677.

Wolfbeis, O. S. 2008. Fiber-optic chemical sensors and biosensors. *Analytical Chemistry* 80: 4269–4283.

Yeo, T. L., M. A. C. Cox, L. F. Boswell et al. 2006. Optical fiber sensors for monitoring ingress of moisture in structural concrete. *Review of Scientific Instruments* 77: 055108.

Yeo, T. L., T. Sun, K. T. V. Grattan et al. 2005. Characterisation of a polymer-coated fiber Bragg grating sensor for relative humidity sensing. *Sensors and Actuators B* 110: 148–156.

Zhang, C. Y. and L. W. Johnson. 2009. Single quantum-dot-Based aptameric nanosensor for cocaine. *Analytical Chemistry* 81: 3051–3055.

19. Industrial Fiber Strain Gauge Sensors

Francisco Araújo
INESC TEC
FiberSensing

Luís Ferreira
INESC TEC
FiberSensing

19.1 Introduction

The impact of critical infrastructures in the economy is unquestionable, and improving their design in order to increase efficiency, safety, security, and reliability is a common challenge for most developed countries. Nowadays, most of the major engineering structures employ innovative materials, advanced structural solutions, and sophisticated construction procedures, being the real-time assessment of the structure performance mandatory since day-one construction to long-term service. In many of the most demanding industrial

Handbook of Optical Sensors. Edited by José Luís Santos and Faramarz Farahi © 2015 CRC Press/Taylor & Francis Group, LLC. ISBN: 9781439866856.

applications, monitoring of structural performance is thus becoming increasingly important in order to guarantee the safety of assets while reducing maintenance and inspection costs. Over the last decades, fiber-optic sensors have become one of the most attractive technologies to implement in large-scale sensing networks (Crossley 2002).

Fiber sensors can be designed such that the measurand interacts with one or several optical parameters of the guided light (intensity, phase, polarization, and wavelength). Independent of the sensor type, the light modulation must be processed into an optical intensity signal at the receiver, which subsequently performs a conversion into an electric signal. In general, the main interest in this type of sensors comes from the fact that the optical fiber itself offers numerous operational benefits. It is electromagnetically passive, so it can operate in high and variable electric field environments (like those typical of the electric power industry) and where there is explosion risk; it is chemically and biologically inert since the basic transduction material (silica) is resistant to most chemical and biological agents; its packaging can be physically small and lightweight. Considering the intrinsic low optical attenuation of the fiber (around 0.2 dB/km), it is possible to attain distributed sensing, that is, determine the measurand as a function of the position along the length of the fiber interrogating from only one end. Also, the optical fiber can be operated over very long transmission lengths, so the sensor can easily be placed kilometers away from the monitoring station. In addition to this, it is also possible to perform multiplexed measurements using large arrays of remote sensors, operated from a single optical source and detection unit, with no active optoelectronic components located in the measurement area, thereby retaining electromagnetic passiveness and environmental resistance (Ansari and Libo 1998, Tao et al. 2000, López-Higuera 2002, Wan and Leung 2007, Gagliardi et al. 2010, Cusano et al. 2011, Pevec and Donlagic 2011, Yang et al. 2007).

There are several mechanisms that can be used to measure strain with optical fibers. Here we consider intrinsic sensing, that is, situations where the measurand interacts with light that keeps propagating in the optical fiber. In such cases, the measurement principle stands invariably in the strain-induced variation of the optical path—that is, the variation of the product nL, where n is the refractive index of the optical fiber core and L is the fiber length between the fixing points (gauge length).

One specific class of fiber-optic intrinsic sensors—fiber Bragg grating (FBG) sensors—is particularly attractive for strain monitoring in industrial applications. Besides all the demonstrations of the high potential associated with Bragg grating technology, the development of high-reliability strain gauges is of particular interest due to the capacity to mitigate well-known deficient performance of conventional electrical technology in industrial hazard applications. This type of sensors must be designed to enable direct deployment in harsh industrial environments, providing a high survivability rate throughout the installation and ensuring a long-term reliability in line with the infrastructure's lifetime.

High-reliability fiber Bragg strain sensors are now commercially available for a wide range of industrial applications, ranging from embedding in glass fiber composite materials for load assessment in wind turbine blades to remote distributed strain mapping in pipelines. This chapter addresses the fiber sensing design and manufacturing considerations looking for reliability in the measurement of strain in industrial applications; performance comparison with the electrical strain gauge; and corrective techniques for the sensors' temperature cross-sensitivity. A brief discussion on optical sensor market position and a guide for future developments will also be presented.

19.2 Design and Manufacturing Considerations

Despite the fact that fiber-optic sensor technology has gathered considerable interest for structural health monitoring over the last few years, many challenges associated with its application in industrial applications are still the scope of intense research. Sensor design issues ranging from temperature compensation to methods that ensure long-term attaching to structure; sensing network architecture and implementation issues such as the lack of standards and installation guidelines and fiber lead cable handling, splicing, and protection in harsh conditions; and instrumentation issues related to technology maturity and cost are critical in broadening the scope of industrial applications.

As fully detailed in other chapters of this book, fiber Bragg sensor is a small-size microstructure (less than 10 mm long) that can be photo-imprinted in photosensitive optical fibers by side exposure to patterned UV laser radiation. Such a microstructure consists of a periodic modulation of the refractive index of the core of the optical fiber that is characterized by a narrowband resonance spectral reflection (Morey et al. 1989). Since this is a deeply embedded device in the optical fiber structure, the resonance behavior strictly follows external actions in the exact proportion as the silica matrix surrounding the microstructure. This results in a localized sensor offering high sensitivity to strain, particularly suitable for quasi-distributed measurements ranging from few centimeters to tens of kilometers. Fiber Bragg strain sensors add to the long recognized advantages of the fiber-optic sensors—for example, immunity to EMI/RFI, remote monitoring, small size and weight, electrical isolation, intrinsically safe operation, high sensitivity, long-term reliability—the inherent multiplexing capability and the ability to provide absolute measurement without the need for referencing. These are two critical qualities that make Bragg grating sensors very attractive for strain monitoring in industrial applications. Multiplexing enables many Bragg sensors to be interrogated using common optoelectronic instrumentation. By sharing the processing optics and electronics, multiplexing reduces the cost per sensor, strongly reduces the overall network complexity, and enhances the robustness of the system. On the other hand, self-referencing refers to the fact that measurements can be made relative to the time the sensor is manufactured and calibrated, and are not jeopardized if the electronic instrumentation is turned off or replaced. In an analogy to resistance strain gauge technology, Bragg grating sensors do not require *bridge balancing*. This feature is invaluable in applications involving measurement over days, months, or years, as is commonly required for industrial applications.

Bragg grating sensors have also some potential drawbacks. The production of the fiber Bragg sensors requires stripping of the polymeric coating from the fiber as it is highly absorbing in the UV region of the spectrum, thus rendering impossible the manufacturing procedure. It is well known that uncoated optical fibers are very susceptible to surface damage (Komachiya et al. 1999, Mauron et al. 2000). Any polymeric residue left on the surface can also have an adverse effect on the mechanical strength of the fiber. Rubbing the surface has almost a catastrophic effect, being compulsory the usage of hot sulfuric acid stripping process to keep unaffected the high mechanical strength of the pristine optical fiber. Alternatively, fiber Bragg sensors can be manufactured by combining the drawing of the optical fiber process with the UV inscription of the grating (Askins et al. 1994) or by exposure through the coating material (Imamura et al. 1998, Chao et al. 1999). Although these processes result in reduced flexibility in the definition of the spectral feature of the sensor, it enables the production of spliceless and high-strength fiber Bragg sensor chains, since the coating is applied after the grating inscription. On the other hand, the refractive index change induced by UV exposure during fabrication evidences temperature-induced decay that must be annealed in order to ensure proper long-term reliability of the manufactured devices (Erdogan et al. 1994, Sennhauser et al. 2000). Moreover, even at the end of optimized manufacturing process, the Bragg sensors are not much different from pristine optical fiber that is intrinsically susceptible to contamination and thus requires protection of the fiber surface, bonding the sensor element to an adequate mechanical transducer, and proper outside packaging and lead cabling.

The overall manufacturing process must then ensure that the strength of the Bragg grating sensing element is kept close to the strength of the pristine optical fiber. The use of polyimide or ormocer recoating just after sensor manufacturing minimizes the risk of damaging of the fiber surface ensuring long-term reliability. In addition, the implementation of accelerated aging procedures allow long lifetime to be achieved by increasing the screening strain for postassembly proof tests. Again the impact of these procedures on the long-term reliability of the packaged devices must be accurately assessed and controlled during the manufacturing process. This is particularly important in strain sensors designed to be embedded directly into composite structures, since afterward any repair of a malfunction sensor is virtually impossible.

19.3 Performance Comparison with the Electrical Strain Gauge

Electrical strain gauges are the most cost-effective solution for the great majority of strain sensing applications. It is a fully matured technology with more than 50 years of industrial effective usage. However, fiber-optic sensors, in particular FBGs, have become an increasingly attractive alternative for niche applications in which electrical strain gauges cannot be employed due to technical constraints. Strain measurement in large electrical machines or in large-scale distributed systems are examples in which the unique advantages of fiber-optic systems can be exploited to provide more effective monitoring solutions.

It is well known that electrical strain gauges are based on the variation of electrical resistivity and geometry of thin conductors when subjected to strain. As a conductor is stretched, its resistivity and length increase and its cross section decreases, resulting in a small increase in resistance. The gauge factor—that is, proportionality constant that relates fractional resistance change to fractional length change (strain)—for electrical strain gauges is usually around 2. Thus, a strain of 1 $\mu\varepsilon$ applied to a 120 Ω strain gauge results in a resistance change of 240 $\mu\Omega$. A Wheatstone balanced bridge is usually employed to provide signal conditioning of electrical strain gauges, in which a constant voltage applied across the bridge results in a voltage drop along the arms of the bridge due to resistance unbalance proportional to the applied strain.

Similar to electrical strain gauges, fiber Bragg sensors are also directly sensitive to strain. The longitudinal strain applied to the fiber transduces to the resonant wavelength through a direct one-to-one geometrical relation, since the resonant wavelength is directly proportional to the periodicity of the Bragg grating. On the other hand, the refractive index of the fiber is dependent on strain due to the photoelastic effect, resulting in a decrease in the unitary geometrical gauge by a factor of 0.22. Thus, the overall gauge factor—that is, proportionality constant that relates fractional wavelength change to fractional length change—for FBGs is 0.78, less than half of the one for strain gauges. A strain of 1 $\mu\varepsilon$ applied to a 1550 nm Bragg grating results in a wavelength change of 1.2 pm. Standard spectroscopic techniques in conjunction with peak detection algorithms are usually exploited to measure accurately such small wavelength shifts.

Next, a comparative analysis of the performance of electrical strain gauges versus FBGs taking into account the parameters with utmost impact on strain monitoring in industrial applications is presented. While static strain measurements are very demanding in terms of accuracy, drift, and temperature cross-sensitivity, for dynamic applications, fatigue, force required to apply strain, and acquisition speed are crucial parameters.

19.3.1 Accuracy

FBGs present a unique performance in terms of accuracy, since strain effect is directly encoded in the spectral content of the resonant peak. In fact, any change of that spectral content can only occur at the microstructure, enabling absolute measurement of strain to be attained. Due to the nature of strain to resistance transduction, electrical strain gauges cannot match Bragg gratings' absolute measurement performance in field applications. Bridge imbalance, residual contact resistance, and cabling interference are issues that make rather difficult the implementation of practical electrical strain gauge systems in many industrial environments.

19.3.2 Networking

Many industrial applications require the implementation of sensing networks comprising tens of strain gauges deployed over wide areas. The resonant spectral feature of FBGs offers intrinsic multiplexing capability over a single fiber lead. In addition, low loss transmission in optical fibers allows flexible implementation of fiber-optic systems to measure strain at hundreds of locations along tens of kilometer range infrastructures. Although multiwire techniques exist to minimize the impact of lead wire resistance on electrical strain gauge applications, when spanning over long distances, the voltage drop along lead wire becomes an outreach obstacle. Nevertheless, the advent of bus-like networking with delocalized signal conditioning and digitalization applied to strain gauge technology has significantly increased its networking capability.

19.3.3 Drift

Long-term applications in harsh environments are specially demanding in terms of drift. Drift is usually associated to nonreversible residual changes of sensor's offset caused by activation reactions highly dependent on temperature. In strain gauges, drift is initiated by oxidation effects in the resistive foil that is mainly triggered at high temperatures. FBGs are deeply embedded in silica matrix that is highly corrosion resistant; however, drift also occurs due to thermal activation of defects responsible for the refractive index modulation that leads to residual wavelength shifts.

High-quality electrical strain gauges and FBGs are subjected to thermal annealing during the manufacturing process, thus preventing drift to occur within the operating temperature range (Erdogan et al. 1994, Sennhauser et al. 2000).

19.3.4 Physical Constraints

Both electrical strain gauges and fiber Bragg strain gauges are directly sensitive to strain and attain high linearity. Fiber Bragg strain gauges can measure high strain levels and provide a resistance to fatigue that is orders of magnitude higher than the one of electrical strain gauges. Optical fibers produced with synthetic fused silica have remarkable strength (65,000 με), being this property critically dependent on the presence of small flaws on the surface of the optical fiber. This value is typically limited in common electrical strain gauges to values in extent of 10,000 με. Fiber Bragg strain gauges can also endure almost limitless fatigue cycles at normal strain levels of ±2500 με, while electrical strain gauges typically fail at 10 M cycles.

In order to transfer strain from the specimen material to the sensor, a force is required, which is usually provided through an adhesive. Electrical strain gauges are made of metallic alloys that require ~2.0 N to be strained to 1000 με, while fiber Bragg sensors are made of silica requiring ~0.8 N to attain the same strain value. The use of low-diameter optical fibers (Ø 50 μm) can further reduce this force to 0.15 N. It is important to point out that although the force necessary to generate strain on an electrical strain gauge is higher, so is its area thus resulting in a lower pressure over the adhesive.

Electrical strain gauges can be installed in curved substrates with radius lower than 1 mm, while Bragg strain gauges manufactured in standard optical fiber withstand minimum bending radius of 25 mm. The advent of bend-insensitive fibers has relaxed this constraint to values lower than 5 mm.

It is well known that fiber Bragg strain gauges are completely immune to electromagnetic fields, being their use in potentially explosive atmospheres also possible without restrictions. Since electrical strain gauges use conductive lead wires, high voltages caused by lightning strikes or partial discharges generate high ignition risk and can lead to the failure of the sensor, as well as of the signal condition and data acquisition hardware.

19.3.5 Temperature Compensation

The resistance of electrical strain gauges is temperature dependent since both the resistivity and dimensions of the metallic foil are temperature dependent.

These effects can be minimized by processing the gauge's material to not only self-compensate its thermal-induced resistance change but also further adjust to match the thermal expansion of the specimen material to which the sensor is bonded. Residual errors on the order of 10 με/°C are common even when using proper temperature-compensated strain gauges. Besides the direct impact of temperature on the strain gauge element, it is also necessary to consider the impact of temperature in the resistance of the lead wires. It is now common practice to use multiwire quarter-bridge arrangements to cancel the impact of lead wire resistance when performing static strain measurements.

Although the thermal expansion of silica is low, the resonant wavelength of Bragg sensors is dependent on temperature mainly due to refractive index dependence on temperature. A temperature variation of 1°C results in a wavelength shift of ~10 pm in a fiber Bragg strain gauge resonant at 1550 nm, corresponding to an apparent strain change of ~8.3 με. Thermal expansion and thermo-optic effects add up, and silica's properties cannot be engineered to cancel them out. However, since many sensors can be daisy-chained in the same optical fiber, active compensation based on simultaneous temperature measurement is usually employed to effectively cancel thermal drift (Morey et al. 1992). When compensation of the thermal expansion of the specimen material is also required, differential compensation techniques can be employed through the use of dummy strain gauges bonded to unstrained substrates. However, such compensation procedures reduce the number of sensors that can be multiplexed in a single fiber and are also prone to thermal gradients. To overcome this limitation, a technique for designing athermal FBG strain gauges is discussed in the next section.

19.3.6 Signal Condition

Strain measurements are intrinsically demanding in terms of signal conditioning, since, besides gauge factor multiplication effect, parts-per-million fractional length change sensitivity is required at microstrain level.

The strain gauge resistance imbalance induced on a Wheatstone bridge typically results in voltage differences on the order of millivolt; thus, in harsh environments, both the sensors and the lead wires are extremely vulnerable to grounding issues and electromagnetic interference. The impact of these effects can be minimized by delocalizing the signal conditioning and digitalization components to the close proximity of the sensor. Electrical strain gauge technology is fully matured with a large number of established providers of standard signal conditioning solutions.

Fiber Bragg sensors codify the strain measurement into optical frequency and thus can withstand long distances (more than 10 km) without the need for delocalized signal conditioning. Although several spectroscopic techniques can be effectively used to measure the strain-induced resonant wavelength shifts, interrogation systems based on tunable lasers are becoming the recurrent option for the implementation of high-end Bragg monitoring applications (Santos et al. 2012). Besides the intrinsic high dynamic range and multiplexing capability, tunable laser systems often employ embedded gas-cell spectral calibration to attain absolute strain measurement.

19.3.7 Field Deployment

As discussed previously, fiber Bragg strain gauges provide many advantages over conventional electrical strain gauges particularly in applications that are exposed to harsh environments. Electrical strain gauges generally do not withstand long life spans and therefore require recurrent replacement in long-term monitoring applications. In addition, lead wires connecting the strain gauge to the signal conditioner unprotected against moisture result in parasitic resistance that leads to measurement errors. To avoid this source of errors, it is mandatory to provide proper protection of strain gauges deployed in field.

Fiber-optic sensing technology is completely immune to electromagnetic interferences and high voltages, enabling its use in applications that require measurements to be performed in close proximity to noise sources. On the other hand, electrical strain gauges require the use of shielded cables and elaborate grounding techniques to protect the sensors, signal conditioning electronics, and acquisition equipment from lightning strikes and partial discharges.

For large-scale applications integrating hundreds of sensors spanning over large distances, optical fiber sensing systems enable the implementation of very competitive monitoring solutions. Tens of sensors can be chained on demand along a single optical fiber, and tens of optical fibers can be effectively multiplexed in a single remote instrument. The installation procedures are substantially simplified, making overall system deployment cost-effective.

19.4 Passive Athermal Fiber Bragg Strain Gauge

Fiber Bragg sensors are particularly suitable for measuring strain. Currently, such a single parameter measurement is difficult to implement, since cross-sensitivity to temperature compels the use of an additional temperature reference. As mentioned, the most common method to overcome cross-sensitivity to temperature while measuring strain with fiber Bragg sensors relies on the use of an additional temperature reference—for example, strain-inactive FBG (Santos et al. 2012). Other methods, based on the use of dual-wavelength FBG (Xu et al. 1994), nonsinusoidal FBG (Brady et al. 1994), FBG written in different diameter fiber (James et al. 1996), FBG and long period grating (Patrick et al. 1996), FBG and in-line fiber etalon (Singh and Sirkis 1996), and FBG pair written in Hi-Bi PANDA fiber (Ferreira et al. 2000), have been demonstrated, but they are often rather complex and difficult to implement in real-world structures. Moreover, besides being not required in all the strain monitoring cases, the measurement of temperature by the referred methods implies the allocation of additional bandwidth to each sensor, therefore limiting the total number of sensors in a given sensing network. The design of a passive athermal fiber Bragg strain gauge is thus of particular interest since it renders optional the measurement of temperature, benefiting large-scale system design and performance. In addition, such a strain gauge can also be designed to further compensate for structural material thermal expansion, enabling load-induced strain components to be discriminated from the thermal-induced strain components.

A passive athermal fiber Bragg strain gauge renders unnecessary the measurement of temperature. In line with this, a particular example of a gauge design for fiber Bragg sensors that enables strain measurements to be performed while canceling temperature sensitivity is discussed here. The design is based on a structure composed of a single material that defines a length ratio that allows the adjustment of the temperature sensitivity of the fiber Bragg sensor to zero, providing athermal operation of the strain gauge. Such assembly can be designed to further compensate for structural material thermal expansion, enabling load-induced strain components to be measured. Such a design enables strain measurements to be performed through a single fiber Bragg sensor while canceling its intrinsic temperature sensitivity, thus extending the range of applications of such components.

19.4.1 System Design

The fiber Bragg strain gauge is characterized by a narrow reflective spectral response in which center wavelength λ_B of the resonance band matches the Bragg condition:

$$\lambda_B = 2n_{eff}\Lambda \tag{19.1}$$

where
- n_{eff} is the effective index of the guided mode
- Λ is the period of the index modulation

The resonance wavelength will vary according to temperature and/or strain changes experienced by the fiber. For a temperature change ΔT, the corresponding wavelength shift is given by

$$\Delta\lambda_B = \lambda_B\left(\frac{1}{\Lambda}\frac{\partial\Lambda}{\partial T}+\frac{1}{n}\frac{\partial n}{\partial T}\right)\Delta T = \lambda_B(\alpha+\xi)\Delta T = \lambda_B\beta_T\Delta T \tag{19.2}$$

where
- α is the fiber coefficient of thermal expansion (CTE)
- ξ is the fiber thermo-optic coefficient, with values of 0.55 and 6.7 ppm/°C, respectively

The wavelength shift induced by a longitudinal strain variation ε is given by

$$\Delta\lambda_B = \lambda_B\left(\frac{1}{\Lambda}\frac{\partial\Lambda}{\partial \varepsilon}+\frac{1}{n}\frac{\partial n}{\partial \varepsilon}\right)\varepsilon = \lambda_B(1-p_e)\varepsilon = \lambda_B\beta_\varepsilon\varepsilon \tag{19.3}$$

where p_e is the photoelastic coefficient of the fiber, with a value of 0.22.

In the last two equations, β_T and β_ε are defined as the temperature and strain sensitivities of the fiber Bragg strain gauge, respectively. The usual approximate values for these two coefficients at 1550 nm are $\beta_T = 7.25 \times 10^{-6}$°$C^{-1}$ and $\beta_\varepsilon = 0.76 \times 10^{-6}\ \mu\varepsilon^{-1}$.

The overall Bragg wavelength shift induced by temperature change and/or strain is then given by

$$\frac{\Delta\lambda_B}{\lambda_B} = \beta_T\Delta T + \beta_\varepsilon\varepsilon \tag{19.4}$$

A possible design for enabling strain measurements while canceling temperature sensitivity—TS—relies on subjecting the fiber Bragg strain gauge to additional temperature-induced strain, $\varepsilon(T)$, according to the following expression:

$$TS = \frac{\Delta\lambda_B / \lambda_B}{\Delta T} = \beta_T + \frac{\beta_\varepsilon \varepsilon(T)}{\Delta T} \tag{19.5}$$

This method enables annulled temperature sensitivity to be attained, being the exact balancing between the intrinsic FBG temperature sensitivity and temperature-induced strain—$\beta_T = -\beta_\varepsilon \varepsilon(T)/\Delta T$—what provides the so-called athermal operation of the strain gauge.

As mentioned, the athermal design relies on subjecting the fiber Bragg strain gauge to additional temperature-induced strain. The simplest method of applying temperature-dependent strain to a fiber Bragg sensor is to attach it to a material with a CTE dissimilar to one of silica. However, this restricts the adjustment of the fiber Bragg strain gauge thermal sensitivity to the set of discrete values that can be obtained employing available materials. A well-known method of attaining a broad range of effective CTEs—including negative CTE values—is to provide a structure incorporating a proper arrangement of two materials with distinct CTEs. A proper design of such a structure can be used for packaging FBGs, allowing for continuous adjustment of the temperature sensitivity. Here, it is analyzed a simpler fiber Bragg strain gauge design that relies on a single material to attain athermal strain measurements in a considerable number of structural health monitoring applications.

The basic structure of such passive athermal fiber Bragg strain gauge is shown in Figure 19.1—it consists of a structure comprising beams and anchoring stands. Each beam has a length $L_2/2$ and is made of a material with CTE, α_B, not attached to the material being monitored.

The anchoring stands are made of the same material as the beams and are attached at a distance of L_1 to the material being monitored. The FBG is attached at its ends to the beams. The anchoring stands are connected through a frame to form a suitable structure to define the distance L_1 between the anchoring stands and to allow simple installation. This frame comprises flexures to minimize the load that is applied to the bonds between the anchoring stands and the material being monitored.

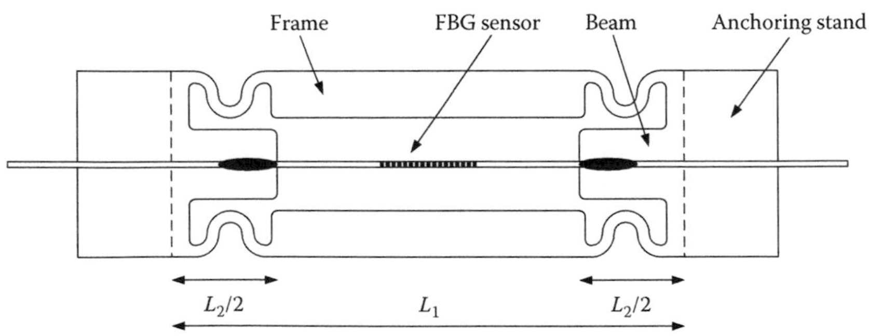

FIGURE 19.1 Schematic representation of a possible athermal fiber Bragg strain gauge.

19.4.2 Athermal Performance

The performance of the athermal strain gauge design in compensating for both the FBG intrinsic temperature sensitivity and the CTE of the material being monitored must also be analyzed. It is straightforward that the athermal design can be adjusted to compensate only for the fiber Bragg strain gauge intrinsic temperature sensitivity. Consider now that the athermal strain gauge is bonded to the material being monitored—for example, composite material, concrete, and steel—at the bottom surface of the anchoring stands. Under this condition, the strain gauge's "effective CTE" becomes equal to one of the materials being monitored, since the material deformation under temperature variations will be integrated by a geometrical change of the distance between the anchoring stands. In this case, and being α_B the CTE of the beams and α_M the CTE of the material being monitored, the following expression can be written for the FBG wavelength shift induced by temperature:

$$\frac{\Delta\lambda_B}{\lambda_B} = \left[\beta_T + \beta_\varepsilon\left(\frac{\alpha_M L_1 - \alpha_B L_2}{L_1 - L_2}\right)\right]\Delta T \tag{19.6}$$

where

L_1 is the distance between the anchoring stands
L_2 is the length of the beams
β_T and β_ε are defined as the temperature and strain sensitivities of the fiber Bragg sensor, respectively

Minimum wavelength drift with temperature can be obtained by balancing the length of the beams to the distance between the anchoring stands, which results in the following condition for the balancing ratio (BR):

$$\text{BR} = \frac{L_2}{L_1} = \frac{\beta_T + \beta_\varepsilon \alpha_M}{\beta_T + \beta_\varepsilon \alpha_B} \tag{19.7}$$

From this ratio, it is straightforward to state that minimum length design is obtained for the maximum value of the CTE of the beams α_B. It should also be emphasized that this design restricts the maximum fiber Bragg strain gauge length to $L_1 - L_2$; thus, for a given sensor length, it is always possible to calculate the lengths L_1 and L_2 that fulfill both conditions. Taking into account these constrains, aluminum is a good option for building the strain gauge since it presents several beneficial properties for manufacturing the athermal strain gauge—for example, high Young's modulus, high CTE ratio, out-of-the-shelf availability in a broad dimensional range, and easy material processing. Figure 19.2 shows the strain test setup of the athermal strain gauge under temperature control.

Figure 19.3 shows the performance of an athermal fiber Bragg strain gauge compared to a standard fiber Bragg strain gauge for a temperature range compatible with most structural health monitoring applications. The athermal strain gauge considered here was built in aluminum and optimized for stress measurements in composite materials according to the discussed design. Data in Figure 19.3 also compare the strain shift of an unbonded bare FBG to one of the standard fiber Bragg strain gauges bonded to a composite material.

It is possible to observe that the standard fiber Bragg strain gauge presents additional strain shift induced by the CTE of the material. In this case, the athermal FBG strain gauge provides compensation for both the FBG thermal sensitivity and the CTE of the material.

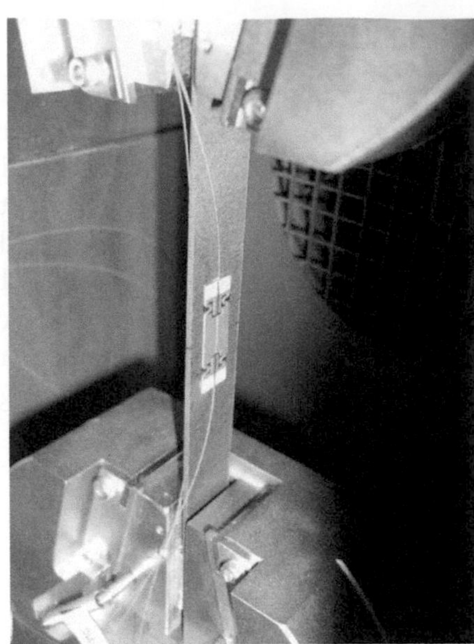

FIGURE 19.2 Strain test of the athermal fiber Bragg strain gauge under temperature control.

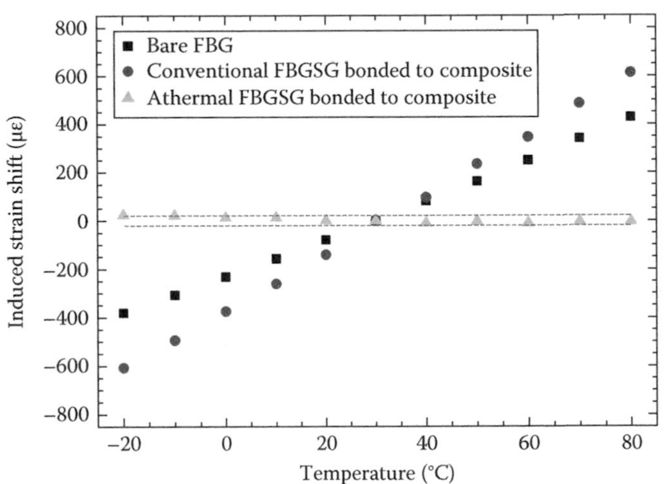

FIGURE 19.3 Performance of the athermal fiber Bragg strain gauge bonded to a carbon composite structure.

19.4.3 Strain Gauge Performance

It is also important to discuss the strain measurement performance of the athermal strain gauge design that compensates for both the FBG intrinsic temperature sensitivity and the CTE of the material being monitored. For a strain α_M applied to the material being monitored, the fiber Bragg sensor will be subjected to a strain given by

$$\varepsilon_{FBG} = \varepsilon_M \frac{L_1}{L_1 - L_2} \tag{19.8}$$

The Bragg wavelength shift associated with this strain is, therefore, given by

$$\frac{\Delta\lambda_B}{\lambda_B} = \beta_\varepsilon \frac{L_1}{L_1 - L_2} \tag{19.9}$$

The mentioned strain will arise only from load applied to the material being monitored, since any temperature-induced deformation of the material will be effectively cancelled by the athermal design.

19.5 Fiber Bragg Sensors' Market Position

Recent market studies from different sources confirm the growing impact of fiber-optic sensors, in general, and of FBGs in particular, on multiple areas of monitoring (Méndez 2007). It is difficult to refer a figure for the total sensing market size, but \$50 bn is often pointed out as a conservative estimation. Of course, this number includes every type of sensing devices, from the costly monitoring systems that equip satellites down to extremely low-cost thermometers and accelerometers that are massively present in car industry, mobile phones, etc. It is expected that by 2015/2016, fiber-optic sensors' market size will reach \$1 bn. Even if this number represents only 2% of the total sensors' market, it is highly relevant as fiber-optic sensing evolution and consolidation has been done essentially by addressing needs that could not be satisfied by conventional technologies. In fact, only very recently, fiber-optic sensors started to achieve market position by replacing other technologies not only because of their better performance but also due to the fact that they have become price competitive.

The expected market growth of fiber-optic sensors is essentially driven by distributed sensors, being FBGs included in this group that is dominated by Raman scattering–based sensing (Brillouin is marginal). By 2015/2016, distributed sensing will be responsible for 80% of the total fiber-optic sensing market. From these, 30%, that is, ~\$300 mn, will be FBG-based systems. These numbers may vary significantly from market study to market study, but all of them indicate that \$300 mn for the market size of FBG sensing systems is a rather conservative number, given the fact that China and other regions of Asia are ignored in almost all forecasts.

One of the reasons why FBGs are becoming increasingly successful in monitoring is because of their recognized appropriateness to measure a multitude of parameters in different fields, such as oil and gas, energy, industry, aerospace, civil infrastructures, and transportation. Among these, civil engineering (dams, bridges, tunnels, buildings, etc.) is usually the sector identified with structural health monitoring applications, where strain (or stress) is one of the most important parameters to be measured. But structural health monitoring is also important in other sectors: energy (e.g., wind turbines), oil and gas (e.g., pipelines), aerospace (e.g., airplanes), transportation (e.g., railroad tracks). Even if FBGs can play a major role in these cases, cost is still a decisive factor: a common rule of thumb assumes that the cost of a complete monitoring system to a new structure should add no more than 1% to the total construction cost (Graver et al. 2004).

In some situations, this can be a rather difficult challenge, since a complete monitoring system includes sensors, instrumentation, cabling, accessories, and installation service. Designing high-performance athermal FBG strain sensors becomes therefore particularly important as they can contribute to dramatically reduce the number of temperature sensors used in large sensing networks.

The word *performance* must be yet associated with another important one: *reliability*. This is because over the last years, besides cost, one of the most significant barriers to FBG market penetration has been the dissemination of strain sensors with extremely poor reliability. Proper development, prototyping, test, and production procedures, ruled by strict quality management systems, must be followed to ensure the delivery of highly reliable products.

All these aspects, of course, must be complemented by a definition of standards for fiber Bragg sensor technology, something that is finally being done through COST actions sponsored by the European Union, a topic addressed in Chapter 20. This will definitely contribute to the market growth of FBG monitoring, by turning compatible systems produced by different manufacturers.

The future for FBG strain sensors is therefore bright, and there is no doubt that these devices will be extensively used in structural health monitoring systems for critical assets.

References

Ansari, F. and Y. Libo. 1998. Mechanics of bond and interface shear transfer in optical fiber sensors. *Journal of Engineering Mechanics* 124: 385–394.

Askins, C. G., M. A. Putnam, G. M. Williams et al. 1994. Stepped-wavelength optical-fiber Bragg gratings arrays fabricated in line on a draw tower. *Optics Letters* 19: 147–149.

Brady, G. P., K. Kalli, D. J. Webb et al. 1994. Recent developments in optical fiber sensing using fiber Bragg gratings. *Fiber Optic and Laser Sensors XIV*, Denver, CO, *SPIE*, Vol. 2839, pp. 8–19.

Chao, L., L. Reekie, and M. Ibsen. 1999. Grating writing through fibre coating at 244 and 248 nm. *Electronics Letters* 35: 924–926.

Crossley, S. D. 2002. The commercialization of fibre optic sensors. In *Handbook of Optical Fibre Sensing Technology*, J. M. López-Higuera (ed.). John Wiley & Sons, Chichester, U.K.

Cusano, A., A. Cutolo, and J. Albert (eds.). 2011. *Fiber Bragg Grating Sensors: Recent Advancements, Industrial Applications and Market Exploitation*. Bentham, Oak Park, IL.

Erdogan, T., V. Mizrahi, P. J. Lemaire et al. 1994. Decay of ultraviolet-induced fiber Bragg gratings. *Journal Applied Physics* 76: 73–80.

Ferreira, L. A., F. M. Araújo, J. L. Santos et al. 2000. Simultaneous measurement of strain and temperature using interferometrically interrogated fiber Bragg grating sensors. *Optical Engineering* 39: 2226–2234.

Gagliardi, G., M. Salza, S. Avino et al. 2010. Probing the ultimate limit of fiber-optic strain sensing. *Science* 19: 1081–1084.

Graver, T., D. Inaudi, and J. Doornink. 2004. Growing market acceptance for fiber-optic solutions in civil structures. *Fiber Optic Sensor Technology and Applications III, SPIE*, Vol. 5589, pp. 44–55.

Imamura, K., T. Nakai, K. Moriura et al. 1998. Mechanical strength characteristics of tin-codoped germanosilicate fibre Bragg grating by writing through UV-transparent coating. *Electronics Letters* 34: 1016–1017.

James, S. W., M. L. Dockney, and R. P. Tatam. 1996. Simultaneous independent temperature and strain measurement using in-fiber Bragg grating sensors. *Electronics Letters* 32: 1133–1134.

Komachiya, M., R. Minamitani, T. Fumino et al. 1999. Proof-testing and probabilistic lifetime estimation of glass fibers for sensor applications. *Applied Optics* 38: 2767–2774.

López-Higuera, J. M. (ed.). 2002. *Handbook of Optical Fiber Sensing Technology*. John Wiley & Sons, Chichester, U.K.

Mauron, P., Ph. M. Nellen, and U. Sennhauser. 2000. High-strength proof-test parameters for fibre Bragg grating sensors. *SPIE* 4215: 183–190.

Méndez, A. 2007. Fiber Bragg grating sensors: A market overview. *Third European Workshop on Optical Fibre Sensors (EWOFS2007)*, Napoli, Italy, *SPIE*, Vol. 6619, pp. 661905.

Morey, W. W., G. Meltz, and W. H. Glenn. 1989. Fiber optic Bragg grating sensors. *Proceedings Fiber Optic and Laser Sensors VII*, Boston, MA, *SPIE*, Vol. 1169, pp. 98–101.

Morey, W. W., G. Meltz, and J. M. Weiss. 1992. Evaluation of a fiber Bragg grating hydrostatic pressure sensor. *Eighth International Conference on Optical Fiber Sensors (OFS 8)*, Monterey, CA.

Patrick, H., G. M. Williams, A. D. Kersey et al. 1996. Hybrid fiber Bragg grating/long period fiber grating sensor for strain/temperature discrimination. *Photonic Technology Letters* 8: 1223–1225.

Pevec, S. and D. Donlagic. 2011. All-fiber long-active-length Fabry-Perot strain sensor. *Optics Express* 19: 15641–15651.

Santos, J. L., L. A. Ferreira, and F. M. Araújo. 2012. Fiber Bragg grating interrogation systems. In *Fiber Bragg Grating Sensors: Recent Advancements, Industrial Applications and Market Exploitation*, A. Cusano, A. Cutolo, and J. Albert (eds.). Bentham, London, U.K., pp. 78–98.

Sennhauser, U., A. Frank, P. Mauron et al. 2000. Reliability of optical fiber Bragg grating sensors at elevated temperature. *IRPS-CS-41 Proceedings of the International Reliability Physics Symposium*, pp. 264–269.

Singh, H. and J. Sirkis. 1996. Simultaneous measurement of strain and temperature using optical fiber sensors: Two novel configuration. *Eleventh International Conference on Optical Fiber Sensors (OFS 8)*, Sapporo, Japan, pp. 108–111.

Tao, X. M., L. Q. Tang, W. C. Du et al. 2000. Internal strain measurement by fiber Bragg grating sensors in textile composites. *Composites Science and Technology* 60: 657–669.

Wan, K. T. and C. K. Y. Leung. 2007. Fiber optic sensor for the monitoring of mixed mode cracks in structures. *Sensors and Actuators A* 135: 370–380.

Xu, M. G., J.-L. Archambault, L. Reekie et al. 1994. Discrimination between strain and temperature effects using dual/wavelength fiber grating sensors. *Electronics Letters* 30: 1085–1087.

Yang, X. F., S. J. Luo, Z. H. Chen et al. 2007. Fiber Bragg grating strain sensor based on fiber laser. *Optics Communications* 271: 203–206.

20. Standardization and Its Impact on Measurement Reliability

Wolfgang R. Habel
BAM Federal Institute for Materials Research and Testing

Handbook of Optical Sensors. Edited by José Luís Santos and Faramarz Farahi © 2015 CRC Press/Taylor & Francis Group, LLC. ISBN: 9781439866856.

20.1 Expectations of Fiber-Optic Sensor Standard

When users want to get a measurement task solved by using fiber-optic sensors, they first need a comprehensive overview of basic characteristics, the terminology used in this field, and some information on how to design fiber-optic sensor networks. Second, they generally expect optimum performance from sensor systems available on the market. A complete measurement system consists of the sensing element, power supply and recording device(s), and some accessories. Unfortunately, there is often a lack in the performance description of the system components. The weakest point is, however, the nonavailability of established and standardized application methodologies for fiber-optic sensors and finally testing methods to prove the performance of differently applied sensors. In some cases, users prefer to design the sensor system by themselves; they purchase components of the fiber-optic sensor system, similar to resistance strain gauges (RSGs) and corresponding devices, and, here and there, they also rent measuring equipment. For an appropriate selection of monitoring components, and to be sure to create a reliable measurement system, confirmed information about the component's characteristics, reproducibility, long-term stability, expected drifts and creep, parameter limits, and eventually the expected uncertainty of the results measured is needed. The design work for a sensor system and the choice of an appropriate application procedure must follow specific demands according to the measurement task. Engineers want and have to find optimum answers with regard to all components of a sensor system, which also includes the selection of an appropriate sensing method out of a multitude of possibilities.

In order to create high-quality sensor systems, all involved cooperating partners need a well-founded overview of fiber-optic sensor technologies, available components, and recording systems. They have to analyze the measurement task and the influencing operational and environmental conditions. Depending on these prerequisites, the appropriate sensing system consisting of the optical source, optical fiber with a specific coating, sensing element itself (extrinsic or intrinsic type), some components like connectors, beam splitters, and finally a recording device with appropriate specifications can be selected. Standards and guidelines can be very helpful to find all this information summarized. Except in very few handouts or company-related guidelines, there are not yet standards that generally summarize the wanted information for design, application, and operation of optical sensor systems. Such existing guidelines are, for example, the gyroscope and the DTS guideline (see Section 20.7.4). Hence, the most appropriate sensor system with an acceptable measurement uncertainty is sometimes not created, and the customers are disappointed with the results achieved. More guidelines and standards are therefore increasingly requested by different groups with different expectations.

20.1.1 Customers

They want to find basic (not really scientific) information about functional principles and basic specifications. In the majority of cases, they are interested in cheap solutions. Only if the new fiber-optic sensor technology has advantages over the conventional one, they are able or willing to pay an appropriate price. To have confidence in any discussion with fiber-optic sensor experts or consultants, customers should have a guideline —or at least a checklist—containing all important aspects.

20.1.2 Consultant Engineers

First, they need an overview of available sensing methods and corresponding components to design appropriate state-of-the-art fiber-optic sensor systems. Customers would be well advised, if they do not receive only one proposal for a measurement system structure or an offer from only one company. Consulting engineers should be able to compare different offers to find out the most effective sensor technology (sensor system). Such systems should be described according to standard rules. High-quality systems, for example, for long-term measurements, require high-performance components and facilities; measurement tasks with low-resolution requirements or for only short-term measurements (e.g., without disconnecting cables) tolerate systems with lower performance. Consulting engineers have to estimate and recommend the system that is the most effective one for the user's task.

20.1.3 Innovative Companies, Skilled Personnel

They must have clear performance specifications for all components of a fiber-optic sensor system. This requires the correct use of performance-determining quantities as well as static and dynamic specifications. This must be done in conformity with available international standards and must use appropriate vocabulary for the description of general terms in metrology (including the use of SI units). The correct terminology ensures that the interdisciplinary community of, for example, physicists, fiber-optic experts, mechanical, and civil engineers understand each other (Habel 2005).

20.1.4 Manufacturers

Companies that produce and deliver sensors or sensor components have to provide validated products. This means that they have to confirm by examination and prove that the objectives of the particular requirements for a specific intended use are fulfilled.

Validation of products as well as measurement methods should therefore be carried out according to the relevant standard ISO/IEC 17025/2000 of the International Standardization Organization (ISO) (Standard EN ISO/IEC 17025:2000, 1999). In this case, for a specific intended use, the sensor or the measurement system should work as reliably as has been requested. Using standards and guidelines, the confidence of the user community into the fiber-optic sensor technology will increase. Standards generally facilitate the use of technology. As a competent summary of valuable technical recommendations, they contain all details for the specification and use of a sensor system.

Standards and guidelines have undoubtedly a number of positive effects; custom products manufactured according to standards are validated and therefore more reliable, and eventually cheaper for the users. Standards on the other hand must not limit the necessary variability of the system components in an emerging technology.

20.2 Importance of Standards with Regard to Measurement Reliability

There are numerous fiber-optic sensor experts who have sufficient experience to design, install, and operate fiber-optic sensor systems. However, in the interest of a broad dissemination of this outstanding technology, guidelines or standards for skilled personnel should therefore be available to provide recommendations, and particularly knowledge, which validation and application procedures are mandatory to ensure reliable measurements.

There are four major types of standards:

1. Fundamental standards that define terminology, signs, symbols, and basic conventions.
2. Test methods as well as data analysis standards that define measurement characteristics (temperature, pressure, physical, and chemical measurands).
3. Organization standards that describe company-related procedures such as quality management system, maintenance procedures, and product or logistic management.
4. Specification standards, which in the current fiber-optic sensor development stage is the most relevant standard type, since it comprises the particular characteristics of a product, its performance threshold such as functional parameters (measurement accuracy, stability), interface and interchangeability, measures enabling cost reduction, environmental protection, and overall health and safety. This standard type will have a strong influence on the reliability of a measurement system.

Concerning sensor systems, all aspects should be considered in guidelines: characteristic features of sensor components, characteristic features of interrogation systems, and application and service aspects. If the component specifications meet the requirements of practical applications, all devices, cables, and connectors will work reliably. Standards should also define fiber sensor-specific validation procedures, which deliver the largest contribution to the user's confidence in fiber-optic sensing systems. Reliability-relevant aspects concerning mainly sensor characterization and methodologies necessary for the validation of fiber-optic sensor systems are described later in detail. Because FBG strain and temperature sensors are very often used for *structural health monitoring* (SHM) purposes, some of the relevant aspects of fiber-optic sensor applications are exemplified for FBG sensors. Most of these aspects can certainly be transferred to other sensor types and systems.

Generally, a distinction has to be made between the sensor signal coming from the sensing element (FBG, Fabry–Perot [FP] interferometer, distributed fiber-optic sensor) itself and the sensor signal recorded from the applied sensor. The characteristics of the unapplied sensor will more or less differ from the attached one. Moreover, it must be aimed to distinguish between measurement signal-relevant information that comes from the sensor, including surrounding effects, and the contribution from the structure to be evaluated. In almost all cases, it is not possible to discriminate between creep of the structure (measuring object) and creep of the adhesive, for example, used for bonding of the FBG sensor. Thus, the customer is only able to record the behavior of the structure from the measurement result with a certain sensor-related uncertainty. The largest challenge for fiber-optic sensor experts is therefore to reduce application-related influences.

Guidelines should recommend methods to test and evaluate the bonding behavior (creep) of applied sensors. Not only commissioning tests must be possible but also test of applied sensors, which have already been in use for a long period of time. By now, there is a lack of validation methods for this purpose.

Specific aspects have to be considered especially when FBG sensors are used, for example, possible influences of unwanted mechanical perturbations like transverse pressure or bending of the sensing element. Transverse pressure into the fiber at the location of the grating or any mechanically induced nonhomogeneous deformation of the grating area might change the spectral characteristics of the FBG element. This spectral change can lead to faulty measurement results if the interrogation system cannot tolerate such influences in the grating spectra. The possibility of the appearance of transverse influence must be estimated, especially when FBG sensors are embedded in nonhomogeneous materials like reinforced composites or rough materials such as concrete and are subjected to thermally induced material contraction.

Other points concern hydrogen-loaded FBG sensors. A decay of the reflectance can be caused over time by out-diffusion of hydrogen, or applied and permanently strained fibers can be subjected to stress relaxation effects, which lead to a wavelength shift. Finally, decay of the refractive index modulation could occur over time. All these characteristic features have to be considered with regard to long-term measurements. Guidelines will help here to avoid the underestimation of these effects.

If strain sensors are to measure not only tensile but also compressive strain, they have to be pretensioned. This procedure is not trivial during application under field conditions because the pretension must be held as long as the curing or hardening process is incomplete. Sometimes, special facilities are needed to introduce and hold strain in FBG during the manufacturing of large composite components in a factory, for example, rotor blades of wind turbines. Prefabricated sensor patches could be used alternatively; however, the patches must hold the pretension of the sensing element during the manufacturing of the patch and later on during potential heat treatment (tempering). In this case, the patch or the adhesive could shrink, and the pretension of the fixed grating is reduced or lost. The appropriate methodology for the application of continuously fixed FBG sensor depends on technological conditions.

Apart from FBG sensor application procedures, the arrangement of the wiring of the sensors must not be underestimated. Under harsh mechanical or environmental conditions, splices inside the material to be assessed and ingress/egress points are critical zones. Splices should also be embedded just like the fibers. This is usually possible

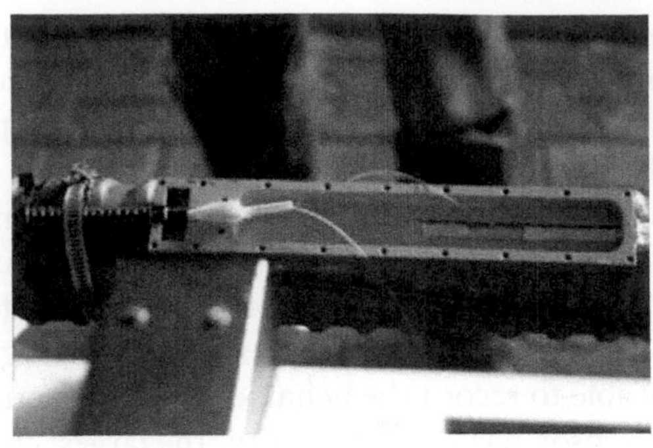

(a)

(b)

FIGURE 20.1 Critical points are splice joints inside of tight tubes (a) and cable egress points (b), for example, in heavy steel anchors. (Photo courtesy of BAM, Berlin, Germany; From Habel, W.R., Reliable use of fiber optic sensors, in *Encyclopedia of Structural Health Monitoring*, Boller, C. et al., eds., John Wiley & Sons, Chichester, U.K., 2009.)

in huge composite structures with sufficient thickness. However, in smart structures or when only very little space is available, other methods have to be developed. Figure 20.1a shows as an example of a solution to protect splice joints inside a steel anchor tube by storing them in a very small box.

If components with integrated fiber sensor have to be cut during the manufacturing process, such as usual on pultruded composite profiles or rods, special connecting technologies must be used to connect the embedded but cut fiber with a cable link to the interrogation device. Such challenging connecting technologies are being developed; guidelines or standards are therefore not yet available. Another critical area is the ingress/egress point of sensor fiber–equipped components. Depending on the structural component and the probable use, some rules should be observed to ensure long-term stable link to the applied sensor. Figure 20.1b shows an example of the ingress/egress point of a prefabricated anchor equipped with FBG arrays and protected for the process when the anchor is introduced into the bore hole.

On the basis of the manifold experiences of fiber-optic sensor experts in creating ruggedized packaging, general guidelines are as follows:

- Preparation of the area where the sensor is to be installed
- Connecting fiber ends and protecting splice areas very close to the measuring object (under field condition)
- Supervisory procedure to check the appropriateness of the sensor installation
- Protection of sensing area
- Robustness/aging of, for example, connectors when they are frequently opened
- Mechanical and thermal hysteresis of embedded or applied sensor fibers, for example, containing FBG sensors
- Identification of zero-point changes when devices have to be disconnected or leading cables have to be cut and exchanged
- Packaging and protection of the ingress/egress areas to make sure that embedded or attached fiber-optic sensors are fully protected over the whole service life of the structure without jeopardizing data integrity

- Sensor reaction to mechanical or thermal impacts (shocks) during operation
- Sensor behavior/aging under repeated vibration
- Sensor behavior (including cabling) under harsh climate conditions
- Durability of sensor-related materials under thermal, chemical, and mechanical conditions

Additionally, recommendations and advice for the repair of applied components of the sensor system should be given in the guidelines. The demand on exchangeability of sensors or components clearly determines the choice of the measurement and application method. Exchange and repair of embedded sensors are much more difficult than those of surface-applied sensors. Guidelines should also give recommendations for some other measurement task–related critical aspects that can affect reliability and/or stability.

20.3 Design Aspects for Reliable Fiber-Optic Sensor Systems in Standards

This section and the following ones describe important aspects that have to be considered (in standards and/or guidelines) when a measurement system is designed, sensors are applied onto or in measurement objects, and the measuring system is in service over a number of years. To solve a measurement task by using a fiber-optic sensor, first, the appropriate sensor type with specific performance features has to be selected from the different types that are available and offered. The optimum design of the fiber-optic sensor systems comprises a number of components that have to work reliably together under on-site conditions. Figure 20.2 shows the hardware structure of a fiber-optic sensor system with the components that have to be well coordinated.

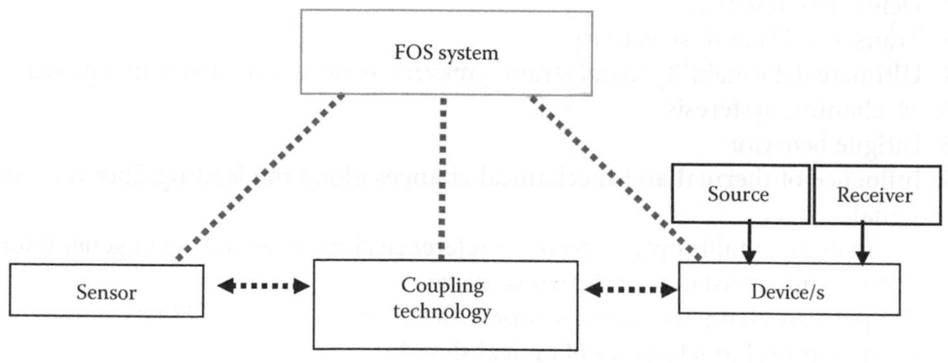

Sensitivity
- Characteristic function
- Mech./thermal hysteresis
- Drift of the measurement signal
- Temperature influence
- Sensor reaction to the measuring object
- Stability of materials

Splice or connector
- Compatible to standardized components
- Line neutrality
- Allowed number of disconnections
- Robustness for on-site use

Parameter stability
- Reproducibility
- Vibration resistance
- Tolerant against perturbations

FIGURE 20.2 Main components of a fiber-optic sensor system.

Behind these physical components, numerous parameters that influence the performance (reliability) of the sensor signal and the stability of the system components are hidden. Most of the commercially available fiber-optic components (connectors, connecting cables, FBG elements, and interrogating devices) are still validated according to standard test procedures developed for data-communication purposes or according to the individual manufacturer's own rules. At least functional tests are carried out. However, the optical measurement equipment is often sensitive to ambient temperature and other perturbing parameters, for example, relative humidity variations. For example, an optical spectrum analyzer occasionally used for recording the FBG signal can have a temperature sensitivity K_T of about 10 pm/K, which means it corresponds to the same temperature sensitivity of an FBG. If such a readout device is used to record a signal from an FBG sensor on-site, the measurement uncertainty could exceed the acceptable level. On the other hand, the calibration curve for the sensing element and/or the completed sensor sample must be available. If the equipment has to be replaced later by another one with different temperature sensitivity, the measured strain or temperature signal does not represent the previous sensor behavior.

In order to get reliable measurement results, the influence of overall climatic conditions on all system component specifications should be considered. Primarily, the component's behavior under reference climate conditions must be understood. The following list provides an example of some of the important technical features and characteristic quantities that have to be proven for deformation sensors at least at reference climate:

1. Optical characteristics (i.e., attenuation, reflection coefficient, and pulse characteristics, e.g., for FBG)
2. Deformation sensitivity
3. Transverse (lateral) sensitivity
4. Ultimate deformability (axial strain, pressure, bending) of fiber sensing area
5. Mechanical hysteresis
6. Fatigue behavior
7. Influence of thermal and mechanical changes along the leading fiber on a sensor signal
8. Ultimate acceptable optical power at reference climate, as well as in general terms
9. Temperature resistance of the whole system
10. Temperature characteristics of sensor material [$\alpha_{\text{Sensor}}(T)$, $\alpha_{\text{Sensor}}(T)\psi -\leftarrow \alpha_{\text{Material}}(T)$]
11. Temperature characteristics of optical signal
12. Influence of local temperature changes on sensor signal
13. Temperature characteristics of sensitivity to deformation and thermal hysteresis

If sensor systems have to work under very different environmental conditions, performance should also be proven for the limits of intended use. The IEC 61757 standard for fiber-optic sensors (IEC 61757-1 2011), the VDI/VDE 2660 guideline (VDI/VDE 2660 2010), and specific sector standards propose few more parameters and test procedures. Some selected aspects for the main components of an FBG strain sensor system are described in the following.

20.3.1 Optical Source, Detector, Modulation, and Demodulation

It can certainly be assumed that the optical source delivers a signal with sufficiently stable spectral and intensity characteristics as is needed for the respective modulation technique. Depending on the sensor signal resolution requirements, devices with different interrogation methods are used. For example, a spectroscopic method using a charged coupled device spectrometer provides a simple method with fairly good resolution. In contrast to this, devices that use interrogation methods on the basis of tunable filters, fiber interferometers, or planar integrated optical chips enable very high signal resolution. Stability and reproducibility of the wavelength measurement on FBG sensors decisively determine the uncertainty of the sensor system. This problem can be understood by considering two aspects.

The first aspect concerns parameters of the FBG that define the Bragg wavelength. In order to measure strain or temperature, it is necessary to know the setup wavelength of the grating. The key question is which part of the spectrum should be used for the center wavelength calculation. Figure 20.3 highlights this problem. Using the definition that the width of a grating peak is defined by the distance between the two minimum points in the peak, this calculated Bragg wavelength λ_B differs from another estimation of the wavelength that is based on the –3 dB level definition. In those cases where the main peak has another side peak, for example, due to birefringence effects, or has split into asymmetric peaks, the spectral shape has an enormous influence on the λ_B value, and the –3 dB definition must be modified.

In between, there are recommendations concerning peak form and stability of the FBG spectra, when the sensor is strained. In the first guideline for such fiber-optic strain

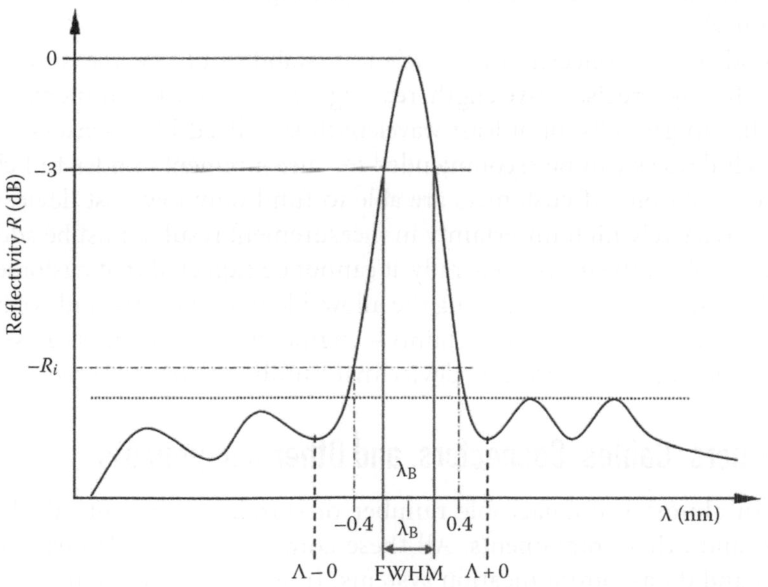

FIGURE 20.3 Basic spectral parameters and values of an FBG peak wavelength (Fernandez et al. 2004) needed to calculate/define the Bragg wavelength from the spectral characteristics. FWHM (full width at half maximum) describes the difference between the two independent values (here: wavelength) at which the dependent variable is equal to half (–3 dB) of its maximum.

sensors based on fiber Bragg gratings (FBGs; Fernandez et al. 2004), it was defined that maximum strainability is reached when the characteristic of the strain sensor under the appropriate strain deviates by more than 5% from the (elsewhere) defined sensor characteristic; or the first subsidiary maxima in the spectrum do not lie at least 5 dB below the main peak, or the spectrum shows a structure no longer amenable to evaluation (failure criteria). However, there are no recommendations in the existing sensor standards or guidelines which method for calculating the peak wavelength should be used. Therefore, it is urgently recommended (Fernandez et al. 2004) that the measured values obtained through tests must be unambiguously specified. It should be clearly stated whether the value is an original indication of a measuring device, an uncorrected or corrected measured result, or the result of averaging several measured values. If the result was calculated using an indicated value, this must be noted, together with the relevant calculation rule.

Another important remark concerning the true signal response from an FBG sensor: it is very likely that the Bragg wavelength λ_B of an FBG designed, manufactured, and delivered differs from that of the grating after application. In the VDI/VDE 2660 guideline, it is therefore proposed to denote the measurement result of the zero-point measurement after the installation of the FBG sensor (without loading) as "reference wavelength λ_0." The reference wavelength λ_B does not necessarily have the same value as the "delivered" Bragg wavelength.

Generally, to get a good signal-to-noise ratio, a broad spectrum is desired. To achieve good peak identification, on the other hand, a narrow reflection spectrum with a sharp peak is recommended. In any case, peaks with flat or nonuniform tops, asymmetric peaks, and large side lobes must be avoided. Thus, the precision in the estimation of the grating wavelength strongly depends on the grating characteristics (mainly strength and uniformity).

The second aspect concerns the wavelength stability of the measurement device. Devices for highly precise wavelength reading use standard commercially available etalons similar to gas cells, or at least wavelength-stabilized FBG sensors. It is obvious that only such devices can be recommended for measurements under full climatic and temperature conditions. If customers are able to fund only low-cost devices for a specific task, the relatively high uncertainty in measurement results must be addressed and must be acceptable to the users. Generally, it cannot be expected that customers know or understand all these aspects. Following the allowable uncertainty—and considering the achievable real system-inherent uncertainty—engineering consultants can seriously recommend the measuring method and device that fulfills the customer requirements best.

20.3.2 Fibers, Cables, Connectors, and Other Components

There are an almost unmanageable number of standards for optical fibers, cables, connectors, and other components. All these components are also used in telecommunication and data-communication systems. International standards can therefore be found in the International Standard Classification, nr. 33.180, and on the web, for example (US and International Fiber Optic Standards). To enhance the reliability of components, some investigations were made within the COST 270 action, for example, the connector group defined a reliability qualification test that resulted in the IEC standard 62005-9-2. In any case, when components developed for telecommunication purposes are used in sensor systems, their behavior under specific environmental conditions has to be verified.

20.3.3 Sensing Element

From the measurement reliability point of view, the most challenging part of a fiber-optic sensor system is the signal response from the sensing element itself. In case of strain sensors, the question is how much of strain is durably transferred from the structure to be evaluated into the sensing element. Depending on the method of sensor application (embedded, clamped, surface-glued, and welded) and the type of sensor (sensor fibers without or with gratings inside, sensor patches, and rods), the interaction between the measurand and the sensing element is influenced by numerous factors. From the author's point of view, the appropriate application of sensors, especially FBG strain sensors, is very complex and quite difficult. Hence, some application aspects related to the reliability of the measurement result that should definitely be the content of a guideline or standard are considered for FBG strain sensors in the following section. Almost every expert team uses a different method of attachment or embedment of FBG sensors. Coatings and adhesives with different specifications even for similar measurement tasks are used. This very complex part of the measurement system brings the largest uncertainty—actually, this could even be considered a skills-based uncertainty. The use of a new validation methodology for the evaluation of strain transfer of surface-applied strain sensor (described later in this chapter) revealed that even not all of the commercially offered strain sensors represent the true strain value of a measurement object. On the other hand, it must be realized that there is still a lack of well-validated methods for the embedment of strain sensors. Moreover, only few companies offer application expertise that includes a warranty claim. For example, the embedment of FBG strain sensors in rotor blades of helicopters or wind turbines is, by now, still often carried out by research institutions or research departments of companies according to their best knowledge because generally accepted technical rules are not yet available. However, the user community interested in fiber-optic sensors promotes and supports more and more research and development (R&D) activities in this area to close the gap in the expert's knowledge and to push the development of validation methods or completion of corresponding guidelines and standards.

20.4 Specific Reliability Issues for the Sensing Components of Fiber-Optic Sensor Systems

In order to ensure high-quality results with fiber-optic sensor systems, the most important prerequisite is that all involved project partners have carefully analyzed the measurement task and the influencing operating and environmental conditions and finally have chosen appropriate system components. This include the optical source, optical fiber with a specific coating, appropriate sensing element itself (extrinsic or intrinsic sensor type), some components like connectors, beam splitters, and finally an interrogator with appropriate specifications. Standard or guidelines will certainly help here. Not to forget, entirely reliable data sheets and specifications for all components must be available.

Commonly, application on-site is carried out under harsh conditions so that some technical inadequacies occur from time to time. Even if they have considered all design aspects and made the installation precisely, there are always few imponderables that influence the reliability and long-term stability. Unfortunately, for applied or embedded sensors, it is not possible to eliminate all disturbing influences on the optical signal. In the following, the main sources for uncertain measurement results will be considered (see also (Habel 2009)). Because the most requested short-gauge length fiber-optic strain sensor type is the FBG-based strain sensor, the special case of FBG-based sensor is considered more frequently.

20.4.1 Strain Transfer Issues

Whatever type of mechanical sensor and fixing method (embedded, clamped, continuously surface-glued, or welded) are used, enduring reliable strain transfer from the measurement object into the sensing element must be ensured over the complete period of operation. Irrespective of the mode of sensor fixation, the sensitive fiber and/or the fiber-optic sensing element (e.g., FBG) has always to be bonded to a support component or to a fixing element, and often, this support component is separately fixed at the measuring object. In order to avoid uncertain measurement results for the expected operating conditions, the following aspects have to be considered:

- Physical–mechanical properties of fixing components (substrate, adhesives, protecting layers/coating), material combination sensing element/measuring object, and covering must be known and characterized for all ambient conditions.
- Proof of sensor position accuracy.
- Thermally induced stress loading of the sensing element during component manufacturing or application.
- Sensitivity of all components (sensing element, leading fiber, connectors and splices, optoelectronic components) to mechanical and thermal influences.
- Attacks from aggressive media.
- Chemical and physical interactions of materials close to the sensor.

From the user's perspective, the appropriate strain transfer from the sensor to the measurement object is the challenge because it brings uncertainty even if the single components have excellent specification. Few selected examples shall show how hidden problems reduce the reliability of a sensor system.

20.4.1.1 Embedded Sensors

In some cases, embedment of fiber-optic strain sensors to measure deformation of structure components is rather recommended than surface application. This applies to wind turbine blades or airplane wings, because surface-glued fiber-optic sensors produce an unsymmetrical interface structure (Schlüter 2007) and can also be damaged when located on surfaces. On the other hand, operation of embedded fiber-optic sensors can become a challenge, particularly when nonhomogeneous and/or orthotropic materials are used such as composite materials or layered structural components with nonsmoothly located layers. It must be ensured that the sensitive area, for example, the FBG

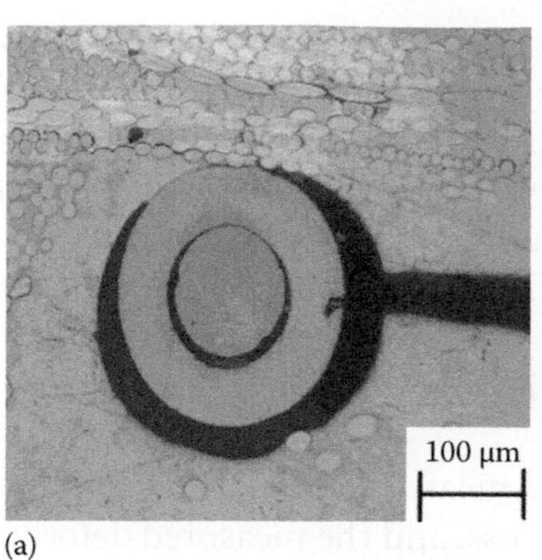

(a)

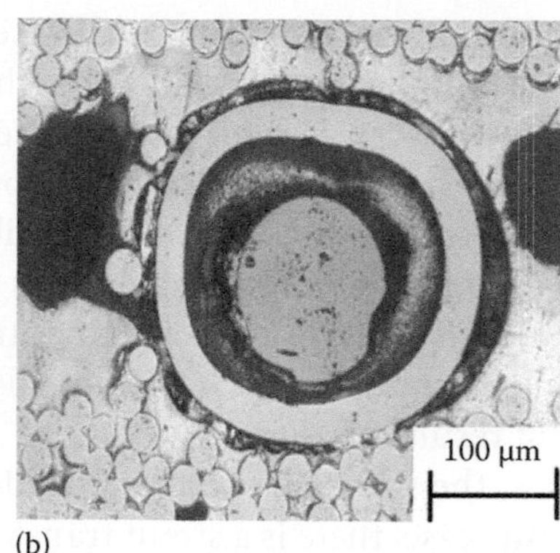

(b)

FIGURE 20.4 Strain transfer problems with embedded fiber-optic strain sensors because of inappropriate choice of fiber coating and wrong method to integrate FBG strain sensors into reinforced epoxy resin components (small fibers around the optical fiber are the reinforcing glass fibers; (a) one-layer high-temperature acrylate coating; (b) two-layer standard acrylate coating). (From Schlüter, V., Strain *transfer characterization of surface applied fiber* Bragg grating sensors, *Young Stress Analyst Competition, British Society for Strain Measurement* [BSSM], 2007, pp. 34–38.)

element, is surrounded on all sides by the bonding or the hosting material, and not bent by local inhomogeneities. Depending on the material used, in particular in reinforced materials, that might be difficult. Another problem arises when high temperatures lead to a local deformation of coating materials, or during cooling-down phase, shrinkage effects occur. Figure 20.4 shows the micrograph of a fiber-optic sensor element embedded in a glass fiber–reinforced sample.

Apart from the dimension of the fiber-optic sensor element, which is commonly a quite large inclusion in fiber-reinforced materials, two critical problems can arise that will seriously affect the strain transfer:

- Deformation of the fiber coating, which represents the bonding—the strain transfer—interface due to increased thermal strain during curing (Figure 20.4b)
- Delamination of the sensing fiber due to shrinkage of the material to be evaluated (Figure 20.4a)

Another critical influence on the reliability of embedded strain sensors comes from the stress/strain conditions in the interface area between sensing element and structural material under operating loads (shear stress development). Commonly, there are two boundary layers for embedded fiber-optic sensors: (1) the interface between the structural material and the fiber coating and (2) the interface between the optical fiber and its coating. Because any deformation of the structural material must be transferred into the optical fiber, the mechanical and physicochemical properties (coating thickness, Young's modulus, and viscoelastic properties) of selected and/or existing materials at the interface profoundly influence both the static and dynamic performance of the sensor.

Assuming that an embedded fiber-optic strain sensor has completely bonded along the fiber length with the material, elastic stress transfer will be the dominant mechanism at the interface up to a definite stress level. The elastic shear stress distribution along the optical fiber is constant, with exception at its ends if the fiber ends inside the matrix material. If the sensor is located in the area along the fiber where the shear stress is constant, it will work without systematic error. In this case, there are no irregularities in shear stress distribution. However, a more realistic situation is that the sensor fiber will have a nonconstant shear stress distribution along the sensing length caused by reduced bond strength due to irregularities in the sensor surface or in the matrix material to be evaluated. This leads to a premature debonding of the sensor from the matrix, and its function could fail early. Although the sensor delivers any signals, no correct measurement result is achieved.

Sometimes, the fixed part of an embedded measuring area along a fiber is not long enough. In this case, there is a strain transfer loss, and the measured deformation value is wrong. The same strain transfer loss can occur when the coating material of the sensing area does not satisfy the requirements regarding temperature loads. If it becomes too stiff at lower temperatures, it cannot follow the deformation introduced; if it is too pliant at higher temperatures, it creeps indeterminately.

And finally, every continuously bonded fiber-optic strain sensor will start to debond at a certain level of shear stress in the interface. This detrimentally affects the transfer of the measurand into the FBG strain sensor and results in erroneous readings. Only in few cases, a sudden loss of the sensor function is detectable. The more critical case is the gradual process of increasing creep and delamination from the measuring object, so that ongoing erroneous measurement results cannot be identified. There is no chance, by now, to check the full function of embedded sensors. Therefore, there is an urgent need to have guidelines that propose and help to estimate the actual range of deformation, which does not lead to debonding and reliably transfers the strain. This estimation should definitely provide the bond strength value for which the strain transfer mechanism fails (performance limit of the sensor). Such estimation can be carried out using a standard testing approach developed for maximum load identification of reinforced composite materials. The so-called push-in testing facility (Kalinka 1997) allows characterizing the interface bonding behavior of embedded fiber-optic deformation sensors. This method can therefore be used to determine and approve the appropriate sensor coating for a specific measurement task. Apart from the experimental proof of the coating's bond strength at the fiber surface, this test method delivers characteristic behavior of the shear stress development at the interface coating/structure. Details and examples of differently bonding sensor elements are described in Section 20.6.2.

20.4.1.2 Surface-Attached Sensors

Strain transfer into surface-attached FBG sensors is generally more difficult because of the unsymmetrical interface structure. A reliable and reproducible bonding of the sensor fiber to the flat surface is not easy to reach. The strain transfer into embedded sensor fibers, especially FBG sensors, is considered to be much easier. Also, the risk of debonding the embedded sensor elements under combined environmental and dynamic loads is less. Although surface-attached sensor fibers can visually be evaluated how the sensor is fixed to the surface and is often seen as good, extensive investigations revealed that bonding defects and thermal influences can lead to erroneous signal readings. Alternatively, prefabricated fiber-optic sensor patches are also used for strain measurements at surfaces; however, patches are quite complex. The patch itself or the adhesive used could shrink. Test series revealed that few commercially offered strain patches show aging effects after thermal cycling resulting in an inacceptable measurement result (Habel et al. 2011a, Schukar et al. 2012). Few examples of weak and unreliable strain transfer are given in Section 20.6.1.2.

Considering the fixing length of installed fiber-optic sensors, it is worthy to expand the discussion and illustrate the problem by an example. This also shows how important such aspects are to become part of guidelines for fiber-optic strain sensors. Due to the shear stress distribution over a strained fixed fiber and its decrease at the end of the fixed area, the length of the fixed fiber must be larger than the length of measuring fiber section. Figure 20.5 impressively shows what happens when the fixing length is too small. In a tensile test of sensor-equipped samples (FBG with a grating length of 6 mm) was revealed that even a fixing length of 30 mm was too small. The relative strain transfer factor under room temperature condition was for the fixing length of 30 mm only $k_{RT} = 0.938$ (the correct factor is 1.0). This too small k-factor deteriorates for increasing temperatures (80°C and 100°C) to an amount of $k_{80} = 0.902$ and $k_{100} = 0.601$. It is obvious that, in this case, the interface zone softens and the applied sensor transfers definitely too less strain values. A sensor guideline should give recommendations what fixing length has to be chosen. On the other hand, it must be considered that there is an upper limit for the fixing length when the applied sensor has to operate also under low temperatures. If the fixing length is too large, a number of temperature effects (e.g., stiffening, contraction) hinder the sensor to follow the deformation of the hosting material.

20.4.2 Sensor Design Issues

Mostly, standard types of sensing elements are used to design the desired strain sensor. If there are special requirements, modified fiber and coating types must be used. One aspect concerns the thermal stability of the fiber coating or even the fiber itself (for temperatures above 500°C). Another aspect concerns the signal stability under mechanical influences. Signal stability can be reduced by microbending effects resulting in a local increase in attenuation. Or, when FBG sensors are used, transverse pressure or local bending of the grating might change its spectral characteristics. This spectral change can lead to faulty measurement results if no fault-tolerant interrogation system is used (Dyer 2004). Such possibly appearing transverse influences must be estimated, when FBG sensors are embedded in nonhomogeneous materials, like reinforced composites or rough materials such as concrete, and are subjected to thermally induced material contraction.

One fundamental requirement for strain sensors is that the temperature-correlated strain has to be identified and considered. Strain sensors should be designed as temperature-invariant ones; at least, it must be possible to discriminate between load-induced and temperature-correlated strain. A clear distinction of these reliability-related aspects to be considered is made in the VDI/VDE 2660 guideline (VDI/VDE 2660 2010). Other points that have also to be considered concern the decay of the reflectance caused over time by out-diffusion of hydrogen. Or, permanently strained fibers for high-precision measurements can be subjected to stress relaxation effects, which lead to a wavelength shift over time. Finally, decay of the refractive index modulation could be occurring over time. All these characteristic features have to be considered when sensors are designed for special demands and long-term measurements.

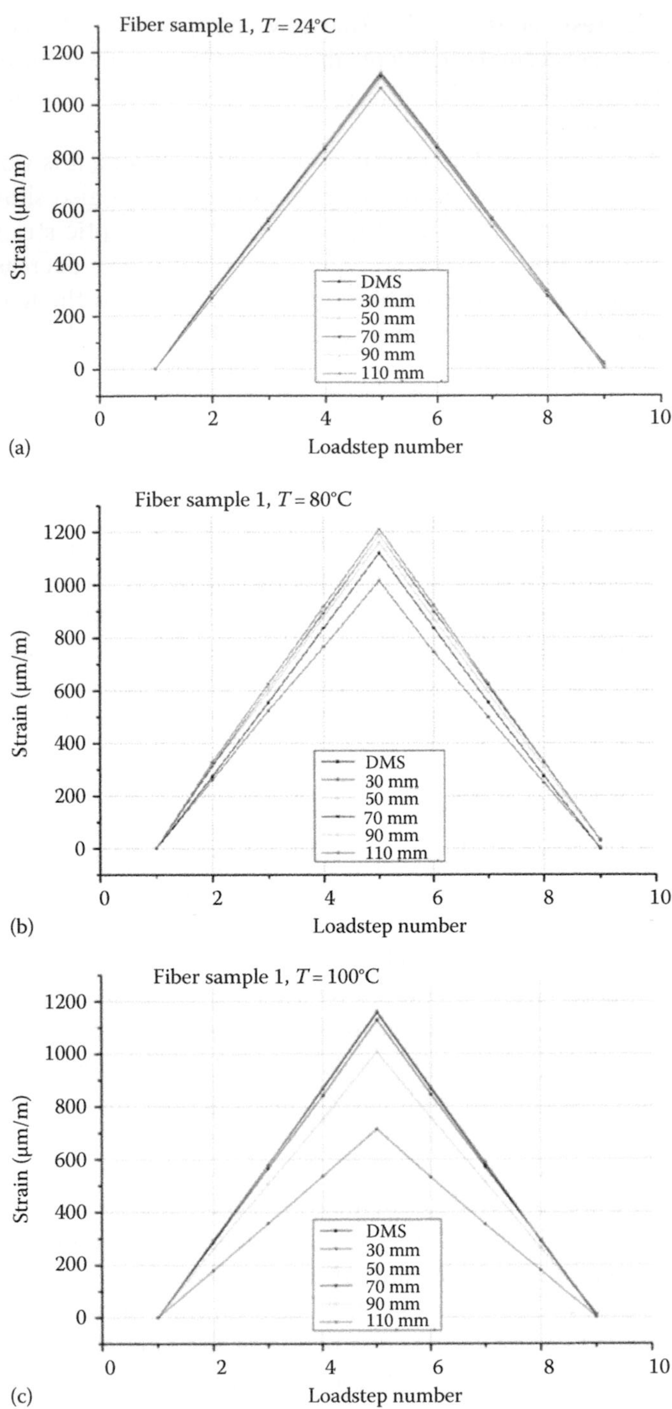

FIGURE 20.5 Strain transfer loss of surface-attached sensor fibers during tensile loading of the test samples depending on the fixing length of the sensor (parameter in pictures in mm) and operating temperature (From Schilder, C. et al. *Measurement and Science Technology* 24:094005, and BAM internal status report, German Research Cooperation Project FZK 01FS10031 on Strain Sensor Guideline Development, HBM GmbH Darmstadt, BAM, Berlin, Germany, 2011): (a) room temperature; (b) 80°C; (c) 100°C. DMS—resistance strain gauge.

The optimal sensor design would be the installed sensor that is able to indicate a malfunction or imminent failure of sensor function. At least, sensors should allow remote checking whether the current parameters are still complied with the calibrated ones. This objective can be considered as one of the main current research topics (see Section 20.8).

20.4.3 Evaluation and Calibration Issues

Basically, all equipment used for on-site measurements (sources, interrogators, and sensors) including equipment for subsidiary measurements, such as environmental conditions, shall be validated and calibrated before being installed and put into service. However, everybody must make it clear that applied or embedded sensor elements cannot be recalibrated after installation without removing them from the measurement object. It is extremely difficult to get validated data from them after a certain time of operation because it is not possible to reveal whether a change in the sensors signal is caused by a change in the structure, a change in the attachment, or a change in the sensor itself due to environmental and mechanical conditions. This basic problem leads always to a certain amount of uncertainty, which cannot be exactly defined. In order to reduce this uncertainty, two ways are imaginable: (1) sensors are designed to allow evaluating the function of the sensitive element, and (2) facilities are available, which allow recalibrating installed sensors on-site. By now, neither of the two methods has thus far been initiated. The only way, by now, is the prediction of the long-term sensor function of applied and/or embedded sensors using test samples investigated in validation and calibration facilities in labs. For surface-applied strain sensors, the newly developed *calibration of fiber-optic sensors* (KALFOS) facility can be used (Habel et al. 2011b). This facility allows simulating long-term operating conditions, testing and evaluating the bonding behavior, and investigating creep and delamination effects of applied sensors. Embedded sensors, on the other hand, can only be evaluated by now by investigating representative test samples in single-fiber push-out test facilities (see Section 20.6.2) and by radiographic methods (Section 20.6.3).

20.5 Importance of Sensor System Validation

Users get appropriate measurement systems only when they use validated components of a fiber-optic sensor system and validated application procedures. Users benefit then from validation because they get assured information about the performance and limitations of the sensor system. Validation places high demands on manufacturers and providers; it is defined in the ISO/IEC standard 17025/2000 (Standard EN ISO/IEC 17025:2000 1999) of the ISO as "Validation is the confirmation by examination and the provision of objective evidence that the particular requirements for a specific intended use are fulfilled." It means that validated systems enable consulting engineers, suppliers, and users to evaluate suitability and reliability of the measurement system for the specific use. The key questions of accuracy, long-term stability, and reliable function under environmental influences to be proven by validation depend on the client's needs. Validation is always a balance between cost, risks, and technical possibilities.

Strictly speaking, validation of fiber-optic sensor system components can only be carried out by institutions that fulfill the requirements for technical competence in the field of fiber-optic sensor technology. The testing institution should preferably be recognized by accreditation bodies. Validation can also be carried out by third-party laboratories (unbiased institutions) that are able to bring in expertise and have the special equipment meeting the requirements concerning the uncertainty of measurement and traceability to etalons or national standards.

Common procedures for the validation of the sensor system components comprise

- Sensor-related characteristics (measurement range, resolution, and sensitivity are the most important)
- Measurement uncertainty, repeatability, and reproducibility of data from components of the measurement system
- Proof of stability of the system components' characteristics

Apart from the stability of the sensor system or the reliability of the measurement method, the uncertainty estimation of the measurement results has particular significance. Validation enables getting data that are important for the evaluation of the measurement uncertainty. Basically, this information can only be achieved under certain conditions:

- All equipment used for tests and or calibrations (including equipment for subsidiary measurements, such as environmental conditions) shall be calibrated before being put into service.
- Equipment shall be operated by personnel with sufficient competence in the field of fiber-optic sensors.
- The calibration procedure has to ensure that all measurements are traceable to the International Systems of Unit (SI).

The link to SI units may be achieved by reference to measurement standards and high-quality measurement instruments. A calibration laboratory or an unbiased competence center establishes traceability to the SI by means of an unbroken chain of calibrations or comparisons linking the available standards to relevant primary standards of the ISO units of measurement (e.g., national measurement standards). A short remark concerning traceability seems to be necessary (compare it also with EN ISO/IEC 17025 (Standard EN ISO/IEC 17025:2000 1999) and ISO 10012 (EN ISO 10012 2003)). It is clear that the resolution of measurement results achievable with relevant measurement standards (including its repeatability) must approximately be one order better than those of the measurement systems to be evaluated.

It should be underlined once more that there is a big difference between a virgin fiber-optic strain sensor validated and calibrated by the manufacturer, and the same strain sensor installed, embedded, or applied to a surface. Depending on the type of application or integration, there is a multiplicity of critical issues during installation that might lead to inadequate monitoring and finally to possibly catastrophic misinterpretation of the measurement results. In between, procedures for the validation and calibration of test coupons with surface-applied strain sensors are available (see Section 20.6).

However, it is extremely difficult to get validated data for embedded sensors even when test coupons are in use in laboratory. It is not possible precisely to discriminate between the sensor signal response and the real loading coming from the measuring object. This basic problem depends on the (long-term) quality of the sensor interface and leads to a certain amount of uncertainty. In any case, recommendations for application procedures in guidelines or standards based on fully established expertise will minimize uncertainty.

Not only users but also manufacturers and sales agencies benefit from validation and corresponding guidelines because damage can then be prevented during the production stage. Performance gaps in the measurement system can be displayed and removed by systematic optimization. In case of damage or if measurement results are not accepted by the user, the supplier is able to prove that he has complied with his obligation to exercise due care (product liability).

20.6 Experimental Methods to Characterize Measurement Reliability and to Confirm Recommendations in Guidelines

Because an important influence on the reliability of mechanical (deformation/strain) sensors comes from the stress/strain conditions in the interface area between the sensing element and the material or structure to be measured or monitored, the strain transfer quality must be validated. Commonly, there are at least two boundary layers: (1) the interface between the structural material and the fiber coating, and (2) the interface between the optical fiber and its coating. Because deformation of the structural material must be transferred into the fiber-optic sensor, the mechanical and physicochemical properties of selected and/or existing materials at the interface determine the performance of the sensor. A debonded interface of a continuously attached sensor fiber (Figure 20.4b) or changes in the interface's behavior due to temperature effects or aging of the involved materials would detrimentally affect the transfer of the measurand to the sensing element and result in erroneous readings. For example, fatigue tests with acrylate-coated FBG embedded in composite samples of wind turbine rotor blades revealed a continuously increasing aging effect in the FBG signal (150 μm/m loss after 1,600,000 cycles) for exactly the same deformation values. Some more results are described in Krebber et al. (2005).

Another systematic influence on the measurement signal of continuously fixed fiber-optic sensors comes from the properties of the coating. Coating thickness and Young's modulus have an influence on both the static and dynamic responses of the sensor. In every application (embedment or surface application), the interfacial mechanics has to be well understood. Some efforts have been done to model the interface conditions of installed deformation sensors (Luyckx et al. 2010); however, only an experimental test facility enables to fully investigate the strain transfer behavior of surface-attached deformation sensors. As described already, experimental facilities to fully verify models with all the correlating effects for embedded deformation sensors must be developed.

In the next section, some approaches to validate the strain transfer behavior of deformation sensors, especially FBG, are presented. The most progress has been made with the validation of surface-applied sensors. In this case, imaging and visualization techniques can be used when the sensor arrangement is visible to, for example, a camera. The validation and/or calibration of embedded—that is, hidden—sensors are much more difficult. Only indirect methods or some tricky measurement arrangements are possible.

20.6.1 Validation Facility KALFOS for Surface-Applied Sensors

Guidelines for the characterization of strain sensors require, among other data, the statement how to determine the strain sensitivity factor (*k*-factor). Therefore, the exact knowledge of the strain transfer from a real component under load into surface-applied strain sensors is of essential importance for structure deformation monitoring. Commonly, strain transfer performance of surface-attached sensors is determined with special facilities that refer strain deformation to high-precisely fixed strain sensors. Or, the sensing elements to be evaluated are installed on a steel rod specimen with a square cross section, which is installed in a stiff mechanical structure with four supporting points. The strain information from the attached sensor is compared with the precisely measured and calculated deformation in the bending line (Roths et al. 2010). Those methods allow calibrating strain sensors with high precision. However, investigation of possible local creep effects or evaluation of the different materials in the sensing area that influence the long-term behavior is not possible. Such information is urgently needed when sensor reliability has top priority, such as in sensor systems installed in safety-relevant structures. Safety-relevant measurement systems must provide trustworthy signal responses. Therefore, it is needed to know whether a change in the sensors' signal is caused by a change in the structure, a change in the attachment, or a change in the sensor itself due to environmental and/or mechanical influences. For the purpose of revealing weaknesses for surface-attached strain sensors, a complex validation facility KALFOS for the characterization of surface-applied sensors was developed at the BAM Federal Institute for Materials Research and Testing in Germany.

20.6.1.1 Description

The facility KALFOS is based on a physically and application-independent referencing method and allows qualifying the application of sensors, determining relative movements between sensor and specimen, and therefore the validation and calibration of surface-attached strain sensors. It also enables exploration and investigation of the strain transfer mechanism including all influences affecting the strain measurement. The methodology used helps to reveal weak points in the strain transfer because the common method to validate glued sensors with another but very precise glued sensor, which are applied in the same manner, does not tell anything about the application reliability. In other words, for reliable information about the strain state of the specimen, noncontact unbiased reference technologies have to be used.

The KALFOS facility consists of a load-bearing structure that allows introducing a mechanical stress to a host structure and a temperature chamber (Figure 20.6). The temperature chamber, which is mounted on a separate movable carrier and therefore mechanically isolated from the load-bearing structure, allows combined temperature/load cycling in a temperature range from −60°C to +100°C. The specimens can be loaded up to 20 kN

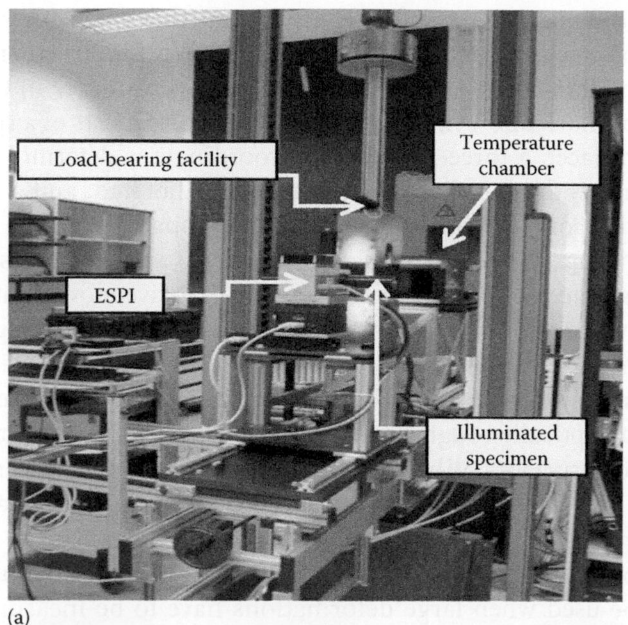

(a)

(b)

FIGURE 20.6 (a) KALFOS testing facility with ESPI reference system and temperature chamber (behind the jaws); (b) (i) tensile test specimen, (ii) clamping jaws, (iii) ESPI system. (Photos by BAM, Berlin, Germany.)

tensile force. That way, the tests can be carried out under fully simulated and reproducible environmental conditions. In order to compare the strain information provided by the sensor with that of the host material under loading conditions, two noncontact optical systems are used: (1) digital image correlation (DIC) technique and (2) electronic speckle pattern interferometer (ESPI) system. These noncontact validation techniques allow referencing the specimen deformation and visualizing the relative movement between the sensor and the strained specimen and are therefore not affected by any attachment problems, for example, by the adhesive used for commonly chosen RSGs as "reference" sensors. Using these noncontact optical systems, the strain in all involved materials, the specimen, the adhesive, and the fiber-optic sensor, can objectively be analyzed.

DIC is a three-dimensional measurement technique that determines the coordinates of points on an object when images are taken at different positions. Therefore, a random black-and-white pattern has to be applied to the specimen surface in the desired measurement area. The black-and-white pattern is divided into facets of a specified number of pixels. For each facet, a three-dimensional coordinate is determined. Before starting the measurement, the camera system has to be calibrated, and a reference image is taken. Every image of the deformed specimen is compared with the reference image taken at the beginning. That way, a displacement vector and the corresponding strains on the specimen surface can be obtained. A possible relative movement between specimen and fiber can be detected with a resolution individually depending on the number and the size of the facets and will easily be depicted in a false-color image. The ESPI system is based on the physical principle of speckles, which can be observed when laser light is reflected by optically rough surfaces. The speckles form a characteristic "finger print" of illuminated surfaces. When this surface undergoes a displacement, the speckles displace then, too. It is possible to track the speckles and to calculate the materials strain values from this speckle displacement.

More details of the optic reference systems are described in Schukar et al. (2011, 2012). DIC can be used when large deformations have to be measured, up to 100% strain increase. The resolution of the DIC system depends on the number and size of facets, and is in the range of 2 μm (in displacement). The ESPI system can be used when very small deformations, for example, in the range of 100 nm, have to be detected. The resolution of the ESPI system under perfect alignment conditions is up to 30 nm in deformation. The ESPI system embedded in the KALFOS facility enables a deformation resolution of 100 nm. The maximum strain range is limited to the number of fringes in the measurement field that the system is able to differentiate. The usual number of fringes for this system is up to 20, which is good to handle and calculate. The measurement fields for the ESPI system are very small compared to DIC. Additionally, to guarantee a certain number of fringes for the deformation calculation of each measurement step, the ESPI sensor has to be moved along to the movement of the specimen itself. A special ESPI holder was designed that allows the movement of the cameras in three directions (x, y, and z) to optimally locate the ESPI in front of the measurement target. Additionally, the carrier construction allows the ESPI cameras to move along with the deformation of the host material in order to keep the measurement field in its focus. Because of the ESPI's extreme sensitivity with respect to dynamic vibrations, the whole facility is founded on damped floor. The system control panels and all measurement units are mechanically separated from the testing facility itself. The ESPI system and the load-bearing facility are controlled by complex process software especially developed for this purpose.

20.6.1.2 Experimental Results

The following examples impressively show how important the recommendation in guidelines is to verify the strain transfer of an applied sensor. Standards and guidelines can give important hints as to which aspects have to be considered. However, the correct characteristics can only be approved by exemplary experimental investigations. Following, significant sources of measurement errors are demonstrated using the KALFOS facility.

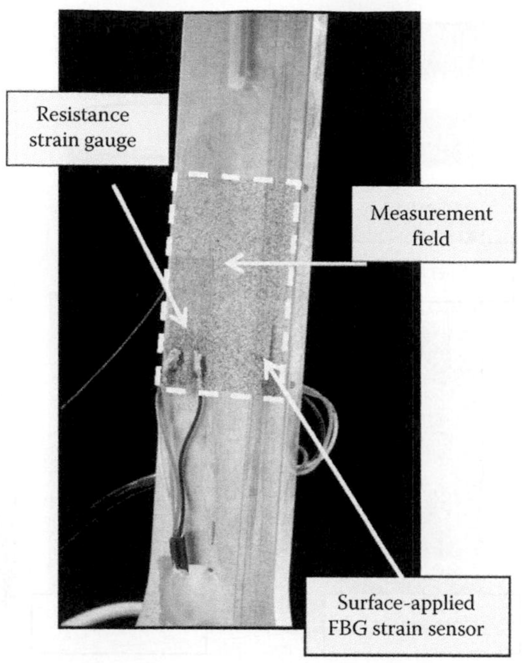

FIGURE 20.7 Tensile test coupon with surface-applied FBG sensor and resistance strain gauge sensors. (Photo by BAM, Berlin, Germany.)

20.6.1.2.1 Inappropriate Attachment of Fiber-Optic Strain Sensors onto Surfaces In order to investigate strain transfer effects, test specimens as shown in Figure 20.7 were manufactured. Usually, they contain one surface-applied FBG strain sensor on the front side, and additional RSGs on the front and back. In this case, the two sensor types, FBG and RSG sensors, were referenced with DIC. Therefore, a random black-and-white pattern had to be sprayed onto measurement field of the specimen (see Figure 20.7). The measurement area consisting of this black-and-white pattern was then divided into single facets measuring 15 × 15 pixels. The deformation of the facets with increasing and decreasing load was then evaluated by comparison with a reference image while the specimen was unloaded. From the measured deformations, the strain of the specimen, of the sensors, and in the adhesive could then be calculated.

It can never be excluded that a fiber-optic sensor has been applied inappropriately or, in the worst case, that the used adhesive or the chosen fiber coating will change their mechanical characteristics during the operation time under harsh environmental conditions. This must be avoided when the sensor is to monitor the health state and provide data for the evaluation of the structure's integrity or residual life. If high-precision measurement is required, long-term stable measurement results over the expected sensor lifetime and under operational conditions must be ensured. In other words, wrong application procedures must be revealed and prevented. Figure 20.8 shows a typical deformation response of an FBG sensor inappropriately applied. The indicated deformation measured by the FBG sensor is seen to lag behind the actual deformation of the specimen measured with DIC for growing displacement (different gray scaling inside the circle). Comparing all strain values by using the DIC system including deformation of the adhesive, it follows that the sensor signal does not match with the strain

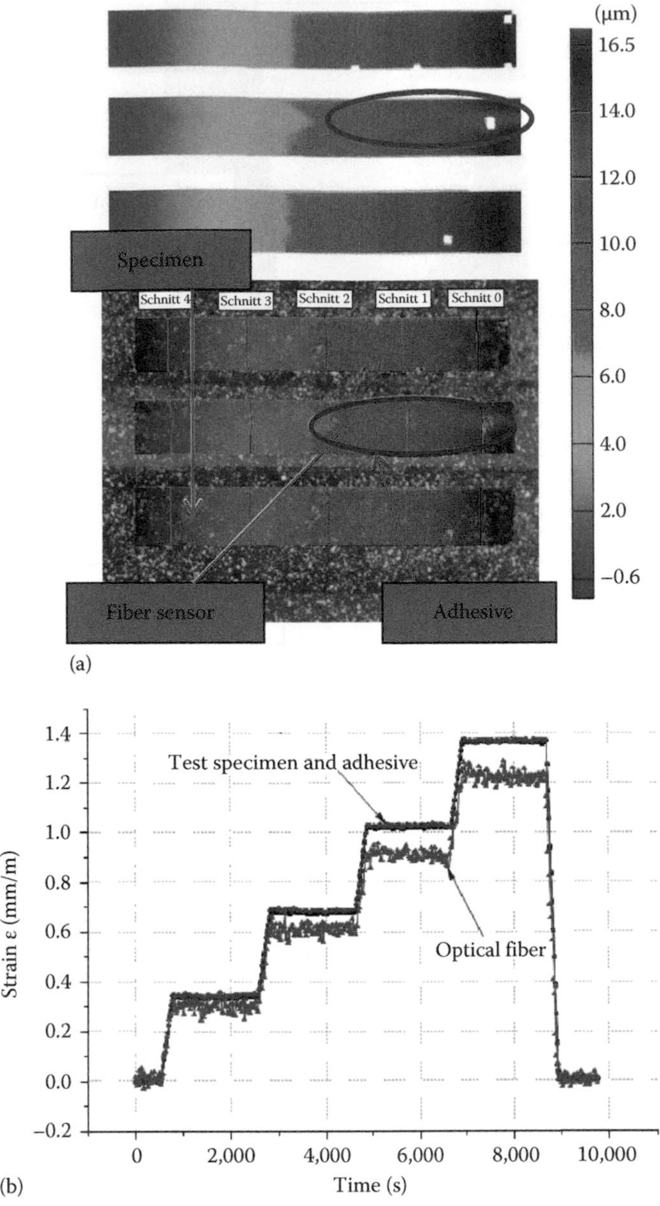

FIGURE 20.8 (a) Visualization of the sensor components' deformation during tensile strain test. The sensor lags behind the deformation of the specimen at higher displacement values; (b) strain response for the substrate deformation (by DIC) and from the surface-applied sensor. The sensor works with a too less gauge factor and does not represent the actual deformation.

of the substrate (Figure 20.8b). From these results, a strain sensitivity factor of only 0.6 was determined. In other words, the sensor signal may hypothesize that the load on the specimen is less; however, in reality, the load might already be over its structural limits (usually, when the sensor is appropriately and reproducibly applied, the gauge factor is then determined to be in the range of 0.7–0.8; for unapplied bare FBG sensors, the gauge factor is 0.78). Such unreliable sensor arrangements must be excluded by the validation of the sensor performance before and after application. This validation methodology can also be applied to other types of surface-mounted strain sensors such as RSGs.

20.6.1.2.2 Fatigue Effects Leading to Slowly Changing Erroneous Measurement Results Strained sensors are increasingly subjected to aging caused by thermal, environmental, or mechanical stresses. If applied strain sensors are dynamically stressed, aging effects should be investigated to validate the sensors for long-term measurements under dynamic loads. Such aging test could be very important to find out weak sensor design samples with the risk of early failure. Figure 20.9 shows aging test results obtained from fiber-optic strain patches attached to a surface of a carbon fiber–reinforced composite carrier (thickness 6 mm). Flexural fatigue tests were carried out with a cycling frequency of 0.5 Hz. The first 50,000 load cycles were carried out with an elongation of ±1 mm/m; after that, the elongation was increased every 10,000 load cycles by ±1 mm/m. One patch (see picture in Figure 6.4) showed critical bonding behavior. Comparing the bonding behavior of fiber-optic strain patches with RSGs, it can generally be stated from our investigations that RSG failed significantly earlier than the fiber-optic ones.

Another important aspect—and therefore a necessary part of guidelines—concerns thermally induced influences not only on the optical (strain) signal but also on accompanying materials like coatings, adhesives, and special interface materials. Such influences may lead to temperature-correlated strain effects. In addition to the material durability, the sensor characteristics must be specified and validated for the desired operating temperature range. Strain sensors indicated for use in an extended temperature range should certainly be resistant to temperature stresses in this range and must not show aging effects. Figure 20.10 shows that this is not always the case. Commercially available fiber-optic patches provided by several companies and specified by the manufacturers for an operating temperature range of up to 80°C were tested. Strain cycling tests were carried out for the maximum indicated temperature of 80°C. In the worst case, one of

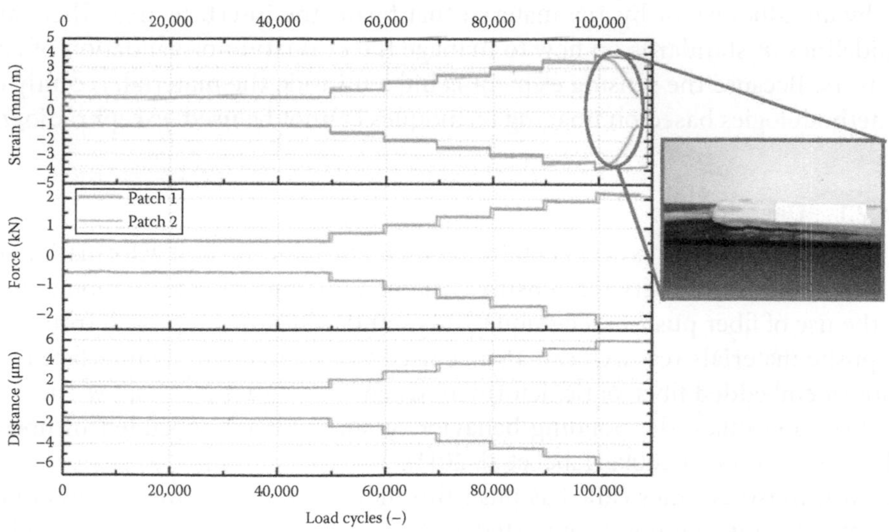

FIGURE 20.9 Beginning delamination of a sensor patch due to dynamic loading with strain cycles of ±4 mm/m. (Photo by BAM, Berlin, Germany.)

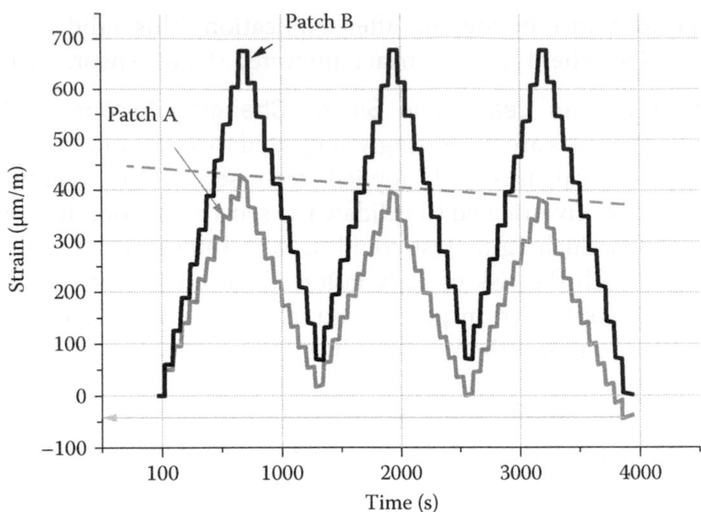

FIGURE 20.10 Fatigue test of fiber-optic strain sensor patches at a temperature of 80°C. The inclined dashed line (red) shows that the strain response of patch A is not stable after cycling; the strain was not correctly transferred into the sensor element.

the FBG strain patch types (Sample A) decayed at 80°C after a few cycles at a strain of about 400 µm/m. The measured strain after cycling was −49 µm/m lower than at the beginning of the test. This means that no proper strain transfer is possible with such a product because of its instability in the specified temperature range.

20.6.2 Push-In Test Facility to Estimate Bonding Behavior of Embedded Sensors

Embedded fiber-optic strain sensors are surrounded on all sides by the measuring object, by an adhesive, or by any material that forms the interface area. There are not any guidelines or standards on how to manage the evaluation or validation of embedded sensors. Because the sensing element is hidden inside the material, recently developed methodologies based on imaging techniques cannot be used, except radiographic methods.

20.6.2.1 Description

There is one possibility to experimentally simulate the bonding and deformation behavior in the interface zone between sensing element and material deformation to be measured: the use of fiber push-in technique. This standard testing approach from the field of composite materials research can also be used to characterize the interface bonding behavior of embedded fiber-optic deformation sensors. Figure 20.11 shows such a testing facility to evaluate the bonding behavior of small fibers embedded in fiber-reinforced materials (Habel 2009, Habel et al. 2011).

The push-in (sometimes called as push-through) testing machine consists of a very stiff beam, which is one-point supported at the middle of the beam. At one end of the beam, a high-precision stepping motor is positioned, which drives the beam in the vertical direction (upward). At its other end, the indenter needle with a diameter little less

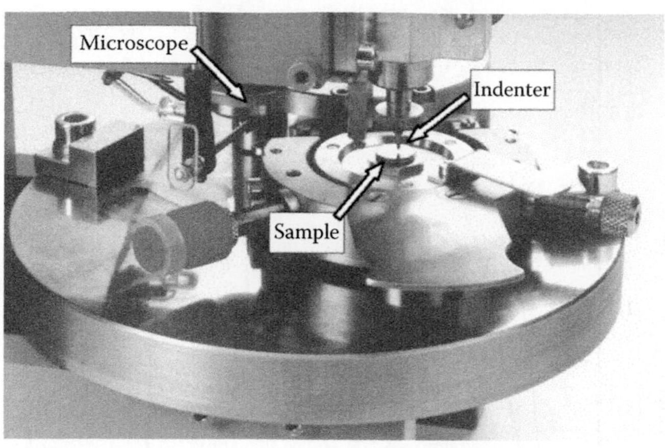

FIGURE 20.11 Push-in test facility for the evaluation of the bonding behavior of fibers embedded in materials. (Photo by BAM, Berlin, Germany.)

than the fiber diameter is fixed at the beam. The indenter introduces the force into the test sample. A microscope is available to position the sample exactly below the indenter needle. The testing procedure is computer controlled, and the force introduced into the fiber as well as the displacement of the end face of the fiber due to the pushing force is automatically measured and recorded. This method is primarily used to determine the appropriate sensor coating for a specific measurement task or, on the other hand, to approve the bond strength of the coated sensor fiber for the expected deformation. Apart from this experimental proof, and also from the proof of bond strength of the coating at the fiber surface, this test method provides also the characteristic behavior of the shear stress development in the border zone coating/structure. Further research shall allow correlating these test procedures with other interfacial parameters such as coefficient of friction, thermal residual radial stress, and coating stiffness.

20.6.2.2 Examples of Experimentally Achieved Results

Several test samples with different sensor fiber coatings were manufactured and then tested in the push-in test facility to find out which coating shows best bonding behavior to a chosen material. One of the test series was done with a composite component on the basis of epoxy resin. One important prerequisite was its appropriateness to survive curing temperatures in the range of up to 120°C. Figure 20.12 shows exemplarily the bonding behavior of three fibers protected with different high-temperature acrylate coatings. It can clearly be distinguished between appropriate and weak strain transfer from the deformed material to the embedded sensor fiber. It shows that this method enables an estimation as to which coating is able to survive the embedment procedure and remain its material characteristics for long-term strain monitoring over a wide range of material deformation.

In some cases, the bonding behavior of the used coating to the glass fiber is worse than that to the surrounding material. For strain sensors with high demands on static and dynamic deformability, it is recommended to approve that the bond strength of the fiber coating is sufficiently high. Figure 20.13 shows a tested fiber sample with insufficiently stable coating. It can clearly be seen that the fiber slipped already along the coating during strain simulation before it debonds from the material to be measured.

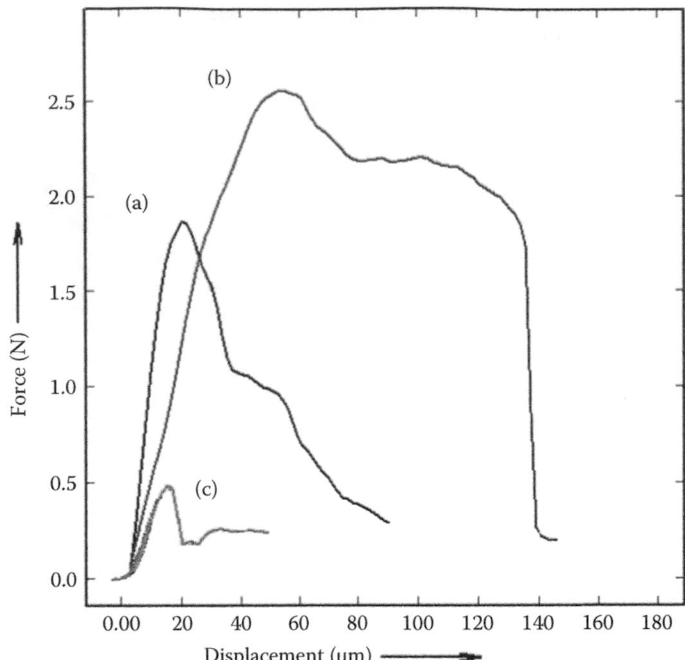

FIGURE 20.12 Experimentally estimated strain transfer ability for optical sensor fibers with high-temperature acrylate coatings embedded in glass fiber–reinforced epoxy material. Curve (a) shows a quite good shear stress behavior; curve (b) shows a strain transfer behavior with higher bond strength but slightly nonlinear sensor response; and curve (c) represents an inappropriate coating with very low bond strength.

20.6.3 Radiographic Method to Reference Deformation Measurements of Embedded Sensors

In order to evaluate the deformation behavior of sensors embedded in materials, radiographic methods can also be used. This methodology was used to reference deformation measurements with embedded flexible interferometer sensors at a very early age of curing cementitious materials. A special silicon mold was used to investigate the deformation process of green cement paste inside the specimen (Figure 20.14). The applicability of this method depends on the object's dimensions. It is difficult to make visible very tiny embedded fiber-optic sensors; however, the contrast can be enhanced by the interspersion of small metallic spheres or particles into the mineral building material; certainly, such inclusions must not lead to material damage. In this way, deformation of hydrating cement paste could be compared with the deformation of embedded fiber-optic FP interferometer sensors (Schuler et al. 2008, 2009). This x-ray tomography technology was accepted to validate sensor signals recorded from embedded flexible FP sensors during the hardening process of building materials. Moreover, the visualization of deformation in the inner part of materials using digital radiography enables the investigation of possible reactions to the material due to embedded sensors and makes irregularities and inhomogeneities in the structural material visible.

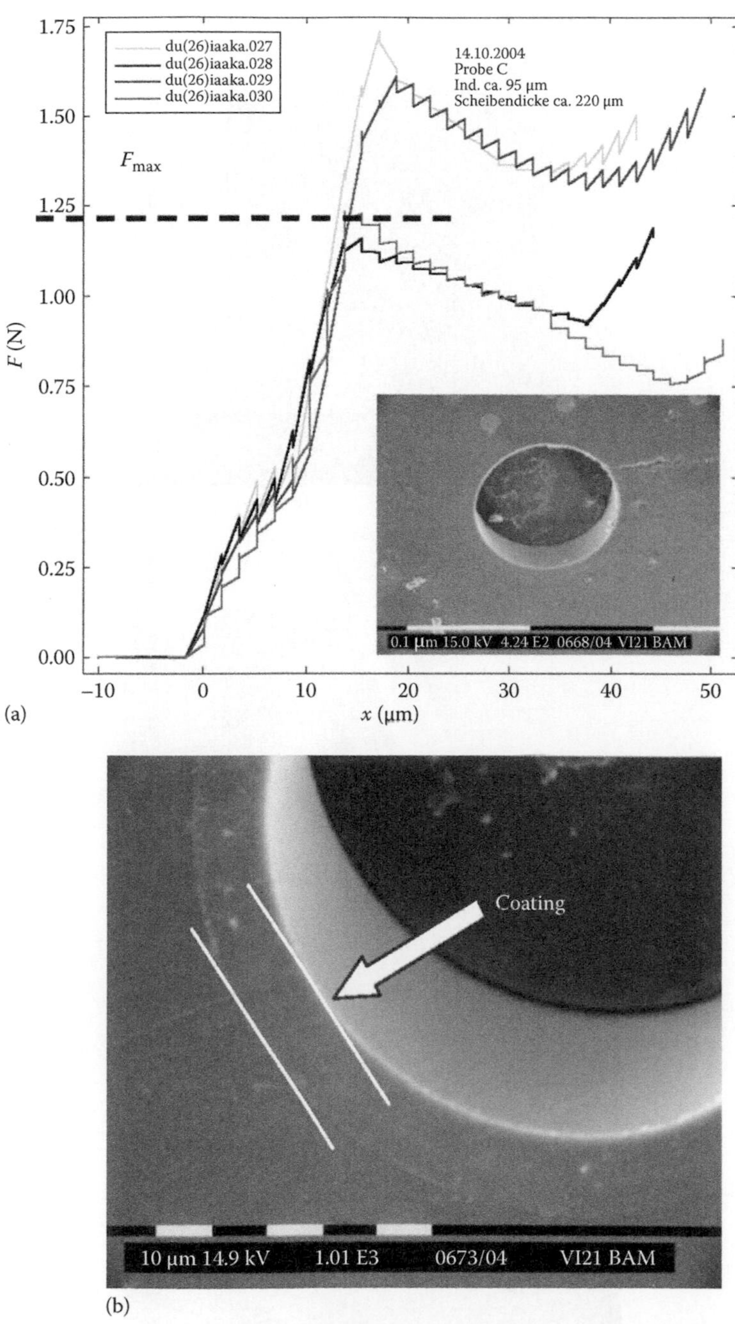

FIGURE 20.13 (a) Strain transfer characteristic of polyimide-coated optical fibers embedded in typical glass fiber–reinforced epoxy resin used for composite components (abscissa: displacement of the indenter, and ordinate: force at the indenter); (b) microphotograph shows debonding of the coating from the optical fiber beyond a certain amount of introduced material deformation.

(a)

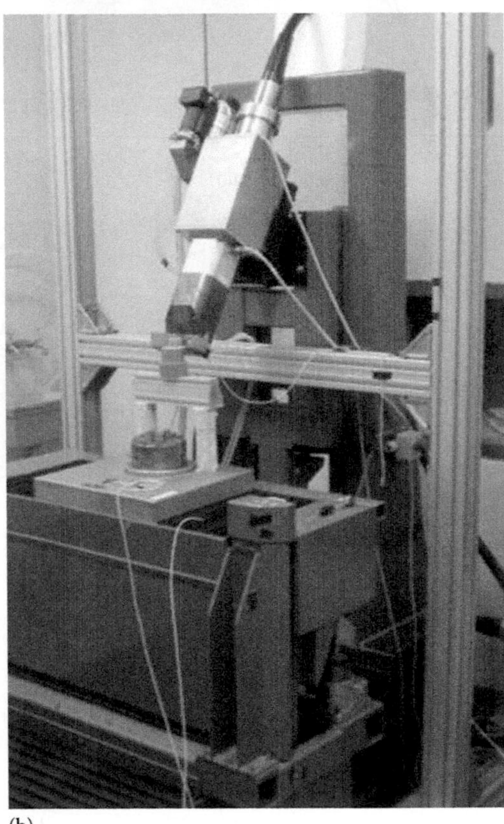

(b)

FIGURE 20.14 (a) Mold with fixed Fabry–Perot sensor before casting; (b) computer tomography (CT) facility with prepared and positioned test sample.

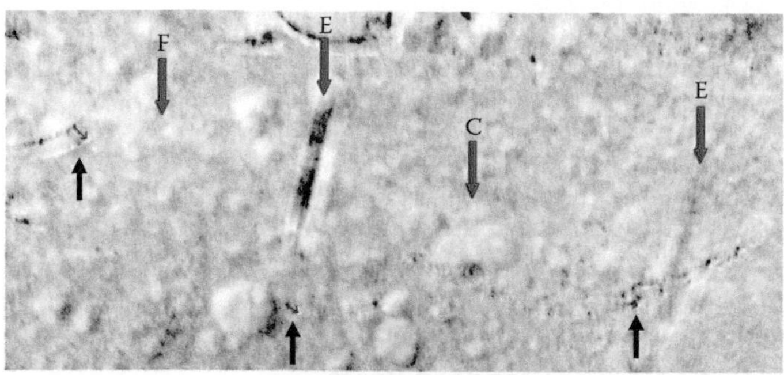

FIGURE 20.15 Summarized deformation of a Portland cement paste during the first 20 h of hydration. Light arrows show components of the embedded extrinsic Fabry–Pérot interferometric sensor: leading fiber (F), capillary (C), and displacement measurement eyelets (E). The tiny black arrows point to the displacement vectors that quantify the deformation of the matrix material during the first 20 h of hydration process.

Figure 20.15 shows in an x-ray picture the summarized deformation of a part of the inner sample matrix during the first 20 h of curing. In order to come to a broader insight into materials and to standardize this validation method, some research is needed in this subject.

20.7 International Activities in the Area of Fiber-Optic Sensor Guideline Development

Development of standards and guidelines for fiber-optic sensor's performance specification and relevant testing procedures has been discussed in the scientific community as well as in the industry since the mid-1990s of the last century. Very global standards for the use of fiber-optic components in data communication and telecommunication have been available for more than 20 years. Guidelines or substantial standards for fiber-optic *sensors* are still an exception. The first standard draft on generic specification of fiber-optic sensors—IEC 61757—has been published in 1995; the first Working Draft P952/D24 for a specific type of a fiber sensor—the fiber-optic gyroscope—has been published in December 1996. The confirmed IEC standard 61757, part 1: "Generic standard" was published shortly after confirmation of the Final Draft International standard (FDIS) document (see Section 20.7.1) in early spring 2012. The gyro standard draft is meanwhile a basis for the gyro production (IEC 61757-1 2011, IEEE 952-1997 Standard 1998).

The most important prerequisite when a new measurement technology is to be launched is the availability of a consistent (standardized) terminology. A lot of terms and descriptions for the fiber-optic sensor technology can be used from fiber-optic communication standards and from the generic standard 61757. However, there are specific issues associated with characterization, validation, and application of specific fiber-optic sensing systems. For instance, standards for fiber-optic sensors have to cover characteristic details related to the respective physical sensor mechanism, to the sensor

Chapter 20

response to different measurands, to application, and finally to specific perturbing influences coming from environmental situation. These sensor-specific issues mark other aspects in fiber-optic sensor technology; however, they were not yet fully considered in existing fiber-optic systems guidelines or standards.

Subsequently, a short overview of activities in the field of fiber-optic standardization including activities to promote the standardization for fiber-optic sensors is given.

20.7.1 Revision of the International Generic Fiber-Optic Sensor Standard Draft

A first step on the way to a specific fiber sensor guideline was done in 1995 with the standard draft IEC 61757-1:1995. This draft was revised in 1998 and then published as the first edition of the generic standard draft IEC 61757-1 Ed. 1.0 b:1998 (Habel 2009). Because this draft had to be revised in 2010 (or rejected), revision of the first edition was started in 2010 by the subcommittee SC 86C of the IEC. During three meetings, significant changes and extension have been discussed and done—the last during the commission meeting in San Louis Obispo/CA/USA in March 2011—and the new document was circulating for revision by international experts until summer 2011. In October 2011, the IEC standard 61757 was presented at the International IEC congress in Australia as Committee Draft for Vote. In the beginning of January 2012, it was in the FDIS stage, and the publication was shortly after confirmation in early spring 2012.

In connection with the revision of the IEC 61757, activities in standardization of fiber-optic sensors have been expanded. In January 2012, a New Work Item Proposal (NP) was communicated to the members of the SC 86C with the objective to integrate further guidelines into the IEC standards and regulations family. The next parts of the fiber-optic sensor family contain IEC 61757, part 2: "Strain sensors based on Fibre Bragg Gratings," and IEC 61757, part 3: "Distributed Temperature Sensing." Relevant documents for these activities are the VDI/VDE guideline (see Section 20.7.3) and the SEAFOM document (see Section 20.7.4).

20.7.2 COST-299 "Guideline for the Use of Fiber-Optic Sensors"

The European COST Action 299 "FIDES" (*Optical Fibres for New Challenges Facing the Information Society*), established in 2005 for a period of 4 years (until January 2010) consolidated a respectable number of fiber sensor experts from all over Europe, from very different topics and application fields. This platform provided an excellent basis for the discussion of terms and very important details for the appropriate use of fiber-optic sensors. Working Group 4 of this action: *New Challenges in Fiber Optic Sensors* dealt with the development of guidelines for preferably long-gauge-length sensors. In order to support an appropriate design and use of fiber-optic sensor systems, undoubtedly clear definitions of sensor-specific terms and clear information on sensor characteristics were developed. An international expert group formulated basic terms and described basic sensor categories. It is not possible to cover all different aspects of fiber-optic sensors in one standard or a set of harmonized standards. The high complexity of fiber-optic sensors requires a number of different guidelines for specific sensors. The structure of the COST 299 guidelines is as follows:

1. Introduction
2. General Terms
   This category describes terms affecting all or most fiber-optic sensors: type of used fiber, type of sensor, distance range, measurement range, wavelength of operation, characteristic wavelength at reference temperature (FBGs), gauge factor/scale factor at reference conditions, true value.
3. Functionality
   This category describes terms useful while working with fiber-optic sensors: fatigue, life expectancy/lifetime, durability, failure criteria, gauge length, sampling interval, optical power dynamic range (optical budget), warm-up time, measuring time, updating time, limiting conditions.
4. Response Characteristics
   This category describes terms that give correlation between output quantities of a measurement system and the corresponding quantitative characteristics of the measurand: resolution, spatial resolution, measurement dynamic range, scale factor, responsivity, linearity, drift, cross-sensitivity, full scale.
5. Quantities of Random Nature
   This category describes terms specifying unpredictable variations in measurement results affecting system reliability: accuracy, location accuracy, distance accuracy, precision, repeatability, reproducibility, uncertainty of measurement, bias, noise, stability.
6. Optical Safety-Related Quantities
   These guidelines refer to the safe use of fiber-optic sensors with laser-based interrogators.
7. Sources, References
   The spirit and purpose of these guidelines is illustrated here by the example term "resolution" for a distributed sensor. Not only users of fiber-optic distributed sensors are confused by important terms like resolution, but even experts have different views what this term actually means and how to precisely define and measure it. The terms defined in the guideline are the outcome of extensive and engaged discussions to provide generally acknowledged understanding of important fiber-optic sensor terms. When adopted, the proposed definitions will facilitate a more objective and precise comparison of the characteristics of different devices.

According to the given definition (Figure 20.16), *resolution* represents the smallest change in the measurand, meaningfully detectable by the measurement system. The resolution is limited by either the instrument readout or precision (σ), whichever is bigger. Next, the following derivatives definitions are presented: *Spatial resolution* is specified for a fiber by the minimum distance between two step transitions of the measurand of at least 20 times its resolution. *Measuring spatial resolution* results in the minimum distance over which the system is able to indicate the value of the measurand within the specified uncertainty. *Detection spatial resolution* results in the minimum distance that generates readings that are within 10% of the measurand transition amplitude. The guideline is available on the website of the COST 299 action (COST 299 2010).

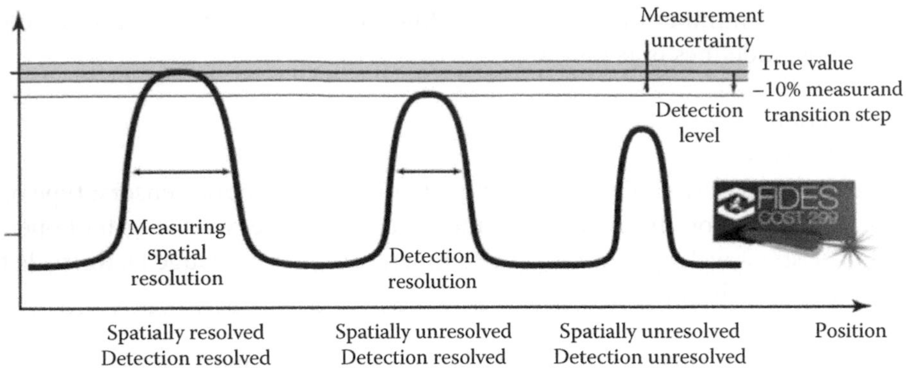

FIGURE 20.16 Definition of the term "resolution."

20.7.3 VDI/VDE 2660 Guideline: "Optical Strain Sensor Based on Fibre Bragg Grating-Fundamentals, Characteristics and Sensor Testing"

The VDI/VDE 2660 guideline is the first fiber-optic sensor standard that focuses on specific fiber-optic strain sensors based on FBGs. This guideline (VDI/VDE 2660, 2010) was developed in the German expert committee 2.17, *Fibre Optic Sensors*, within the VDI—*The Association of German Engineers* as a cooperative work between two research institutes (IPHT Jena and BAM Berlin) and three small and medium companies (AOS Dresden, FBGS Jena, and HBM Darmstadt). This guideline was published in July 2010 in a bilingual version (English/German). The coversheet of this guideline is shown in Figure 20.17.

Background of this activity was the permanent discussion of application-oriented aspects in the measurement technology community when fiber-optic strain sensors are used on-sites, especially the following question: How to avoid unreliable measurement results? The VDI/VDE 2660 guideline was drawn up with the objective to provide directive technical and scientific procedural documentation and decision-making aids for manufacturers, distributors, and users of FBG strain sensors. It treats significant technical questions from scientific and economic standpoints and formulates assessment and evaluation criteria that facilitate the technology transfer of new products. The guideline covers all quality assurance aspects arising during manufacturing, testing, storing, and finally the use of FBG strain sensors in different fields.

The guideline covers the most important terms and aspects for manufacturers and users alike; its terminology is consistent with the COST 299 guideline described in Section 20.7.2. This was initially a German VDI/VDE guideline and is being integrated into the IEC fiber-optic sensor standard work.

Important issues addressed in the guideline are

- Terms and definitions
- Design-specific features and characteristics to be measured
- Recurring specifications for characteristics testing
- Test devices requirements for use in test facilities and on-site
- Statement of the measurement results for investigated characteristics

ICS 17.180.99, 33.180.10

VDI/VDE-RICHTLINIEN

Juli 2010
July 2010

VEREIN DEUTSCHER INGENIEURE

VERBAND DER ELEKTROTECHNIK ELEKTRONIK INFORMATIONSTECHNIK

Experimentelle Strukturanalyse
Optischer Dehnungssensor basierend auf Faser-Bragg-Gitter
Grundlagen, Kenngrößen und deren Prüfung

Experimental stress analysis
Optical strain sensor based on fibre Bragg grating
Fundamentals, characteristics and sensor testing

VDI/VDE 2660

Blatt 1 / Part 1

Ausg. deutsch/englisch
Issue German/English

Die deutsche Version dieser Richtlinie ist verbindlich.

The German version of this guideline shall be taken as authoritative. No guarantee can be given with respect to the English translation.

Inhalt Seite

Contents Page

VDI/VDE-Gesellschaft Mess- und Automatisierungstechnik (GMA)
Fachbereich Prozessmesstechnik und Strukturanalyse

VDI/VDE-Handbuch Prozessmesstechnik und Strukturanalyse
VDI/VDE-Handbuch Mikro- und Feinwerktechnik
VDI/VDE-Handbuch Optische Technologien

Frühere Ausgabe: VDI/VDE 2660:2009-05 Entwurf, deutsch
Former edition: VDI/VDE 2660:2009-05 Draft, in German only

VDI-Handbuch - Stand 2011-11

FIGURE 20.17 Coversheet of the VDI/VDE 2660 guideline (with part of the structure).

The discussed optical strain sensors (OSSs) based on FBG sensors are divided into three different engineering designs: (1) as a fiber with one or several sensors along its length (FBG array), (2) as extensometer sensor whereas the fiber ends of the grating are fixed at two points with definite distance, and (3) as an FBG patch, where the grating is embedded in a protective layer or directly fixed to a carrying frame. Commonly, the planar patch is fixed to the measured object. These three design versions are considered and basic terms are defined, such as gauge length, strain, reference strain, wavelength reference, stability criteria for the OSS, and zero-point-related and nonzero-point-related measurement.

An important point of the guideline concerns design-specific features that affect the characteristics of the sensor. Significant parameters and terms are therefore considered and defined: for example, Bragg wavelength, spectral width, reflectivity, strain sensitivity (k-factor characteristics), ultimate strain for room temperature, fatigue behavior, minimum bending radius, response to temperature changes, and temperature-correlated strain. It is well known that the reproducibility of measurements using FBG sensor systems depends not only on the quality of the Bragg grating spectra but also on the stability and reproducibility of the wavelength measurement with the employed recording devices. One aspect concerns therefore parameters of the FBG that define the Bragg wavelength. It is necessary to know the setup wavelength of the grating. The determination of the Bragg wavelength using a full width at half maximum (FWHM) algorithm can differ from a Bragg wavelength determined by a centroid calculation algorithm. This is due to the influence of the grating characteristics (mainly uniformity of the spectral shape) and the sampling density of the FBG spectrum, as well as on the uncertainty of the curve fit algorithm for the determination of the Bragg wavelength (Dyer 2004, Fernandez et al. 2004). The guideline recommends which parameters and terms have to be specified in order to record reliable measurement results.

Another important section of the guideline focuses on the requirements for certified testing facilities used by manufactures to characterize FBG-based sensors. In order to ensure reliable measurements with such sensors, requirements on their recording devices are also defined, such as requirements on the estimation of the Bragg wavelength, spectral width, reflectivity, strain sensitivity, and the other previously defined terms. Recommendations are also given on how to state the measured results. Therefore, the guideline refers to the GUM (ISO/IEC Guide 2008) as well as to the corresponding standard (Standard EN ISO/IEC 17025:2000, 1999), which ensures the use of calibrated and validated testing facilities. Finally, the guideline presents a model of a data sheet, which lists a number of functional characteristics that have to be specified, such as Bragg wavelength, k-factor characteristics, response to temperature changes, spectral width, and reflectivity. The model data sheet also recommends a number of other important specific parameters and design specifications that provide comprehensive information for designers and users of such sensor systems. In any case, manufacturers and users who observe the recommendations of this guideline are allowed to sell their products with the cachet: *Qualified according to guideline VDI/VDE 2660.*

All aspects that are related to commercially applied sensors (preparation of the measurement zone, selection of fixing materials, application or installation procedures, and so on) are not considered in this guideline. These aspects are discussed in Habel (2009) and will become the basis for another guideline that comprises application and operation aspects of fiber-optic sensors.

20.7.4 Subsea Fiber-Optic Monitoring (SEAFOM) Guidelines

The SEAFOM-MSP-01 guideline *Measurement Specification for Distributed Temperature Sensing* published in March 2010 (SEAFOM-MSP-01 2010) focuses on distributed temperature sensing (DTS). This document was written by and on an initiative of the SEAFOM Measurement Specification Working Group. SEAFOM™ is an international joint industry project launched in June 2006 and is promoting the growth of fiber-optic sensing in subsea applications. The members of the SEFOM group come mainly from oil and gas industry and focuses on international standardization of subsea optical monitoring and sensing functional and test parameters. The SEAFOM-MSP-01 document is to be used as a guide to enable the characterization of performance of any DTS as defined by the measurement parameters. It describes a harmonized set of DTS performance testing procedures, for example, measurement practices such as test setups, test procedures, and calculation methods. Measurement performance parameters such as calibration error, spatial resolution, repeatability of temperature measurements, spatial temperature resolution, environmental temperature stability, and warm-up time are specified. Additionally, supporting parameters that support the definition of the measurement specification and their associated test procedures are defined: temperature sample point, temperature trace, distance range, total fiber length, sampling interval, start-up time, temperature disturbance width, and some more. This guideline is accompanied by another one: SEAFOM-TQP-01 *Functional Design and Test Requirements for an Optical Feedthrough System used in Subsea Xmas Tree (XT) Installations*, published in February 2011 (SEAFOM-TQP-01 2011). This guidance note has been prepared under the auspices of the SEAFOM joint industry group and the Feedthrough System Working Group. This document specifically addresses the interface between the subsea environment of the XT and the downhole or reservoir environment for the transmission of optical data through the XT.

20.7.5 Other Fiber-Optic Guidelines Activities

First activities in Europe to develop guidelines for use of the most requested fiber-optic sensor type—the FBG strain sensor—have been launched in 2003. Companies were interested to get guidelines for measurement of strain profiles and for vibration measurements on/in large structures to assess their integrity. In July 2004, a very effective European group formed from 13 institutions proposed the collective EU research project *Development of Guidelines and Standards for Reliable Commercial Use of Fibre Bragg Grating Sensors* (DoGStaR), which addressed the issue of the lack of recognized standards for FBGs and FBG measurement systems. Although the first stage of submission and evaluation was successful, a number of small and medium companies originally highly motivated were not able to join this research project because of unfavorable prerequisites for subsidies. Eventually, this early activity to push specific standardization activities in Europe was not successful.

First successful European standardization activities in the field of fiber-optic sensors have been carried out within the European COST 299 action, which ended in January 2010. These European activities were reorganized and extended within the new ICT COST TD 1001 action *Novel and Reliable Optical Fibre Sensor Systems for Future Security and Safety Applications* (European COST TD1001 2010). The objective was to facilitate and promote end-user adoption of state-of-the-art fiber-optic sensor systems for reliable use in safety- and security-relevant applications in society, through the establishment of standardized characterization and application procedures, which required a highly interdisciplinary and strongly coordinated European-wide approach. In the working group WG 3 "Sensor Characterization & Onsite Evaluation," the key requirement of companies that are interested to provide high-quality products manufactured, tested, specified, and validated according to international guidelines was discussed. There were four study groups in the WG 3: Sensor interfaces, test procedures, on-site implementation and validation, and standardization. Some of the related special aspects were validation of fiber-optic sensors for long-term use and validation of interrogation systems for reliable data recording under adverse conditions. This international COST platform was open for all experts from industry, research and consulting who were interested to collaborate. There were also dissemination activities, technology transfer discussion, and training activities, to ensure that the proposed standards discussions are widely accepted and strongly implemented.

Early activities in developing guidelines for implementing SHM methods in civil engineering were also done in the ISIS Canada Research Network (Intelligent Sensing for Innovative Structures). It focused on guidelines for fiber-reinforced polymers containing integrated fiber-optic sensors. The first manual that gives a brief introduction on how to select fiber-optic sensor technologies was published in 2001 (Tennyson 2001). It provides a number of specifications and instructions on handling and installation, and considers application-related aspects for fiber-optic sensor systems. However, these recommendations are not considered as generally approved guidelines or technical standards for fiber-optic sensors. Nevertheless, this manual was an important step toward developing guidelines for the handling of fiber-optic sensors on-site.

Research groups in the US standardization organization National Institute of Standards and Technology in Boulder, Colorado, USA, have been involved in research into the behavior of FBG sensors and associated devices. The basic metrology considerations developed by them provided valuable input for the definition of system component specifications (Dyer 2004). Preliminary specifications have been developed for FBG sensors, FBG interrogators, and interferometric sensing systems. These activities were mainly driven by companies that offer devices and systems or those that want to use fiber-optic sensor technology. The US Optoelectronics Industry Development Association provides an effective platform to overcome the barriers to generate a robust fiber-sensor market environment. This objective includes the development of standards and guidelines. NASA utilizes fiber-optic components for the space flight sensor systems and is therefore obliged to validate the components of these systems. They use test procedures for materials validation as well as special test programs developed by the American Society for Testing and Materials (ASTM international) to validate sensor functions under specific space-typical requirements (Ott 2005).

In this case of a very specific application area, the testing parameters must be adjusted for each component to simulate the environmental conditions or the worst case to be expected. Although there is no document cited comparable to a standard for the validation of fiber-optic sensors, the described test program can be considered as a discussion basis along the way to fiber-optic sensor guidelines. The ASTM activities in two standard committees, F25 on Ships and Marine Technology and E13.09 on Fiber Optics, Waveguides and Optical Sources, include fiber-optic sensor elements, however, only as a possible option for, for example, pressure transducers or fiber-optic position switches (American Society for Testing and Materials 2011).

Another international (industry-driven) standardization activity is done in the Society of Automotive Engineers (SAE), in the Technical Committee AS-3: *Fiber Optics and Applied Photonics* (SAE Society 2011). The fiber-optic sensors subtask group is composed of developers, manufacturers, and users of fiber-optic sensors for aerospace applications. The purpose is threefold: first, the development of a set of standards that will define the function, installation, operating parameter ranges, and interface requirements for fiber-optic sensor applications; second, elaboration of standards that apply to fiber-optic sensors used in aerospace; and third, generating interface standards. The interface standards will cover electronic and optical interface methods, and standard requirements for interface devices such as interrogators. In 2011, well-prepared drafts of guidelines are under revision, for example, Aerospace Resource Document ARD 040711, *Fiber Optic Coupled Sensors for Aerospace Applications,* and Aerospace Information Report AIR6031, *Fiber Optic Cleaning*, which contains information relating to available cleaning materials, their pros and cons, and recommended practices.

Under the aegis of the "International Union of Laboratories and Experts in Construction Materials, Systems, and Structures" (RILEM), the Technical Committee "Fiber-Optic Sensors" (TC-OFS) is also dealing with guidelines and standards for the use of fiber-optic sensor technologies in civil engineering. The general objective of this RILEM TC-OFS committee is to promote the proper use of fiber-optic sensors in civil engineering applications such that their advantages can be fully exploited. A state-of-the-art report is under development. On this basis, application guidelines for fiber-optic sensors in civil engineering will provide expertise on the most important questions concerning reliability and stability of such sensor systems. More details can be found on the website of the TC-OFS (RILEM 2011). The activities of this Technical Committee are—according to the general intentions of the RILEM organization—limited to sensing needs that are of direct relevance to civil engineering. However, it is expected that these guideline activities will serve as a model for other application areas, for example, composite materials monitoring, monitoring of industrial plants with specific risks, and evaluation of new materials. Except for the RILEM TC-OFS activities chaired by an expert from the Hong Kong University of Science & Technology, any noticeable standardization activities in China, Japan, or other Asian countries are not known.

Beginning with 2010, some more R&D activities have been started in Germany within a research program partially supported by the German Federal Ministry of Economics and Technology. This program called "Transfer of R&D Results by Standards and Standardization" supports German research organizations and experienced SMEs to work together with the objective to make results of industrial research ready for integration into standards or to develop standards based on application-oriented R&D activities. The challenge is to transfer high-technology results into products that can be better launched into the global market. On the other hand, these activities will prepare and initiate also new standards on the national (e.g., DIN), European (CEN/CENELEC), and international (ISO/IEC) scales. These projects led by experts of the BAM Federal Institute for Materials Research and Testing in Berlin and cooperated with the German Companies GESO GmbH Jena and HBM GmbH Darmstadt deal, on the one hand, with the development of composite-embeddable FBG strain sensors as well as of validation methods and test apparatus for the characterization of applied and embedded fiber-optic strain sensors. On the other hand, standards for the use of fiber-optic sensors in the offshore environment are to be developed. In the offshore project, the Fraunhofer Institute for Wind Energy and Energy System Technology IWES in Bremerhaven is an experienced project partner. In these projects, special emphasis is given to the optimization and validation of a reliable sensor function as well as the development of appropriate methods of application and installation under extremely harsh environmental conditions.

20.8 Summary and Outlook

As the most noteworthy outcome of the activities in the last decade in the standardization of fiber-optic sensors, the frequently requested fiber-optic sensor standards documents 952-1997 IEEE gyro standard, the generic fiber-optic sensor standard IEC 61757, the guidelines for distributed temperature sensors (DTS—SEAFOM-MSP-01), and the guideline for fiber-optic strain sensors based on FBG (VDI/VDE 2660) can be considered. These documents have model character for establishing further competent international fiber-optic sensor standards and guidelines. In this sense, there are some activities in international expert and research groups where open questions concerning reliability, reproducibility, stability, and technical rules are addressed. Scientists and interested engineers with hands-on expertise are aiming to establish well-researched guidelines and recommendations on how to handle the sensor system outside the laboratory or manufacturing environment, the type of adhesion or bonding material to be chosen, and finally, which influences resulting from application might perturb or possibly damage the sensor signal. Such activities are the prerequisites of serious standards and create confidence in the fiber-optic sensor's user community.

From the author's point of view, two main lines are expected in the near future: first, developing a set of further guidelines for different FBG-based sensors (acoustic waves, temperature, pressure, etc.) because this sensor type is part of very different sensor systems. Second, guidelines or standards that cover all important aspects when fiber-optic sensors are to be applied: use of appropriate system components, methods to validate application procedures, and on-site test methods to evaluate the full function of applied sensors after being in operation for long time. For this purpose, research has to be carried out to establish validation methodologies for embedded fiber-optic sensors. There is already extensive knowledge of the materials properties, interaction of specific sensor materials with environment, and characterization of the sensor behavior under specific operational conditions. On the other hand, facilities from the field of materials characterization can be adapted to the sensor-specific demands. All this knowledge will be used if necessary for developing standards and corresponding recommendations.

References

American Society for Testing and Materials (ASTM). http://www.astm.org/. (accessed June 19, 2014).

COST 299-Guideline for the use of fibre optic sensors. Document of the European Cost 299 Action. September 2009. (www.bam.de, Division 8.6).

Dyer, S. D. 2004. Key metrology considerations for fiber Bragg grating sensors. *Conference on Smart Structures and Material*, San Diego, CA. *SPIE*, Vol. 5384, pp. 181–189.

EN ISO 10012. 2003. Measurement management systems—Requirements for measurement processes and measuring equipment. http://www.iso.org/iso/catalogue-detail? csnumber=26033 (accessed June 19, 2014).

European COST TD1001 Action. Novel and reliable optical fibre sensor systems for future security and safety applications. http://www.ul.ie/td1001. (accessed June 19, 2014).

Fernandez, A. F., A. Gusarov, F. Berghmans et al. 2004. Round-robin for fiber Bragg grating metrology during COST270 action. *Proceedings of the SPIE* 5465:210–216.

Habel, W. R. 2005. Fiber optic sensors for deformation measurements: Criteria and method to put them to the best possible use. *Proceedings of the SPIE* 5384:158–168.

Habel, W. R. 2009. Reliable use of fiber optic sensors. In: *Encyclopedia of Structural Health Monitoring*, Boller, C. et al. (eds.). Berlin, Germany: John Wiley & Sons, Cambridge, UK.

Habel, W. R., V. G. Schlüter, and V. V. Tkachenko. 2011a. How do application-related issues influence the reliability of fiber optic strain measurements? *IEEE Sensors*, Limerick, Ireland.

Habel, W. R., V. G. Schukar, and N. Kusche. 2011b. Calibration facility for evidence of reliable measurements with surface-attached fiber optic and electrical strain sensors. *IEEE Sensors*, Limerick, Ireland.

IEC 61757-1 Ed. 2.0 b:2011. Fibre optic sensors—Part 1: Generic specification. Committee Draft for Vote (CDV).

IEEE 952-1997 Standard specification format guide and test procedure for single-axis interferometric fiber optic gyros. Issue Date: 1998.

ISO/IEC Guide 98-3:2008. Uncertainty of measurement—Part 3: Guide to the expression of uncertainty in measurement (GUM:1995).

Kalinka, G., A. Leistner, and A. Hampe. 1997. Characterization of the fibre/-matrix interface in reinforced polymers by the push-in technique. *Composite Science and Technology* 57:845–851.

Krebber, K., W. R. Habel, T. Gutmann et al. 2005. Fibre Bragg grating sensors for monitoring of wind turbine blades. *Proceedings of the SPIE* 5855:1036–1039.

Kusche, N. and C. Schilder. 2011. BAM-internal status report. *German Research Cooperation Project FZK 01FS10031 on Strain Sensor Guideline Development.* HBM GmbH Darmstadt, BAM Berlin.

Luyckx, G., E. Voet, W. De Waele et al. 2010. Multi-axial strain transfer from laminated CFRP composites to embedded Bragg sensor: I. Parametric study. *Smart Materials Structures* 19:105017.

Ott, M. N. 2005. Validation of commercial fiber optic components for aerospace environments. *Proceedings of the SPIE* 5758:427–439.

RILEM Technical Committee. Optical fibre sensors. http://www.rilem.net/gene/main.php?base=8750&gp_id=221. (accessed June 19, 2014).

Roths, J., A. Wilfert, P. Kratzer et al. 2010. Strain calibration of optical FBG-based strain sensors. *4th EWOFS Porto/Portugal. SPIE* vol. 7653, paper 76530F-1.

SAE Society of Automotive Engineers Committee. AS-3 fiber optics and applied photonics committee. http://www.sae.org/servlets/works/committeeHome.do?comtID=TEAAS3. (accessed June 19, 2014).

Schlüter, V. 2007. Strain transfer characterization of surface applied fiber Bragg grating sensors. *Young Stress Analyst Competition.* British Society for Strain Measurement (BSSM), pp. 34–38.

Schlüter, V. G. 2010. Development of an experimentally-based evaluation method for optimisation and characterisation of the strain transfer of surface-applied fibre Bragg grating sensor. (Original in German). Dissertation TU Berlin 2009. Published at BAM-Dissertationsreihe, vol. 56. 2010, http://www.bam.de/de/service/publikationen/publikationen_medien/dissertationen/diss_56_vt.pdf.

Schukar, V. G., D. Kadoke, N. Kusche et al. 2012. Validation and qualification of surface applied fibre optic strain sensors with application independent optical techniques. *Measurement and Science Technology* 23: 085601.

Schlüter, V. G., N. Kusche, and W. R. Habel. 2010. How reliable do fibre Bragg grating patches perform as strain sensors? *4th EWOFS*, Porto, Portugal. SPIE-vol. 7653, paper 76533N-1.

Schilder, C., N. Kusche, V.G. Schukar et al. (2013). Experimental qualification by extensive evaluation of fibre optic strain sensors. *Measurement and Science Technology* 24:094005.

Schuler, S., W. R. Habel, and B. Hillemeier. 2008. Embedded fibre optic micro strain sensors for assessment of shrinkage at very early ages. *International Conference on Microdurability*, Nanjing, China.

Schuler, S., B. Hillemeier, M. Fuhrland et al. 2009. Investigations of durability parameters of concrete by means of embedded flexible fiber-optic Fabry–Perot interferometers (in German). *Tm—Technisches Messen* (Oldenbourg-Verlag München) 76:517–526.

SEAFOM-MSP-01. 2010. Measurement specification for distributed temperature sensing (DTS). Document published March 2010 by the SEAFOM Measurement Specification Working Group. http://www.seafom.com. (accessed June 19, 2014).

SEAFOM-TQP-01. 2011. Functional design and test requirements for an optical feedthrough system used in subsea Xmas Tree (XT) installations. Document No: SEAFOM-TQP-01, published February 2011 by SEAFOM joint industry group and the Feedthrough System Working Group. http://www.seafom.com/. (accessed June 19, 2014).

Standard EN ISO/IEC 17025:2000 (trilingual version). 2009. General Requirements for the Competence Q2 of Testing and Calibration Laboratories. ISO/IEC 17025:1999.

Tennyson, R. (ed.). 2001. Installation, use and repair of fibre optic sensors. In Design Manual ISIS-M02-00, Canada, Spring 2001 and Civionics Specification. Design Manual No. 6 (Chapter 2: Specifications for fibre optic sensors (FOS)). ISIS Canada Research Network, October 2004.

US and International Fiber Optic Standards, EIA-TIA Fiber Optic Standards. http://www.thefoa.org/tech/standards.htm.

VDI/VDE 2660 Guideline. 2010. *Experimental Stress Analysis—Optical Strain Sensor based on fibre Bragg grating; Basics, Characteristics and its Testing*. Beuth-Verlag, Berlin, Germany. http://www.beuth.de (developed by VDI expert committee 2.17 “Fiber Optic Sensors”. http://www.vdi.de/46447.0.html.

IV

The Dynamic Field of Optical Sensing

21. Optical Sensors

Final Thoughts

José Luís Santos
University of Porto

Faramarz Farahi
University of North Carolina at Charlotte

Handbook of Optical Sensors. Edited by José Luís Santos and Faramarz Farahi © 2015 CRC Press/Taylor & Francis Group, LLC. ISBN: 9781439866856.

21.1 Introduction

Previous chapters have covered a broad range of technologies and applications in the general field of optical sensing. Section I is dedicated to an introduction to optical sensing with Chapters 1 and 2 on the fundamentals of optical sensors and the principles of optical metrology. Section II addresses optical measurement principles and techniques, which, after Chapter 3, focused on the role of optical waveguides in sensing. Chapters 4 through 7 detail how this functionality can be achieved via intensity, interferometric, fluorescence, and plasmonic approaches, followed by Chapters 8 through 10, which focused on wavefront sensing and adaptive optics, multiphoton microscopy, and imaging based on optical coherence tomography. Section III is dedicated specifically to the field of optical fiber sensing. In Chapter 11, a historic overview of the subject is presented, followed by one on different types of optical fibers (Chapter 12), light propagating principles, fabrication techniques, and their main characteristics. Chapters 13 and 14 have focused on fiber sensing based on light intensity and phase modulation, followed by Chapters 15 and 16 describing multipoint measurements and distributed sensing. Due to importance of fiber Bragg grating (FBGs) in the field of optical sensors, Chapter 17 is dedicated to this topic followed by Chapter 18 on chemical sensors. Section III ends with two chapters (Chapters 19 and 20) describing some applications of fiber sensors and issues related to standardization and its impact on measurement reliability.

In the rest of this chapter, and before wrapping up our discussions, we would like to provide a summary of trends in optical sensing and some of authors' perspectives on the progress of this field.

21.2 Trends in Optical Sensing

Review of progress in the field of waveguide sensors in the past decade shows that there are some noteworthy achievements with long-term consequences on its development, which will be briefly discussed.

21.2.1 Sensing Platforms

Optics is a branch of science that seems naturally tailored for sensing and of course so many other fields. Optical-based sensors could be designed to have high sensitivity, have minimal cross-talk, have small sizes, and be lightweight. In addition, they can be designed for remote operation and interrogation of signals, therefore offering electrically passive sensing with immunity from electromagnetic interference, resistant to harsh thermal and chemical environments. Optical waveguides are uniquely attractive since they provide added flexibility in some specific applications. In most cases, integrated optics–based sensors can be designed for many in situ sensing applications with some signal conditioning performing on the same substrate, but light coupling in/out is a challenge, limiting its practical use to some controlled environments. However, this limitation can be overcome by combining the planar sensing head with lead fibers that provide illumination and a controlled propagating path for the light having its characteristics modulated by the targeted measurand. This hybrid approach is effective, and many developments have been reported on its utilization, particularly for biochemical sensing (Pruneri et al. 2009).

Surely, the possibility to have all-fiber sensing configurations is in general the preferred option since light propagates in the fiber without the need to encounter other optical elements. The potential performance enhancement associated with this intrinsic sensing principle was a motivating factor to work on a variety of fiber sensors supported by standard optical fibers with degrees of freedom anchored almost exclusively on fiber core and cladding dimensions and doping of the silica core. Notable exceptions to this constraint were the birefringent optical fibers that constituted a platform for many important sensing developments since the pioneer work of Eickhoff (1981) and the bi-core fiber (also known as dual-core fiber or twin-core fiber), proposed in 1981 by Meltz and Snitzer in a US patent entitled *Fiber Optic Strain Sensor* (Meltz and Snitzer 1981), and further discussed in a journal article 2 years later (Meltz et al. 1983).

A major breakthrough occurred with the advent of microstructured optical fibers (MOF), which have been discussed in detail in Chapters 11 and 12. Since then, the design freedom offered by these fibers has been widely explored and demonstrated in a wide range of sensing configurations for the measurement of many physical, chemical, and biochemical parameters. It can confidently be stated that microstructured fibers will play a central role in the development of many advanced optical fiber sensors. If we choose a word to characterize these fibers, probably the best choice would be *flexibility*. Indeed, what is feasible with these fibers cannot be well described within the context of standard fibers. Surely, standard fibers will play an essential role, but more in illumination of the sensing system and transmission of optical signal over long distances. With microstructured fibers, the freedom to draw a cross section with almost any pattern allows tailoring the distribution of optical field to optimize the interaction with the measurand, particularly when we are dealing with chemical and biological species with access to the internal fiber void spaces. Additionally, these regions may be filled with specific materials to increase the effect of measurand on the optical field. Other structures such as Bragg gratings or layouts supporting plasmonic behavior can be added to these fibers. In hollow-core photonic crystal fibers, the relatively easy access of fluids to the fiber core allowing high overlap with the optical field turns these fibers remarkably suitable for environment and biochemical sensing. Other possibilities are around the corner, such as the controlled conduction of particles along the hollow core guided by the optical field, atom interferometers, etc., opening a new door to a completely new range of ultrasensitive sensing devices with applications ranging from gravitational wave detection to ultrasensitive rotation and magnetic sensing (Vorrath et al. 2010, Kayani et al. 2012, Pinto and Lopez-Amo 2012).

21.2.2 Wavelength Encoding

A sensing head can be configured to encode the measurand effect into the modulation of the amplitude of the optical field, its polarization state, its phase, frequency, or wavelength. This mapping is the basis of an established classification of optical sensors. Initially, intensity encoding was the preferred choice, and over the years, techniques relying on polarization and phase modulation have been developed. High-sensitive interferometric sensors have been often limited to laboratory demonstrations with a few exceptions, most notably the fiber-optic gyroscope. Soon it was recognized that optical sensors based on wavelength encoding have interesting characteristics since the measurand action is encoded into an absolute parameter, the light wavelength (or optical frequency in the case of Doppler effect sensors).

Invention of fiber Bragg grating (FBG) was a major breakthrough that brought, for the first time, truly wavelength encoded sensors As will be discussed in the following section, these devices will play a significant role in the future of fiber sensors, as it is true for DFB fiber laser sensors, structures that are related to FBGs as will be outlined in Section 21.2.2.2.

The consideration of the phenomena of plasmonic resonance for optical sensing is not new, starting first within the context of bulk optics and later moving to the domain of optical fiber platforms, which are particularly suited for spectral interrogation. This wavelength-encoded characteristic, combined with the progresses that have been reported in the field, points out the growing relevance plasmonics will have in optical sensing. Section 21.2.2.3 details this topic.

21.2.2.1 Bragg Grating Sensors

The basic principle of FBG sensors is discussed thoroughly in Chapter 17. The impacts of FBGs on optical communications, optical signal processing, and optical sensing have been significant. New techniques have been developed to fabricate these structures (and the related long period fiber gratings—LPGs) in microstructured fibers. Quite successful has been the femtosecond laser writing technique that permits the fabrication of FBGs in no-doped silica fibers (Koutsides et al. 2011). Even in the standard form, these gratings in the microstructured fiber environment provide a range of sensing functionalities that far exceed what is possible with FBGs in conventional fibers.

A recent development in FBG technology allows fabrication of devices with dimensions much smaller than what is typical for standard FBGs where the length is in the millimeters range and the thickness up to approximately 100 μm. For long, these dimensions have been considered to impose a limitation on their applications when highly localized measurements are needed. The possibility to scale down these dimensions arose with the progress in the fabrication of submicrometric optical wires, which allow for low-loss evanescent waveguiding (Tong et al. 2003). These are known as optical microfibers, where light traveling along such a fiber is tightly confined to the fiber core owing to the large refractive index contrast between the core and air, while a large fraction of the guided light can propagate outside the fiber as an evanescent wave, which makes it highly sensitive to the ambient medium. The small size of a microfiber also provides excellent flexibility and convenient configurability, compatible with easy manipulation of microfiber-based devices with a complex topology, reasons that justify the extensive investigation of these structures for the fabrication of miniature optical components and photonics integration.

When compared with standard FBGs, the fabrication of these devices in microfibers may offer advantages such as compactness, strong near-field interaction with surrounding materials, and high resistance to mechanical and thermal shocks. An interesting example is the optical fiber tip micrograting thermometer (Feng et al. 2011) where an ultrasmall grating is machined onto a photosensitive fiber tip by focused ion beam milling, a suitable technique for nanofabrication due to its small and controllable spot size. Figure 21.1 shows an SEM image of the nanostructured sensor, showing a corrugated grating with 11 notches (depth of ~1.6 μm) and a total length of ~12 μm.

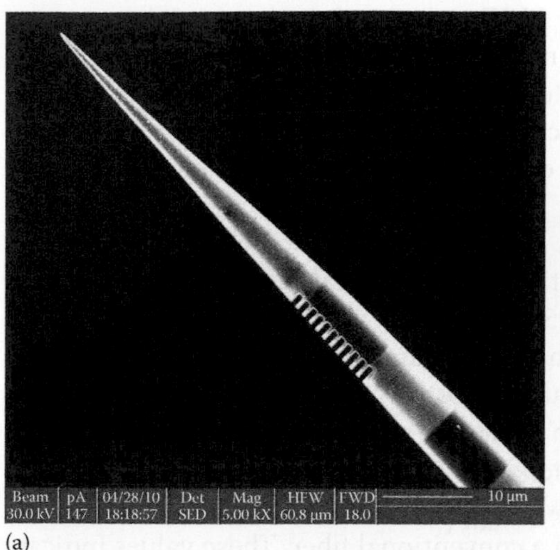

(a)

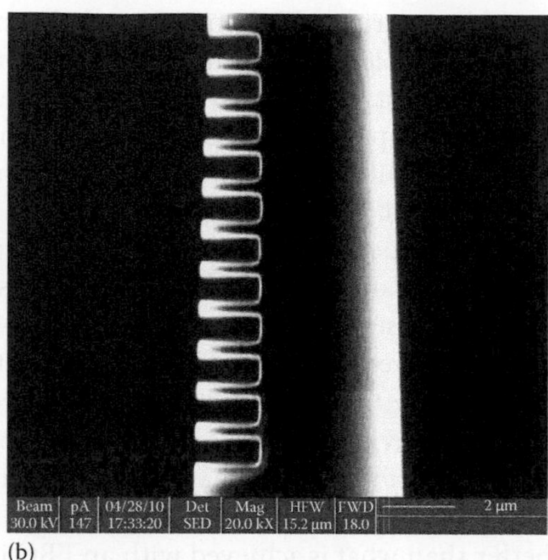

(b)

FIGURE 21.1 SEM picture of a micro FBG machined into a fiber tip (a) and magnified image of the grating region (b). (Reprinted from Feng, J. et al., *IEEE Photon. J.*, 3, 810, 2011. With permission.)

The results reported indicate a refractive index modulation $\Delta n_{mod} > 10^{-1}$ due to high-index contrast in silica–air interface. Such a high value is required for the structure to have resonant behavior with a very short length (typical FBGs have Δn_{mod}~10^{-4}–10^{-3}). This sensor tip demonstrated a temperature sensitivity of ~22 pm/°C, more than double of standard FBGs in the 1550 nm region. More than the sensitivity value, the highest merit of this sensing structure is its ultracompact size. These characteristics open prospects for high-resolution temperature mapping at minuscule spaces, even in the presence of large temperature gradients.

Several techniques have been reported for the fabrication of FBGs in microfibers. The most commonly used technique is to etch a single-mode fiber (SMF) after the FBG has been written in the photosensitive Ge-doped core. Therefore, the grating region experiences only refractive index modulation and not structural perturbations. These structures have been considered for refractive index sensing (Iadicicco et al. 2004) and for the simultaneous measurement of refractive index, temperature, and strain (Lee et al. 2010). Another fabrication approach is focused ion beam milling, which employs accelerated ions to mill nanometer-scale features on microfibers to form corrugated structures; therefore, the index modulation results from changes in the structure. The optical fiber tip micrograting thermometer referenced earlier uses this technique. These gratings are characterized by a high-index contrast that favors higher sensitivity to refractive index. Indeed, a refractive index sensitivity in the range of 660 nm/RIU (at RI ~ 1.39) was demonstrated with a grating length of ~500 μm in a microfiber of 1.8 μm diameter (Liu et al. 2011).

FBGs in microfibers have also been fabricated using other techniques. One of them is the femtosecond laser approach, where the high power of the ultrashort pulses induces permanent structural damage, causing periodical physical deformation on the surface of the microfiber. The work of Fang et al. (2010) reports the fabrication of FBG with this technique in a tapered fabricated microfiber (~2 μm diameter) showing a refractive index sensitivity of ~230 nm/RIU (at RI ~ 1.39). This fabrication approach is particularly suitable to writing FBGs in suspended-core MOF as it was demonstrated for sensitive refractive index (Huy et al. 2007) and strain and temperature (Fernandes et al. 2012) measurements. It has also been demonstrated that FBGs in microfibers can be fabricated by wrapping the tapered fiber on a microstructured rod with an internal channel (Xu et al. 2010) or by laying the microfiber on a platform with pretreated microstructures (Kou et al. 2011a).

The temperature sensitivity of FBG in microfibers is typically on the order of 20 pm/°C (Kou et al. 2011b). The strain sensitivities of these gratings are similar to those found with standard FBGs (~1.2 pm/°C at 1550 nm); however, FBGs in microfibers are very sensitive to force. An FBG in a fiber with 3.5 μm diameter reaches a force sensitivity of ~1900 nm/N (Wieduwilt et al. 2011). Results reported by Luo et al. (2012) indicate a force sensitivity of ~4133 nm/N at 1550 nm for a 58 μm length microfiber FBG fabricated using focused ion beam milling technique in a tapered fiber with ~2 μm diameter (grating with ~100 shallow corrugations), which is more than 3 orders of magnitude better than what is achieved with an FBG in a conventional fiber. These values indicate that with sensitive FBG interrogation techniques, forces as small as nN are measurable. This level of resolution opens new applications such as measuring forces during some surgical procedures (Chung et al. 2012).

Microfiber-based Bragg gratings are an emerging concept in ultrasmall optical fiber sensors with potential to outperform standard FBGs in light of their large evanescent field and compactness (Kou et al. 2012). Because of their small dimensions and extreme sensitivities, microfiber-based Bragg gratings will find many applications, and their study will remain a hot R&D topic in the future.

21.2.2.2 DFB Fiber Laser Sensors

Sensing based on fiber lasers is attractive for three reasons: (1) the lasing wavelength is a function of the measurand action, which is translated into an absolute parameter; (2) the active operation of the sensor, in principle, offers high signal-to-noise ratio, an advantage for high-resolution measurement; and (3) its compatibility with wavelength multiplexing techniques. Initially, the, demonstration of this sensing approach based on a length of doped fiber between two reflectors was compromised by the presence of large number of modes and high level of phase noise. Therefore, the concept remained as an academic interest until 1992 when Ball proposed the first in-fiber laser formed by arranging two FBGs on either sides of a short length of erbium-doped fiber (Ball and Glenn 1992), a layout that later became known as distributed Bragg reflector (DBR) fiber laser. Soon after this, a number of potential sensing applications were identified that exploited the high strain sensitivity of such devices (on the order of 10^{-14}), wide bandwidth (from DC to several megahertz), and relatively high optical output (>−10 dBm) of such devices, together with the ability to be wavelength division multiplexed along a single fiber to form sensor arrays.

Some early problems associated with severe burst mode operation and mode hopping were resolved, where the two FBGs of the DBR laser merged into a single one with a π-phase shift at the center, resulting into a configuration known as distributed feedback (DFB) laser, which shows robust single longitudinal mode operation, and soon were applied as sensors (Kringlebotn et al. 1996). These laser sensors have many advantages in common with passive FBG sensors, such as compact in-fiber design, a highly localized

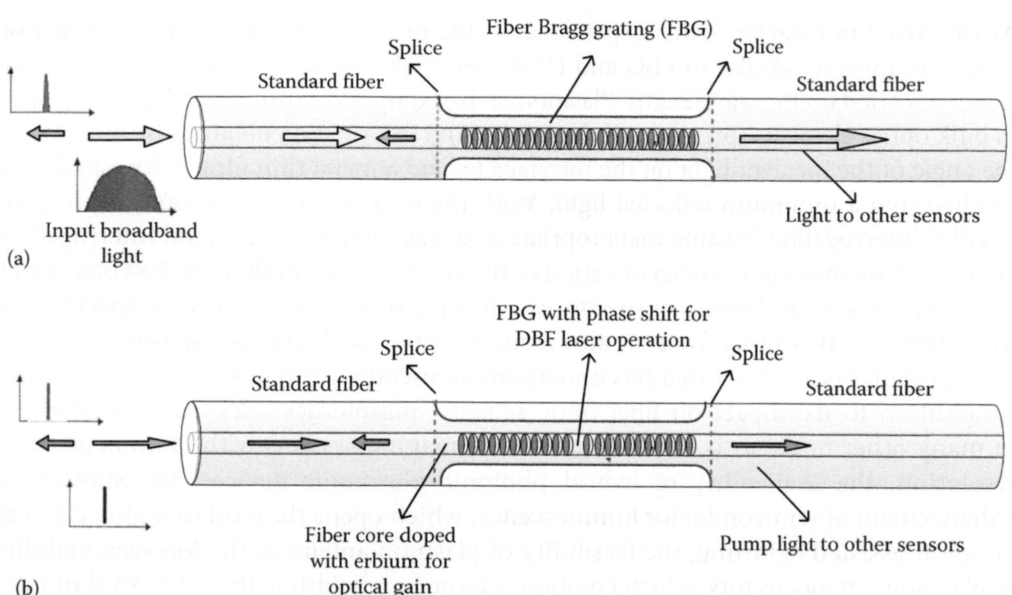

FIGURE 21.2 Configuration for refractive index sensing based on a standard FBG sensor (a) and on a DFB fiber laser sensor (b).

sensitive region, and wavelength multiplexing capabilities. Both sensor types are based on interrogation of a shift in the FBG resonance frequency. The reflection bandwidth of a passive FBG may typically be on the order of 10 GHz, while the emission linewidth of fiber DFB lasers is in the range of 1–10 kHz. This difference will allow for an improvement of sensor sensitivity by orders of magnitude. In 2008, Cranch et al. reported that DFB fiber lasers can resolve effective length change of less than 0.76 fm/Hz$^{1/2}$ at 2 kHz, which is roughly the estimated value for the size of a proton! This was achieved based on an interferometric technique to decode the wavelength shift of the laser sensor and minimization of noise sources (Cranch et al. 2008). Such results indicate that fiber laser sensors are capable of achieving fundamentally limited strain resolution (Miller et al. 2012). These characteristics provide opportunities to develop very sensitive temperature measurement, as well as measurement of change in the refractive index of external medium by decreasing the fiber diameter in the sensor region, as shown in Figure 21.2.

Wavelength encoding, small size, narrow linewidth, large signal-to-noise ratio, resilience to environmental perturbations, ease of multiplexing, as well as flexibility to choosing the wavelengths are characteristics that make DFB laser sensors appealing in a broad range of applications where compactness, high resolution, and multipoint sensing are in demand. These attributes, as well as the increasingly availability of DBF lasers in a wide range of wavelengths, will certainly induce the acceleration of application of this sensing technology in a large range of applications (Zeller et al. 2010).

21.2.2.3 Optical Sensing Based on Plasmonics

The resonant interaction of light with free electrons in metals originates loss bands in the light spectrum with a central wavelength that depends on the characteristics of the metal as well as of the surrounding medium, particularly its refractive index.

As discussed in Chapter 7, this dependence is the principle of optical sensing based on plasmonics where, similar to FBG and DFB fiber laser sensors, the measurand information is encoded on the wavelength. Plasmonics-based optical sensors were first developed in bulk optics elements (prisms) and the measurand information obtained by monitoring the angle of the incident light on the interface (where a metal thin film is deposited) that resulted into a minimum reflected light. With the introduction of the optical fiber, the angular interrogation became inappropriate and was replaced by spectral interrogation, either directly through the determination of the central wavelength of the loss band or by monitoring the optical power at specific wavelengths on the two sides of the spectral loss band, therefore benefiting from the advantages of wavelength-encoded sensors.

Plasmonics is a subject that has undergone tremendous progress in the last decade. In addition to its impact on fiber-optic sensing, plasmonics offers new possibilities in many other fields, as is the case of optical imaging systems with nanometer-scale resolution, the availability of hybrid photonic–plasmonic devices, the substantial enhancement of semiconductor luminescence, which opens the road to highly efficient plasmon-assisted lightning, the feasibility of plasmon antennas, the foreseen viability of plasmonic nanocircuits, which combine a large bandwidth with a high level of integration, and many others.

In optics, two types of surface plasmon resonances (SPRs) need to be considered: (1) propagating surface plasmon polaritons (SPPs) and (2) nonpropagating localized SPRs (LSPRs). In both cases, evanescent electromagnetic fields are confined. For SPPs, the plasmons propagate tens to hundreds micrometers along the metal surface with an associated electric field that decays exponentially from the surface (normal to the dielectric–metal interface). Change of the refractive index of the medium above the metal shifts the plasmon resonance condition. For LSPRs, plasmons are resonantly excited in metal nanoparticles and around nanoholes or nanowells in thin metal films. The spectrum and the magnitude of the LSPR depend on the size, shape, composition, and local dielectric environment.

The fundamental reason for the growing importance of plasmonics lies on the interaction process between electromagnetic radiation and conduction electrons at metallic interfaces or in small metallic nanostructures, leading to an enhanced near-field of subwavelength dimension. Physically, this derives from the fact the mode coupling between photons and electrons implies the relevant length scale is no longer the light wavelength but a much smaller effective value essentially determined by the electron dynamics in the metal, with the light energy carried as a package of electron oscillations. These hybrid electromagnetic modes (surface plasmon resonances or simply *plasmons*) have properties that enable two unique characteristics: (1) the concentrated optical field in the metal surface means that if light is first coupled to a plasmon, any further interaction (e.g., with adjacent molecules or with a nonlinear material) will be much more effective; (2) the concentrated optical field permits to confine light to subwavelength spaces well below the usual diffraction limit. These remarkable characteristics make plasmonics particularly important in nanophotonics. It allows bridging the gap between the nanoworld and the optical world, thereby enabling the powerful tools of optics to be more efficiently applied in nanoscience and nanotechnology (Atwater 2007).

Independent of the application, the incorporation of surface plasmons in any kind of device architecture requires the consideration of three steps, illustrated in Figure 21.3: (1) the coupling of the input light into plasmons; (2) control of the plasmons in the device

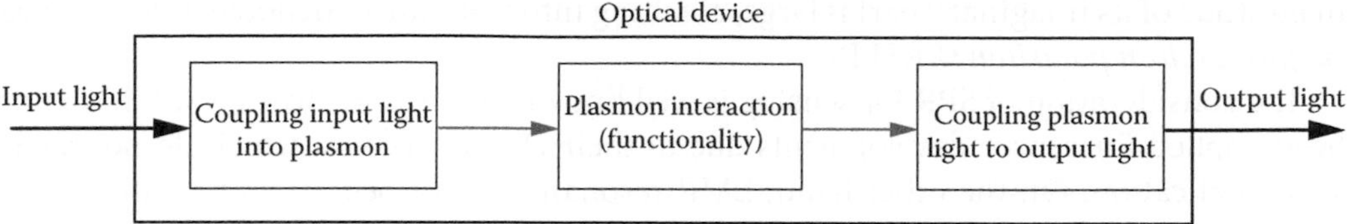

FIGURE 21.3 General structure of an optical device incorporating a plasmonic functionality.

dependent on the target application; (3) coupling the light out of the plasmon to the output channel that propagates light to the photodetection, amplification, and processing unit. The first and the third steps are essentially application independent. For sensing, the crucial factor is to produce a strong interaction of the plasmon with the nearby environment.

It should be mentioned that in the configuration where a thin film is deposited onto a length of the fiber core (no cladding), or on a tapered region, which is thin enough for the metal layer be within the evanescent field range of the light that propagates in the core, three different types of electromagnetic resonances are possible depending on the electric permittivity of the thin film, fiber core, and the surrounding medium, as shown in Table 21.1 (Yang and Sambles 1997). ε_{core} is the electric permittivity of the fiber core, $\varepsilon_{Layer} = \varepsilon_{r_{Layer}} + i\varepsilon_{i_{Layer}}$ the complex permittivity of the thin film (if the film is structured in layers of different materials, the permittivity is an effective value), and ε_{medium} the electric permittivity of the external medium (normally a dielectric).

The first row of Table 21.1 refers to the cases when the real part of the thin-film permittivity is negative and higher in magnitude than both its own imaginary part and the permittivity of the material surrounding the thin film (external medium and fiber core). In this case, the thin film is metallic, and the electromagnetic resonance is known as *surface plasmon resonance*. The second row addresses the cases where the real part of the thin-film permittivity is positive and higher in magnitude than both its own imaginary part and the permittivity of the material surrounding the thin film, producing resonances known as *lossy mode resonance* (LMR). Finally, the third row deals with arrangements when the real part of the thin-film permittivity is close to zero, while the

Table 21.1 Electromagnetic Resonances in an Optical Waveguide with a Thin-Film Overlay

Electric Permittivity Conditions	Electromagnetic Resonance
$\varepsilon_{r_{Layer}} < 0$ $\left\|\varepsilon_{r_{Layer}}\right\| > \left\|\varepsilon_{i_{Layer}}\right\|, \varepsilon_{medium}, \varepsilon_{core}$	Surface plasmon resonance
$\varepsilon_{r_{Layer}} > 0$ $\varepsilon_{r_{Layer}} > \left\|\varepsilon_{i_{Layer}}\right\|, \varepsilon_{medium}, \varepsilon_{core}$	Lossy mode resonance
$\varepsilon_{r_{Layer}} \approx 0$ $\left\|\varepsilon_{i_{Layer}}\right\| \gg 0$	Long-range surface exciton polariton

magnitude of its imaginary part is large, resulting into resonances denoted as *long-range surface exciton polariton* (LRSEP).

The consideration of SPR for sensing is well known. The resonances LRSEP have not been applied for sensor developments due to their characteristics, which do not favor such application. On the other hand, LMR resonances have been used by Villar et al. (2010) and Hernáez et al. (2010), for refractive index measurement and for other parameters such as humidity and pH (Socorro et al. 2012). They show resemblances with SPR resonances, with some particular properties that favor in some situations their application in sensing, as is the case of polarization independence, feature that does not happen with SPR sensors.

The progress of optical sensing based on the SPR phenomenon will benefit substantially from advances in nanofabrication technologies. They have led to the realization of metal nanostructures composed of nanoparticles, nanoholes, nanorods, and other components with precisely controlled shapes, sizes, and spatial distributions. This fabrication control, in combination with advances in theory and the emergence of quantitative electromagnetic modeling tools, has provided a better understanding of the optical properties of isolated and electromagnetically coupled nanostructures of different sizes and shapes. These properties may result into a substantial enhancement of the performance of sensors based on these plasmonic structures.

21.2.3 Sensing Based on Surface-Enhanced Raman Scattering

Raman spectroscopy is a powerful technique for identifying molecular species. The unique vibrational frequencies of different molecules are in the origin of their Raman spectrum signature. Unfortunately, Raman scattering is an exceedingly weak phenomenon: only about 1 in 10^{12} photons incident on a molecule undergoes Raman scattering. However, when a molecule is in close proximity to a metallic nanostructure, the Raman scattering can be enhanced by a factor of 10^6–10^{16} due to a phenomenon known as *surface-enhanced Raman scattering* (SERS), which has emerged as an extremely sensitive and selective technique for identifying molecular species (Stiles et al. 2008). The discovery of this phenomenon dates back to the 1970s of last century (Albrecht and Creighton 1977), and since then, this topic has grown dramatically as an analytical tool for sensitive and selective detection of molecules adsorbed on noble metal nanostructures and, consequently, as a mechanism that supports highly sensitive chemical/biochemical optical sensing.

In the context of molecules, Raman scattering describes the inelastic scattering process between a photon and a molecule, mediated by a fundamental vibrational or rotational mode of the molecule. Due to energy exchange between the scattering agents, the incoming photon of energy $h\upsilon_p$ is shifted in energy by the characteristic energy of vibration $h\upsilon_m$. These shifts can be in both directions, depending on whether the molecule in question is in its vibrational ground state or in an excited state. In the first case, the photon loses energy by excitation of a vibrational mode (Stokes scattering), while in the other situation, the photon gains energy by de-excitation of such a mode (anti-Stokes). In general, the photons involved in Raman transitions are not in resonance with the molecule, and the excitation takes place via virtual levels. No absorption or emission of photons is involved, and the transition is a pure scattering process. This is true even in the case where the incoming photon is in resonance with an electronic transition. This resonant Raman scattering is stronger than normal Raman scattering, but its efficiency is still much weaker than that of fluorescent transitions. Typical Raman scattering cross sections σ_{RS} are usually more than 10 orders of magnitude smaller than those of a fluorescent process, with values in the range, 10^{-31} cm^2/molecule $\leq \sigma_{RS} \leq 10^{-29}$ cm^2/molecule, depending on whether the scattering is nonresonant or resonant (Maier 2007).

The optical power of the scattered beam, P_S, can be expressed as

$$P_S(\nu_S) = N\sigma_{RS}I(\upsilon_p) \tag{21.1}$$

where

N is the number of Stokes scatters within the excitation region

$I(\upsilon_p)$ is the intensity of the excitation beam (ν_S is the frequency of the scattered light)

SERS describes the enhancement of this process, accomplished by placing the Raman-active molecules within the near field of a metallic nanostructure, which can be specifically designed nanoparticles ensembles or the topography of a roughened surface.

The enhancement of the scattered optical power is due to two effects. One is associated with the modification of the cross section due to a change of the environment of the molecule. This change results into $\sigma_{SERS} > \sigma_{RS}$, which is the chemical or electronic contribution to the Raman enhancement. Such enhancement factor in general does not exceed 100. The other effect is substantially more important. It is quite often identified as the electromagnetic effect and has two components: increase in the electromagnetic field due to excitations of localized surface plasmons and a crowding of the electric field lines at the metal interface (lightning rod effect). Of these two components, only the plasmon resonance shows strong frequency dependence, while the lightning rod effect is due to purely geometric phenomenon of field line crowding and the accompanying enhancement near sharp metallic features. Both contribute to the enhancement of the incident and scattered optical fields, and this enhancement can be expressed as (known as the electromagnetic enhancement factor)

$$L(\upsilon) \equiv \frac{E_{loc}(\upsilon)}{E_0} \tag{21.2}$$

where

$E_{loc}(\upsilon)$ is the magnitude of the local field amplitude at the Raman-active location

E_0 is the amplitude of the unperturbed incident field

The enhancement of the incoming field means an effective intensity of the excitation beam given by $L^2(\upsilon_p)I(\upsilon_p)$. On the other hand, the electromagnetic enhancement of the scattered field is expressed by multiplying the normal scattered power by $L^2(\upsilon_S)$. Therefore, the total power of the Stokes beam under SERS conditions is

$$P_S(\upsilon_S)_{SERS} = N\sigma_{SERS}L^2(\upsilon_p)L^2(\upsilon_S)I(\upsilon_p) \tag{21.3}$$

Considering that the difference in frequency $\Delta\upsilon = \upsilon_p - \upsilon_S$ between the incident and scattered photons is, in general, much smaller than the linewidth of a localized surface plasmon mode, then $L(\upsilon_p) \approx L(\upsilon_S)$. Therefore, combining Equations 21.1 and 21.3, the enhancement factor of the optical power of the Stokes beam (Kneipp et al. 2006) can be derived as

$$R \equiv \frac{P_S(\upsilon_S)_{SERS}}{P_S(\upsilon_S)} = \frac{\sigma_{SERS}}{\sigma_{RS}}\left(\frac{E_{loc}}{E_0}\right)^4 = \frac{\sigma_{SERS}}{\sigma_{RS}} L^4(\upsilon_S) \tag{21.4}$$

This equation shows that the electromagnetic contribution to the total SERS enhancement is proportional to the fourth power of the field (electromagnetic) enhancement factor. As a result, enhancements of Raman scattering signals by a factor of 10^{16} have been achieved, associated with substrates based on chemically synthesized noble metal nanoparticles and colloidal systems (Culha et al. 2012), showing SERS can have the sensitivity sufficient for the detection of analytes down to single molecules (Lim et al. 2010).

A critical factor of the SERS mechanism is the distance dependence. The electromagnetic effect predicts that SERS does not require the measurand to be in direct contact with the surface but within a certain sensing volume. From a practical perspective, there are certain experiments, such as those involving surface-immobilized biological molecules, in which direct contact between the adsorbent of interest and the surface is not possible because the surface is modified with a capture layer for specificity or biocompatibility. Because the field enhancement around a small metal sphere decays with r^{-3}, where r is the radial distance from the surface of the sphere, E_{loc}^4 dependence implies that the overall distance dependence should scale with r^{-12}. Taking into account that the increased surface area scales with r^2 as one considers shells of molecules at an increased distance from the nanoparticle, one should observe the r^{-10} distance dependence:

$$P_S(\upsilon_S, r)_{SERS} = P_S(\upsilon_S, 0)_{SERS}\left(1 + \frac{r}{a}\right)^{-10} \tag{21.5}$$

where a is the average size of the field-enhancing features on the surface of the sphere. Values for this parameter depend on the materials used, but they are typically in the range of 10–20 nm. To illustrate this characteristic, for Al_2O_3 multilayers onto Ag film over nanosphere (AgFON), it comes out $a = 12$ nm, and the distance at which the SERS intensity decreases by a factor of 10 is 2.8 nm.

It is important to point out that the E_{loc}^4 enhancement approximation predicts that the best spectral location of the LSPR for maximum electromagnetic enhancement is coincident with the laser excitation wavelength, which would lead to maximum enhancement of the incident field intensity at the nanoparticle surface. In practice, this is not the case considering it is necessary to achieve electromagnetic enhancement of both the incident field and the radiated field, which, for Raman scattering, happens at different wavelengths. Experimental studies indicate that to achieve maximum electromagnetic enhancement, the frequency of the incident light shall have a value higher than the one where maximum LSPR extinction occurs, with the exact value depending on the geometry/materials involved (Willets and Duyne 2007).

The level of sensitivity achieved with SERS is particularly useful for the detection of a small number of analyte molecules normally encountered in a single cell. Apart from the sensitivity, the inherent attributes of Raman are retained in SERS. Thus, it distinguishes vibrational signatures of molecular bonds, enabling label-free positive identification of analytes in complex cellular environments. Analyte detection can be multiplexed without spectral overlap because SERS provides spectra with narrow bandwidth. It is also very sensitive to slight changes in the orientation and structure of the molecules, allowing for structural elucidation. These characteristics, coupled with the weak Raman scattering of water, make SERS an ideal technique for analyzing complex biological samples that require little or no sample preparation. Importantly, SERS is achieved using a wide range of excitation frequencies, allowing for the selection of less energetic excitation (NIR to red) in order to reduce photodamage and background autofluorescence. Additionally, the analyte detection takes place at a close proximity to the SERS-active metal surface, thereby further reducing background autofluorescence through quenching. Therefore, SERS addresses most of the challenges of fluorescence detection for biological applications while providing comparable sensitivity.

The realization that localized plasmons play a crucial role in the Raman enhancement of molecules at a metal surface has triggered a great amount of research into the design and fabrication of SERS substrates with controlled surface structure optimized for field enhancement. Topographies based on closely spaced nanoparticles (mimicking a surface with controlled, regular roughness), specially shaped nanostructures, or nanovoids have been analyzed for their effectiveness as SERS substrates. The motivation to consider fiber platform is associated with the attempts to transfer the technology from the laboratory to the production line, clinic, or field, which, with the standard approach, have been frustrated by the lack of robust affordable substrates and the complexity of interfacing between sample and spectrometer. Prompted by the success of optical fiber systems for implementing normal Raman scattering spectroscopy in remote locations and biomedical applications, attention has been shifting to the development of SERS-active optical fiber systems (Stoddart and White 2009).

Optical fiber SERS sensors have generated steady interest as they are seen as a potentially robust means of extending SERS into practical applications. Moreover, optical fiber SERS sensors lend themselves to single-ended measurement geometries, which are particularly attractive for minimally intrusive monitoring, such as in vivo biosensing and biomedical applications. Certainly, optical fiber SERS sensors do have some drawbacks. Some problems, such as interference due to ambient light and limited stability of the sensor, are not unique to optical fibers. Background absorption, fluorescence, and Raman scattering from the fiber itself present a more specific challenge, which will be discussed later. As with any novel technology, costs can be high at the prototyping level. However, the recent history of key components such as SMF and laser diodes shows that dramatic cost reductions can be achieved through economies of scale.

Optical fiber SERS sensing has, to date, largely focused on point sensing. Nevertheless, it is clear that SERS imaging probes and distributed sensors offer another potentially powerful extension of the technique. In the context of the broader problem of integrating an optical fiber platform with a nanostructured metal transducer, many of the related techniques that rely on plasmon resonance effects, such as surface-enhanced resonance Raman scattering (Smith 2008), surface-enhanced hyper-Raman scattering

(Valley et al. 2010), and surface-enhanced infrared absorption spectroscopy (Osawa 2001) could also be applied.

A number of optical-fiber-based SERS configurations have been researched. The so-called optrode design aims to exploit the potential for full alignment of samples with optical fibers, by using a single fiber to carry both the excitation light and the back-scattered SERS signal. The basic idea of a SERS optrode is that the SERS substrate be fabricated on the *active* tip of the sensor fiber, thereby ensuring perfect overlap between the excitation and collection fields. The optrode design requires that the SERS substrate be reasonably transparent, as scattered light arising at the sampling interface should be able to pass back through the substrate to be captured by the optical fiber. The optrode design minimizes the required number of optical elements, eliminates the need for optical alignment with the sample, and avoids free space light propagation. It also reduces the sensitivity of the measurement to scattering or absorption by suspended particulate matter.

Different techniques have been used to produce effective SERS-active optrodes. These include the immobilization of colloidal silver particles, slow evaporation of metal island films, vacuum deposition of metal films over alumina and diamond particles, or on fiber tips roughened by nanoimprint lithography or mechanical means (Viets and Hill 1998).

The signal obtained from SERS optrodes can be increased by appropriate engineering of the tip geometry (Viets and Hill 2001). It has been shown that the SERS intensity for different tip coatings could be increased by as much as a factor of 20 by angle-polishing the fiber tip. The optimal tip angle was between 40° and 60°, depending on the type of coating. Conically etched tips also gave an increased SERS intensity, which was attributed to the enlarged surface area, multiple internal reflections of the laser light, and excitation of delocalized plasmons.

The single-fiber-optrode configuration is the simplest configuration used for SERS sensors. Compared with multiple-fiber arrangements, the compact size of the single-fiber optrode is well suited to in situ and in vivo applications. The main disadvantage with SERS optrodes is the background that is generated by the laser excitation within the fiber core. In general, this requires short lengths of fiber to minimize this effect.

While end-coupled sensors rely on a direct excitation of the LSPR by the propagating electromagnetic field, excitation can also occur via evanescent field that arises owing to total internal reflection at the boundary between the optical fiber core and cladding (Stoddart and White 2009). This is an intrinsic sensing modality, as the waveguide structure plays a key role in determining the strength of the interaction with the analyte. Evanescent field excitation has proved attractive for SERS, possibly owing to the similarity with the well-known attenuated total reflection and SPR sensing modalities. At a practical level, the large increase in the available sensing area is expected to increase sensitivity. It is important to note that signal photons generated in the cladding can only be coupled back into propagating modes of the waveguide by the same evanescent field mechanism. This is a relatively inefficient process, with typical capture efficiency on the order of 1% for light sources that are concentrated in a thin layer at the core–cladding interface. A strong evanescent field interaction with metallic nanoparticles also implies large transmission loss due to absorption (Gu et al. 2008).

The MOF brought an important development for optical-fiber-based SERS sensors (Gu et al. 2008). Both light guidance mechanisms of these fibers (through total internal reflection or by the photonic bandgap effect) have an associated electromagnetic field component in the cladding region, which may be used to perform SERS measurements, for example, in the air holes surrounding a solid-core MOF (Amezcua-Correa et al. 2007). Silver nanoparticles were introduced within the MOF template from organometallic precursors using a high-pressure chemical deposition technique. A large SERS response was detected when analyte molecules infiltrated the structure, which was attributed to the long electromagnetic interaction lengths of the optical-fiber-guided modes. Despite the relatively small fraction of the overall power propagating in air in the fiber studied, the deposited silver nanoparticles increased the fiber attenuation to ~2 dB/cm.

The application of solid-core MOF for SERS is still at an early stage of development, but it is already obvious that these fibers form another interesting new category of sensor platforms. The large evanescent field overlap in these fibers, together with localized regions of high field intensity near the surface of the core, leads to more efficient capture of scattered light, which could lead to future SERS sensor development.

Hollow-core PCF fibers have attracted most attention as platforms for SERS sensing (Yang et al. 2011). In a SERS sensor of this type, the sample solution usually goes into the core (and possibly cladding) channel(s) via the capillary effect. As the excitation light interacts directly with the sample solution inside the central core, where the electromagnetic field is strongest, most of the light is utilized for the generation of the SERS signal. In addition, as light is confined inside the hollow core, the interaction between light and silica can be extremely small, which greatly reduces the interference from the fiber background. Consequently, better signal-to-noise ratio is expected as it has been shown for the ultrasensitive detection of cancer proteins in an extremely low sample volume (Dinish et al. 2012).

The integration of SERS systems with optical fiber technology is a challenging but potentially rewarding endeavor that has just begun. It is expected that in the near future, the SERS sensors benefiting from the advances in the fiber technology and the field of metamaterials attract many interests and find many new applications.

21.2.4 Distributed Sensing

Distributed sensing associated with the phenomenon of light scattering in a medium, particularly in optical fibers, has been extensively discussed in Chapter 16. The beauty of this approach is that a spatial resolution approaching centimeter scale over tens of kilometer range could be achieved with the bare (standard) fiber itself acting as a sensing element without any special modification. The targeted measurands for distributed sensing have been strain and temperature, and refractive index of the surrounding medium can also be a primary parameter if the evanescent field of the guided core mode is allowed to expand into this medium. This means the field of biochemical sensing is also a future target application for distributed sensors.

Raman-based distributed temperature measurement systems were the first to be developed. Raman techniques are intensity based and offer temperature resolutions of ~1°C with a spatial resolution of 50 cm and a measurement length up to 2 km (Yokogawa 2012), very useful for applications where there is a need for distributed temperature monitoring of large structures, such as gas and oil tanks (where a local temperature change can be an indication of leakage), and for fire detection in large buildings.

The phenomenon of Rayleigh scattering has been explored only recently for distributed sensing systems. Surely, the *optical time-domain reflectometry* (OTDR) technique has for long been a valuable tool in the field of optical communication for the detection of points of loss, but rarely has been used for sensing (Frazão et al. 2005). The development of systems based on the concept of *optical frequency-domain reflectometry* (OFDR) has provided a new opportunity to explore Rayleigh scattering for sensing (Soller et al. 2005).

OFDR systems are able to measure distributed Rayleigh backscatter with high spatial resolution along a single-mode optical fiber and records a *finger-print*-scattered pattern as a function of length. Although the pattern is random and statistically distributed, it is a stable and invariant property of a fiber segment. External perturbations (strain and temperature) on the fiber result in a shift or a change in *periodicity* of this *finger print*, and using suitable algorithms, the magnitude and location of perturbations can be recovered. Temperature and strain resolutions on the order of a fraction of 1°C and few microstrains can be achieved with millimeter spatial resolution and relatively short measurable length (tens of meters (Kreger et al. 2006)), or up to several hundred meters with reduced spatial resolution (Duncan et al. 2007, Luna Technologies 2012). This technology fills a gap in sensitivity and spatial resolution between FBG-based sensing, which is well suited for distributed sensing over short lengths or at discrete points and Brillouin scattering technology, which is ideally suited to sensing over very long lengths at a more reduced spatial resolution.

Distributed sensing has been focused in stationary measurements, but recently, dynamic distributed sensing has been explored using Brillouin scattering, as will be seen soon. Concerning Rayleigh backscattering, the research of Zhou et al. (2012) indicates the feasibility of OFDR systems for distributed dynamic sensing. Indeed, by determining the spectral shift of the Rayleigh backscatter of the vibrated state with respect to that of the nonvibrated state, dynamic strain information in the frequency range 0–32 Hz could be obtained with 10 cm spatial resolution over 17 m sensing length.

Since the pioneering work of Horiguchi and Tateda in 1989 (Horiguchi and Tateda 1989), distributed sensing in optical fibers based on Brillouin scattering was explored following a technique similar to the Rayleigh-based OTDR, known as *Brillouin optical time-domain reflectometry*, and later evolved to an approach that takes advantage of the stimulated version of Brillouin scattering (*Brillouin optical time-domain analysis*: BOTDA). Several reasons explain the successes obtained with Brillouin distributed measurement, but three core ones can be identified (Thévenaz 2010): (1) it involves the use of standard low-loss single-mode optical fibers offering several tens of kilometers of distance range and a compatibility with telecommunication components; (2) it is a frequency-based technique, inherently accurate and stable on the long term; (3) since the information is not a consequence of the background thermal activation, experimental solutions based on stimulated Brillouin scattering (SBS) can be exploited, leading to a much greater intensity of the scattering mechanism and, consequently, a more acceptable signal-to-noise ratio.

Details about Brillouin scattering, including theoretical background, configurations, performance, and applications, can be found in Chapter 16. The relevant point to indicate here is what may be called the *classical* BOTDA technique routinely achieves 1 m spatial resolution up to distances on the order of 30 km. The word *classical* is justified considering in recent years novel developments brought the technique to new qualitative performance levels. It is the case of the *Brillouin echo-distributed sensing* (BEDS), associated to the preexcitation of the acoustic wave through the presence of a continuous background pump. Several variants of BEDS (such as dark and π-phase pump pulses) permitted to bring the spatial resolution down to some centimeters, in measurement lengths of kilometers, keeping the determination accuracy of the measurand-induced Brillouin shift comparable to what is obtained with the standard BOTDA technique (Brown et al. 2007, Foaleng et al. 2009).

Another important progress was the demonstration of the concept of *Brillouin dynamic grating* (BDG) in polarization maintaining fibers (Song et al. 2008). In these fibers, the acoustic wave generated during the process of SBS by the pump wave in one polarization is used to reflect an orthogonally polarized wave (probe wave) at a different optical frequency from the pump. The frequency separation between the pump and the probe waves is determined by the birefringence of the fiber and, normally, lies in the range of several tens of GHz. This difference depends on the targeted measurand, feature that is the basis of using this concept for distributed sensing, as demonstrated by Dong et al. (2009), achieving an order-of-magnitude higher sensitivity in the measurement of temperature compared to what has been obtained with standard Brillouin sensors. This approach has also impact on the enhancement of the spatial resolution associated with the conventional BOTDA technique, with demonstration of a 1 cm spatial resolution, which is the best result ever reported using the concept of Brillouin distributed sensing (Song et al. 2009). This value is close to the better figure achieved so far for the spatial resolution of a fiber-optic-based distributed sensing system, obtained using the technique of the synthesis of the optical-coherence function, however, only applicable over a reduced fiber length (Hotate and He 2006).

This evolution track indicates from the three mechanisms that support distributed sensing (Rayleigh, Brillouin, and Raman), Brillouin systems promise major progresses, as demonstrated by the novel techniques BEDS and BDG *sensing*, as well as by the following state-of-the-art developments.

In BOTDA systems, to achieve high-strain/temperature resolution over a wide dynamic range, the optical frequency of either the pump or probe wave needs to be swept across 100–200 MHz to determine the Brillouin frequency shift (BFS, ν_B) along the sensing fiber. Such a large frequency scan results in a fairly slow process, which, together with the need for averaging, limits the BOTDA method to the quasi-static measurement. To overcome this limitation, Brillouin sensing based in correlation domain techniques have been demonstrated to be amenable to dynamic measurement and reported results show recording of 200 Hz strain signal with 10 cm spatial resolution over 20 m measurement range (Kwang and Hotate 2007).

Recent progress pointed out that, in principle, it is just needed a small modification of the classical BOTDA setup to get the dynamic measurement functionality. This has been recently emphasized by Peled et al. (2011) showing that by using a pump pulse of a fixed optical frequency and a variable optical frequency CW probe wave, instead of a swept frequency pump pulse and a CW probe, dynamic measurement is possible with the BOTDA technique. The time evolution of the probe frequency is designed in such a way that when the probe wave meets the counterpropagating pump pulse at a certain location along the fiber, the optical frequency difference between these two waves coincides, as close as possible, with the middle of the slope of the Brillouin gain spectrum (BGS). Any fast change in the local strain or temperature will shift the BGS and, consequently, will be translated into gain variations of the probe wave. With this technique, the whole length of the fiber can be interrogated with a single pulse, or a few if averaging is required.

The concept was illustrated by the authors considering a fiber length of 85 m in a layout shown in Figure 21.4a. Five sections of SMF fiber and two patch cords were considered. Two 1 m sections were mounted on manually stretching stages, making it possible to adjust their static BFSs, and audio speakers were physically attached to these two sections in order to induce fast strain variations of various frequencies and amplitudes. To demonstrate Brillouin sensing over a nonuniform fiber (different average strain values along the fiber), the two 1 m sections (I and II in the figure) were stretched to different strain values, and audio signals of frequency 180 and 320 Hz were applied to sections I and II, respectively.

The results shown in Figure 21.4b clearly illustrate the feasibility of the proposed new technique for distributed dynamic strain measurement and, interestingly, its compatibility with BEDS and BDG for the enhancement of spatial resolution. The probe waveform can be adaptively modified to follow slow changing strain/temperature conditions. The dynamic range of the allowable vibrations is limited to the linear

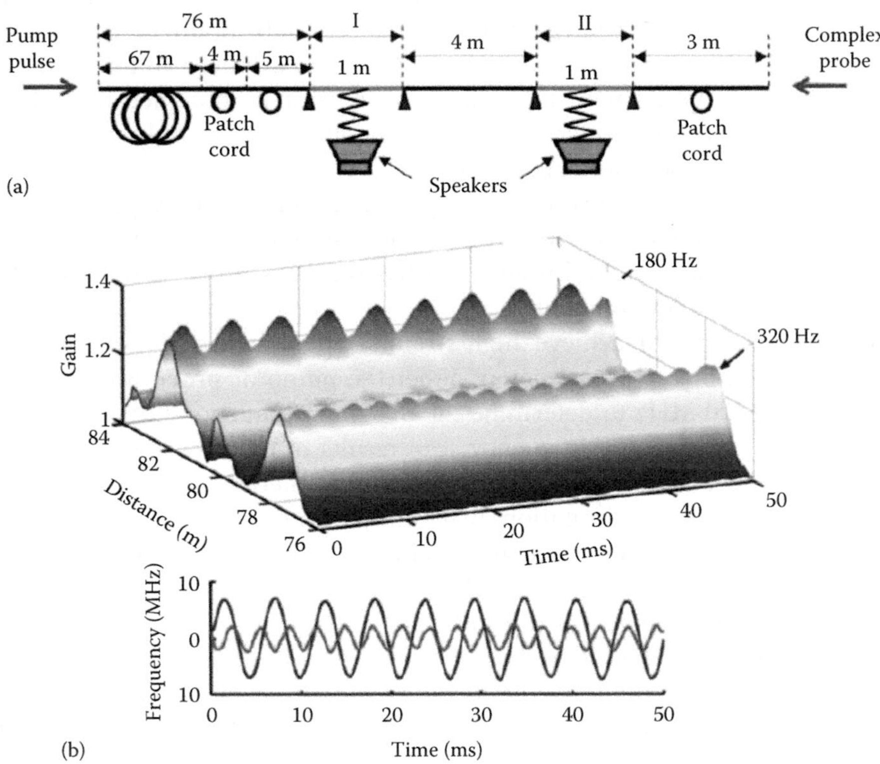

FIGURE 21.4 Fiber layout to demonstrate dynamic strain Brillouin sensing (a) and strain-induced gain vibrations at 180 and 320 Hz induced by the audio speakers at sections I and II, showing also the conversion of the gain variations into frequency values (b). (Adapted from Peled, Y. et al., *Opt. Express*, 19, 19845, 2011. With permission.)

frequency range of the BGS slope, which means for 10 ns pump pulse, the amplitude of the strain vibration must not exceed ~600 με. An upgraded version of this technique was later proposed by the same authors, which permitted distributed Brillouin dynamic strain sensing over 100 m long fiber with a resolution of 5 με and spatial resolution of 1.3 m (Peled et al. 2012).

Another direction of research concerns with extending the length of fiber for distributed sensing without substantial degradation of the measurement performance. A distinctive key factor in Brillouin-based distributed sensing is the possibility to measure along many kilometers of fiber with high spatial resolution. But the challenge is to achieve submeter spatial resolution over fiber lengths of several tens of kilometers, desirably more than 100 km. The issues associated with such challenge are twofold: one associated with fiber attenuation that inevitably reduces the Brillouin gain along the fiber, leading to a contrast loss and an increase in the uncertainty in the far end, making measurement over fiber lengths in excess of 50 km very difficult using BOTDA, and the other is the spatial resolution, which in classical BOTDA is determined by the time width of the pump pulse, resulting into a spatial resolution of ~1 m with a pulse duration of 10 ns. Shorter pulses lead to better resolutions. This approach has an intrinsic limitation, associated with the fact that for shorter pulses, the observed BGS broadens. This broadening leads to larger uncertainties in determination of the Brillouin shift. A natural limit can be found when pump pulse and natural Brillouin gain spectra show comparable bandwidths, corresponding to pump pulses with a duration of ~10 ns, or ~1 m spatial resolution. This is known as the resolution-uncertainty trade-off (Naruse and Tateda 1999). In long-range systems, this trade-off is worsened by self-phase modulation, which introduces an extra spectral broadening of the pulses as they travel along the fiber.

The research of Vinuesa et al. (2012) addressed these two issues. To avoid the physical limitations associated with pump pulse shortening, the authors considered the use of the differential pulse-width pair technique, first proposed by Li et al. (2008). It allows increasing the resolution of BOTDA fiber sensors without broadening the gain spectrum and avoiding self-phase modulation. Its working principle is based on the subtraction between gain traces obtained with slightly different pulse widths. The spatial resolution is then given by the differential width between the pulses, while the broadening in the gain remains bounded since the pulses used are always much longer than the phonon lifetime (typically four to six times larger). The effect of self-phase modulation could also be ignored for the typical power levels used.

To overcome the attenuation issue, a low-gain distributed Raman amplification along the sensing fiber was implemented, therefore providing a process of maintaining the power level of the pump pulse along the fiber, and hence the Brillouin gain contrast. The effect of the relative intensity noise (RIN) transfer from the Raman pumps to the Brillouin probe was attenuated using a novel RIN de-noising technique. The combination of all these techniques permitted the authors to report 0.5 m spatial resolution over a range of 100 km of hot spots with the temperature uncertainty on the order of 2.9°C.

Finally, as described in this chapter, there are two approaches to distributed measurement; one is distributed point measurement (multiplexing), and another relies on scattering. Until recently, it was thought that these two approaches had very little in common and each offered distinct methodologies and operational principles and techniques. As a matter of fact, that has not to be the case, as demonstrated by the research *random-access distributed fiber sensing* (Zadok et al. 2012). Applying a new radar-inspired technique, each fiber segment is selectively addressed as a distinct sensing element. The concept is illustrated in Figure 21.5a.

Similar to matched-filtering processing in radars, both Brillouin pump and signal waves are jointly phase-modulated by a common binary pseudorandom bit sequence (PRBS), whose symbol duration, T, is much shorter than the silica phonon lifetime ($\tau \sim 5$ ns). The modulation phase within each symbol assumes a value of either 0 or π with equal

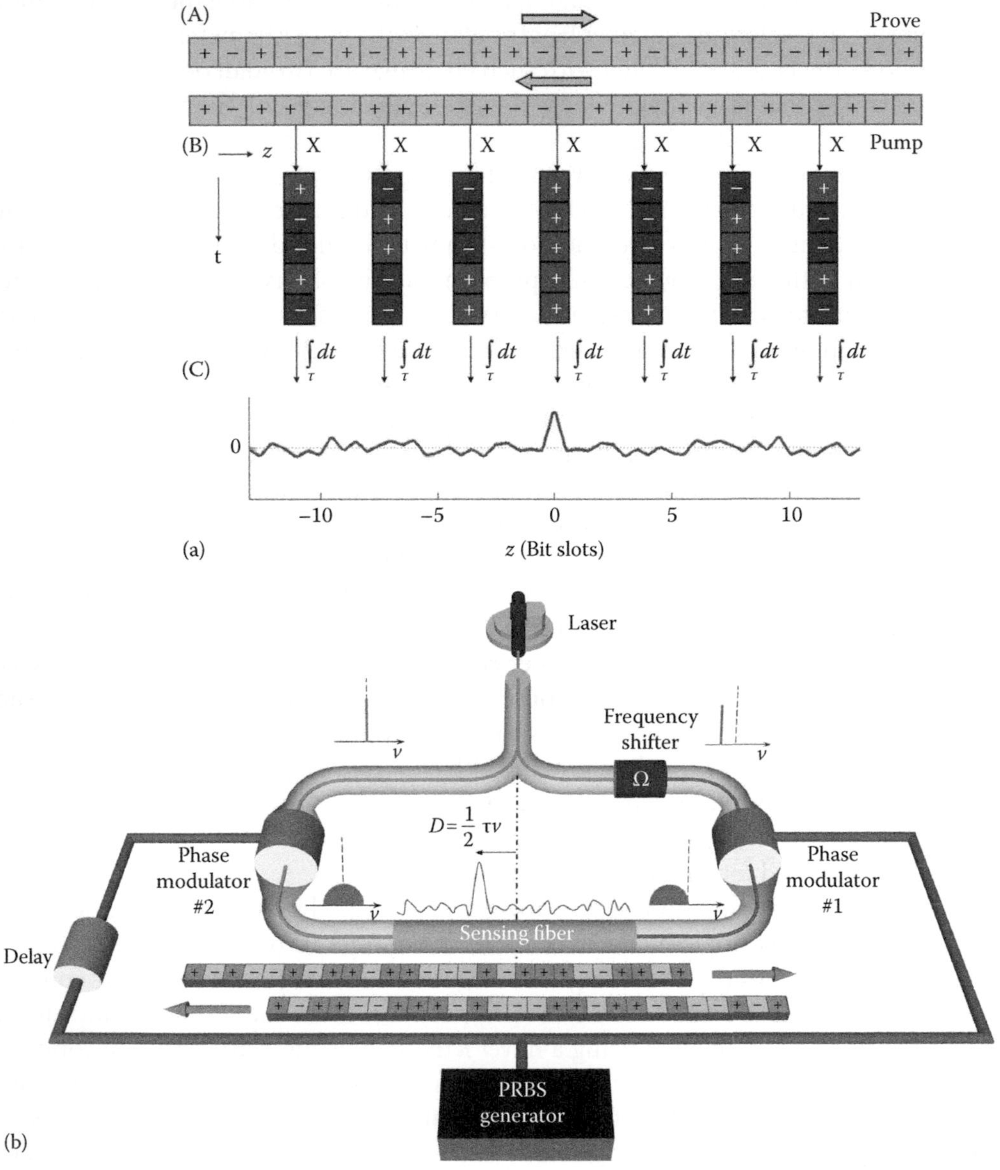

FIGURE 21.5 Principle of random-access distributed fiber-optic sensing using stimulated Brillouin scattering (a) and experimental layout (b). (Adapted from Zadok, A., Antman, Y., Primerov, N. et al.: Random-access distributed fiber sensing. *Laser Photon. Rev.* 2012. 2012. 1–5. Copyright Wiley-VCH Verlag GmbH & Co. KGaA. With permission.)

probabilities. The instantaneous driving force for the SBS acoustic field generation is proportional to the product of the pump wave envelope and the complex conjugate of the signal wave envelope. Therefore, it is necessary to distinguish between the dynamics of the SBS-induced acoustic field in two different regions. In the vicinity of the fiber center (assuming the phase modulation synchronization shown in Figure 21.5aA), the pump and signal are correlated (Figure 21.5aB); hence their phase difference is constant, and the driving force for the acoustic field generation keeps a steady nonzero value. Consequently, the acoustic field is allowed to build up to its steady-state value and permits the interrogation of the local Brillouin shift. In all other locations, the driving force for the acoustic field is randomly alternating in sign; thus, considering $T \ll \tau$, the acoustic signal amplitude thus averages out to a zero expectation value, and the SBS interaction outside the correlation peak is largely inhibited (Figure 21.5aC). Off-center fiber segments are interrogated introducing a specific phase shift between the PRBSs that modulate the Brillouin pump and probe waves.

The setup implemented by the authors to test this random-access distributed sensing concept is shown in Figure 21.5b. Both pump and signal waves are drawn from the output of a single monochromatic laser. The signal wave is offset in frequency by Ω, which is on the order of the BFS in the fiber under test. Both waves are comodulated by a common PRBS phase code and launched from opposite ends of the fiber. Careful timing of the two-phase modulators provides a scan of the code correlation peak across the fiber and allows for an arbitrary addressing of a specific fiber segment, and by varying Ω, in each segment, the BGS is reconstructed. Random-access monitoring of 9 mm-long sections at arbitrary locations along 200 m of fiber and temperature measurement with accuracy of ±0.5°C has been reported.

There is no fundamental reason limiting the applicability of this technique to relatively short fiber lengths. Indeed, it can be scaled to have an effective random access of centimeter-long segments of fiber over several kilometers, offering possibilities similar to what is found in biological nervous systems.

21.3 Novel Paths for Optical Sensing

In previous sections, some of the most promising sensing concepts, methodologies, and technologies have been noted, and their potential impact on future trends in the field of fiber-optic sensor have been discussed. Here, the focus will be on more exploratory, sometimes even speculative, ideas that are being considered in the general field of wave propagation and their potential implications in optical sensors. We begin this part with metamaterials, the engineered materials with properties unmatched in nature. The use of metamaterials in optical sensors has already begun, and its long-term impact would be immense. In Section 21.3.1, we discuss how metamaterials could enhance plasmonics-based sensors. In Section 21.3.2, the topics of slow and fast light will be described, focusing on the potential of these phenomena for high-performance optical sensing. In Section 21.3.3, optical quantum sensing will be viewed at a glance, but with sufficient detail to realize how quantum states of light, such as squeezed states (Section 21.3.3.2) or entangled states (Section 21.3.3.3), can be used to perform sensing with sensitivities far exceeding what is possible classically and also exceeds what was once thought to be possible within quantum mechanics.

21.3.1 Metamaterials and Optical Sensing

For long, it was known that the electromagnetic response of a material can be modified via periodic variations of its structure and composition. However, recent technological advances permit these concepts to be applied in very small dimensions with high accuracy and control to develop materials/media with interesting properties. A well-known example in the field of fiber optic is photonic crystals, dielectric materials with a periodic modulation of their (real) refractive index ($n = \sqrt{\varepsilon}$), achieved via the inclusion of scattering elements such as holes of different dielectric constant into the embedding host. In these structures, both the size and the periodicity of index modulations are on the order of the light wavelength in the material. In this way, the dispersion relation for electromagnetic waves propagating through an artificial crystal can be engineered, and bandgaps in frequency space may be established to inhibit propagation. This approach permits photonic structures with unconventional properties to be designed. As we will see in the following section, artificial materials with controlled photonics response provide new opportunities that were not imaginable until recently.

21.3.1.1 Metamaterials

In contrast to photonics materials, the metamaterial concept is grounded on the fact that both the size and periodicity of the scattering elements are significantly smaller than λ. Therefore, they can in a sense be viewed as microscopic building blocks of an artificial material, in analogy to atoms in conventional materials found in nature (Figure 21.6). Using the same rational and transitioning from the microscopic to the macroscopic form of Maxwell equations, the electromagnetic response of metamaterials can be described via both effective electric permittivity (ε_{eff}) and magnetic permeability (μ_{eff}). Since on the subwavelength scale, the electric and magnetic fields are essentially decoupled, ε_{eff} and μ_{eff} can often be controlled independently by the use of appropriately shaped scatters.

A pure intellectual exercise of Victor Veselago of the Moscow Institute of Physics and Technology, who, in the 1960s, understood that negative values for the electric permittivity and magnetic permeability were allowed by Maxwell equations, led him to ask the question: *what physical sense can be attributed to the negative values of* ε *and* μ? It was by considering this question that the idea of metamaterials and the notion of negative refractive index came about as expressed in his research paper entitled *The Electrodynamics of Substances with Simultaneously Negative Values of* ε *and* μ (Veselago 1968). Initial responses to this paper were far from enthusiastic, and the idea was essentially ignored as an interesting but quirky idea for 30 years, but eventually investigation into metamaterials, and their design and fabrication inevitably began to take shape 30 years later.

Negative refractive index does not change Snell's law, but it does change the refraction rule. When a light encounters a negative refractive index surface, rather than being refracted at an angle to the right of the surface normal, it is refracted to the left. It is for this reason that negative refractive index metamaterials are sometimes referred as left-handed materials. A material with negative refractive index shows also other strange properties (Beech 2012). For example, while in a right-handed material, the individual waves, the wave profile, and the energy carried by it all propagate in the same direction,

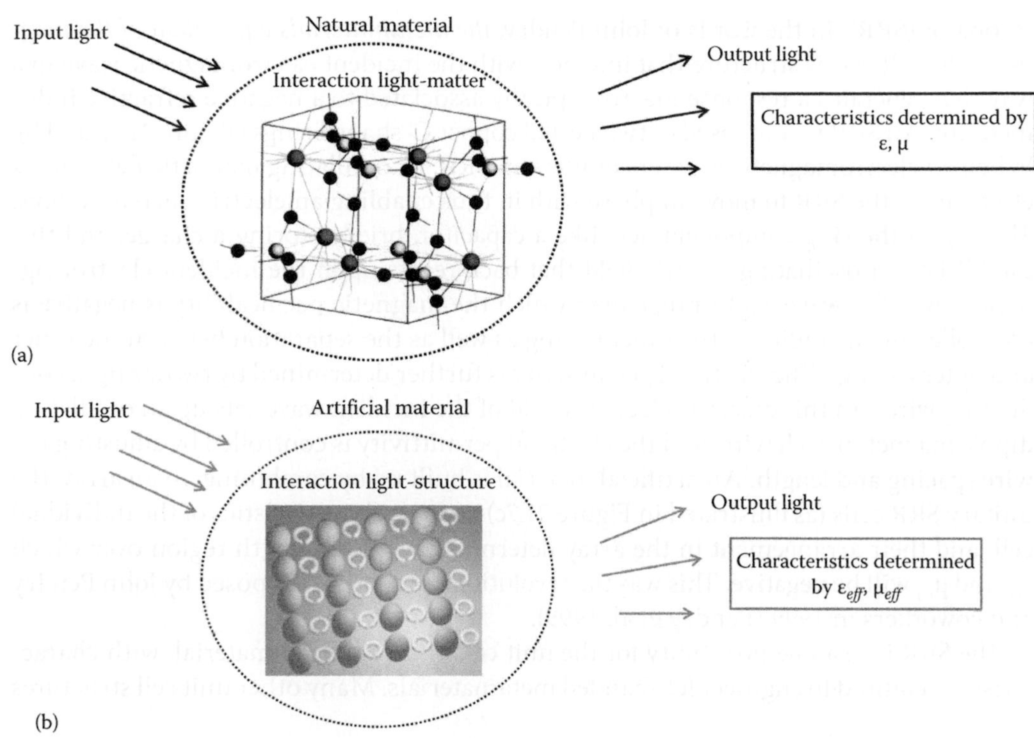

FIGURE 21.6 Analogy between light–matter interaction in a natural material, macroscopically modeled by the electric permittivity (ε) and magnetic permeability (μ) (a) and light–artificial material (*metamaterial*) interaction with macroscopic effects associated with effective values for the permittivity (ε_{eff}) and permeability (μ_{eff}) (b).

although the phase and group velocities may be different, in a negative refractive index medium, it turns out that the individual waves move in the opposite direction to the wave packet profile and the energy. The phase and the group velocity have different signs and move in opposite directions from each other. Surprisingly, we have the situation that the individual waves (that interfere to make up the wave packet) travel backward, while the wave packet itself and the energy associated with it move forward in the medium. As pointed out by John Pendry, the strangeness of the situation can be understood by noting that if a flashlight emitting a continuous beam of light was embedded in an optical metamaterial with a negative refractive index, then if one could follow the individual waves, they would appear to be moving backward from the metamaterial into the flashlight, as if one were seeing a movie running backward (Pendry et al. 1999). The energy and wave packet, however, travel away from the flashlight into the metamaterial. Another oddity associated with left-handed metamaterials is that the Doppler effect will behave in the opposite manner to what is normally experienced, that is, light emitted by a moving object approaching the observer will experience a red shift, and when it passes away, a blue shift would be registered.

The identification of the peculiar characteristics of left-handed materials adds extra motivation to address the problem of how to engineer the materials to exhibit metamaterial effects, specifically to design a metamaterial with negative refractive index. This brings us to the structure that is of central importance in this field: the split-ring resonator (SRR), in the words of John Pendry, *the metamaterials equivalent of the magnetic atom*. It is this structure that interacts with the incident electromagnetic wave in a way that generates a response macroscopically associated to a negative refractive index medium. An SRR is composed of two nested copper C-shaped rings (Figure 21.7a and b). When an electromagnetic wave encounters an SRR, its oscillating magnetic field causes electrons in the SRR to move in phase with it, thus enabling an electric current to flow. The gap in the ring component acts like a capacitor, briefly storing a charge, and this establishes an oscillating electric field that back-reacts upon the incident electromagnetic wave. The wavelength range over which the magnetic permeability is negative is controlled by the radius of the inner C-ring as well as the separation between the inner and outer C-rings. The electrical permittivity is further determined by two straight conducting wires. In this case, the electrical field of the incident wave sets up an oscillating dipole magnet in each wire, and the electrical permittivity is controlled by adjusting the wire spacing and length. An artificial material is built up by combining, in an array, the unitary SRR cells (as illustrated in Figure 21.7c) and the characteristics of the individual cell, and their arrangement in the array determines the wavelength region over which ε_{eff} and μ_{eff} will be negative. This was the revolutionary concept proposed by John Pendry and coworkers in 1999 (Pendry et al. 1999).

The SRR is just one possibility for the unit cell of an artificial material, with characteristics required to engineer left-handed metamaterials. Many other unit cell structures

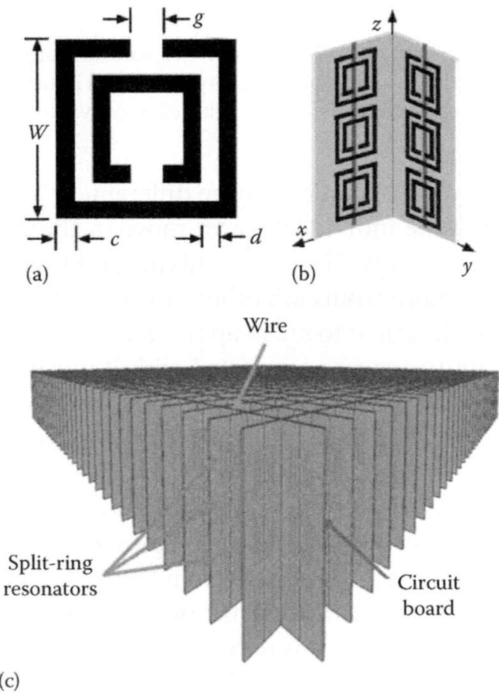

FIGURE 21.7 Schematic of a single split-ring resonator (SRR) showing the geometric parameters that determine its characterization (a) and the unit cell of the metamaterial with left-handed properties, formed by six copper SRRs and two wire strips on two thin fiberglass boards in a 90° geometry (b); artistic layout of the left-handed metamaterial array (c). (a and b: From Shelby, R.A., et al., *Appl. Phys. Lett.*, 78, 489, 2001. With permission; c: Courtesy of Wikipedia Commons.org.)

are possible permitting freedom to tailor the pair (ε_{eff}, μ_{eff}) as needed. For example, considering *I*-shaped unit cells in a closely packed grid (Choi et al. 2011), a positive peak refractive index of 38.6 along with a low-frequency quasi-static value over 20 was experimentally realized for a metamaterial operating at terahertz frequencies. The authors suggest that refractive indices of several hundreds might eventually be achieved by reducing the spacing between *I*-shaped elements and embedding them in a higher refractive index substrate.

Figure 21.8 permits to realize the amplitude of this field by illustrating the real (ε, μ) plane, that is, considering real values for the electric permittivity and for the magnetic permeability, therefore not considering losses (when dealing with metamaterials, in rigor, we have the (ε_{eff}, μ_{eff}) plan).

To situate the analysis of such plane, it can be observed that the propagation of electromagnetic waves in vacuum is represented by a single point and that most of the classical optics just corresponds to a small portion of the $\mu = 1$ line (nonmagnetic materials), which roughly can be located in the intervals [$\mu = 1$, $\varepsilon \in (1,4)$] (dielectric optics) and [$\mu = 1$, $\varepsilon \in (-10,-0.1)$] (plasmonics, for the wavelength region around 632 nm; aluminum is a famous exception, with a real part of the dielectric constant on the order of −55) (Weber 2003). This means that there is plenty of room for many revolutionary ideas in optics.

In general, the four quadrants of the plane identify regions with distinct characteristics for electromagnetic wave propagation. When ($\varepsilon > 0$, $\mu > 0$), the refractive index is positive, and we have a propagating field in materials with right-handed properties. Values of $\mu > 1$ are associated with magnetic materials, available in the nature or synthesized following the metamaterial approach. The third quadrant is associated with materials where ($\varepsilon < 0$, $\mu < 0$), also compatible with electromagnetic field propagation, now with the left-handed characteristics.

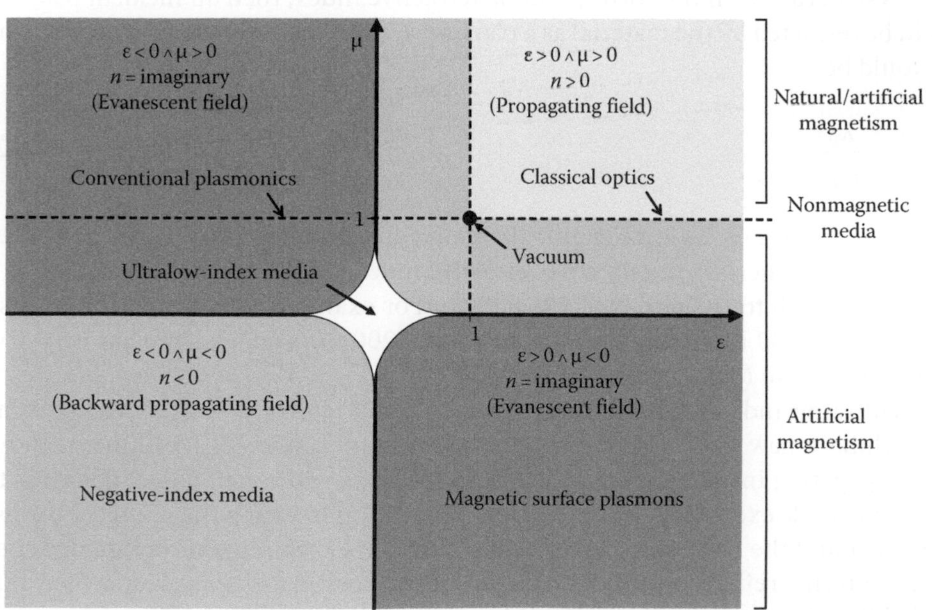

FIGURE 21.8 Schematic of a single (ε, μ) plane with some properties shown in each of the quadrants.

From Maxwell equations, we have, $n^2 = \varepsilon\mu \Rightarrow n = \pm\sqrt{\varepsilon\mu}$, and by physical reasons, Veselago showed that we need to have (Veselago 1968)

$$n = \begin{cases} +\sqrt{\varepsilon\mu} \Leftarrow & \varepsilon > 0 \text{ and } \mu > 0 \\ -\sqrt{\varepsilon\mu} \Leftarrow & \varepsilon < 0 \text{ and } \mu < 0 \end{cases} \tag{21.6}$$

Therefore, for ($\varepsilon < 0$, $\mu < 0$), the refractive index is negative.

The second quadrant involves ($\varepsilon < 0$, $\mu > 0$), from where n takes imaginary values and propagation is not allowed (besides evanescent field). The line $\mu = 1$ includes the region where materials support conventional plasmonics, most notably gold and silver at optical frequencies, which have specific values for ε that are wavelength dependent but essentially determined by their intrinsic atomic structures. Finally, the fourth quadrant is characterized by ($\varepsilon > 0$, $\mu < 0$), again n is imaginary like second quadrant. While standard surface plasmons originate from the collective resonance of electronic density waves in a system consisting of metallic building blocks, the symmetry of Maxwell equations implies that there is a magnetic counterpart of this excitation exhibited in magnetic systems, known as magnetic surface plasmons, which occurs with artificial materials located in this quadrant, showing positive electric permittivity and negative magnetic permeability.

Finally, the central region of the (ε, μ) plane shown in Figure 21.8 corresponds to $\varepsilon < 1$ and $\mu < 1$, known as the region of ultralow-index media. For these materials, the real part of the effective refractive index is smaller than 1 and the phase velocity exceeds that of light in vacuum. The value of the refractive index can be calculated from Equation 21.6, but it is possible to evaluate it in a more operational way. Keeping in mind that a subwavelength structure with an effective refractive index should refract light like a homogeneous structure with the corresponding refractive index, then an incident plane wave should be refracted by the material as a plane wave, and the wavelength inside the material would be

$$\lambda_{mat} = \frac{\lambda_0}{n_{eff}} \tag{21.7}$$

where λ_0 is the free space wavelength. Therefore, determining the wavelength inside the material (metamaterial), the effective refractive index is obtained by the ratio of the free space wavelength to the refracted wavelength. For example, a metamaterial consisting of silver rods of 30 nm diameter arranged into a 200 nm square array has $n_{eff} = 0.62 + 0.024i$ at $\lambda_0 = 1\ \mu$m (Schwartz and Piestun 2003).

This ultralow index medium, also known as ULIMs (ultralow refractive-index materials) exhibit a new optical property, *total external reflection* (TER). This occurs when light propagating in vacuum is incident on a medium with a refractive index less than unity at an angle exceeding the critical angle, situation in which the refracted waves are evanescent and the reflectivity is very close to unity. Conventional waveguides operate by total internal reflection, where the index of the core material is greater than that of the cladding. On the other hand, ULIMs open up the possibility of hollow-core waveguides based on TER because the cladding (metamaterial) refractive index (real part) is smaller than the core refractive index (that can be an open channel filled with a gas whose concentration is the targeted measurand, offering new possibilities for gas sensing). The limitation may be the inherent loss of metamaterials with ultralow index, but with proper design, the attenuation of propagating waves in some metamaterials is much lower than in metals at optical frequencies. Indeed, while in metals, at optical wavelengths, the imaginary part of the refractive index dominates, in metamaterials, it can be more than one to two orders of magnitude smaller than the real part, as the earlier example of silver rods illustrates.

Dielectric photonic bandgap materials can also refract incident light in vacuum away from the surface normal for frequencies close to the bandgap. Thus, one could assign them a refractive index less than unity (and even less than zero) that will be consistent with Snell's law (Notomi 2000). However, this effect differs from the reflection from ULIM considering (1) the photonic bandgap materials do not, in principle, have the same angle-dependent reflectivity and transmissivity predicted by Fresnel formulas; (2) their unusual values of refractive index apply only to narrow bandwidths; and (3) they require a bandgap.

Therefore, the flexibility associated with ULIM structures is substantially larger compared with photonic bandgap materials, opening new opportunities for device design and applications such as in optical sensing (Schwartz and Piestun 2005).

It should be noted that ultralow index media and their properties are different from those involved with superluminal wave propagation where electromagnetic waves appear to propagate with a group velocity faster than the speed of light in vacuum, an effect related with many wave interaction mechanisms, including anomalous dispersion, evanescent propagation, and wave interference (Withayachumnankul et al. 2010).

The focus of this book is optical sensing; hence, we will briefly discuss some demonstrated examples of combining plasmonics and metamaterials for sensing applications.

21.3.1.2 Metamaterial-Enhanced Plasmonics

As discussed in Section 21.2.2.3, most practical SPR biosensors use SPPs, which are excited on continuous Au or Ag films in the attenuated total internal reflection geometry. Owing to the resonant photon SPP coupling conditions, this technique provides an extremely small detection limit. However, the SPP-based approach still needs an improvement in sensitivity for the detection of small analytes and does not always satisfy requirements of biotechnology, advancing toward new nanoscale designs and selective chemical and biochemical nanoarchitectures (Prasad 2003). LSPRs of metallic nanostructures seem much more suitable to match these new trends, as well as to bring new functionalities, such as spectral tunability and strong enhancement of the local electric field. However, LSPR-based sensors are known to provide at least an order of magnitude lower sensing response to refractive-index change compared to SPPs, as well as 10 times smaller probe depth, making them applicable for only a limited number of biological species (Anker et al. 2008).

These limitations are being overcome by newly emerging plasmonic metamaterials capable of supporting similar or more sensitive guiding modes than SPPs (Kabashin et al. 2009). In this work, an assembly of Au nanorods electrochemically grown into a substrate-supported, thin-film porous aluminum oxide. The final structure represents an array of parallel nanorods occupying an area of up to 2 cm^2. The lateral size and separations between the nanorods are much smaller than the wavelength of light used in the experiments, so only average values of nanorod assembly parameters are important, and individual nanorod size deviations have no influence on the optical properties that are well described by an effective medium model.

The authors showed that this metamaterial structure works similarly to a conventional SPP-based sensor, exhibiting a red shift of the resonance in response to an increase in the refractive index. Furthermore, a change of the refractive index of 10^{-4} RIU causes a shift of the resonance by 3.2 nm. The estimated sensitivity of 32,000 nm per RIU exceeds the sensitivity of localized plasmon-based schemes by two orders of magnitude (Anker et al. 2008). Also, the sensitivity of the nanorod metamaterial-based sensing system exceeds the one achievable with SPP-based sensors (with spectral interrogation) by a factor of 10.

The enhancement factor associated with the metamaterial-based structure is impressive, providing a glimpse of what may be possible with metamaterials supported optical sensing. The origin of such high sensitivity gain is associated with the fact that in nanorod arrays, the sensed substance is incorporated between the initially bare rods; therefore, the waveguide mode provides a better overlap between the sensing field and the sensed substance compared with what happens in standard SPR sensors. Furthermore, the effective dielectric constant, ε_{eff}, of the metamaterial strongly depends on the dielectric constant of the tested medium as a result of modification of the plasmon interaction in the nanorod array, thus leading to modification of resonant conditions of the guided-mode excitation caused by the sensed analyte (Wurtz et al. 2008). This means that the dielectric constant seen by the analyte is no more that of the host material (gold), but a different value that depends not only on the material but also on the shape/dimensions of the rods and their spatial arrangement, parameters that can be adjusted to tune ε_{eff} to a value that turns maximum the sensitivity of the structure to analyte changes.

In addition, the discontinuous porous nanotexture of the nanorod matrix enables the implementation of new sensing geometries, not feasible with conventional film-based SPR. Indeed, by functionalizing the nanorods and immobilizing a receptor on their surface, one can follow the binding of a selective analyte with the receptor inside the nanorod matrix. The considerably increased surface area given by the nanoporous texture of the metamaterial significantly increases the amount of biomaterial that can be incorporated into the matrix within the available probe depth, maximizing the *biological* sensitivity of the system. Furthermore, the distance between the nanorods can be selected to match the size of biological species of interest, giving access to a further size selectivity option that is important for many situations in immunoassays and virus and protein detection (Kabashin et al. 2009).

Another interesting example of the application of metamaterials for optical sensing is the use of such engineered material as perfect absorber coupled to an LSPR configuration as shown in Figure 21.9 (Liu et al. 2010).

The diameter and thickness of the gold disks are 352 and 20 nm, respectively. The periods in the x- and y-directions are 600 nm. The thickness of the MgF_2 spacer is 30 nm, and the thickness of the gold mirror is 200 nm. The whole structure resides on a glass subtract. Because of the presence of the gold mirror, the transmittance of the structure is totally eliminated across the entire near-infrared frequency regime. Additionally, the presence of the gold disks in the array induces a residual reflectivity of ~0.28% at 1.6 μm,

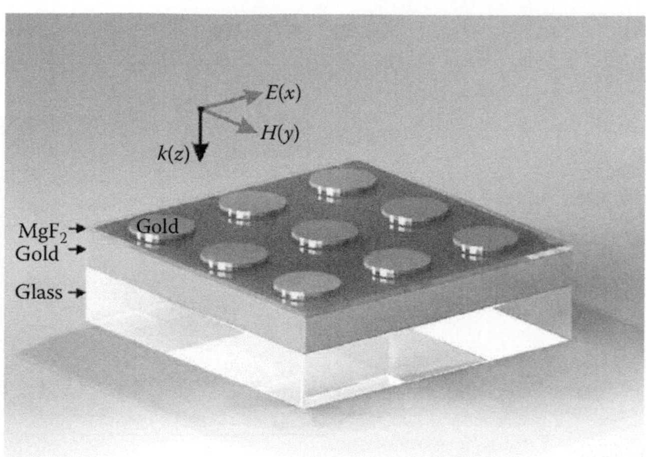

FIGURE 21.9 Schematic of the perfect absorber structure. (Reprinted with permission from Liu, N., Mesch, M., Weiss, T. et al., Infrared perfect absorber and its application as plasmonic sensor, *Sensor Nano Lett.*, 10, 2342. Copyright 2010 American Chemical Society.)

turning this structure like a perfect absorber. It was also found that this characteristic holds for a wide range of incident angles.

The sensing principle relies on the fact that zero reflectance (i.e., perfect impedance matching) occurs only for a certain refractive index of the surrounding medium. The variation of the refractive index of the surrounding medium gives rise to nonzero reflectance (i.e., nonperfect absorbance) and therefore allows for the extremely sensitive detection of the intensity change in reflectance at a fixed light wavelength.

This type of sensing layouts shows the feasibility of refractive index sensing based on straightforward reflectance measurements using a single-wavelength light source. So far, the best plasmonic sensors have been mostly synthesized by chemical methods because gold or silver nanostructures obtained from chemical synthesis can be single crystalline. The introduction of the perfect absorber sensors discards this restriction since the intrinsic losses in the metal are essential to achieve perfect absorption. As a result, many nanofabrication technologies such as electron-beam lithography, mask colloidal lithography, interference lithography, nanosphere lithography, nanoimprint lithography, and focused ion beam writing can be widely applied to manufacture ultrasensitive plasmonic sensors.

The examples described dealt with planar substrates, and further steps are needed to couple metamaterial structures with optical fibers. The task is not trivial, but recently, there has been substantial progress in this direction, as illustrated in Figure 21.10 (Smythe et al. 2009). The integration of metallic nanopattern onto a fiber facet was achieved using a technique that harness the high resolution and geometric versatility of e-beam lithography and the topographical adaptability of polymer-based *soft* nanofabrication methods to transfer arbitrary metallic nanopatterns to various unconventional substrates, particularly optical fibers.

The transferred pattern is the same as the original created by e-beam lithography, shown in Figure 21.10d, which consists of 100 μm × 100 μm array of gold nanorods, each approximately 40 nm tall, 100 nm long, and 30 nm wide. The structures are separated

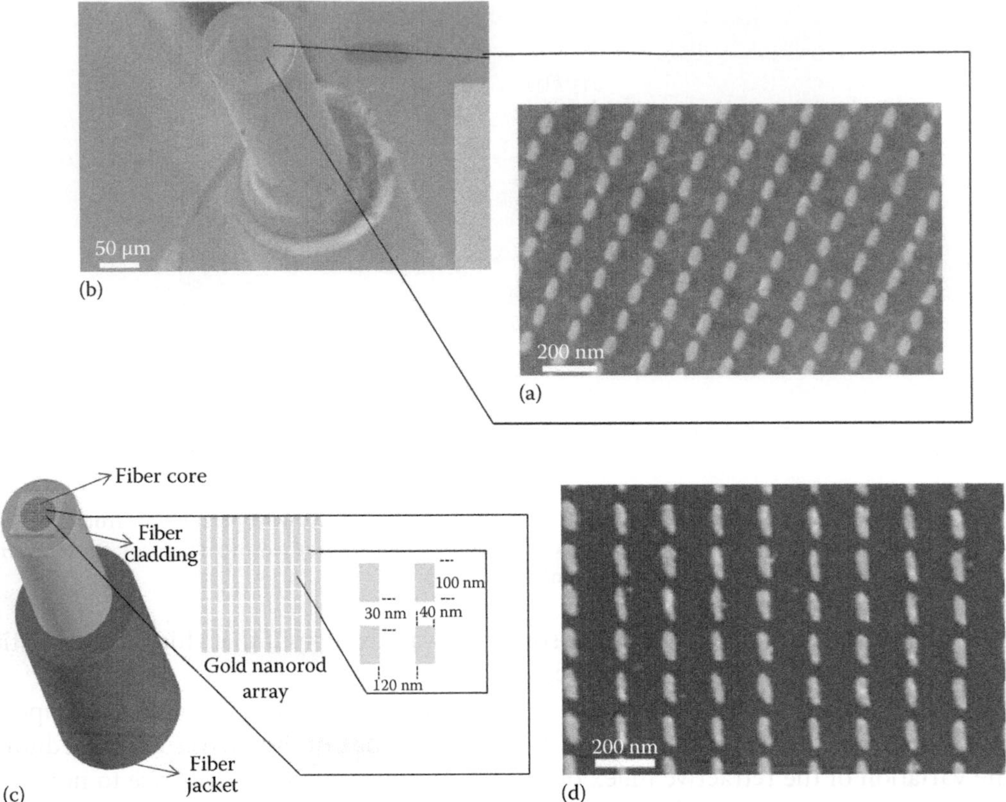

FIGURE 21.10 SEM micrograph showing a section of a gold nanorod array transferred to the facet of an optical fiber (a); SEM image of a transferred nanorod array on the facet of an optical fiber (b); schematic illustrating the arrays on the fiber facet (c); image of the nanorods on a silicon substrate before the transfer (d). The images of the transferred array (a) are of lower resolution than those of the arrays on silicon shown in (d). (Reprinted with permission from Smythe, E.J., Dickey, M.D., Whitesides, G.M. et al., A technique to transfer metallic nanoscale patterns to small and non-planar surfaces, *ACS Nano*, 3, 59. Copyright 2009 American Chemical Society.)

by gaps of approximately 30 nm along their longitudinal axis and 150 nm along their short axis. The nanorod array transferred to the facet of an optical fiber is shown in Figure 21.10a. The circle in the middle of the facet is the fiber core (62.5 μm in diameter), and the region surrounding it is the cladding (125 μm in diameter). In principle, the shapes and spacing of the transferred structures are limited only by the techniques used to form them: electron-beam lithography should allow for large areas of transferred patterns with structures and spacings of 10 nm and smaller.

The array geometry allows to tailor its (ε_{eff}, μ_{eff}) pair to values that optimize the surface plasmonic resonance for high sensitive detection of analytes or for optimized operation of SERS applied to an optical fiber. Recently, the advantage of fiber implementation was explored in a multiplexed fiber-based SERS system using gold nanovoid arrays (Chang et al. 2012).

In a remarkable work, Kostovski and collaborators showed a combination of biological templates with nanoscale replication, and optical fibers provides high-performance optical fiber SERS sensors (Kostovski et al. 2009). In this research, the SERS nanostructures are derived from a biological source, the cicada wing, where the antireflection structures on the surface of the wings are used as nanostructured templates. Photonic structures in biological organisms are often startling in their complexity and accomplishment, as demonstrated by the examples of the multilayer structures on the wings of *Morpho* butterflies, and the nano-protuberance antireflection arrays found on the corneas of butterflies and on the transparent wings of some hawk moths and termites. In the present case, the nanostructure on a cicada wing consists of a dense two-dimensional array of pillars that have separations, diameters, and heights of approximately 50, 110, and 200 nm, respectively. These pillars have approximate hexagonal distributions within microscale domains. The authors used nanoimprint lithography to integrate the cicada nanostructure onto an optical fiber platform, with the transferred nanoarrays coated with silver to make them SERS active. Thiophenol and rhodamine 6G were used as test analytes, using strong and well-defined SERS spectra collected with direct endface illumination and through fiber interrogation.

21.3.2 Sensing with Fast and Slow Light

The ability to control the group velocity of light propagating through an optical medium has been the focus of intense research. Slow/fast light refers to a large reduction/increase in the group velocity of light. If the dispersion relation of the refractive index is such that the index changes rapidly over a small range of frequencies, then the group velocity might be very low or very high, a behavior associated with the phenomenon of slow and fast light, respectively (Boyd and Gauthier 2002, Boyd et al. 2010). The early demonstrations of slow and fast light were in atomic media at low temperatures, but developments of the last decade demonstrated the possibility to have a wide range of group velocities in solid-state materials at room temperature, which are more suitable for practical applications (Boyd 2009).

In the following section, some applications of slow/fast light in optical sensing will be briefly discussed.

21.3.2.1 Fast- and Slow-Light Phenomena

The physical mechanisms that generate slow/fast light create narrow spectral regions with high dispersion, and they are generally grouped into two categories (Khurgin and Tucker 2008): material dispersion and waveguide dispersion. Material dispersion mechanisms, such as electromagnetically induced transparency (EIT), coherent population oscillation, and various four-wave mixing schemes, produce a rapid change in refractive index as a function of optical frequency; therefore, it involves modification of the temporal component of a propagating wave. The generation of slow/fast light in this way is identified as *material slow/fast light*. Waveguide dispersion mechanisms that appear in photonic crystals, in coupled resonator optical waveguides, in FBGs, and in other microresonator structures modify the spatial component (k-vector) of a propagating wave, allowing slow/fast light, identified as *structural slow/fast light*.

21.3.2.1.1 Material Slow/Fast Light This type of slow/fast light is associated with situations in which the velocity of light pulses can be fully described in terms of a spatially uniform but frequency-dependent refractive index n of the material. Under these circumstances, the group velocity of a light pulse is defined as

$$v_g = \frac{c}{n_g} \tag{21.8}$$

where
c is the vacuum light velocity
the group index, n_g, is given by

$$n_g = n + \omega \frac{dn}{d\omega} \tag{21.9}$$

where ω is the angular frequency of the carrier wave of the light field. Values of v_g substantially different from c rely on the dominance of the second term in Equation 21.9. This is resulted from the frequency dependence of the refractive index; hence, it is normally associated with the resonant or near-resonant response of material systems (Boyd 2011).

For a dispersive optical medium, without appreciable gain or loss in the spectral region of interest, it is possible to apply the Poynting vector formalism, resulting for the energy density u as (Boyd 2011)

$$u = \frac{1}{2} n n_g \varepsilon_0 |E|^2 \tag{21.10}$$

where
E is the amplitude of the optical field
ε_0 is the vacuum electrical permittivity

This equation shows no increase in field strength within the medium even though the energy density has increased. The energy flow per unit area (normal to the propagation direction) and unit time, that is, the intensity, is given by

$$I = \frac{1}{2} n c \varepsilon_0 |E|^2 \tag{21.11}$$

and consequently

$$I = u v_g = u \frac{c}{n_g} \tag{21.12}$$

These results show that when a pulse enters a slow-light medium, it becomes spatially compressed by a factor equal to the group index, increasing its peak energy by the same factor. It is important to note that pulse becomes compressed in space but not in time, indicating the pulse intensity remains constant, as dictated by energy conservation considering the power flow of an optical wave through each transverse plan cannot increase when propagating through a passive medium.

21.3.2.1.2 Structural Slow/Fast Light A qualitatively different mechanism to generate slow/fast light involves structures containing periodic patterns with periodicities on the scale of the optical wavelength, which substantially modify light propagation conditions. A typical example of such behavior occurs in photonic crystals, formed by a periodic modulation of the local dielectric constant in either one, two, or three dimensions. Introduction of a defect in a periodic structure, such as a resonator or a waveguide causes such an effect. For example, an FBG, which can be thought of as a one-dimensional photonic crystal (Boyd 2011), supports a multiwave interference (Figure 21.11a), which can generate slow and fast light. The dependence of the normalized group delay (equal to the dependency of the normalized stored energy) of an FBG structure on the detuning of the incident light field from the Bragg frequency ω_B is shown in Figure 21.11b for $\kappa L = 4$ (Winful 2002, 2006). L is the FBG length and κ is the coupling strength given by $\kappa = n_0 n_1 \omega_B / 2c$, where n_1 is the amplitude of the index modulation of the FBG, n_0 the background refractive index, and c the speed of light in vacuum.

Careful examination of this figure reveals that an FBG has regions of large group delay (higher values of stored energy) outside the main reflective window (bandgap) of the structure, which are spectral regions of slow-light generation, while inside the bandgap, the group delay and the stored energy have lower values, allowing fast-light behavior. In the context of slow light for sensing, it has been demonstrated that FBGs has promising potential, as will be discussed in Section 21.3.2.2.

For structural based slow light, unlike the case of material slow light, there is a true increase in electric field strength within the structure, which can lead to enhancement of nonlinear optical interactions (Monat et al. 2010). Defining *the slow-down factor* Ψ as the ratio of the effective group index of the photonic structure and its mean refractive index,

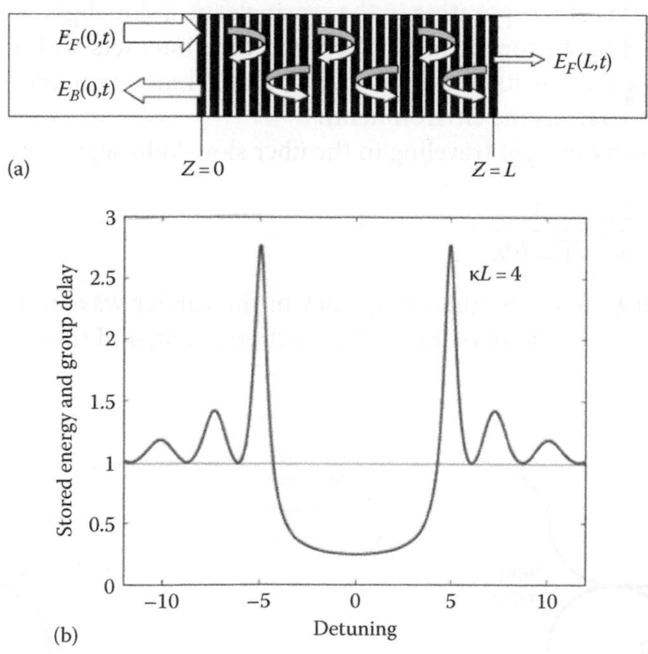

FIGURE 21.11 (a) Illustration of multiwave generation and interference in a fiber Bragg grating; (b) dependence of the normalized group delay and normalized stored energy density of an FBG structure on the detuning of the incident light frequency from the Bragg frequency. (Reproduced from Winful, H.G., *New J. Phys.*, 8, 101, 2006. With permission.)

it has been observed that linear optical properties, such as optical attenuation and phase accumulation, tend to scale linearly with Ψ, demonstrating that the dominant effect of structuring a material is to increase the effective path length that light undergoes when passing through it. Additionally, it has been pointed out that for large values of Ψ (typically larger than 30–50), there is a contribution to the attenuation that scales quadratically with Ψ, as a consequence of multiple scattering within the waveguide. Therefore, an important research area in the future would be to design and fabricate photonic crystals, particularly optical fiber based, in a way to minimize these sources of loss (O'Faolin et al. 2010).

In summary, there are substantial differences between the two approaches to generate slow/fast light. For material slow/fast light, the energy density u, the group velocity v_g, and the intensity I obey the relation given by Equation 21.12 ($I = uv_g$), while there is no such analogous relation for the case of structural slow/fast light. As a consequence of this difference, for the case of slow light, there is no enhancement of the electric field strength within a medium that supports material slow light, whereas there is an increase for the case of structural slow light. Therefore, nonlinear optical processes are inherently enhanced when considering structural slow light, but not when dealing with material slow light. Another important difference is related to linear absorption. It has been experimentally confirmed that linear absorption scales with the group index in structural slow/fast light, while no such result is obtained for material slow/fast light (Dicaire et al. 2012).

21.3.2.2 Slow Light for Sensing

The impact of the slow-light phenomenon in optical sensing can be seen in a waveguide interferometer, like a fiber Mach–Zehnder interferometer. Figure 21.12 shows an interferometer of this type, incorporating in the sensing arm, a slow-light segment where the evanescent field of light propagating in the waveguide interacts with the environment, hence being influenced by its refractive index (the reference arm and the input/output couplers are isolated from the environment).

The group velocity of light traveling in the fiber slow-light segment is given by

$$v_g \equiv \frac{\partial\omega}{\partial\beta_{eff}} = -\frac{2\pi c}{\lambda^2}\frac{1}{\partial\beta_{eff}/\partial\lambda} \tag{21.13}$$

where ω, β_{eff}, and λ are the angular frequency of the carrier wave of the light field, the effective propagation constant of the light in such segment, and the vacuum light wavelength, respectively.

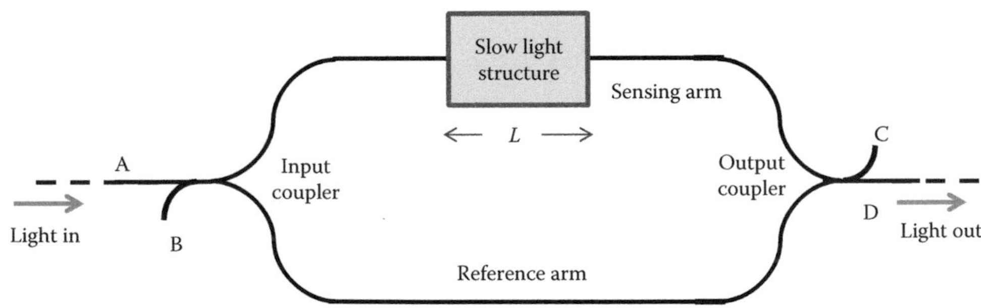

FIGURE 21.12 Fiber Mach–Zehnder interferometer incorporating a slow-light region for phase/spectral sensitivity enhancement.

The sensitivity S is defined as the ratio of the phase change $\partial\phi_s$ of light when it propagates in the slow-light region (assumed to be the sensing element of the interferometer sensing arm) and the change ∂n_{eff} of the effective refractive index of the waveguide mode, associated to the change of the environmental refractive index ∂n_{mea}. On the other hand, if S_{mea} is the phase sensitivity to variations of n_{mea}, then

$$S_{mea} \equiv \frac{\partial\phi_s}{\partial n_{mea}} = \frac{\partial n_{eff}}{\partial n_{mea}} S \tag{21.14}$$

Considering $\beta_{eff} = \omega n_{eff}/c$ and $\phi_s = \beta_{eff}L$, with L the geometric length of the slow-light region, then Equation 21.14 becomes

$$S_{mea} = \frac{1}{v_g}\left(\frac{\omega L}{n_{eff}}\right)\left(\frac{\partial n_{eff}}{\partial n_{mea}}\right) \tag{21.15}$$

This equation shows the relation between the sensitivity of the interferometer to detect phase changes and the slowness of the light. As light travels more slowly (smaller v_g), the interferometer becomes more sensitive to the changes of n_{eff} if other parameters are kept constant and, therefore, to the changes of the environmental refractive index n_{mea}. The result given by Equation 21.15 was derived independently of the physical mechanisms determining n_{eff}, meaning it shall be valid independent of the origin of the slow light (material or structural based).

The sensitivity enhancement associated with structural slow light in an FBG has been demonstrated (Wen et al. 2012). It is known that these devices, intrinsically wavelength sensitive to strain, temperature, and refractive index (when the fiber is tapered or polished to allow the evanescent interaction of the core mode light field with the surrounding medium), as well as to other measurands through adequate interfacing, offer resolutions in a wide range depending on the interrogation approach considered. The utilization of a processing interferometer to convert the measurand-induced Bragg wavelength shift into a phase variation provides very high resolutions in the readout of standard FBG sensors. Indeed, the use of an unbalanced fiber Mach–Zehnder interferometer to interrogate a passive FBG-based strain sensor enables a dynamic strain resolution of $600\,\text{p}\varepsilon/\sqrt{\text{Hz}}$ (Kersey et al. 1992). More complex signal processing techniques, such as the Pound–Drever–Hall technique, provide better resolutions, as demonstrated by the research of Gatti et al. (2008), where this technique is applied to a strain sensor based on a π-phase-shifted FBG, resulting into a minimum detectable dynamic strain of $5\,\text{p}\varepsilon/\sqrt{\text{Hz}}$ for frequencies larger than 100 kHz. The research of Wen and coworkers showed that this value can be improved significantly with slow-light-based FBG strain sensor, as will be described next.

Different physical mechanisms have been investigated to generate slow light in optical fibers, namely, EIT (Slepkov et al. 2008), stimulated Brillouin (SBS) (Song et al. 2005), and FBGs (Mok et al. 2006). EIT is the most promising method for generating extremely slow light but its implementation is complex; on the other hand, SBS is a more practical technique but has modest velocity reductions limited by the SBS gain.

Both approaches require supplying high optical power to the fiber, and for the SBS, long lengths of fiber are also needed.

These constraints are removed in FBGs because it relies on a passive resonant effect and does not need an optical pump. In addition, FBG is a very short and an inexpensive component. As shown in Figure 21.11, an FBG structure supports slow light on the edges of the spectral stop band, as happens in general on the edges of the spectral bandgap of photonic crystals. To maximize the group index attainable at certain band-edge frequencies in an FBG, the index modulation should be as large as possible and the loss as low as feasible (Wen et al. 2012). For a given loss, there is also a grating length that maximizes the group index. Also, apodization can significantly increase the group index. These specific considerations explain why slow-light phenomenon has not been observed before, since a typical FBG has a weak index modulation, their length was too short or not optimized, no apodization was involved, or a combination of these factors.

To quantify the impact of the FBG slow-light behavior, the authors placed the FBG in one arm of a fiber Mach–Zehnder interferometer illuminated by a slow-light wavelength, a configuration similar to Figure 21.12. The interferometer can be geometrically balanced to reduce the impact of laser phase noise and makes the interferometer more stable against external perturbations such as temperature changes.

The optical power P_{out} at the output of the interferometer is given by (Wen et al. 2012)

$$P_{out} = P_o\left[(1-\eta)^2 T_1 + \eta^2 T_2 - 2\eta(1-\eta)\sqrt{T_1T_2}\cos(\Delta\phi)\right] \tag{21.16}$$

where

- P_o is the optical power injected into the input coupler
- η is the coupler's coupling coefficient
- $\Delta\phi$ is the phase difference between the light propagating in the two arms
- T_1, T_2 are the power transmission coefficients of the upper and lower arms of the couplers, respectively

If the interferometer is in quadrature [$\Delta\phi = (2m+1)\pi/2$, m integer] and if the dynamic strain-induced phase perturbation applied to the FBG is $\delta\phi \ll 1$, then Equation 21.16 becomes

$$P_{out}(\delta\phi) \approx 2P_o\eta(1-\eta)\sqrt{T_1T_2}\cos(\delta\phi) \tag{21.17}$$

The strain sensitivity is given by

$$S \equiv \frac{1}{P_o}\frac{dP_{out}}{d\varepsilon} = \frac{1}{P_o}\frac{dP_{out}}{d\phi}\frac{d\phi}{d\lambda}\frac{d\lambda}{d\varepsilon} \tag{21.18}$$

The factor $dP_{out}/d\phi$ is obtained from (21.17). On the other hand, following the argument associated with Equation 21.13, it comes out

$$\frac{d\phi}{d\lambda} = n_g \frac{L}{c}\frac{d\omega}{d\lambda} \tag{21.19}$$

For silica fibers, $d\lambda/d\varepsilon = -0.79\lambda_B$, hence we obtain

$$S = 3.16\pi\eta(1-\eta)\sqrt{T_1}\, n_g L \frac{\lambda_B}{\lambda^2} \tag{21.20}$$

where it is assumed that the reference arm is made essentially lossless ($T_2 = 1$). To maximize the sensitivity, it is required to maximize the product $n_g\sqrt{T_1}$, which can be considered the figure of merit of the interferometric slow-light sensing structure, as well as to consider $\eta = 0.5$. This implies that the first FBG slow-light peak (closest to the band edge) does not necessarily produce the highest sensitivity because its transmission is usually small. Therefore, the best choices are typically the second or third peaks.

Wen et al. compared the sensitivity performance of the FBG slow-light scheme proposed by them with the one obtained with interferometric interrogation of a standard FBG, for the same FBG length (2 cm) and index modulation amplitude (Wen et al. 2012). The results are shown in Figure 21.13. For the conventional sensor, the sensitivity is maximum when the grating is very weak and decreasing when the index modulation increases as expected. In contrast, the sensitivity of slow-light sensor increases steadily when Δn increases, at a faster rate for larger Δn. As shown by the authors, in the lossless limit, it can be as much as five orders of magnitude more sensitive. Considering losses of 1 m^{-1}, the sensitivity enhancement factor is ~260, an improvement that comes

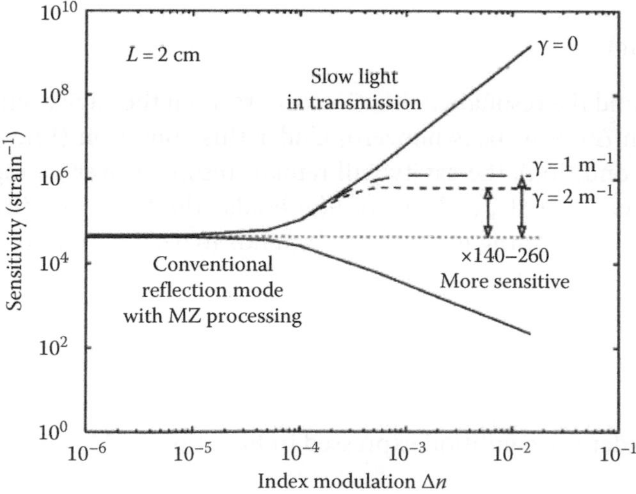

FIGURE 21.13 Strain sensitivity dependence on index modulation amplitude of a slow-light FBG sensor operated in transmission with interferometric interrogation. (Reproduced with permission from Wen, H., Terrel, M., Fan, S. et al., Sensing with slow light in fiber Bragg gratings, *IEEE Sens. J.*, 12, 156, Copyright 2012 IEEE.)

together with the use of a more stable layout for the interferometric interrogation (balanced interferometer).

This study clearly demonstrates the potential for highly sensitive optical sensing utilizing slow light. Wen et al. used an apodized strong FBG with a length of 1.2 cm and obtained a group index of ~127, the largest value associated with an FBG to date. A dynamic (25 kHz) strain sensitivity of 3.14 × 10^5 strain^{-1} and a strain resolution of $\sim 880\,\mathrm{f}\varepsilon/\sqrt{\mathrm{Hz}}$ were also obtained, which is ~700 times better than the value reported using the conventional approach of interferometric interrogation of passive FBGs (Kersey et al. 1992) and a factor ~6 times better than the value obtained with FBG interrogation based on Pound–Drever–Hall processing techniques (Gatti et al. 2008).

21.3.2.3 Fast Light for Sensing

As mentioned before, fast light refers to the phenomenon of the group velocity of light in a medium exceeding the vacuum speed of light (Withayachumnankul et al. 2010). This n_g superluminal phenomenon provides opportunities for high-performance optical sensing as briefly discussed later (Shahriar and Salit 2008).

Let us assume a Fabry–Pérot cavity (FPC) of length L tuned to resonance frequency υ_0, corresponding to wavelength λ_0. When the frequency is changed from this value, the wavelength also changes, so that the resonance condition is no longer fulfilled, leading to dephasing and a drop in the intracavity light intensity and in the cavity transmission. When a medium with a frequency-dependent refractive index, $n(v)$, is placed inside the FPC, the wavelength of the light with frequency v is given by

$$\lambda = \frac{c}{\upsilon n(\upsilon)} \tag{21.21}$$

If $n(v)$ is chosen to depend on frequency in a manner so that

$$vn(v) = \text{constant} \tag{21.22}$$

for v around υ_0, and if a resonance situation occurs, then the cavity will be kept in resonance even when $\Delta\upsilon \equiv \upsilon - \upsilon_0$ is nonzero. Under this condition (known as white light cavity—WLC—condition), the cavity will remain resonant over a range of frequencies for which this frequency dependence of $n(v)$ holds. The WLC condition requires that $n(v)$ decreases with increasing v, which corresponds to negative (anomalous) dispersion. The group index is defined as

$$n_g \equiv \frac{\partial(n\upsilon)}{\partial\upsilon} \tag{21.23}$$

which means, under the condition expressed in Equation 21.22,

$$n_g = 0 \Rightarrow \text{Group velocity} \rightarrow \infty \tag{21.24}$$

Because condition expressed by Equation 21.22 can only hold for a finite bandwidth for a real medium, the front edge of the wave packet can never propagate faster than the free space velocity of light, so there is no violation of special relativity (Dogariu et al. 2001).

The result expressed in Equation 21.24 indicates that a medium with negative dispersion can have an exceedingly large group velocity. However, there is another aspect of this phenomenon that is very important for sensing.

Let us consider again a Fabry–Pérot interferometer in vacuum with length L and tuned to resonance frequency υ_0, corresponding to a wavelength λ_0, that is,

$$L = L_0 = m\frac{\lambda_0}{2} = m\frac{c}{2\upsilon_0} \tag{21.25}$$

where m is an integer. When the cavity length changes by ΔL, the cavity is no longer in a resonant condition. It can be restored to resonance by changing the optical frequency by

$$\Delta\upsilon = \frac{\Delta L}{\left(\partial L/\partial\upsilon\right)_{L_o}} = S_{vacuum}\Delta L \tag{21.26}$$

where the sensitivity S_{vacuum} is expressed as

$$S_{vacuum} \equiv \frac{\Delta\upsilon}{\Delta L} = \frac{1}{\left(\partial L/\partial\upsilon\right)_{L_o}} = -\frac{2\upsilon_o^2}{mc} \tag{21.27}$$

Now, it is assumed that a medium with frequency dependent index, $n(\upsilon)$, is placed inside the cavity. Therefore, in resonance at frequency υ_0, we have

$$L = m\frac{\lambda}{2} = m\frac{c}{2\upsilon_o n\left(\upsilon_o\right)} \tag{21.28}$$

Again, if the cavity length changes by ΔL, the cavity is no longer in resonance. However, if condition (21.22) is satisfied, the wavelength ($\lambda = c/\upsilon n[\upsilon]$) remains unchanged. This implies infinite sensitivity to mirror displacement ΔL, or one could say the value of L for resonance remains unchanged when v changes.

In practice, due to the fact that $n(\upsilon)$ is linear with v only over a finite range, the value of S_{Medium} remains finite and scales inversely to the group index, n_g (Shahriar et al. 2007, Pati et al. 2008)

$$S_{Medium} = \frac{S_{vacuum}}{n_g} = \xi S_{vacuum} \tag{21.29}$$

where ξ is called the superluminal enhancement factor, designation connected with the fact that the group velocity of light exceeds the free space velocity of light. Enhancement factors as large as 10^7 can be obtained under realistic conditions (Yum et al. 2010).

To assess the impact of this enhancement phenomenon, let us assume that we would like to measure the optical frequency shift of the interferometric transmitted spectral peak with a resolution of $\delta\upsilon_{min} = 1$ kHz associated with an interferometric cavity that in vacuum has $S_{vacuum} = 1$ MHz/mm, thus allowing a minimum measurable displacement of

$\delta L_{min} = 1$ μm. If a negative dispersion medium was inserted in the cavity such that $n_g = 0.001$, and if the same $\delta\upsilon_{min} = 1$ kHz was possible, it would result in a value $\delta L_{min} = 1$ nm.

This enhancement in the minimum detectable interferometric length change can be used to design highly sensitive interferometric structures for the sensing of rotation, acceleration, vibration, and other parameters. However, we note that enhanced sensitivity provided by the WLC does not lead to a corresponding improvement in the minimum measurable phase shift. This is because the accuracy of measuring the resonance wavelength depends on the cavity linewidth, which is broadened by the WLC effect. In fact, the WLC is broadened by a factor that essentially matches the enhancement in sensitivity (Shahriar et al. 2007). This statement is drastically altered if we considered a situation where, instead of the cavity being fed by an external laser, the dispersive cavity itself contained an active gain medium. The laser linewidth would be determined by several mechanisms and would be substantially narrower than that of the cold cavity, thus allowing one to exploit the sensitivity enhancement of the WLC concept. Indeed, studies performed on such superluminal lasers indicate displacement sensitivity of nearly five orders of magnitude better compared with the situation of no WLC effect (Yum et al. 2010). Recent developments on superluminal ring fiber lasers supported by Brillouin gain suggest possibility of values for ΔL_{min} nearly eight orders of magnitudes smaller than the values achieved with conventional fiber lasers with the same cavity length (Kotlicki et al. 2012).

The sensitivity enhancement associated with the operation of interferometric sensors incorporating the WLC concept is very promising for optical sensors (Yum et al. 2010). One of the most ambitious projects is the upgraded version of the *Laser Interferometer Gravitational-Wave Observatory* (LIGO), a scientific collaboration of the California Institute of Technology and the Massachusetts Institute of Technology, established in 1992, with infrastructures located at Hanford, Washington, and Livingston, Louisiana, essentially two Michelson interferometers with 4 km long arms where optical power at 100 kW level circulates. The upgraded system, known as *Advanced LIGO*, will include better mirrors and other optical components, high power lasers, additional components for power and signal recycling, and, eventually, the incorporation of novel technology supported by concepts such as squeezed light, fast light, and others (Matson 2009). Studies involving the application of fast light for such a system suggest the possibility of a sensitivity enhancement factor on the order of 10^5 in the interferometric detection of gravitational wave–induced strain variations (Shahriar and Salit 2008). Such improvement would permit the opening of new observational fields for gravitational waves and the design of more compact interferometric systems, facilitating its calibration and ease of operation.

21.3.3 Optical Sensing and Quantum Mechanics

The advent of lasers, waveguides, and many other optical structures and devices permits light properties from the intensity, polarization, phase, and frequency to be well controlled. Understanding the operations of these devices and properly engineering them for specific applications require a deep understanding of quantum mechanical principles. Some other quantum mechanical principles that have not been explored in optics, such as nonlocal correlations, could soon become important for two reasons. One is due to constant desire for miniaturization, which means that when sizes approach nanometer scale, the interaction rules derived from quantum mechanics need to be considered. The second reason is not based on the need but instead on the opportunities that quantum mechanics principles provide (Milburn 1997, Dowling and Milburn 2003).

Quantum technology is supported by the fundamental principles of quantum mechanics, but as a technology, it requires a set of auxiliary tools necessary to build up functional quantum-based systems. The ability to incorporate the control elements into these systems is crucial. It has been shown that classical control theory is inadequate to handle quantum systems; therefore, the development of general principles of control quantum theory is essential for the future of quantum technology (Dong and Peterson 2011). A control system requires the ability to feel its surrounding, which means the need of sensing elements, the so-called quantum sensors, the main focus of our discussion here (Giovannetti et al. 2011). Communication between quantum-based systems (or subsystems) requires new principles of operation and new protocols that might incorporate, as fundamental components, quantum key distribution and error correction algorithms. This field is in its infancy and requires a deep understanding of the quantum communication complexity (Gisin and Thew 2007).

On the core of the quantum technology development was the discovery that quantum mechanics, under certain conditions, enables exponentially more efficient algorithms compared with what can be implemented in a classical computer (Ladd et al. 2010). Building a quantum computer is a landmark of quantum technology, requiring the ability to manipulate quantum-entangled states for millions of subcomponents, involving the coherent interplay of sensing and measurement (quantum sensors and quantum metrology), error correction (quantum control), and interconnection (quantum communication).

In the following sections, we will briefly address the measurement resolution limits imposed by quantum mechanics in classical sensing systems, and their improvement through utilization of squeezed light. Finally, *quantum mechanical devices that can* feel *the quantum state of the system exploiting quantum correlations such as quantum entanglement* will be discussed as the basis of quantum sensors.

21.3.3.1 Resolution Limit of Classical Sensing Systems

In classical optical sensing systems, the ultimate resolution is limited by the fact that light intensity cannot be measured with infinite precision in view of its fluctuations around some average value. To quantify this effect, we can start from the Heisenberg uncertainty principle, namely, $\Delta E \Delta t \geq h/2\pi$, where ΔE and Δt are the uncertainty in the measured energy and the uncertainty in the time window the measurement is performed, with h the Planck constant. If n is the average number of photons in the light field (we have assumed monochromatic field, i.e., $\Delta\omega = 0$), then $E = n\omega h/2\pi \Rightarrow \Delta E = \omega h \Delta n/2\pi$, where ω is the light angular frequency and Δn is the fluctuation in the detected photon number. At a specific point in space, the light phase is given by $\phi = \omega t + \text{constant} \Rightarrow \Delta\phi = \omega\Delta t$, with $\Delta\phi$ the fluctuation of detected optical phase. The insertion of these relations into the energy–time uncertainty condition gives

$$\Delta n \Delta\phi \geq 1 \tag{21.30}$$

known as the Heisenberg number–phase uncertainty relation.

It is known that a single-frequency laser light field is well approximated by a coherent state, commonly represented by $|\alpha\rangle$, with $\alpha = |\alpha|e^{i\phi}$ proportional to the electric field amplitude. Therefore, $|\alpha|^2$ is proportional to n and thus to the intensity of the light field. The intensity can be expressed as $I = nI_{sp}$, where I_{sp} is the intensity associated to a single photon given by $I_{sp} = h\omega/(2\pi\varepsilon_0\eta)$, where ε_0 is the vacuum electric permittivity and η is the mode volume of the electromagnetic field (Scully and Zubairy 1997). The statistics of the light intensity fluctuations follows a Poisson distribution meaning $\Delta n = \sqrt{n}$; also, for a coherent state light field, $\Delta\phi = 1/\sqrt{n}$, which means the equality is considered in Equation 21.30, indicating such light state is a minimum uncertainty state, with the relevant relations summarized as (Dowling 2008)

$$\begin{cases} \Delta n \Delta\phi_{sn} = 1 \\ \Delta n = \sqrt{n} \\ \Delta\phi_{sn} = \dfrac{1}{\sqrt{n}} \end{cases} \tag{21.31}$$

The third equation of this group can be expressed in terms of intensities in the form

$$\Delta\phi_{sn} = \left(\frac{I_{sp}}{I}\right)^{1/2} \tag{21.32}$$

known as the shot-noise limit for the determination of the field light phase. This phase uncertainty can be expressed as an equivalent length uncertainly $\Delta\ell_{sn}$ through the spatial phase relation $\Delta\phi_{sn} = k\Delta\ell_{sn}$, where the wavevector is $k = 2\pi/\lambda$, with λ the light wavelength in the medium. The third equation of (21.31) leads to $\Delta\ell_{sn} = \lambda/(2\pi\sqrt{n})$, which is the minimum length change that can be measured in a shot-noise-limited interferometric sensing system. In the LIGO equipment mentioned in Section 21.3.2.3, the optical power circulating in the interferometer is on the order of 100 kW with a wavelength of $\lambda \approx 1$ μm, which corresponds to a mean photon number of $n \approx 10^{24}$. Therefore, the minimum detectable displacement of the interferometric mirrors limited by shot noise is $\Delta\ell_{sn} \approx (1/2\pi)\times 10^{-18}$ m, roughly 1000 times smaller than the diameter of a proton (Dowling 2008).

21.3.3.2 Squeezed Light and Sensing

The shot-noise phase uncertainty expressed by Equation 21.32 results from the photon-number fluctuations associated with a light field. In a classical electromagnetic field perspective, the field amplitude and phase can be measured simultaneously with infinite precision, meaning with proper design and experimental effort, it would be possible to reach smaller and smaller minimum detectable phases (and detectable lengths) in interferometric sensing systems (the same would happens with sensing based on different modulation approaches, such as those supported by intensity light modulation—intensity sensors—where now the relevant phenomenon would be the intensity fluctuations). The quantification of the light field implied such picture was no more valid, and for long, it was assumed that the shot-noise effect determined the ultimate resolution as imposed by quantum mechanics. The situation changed in 1981 with a breakthrough paper of Carlton Caves, where the idea of using nonclassical states of light was proposed to improve the resolution of optical measurement systems to even below the shot-noise limit (Caves 1981). These nonclassical states were identified as squeezed states of light, and to understand why they allow better resolutions, the diagram shown in Figure 21.14 is useful (Dowling 2008).

The figure gives a phase-space diagram where fluctuations along the radial direction correspond to intensity fluctuations and those in the angular direction are associated with phase fluctuations. A light field in classical view has no fluctuations; therefore, it is represented by a point in the phase space. A coherent light state is shown by a disk, showing intensity and phase fluctuations of the same amplitude as given by Equation 21.31. If the area of this disk is A, Equations 21.30 and 21.31 indicate that any light quantum state must have an area greater or equal to A; also, any minimum uncertainty light state must be represented in the phase space by a region of area A. In other words, while for the coherent state the three relations of Equation 21.31 are fulfilled, minimum uncertainly light states are possible by relaxing the last two relations and keeping valid the first one.

The consequence of this argument is that it is possible to decrease $\Delta\phi$ when it is permitted to increase Δn, so that the product $\Delta n\Delta\phi$ remains unitary and the area of the enclosed uncertainty region remains equal to A therefore maintaining a minimum uncertainty light state. In the phase space, this is represented by squeezing the disk associated to the coherent state into an ellipse with the larger dimension along the radial direction, equivalent to saying the phase uncertainty is decreased at the expense of increasing the light intensity uncertainty.

A simple argument permits to estimate how much squeeze can be performed on a light state. It is based on the question of what would be the largest uncertainty it can be

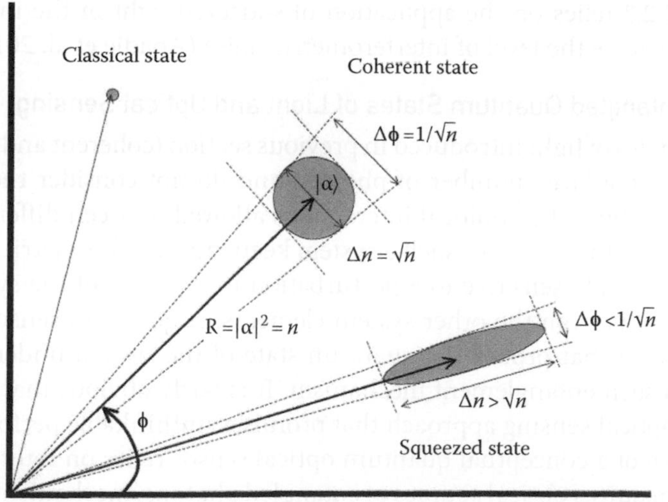

FIGURE 21.14 Phase-space diagram showing intensity and phase uncertainties of classical, coherent, and squeezed light states (fluctuations in the radial direction correspond to intensity fluctuations and those in the angular direction to phase fluctuations). (Adapted from Dowling, J.P., *Contemp. Phys.*, 49, 125, 2008. With permission.)

produced in the intensity of a light state with a mean photon number n while keeping the minimum uncertainty state condition, that is, the area of the ellipse remains equal to A. It is fairly obvious that the energy fluctuations of the light state cannot exceed its average energy, or at maximum $\Delta n = n$. When this relation is substituted into the first equation of the set (21.31), we obtain

$$\Delta\phi_{HL} = \frac{1}{n} = \frac{I_{sp}}{I} \tag{21.33}$$

This relation is known as the Heisenberg limit, designation justified by the fact that Equation 21.33 gives the minimum phase uncertainty possible for light with an average photon number n (Konwar and Saikia 2012).

In an interferometer, this phase uncertainty is converted into a displacement uncertainty given by $\Delta\ell_{HL} = \lambda/(2\pi n)$. In the example of LIGO, for an average number of photons $n = 10^{24}$ circulating in the interferometer and $\lambda = 1$ μm, the minimum measurable change in length would be $\Delta\ell_{HL} \approx (1/2\pi)\times 10^{-30}$ m, approaching the Planck length limit (10^{-35} m) where the classical notion of space breaks down (Dowling 2008). This improvement of 12 orders of magnitude in the interferometric phase resolution relative to the one determined by shot noise is not expectable to be reached in the laboratory in the near future due to technological limitations.

Squeezed light can be generated from light in a coherent state by using certain optical nonlinear interactions, such as optical parametric amplification and frequency doubling. The Kerr nonlinearity in optical fibers also allows the generation of squeezed light, as well as semiconductor lasers when operated with a stabilized pump current. Gradually, the achievable noise reduction through light squeezing is improving, being reported values of 10 dB (Vahlbruch et al. 2008) and 12.3 dB at 1550 nm (Mehmet et al. 2011), in a trend that surely will accelerate in the near future. The *Advanced LIGO Project* discussed in Section 21.3.2.3 relies on the application of squeezed light in the interferometer to substantially decrease the level of interferometric noise (Abadie et al. 2011).

21.3.3.3 Entangled Quantum States of Light and Optical Sensing

The quantum states of light introduced in previous section (coherent and squeezed states of light) deal with a large number of photons and do not consider the entanglement phenomena describing the nonlocal interactions allowed between different elements of a quantum system. In some way, such a system keeps an interferometric pattern instantaneously and strongly sensitive to a perturbation in any part of the system, independent of how far it is from the other system elements. A quantum sensor is a quantum mechanical system that probes the quantum state of the system under measurement, and it relies on such entanglement mechanism. It is fairly obvious that we are facing a radically new optical sensing approach that promises unthinkable performance.

The operation of a conceptual quantum optical sensor relies on interferometric configuration and requires optical sources of entangled photons, mechanisms of measurand interaction with the entangled photons set, the conversion of the measurand-modulated photon-entangled state into an optical intensity signal, the identification of factors that degrade the photon-entangled condition with the consequent degradation of sensor performance, and many others (Kapale et al. 2005).

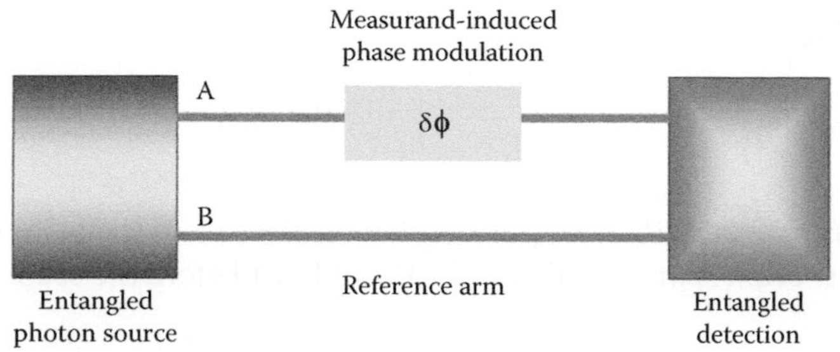

FIGURE 21.15 Mach–Zehnder interferometer with entangled optical quantum states. (Adapted from Dowling, J.P., *Contemp. Phys.*, 49, 125, 2008. With permission.)

The following analysis is based on a Mach–Zehnder interferometric configuration shown in Figure 21.15, a two-mode (two paths) layout that deals with an entangled photon state identified in the literature as N00N (Dowling 2008). This belongs to the class of the *Schrodinger Cat States*, first presented by Barry Sanders in 1989 in the context of his studies of quantum decoherence (Sanders 1989) and rediscovered in 2000 by Jonathan Dowling and coworkers when investigating quantum lithography (Boto et al. 2000). In the *N00N* state, a fixed number (*N*) of photons is either all in the upper mode (path A) or all in the lower mode (path B), but intrinsically, it is not possible to say in which path they are. The state of all photons in mode A and none in mode B is expressed as $|\text{up}\rangle = |N\rangle_A|0\rangle_B$, while the state of all in mode B and none in mode A is written as $|\text{down}\rangle = |0\rangle_A|N\rangle_B$, with this notation indicating a product state of *N* photons either in A or in B (but not in both). In Figure 21.15, the *entangled photon source* is supposed to produce superpositions of these two states in the following form:

$$|N00N\rangle = |\text{up}\rangle + |\text{down}\rangle = |N\rangle_A|0\rangle_B + |0\rangle_A|N\rangle_B \tag{21.34}$$

where, for convenience, the $1/\sqrt{2}$ normalization constant has been removed. It has been shown that the $|N00N\rangle$ superposition state maintains quantum entanglement between the states $|\text{up}\rangle$ and $|\text{down}\rangle$ with its intrinsic nonlocality, meaning a measurement of the *N* photons in mode A collapses, instantaneously, the photon number in mode B to zero, even if mode A is associated to a light propagation path on Earth and mode B to a path at another side of the galaxy!

In Figure 21.15, it is shown that the measurand region is located in path A, meaning the measurand action introduces a phase shift δϕ in a classical monochromatic light field that goes through that path. This outcome happens also with quantum mechanics coherent states, but not in states like $|N00N\rangle$. Indeed, it can be shown that the introduced phase shift is directly proportional to *N*, the number of photons, a radically different behavior from what is found in coherent states where there is no *n*-dependence of the measurand-induced phase shift, where *n* is the average number of photons in the interaction region. Formally, this can be expressed by the diverse output of the phase shift operator $\widehat{U} \equiv \exp(i\delta\phi\widehat{n})$ for different photon states at its input ($\widehat{n}$ is the photon-number operator). Therefore, for the coherent state $|\alpha\rangle$ and for a quantum state of *N* photons, $|\text{N}\rangle$, we have (Gerry and Knight 2005)

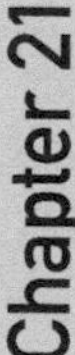

$$\widehat{U}_{\delta\phi}\left|\alpha\right\rangle = e^{i\delta\phi}\left|\alpha\right\rangle \tag{21.35}$$

$$\widehat{U}_{\delta\phi}\left|N\right\rangle = e^{i(N\delta\phi)}\left|N\right\rangle \tag{21.36}$$

The result (21.36) indicates that the state $|N00N\rangle$ evolves from the form given by (21.34) just after the first *beam splitter* in Figure 21.15, and just before the second *beam splitter*, it becomes

$$\left|N00N\right\rangle_{After} = e^{i(N\delta\phi)}\left|N\right\rangle_A\left|0\right\rangle_B + \left|0\right\rangle_A\left|N\right\rangle_B \tag{21.37}$$

This superposition state collapses in the entangled detection block producing a light intensity signal given by

$$I_{out|Entangled}\left(\delta\phi\right) = I_o\left[1 + V_{|Entangled}\cos\left(N\delta\phi\right)\right] \tag{21.38}$$

where

I_o is proportional to the optical intensity injected into the interferometer

$V_{|Entangled}$ is the fringe visibility. This result contrasts with the one obtained with classical interferometry,

$$I_{out|Classical}\left(\delta\phi\right) = I_o\left[1 + V_{|Classical}\cos\left(\delta\phi\right)\right] \tag{21.39}$$

The comparison of results (21.38) and (21.39) shows a sensitivity improvement of the entangled interferometric approach by a factor of N relative to the classical interferometry (assuming equal visibility in both cases). Hence, if $\Delta\phi_{sn}$ is the shot-noise limited phase resolution of the interferometer in the classical structure, the interferometric entangled configuration provides factor of N improvement,

$$\Delta\phi_{Entangled} = \frac{\Delta\phi_{sn}}{N} \tag{21.40}$$

This is in line with the Heisenberg limit expressed by Equation 21.33, and it is shown that this method provides better phase resolution than squeezed states of light (Zwierz et al. 2012). The result (21.38) can be interpreted as the effective wavelength ($\lambda_{Entangled}$) of N entangled photons in the $N00N$ state is N times smaller than the wavelength associated to each individual photon (λ_{ph}):

$$\lambda_{Entangled} = \frac{\lambda_{ph}}{N} \tag{21.41}$$

Such a short effective wavelength provides new possibilities in optical imaging with spatial resolution limits well below the Rayleigh limit (Giovannetti et al. 2009).

The result (21.40) assumes equal interferometric visibility for the entangled and classical interferometric approaches ($V_{|Entangled} = V_{|Classical}$). The visibility depends on several factors, and one of them is the balance of the optical power in the two interferometer arms. If this does not happens, the situation changes. Admitting the phase shift $\delta\phi$ in the measurand region is accompanied by an excess loss represented by γ (the rate at which the photons are absorbed in the region, equal to gL, where g is the loss per unit length and L the distance traveled through the lossy region), then for the coherent state $|\alpha\rangle$ and for the quantum number state $|N\rangle$, it comes out (Gerry and Knight 2005) as

$$|\alpha\rangle \rightarrow e^{-\gamma} e^{i\delta\phi} |\alpha\rangle \tag{21.42}$$

$$|N\rangle \rightarrow e^{-N\gamma} e^{i(N\delta\phi)} |N\rangle \tag{21.43}$$

The exponential dependence of the loss in the coherent (classical) state is known as the Beer's law for optical absorption. Equation 21.43 shows that for the N-photon-number state, there is a much more pronounced effect, sometimes identified as *super-Beer's law.* This means a large dependence on the loss of the visibility of interferometers operating with entangled photons compared with the standard interferometric layout, indicating a rapid degradation of the phase resolution $\Delta\phi_{Entangled}$ when the loss in the measurement region increases even by a small amount, very soon reaching a point where no performance advantage is obtained relative to the classical approach. This behavior is deeply related with the fragility of $N00N$ entangled quantum states, indicating the relevance of searching specific types of photon-number-entangled states more robust to environment loss (Joo et al. 2011).

21.4 Final Remarks

In a broadly defined term, optical sensing including optical metrology and imaging is a dynamic field of research that utilizes a large range of classical and novel concepts. The broad reach of this field and its dynamic nature make it challenging to offer a comprehensive review of the state-of-the art of this field. Given this fact, this book presents 20 chapters prepared by recognized experts in this field and finishes with this chapter. The first two chapters deal with the fundamentals of optical sensors and principles of optical metrology, chapter 3 provides a general view on optical sensing based on waveguides, followed by four chapters (4–7) with detail discussions on sensors based on intensity measurement, interferometric phase measurement, fluorescence detection and plasmonic approach. There are three chapters (8–10) on wavefront sensing and adaptive optics, multiphoton microscopy and imaging based on optical coherence tomography. Optical sensors relying on optical fiber are becoming a widely accepted form of waveguide sensors, hence a few chapters are dedicated to this field. After Chapter 11 that presents a historical overview of fiber optic based sensors, and Chapter 12 on optical fibers, there are two chapters describing optical fiber sensors using intensity modulation and phase modulation (13–14). Some unique advantages of optical fiber sensors are fully described in chapters dedicated to multi-point and distributed sensing (15–16). Due to importance of fiber Bragg gratings in the field of fiber sensing, Chapter 17 is dedicated to this technology, followed by Chapters 18 and 19 that provide some examples of applications of fiber sensors for detection of chemicals and physical parameters and finally a Chapter 20 with focus on the *standardization*, an issue that should be in the forefront of this field.

This final chapter was intended to be authors' outlook of the field and their assessment of possible new trends and opportunities. This is not intended to be a review of what have been achieved so far and a review of topics presented in other chapters.

The chapter began with wavelength-encoded sensors, such as FBGs, DFB fiber laser sensors, and plasmonics-based sensors (including the specific but highly important subject of optical sensing supported by SERS), and discussion was expanded to MOF, sensor multiplexing, and distributed sensing, with no counterpart in any other sensing technology. Finally, we have attempted to identify, qualitatively, new development paths for optical sensing. Such exercise is always risky when it deals with scientific progresses and discoveries. It is often the case that unexpected discoveries open new frontiers for science and technological progress in a way that was previously unthinkable. But it is evident that optical sensing has a huge unexplored space to expand its frontiers, integrating new concepts, methodologies, and technologies to its already vast portfolio of knowledge. Integration of metamaterials in sensing structures is imminent, promising substantial improvement in the performance of optical sensors and new opportunities for new sensors. The impact of metamaterials on plasmonic-based optical sensors has already been significant. The progress of the metamaterial science and technology opens the opportunity to design artificial materials characterized by effective (ε_{eff}, μ_{eff}) pairs located in regions of the (ε, μ) plan associated with optical properties compatible with ultrahigh performance optical sensing. Further novelty paths will surely also derive from the combination of metamaterials with the phenomena of slow and fast light.

Although proper design and implementation of metamaterials require knowledge of atomic and molecular science, metamaterials are essentially rooted in classical electromagnetism governed by Maxwell equations. Based on such analysis, an artificial material could be designed with required effective values for its electric permittivity and magnetic permeability, that is, the pair (ε_{eff}, μ_{eff}). A radically different approach is needed when dealing with quantum sensors where nonlocal correlations are essential. The phenomenon of photon entanglement when applied to optical sensing would bring measurement possibilities considered impossible a few years ago, not only because of orders of magnitude improved resolution, but also because the nonlocal feature of these sensors permits obtaining information over an extended space almost instantaneously.

Control and manipulation of light have played important roles in the evolution of mankind and in technological progress. Nimrud lens, the oldest known lens artifact dating back to ancient Assyria over 3000 years ago, and the earliest written records of lenses in Ancient Greece show that attempts to correct for vision loss go back to the early stage of civilization. Later on, telescopes allowed human kind to expand its horizon and to explore the astronomical world, and microscope allowed the exploration of things at microscale. In a broad perspective, all of these are optical sensors, although only some 50 years ago, the word *sensor* was used as we understand it today. Since then, the field of optical sensing has experienced rapid progress and has made significant impacts on the society and people's lives. It is virtually impossible to grasp how this field will evolve in the future, but sure it will be fascinating to follow its progress and the way it will shape the comprehension of our interaction with the natural world.

References

Abadie, J., B. P. Abbott, R. Abbott et al. 2011. A gravitational wave observatory operating beyond the quantum shot-noise limit: Squeezed light in action. arXiv:1109.2295v1 [quant-ph].

Albrecht, M. G. and J. A. Creighton. 1977. Anomalously intense Raman spectra of pyridine at a silver electrode. *Journal of America Chemical Society* 99: 5215–5217.

Amezcua-Correa, A., J. Yang, C. Finlayson et al. 2007. Surface-enhanced Raman scattering using microstructured optical fiber substrates. *Advanced Functional Materials* 17: 2024–2030.

Anker, J. N., W. P. Hall, O. Lyandres et al. 2008. Biosensing with plasmonic nanosensors. *Nature Materials* 7: 442–453.

Atwater, H. A. 2007. The promise of plasmonics. *Scientific American* 296: 56–63.

Ball, G. A. and W. H. Glenn. 1992. Design of a single-mode linear-cavity erbium fibre laser utilizing Bragg reflectors. *Journal of Lightwave Technology* 10: 1338–1343.

Barnes, W. L., A. Dereux, and T. W. Ebbesen. 2003. Surface plasmon subwavelength optics. *Nature* 424: 824–830.

Beech, M. 2012. *The Physics of Invisibility: A Story of Light and Deception*. Springer, New York.

Boto, A. N., P. Kok, D. S. Abrams et al. 2000. Quantum interferometric optical lithography: Exploiting entanglement to beat the diffraction limit. *Physics Review Letters* 85: 2733–2736.

Boyd, R. W. 2009. Slow and fast light: Fundamentals and applications. *Journal of Modern Optics* 56: 1908–1915.

Boyd, R. W. 2011. Material slow light and structural slow light: Similarities and differences for nonlinear optics. *Journal of Optical Society of America* 28: A38–A44.

Boyd, R. W. and D. J. Gauthier. 2002. Slow and fast light. *Progress in Optics* 43: 497–530.

Boyd, R. W., O. Hess, C. Denz et al. 2010. Slow lights. *Journal of Optics* 12: 100301.

Brown, A. W., B. G. Colpitts, K. Brown et al. 2007. Dark-pulse Brillouin optical time-domain sensor with 20-mm spatial resolution. *Journal of Lightwave Technology* 25: 381–386.

Caves, C. M. 1981. Quantum-mechanical noise in an interferometer. *Physical Review D* 23: 1693–1708.

Chang, S., J. Nyagilo, J. Wu et al. 2012. Optical fiber-based surface-enhanced Raman scattering sensor using Au nanovoid arrays. *Plasmonics* 7: 501–508.

Choi, M., S. H. Lee, Y. Kim et al. 2011. A terahertz metamaterial with unnaturally high refractive index. *Nature* 470: 369–373.

Chung, K. M., Z. Liu, C. Lu et al. 2012. Highly sensitive compact force sensor based on microfiber Bragg grating. *IEEE Photonics Technology Letters* 24: 700–702.

Cranch, G. A., G. M. H. Flockhart, and C. K. Kirendall. 2008. Distributed feedback fiber laser strain sensors. *IEEE Sensors* 8: 1161–1172.

Cullha, M., B. Cullum, N. Lavrik et al. 2012. Surface-enhanced Raman scattering as an emerging characterization and detection technique. *Journal of Nanotechnology* 2012: 971380.

Dicaire, I., A. De Rossi, S. Combrié et al. 2012. Probing molecular absorption under slow-light propagation using a photonic crystal waveguide. *Optics Letters* 37: 4934–4936.

Dinish, U. S., C. Y. Fu, K. S. Soh et al. 2012. Highly sensitive SERS detection of cancer proteins in low sample volume using hollow core photonic crystal fiber. *Biosensors and Bioelectronics* 33: 293–298.

Dogariu, A., A. Kuzmich, and L. J. Wang. 2001. Transparent anomalous dispersion and superluminal light pulse propagation at a negative group velocity. *Physics Review A* 63: 053806–053812.

Dong, D. and I. R. Peterson. 2010. Quantum control theory and applications: A survey. *Control Theory and Applications*, IET, 4, 2651–2671.

Dong, Y., X. Bao, and L. Chen. 2009. Distributed temperature sensing based on birefringence effect on transient Brillouin grating in a polarization maintaining photonic crystal fiber. *Optics Letters* 34: 2590–2592.

Dowling, J. P. 2008. Quantum optical metrology—The lowdown on high-N00N states. *Contemporary Physics* 49: 125–143.

Dowling, J. P. and G. J. Milburn. 2003. Quantum technology: The second quantum revolution. *Philosophical Transactions of the Royal Society A* 361: 1655–1674.

Duncan, R. G., B. J. Soller, D. K. Gifford et al. 2007. OFDR-based distributed sensing and fault detection for single- and multi-mode avionics fiber-optics. *Joint Conference on Aging Aircraft*, Palm Springs, CA, June 16–19, 2007.

Eickhoff, W. 1981. Temperature sensing by mode-mode interference in birefringent optical fibers. *Optics Letters* 6: 204–206.

Fang, X., C. R. Liao, and D. N. Wang. 2010. Femtosecond laser fabricated fiber Bragg grating in microfiber for refractive index sensing. *Optics Letters* 35: 1007–1009.

Feng, J., M. Ding, J. Kou et al. 2011. An optical fiber tip micrograting thermometer. *IEEE Photonics Journal* 3: 810–814.

Fernandes, L. A., M. Becker, O. Frazão et al. 2012. Temperature and strain sensing with femtosecond laser written Bragg gratings in defect and non-defect suspended-silica-core fibers. *IEEE Photonics Technology Letters* 24: 554–556.

Foaleng, M. S., J. C. Beugnot, and L. Thévenaz. 2009. Optimized configuration for high resolution distributed sensing using Brillouin echoes. *Proceedings of the 20th International Conference on Optical Fiber Sensors*, Edinburgh, Scotland, October 5–9, 2009.

Frazão, O., R. Falate, J. M. Baptista et al. 2005. Optical bend sensor based on a long-period fibre grating monitored by an optical time-domain reflectometer. *Optical Engineering Letters* 44: 110502.

Gatti, D., G. Galzerano, D. Janner et al. 2008. Fiber strain sensor based on a π-phase-shifted Bragg grating and the Pound–Drever–Hall technique. *Optics Express* 16: 1945–1950.

Gerry, C. C. and P. Knight. 2005. *Introduction to Quantum Optics*. Cambridge University Press, Cambridge, U.K.

Giovannetti, V., S. Lloyd, and L. Maccone. 2011. Advances in optical metrology. *Nature Photonics* 5: 222–229.

Giovannetti, V., S. Lloyd, L. Maccone et al. 2009. Sub-Rayleigh-diffraction-bound quantum imaging. *Physical Review A* 79: 013827.

Gisin, N. and R. Thew. 2007. Quantum communication. *Nature Photonics* 1: 165–171.

Gu, C., C. Shi, H. Yan et al. 2008. Recent advances in fiber SERS sensors. *Proceedings of SPIE* 7056: 70560H.1–70560H.12.

Hernáez, M., I. D. Villar, C. R. Zamarreño et al. 2010. Optical fiber refractometers based on lossy mode resonances supported by TiO_2 coatings. *Applied Optics* 49: 3980–3985.

Horiguchi, T. and M. Tateda. 1989. Optical-fiber-attenuation investigation using stimulated Brillouin scattering between a pulse and a continuous wave. *Optics Letters* 14: 408–410.

Hotate, K. and Z. He. 2006. Synthesis of optical-coherence function and its applications in distributed and multiplexed optical sensing. *Journal of Lightwave Technology* 24: 2541–2557.

Huy, P. M. C., G. Laffont, V. Dewynter et al. 2007. Three-hole microstructured optical fiber for efficient fiber Bragg grating refractometer. *Optics Letters* 32: 2390–2392.

Iadicicco, A., A. Cusano, A. Cutolo et al. 2004. Thinned fiber Bragg grating as high sensitivity refractive index sensor. *IEEE Photonics Technology Letters* 16: 1149–1151.

Joo, J., W. J. Munro, and T. P. Spiller. 2011. Quantum metrology with entangled coherent states. *Physics Review Letters* 107, 083601.

Kabashin, A. V., P. Evans, S. Pastkovsky et al. 2009. Plasmonic nanorod metamaterials for biosensing. *Nature Materials* 8: 867–871.

Kapale, K. T., L. D. Didomenico, H. Lee et al. 2005. Quantum interferometric sensors. *Concepts of Physics* 2: 225–240.

Kayani, A. A., K. Khoshmanesh, S. A. Ward et al. 2012. Optofluidics incorporating actively controlled micro- and nano-particles. *Biomicrofluidics* 6: 031501.

Kersey, A. D., T. A. Berkoff, and W. W. Morey. 1992. High resolution fibre-grating based strain sensor with interferometric wavelength-shift detection. *Electronics Letters* 28: 136–138.

Khurgin, J. B. and R. S. Tucker (eds.). 2008. *Slow Light: Science and Applications*. CRC, Boca Raton, FL.

Kneipp, K., M. Moskovits, and H. Kneipp (eds.). 2006. *Surface-Enhanced Raman Scattering: Physics and Applications*. Springer, Berlin, Germany.

Konwar, H. and J. Saikia. 2012. On the nature of vacuum fluctuations and squeezed state of light. *Archives of Physics Research* 3: 232–238.

Kostovski, G., D. J. White, A. Mitchell et al. 2009. Nanoimprinted optical fibres: Biotemplated nanostructures for SERS sensing. *Biosensors and Bioelectronics* 24: 1531–1535.

Kotlicki, O., J. Scheuer, and M. S. Shahriar. 2012. A Brillouin fast light fiber laser sensor. *Proceedings of the 22nd International Conference on Optical Fiber Sensors*, Beijing, China, October 15–19, 2012.

Kou, J., M. Ding, J. Feng et al. 2012. Microfiber-based Bragg gratings for sensing applications: A review. *Sensors* 12: 8861–8876.

Kou, J. L., Z. D. Huang, G. Zhu et al. 2011a. Wave guiding properties and sensitivity of D-shaped optical fiber microwire devices. *Applied Physics B: Laser and Optics* 102: 615–619.

Kou, J. L., S. J. Qiu, F. Xu et al. 2011b. Demonstration of a compact temperature sensor based on first-order Bragg grating in a tapered fiber probe. *Optics Express* 19: 18452–18457.

Koutsides, C., K. Kalli, D. J. Webb et al. 2011. Characterizing femtosecond laser inscribed Bragg grating spectra. *Optics Express* 19: 342–352.

Kreger, S. T., D. K. Gifford, M. E. Froggatt et al. 2006. High resolution distributed strain or temperature measurements in single- and multi-mode fiber using swept-wavelength interferometry. *Proceedings of the 18th International Conference on Optical Fiber Sensors*, Cancún, Mexico, October 23–27, 2006.

Kringlebotn, J. T., W. H. Loh, and R. I. Laming. 1996. Polarimetric Er^{3+}-doped fiber distributed-feedback laser sensor for differential pressure and force measurements. *Optics Letters* 21: 1869–1871.

Kwang, Y. S. and K. Hotate. 2007. Distributed fiber strain sensor with 1 kHz sampling rate based on Brillouin optical correlation domain analysis. *IEEE Photonics Technology Letters* 19: 1928–1930.

Ladd, T. D., F. Jelezko, R. Laflamme et al. 2010. Quantum computers. *Nature* 464: 45–53.

Lee, S. M., S. S. Saini, and M. Y. Jeong. 2010. Simultaneous measurement of refractive index, temperature and strain using etched-core fiber Bragg grating sensors. *IEEE Photonics Technology Letters* 22: 1431–1433.

Li, W., X. Bao, Y. Li et al. 2008. Differential pulse-width pair BOTDA for high spatial resolution sensing. *Optics Express* 16: 21616–21625.

Lim, D. K., K. S. Jeon, H. M. Kim et al. 2010. Nanogap-engineerable Raman-active nanodumbbells for single-molecule detection. *Nature Materials* 9: 60–67.

Liu, N., M. Mesch, T. Weiss et al. 2010. Infrared perfect absorber and its application as plasmonic sensor. *Sensor Nano Letters* 10: 2342–2348.

Liu, Y., C. Meng, A. P. Zhang et al. 2011. Compact microfiber Bragg gratings with high-index contrast. *Optics Letters* 36: 3115–3117.

Luna Technologies. 2012. http://www.lunatechnologies.com/applications/OFDR-Based-Distributed-Sensing.pdf.

Luo, W., F. Xu, and Y. Lu. 2012. Ultra-small microfiber Bragg grating force sensor with greater sensitivity. *Proceedings of the 22nd International Conference on Optical Fiber Sensors*, Beijing, China, October 15–19, 2012.

Maier, S.A. 2007. *Plasmonics: Fundamentals and Applications*. Springer 2007.

Matson, J. 2009. Listening for gravitational waves, silence becomes meaningful. *Scientific American*, August 26.

Mehmet, M., S. Ast, T. Eberle et al. 2011. Squeezed light at 1550 nm with a quantum noise reduction of 12.3 dB. *Optics Express* 19: 25763–25772.

Meltz, G., J. R. Dunphy, W. W. Morey et al. 1983. Cross-talk fiber-optic temperature sensor. *Applied Optics* 22: 464–477.

Meltz, G. and E. Snitzer. 1981. Fiber optic strain sensor. US Patent No. 4295738.

Milburn, G. J. 1997. *Schrodinger's Machines*. W. H. Freeman & Co, New York.

Miller, G. A., G. A. Crench, and C. K. Kirkendall. 2012. High-performance sensing using fiber lasers. *Optics and Photonics News* 23: 30–36.

Mok, J. T., C. M. Sterke, I. C. M. Littler et al. 2006. Dispersionless slow light using gap solitons. *Nature Physics* 21: 775–780.

Monat, C., M. de Sterke, and B. J. Eggleton. 2010. Slow light enhanced nonlinear optics in periodic structures. *Journal of Optics* 12: 104003.

Naruse, H. and M. Tateda. 1999. Trade-off between the spatial and the frequency resolutions in measuring the power spectrum of the Brillouin backscattered in an optical fiber. *Applied Optics* 38: 6516–6521.

Notomi, M. 2000. Theory of light propagation in strongly modulated photonic crystals: Refraction-like behaviour in the vicinity of the photonic bandgap. *Physics Review B* 62: 10696–10705.

O'Faolin, L., S. A. Schulz, D. M. Beggs et al. 2010. Loss engineered slow light waveguides. *Optics Express* 18: 27627–27638.

Osawa, M. 2001. Surface-enhanced infrared absorption. *Topics in Applied Physics* 81: 163–187.

Pati, G. S., M. Salit, K. Salit et al. 2008. Demonstration of displacement-measurement-sensitivity proportional to inverse group-index of intra-cavity medium in a ring resonator. *Optics Communications* 281: 4931–4935.

Peled, Y., A. Motil, and M. Tur. 2012. Fast Brillouin optical time domain analysis for dynamic measurement. *Optics Express* 20: 8584–8591.

Peled, Y., A. Motil, L. Yaron et al. 2011. Slope-assisted fast distributed sensing in optical fibers with arbitrary Brillouin profile. *Optics Express* 19: 19845–19854.

Pendry, J. B., A. J. Holden, D. J. Robbins et al. 1999. Magnetism from conductors and enhanced nonlinear phenomena. *IEEE Transactions on Microwave Theory and Technology* 47: 2075–2084.

Pinto, A. M. R. and M. Lopez-Amo. 2012. Photonic crystal fibers for sensing applications. *Journal of Sensors* 2012: 598178.

Prasad, P. N. 2003. *Introduction to Biophotonics*. Wiley-Interscience, Hoboken, NJ.

Pruneri, V., C. Riziotis, P. G. R. Smith et al. 2009. Fiber and integrated waveguide-based optical sensors. *Journal of Sensors* 2009: 171748.

Sanders, B. C. 1989. Quantum dynamics of the nonlinear rotator and the effects of continual spin measurement. *Physics Review A* 40: 2417–2427.

Schwartz, B. and R. Piestun. 2005. A new path: Ultralow-index metamaterials present new possibilities for controlling light propagation. *SPIE's OE Magazine* 5: 30–32, January.

Schwartz, B. T. and R. Piestun. 2003. Total external reflection from metamaterials with ultralow refractive index. *Journal of Optical Society of America* 20: 2448–2453.

Scully, M. O. and M. S. Zubairy. 1997. *Quantum Optics*. Cambridge University Press, Cambridge, U.K.

Shahriar, M. S., G. S. Pati, R. Tripathi et al. 2007. Ultrahigh enhancement in absolute rotation sensing using fast and slow light. *Physical Review A* 75: 053807.

Shahriar, M. S. and M. Salit. 2008. Application of fast-light in gravitational wave detection with interferometers and resonators. *Journal of Modern Optics* 55: 3133–3147.

Shelby, R. A., D. R. Smith, S. C. Nemat-Nasser et al. 2001. Microwave transmission through a two-dimensional, isotropic, left-handed metamaterial. *Applied Physics Letters* 78: 489–491.

Slepkov, A. D., A. R. Bhagwat, V. Venkataraman et al. 2008. Generation of large alkali vapor densities inside base hollow-core photonic band-gap fibers. *Optics Express* 16: 18976–18983.

Smith, W. E. 2008. Practical understanding and use of surface enhanced Raman scattering/surface enhanced resonance Raman scattering in chemical and biological analysis. *Chemical Society Review* 37: 955–964.

Smythe, E. J., M. D. Dickey, G. M. Whitesides et al. 2009. A technique to transfer metallic nanoscale patterns to small and non-planar surfaces. *ACS Nano* 3: 59–65.

Socorro, A. B., I. D. Villar, C. R. Zamarreño et al. 2012. Tapered single-mode optical fiber pH sensor based on lossy mode resonances generated by a polymeric thin-film. *IEEE Sensors Journal* 12: 2598–2603.

Soller, B., D. Gifford, M. Wolfe et al. 2005. High resolution optical frequency domain reflectometry for characterization of components and assemblies. *Optics Express* 13: 666–674.

Song, K. Y., S. Chin, N. Primerov et al. 2009. Time-domain distributed sensor with 1 cm spatial resolution based on Brillouin dynamic grating. *Proceedings of the 20th International Conference on Optical Fiber Sensors*, Edinburgh, Scotland, October 5–9, 2009.

Song, K. Y., M. Gonzalo-Herraez, and L. Thévenaz. 2005. Optically controllable slow and fast light in optical fibers using stimulated Brillouin scattering. *Applied Physics Letters* 87: 081113.

Song, K. Y., W. Zou, Z. He et al. 2008. All-optical dynamic grating generation based on Brillouin scattering in polarization-maintaining fiber. *Optics Letters* 33: 926–928.

Stiles, P. L., J. A. Dieringer, N. C. Shah et al. 2008. Surface-enhanced Raman spectroscopy. *Annual Review of Analytical Chemistry* 1: 601–626.

Stoddart, P. R. and D. J. White. 2009. Optical fiber SERS sensors. *Analytical and Bioanalytical Chemistry* 394: 1761–1774.

Thévenaz, L. 2010. Brillouin distributed time-domain sensing in optical fibers: State of the art and perspectives. *Frontiers of Optoelectronics in China* 3: 13–21.

Tong, L., R. R. Gattass, J. B. Ashcom et al. 2003. Subwavelength-diameter silica wires for low-loss optical wave guiding. *Nature* 426: 816–819.

Vahlbruch, H., M. Mehmet, S. Chelkowski et al. 2008. Observation of squeezed light with 10-dB quantum noise reduction. *Physics Review Letters* 100: 033602.

Valley, N., L. Jensen, J. Autschbach et al. 2010. Theoretical studies of surface enhanced hyper-Raman spectroscopy: The chemical enhancement mechanism. *Journal of Chemical Physics* 133: 054103.

Veselago, V. G. 1968. The electrodynamics of substances with simultaneously negative values of ε and μ. *Soviet Physics Uspekhi* 10: 509–513.

Viets, C. and W. Hill. 1998. Comparison of fibre-optic SERS sensors with differently prepared tips. *Sensors and Actuators B: Chemical* 51: 92–99.

Viets, C. and W. Hill. 2001. Fibre-optic SERS sensors with conically etched tips. *Journal of Molecular Structure* 563–564: 163–166.

Villar, I. D., C. R. Zamarreño, M. Hernáez et al. 2010. Generation of lossy mode resonances with absorbing thin-films. *Journal of Lightwave Technology* 28: 3351–3357.

Vinuesa, X. A., S. M. Lopez, P. Corredera et al. 2012. Raman-assisted Brillouin optical time-domain analysis with sub-meter resolution over 100 km. *Optics Express* 20: 12147–12154.

Vorrath, S., S. A. Moller, P. Windpassinger et al. 2010. Efficient guiding of cold atoms through a photonic band gap fiber. *New Journal of Physics* 12, 123015.

Weber, M. J. 2003. *Handbook of Optical Materials*. CRC Press, Boca Raton, FL.

Wen, H., M. Terrel, S. Fan et al. 2012. Sensing with slow light in fiber Bragg gratings. *IEEE Sensors Journal* 12: 156–163.

Wieduwilt, T., S. Bruckner, and H. Bartelt. 2011. High force measurement sensitivity with fiber Bragg gratings fabricated in uniform-waist fiber tapers. *Measurement Science and Technology* 22: 075201.

Willets, K. A. and R. P. V. Duyne. 2007. Localized surface plasmon resonance spectroscopy and sensing. *Annual Review of Analytical Chemistry* 1: 601–626.

Winful, H. G. 2002. Energy storage in superluminal barrier tunneling: Origin of the "Hartman effect". *Optics Express* 10: 1491–1496.

Winful, H. G. 2006. The meaning of group delay in barrier tunnelling: A re-examination of superluminal group velocities. *New Journal of Physics* 8: 101–103.

Withayachumnankul, W., B. M. Fischer, B. Ferguson et al. 2010. A systemized view of superluminal wave propagation. *Proceedings of the IEEE* 98(10): 1775–1786.

Wood, R. W. 1902. On a remarkable case of uneven distribution of light in a diffraction grating spectrum. *Proceedings of the Physical Society of London* 18: 269–275.

Wurtz, G. A., W. Dickson, D. O'Connor et al. 2008. Guided plasmonics modes in nanorod assemblies: Strong electromagnetic coupling regime. *Optics Express* 16: 7460–7470.

Xu, F., G. Brambilla, J. Feng et al. 2010. A microfiber Bragg grating based on a microstructured rod: A proposal. *IEEE Photonics Technology Letters* 22: 218–220.

Yang, F. and J. R. Sambles. 1997. Determination of the optical permittivity and thickness of absorbing films using long range modes. *Journal of Modern Optics* 44: 1155–1163.

Yang, X., C. Shi, R. Newhouse et al. 2011. Hollow-core photonic crystal fibers for surface-enhanced Raman scattering probes. *International Journal of Optics* 2011: 754610.

Yokogawa. 2012. http://www.koreayokogawa.com/product/view.php?idx=205&Ltm=13&cid1=&cid2=&mode=&keyfield=&key=&page=.

Yum, H. N., M. Salit, J. Yablon et al. 2010. Superluminal ring laser for hypersensitive sensing. *Optics Express* 18: 17658–17665.

Zadok, A., Y. Antman, N. Primerov et al. 2012. Random-access distributed fiber sensing. *Laser Photonics Review* 2012: 1–5.

Zeller, W., L. Naehle, P. Fuchs et al. 2010. DFB lasers between 760 nm and 16 μm for sensing applications. *Sensors* 10: 2492–2510.

Zhou, D. P., Z. Qin, W. Li et al. 2012. Distributed vibration sensing with time-resolved optical frequency-domain reflectometry. *Optics Express* 20: 13138–13145.

Zwierz, M., C. A. Pérez-Delgado, and P. Kok. 2012. Ultimate limits to quantum metrology and the meaning of the Heisenberg limit. *Physical Review A* 85, 042112.

Index

D

E

F

G

H

J

K

L

M

N

O

P

Q

R

S

T

U

V

W

X

Y